Birkhäuser

***Modeling and Simulation in Science, Engineering and Technology***



For further volumes:
http://www.springer.com/series/4960

Enzo Tonti

# The Mathematical Structure of Classical and Relativistic Physics

## A General Classification Diagram

Birkhäuser

Enzo Tonti
Department of Engineering and Architecture
University of Trieste
Trieste, Italy

ISSN 2164-3679 ISSN 2164-3725 (electronic)
ISBN 978-1-4939-4232-9 ISBN 978-1-4614-7422-7 (eBook)
DOI 10.1007/978-1-4614-7422-7
Springer New York Heidelberg Dordrecht London

Mathematics Subject Classification (2010): 55-XX, 57-XX, 68-XX, 70-XX, 74-XX, 76-XX, 78-XX, 80-XX, 83-XX



Printed on acid-free paper

Springer is part of Springer Science+Business Media (www.birkhauser-science.com)

# Acknowledgements

Jesus said to the twelve, "Do you also wish to go away?" Simon Peter answered him, "Lord, to whom shall we go? You have the words of eternal life; and we have believed, and have come to know, that you are the Holy One of God."

[Gospel of John 6:67]

Realized with the contribution of

Fondazione

FONDAZIONE CRTRIESTE

# Preface

Why do we have analogies in physics?

This question has been the leitmotif of the author's research for about half a century. Six physical theories are considered: particle mechanics, electromagnetism, mechanics of deformable solids, fluid mechanics, gravitational field and heat conduction. The reasons for the analogies have been discovered, and the main purpose of this book is to make them known.

Usually, analogies are revealed based on the similarities of equations of various physical theories. As the present research developed, it was found that, instead of starting from the equations, the first step must be an analysis of the physical variables that compose them, more precisely, an analysis of the global variables from which the field variables follow as their density and rates. The reason for analogies has been localized in the fact that global variables have a natural association with the four so-called space elements, i.e. points, lines, surface and volumes, and with the two time elements, i.e. time instants and time intervals. In this association, a fundamental role is played by the notion of the orientation of a space and time element. It will be shown that there are two kinds of orientation, inner and outer. Since each of the four space elements can have two possible orientations, it follows that we must consider *eight* distinct space elements and *four* distinct time elements.

This discovery leads to the construction, for the first time, of a classification diagram of these eight oriented space elements and, due to the correspondence of global variables with space and time elements, the construction of a corresponding classification diagram for the global variables of the physical theories.

It follows that the mathematical structure underlying the classification diagram is of a geometric nature, specifically of a topological nature. This diagram brings to light the existence of a mathematical structure that is common to various branches of physics.

However, we emphasize that this is a book about physics, not mathematics. It begins with an analysis of the operational definition of physical variables, with their measurement process, and not with a mathematical theory as the theory of affine spaces or that of differential forms.

The book offers an interdisciplinary approach to the various branches of classical and relativistic physics. To this end, an attempt has been made to write the book with Ph.D. students in mind.

At this point, we would like to quote Cornelius Lanczos from the preface of his book *The Variational Principles of Mechanics*: "Many of the scientific treatises of today are formulated in a half-mystical language, as though to impress the reader with the uncomfortable feeling that he is in the permanent presence of a superman. The present book is conceived in a humble spirit and is written for humble people."

To help the reader, we have made copious citations of books and articles, indicating also the page of the cited books for easy referencing and comparing of statements and formulae by the reader.

**Epigraph**

The author is grateful to Prof. Nicola Bellomo, who proposed publishing the book with Birkhäuser; to Prof. Marc Gerritsma, Prof. Piero Villaggio, Dr. Alberto Favaro, and Prof. Salvatore Noè, who revised some of this book's chapters. The author is grateful to Prof. Stefan Kurz for his pertinent remarks.

Heartfelt appreciation goes to Prof. Friedrich Hehl for his numerous and in-depth comments, which helped to improve many details and some conceptual matters.

Not all suggestions made by these reviewers were accepted, and the author retains sole responsibility for any flaws or faulty judgments expressed in the book.

The author received considerable and crucial assistance in preparing the book from his collaborator Federica Zarantonello. Her enthusiasm for the subject, attention to my explanations on the blackboard, relevant comments, remarks, and corrections, and passion for books were instrumental in bringing the book to a conclusion.

Special thanks go to the *Fondazione Cassa di Risparmio di Trieste*, which provided a generous grant for the preparation of the book.

Lastly, I am grateful to my wife, Pia, who in the 50 years of our marriage has been ever generous with her affection and allowed me to devote myself entirely to studying and teaching. Her dedication to our family and consideration for my work was essential in producing that rarity, a successful marriage.

Trieste, Italy Enzo Tonti

# Contents

1 **Introduction** ... 1
1.1 Aim of the Book ... 1
1.2 Analogies in Physics ... 9
1.3 Role of a Classification ... 13
1.4 Role of Geometry in Physics ... 14
1.4.1 Topological Concepts ... 15
1.5 Algebraic Formulation of Fields ... 17
1.6 Summary ... 17
1.6.1 Notation Used in Book ... 18

**Part I Analysis of Variables and Equations**

2 **Terminology Revisited** ... 21
2.1 Why Do We Need Proper Notation and Terminology? ... 21
2.2 Many Meanings of $\Delta$ ... 22
2.3 Use and Misuse of Symbol 'd' ... 23
2.4 Rates and Derivatives ... 26
2.5 Definite and Indefinite Integrals ... 27
2.6 Multiplication Symbols $\times$ and $\wedge$ ... 27
2.7 Many Meanings of "Force" ... 30
2.8 Many Meanings of "Flux" ... 30
2.9 Many Kinds of Equality ... 32
2.10 Irrotational and Solenoidal Vector Fields ... 34
2.11 Reversal of Motion Is Not Time Reversal ... 35

3 **Space and Time Elements and Their Orientation** ... 39
3.1 Space Elements ... 39
3.2 Orientation of Space Elements ... 40
3.3 Combinatorial Side of Orientation ... 41
3.4 Orientation in Geometry ... 42

3.5 Two Kinds of Orientation: Inner and Outer ... 44
3.5.1 Inner and Outer Orientation of Lines ... 45
3.5.2 Inner and Outer Orientation of Surfaces ... 45
3.5.3 Inner and Outer Orientation of Volumes ... 47
3.5.4 Inner Orientation of Points ... 49
3.5.5 Outer Orientation of Points ... 51
3.6 Role of Space of Immersion ... 53
3.6.1 Induced Orientation ... 54
3.6.2 Holes ... 54
3.7 Historical Note on Orientation ... 56
3.8 Time Elements ... 57
3.8.1 Primal Time Elements and Their Orientation ... 58
3.8.2 Dual Time Elements and Their Orientation ... 59

**4 Cell Complexes** ... 63
4.1 Coordinate Systems and Cell Complexes ... 63
4.1.1 Coordinate Cell Complexes ... 65
4.1.2 Families of Cells ... 66
4.1.3 Simplicial Cell Complexes ... 67
4.2 Dual Cell Complexes ... 69
4.2.1 Duals of Simplicial Complexes ... 72
4.2.2 Voronoi Dual ... 72
4.2.3 Barycentric Dual ... 74
4.3 Inner and Outer Oriented Cell Complexes ... 75
4.3.1 Outer Orientation ... 76
4.4 Role of Dual Complex in Mathematics ... 78
4.5 Role of Dual Complex in Physics ... 79
4.6 Global Variables and Computational Physics ... 81
4.6.1 Finite Difference Method Reinterpreted ... 82
4.7 Classification Diagram for Space Elements ... 84
4.8 Classification of Time Elements ... 85

**5 Analysis of Physical Variables** ... 89
5.1 Role of Mathematics in Physics ... 89
5.2 Material and Spatial Descriptions ... 90
5.2.1 Material Description ... 92
5.2.2 Spatial Description ... 93
5.2.3 Material and Spatial Descriptions: An Overview ... 94
5.3 Physical Quantities ... 96
5.3.1 Physical Constants ... 96
5.3.2 Physical Parameters ... 96
5.3.3 Physical Variables ... 97

5.4 Configuration, Source and Energy Variables ... 99
5.4.1 Source Variables ... 100
5.4.2 Configuration Variables ... 101
5.4.3 Energy Variables ... 102
5.5 Fundamental Problem of a Physical Theory ... 103
5.6 Set Functions ... 104
5.7 Global Variables and Field Variables ... 105
5.8 Global Variables ... 106
5.8.1 Extension of the Notion of Density ... 108
5.8.2 Global Variables in Space ... 109
5.8.3 Continuity of Global Variables in Space ... 112
5.8.4 Space Association ... 113
5.9 Genesis of + and – Signs ... 118
5.10 Sign of Physical Variables ... 118
5.11 Oddness Principle ... 122
5.11.1 Global Variables in Time ... 125
5.12 Time Association ... 126
5.12.1 Primal or Dual Time Elements? ... 130
5.12.2 Global Variables in Space and Time ... 133
5.13 Field Variables: Inherited Association ... 135
5.14 How to Find the Space and Time Association ... 137
5.15 Physical Variables Can Be Grouped into Families ... 139
5.16 Identification Criteria ... 141
5.16.1 Possible Ambiguities ... 143
5.17 Differential Formulation and Orientation ... 145
5.18 What Vector Calculus Ignores ... 147
5.18.1 Energy as a Potential of Constitutive Equations ... 148
5.19 Conjugated Variables ... 149
5.20 Phase, Angular Frequency and Wave Vector ... 152

**6 Analysis of Physical Equations** ... 155
6.1 Introduction ... 155
6.2 Phenomenological Equations ... 160
6.3 Constitutive Equations ... 160
6.3.1 Reversible and Irreversible Constitutive Equations ... 161
6.3.2 Interaction Equations ... 165
6.4 Topological Equations ... 168
6.4.1 Balance Equations ... 170
6.4.2 Circuital Equations ... 171
6.4.3 Space Differences ... 173
6.5 Invariance of Equations Under Inversion of Orientation ... 174
6.6 Defining Equations ... 176
6.7 Equations of Behaviour ... 177

**7 Algebraic Topology** .... 181
7.1 Why Algebraic Topology? .... 181
7.2 Topology and Algebraic Topology .... 184
7.3 Role of Cell Complexes .... 185
7.3.1 Faces of a $p$-Cell and Boundary .... 185
7.3.2 Cofaces of a $p$-Cell and Coboundary .... 185
7.3.3 Incidence Numbers and Incidence Matrices .... 186
7.4 The Notion of a Chain .... 190
7.4.1 Boundary of a Line, Surface and Volume .... 193
7.4.2 Boundary of a Chain .... 193
7.4.3 The Boundary of a Boundary Is a Null Chain .... 194
7.5 Notion of Discrete Form .... 196
7.5.1 Value of a Discrete Form on a Chain .... 198
7.5.2 Coboundary of a Discrete $p$-Form .... 200
7.5.3 Coboundary of a Discrete 0-Form .... 200
7.5.4 Coboundary of a Discrete 1-Form .... 202
7.5.5 Coboundary of a Discrete 2-Form .... 203
7.5.6 General Definition of Coboundary Process .... 204
7.5.7 Discrete Version of Stokes' Theorem .... 204
7.5.8 The Coboundary of the Coboundary Vanishes .... 207
7.5.9 Chains or Discrete Forms in Physics? .... 208
7.6 Coboundary Process in Two Dimensions .... 209
7.6.1 Coboundary Process on Primal Complex .... 209
7.6.2 Coboundary Process on Dual Complex .... 211
7.7 The Wonderful Role of the Coboundary Process in Physics .... 212
7.7.1 Algebraic Analogue of Derivative .... 213
7.7.2 Algebraic Analogue of Gradient .... 213
7.7.3 Algebraic Analogue of the 'Curl' .... 214
7.7.4 Algebraic Analogue of Divergence .... 215
7.8 Examples of Coboundary Process in Physics .... 216
7.8.1 Gauss's Law of Electrostatics .... 217
7.8.2 Equilibrium Law of Continuum Mechanics .... 218

**8 Birth of Classification Diagrams** .... 221
8.1 Classification Diagram of Physical Variables .... 221
8.2 Statics of Strings .... 224
8.3 A Time Diagram .... 226
8.4 How to Combine Space and Time .... 227
8.4.1 Transversal Vibrations of Strings .... 227
8.5 The Structure of the Diagrams .... 229
8.6 What the Diagram Shows .... 230
8.6.1 Global Variables Versus Field Variables .... 230
8.6.2 Reversible and Irreversible Links .... 231

8.6.3 Topological Relations ... 231
8.6.4 Scalar and Vectorial Theories ... 232
8.6.5 Exterior Differential Forms ... 233
8.6.6 Material Parameters ... 233
8.6.7 Configuration Variables are Material Dependent ... 233
8.6.8 Tensorial Nature of Field Functions ... 235
8.6.9 Composing the Equations ... 237

**Part II Analysis of Physical Theories**

**9 Particle Dynamics** ... 241
9.1 Fundamental Problem ... 241
9.2 Source Variables ... 242
9.2.1 Force ... 242
9.2.2 Impulse ... 242
9.2.3 Momentum ... 243
9.2.4 Force from Momentum or Vice Versa? ... 244
9.3 Configuration Variables ... 247
9.3.1 Radius Vector ... 247
9.3.2 Displacement ... 248
9.3.3 Velocity ... 248
9.3.4 Acceleration ... 249
9.4 Constitutive Laws ... 250
9.4.1 Momentum–Velocity Relation ... 250
9.4.2 Force–Radius Vector Relation ... 252
9.4.3 Force–Velocity Relation ... 252
9.4.4 Classification of Forces ... 252
9.5 Energy Variables ... 253
9.5.1 Power, Work, Kinetic Energy ... 253
9.5.2 Potential Energy ... 255
9.5.3 Potential Energy of a Particle in a Force Field ... 257
9.5.4 Potential Energy of a System ... 259
9.5.5 Kinetic Energy and Kinetic Co-energy ... 262
9.5.6 Lagrangian and Hamiltonian ... 264

**10 Electromagnetism** ... 273
10.1 Fundamental Problem ... 273
10.2 From Field Variables to Global Variables ... 273
10.3 Source Variables: Space and Time Classification ... 276
10.3.1 Electric Charge Content ... 276
10.3.2 Electric Flux ... 277
10.3.3 Birth of Electric Displacement ... 279
10.3.4 Critical Remarks ... 281
10.3.5 Electric Charge Flow and Electric Current ... 281

10.3.6 Birth of Electric Current Density Vector ............. 282
10.3.7 Electric Vector Potential ............. 283
10.3.8 Magnetic Field Strength ............. 284
10.3.9 Magnetomotive Force ............. 286
10.3.10 Scalar Magnetic Potential ............. 287
10.4 Configuration Variables: Space and Time Classification ............. 289
10.4.1 Electric Field Strength ............. 289
10.4.2 Electromotive Force and Its Impulse ............. 290
10.4.3 Electric Potential ............. 291
10.4.4 Magnetic Flux ............. 291
10.4.5 Magnetic Flux Density ............. 293
10.4.6 Summary of Physical Variables of Electromagnetism ... 294
10.5 Field Laws ............. 295
10.5.1 Gauss's Law ............. 296
10.5.2 Gauss's Law for Magnetism ............. 297
10.5.3 Faraday's Electromagnetic Induction Law ............. 299
10.5.4 Ampère–Maxwell Law ............. 300
10.6 Space-Time Representation of Maxwell's Equations ............. 301
10.6.1 Visualizing Space-Time ............. 303
10.6.2 Geometric View of Maxwell's Equations ............. 304
10.7 Algebraic Formulation ............. 306
10.8 Classification Diagrams of Electromagnetism ............. 307

**11 Mechanics of Deformable Solids** ............. 325
11.1 Introduction ............. 325
11.2 Fundamental Problem ............. 325
11.3 Source Variables ............. 326
11.3.1 Impulse of Volume Forces ............. 326
11.3.2 Impulse of Surface Forces ............. 327
11.3.3 Momentum ............. 327
11.3.4 Momentum Content, Flow and Production ............. 328
11.3.5 Stress Vector, Stress Tensor, Pressure ............. 328
11.4 Configuration Variables ............. 330
11.4.1 Initial Position Vector ............. 331
11.4.2 Relative Position Vector ............. 331
11.4.3 Position Vector ............. 331
11.4.4 Displacement ............. 331
11.4.5 Relative Displacement ............. 333
11.4.6 Displacement Gradient ............. 334
11.4.7 Strain Tensor ............. 334
11.4.8 Velocity ............. 335
11.5 Field Laws ............. 335
11.6 Rod Traction ............. 335

11.7 Vibrations in One Dimension ... 339
11.8 Classification Diagrams of Deformable Solids ... 339

**12 Mechanics of Fluids** ... 355
12.1 Particles and Points ... 355
12.2 Some Peculiarities of the Fluid Field ... 356
12.3 Fundamental Problem ... 357
12.4 Fluids and Flows ... 358
12.4.1 Kinds of Fluids ... 358
12.4.2 Kind of Flows ... 359
12.5 Variables Used in the Steady Motion of a Perfect Fluid ... 360
12.6 Source Variables ... 360
12.6.1 Volume and Surface Forces and Their Impulses ... 361
12.6.2 Mass Content and Mass Density ... 361
12.6.3 Mass Flow, Mass Current and Mass Current Density ... 362
12.6.4 Stream Function ... 362
12.6.5 Stream Vector ... 363
12.6.6 Mass Production, Mass Source ... 364
12.6.7 Momentum Content and Momentum Density ... 364
12.6.8 Momentum Flow and Momentum Current ... 364
12.6.9 Momentum Production and Momentum Source ... 364
12.7 Configuration Variables ... 365
12.7.1 Line Integral of Velocity and Velocity Potential ... 365
12.7.2 Vortex Flux and Vorticity ... 367
12.7.3 Relative Velocity and Velocity Gradient ... 367
12.7.4 Strain Rate Tensor and Volume Dilatation Rate ... 368
12.7.5 Global Form of Mass Balance ... 369
12.7.6 Global Form of Momentum Balance ... 370
12.8 Constitutive Laws ... 372
12.9 Classification Diagrams of Fluid Dynamics ... 372

**13 Other Physical Theories** ... 385
13.1 Equilibrium Thermodynamics ... 385
13.2 Non-equilibrium Thermodynamics ... 386
13.2.1 Internal Energy ... 386
13.3 Thermal Conduction ... 387
13.3.1 Fundamental Problem ... 387
13.3.2 Source Variables: Space and Time Classification ... 387
13.3.3 Configuration Variables: Space and Time Classifications ... 388
13.4 Gravitational Field ... 389
13.4.1 Relativistic Gravitation ... 392
13.5 Quantum Mechanics ... 392

**Part III Advanced Analysis**

**14 General Structure of the Diagrams** ... 415
14.1 Introduction ... 415

**15 The Mathematical Structure** ... 429
15.1 Introduction ... 429
15.1.1 Discovery of Adjointness ... 431
15.1.2 Topological Equivalent of Algebraic Formulation ... 433
15.2 From Differential Operators to Algebraic Operators ... 436
15.2.1 Differential Operators: Some Specifications ... 437
15.2.2 Algebraic Equivalent of Differential Formulation ... 439
15.3 Physics Needs Couples of Vector Spaces ... 440
15.3.1 Bilinear Forms as Scalar Products ... 442
15.3.2 From Transposed Matrix to Adjoint Operator ... 444
15.3.3 Inhomogeneous Boundary Condition: Convex Set ... 445
15.3.4 Adjoint Operator ... 446
15.3.5 Further Extension of Notion of Adjoint Operator ... 447
15.3.6 Role of Boundary Conditions ... 448
15.3.7 Symmetric and Self-Adjoint Operators ... 449
15.3.8 Operators at the Same Level Are Mutually Adjoint ... 450
15.3.9 Formal Operator 'curl' is Self-Adjoint ... 452
15.4 The Three Kinds of Partial Differential Equations ... 455

**A Affine Vector Fields** ... 457
A.1 Affine Fields ... 457
A.1.1 Affine Scalar Field ... 458
A.1.2 Affine Vector Field ... 459

**B Tensorial Notation** ... 467
B.1 Summary of Tensorial Notation Used in This Book ... 467
B.1.1 Generalized Kronecker Delta ... 469
B.1.2 Permutation Symbol ... 470
B.1.3 Main Use of Levi-Civita Pseudotensor ... 472
B.1.4 Vector Components ... 472
B.2 Algebraic and Metric Duals ... 473
B.3 Bivectors ... 474
B.3.1 Exterior Product of Two Vectors ... 475

**C On Observable Quantities** ... 477

**D History of the Diagram** ... 481
D.1 Historical Remarks ... 481

**E List of Physical Variables** ........................................ 485

**F List of Symbols Used in This Book** ................................. 493

**References** ........................................................... 505

**Index** ................................................................ 513

# List of Diagrams


| *Diagram* | *Description* | *Page* |
|---|---|---|
| [AME1] | Analytical mechanics of a particle | 270 |
| [AME2 ] | Analytical mechanics, *Lagrange's and Hamilton's equations* | 271 |
| | Analogies | |
| [ANA1] | Analogy between optics and mechanics | 403 |
| [ANA2] | Analogy between translatory and rotatory motion | 404 |
| [ANA3] | Analogies among different physical theories | 405 |
| | Electromagnetism | |
| [ELE1] | Electrostatics, *vector notation* | 310 |
| [ELE2] | Magnetostatics, *vector notation* | 311 |
| [ELE3] | Electromagnetism, *vector notation* | 312 |
| [ELE4] | Electrostatics, *tensor notation* | 313 |
| [ELE5] | Magnetostatics, *tensor notation* | 314 |
| [ELE6] | Electromagnetism, *space-time formulation* | 315 |
| [ELE7] | Plane electrostatics using complex variables | 316 |
| [ELE8] | Charged particle in an electromagnetic field | 317 |
| [ELE9] | Energetics of the electric field | 318 |
| [ELE10] | Network theory | 319 |
| [ELE12] | Coulomb's law from the differential formulation | 321 |
| [ELE13] | Charged particle in the electromagnetic field | 322 |
| [ELE14] | Electromagnetism: exterior differential forms | 323 |
| | Fluid dynamics | |
| [FLU1] | Perfect fluid, plane motion, *Cartesian notation* | 373 |
| [FLU2] | Plane motion of fluid using complex variables | 374 |
| [FLU3] | Stationary motion of perfect fluid, *vector notation* | 375 |
| [FLU4] | Stationary motion of perfect fluid, *tensor notation* | 376 |

| | | |
|---|---|---|
| | % Fluid dynamics | |
| [FLU5] | Acoustics in fluids | 377 |
| [FLU6] | Viscous fluid, *Cartesian notation* | 378 |
| [FLU7] | Hydraulics: percolation in porous media | 379 |
| [FLU8] | Perfect fluid: Euler equation | 380 |
| [FLU9] | Viscous fluid motion in a pipe: Poiseuille's law | 381 |
| [FLU10] | Fluid motion: convective term | 382 |
| [FLU11] | Soap films (liquid membrane) | 383 |
| | General structure of classification diagram | |
| [GEN1] | Even and odd differential forms in three dimensions | 417 |
| [GEN2] | Mathematical objects in three dimensions, *inner orientation* | 418 |
| [GEN3] | Mathematical objects in three dimensions, *outer orientation* | 419 |
| [GEN4] | From cochains to standard diagram | 420 |
| [GEN5] | Mathematical objects in three dimensions: differential notation | 421 |
| [GEN6] | Standard diagrams and de Rham complexes | 422 |
| [GEN7] | General classification diagram (cochains) | 423 |
| [GEN8] | Even and odd differential forms in space-time | 424 |
| [GEN9] | Mathematical objects in space-time, *inner orientation* | 425 |
| [GEN10] | Mathematical objects in space-time, *outer orientation* | 426 |
| [GEN11] | Space-time diagram | 427 |
| [GEN12] | Standard diagrams for $n = 1, 2, 3, 4$; *tensorial notation* | 428 |
| | Gravitation | |
| [GRA1] | Classical gravitational field | 401 |
| [GRA2] | Relativistic gravitation | 402 |
| | Interactions | |
| [INT3] | Thermoelectricity | 406 |
| | Particle dynamics | |
| [PAR1] | Particle dynamics | 268 |
| [PAR2] | Relativistic particle dynamics | 269 |
| | Quantum mechanics | |
| [QME1] | Quantization process | 407 |
| [QME2] | Spin-zero particles, *space formulation* | 408 |
| [QME3] | Spin-zero particles, *space-time formulation* | 409 |
| [QME4] | Spin-one particles, *space-time formulation* | 410 |
| [QME5] | Bilinear covariants | 411 |
| [QME6] | Elementary particles | 412 |

| Mechanics of deformable solids | | |
|---|---|---|
| [SOL0] | Traction of rods | 340 |
| [SOL1] | Bending of beams | 341 |
| [SOL2] | Torsion of beams | 342 |
| [SOL3] | Shear of beams | 343 |
| [SOL4] | Theory of beams: Bernoulli–Euler | 344 |
| [SOL5] | Theory of beams: Timoshenko theory | 345 |
| [SOL6] | Plane elasticity | 346 |
| [SOL7] | Plates: plane load | 347 |
| [SOL8] | Plates: normal load | 348 |
| [SOL9] | Elastostatics | 349 |
| [SOL10] | Longitudinal vibrations of rods | 350 |
| [SOL11] | Torsional vibrations of rods | 351 |
| [SOL12] | Elastodynamics | 352 |
| [SOL13] | Statics of strings, *small deflections* | 353 |
| [SOL14] | Transverse vibrations of strings | 354 |
| **Thermal conduction** | | |
| [TCO1] | Steady thermal conduction in solids | 397 |
| [TCO2] | Thermal conduction in solids, *entropy representation* | 398 |
| [TCO3] | Thermal conduction in solids, *energy representation* | 399 |
| [TCO4] | Stationary thermal conduction, *algebraic formulation* | 400 |
| **Thermodynamics** | | |
| [THE1] | Thermodynamics, *Legendre's transform* | 394 |
| [THE2] | Thermodynamics, *entropy representation* | 395 |
| [THE3] | Thermodynamics, *energy representation* | 396 |
| **Various topics** | | |
| [ODD] | Oddness principle | 124 |
| [PDE] | Three kinds of partial differential equations | 455 |
| [VIB] | Transverse vibrations of strings, *space + time* | 228 |

# List of Figures

1.1 Cell complex in space (*light lines*) and its dual (*heavy lines*) .... 5
1.2 *Left* Classification of eight oriented space elements. *Right* Corresponding classification of physical variables of a physical theory .... 6
1.3 Generation of space-time diagram from the space diagram ...... 7
1.4 Topological notions at origin of three main differential operators . 15
1.5 The four mathematical formulations of physics: three of them are known .... 18

2.1 (a) The change of $t$ into $-t$ means the reflection of the time axis. (b) Reversal of order of time instants, i.e. reversal of motion .... 35

3.1 The four space elements: points, lines, surfaces, volumes ....... 39
3.2 What is commonly meant by 'oriented' space elements, but there are other kinds of orientation .... 40
3.3 Simplexes of one, two and three dimensions. Recall that the term *simplex* denotes the most simple polygon, i.e. the triangle, and the most simple polyhedron, i.e. the tetrahedron .... 42
3.4 The oriented area of the polygon is the sum of the oriented areas of the triangles with a common vertex .... 43
3.5 The definite integral of a function of one variable has a geometrical representation of the oriented area under the representative curve of the function .... 44
3.6 Orientation of a line: (a) inner orientation; (b) outer orientation .. 45
3.7 Orientation of a surface: (a) inner orientation; (b) outer orientation 45
3.8 Pressure (p)–volume (V) diagram: the orientation of the area is linked with the sign of the work (W). (a) positive area implies a positive work; (b) a negative area implies a negative work ...... 46
3.9 Orientation of a surface and a volume .... 46

3.10 Outer orientation of a surface represented by two colours on two sides, *white* and *black* .......... 47
3.11 Orientation of a volume: (a) inner orientation; (b) outer orientation 47
3.12 The inner orientation of a cube is equivalent to the choice of a right-handed or a left-handed helix .......... 48
3.13 (a) A fountain is a source, whereas a manhole is a sink. (b) A point with an inner orientation means that it is conceived as a source or a sink .......... 49
3.14 (a) The definition of the increment of a function implies that points have been oriented as sinks. (b) The radius vector is always oriented from the origin of a reference system towards a point, and hence the point is oriented as a sink .......... 50
3.15 Orientation of a point: (a) inner orientation; (b) outer orientation . 51
3.16 The inner orientation of a space element induces an outer orientation of the dual element and vice versa .......... 52
3.17 Fluid flow in plane motion .......... 52
3.18 Outer orientation of a line: (a) in three-dimensional space; (b) in a plane; (c) on a line .......... 53
3.19 Outer orientation of a surface element: (a) in three-dimensional space; (b) in a plane .......... 53
3.20 Outer orientation of a point: (a) in three-dimensional space; (b) in a plane; (c) on a line .......... 54
3.21 Relation between inner and outer orientations of coordinate axes: (a) in two dimensions: the outer orientation of one axis coincides with the inner orientation of the other and viceversa; (b) in three dimensions: the inner orientation of a coordinate plane coincides with the outer orientation of the third axis .......... 54
3.22 Picture showing how the inner and outer orientations of a cube in three dimensions are projected in two dimensions and in one dimension .......... 55
3.23 The outer orientation of a space element depends on the dimensions of the embedding space. The windows must be read in rows .......... 55
3.24 The orientation of a space element induces a consistent orientation to its boundary .......... 56
3.25 The propagation of the outer orientation of a volume leads to an opposite orientation of the holes .......... 56
3.26 Two time elements: time instants and intervals .......... 57
3.27 Incidence numbers in a time complex .......... 58
3.28 The two kinds of orientation of time elements using signs .......... 58
3.29 A primal and a dual cell complex on the time axis: four time elements .......... 59

3.30 (a) Outer orientation of surface in space compared with (b) outer orientation of time interval .................... 60
3.31 This is the convention used in books on the strength of materials . 60
3.32 Time intervals endowed with outer orientation ............... 61

4.1 Geometry is the indispensable bridge between physics and mathematics ............................................ 64
4.2 Cartesian cell complexes in two and three dimensions .......... 65
4.3 Cells in main coordinate systems: (a) Cartesian; (b) cylindrical; (c) spherical ............................................. 66
4.4 Families of points, lines, surfaces and volumes of a Cartesian coordinate system in three-dimensional space ................ 66
4.5 Families of points, lines and surfaces of a Cartesian coordinate system in two-dimensional space ......................... 67
4.6 Pascal's triangle gives the number of families of elements in spaces of various dimensions .......................... 68
4.7 A simplicial complex: (a) in two dimensions; (b) in three dimensions ......................................... 69
4.8 Space region containing different materials discretized by a simplicial complex and its dual ....................... 69
4.9 The temperature of a room is measured at its centre ........... 70
4.10 Cartesian cell complexes in two and three dimensions: the dual complex is shown in *thick lines*, the primal complex in *thin lines* . 70
4.11 Cell complex and its dual for plate under normal load .......... 71
4.12 Dual of a cell complex: (a) Voronoi dual; (b) barycentric dual ... 71
4.13 Around each node one can select the dual cell as a tributary region: (a) Voronoi dual; (b) barycentric dual ................ 72
4.14 Circumcentre (*left*) and barycentre (*right*) of a triangle ......... 73
4.15 Region around an electric wire decomposed using a simplicial complex (*left*) and its barycentric dual (*right*) ................. 74
4.16 (a) Domain composed of different materials. (b) Simplicial complex. (c) Its barycentric dual ........................... 74
4.17 The dual of a 1-cell in three dimensions is a face composed of many flat small faces ..................................... 75
4.18 Oriented graph, also called a *directed* graph. (a) The 0-cells (nodes) all have the same orientation: e.g. sinks. (b) The 1-cells (edges or branches) have arbitrary orientations. (c) The 2-cells (loops) have a compatible orientation ......................... 75
4.19 Oriented cell complexes. (a) Oriented Cartesian complex. (b) Oriented simplicial complex ............................. 76
4.20 Oriented cell complexes and their dual: (a) oriented Cartesian complex; (b) oriented simplicial complex and its Voronoi dual ... 77

4.21 Primal and dual cell complexes and their orientation: (**a**) in three dimensions; (**b**) in two dimensions; (**c**) in one dimension... 77
4.22 The inner orientation of an element of the primal complex induces an outer orientation on the corresponding element of the dual ... 77
4.23 (a) The inner orientation of a 0-cell $\overline{\mathbf{P}}$ as a source (*outgoing arrows*) induces on the dual cell $\tilde{\mathbf{S}}$ an outer orientation (*outward arrows*). (b) The inner orientation of the 1-cell $\overline{\mathbf{L}}$ of the primal induces an outer orientation on the 1-cell of the dual ... 78
4.24 A primal and a dual cell complex useful for performing numerical integration and numerical derivation ... 79
4.25 The reason for the introduction of the dual cell complex shown on a stalactite ... 80
4.26 The reinterpretation of the finite difference method ... 83
4.27 In a region of uniform heat, the heat current through a flat surface, for any surface orientation, is proportional to the temperature gradient in the direction normal to the surface ... 83
4.28 The two kinds of space elements: there are eight in total ... 85
4.29 Dual elements in two- and one-dimensional space ... 85
4.30 Primal and dual cell complexes in three dimensions formed by a Cartesian coordinate system. The same diagrams are valid for a coordinate cell complex and its dual ... 86

5.1 Material and spatial descriptions. (a) system description; (b) spatial description ... 91
5.2 Balance equation in spatial formulation ... 94
5.3 Classification of variables of a physical theory ... 102
5.4 Relation between global, integral and field variables. A longer list of field variables which are also global variables is given in Table 5.7 ... 110
5.5 The global variables associated with points are continuous through the separation surface of two media. (a) The gravitational potential is continuous passing from the exterior to the interior of a massive sphere; (b) the electrical potential is continuous passing from the exterior to the interior of a metallic sphere; (c) the phase of a plane wave is continuous through the separation surface between two transparent media; (d) temperature is continuous through the separation surface of different materials; (e) the ray vector in optics is continuous in refraction; (f) the position vector is continuous in the rebound of a particle ... 114
5.6 (a) A polygon with non vanishing area; (b) a polygon with zero area ... 116

5.7 Two different attributes of men: weight and height . . . . . . . . . . . 120
5.8 The sign depends on the viewpoint: two refugees inside the embassy of nation $B$ inside nation $A$ . . . . . . . . . . . . . . . . . . . . . . 121
5.9 (a) The time needed to reach equilibrium $T^e$ is distinct from the registration time $T^r$. (b) The registration time $T^r$ for the position of a particle must be very short . . . . . . . . . . . . . . . . . . . . . . . . . . 128
5.10 The rebounding of a ball against a wall shows that the impulse does not change sign under a reversal of motion. (a) With forward motion the impulse given to the ball points to the left. (b) With backward motion the impulse also points to the left . . . . 132
5.11 The four families of physical variables: some are composed of four members, some of two members and some of a single member . . . . . . . . . . . . . . . . . . . . . . . . . . . . . . . . . . . . . . . . . . . . . . 140
5.12 Energy and energy densities as potentials of reversible constitutive equations . . . . . . . . . . . . . . . . . . . . . . . . . . . . . . . . . . . 150
5.13 The elementary unit of the diagram . . . . . . . . . . . . . . . . . . . . . . . 150
5.14 The three kinds of conjugation, with respect to power, energy and action . . . . . . . . . . . . . . . . . . . . . . . . . . . . . . . . . . . . . . . . . . . . 151

6.1 Hooke's law for a bar: a reversible link . . . . . . . . . . . . . . . . . . . . . 163
6.2 Hooke's law for simple traction: a reversible link . . . . . . . . . . . . . . 163
6.3 Ohm's law for a wire: an irreversible process . . . . . . . . . . . . . . . . . 164
6.4 Ohm's law for a wire: an irreversible phenomenon . . . . . . . . . . . . . 165
6.5 Space and time elements and their boundaries . . . . . . . . . . . . . . . . 168

7.1 Faces and cofaces of 0-, 1-, 2-, 3-cells respectively in a Cartesian complex . . . . . . . . . . . . . . . . . . . . . . . . . . . . . . . . . . . . . . . . . 186
7.2 Incidence numbers between a cell and its faces . . . . . . . . . . . . . . . 189
7.3 Incidence numbers between cells of a cell complex and its dual . . 189
7.4 One-dimensional chains: (a), (b), (c), (d) are different 1-chains on the same complex . . . . . . . . . . . . . . . . . . . . . . . . . . . . . . . . . . . . . . 191
7.5 Pictorial view of boundary process . . . . . . . . . . . . . . . . . . . . . . . . 194
7.6 The boundary of a boundary is a null chain. This pictorial view illustrating a process, similar to fountains, which transfers the coefficient $n$ of the 2-cell in the chain to the 1-cells that form the edge, and from these, in turn, to the 0-cells that form the edge thereof . . . . . . . . . . . . . . . . . . . . . . . . . . . . . . . . . . . . . . . . . . . . . 195
7.7 Computing the boundary of a boundary we obtain systematically zero . . . . . . . . . . . . . . . . . . . . . . . . . . . . . . . . . . . . . . . . . . . . . . . . 195
7.8 The simplest, one-dimensional, cell complex allows one to capture the algebraic root of the generalized Stokes theorem . . . . 205
7.9 (*Left*) A cell complex; (*right*) its dual . . . . . . . . . . . . . . . . . . . . . . . 210
7.10 Coboundary of a discrete 0-form in a two-dimensional complex . 210

7.11 Coboundary process from a discrete 1-form to a discrete 2-form on dual complex ... 211
7.12 The coboundary process in one dimension, here on the time axis, is the discrete analogue of a derivative ... 213

8.1 Classification of space global variables ... 222
8.2 Corresponding classification of field variables ... 223
8.3 Classification of field variables enriched with corresponding relations ... 223
8.4 Equilibrium of a string ... 224
8.5 Differential formulation of statics of strings ... 226
8.6 Classification diagram for particle dynamics ... 227
8.7 Irreversible links are those which connect the boxes of primal time intervals with that of dual time intervals and are represented by *dotted lines*. For the page of each diagram see the index on p. xx ... 232
8.8 The same source gives rise to different configurations depending on the material medium ... 235
8.9 Behaviour of field functions under space and time transformations ... 236
8.10 Possible links between variables of a physical theory for one-, two-, three- and four-dimensional cases ... 237
8.11 The fundamental equation of a theory is that which links the potential with the source of the field ... 238
8.12 Paths which must be followed to compose fundamental equations 238

9.1 The momentum imparted on a body is the time integral of force, starting from rest ... 244
9.2 Calibration of an impulsometer. We shoot $n$ bullets in a box containing sand; the bullets are captured and stopped. The maximum displacement is the measure of the impulse received by the box, and this is equal to the momentum of the bullet before impact ... 246
9.3 (a) Diagram $p - v$ in classical mechanics. (b) Corresponding diagram in relativistic mechanics ... 250
9.4 Classification of forces ... 253
9.5 Force generates work and impulse ... 255
9.6 System formed by two masses and a compressible spring ... 260
9.7 When a bird escapes from its cage, there is not only a mass flow, but also a flow of potential energy (internal and external) .. 260
9.8 Kinetic energy and kinetic co-energy in classical (*left*) and in relativistic (*right*) cases ... 263

10.1 Measure of electric flux on element of surface with outer orientation. $N$ is the normal, which shows the space direction for which the charge collected on the surface is maximum. This fact will be useful for introducing the electric displacement ..... 278
10.2 (a) In a uniform and static electric field, the electric flux associated with a flat surface is invariant under space and time translation. (b) The direction of D is that of maximum $\sigma$ ....... 280
10.3 Torsion balance to measure magnetic field strength. (a) what is significant for the intensity of the magnetic field is the product of the current for the number of loops, hence (2i) N = i (2N); (b) the operation of measurement of H.................... 285
10.4 (a) Compensating coil for measuring $F_m$ and $\check{H}$. (b) The sign of the current in the compensating coil must be such that the magnetic field inside the coil vanishes. (c) This can be tested with a Hall probe ........................................ 287
10.5 The scalar magnetic potential at a point depends on the outer orientation of the point ................................ 289
10.6 By removing the magnetic field a short pulse of current is induced in the coil (the probe). This short pulse implies an electromotive force impulse at the terminal of the coil.......... 292
10.7 (a) The electric field vector E measures the force acting on an electric charge. (b) The magnetic induction vector $\check{B}$ measures the force acting on a linear element of current ............... 294
10.8 The charge induced on the external surface of a metal shell is equal to the charge contained inside the surface ............... 296
10.9 Devices to prove Gauss' law: (*left*) for electrostatics; (*right*) for the magnetic field .......................................... 297
10.10 Gauss's law for magnetism: (a) one possible way to form three loops with a single wire; (b) and (c) two pairs of three loops; (d) the sum of two electromotive forces (hence of their impulses) vanishes ..................................... 298
10.11 Simplest devices for illustrating the four Maxwell equations (*top*) and the three constitutive equations (*bottom*)............. 301
10.12 Different projections of a cube in the plane: (a) central view; (b) lateral view; (c) axonometric view ......................... 303
10.13 Kinematic generation of a cube. (a) Perspective frontal view of cube. (b) Perspective lateral view of cube. (c) Axonometric lateral view of cube ....................................... 303
10.14 Kinematic views of back of a moving train. As the train approaches, its shape expands and shrinks as it moves away. A reversal of motion changes an explosion into an implosion.... 304
10.15 Geometric support for Faraday's law in a two-dimensional space-time ................................................ 304

10.16 (a) A square is formed by two segments. (b) A cube is formed from three squares. (c) A hypercube is formed from one cube and three truncated pyramids ........................ 304
10.17 A hypercube of a four-dimensional space projected into a three-dimensional space and represented in the two-dimensional space of the paper. The *arrows* on the time lines denote the two possible outer orientations of the hypercube .................. 305
10.18 Geometric interpretation of Gauss's law of magnetism ......... 305
10.19 Geometric interpretation of Faraday's law. This is one of the three equations, the one relative to the $z$-axis .................. 306
10.20 The eight scalar equations of the electromagnetic field are balance equations in space-time, one for each of the cubes that compose the four-dimensional hypercube. The figure illustrates the three-dimensional projection of an 'exploded' hypercube of space-time ................................................ 306

11.1 Two particles $\mathscr{P}$ and $\mathscr{Q}$ in the initial configuration occupy the positions P and Q. The same particles, at a later instant, occupy the positions P′ and Q′. The vectors in thin lines are position vectors, whereas those in thick lines are the displacements of the two particles $\mathscr{P}$ and $\mathscr{Q}$ ................................ 330
11.2 The displacement $\eta$ in particle dynamics: (a) corresponds to the displacement $\eta$ in continuum mechanics (b), but in continuum mechanics there is also the *total* displacement $\mathbf{u}$, i.e. the displacement from an initial configuration, taken as the reference configuration, which corresponds to the position vector $\mathbf{r}$ in particle mechanics ........................... 332

12.1 Perfect fluid motion: classification of space global variables..... 360
12.2 Definition of stream function in a plane fluid flow ............. 362
12.3 Velocity of two particles $\mathscr{P}$ and $\mathscr{Q}$, which at the instant $t$ are in the points P and Q. *Thick lines*: velocities; *thin lines*: position vectors ................................................ 368

15.1 Three stages in description of physical theories ............... 430
15.2 Heat flows horizontally only. The defining equations and the equation of balance perform the coboundary process on a one-dimensional cell complex and its dual ...................... 431
15.3 Coboundary process on primal and dual cell complexes in one dimension ............................................... 435
15.4 Primal and dual subdivision of an interval [a,b] ............... 439
15.5 Two couples of spaces placed in duality by non-degenerate bilinear forms. The formulae on the right part will be explained later ..................................................... 441

15.6 (*Left*): linear operator $L : \mathsf{U} \mapsto \mathsf{U}$; (*right*) linear operator $L$ between two isomorphic spaces $L : \mathsf{U} \mapsto \mathsf{V}$ .................. 446
15.7 Straight line segment connecting two elements in a function space .................................. 446
15.8 ............................................ 456

A.1 The gradient of an affine function is orthogonal to the equipotential planes .................................. 458
A.2 The variation of a vector field along a straigth line. (a) The tangential and the normal components vary linearly; (b) The vector in the midpoint of a line segment is the mean of the vectors at its ends .................................. 460
A.3 The sum of the circulations along the boundary of every triangle is equal to the circulation along the boundary of the entire polygon .................................. 461
A.4 In an affine vector field the circulations of the vector along the boundary of all *triangles* of equal area are equal. In another words, the circulation of the vector along a closed line is proportional to the area enclosed. The *small filled circles* denote the application points of vectors, while the line segments, without arrows, denote the vectors .......................... 462
A.5 The curl of an affine vector field (*heavy lines*) describes a uniform vector field .................................. 462
A.6 (a) Flux across a plane surface. (b) Flux across boundary of a cube .................................. 463

B.1 *Left*: A bivector is a parallelogram with an inner orientation; *right*: a couple (mechanics) can be described by a bivector ...... 475
B.2 The exterior product of two vectors generates a bivector. Inverting the order of the two vectors changes the sign of the bivector .................................. 476

D.1 The Roth diagram of electrical circuits ...................... 482
D.2 1966: Branis uses Roth's diagram formed by chains and cochains 483
D.3 1966: Branin's diagram for electromagnetic field .............. 483
D.4 1981: Deschamps' proposed diagram of electromagnetism ...... 484

# List of Tables


| | | |
|---|---|---|
| 1.1 | The 32 combinations of space and time elements | 3 |
| 1.2 | Two classification diagrams for variables of physical theories | 7 |
| 1.3 | Two kinds of equations of a physical field: field equations and constitutive equations | 8 |
| 2.1 | Relation between rates and time derivatives | 28 |
| 2.2 | Definite and indefinite line and time integrals | 29 |
| 2.3 | Modern equivalent of various quantities improperly called 'forces' | 31 |
| 2.4 | Unambiguous terminology for *flux* and its related terms | 32 |
| 2.5 | Reversal of motion and time reversal | 36 |
| 3.1 | The eight oriented space elements | 41 |
| 3.2 | Adjectives referring to motion: the + sign is conventional | 62 |
| 3.3 | Inner and outer orientation of time instants | 62 |
| 4.1 | Summary of symbols and names used to denote space elements | 64 |
| 4.2 | Cells of the same dimension of a coordinate cell complex can be grouped into families | 67 |
| 4.3 | Summary of symbols used to denote space elements endowed with inner or outer orientation | 79 |
| 4.4 | Two possible classification diagrams of time elements | 87 |
| 5.1 | The terminology used by different authors about the two descriptions of continuum mechanics | 92 |
| 5.2 | Passing from a *material* to a *spatial* description global physical variables are split into *content* and *flow* | 94 |
| 5.3 | A classification of physical quantities | 97 |
| 5.4 | Physical parameters | 98 |
| 5.5 | Energy variables of mechanics | 100 |

5.6 The three kinds of variables of all physical theories: configuration, source and energy variables .................. 103
5.7 The two kinds of field variables .......................... 106
5.8 Connection between various classifications used in physics and in engineering ............................................ 112
5.9 Conjugate physical variables: the product *intensive* × *extensive* gives an energy; the product *across* × *through* gives a power .... 113
5.10 Examples of associations of physical variables with space elements .................................................. 114
5.11 Global variables in space ............................................ 115
5.12 Analogies between four physical fields ........................ 117
5.13 Examples of opposite attributes: the + sign is conventional...... 119
5.14 Paradigm to find the association of a physical variable with its space and time element. Filled circles denote space global variables, empty circles time global variables ................. 119
5.15 Procedure to find time association ........................... 130
5.16 Global variables in time associated with time intervals which are increments of other global variables in time associated with instants ................................................................ 134
5.17 The two kinds of field variables as functions of time ........... 134
5.18 Global variables and corresponding field functions ............ 138
5.19 Pseudovectors and pseudoscalars ................................ 147
5.20 Conjugate variables with respect to power are linked by irreversible constitutive equations.......................... 151
5.21 Conjugate variables with respect to energy (and energy density) . 152

6.1 Various kinds of basic equations ............................... 157
6.2 Fundamental equations of the main physical theories .......... 158
6.3 Constitutive equations in the case of affine behaviour of a 'potential'. $A$ is the area of a plane section and $d$ is the distance between two points P and Q ........................... 161
6.4 Constitutive equations ............................................ 166
6.5 Some interaction equations ....................................... 167
6.6 Some defining equations .......................................... 178
6.7 Examples of equations of behaviour............................ 179

7.1 Partition of topology ................................................ 185
7.2 Incidence numbers of a planar cell complex ................... 187
7.3 Incidence numbers of dual of planar cell complex of Fig. 7.2 .... 188
7.4 Incidence numbers of three-dimensional cell complex and its dual 190
7.5 Shopping in a supermarket provides an analogue of chains and cochains ................................................................ 200
7.6 The two steps of the coboundary process performed on primal cell complex ....................................................... 201

7.7 The two steps of the coboundary process performed on dual cell complex . . . 202
7.8 Various forms of Gauss's law of electrostatics . . . 218
7.9 Various forms of equilibrium equation in continuum mechanics . 219

8.1 Material parameters are contained in constitutive equations . . . 234

9.1 Main variables of particle dynamics . . . 242
9.2 Association of mechanical variables with time elements . . . 247
9.3 Notions of kinetic energy and kinetic co-energy in the literature . 264
9.4 Classification diagram for particle dynamics . . . 265
9.5 From field variables to global variables of particle mechanics . . . 266
9.6 A paradigm for the time association . . . 267

10.1 Field variables of electromagnetism and corresponding global variables . . . 276
10.2 Tensorial nature of space-time global variables and of field variables of electromagnetism . . . 288
10.3 Global physical variables expressed as integrals. The global variable on the left side all have the same physical dimension, that of the magnetic flux; those on the right also have the same physical dimension, that of the charge . . . 295
10.4 From the verbal statement to the differential formulation of Gauss's law . . . 297
10.5 From the verbal statement to the differential formulation of Gauss's law of magnetism . . . 298
10.6 From the verbal statement to the differential formulation of Faraday's law . . . 299
10.7 From the verbal statement to the differential formulation of Ampère's law . . . 300
10.8 Various descriptions of Maxwell's equations . . . 302
10.9 Classification diagram for field variables of electromagnetism: *front*: electrostatics; *rear*: magnetostatics . . . 308
10.10 From field variables to global variables of electromagnetism . . . 309

11.1 Fundamental equations of continuum mechanics . . . 326
11.2 From global variables to field functions in the mechanics of deformable solids . . . 327
11.3 Source variables of continuum mechanics . . . 329
11.4 The term *displacement* and its meanings . . . 333
11.5 Configuration variables in the mechanics of deformable solids . . . 336
11.6 From field variables to global variables in continuum mechanics . 337
11.7 Diagram of discrete analysis of the deformation of a rod under axial loading . . . 338

12.1 The two possible associations of velocity in fluid dynamics ..... 357
12.2 The many kinds of fluids and flows ..... 359
12.3 Perfect fluid motion: classification of field variables ..... 361
12.4 Configuration variables in fluid dynamics ..... 366
12.5 From verbal statement to differential formulation via global formulation ..... 370
12.6 From verbal statement to differential formulation via global formulation ..... 371

13.1 First principle of thermodynamics ..... 387
13.2 Steady thermal conduction: classification of space global variables ..... 390
13.3 Steady thermal conduction: classification of field variables ..... 390
13.4 From field variables to global variables in thermal conduction ..... 391

14.1 Link between present classification diagram and de Rham theory 416

15.1 The balance matrix is the transpose of the definition matrix with a minus sign ..... 432
15.2 One dimensional, stationary, fluid flow ..... 434
15.3 Main bilinear forms ..... 445
15.4 Algebraic version of three kinds of boundary value problem ..... 449
15.5 Configuration side of diagrams [FLU1] and [ELE1] ..... 450
15.6 Source side of diagrams [FLU1] and [ELE1] ..... 451
15.7 Configuration side of diagrams [FLU2] and [ELE3] ..... 453
15.8 Source side of diagrams [FLU2] and [ELE3] ..... 454

B.1 The six kinds of scalars ..... 473
B.2 Main formulae involving gradient, curl and divergence ..... 474

# Chapter 1
# Introduction

> ... the aim of mathematical physics is not only to facilitate for the physicist the numerical calculation of certain constants or the integration of certain differential equations. It is besides, it is above all, to reveal to him the hidden harmony of things in making him see them in a new way.
>
> (H. Poincaré, *The Value of Science*, Dover, 1958, p. 79)

## 1.1 Aim of the Book

The purpose of this book is to demonstrate the existence of a mathematical structure that is common to all physical theories of the macrocosm and to explain the origin of this common structure. The starting point of this investigation is the analysis of physical variables under a new profile: we take into consideration all those geometric features that are usually overlooked in physics books.

A detailed analysis of physical variables and of equations of the theories of the macrocosm makes it possible to put forward as evidence a natural association of physical variables with elementary geometric elements, such as points, lines, surface and volumes. This association makes it possible to build a classification diagram of physical variables and equations *that is the same for all theories, both classic and relativistic*.

The association of 'global' variables with space elements is a new perspective in the description of physics and is the *raison d'être* of this book.

The revelation of a common mathematical structure arose from the ardent desire to explain why analogies exist between physical theories that are very different in their physical content. The book gives an answer to this question and explains *why* all these theories have this common structure.

**(a) The role of global variables.** As is obvious, the mathematical description of physics relies on the very existence of quantitative attributes of physical systems, a fact that makes possible the introduction of physical quantities. Since

E. Tonti, *The Mathematical Structure of Classical and Relativistic Physics*, Modeling and Simulation in Science, Engineering and Technology, DOI 10.1007/978-1-4614-7422-7_1, © Springer Science+Business Media New York 2013

these quantities are linked by equations, we are accustomed to detecting analogies between different physical theories through the similarity of the equations that describe their laws. Contrary to this practice, we have realized that to explain the origin of the analogies, we must not start from the similarities of differential equations, but from 'global' physical variables, i.e. those variables that are neither a density nor a rate of other variables (Chap. 5). We must investigate their origins, their operative definitions and the role they play in the corresponding theory. The presentation of a physical theory by starting with global variables, instead of field variables, brings out a simple *topological structure* that the differential formulation is not able to show.

What is remarkable is that, in general, global variables arise *directly* from physical measurements: this fact is in contrast to our practice of deducing such global variables from space and time integration of field variables, a fact that leads us to call them *integral variables*. The use of global variables from the very beginning leads to the *global formulation* and then to the *algebraic formulation* of physical theories, while the use of field variables leads to the *differential formulation*. The global formulation must precede the differential formulation because it does not impose those purely mathematical restrictions on the field functions that are indispensable to perform the derivatives but that are not required on physical grounds.

Global variables are linked with domains such as *volumes* **V**, *surfaces* **S** and *lines* **L**: it is here that topology enters the scene. But we will see that there are also global variables associated with *points* **P**, e.g. temperature and electric potential. We will call **P**, **L**, **S** and **V** space elements, and we will say that a variable is *global in space* when it is not the volume density, surface density, or line density of another variable. Global variables can also be associated with *instants* **I** and *intervals* **T**. We will call **I** and **T** time elements, and we will say that a variable is *global in time* when it is not the rate of another variable.

We will show that this association of global variables with space and time elements is a general property of all physical theories of the macrocosm. This association is intrinsic to the very definition of each physical variable and, when the variable is measurable, is reflected in its measurement process. The field functions, which arise as *densities* of space global variables and as *rates* of time global variables, inherit the association with space and time elements of the corresponding space or time global variable (Chap. 5).

This natural association makes it possible to describe physical theories without using the differential formulation from the outset. A formulation based on algebraic topology emphasizes the topological, geometric and mathematical structures that are common to all physical theories and that the differential formulation leaves in the shadow.

**Remark.** Regarding the microcosm, which is described by quantum mechanics and which replaces observables with operators, the author does not have sufficient knowledge to identify the source variables, nor to identify space and time elements with which these operators are associated. Despite this, the variables and the equations of relativistic quantum mechanics for particles with integer spin, such as those of Klein–Gordon and Proca, find their place in a diagram similar to that of the relativistic formulation of electromagnetism. Moreover, we remark that both Bohm and Schönberg used algebraic topology for quantum mechanics, as do we in this book.[1] We have included in Chap. 13 six tables dealing with quantum mechanics that are formally similar to the classification diagram presented here, with the purpose of inciting theoretical physicists to find their justification.

**(b) The two kinds of orientations of space and time elements.** In the association of global variables with space and time elements, the notion of *orientation* plays a key role. In fact, space and time elements can be equipped with two types of orientation: *inner* or *outer* orientation. It is shown the every global physical variable is associated with a space and a time element, each of which has an inner or an outer orientation (Chap. 3, p. 39).

We will denote the space and time elements endowed with inner orientation by placing a bar over the boldface, uppercase letters, i.e. $\bar{\mathbf{P}}, \bar{\mathbf{L}}, \bar{\mathbf{S}}, \bar{\mathbf{V}}, \bar{\mathbf{I}}, \bar{\mathbf{T}}$, and the space and time elements endowed with outer orientation by placing a tilde over the boldface, uppercase letters, i.e. $\tilde{\mathbf{P}}, \tilde{\mathbf{L}}, \tilde{\mathbf{S}}, \tilde{\mathbf{V}}, \tilde{\mathbf{I}}, \tilde{\mathbf{T}}$.

This association requires consideration of all the possible combinations between the global variables and the oriented space and time elements. As shown in Table 1.1, 32 couples are formed by an oriented space element and an oriented time element.

**Table 1.1** The 32 combinations of space and time elements

| | $\bar{\mathbf{I}}$ | $\bar{\mathbf{T}}$ |
|---|---|---|
| $\bar{\mathbf{P}}$ | $[\bar{\mathbf{I}}, \bar{\mathbf{P}}]$ | $[\bar{\mathbf{T}}, \bar{\mathbf{P}}]$ |
| $\bar{\mathbf{L}}$ | $[\bar{\mathbf{I}}, \bar{\mathbf{L}}]$ | $[\bar{\mathbf{T}}, \bar{\mathbf{L}}]$ |
| $\bar{\mathbf{S}}$ | $[\bar{\mathbf{I}}, \bar{\mathbf{S}}]$ | $[\bar{\mathbf{T}}, \bar{\mathbf{S}}]$ |
| $\bar{\mathbf{V}}$ | $[\bar{\mathbf{I}}, \bar{\mathbf{V}}]$ | $[\bar{\mathbf{T}}, \bar{\mathbf{V}}]$ |

| | $\tilde{\mathbf{I}}$ | $\tilde{\mathbf{T}}$ |
|---|---|---|
| $\tilde{\mathbf{P}}$ | $[\tilde{\mathbf{I}}, \tilde{\mathbf{P}}]$ | $[\tilde{\mathbf{T}}, \tilde{\mathbf{P}}]$ |
| $\tilde{\mathbf{L}}$ | $[\tilde{\mathbf{I}}, \tilde{\mathbf{L}}]$ | $[\tilde{\mathbf{T}}, \tilde{\mathbf{L}}]$ |
| $\tilde{\mathbf{S}}$ | $[\tilde{\mathbf{I}}, \tilde{\mathbf{S}}]$ | $[\tilde{\mathbf{T}}, \tilde{\mathbf{S}}]$ |
| $\tilde{\mathbf{V}}$ | $[\tilde{\mathbf{I}}, \tilde{\mathbf{V}}]$ | $[\tilde{\mathbf{T}}, \tilde{\mathbf{V}}]$ |

First group

| | $\tilde{\mathbf{I}}$ | $\tilde{\mathbf{T}}$ |
|---|---|---|
| $\bar{\mathbf{P}}$ | $[\tilde{\mathbf{I}}, \bar{\mathbf{P}}]$ | $[\tilde{\mathbf{T}}, \bar{\mathbf{P}}]$ |
| $\bar{\mathbf{L}}$ | $[\tilde{\mathbf{I}}, \bar{\mathbf{L}}]$ | $[\tilde{\mathbf{T}}, \bar{\mathbf{L}}]$ |
| $\bar{\mathbf{S}}$ | $[\tilde{\mathbf{I}}, \bar{\mathbf{S}}]$ | $[\tilde{\mathbf{T}}, \bar{\mathbf{S}}]$ |
| $\bar{\mathbf{V}}$ | $[\tilde{\mathbf{I}}, \bar{\mathbf{V}}]$ | $[\tilde{\mathbf{T}}, \bar{\mathbf{V}}]$ |

| | $\bar{\mathbf{I}}$ | $\bar{\mathbf{T}}$ |
|---|---|---|
| $\tilde{\mathbf{P}}$ | $[\bar{\mathbf{I}}, \tilde{\mathbf{P}}]$ | $[\bar{\mathbf{T}}, \tilde{\mathbf{P}}]$ |
| $\tilde{\mathbf{L}}$ | $[\bar{\mathbf{I}}, \tilde{\mathbf{L}}]$ | $[\bar{\mathbf{T}}, \tilde{\mathbf{L}}]$ |
| $\tilde{\mathbf{S}}$ | $[\bar{\mathbf{I}}, \tilde{\mathbf{S}}]$ | $[\bar{\mathbf{T}}, \tilde{\mathbf{S}}]$ |
| $\tilde{\mathbf{V}}$ | $[\bar{\mathbf{I}}, \tilde{\mathbf{V}}]$ | $[\bar{\mathbf{T}}, \tilde{\mathbf{V}}]$ |

Second group

More precisely, these 32 couples can be divided into two groups, each consisting of 16 elements. The 16 couples of the first group are those combinations of time and space elements both of which are endowed with the same kind of orientation, either inner or outer. The 16 couples of the second group are those combinations of time and space elements that are endowed with opposite orientations, one inner the other outer. These two groups are organized in Table 1.2 (p. 7).

[1] Bohm et al. [17]; Schönberg [202].

What is surprising is that *each physical variable of every physical theory (of the macrocosm) can be matched with one of these 32 couples.* This is somewhat similar to the fact that each crystal can be classified in one of the 32 classes of symmetry[2] or that each chemical element can be placed in one of the boxes of Mendeleev's table.

It has been found that the variables of physical theories that can be associated with the elements of the first group, i.e. the one on the left side of Table 1.1, belong to *mechanical theories*, whereas the variables of the physical theories that can be associated with the elements of the second group, on the right side of Table 1.1, belong to *field theories*. In this way we have obtained a classification of global variables that is valid for both the classic and relativistic versions of every theory. This fact reveals that the marriage between physics and mathematics is possible through the intermediation of topology and geometry. This is a natural consequence of the fact that physical phenomena arise in space.

**(c) Role of cell complexes.** The differential formulation, which is based on field variables (i.e. point variables), makes use of coordinate systems. The algebraic formulation, based on global variables, requires a proper reference structure. The need to consider oriented space elements to create the algebraic formulation suggests the need to introduce into the working region of a suitable reference structure whose elements are endowed with spatial extension. A *cell complex*, as defined in algebraic topology (usually in the restricted form of *simplicial complex*), provides the appropriate tool (Chap. 4).

The four kinds of space elements that make up a cell complex, i.e. *vertices*, *edges*, *faces* and *volumes*, can be considered as cells of different dimensions: they have zero, one, two, or three dimensions, respectively. Following the terminology of algebraic topology, we will denote these elements by the terms 0-cells, 1-cells, 2-cells and 3-cells, respectively. In general, we will speak about $p$-dimensional cells or $p$-cells for short. In particular, to recover the traditional differential formulation, one can use a cell complex formed by a coordinate system (p. 65).

Once all the $p$-cells of a complex are endowed with an inner orientation, we obtain a reference structure for those global variables that are associated with space elements with an *inner* orientation. Still lacking is a structure whose $p$-cells correspond to an *outer* orientation. In fact, we need to locate variables that are associated with space elements endowed with an *outer* orientation. The idea now is to use a second complex that is staggered with respect to the first complex (Fig. 1.1). This is the *dual* cell complex whose outer orientation is automatically induced by the inner orientation of the first complex denoted as *primal*.

Note that every vertex of the primal complex, a 0-cell, is contained in a 3-cell of the dual complex, and vice versa. Moreover, every edge of the primal complex, a 1-cell, is crossed by a face of the dual complex, a 2-cell, and vice versa; and so on. In general, every $p$-cell of the primal is contained into (or intersects or contains) a

[2] The coincidence of the number of symmetry classes in crystal classification and of the distinct space-time elements is purely a matter of chance.

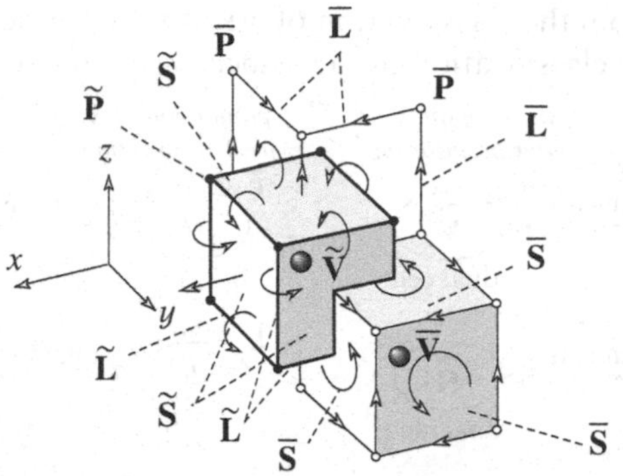


**Fig. 1.1** Cell complex in space (*light lines*) and its dual (*heavy lines*)

$(3-p)$-cell of the dual. Conversely, every $p$-cell of the dual, contains (or intersects or is contained into) a $(3-p)$-cell of the primal. A pair formed of a $p$-cell of one complex and of the corresponding $(3-p)$-cell of the other complex may be called a dual pair. One can view this correspondence by a pair of boxes (Fig. 1.2a). The boxes corresponding to each dual pair are on the same level. This figure represents the four kinds of cells (pieces of oriented space elements) with four elliptic boxes arranged vertically: four boxes for the primal cell complex and four boxes for the dual cell complex.

A great merit of a pair of cell complexes, a primal and a dual, is to enable a classification of the eight oriented space elements, four equipped with an inner orientation, $\overline{\mathbf{P}}, \overline{\mathbf{L}}, \overline{\mathbf{S}}, \overline{\mathbf{V}}$, and four with an outer orientation, $\widetilde{\mathbf{P}}, \widetilde{\mathbf{L}}, \widetilde{\mathbf{S}}, \widetilde{\mathbf{V}}$.

The differential formulation, while making implicit use of a cell complex, that formed by a coordinate system, lacks a support structure for variables associated with an outer orientation. In contrast, in the topological formulation, this role is played brilliantly by the dual complex.

**(d) Classification diagram for physical variables.** If we take into account the association of physical variables with the oriented space elements, then we can use the same classification diagram of space elements as a classification diagram for the physical variables associated with them. In fact, we can store the global variables in the appropriate elliptic boxes (Fig. 1.2b). In this way, the two cell complexes, endowed with the two kinds of orientation, become a *topological frame* for the classification of global variables in every physical theory of the macrocosm.

Since the field variables inherit the same association of the corresponding global variables, it is natural to put them in the same boxes as the relative global variables. This leads to the building of a classification diagram also for field variables and, hence, for the differential formulation of physical theories.

The physical variables of every physical theory can be divided into *configuration, source* and *energy* variables. Analysing the configuration variables we will see that they are associated with space elements endowed with an inner orienta-

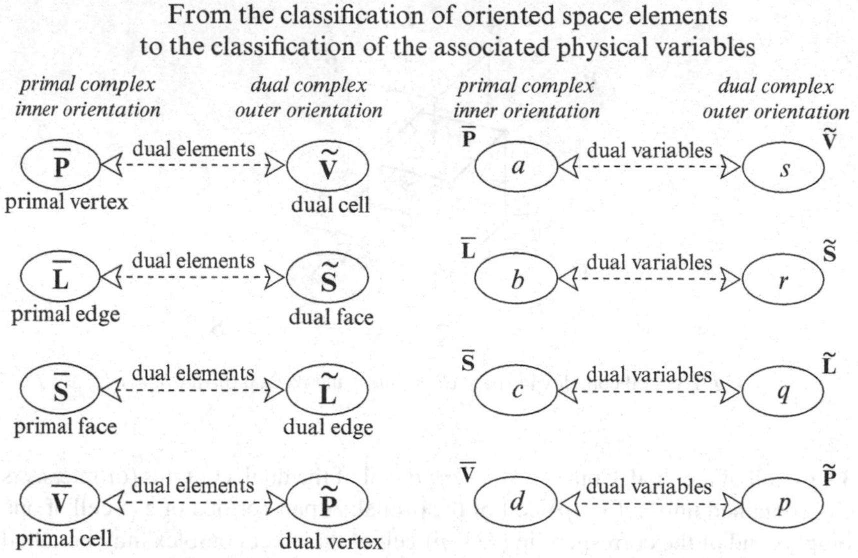


**Fig. 1.2** *Left* Classification of eight oriented space elements. *Right* Corresponding classification of physical variables of a physical theory

tion, whereas the source variables are associated with space elements endowed with an outer orientation.[3] It follows that the *configuration* variables can be associated with the cells of the *primal* complex, while the *source* variables can be associated with the cells of the *dual* complex.

Since global physical variables are associated both with an oriented space element and an oriented time element, it is useful to have a classification diagram that takes into account space and time elements. Such a diagram can be obtained by doubling the diagram of Fig. 1.2b and shifting the diagram to the rear, as shown in Fig. 1.3.[4]

The four different combinations of the oriented space and time elements shown in Table 1.1 can be organized in the two diagrams of Table 1.2.

**(e) Classification of equations.** What about the equations? The equations of every physical theory can be obtained by composing elementary equations of different types. These include *defining* equations, *topological* equations, *phenomenological* equations, and equations of *behaviour*. Since equations are links between the physical variables, they connect the rounded boxes. In the classification

[3] The reason for this astonishing correspondence is not clear, but we take it as an assumption supported by the evidence.

[4] See also Fig. 8.9 on p. 236.

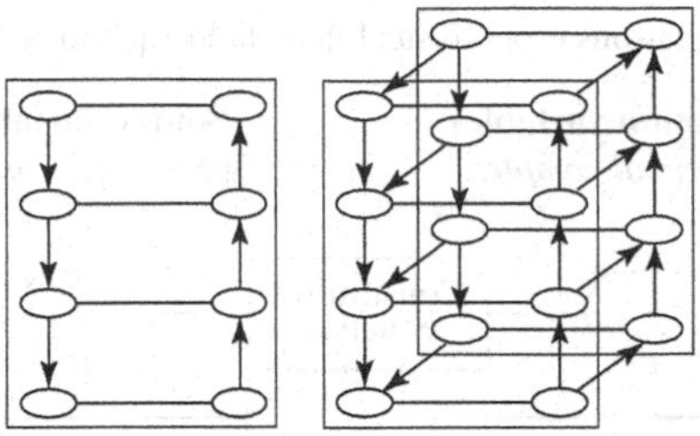

**Fig. 1.3** Generation of space-time diagram from the space diagram

**Table 1.2** Two classification diagrams for variables of physical theories

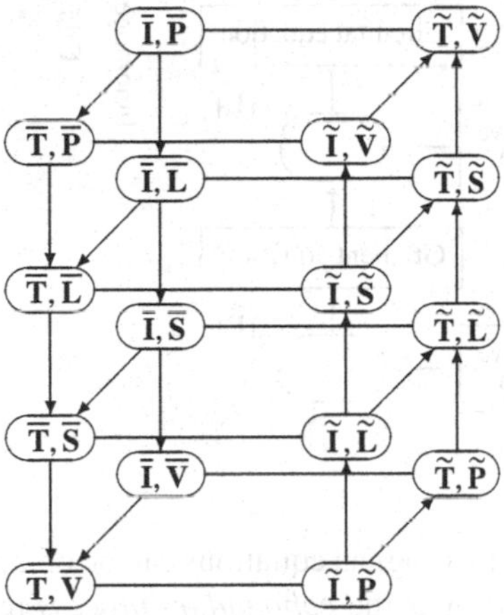


**Mechanical theories**

Particle mechanics, analytical mechanics, mechanics of deformable solids, fluid dynamics

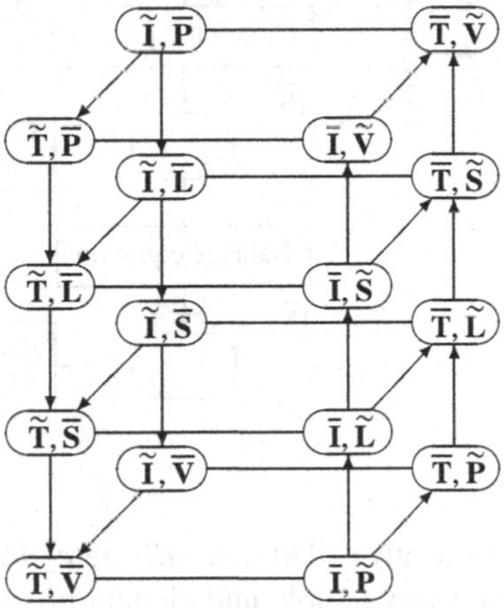


**Field theories**

Electromagnetism, gravitation, irreversible thermodynamics, thermal conduction, diffusion

diagrams, we will represent equations inside rectangular boxes (Table 1.3). We will show that:

- Topological equations connect the physical variables associated with cells of different dimensions of the same complex (primal or dual); they are indicated by arrows;
- Constitutive equations link the physical variables associated with cells in the primal complex to those associated with cells in the dual complex.

The association of global variables of a physical theory with the cells of different dimensions of a cell complex gives rise to a space distribution of global variables known in algebraic topology by the (unfortunate) name of *cochain* but

**Table 1.3** Two kinds of equations of a physical field: field equations and constitutive equations

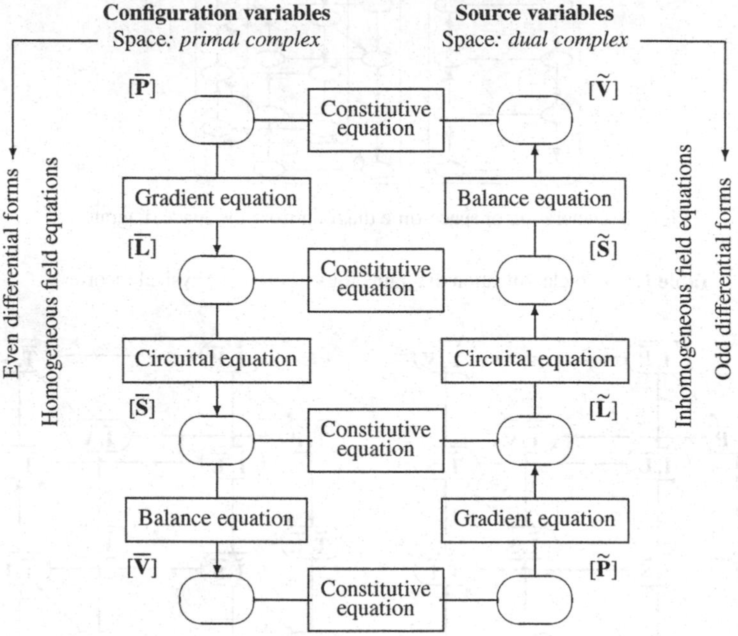


conveniently called today *discrete form*. All topological equations can be obtained from a very simple and elegant process known as the *coboundary process* in algebraic topology. When we deduce field variables from the corresponding global variables, the discrete forms become exterior differential forms and the coboundary process transforms into the *exterior differential* of a form. More precisely, while the configuration variables, associated with space elements endowed with an inner orientation, can be described by exterior differential forms of an even kind, the source variables, associated with space elements endowed with an outer orientation, can be described by differential forms of an odd kind (≡ *twisted* differential forms).

The unifying power of the coboundary process on the discrete forms, that is, the algebraic ancestor of the exterior differential on the exterior differential forms, is manifested in the fact that it generates the three typical differential operators, which give rise to the gradient, curl and divergence.

Concepts like *global variable, oriented space element, cell complex, chain, cochain* and *coboundary operator* are usually ignored in the traditional description of physics, which is based almost exclusively on differential equations.

The differential formulation hides the geometric and topological structures of physical variables and of physical laws because it reduces every physical variable to a field variable, i.e. it deprives physical variables of their geometric content. The differential machinery disregards the geometric description from the outset and in so doing obscures the structure that is common to different physical theories based on the notion of homologous variables. This is why our investigation should not start from field variables, as the differential formulation does, but from *global variables*.

The classification diagram presented in the book reveals a substantial unity of all physical theories of the macrocosm. To stress this unity, we have included in all diagrams, inside a small icon, the frame in which all present theories find their home. This frame was called by Bossavit the Maxwell house because the variables and the equations of electromagnetism fit nicely inside the frame. The present book shows that the frame is not only a house for Maxwell's equations but for all equations of physical theories of the macrocosm.

The classification of variables and equations obtained in this way exhibits a general structure that displays many known properties of physical equations including, for example, the possibility of a variational formulation of the fundamental equation, the existence of reciprocity theorems, the uniqueness of the solution of the fundamental problem of a theory, compatibility conditions, gauge invariance and the possibility of knowing if a constitutive equation is reversible or not.

Another interesting aspect of the common structure is that the classification is applicable both to the algebraic formulation, which is performed on global variables, and to the differential formulation, which utilizes field variables.

## 1.2 Analogies in Physics

As was stated previously, the general framework that we want to set up in this book arises from the question of why analogies exist in physics. It is therefore necessary to understand what an analogy is and exemplify the importance that analogies have had and continue to have today in physics.

In the development of scientific thinking, an essential role has been played by the discovery that physical theories exhibit structural similarities, commonly called *analogies*. Duhem wrote:

> The history of physics shows us that the search for analogies between two distinct categories of phenomena has perhaps been the surest and most fruitful method of all the procedures put into play in the construction of physical theories.[5]

An analogy can be defined as the 'invariance of a relation or statement under changes of the elements involved in it'.[6] The corresponding elements of two

[5] Duhem [60, p. 95].

[6] Rosen [193, Chap. 4].

fields are called *homologous*. Analogy is perhaps the weakest form of invariance: invariance in form.

Analogies are not simply a formal fact; they are a structural fact. Analogies reveal the existence of an underlying structure, though they do not explain why that structure exists.

Scientific activity is strongly based on the use of analogies. We use them, more or less consciously, because they are like roads already traced on the land that we are investigating. Analogies are a fundamental tool for knowledge because they allow us to explore a new field using the established knowledge of another field; we need only find the homologous entities of the two fields.

Analogies are a powerful tool in discovery, learning and teaching, i.e. in the creation and transmission of knowledge.

Analogies have another important merit: when dealing with a set of abstract notions, one may find an analogy with a set of concrete notions; in this way, the abstract notions can be more easily understood by referring to the concrete ones. The understanding of abstract notions is greatly facilitated by a well-chosen analogy! In fact, analogies have played an important role in the development of physics. Often a theory is built using a one-to-one correspondence between its physical variables and those of another theory (homologous variables). For example, the analogy between heat conduction in solids, which is transmitted by contact, and electrostatics, which is deduced from the laws of action at a distance, supported the idea that electromagnetic action was also transmitted by contact.[7]

Similarly, Poisson introduced the idea of electric potential by analogy to the notion of temperature in a thermal field, a subject previously treated by Fourier in his book on heat conduction. Another analogy arises from a comparison of the propagation of waves in a material continuum (solid or fluid) and the propagation of electromagnetic waves in free space. In both cases, we have the typical phenomena of reflection, refraction, interference, diffraction, polarization and others.

In an analogy between two physical phenomena, homologous variables differ in many senses: they have different physical meanings, different physical dimensions, and, in general, different mathematical natures. For example, the homolog of a scalar variable may be a vector variable. The existence of similarities despite these differences may be the reason for the fascination that similarities have on many people.

Since a relation between physical variables is expressed by an equation, it follows that analogies in physics are easily captured by the similarity of equations.

In addion, similarity of the equations in various theories allows us to use the same mathematical formalism. For example, the existence of analogies between field theories is shown by the ubiquitous presence of operators such as 'grad', 'curl' and 'div' and by the equations of Laplace, Poisson and d'Alembert arising from them.

[7] Whittaker [254, Vol. I, p. 241].

Going one step further, we arrive at the formalism of mathematical field theory, which can be used to investigate the possible forms of, for example, field equations (elliptic, parabolic, hyperbolic, linear, or nonlinear), variational principles, invariance properties and conservation laws.

One of the impressive facts underlining the power of analogies in physics and engineering is that they allow for the construction of many mathematical formalisms, such as:

- The formalism of *dynamical systems*;
- The formalism of *generalized network theory*;
- The formalism of *irreversible thermodynamics*;
- The formalism of *mathematical field theory*;
- The formalism of *variational principles*;
- The formalism of the *first quantization*;
- The formalism of the *second quantization*.

We live among formalisms! Mathematics is universally applied because it is the king of formalisms: differential and integral calculus, matrix calculus, vector calculus, operator theory and group theory are all mathematical theories whose application to different fields of science enables great economy of thought, labour and time.

When faced with analogies in physics, two approaches are possible: one is to accept them as a matter of fact and to use them to construct a formalism, the other is to question the reasons for their existence.

This book aims to provide an answer to the latter, i.e.

*What is the reason for analogies between physical theories?*

The answer to this question forms the core of this book. The same question was raised by Richard Feynman[8]: *Why are the equations from different phenomena so similar?* His answer is as follows:

> We might say: 'It is the underlying unity of nature.' But what does that mean? What could such a statement mean? It could mean simply that the equations are similar for different phenomena; but then, of course, we have given no explanation. The 'underlying unity' might mean that everything is made out of the same stuff, and therefore obeys the same equations. That sounds like a good explanation, but let us think. The electrostatic potential, the diffusion of neutrons, heat flow – are we really dealing with the same stuff? Can we really imagine that the electrostatic potential is *physically* identical to the temperature, or to the density of particles? Certainly $\phi$ is not *exactly* the same as the thermal energy of particles. The displacement of a membrane is certainly not like a temperature. Why, then, is there 'an underlying unity'? A closer look at the physics of the various subjects shows, in fact, that the equations are not really identical. The equation we found for neutron diffusion is only an approximation that is good when the distance over which we are looking is large compared with the mean free path. If we look more closely, we would see the individual neutrons running around. Certainly the motion of an individual neutron

[8] Feynman et al. [69, Vol. II; pp. 12–12].

> is a completely different thing from the smooth variation we get from solving the differential equation. The differential equation is an approximation, because we assume that the neutrons are smoothly distributed in space. Is it possible that *this* is the clue? That the thing which is common to all the phenomena is the *space* the framework into which the physics is put?

As we can see, Feynman did not give an answer to the question, even though his claim that *space* is responsible has hit the nail on the head.

Analogies show that alongside the traditional criteria for classifying physical quantities, there is one arising from the fact that some physical quantities are related to lines and others to surfaces. In 1871, Maxwell published an article entitled 'Remarks on the Mathematical Classification of Physical Quantities' [152, p.227], in which he wrote:

> Of the factors which compose it [energy], one is referred to unit of length, and the other to unit of area. This gives what I regard as a very important distinction among vector quantities.

Considering that there are vectors relative to lengths and others relative to areas, he suggested that physical variables could be classified according to their reference to a geometric element, like lines and surfaces. When this is done, analogies make a fourfold distinction of polar–axial and line–surface vectors, a distinction absent in books on vector calculus where with the same vector one performs circulation along a line and flux across a surface.[9] The key point of the present analysis is that global physical variables have a natural association with the four space elements and the two time elements. Once this association has been realized, we can easily see that the *homologous variables of two physical theories are those associated with the same space element*. This fact is at the base of the existence of analogies in physics. Hence, every physical variable can be classified according to the space and time element with which it is associated.

In the same article Maxwell said

> It is only through the progress of science in recent times that we have become acquainted with so large a number of physical quantities that a classification of them is desirable. [...] But the classification which I now refer to is founded on the mathematical or formal analogy of the different quantities, and not on the matter to which they belong.

He called this

> [...] a mathematical classification of quantities. A knowledge of the mathematical classification of quantities is of great use both to the original investigator and to the ordinary student of the science.

Speaking about analogies Maxwell said

> But it is evident that all analogies of this kind depend on principles of a more fundamental nature; and that, if we had a true mathematical classification of quantities, we should be able at once to detect the analogy between any system of quantities presented to us

[9] Post [184, p. 630].

> and other systems of quantities in known sciences, so that we should lose no time in availing ourselves of the mathematical labours of those who had already solved problems essentially the same. [. . .] At the same time, I think that the progress of science, both in the way of discovery and in the way of diffusion, would be greatly aided if more attention were paid in a direct way to the classification of quantities.

In the classification diagram that will be presented in this book, the relations between variables, i.e. the equations, are displayed in an ordered way such that they immediately reveal analogies. We want to show that the classification of physical variables and equations is the key to explaining analogies in physics, and it is the root of many common mathematical properties of physical theories. The structure revealed in this way allows us to obtain a classification diagram of variables and equations that is the same for all physical theories.

In this book we will consider the following six physical theories:

1. Particle mechanics (and part of analytical mechanics),
2. Electromagnetism,
3. Mechanics of deformable solids,
4. Fluid dynamics,
5. Thermal conduction,
6. Gravitation,

and we will unfold the mathematical structure of these physical theories. Although the classification needs the global formulation, the diagrams are written using the differential formulation because, to date, this has been the most used language in physics.

## 1.3 Role of a Classification

What is the role of a classification? Many physicists believe that the value of a classification lies essentially in its ability to predict a *new* property, a *new* fact, a *new* phenomenon: 'if you do not hope to discover something new, what is the interest?'

Let us take as an example the periodic table of chemical elements. It is true that Mendeleev's classification has made possible the prediction of the existence of new chemical elements on the basis of their physical and chemical properties, and it was a big help to discover them. But the value of this classification has not been exhausted at all in the discovery of new chemical elements. If the predictive aspect was the predominant one, how are we to explain the fact that Mendeleev's table is still hanging on the walls in the classrooms of colleges and universities around the world? Perhaps this is done to encourage students to discover some new elements? How do we justify the teaching of the classification of crystals into 32 classes of symmetry? Why do we teach the Linnaean classification of botany? Perhaps to stimulate the discovery of a new type of animal or vegetal organism?

One of the advantages of Mendeleev's table is in the spatial arrangement of the elements within a table, not simply in a list. In fact, an order can be obtained simply by making a list of the chemical elements in increasing atomic weight next to the indication of the physical and chemical characteristics of each element. Such a list would contain all the information necessary to make the combinations to create chemical compounds. But it was soon realized that in such a list, there would be a periodic behaviour of the chemical properties of the elements and the period would be composed of eight elements. This fact suggested organizing the elements in rows of eight elements. In this way, we pass from a one-dimensional list to a two-dimensional table. Much later it was realized that eight is the maximum number of electrons that can stay in the outer orbital of an atom.

Doing a classification means *dividing into classes* according to certain criteria. The primary goal of a classification is to impose order on a set of elements, but the value of a classification depends strongly on the criteria used to classify.

The classification of physical quantities that we introduce in this book is based on the *role* that the variables have in their theory. This leads to a first criterion of dividing the variables into three classes – configuration, source and energy variables. A second criterion concerns the space element with which the variables are associated. Since each physical variable can be classified according to both criteria, this allows the formation of a two-dimensional diagram. A third criterion arises considering the association of a variable with a time element. This leads to a three-dimensional diagram represented in an assonometric view (Table 1.2).

One of the features of the classification diagram is the clear-cut distinction between the *field equations*, such as balance equations, circuital equations and equations for the formation of gradients (vertical links), and the *constitutive equations* which describe both reversible and irreversible processes (horizontal links).

One thing that catches the eye looking at the diagram, as shown in Fig. 8.7 (p. 232), are the irreversible links: these connect a variable located in the left column of the front part of the diagram with another variable located in the right column of the rear part of the diagram.

The diagram gives a systematic procedure to obtain the fundamental equation of a theory: simply combine the equations encountered in the path that goes from the potential to the source.

## 1.4 Role of Geometry in Physics

The mathematical description of physics requires the intermediary role of geometry because all physical phenomena arise in space and physical variables are introduced with reference to lines or surfaces or volumes, and only a few of them refer directly to points. For example, the light flux detected by a photocell depends on the area of the surface and on its position in space, its orientation; a strain-gauge

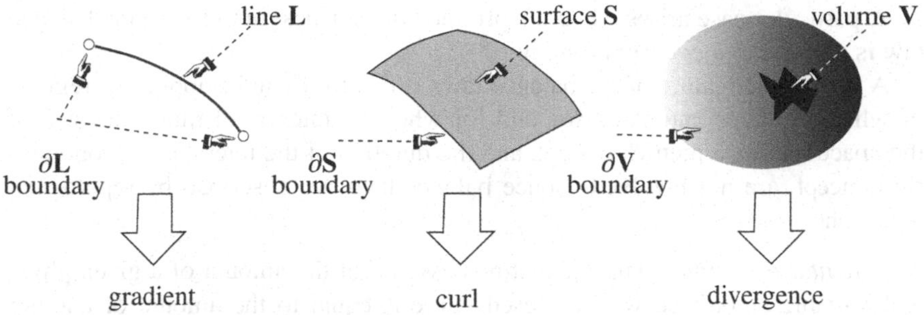


**Fig. 1.4** Topological notions at origin of three main differential operators

measures the variation of the distance between two points by the increase of the electric resistance of a wire.

Geometry enters into physics at two levels: by means of *topological* concepts and *metric* concepts. So Coulomb's law requires the distance between two point charges and is therefore a *metric* law. In contrast, the first law of the electrostatic field says that the sum of the voltages along any closed line is zero. This is a *topological* law because the shape of the closed line and its extension are not involved.

## *1.4.1 Topological Concepts*

Recall that topological properties are those properties of geometric figures which are invariant under continuous deformations without introducing tears and overlaps. Stated in more mathematical language, they are invariant under *homeomorphisms*. These concepts differ from metric concepts because no measures of lengths, areas, volumes or angles are involved.

In physical theories, there are three kinds of topological equations: *balance equations, circuital equations* and *equations forming gradients.*

Figure 1.4 shows the topological ingredients which we use, implicitly or explicitly, in physical theories, i.e. line surfaces, volumes and their boundaries: they are involved in the construction of the three differential operators which are ubiquitous in physics – the gradient, the curl and the divergence.

*Balance equations.* These are the most important physical equations. They play a pivotal role in all physical theories and include the balance of mass, momentum, angular momentum, energy, entropy, electric charge and the number of particles. A balance law states that, given a space region and a time interval, the amount of a given physical variable produced inside the space region in the time interval can be divided into two parts: one part is stored inside the region in the time interval, the other flows across the boundary of the region in the same interval.

A particular case arises when the production vanishes: in this case the balance law is reduced to a *conservation law.*

A distinctive feature of the balance laws is that they can be applied to regions of whatever shape and extension and for whatever interval of time. The size of the space region, a metric concept, and the duration of the interval, a chronometric concept, are not involved. Hence balance laws are described by topological equations.

*Circuital equations.* These equations assert that the amount of a given physical variable associated with a closed curve is equal to the amount of another physical quantity associated with a surface enclosed by this curve. Such equations may express a law or simply the definition of a physical variable. These include André-Marie Ampère circuital law, Faraday's induction law and Kelvin's circulation theorem.

In fluid dynamics, the vortex flux through a surface is *defined* as the velocity's circulation along the boundary of the surface. In electromagnetism, the magnetic flux across a surface is *defined* as the impulse of the electromotive force[10] across the boundary of the surface. A circuital equation is valid for any shape of the surface and for any area; hence, circuital equations are topological equations.

*Equations forming gradients.* These are the third kind of topological equation. When we form the difference in temperature between two points, we define a new variable by means of a simple relation which does not involve metric concepts. Hence the equations forming differences are *topological equations*. In general this is a preliminary step towards defining the average temperature gradient, i.e. dividing this temperature difference by the distance of the two points. The division by a distance introduces a metric attribute and leads to the temperature gradient along the direction of the line connecting the two points.
Besides these three classes of equations there are other topological concepts.

*Connectivity of a region.* The existence, in a region, of closed lines which are not contractible to a point and of closed surfaces which are not contractible to a closed line leads to the concept of multiply connected regions with respect to lines and surfaces, respectively. Hence connectivity is a topological concept. We remark that the projection of a figure, as in axonometry and in perspective, does not respect connectivity because distinct lines in space may have projections which intersect, and it does not respect either the topological properties or the metric properties, as seen in the prospective of a cube.

*Orientation.* The concept of orientation plays a crucial role in the mathematical description of physics. Both the inner orientation and the outer orientation, which we will present in detail in what follows, are used in the description of physical laws. The concept of orientation, which has a combinatorial nature, is a topological concept.

---

[10] See pp. 290 and 291.

## 1.5 Algebraic Formulation of Fields

The use of global variables leads to an *algebraic* (or *discrete* or *direct* or *finite* formulation) of physical theories, which is an alternative to the *differential formulation*. Starting from the algebraic formulation one can easily deduce the differential formulation. This way of presenting physical theories is very useful in teaching because it makes a strong appeal to physical measurements and avoids premature recourse to differential operators whose symbolism is exceedingly abstract for the average university student.

For example, the Faraday electrostatics law, which states that the electric charge $\Psi$ collected on the boundary of a volume is equal to the charge contained inside the volume, can be grasped more easily than the statement $\nabla \cdot \mathbf{D} = \rho$, which states the same thing but in differential formulation.

What is more important is that *the algebraic formulation can be immediately used for computational physics* using a cell complex and its dual in the computational domain. This formulation avoids the discretization of differential equations, which is necessary in the existing numerical methods, because it uses global variables and balance equations in a global form. Hence, from a computational point of view (which is not considered in this book) the direct algebraic formulation avoids all the typical difficulties linked to differentiability such as the use of generalized functions (e.g. Dirac delta function), the splitting of physical laws into differential equations in the regions of regularity and jump conditions across the surfaces of discontinuity.

The numerical method, which is based on a direct algebraic formulation, is called the *cell method*.[11]

## 1.6 Summary

In short, a description of physical theories using global variables as a starting point

1. Shows the link between global variables and the oriented space and time elements;
2. Maintains a close link with physical measurements because global variables are, in general, the variables we measure;
3. Permits a numerical formulation of physical theories from the very outset, i.e. without discretizing the differential equations.

The global formulation (algebraic–topological formulation) which we want to reveal gives an answer to the question of why analogies exist. Moreover, it allows us to deduce

[11] The Web site: `http://discretephysics.dicar.units.it/` has collected a large number of papers dealing with this method.

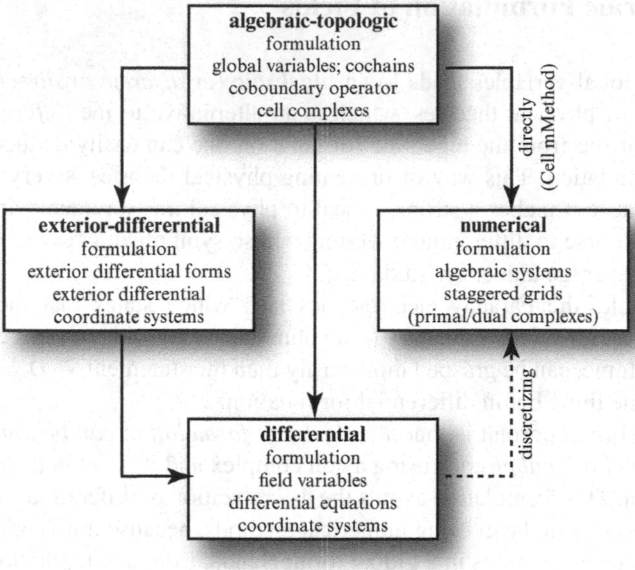


**Fig. 1.5** The four mathematical formulations of physics: three of them are known

- The traditional differential formulation,
- The exterior differential formulation,
- The numerical formulation,

as shown in Fig. 1.5. A further consequence of this association is that balance, circuital and gradient forming equations can be described through a single algebraic process which has inherited an unusual name derived from algebraic topology: the coboundary process. Despite its name, the process is so simple and so intuitive that it can be presented to undergraduate students. Algebraic topology also enables the formulation of physical laws in global form, *in the large*, i.e. considering also multiply connected domains.

### *1.6.1 Notation Used in Book*

For symbols of physical quantities we have followed the indication of the International Union of Pure and Applied Physics (IUPAP). Page 485 contains a list of the physical variables treated in this book. The units used are those of the SI system.

# Part I
# Analysis of Variables and Equations

# Chapter 2
# Terminology Revisited

## 2.1 Why Do We Need Proper Notation and Terminology?

Sometimes the same symbol is used with different meanings. Let us look at the following two formulas:

$$M = \int \rho \, \mathrm{d}V, \qquad W = -\int p \, \mathrm{d}V \,. \tag{2.1}$$

In the first integral, $M$ denotes the mass, $\rho(P)$ the mass density at a point $P$ and the symbol $\mathrm{d}V$ indicates an infinitesimal volume; in the second integral, $W$ denotes the work, $p(V)$ the pressure and $\mathrm{d}V$ indicates an infinitesimal *variation* of the volume. In Sect. 5.2 we will distinguish material descriptions from spatial descriptions: in the first integral, $V$ denotes a fixed *control volume*, typical of a *spatial description*, whereas in the second integral, $V$ denotes a variable volume, typical of a *system description*. We shall denote these two volumes with the symbols $V$ and $\mathcal{V}$ respectively. Hence a more detailed notation is

$$M = \int_V \rho(P) \, \mathrm{d}V, \qquad W = -\int_{\mathcal{V}_1}^{\mathcal{V}_2} p(\mathcal{V}) \, \mathrm{d}\mathcal{V} \,. \tag{2.2}$$

Another example of equivocal notation arises when some authors confuse the *rate* with the *derivative*.

Thus in mechanics of continua we must distinguish between the displacement of a point of a continuum from its position in a reference configuration (total displacement) and the displacement of a point between two arbitrary positions.[1] Hence, in the mechanics of continua, the velocity at an instant $t$ is the time *derivative* of the *total* displacement at the instant $t$, but it is the *rate* of the displacement in a small time interval $\Delta t$ centred around the instant $t$.

---

[1] See p. 331.

E. Tonti, *The Mathematical Structure of Classical and Relativistic Physics*,
Modeling and Simulation in Science, Engineering and Technology,
DOI 10.1007/978-1-4614-7422-7_2, © Springer Science+Business Media New York 2013

Vector calculus does not distinguish line vectors from surface vectors and on the same vector defines the line integral along a line and the flux across a surface.[2] This fact, though mathematically legitimate, can be physically inappropriate. Thus in electrostatics, it is proper to define the line integral of the electric field strength **E** along a line and not the flux of the same vector across a surface; it is proper to define the flux of the electric displacement **D** across a surface and not the line integral of the same vector along a line. In fluid dynamics, it is proper to define the line integral of the velocity **v** along a line and not the flux of the velocity across a surface; it is proper to define the flux of the mass current density $\mathbf{q} = \rho\mathbf{v}$ across a surface and not the line integral of the same vector along a line.

Vector calculus does not usually make a distinction between *polar* vectors and *pseudovectors* (≡ axial vectors) or between *true scalars* and *pseudoscalars*.[3] So the voltage $E$, which is the line integral of the polar vector **E**, is a true scalar, while the magnetic voltage $F_{\mathrm{m}}$, which is the line integral of the pseudovector **H**, is a pseudoscalar.[4] It follows that the electric potential $\phi$ is a true scalar, whereas the magnetic scalar potential $\phi_{\mathrm{m}}$ is a pseudoscalar.

In vector calculus, it is not stated that the unit vector normal to a surface in some cases is a pseudovector, whereas in other cases it is a polar vector. In fact, when the surface is endowed with *inner* orientation, the normal is a pseudovector; in contrast, when the surface is endowed with *outer* orientation, the normal is a polar vector. Thus, recalling that in electromagnetism the vector **B** is a pseudovector whereas **D** is a polar vector, and recalling that the magnetic flux $\mathbf{B}\cdot\mathbf{n}\,\mathrm{d}S$ and the electric flux $\mathbf{D}\cdot\mathbf{n}\,\mathrm{d}S$ are true scalars, it follows that in the first case **n** is a *pseudovector*, whereas in the second case **n** is a *polar* vector.

In this book, we distinguish pseudovectors and pseudoscalars by placing a mark upon the letter: thus, $\check{\mathbf{B}}, \check{\mathbf{H}}, \check{\omega}, \check{\phi}_{\mathrm{m}}$.

## 2.2 Many Meanings of $\Delta$

The symbol $\Delta$ is used with many different meanings in mathematics: it can denote, for example, the Laplacian of a function, a space or a time increment, the determinant of a base in a vector space, or the area of a triangle. The following formulae show some possible meanings:

---

[2] Post [184, p. 630].

[3] See p. 145.

[4] Recall that a scalar is called *axial* or *pseudoscalar* when, by definition, it changes sign when we pass from a right-handed to a left-handed screw, i.e. when we change the inner orientation of space.

$$\begin{cases} -\epsilon \Delta\phi = \sigma \qquad \Delta u = u(\mathbf{B}) - u(\mathbf{A}) \qquad \Delta f = f(t^+) - f(t^-)\,; \\ \Delta = \frac{1}{2}\begin{vmatrix} 1 & x_1 & y_1 \\ 1 & x_2 & y_2 \\ 1 & x_3 & y_3 \end{vmatrix} \qquad \Delta = \det(\mathbf{e}_1 \wedge \mathbf{e}_2 \wedge \mathbf{e}_3)\,. \end{cases} \tag{2.3}$$

These facts make the symbol $\Delta$ ambiguous. When a physical variable depends on time and space coordinates, say temperature $T(t, x, y, z)$, the symbol $\Delta$ is ambiguous because it does not specify if it refers to time or space coordinates or both. We use the rule that *whenever there is a conceptual difference, there must also be a difference in the notation.* To avoid this ambiguity, which we will see implies an erroneous association with space and time elements of a physical variable, we will use the two notations $\Delta_t$ and $\Delta_s$:

$$\begin{array}{ll} \Delta_t T(t,\mathbf{x}) \stackrel{\text{def}}{=} T(t_2,\mathbf{x}) - T(t_1,\mathbf{x}), & \Delta_s T(t,\mathbf{x}) \stackrel{\text{def}}{=} T(t,\mathbf{x}_2) - T(t,\mathbf{x}_1), \\ \mathrm{d}_t T(t,\mathbf{x}) \stackrel{\text{def}}{=} T(t+\mathrm{d}t,\mathbf{x}) - T(t,\mathbf{x}), & \mathrm{d}_s T(t,\mathbf{x}) \stackrel{\text{def}}{=} \\ & T(t,\mathbf{x}+\mathrm{d}\mathbf{x}) - T(t,\mathbf{x}). \end{array} \tag{2.4}$$

The index $t$ refers to a time variation, whereas the index $s$ refers to a space variation. In this book we will use the symbol $\Delta$ with the following meanings:

$\Delta$ for the increment of a function of one variable, in particular:

$\Delta_t$ for a *time* increment of a function of time and space variables,
$\Delta_s$ for a *space* increment of a function of time and space variables;

$\Delta$ for the determinant of the transition matrix in the change of base vectors.

## 2.3 Use and Misuse of Symbol 'd'

Much confusion arises from an improper use of the symbol of differentiation. Let us analyse the notation used in mathematics to denote the integral. The integral of a function of one variable is denoted by

$$F(t) = \int_0^t f(\tau)\,\mathrm{d}\tau, \qquad G(x) = \int_0^x g(\xi)\,\mathrm{d}\xi, \qquad H = \int_0^1 h(s)\,\mathrm{d}s, \tag{2.5}$$

with $0 \le \tau \le t$, $0 \le \xi \le x$ and $0 \le s \le 1$. In this notation the symbol 'd' means the differential of a variable. The notation is well chosen because if we want to change the variable by putting $\xi(\eta)$ we can write $\mathrm{d}\xi = (\mathrm{d}\xi/\,\mathrm{d}\eta)\,\mathrm{d}\eta$.

What happens if we consider line, surface and volume integrals? We can ask ourselves what the symbols $\mathrm{d}L$, $\mathrm{d}S$, $\mathrm{d}V$ mean in the notations

$$A = \int_L f(P)\, \mathrm{d}L, \qquad B = \int_S g(P)\, \mathrm{d}S, \qquad C = \int_V h(P)\, \mathrm{d}V\,. \tag{2.6}$$

*The symbol 'd' here simply denotes an infinitesimal length, area and volume,* not the differential of a variable. It is precisely this notation which leads many authors to use the symbol 'd' to denote an infinitesimal quantity *even if it is not the variation of another quantity*. Thus, it is customary to write

$$M = \int \mathrm{d}m \qquad \textit{which is the continuous analogue of} \qquad M = \sum_{k=1}^{N} m_k, \tag{2.7}$$

where the notation d$m$ does not mean, of course, an infinitesimal *increment* of mass, but only an infinitesimal mass.

**Remark.** We may also ask ourselves whether the notation d$L$, d$S$, d$V$ is convenient from a mathematical point of view. Let us consider the element d$S$, which is usually expressed in a Cartesian plane as $\mathrm{d}S = \mathrm{d}x\, \mathrm{d}y$. What happens if we want to pass from Cartesian to polar coordinates? Since the transformation is given by $x = \rho \cos\theta$, $y = \rho \sin\theta$, a formal substitution of $\mathrm{d}x\, \mathrm{d}y$ in terms of $\mathrm{d}\rho\, \mathrm{d}\theta$ *does not* give the correct areal element, i.e. $\rho\, \mathrm{d}\rho\, \mathrm{d}\theta$. Things change if we write

$$\mathrm{d}S = \mathrm{d}x \wedge \mathrm{d}y, \tag{2.8}$$

i.e. using the symbol $\wedge$ as *exterior product* $x$ and $y$ are coordinate function 0-forms, associated with a point, and therefore $dx$ and $dy$ are functions associated with line elements. The operation $\wedge$ can therefore also be interpreted as the wedge product between differentials and must not be confused with the cross product, as happens in the Italian and French literature. In this case, since $\mathrm{d}\rho \wedge \mathrm{d}\rho \equiv 0$, $\mathrm{d}\theta \wedge \mathrm{d}\theta \equiv 0$ and $\mathrm{d}\rho \wedge \mathrm{d}\theta \equiv -\mathrm{d}\theta \wedge \mathrm{d}\rho$, we obtain the correct result in the form

$$\mathrm{d}S = \rho\, \mathrm{d}\rho \wedge \mathrm{d}\theta\,. \tag{2.9}$$

In physics the symbol 'd' is often used to denote an infinitesimal quantity. For example, in defining the electric field strength **E**, some authors use the notation

$$\mathbf{E} \stackrel{\text{def}}{=} \lim_{\mathrm{d}q \to 0} \frac{\mathrm{d}\mathbf{F}}{\mathrm{d}q} \qquad \textit{instead of} \qquad \mathbf{E} \stackrel{\text{def}}{=} \lim_{q \to 0} \frac{\mathbf{F}}{q}\,. \tag{2.10}$$

To this improper notation we must add that the limit in the second formula 2.10 has no physical meaning because the charge is discrete and the smallest charge known is the charge of the electron. It is better to define the electric field strength as $\mathbf{E} \stackrel{\text{def}}{=} \mathbf{F}/q$, with the specification that $q$ must be *sufficiently small* so that it does not alter the position of the charges that generate the field. Fortunately, this is what is commonly done.[5]

---

[5] Jackson [103, p. 23], Pauli [174, p. 14], Schelkunoff [201, p. 8], Becker [11, p. 61], Lorrain et al. [143, p. 45], Fleury and Mathieu [71, p. 30], Bettini [13, p. 21], Hallen [87, p. 2], Sears and Zemansky [208, p. 431], Fouillé [72, p. 38], Novozilov and Jappa [169, p. 15], Akhiezer [5, p. 13].

The reason why many authors write $\mathrm{d}q \to 0$ instead of $q \to 0$ is because they believe that an infinitesimal quantity must be preceded by the symbol 'd'. This is unnecessary because in mathematics infinitesimal quantities like $\epsilon$ and $\eta$ are used without any such symbol, with the exception of $\mathrm{d}L$, $\mathrm{d}S$, $\mathrm{d}V$ *used in integration*.

The practice of introducing the symbol 'd' when it is not necessary leads to the introduction of other symbols, say $\delta$, $\mathrm{d}^*$, $\bar{\mathrm{d}}$, đ and similar alterations of the letter 'd' to denote quantities which are not an exact differential. This is, for instance, the case in the first law of thermodynamics, which is written in one of the following forms:[6]

$$\begin{array}{lll} \mathrm{d}u = \mathrm{d}'q + \mathrm{d}'w, & \mathrm{d}U = \mathrm{d}Q + \mathrm{d}W, & \mathrm{d}E = \varDelta W + \varDelta Q, \\ \mathrm{d}u = \mathrm{d}q + \mathrm{d}w, & \mathrm{d}U = đq - p\,\mathrm{d}V, & \mathrm{d}U = \delta q + \delta w, \end{array} \tag{2.11}$$

instead of the simpler (and clearer) notation[7]

$$\boxed{\varDelta_t U = Q + W \qquad \text{and} \qquad \mathrm{d}_t u = q + w\,.} \tag{2.12}$$

To denote the entropy differential, three notations are used

$$\underbrace{\mathrm{d}S = \frac{\mathrm{d}Q}{T},}_{\textit{incorrect}} \qquad \underbrace{\mathrm{d}S = \frac{\mathrm{d}'Q}{T} = \frac{\bar{\mathrm{d}}Q}{T} = \frac{đQ}{T} = \frac{\delta Q}{T},}_{\textit{correct but redundant}} \qquad \underbrace{\mathrm{d}S = \frac{q}{T}\,.}_{\textit{correct}} \tag{2.13}$$

The first notation is incorrect because the symbol 'd' has two different meanings in the two members of the same equation.[8] The second notation is correct but redundant because the apostrophe, the overbar, the slash or the $\delta$ has the purpose of correcting the inappropriate symbol 'd' which precedes the heat $Q$. Lastly, the third notation is the correct one.[9]

To denote a finite *difference* of a quantity $Q$, it is customary to use the symbol $\varDelta Q$, and to denote an infinitesimal *difference*, the symbol $\mathrm{d}Q$ is used. The letters $\varDelta$ and d are simply the initials of the words difference and differential respectively. To avoid ambiguities, we recommend denoting an infinitesimal amount of a quantity $Q$ by the symbol $q$ in lower case, instead of $\mathrm{d}Q$ or $\delta Q$.

---

6 Sears [207, p. 45], Callen [33, p. 19], Perucca [176, Vol. I, p. 634], Paterson [172, p. 116], Moore [164, p. 28].

7 Guggenheim [86, Ch. 1, Sect. 12], Fleury and Mathieu [71, Vol. II, p. 95], Fermi [67, Ch. 2, Sect. 3], Dugdale [59, p. 20]. The subscript $t$ is not essential: it is added here to comply with our decision to distinguish between time increments and space increments in this book.

8 Callen [33, p. 32], de Groot and Mazur [49, p. 20], Sommerfeld [216, p. 32].

9 Guggenheim [86, p. 46].

## 2.4 Rates and Derivatives

Recall that the term *rate* denotes the *ratio* of a physical quantity is referred to a time interval and the duration of the interval. This is the meaning we give to the term in everyday language, e.g. when we speak of a reaction rate or of the annual inflation rate. A rate is a time *density*. The term *derivative* denotes the limit of the ratio between two *increments*, not simply between two quantities. Some authors do not distinguish between *rates* and *derivatives*. For example, in particle mechanics we must distinguish between the radius vector $\mathbf{r}(t)$, which depends on time instants, and the displacement $\mathbf{u} = \mathbf{r}(t^+) - \mathbf{r}(t^-)$, which depends on intervals. Thus it is sometimes said that velocity is a *derivative* of the displacement, whereas it is actually the *rate* of displacement and a *derivative* of the radius vector. In fact, we write

$$\mathbf{v}(t) = \lim_{T\to 0} \frac{\mathbf{u}}{T} = \lim_{\Delta t\to 0} \frac{\Delta\mathbf{r}(t)}{\Delta t} = \frac{\mathrm{d}\mathbf{r}(t)}{\mathrm{d}t}, \tag{2.14}$$

where $T = t^+ - t^-$ denotes the time interval. Some authors,[10] after stating that power is the rate of work (which is correct), say that power is the *derivative* of work (incorrect), i.e.

$$P(t) = \frac{\mathrm{d}W}{\mathrm{d}t} = \dot{W} \qquad \textit{should not be used.} \tag{2.15}$$

One cannot take the derivative of a variable which depends on a time interval, like the work $W$, but only of a variable that is a function of a time instant. Some international agencies dealing with the nomenclature[11] recommend denoting the rate of a quantity with a dot above the symbol, but since Newton, the dot over the letter has denoted a time derivative, not a rate!

$$\dot{W},\ \dot{Q},\ \dot{N} \qquad \textit{should not be used.} \tag{2.16}$$

Some authors write[12]

$$\mathbf{F} = \lim_{\Delta m\to 0} \frac{\Delta\mathbf{F}}{\Delta m} \qquad \text{instead of} \qquad \mathbf{f} = \lim_{m\to 0} \frac{\mathbf{F}}{m}. \tag{2.17}$$

The first notation is improper not only because the symbol $\Delta$ is superfluous but also because we use the same letter to denote the force **F** and the specific force **f**. As an example of this improper notation we present the following relation:[13]

---

[10] Saletan and Cromer [199, p. 10], Mansfield and O'Sullivan [150, p. 102].

[11] Nomenclature in the *Journal of Heat Transfer*, Vol. 121, No. 4, pp. 770–773, November 1999.

[12] Loitsyanskii [142, p. 73].

[13] Tolman [225, p. 246].

$$\mathbf{v} = \lim_{t \to 0} \frac{\mathbf{l}}{t} = \frac{\mathrm{d}\mathbf{l}}{\mathrm{d}t} \qquad \textit{should not be used.} \tag{2.18}$$

The vector **l** used here is the displacement, and the fact that it becomes small does not authorize us to use d**l** because the displacement is already the increment of the radius vector. Moreover, the time $t$ in the first fraction is a duration, better denoted by $T$ or by $\tau$, and must not be confused with a time instant $t$ used in the second fraction.

Table 2.1 illustrates the relation between time derivatives, mean rate, primitive functions and impulses for many physical quantities.

## 2.5 Definite and Indefinite Integrals

The integral sign is often used without an explicit distinction between *definite* and *indefinite* integrals. For example, some authors use the same letter, say $S$, to denote both the definite and the indefinite time integral of Lagrange functions[14]: the definite integral gives the action, while the indefinite integral gives Hamilton's principal function. Another example: the impulse of a force can be defined as the *definite* time integral of force, while momentum is the *indefinite* time integral of force. The latter definition seems to be absent in books on mechanics where momentum is improperly *defined* as the product of the mass times velocity, i.e. by a constitutive law.[15]

Table 2.2 shows that a careful distinction between definite and indefinite space and time integrals permits one to make clear the meaning of some physical variables and of their mutual relations.

## 2.6 Multiplication Symbols × and ∧

Gibbs denoted the *vector product* of two vectors **u** and **v** by the symbol $\mathbf{w} \overset{\text{def}}{=} \mathbf{u} \times \mathbf{v}$. In the Italian and French literature, the vector product was denoted (and many authors still use it today) with the symbol ∧, i.e. $\mathbf{w} \overset{\text{def}}{=} \mathbf{u} \wedge \mathbf{v}$. With the introduction of the notion of bivector,[16] the latter symbol was used to denote the *bivector* formed by two vectors, i.e. $\mathbf{B} \overset{\text{def}}{=} \mathbf{u} \wedge \mathbf{v}$. This is a new geometric entity completely different from the vector product: only in the three-dimensional space is the bivector the complement of the vector product.

[14] Kompaneyets [115, p. 85; p. 244], Landau and Lifshitz [123, p. 138,139].

[15] See Chap. 10, p. 241.

[16] For the notion of *bivector*, which is not very well known, see Appendix B, p. 467.

**Table 2.1** Relation between rates and time derivatives

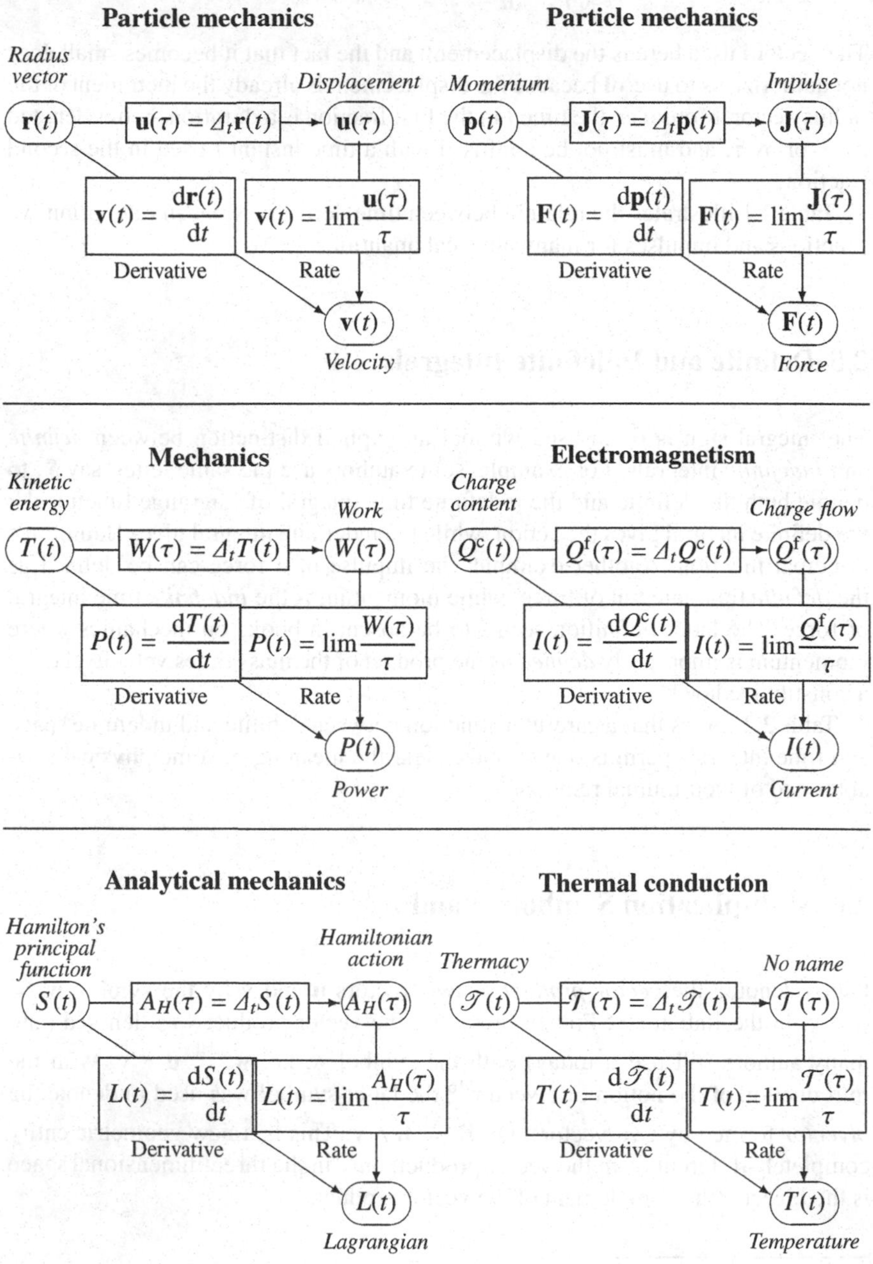


The symbol $\tau$ denotes the duration of a time interval.

**Table 2.2** Definite and indefinite line and time integrals

This table aims to show that a systematic distinction between definite and indefinite integrals both on line and on time clarifies the link between many physical variables.

| | | | |
|---|---|---|---|
| $\mathbf{F}$ Force | $W^* \stackrel{\text{def}}{=} \int_{\mathbf{A}}^{\mathbf{B}} \mathbf{F}\cdot\mathbf{t}\,\mathrm{d}L$ | Virtual work | $W^* = V(\mathbf{A}) - V(\mathbf{B})$ *(note the order)* |
| | $V \stackrel{\text{def}}{=} \int_{\mathbf{P}}^{\mathbf{O}} \mathbf{F}\cdot\mathbf{t}\,\mathrm{d}L$ | Potential energy | $\mathbf{O}$ reference point; $\mathbf{F} = -\operatorname{grad} V$ |
| $\mathbf{E} \stackrel{\text{def}}{=} \dfrac{\mathbf{F}}{q}$ Electric field strength | $E \stackrel{\text{def}}{=} \int_{\mathbf{A}}^{\mathbf{B}} \mathbf{E}\cdot\mathbf{t}\,\mathrm{d}L$ | Voltage | $E = \phi(\mathbf{A}) - \phi(\mathbf{B})$ *(note the order)* |
| | $\phi \stackrel{\text{def}}{=} \int_{\mathbf{P}}^{\infty} \mathbf{E}\cdot\mathbf{t}\,\mathrm{d}L$ | Electric potential | $\mathbf{E} = -\operatorname{grad} \phi$ |
| $\mathbf{g} \stackrel{\text{def}}{=} \dfrac{\mathbf{F}}{m}$ Acceleration of gravity | $V_{\mathrm{g}} \stackrel{\text{def}}{=} \int_{\mathbf{A}}^{\mathbf{B}} \mathbf{g}\cdot\mathbf{t}\,\mathrm{d}L$ | Gravitational potential difference | $V_{\mathrm{g}} = U_{\mathrm{g}}(\mathbf{A}) - U_{\mathrm{g}}(\mathbf{B})$ *(note the order)* |
| | $U_{\mathrm{g}} \stackrel{\text{def}}{=} \int_{\mathbf{P}}^{\infty} \mathbf{g}\cdot\mathbf{t}\,\mathrm{d}L$ | Gravitational potential | $\mathbf{g} = -\operatorname{grad} U_{\mathrm{g}}$ |
| $\mathbf{F}$ Force | $\mathbf{J}(t^-, t^+) \stackrel{\text{def}}{=} \int_{t^-}^{t^+} \mathbf{F}(t)\,\mathrm{d}t$ | Impulse | $\mathbf{J}(t^-, t^+) = \mathbf{p}(t^+) - \mathbf{p}(t^-)$ |
| | $\mathbf{p}(t) \stackrel{\text{def}}{=} \int_{0}^{t} \mathbf{F}(t')\,\mathrm{d}t'$ | Momentum | $\mathbf{F}(t) = \dfrac{\mathrm{d}\mathbf{p}(t)}{\mathrm{d}t}$ |
| $P$ Power | $W(t^-, t^+) \stackrel{\text{def}}{=} \int_{t^-}^{t^+} P(t)\,\mathrm{d}t$ | Work | $W(t^-, t^+) = T(t^+) - T(t^-)$ |
| | $T(t) \stackrel{\text{def}}{=} \int_{0}^{t} P(t')\,\mathrm{d}t'$ | Kinetic energy | $P(t) = \dfrac{\mathrm{d}T(t)}{\mathrm{d}t}$ |

$$W(t^-, t^+) \stackrel{\text{def}}{=} \int_{t^-}^{t^+} \mathbf{F}(t)\cdot \mathrm{d}\mathbf{r}(t) = \int_{t^-}^{t^+} \mathbf{F}(t)\cdot\mathbf{v}(t)\,\mathrm{d}t = \int_{t^-}^{t^+} P(t)\,\mathrm{d}t$$

To increase the confusion, some authors call the vector product of two vectors an *exterior* or *outer* or *cross* product; others call the *exterior* or *outer* or *wedge* product of the two vectors the bivector formed by them. As we can see, the terms *exterior* and *outer* products in the two cases give rise to different entities.[17]

The vector product is possible only in three-dimensional space and makes use of the screw rule; hence, the vector product of two polar vectors gives a *pseudovector*. In contrast, the exterior product of two polar vectors is a bivector; it is valid in a space of two or more dimensions and the screw rule is not involved.

## 2.7 Many Meanings of "Force"

Some physical variables are improperly denoted as *forces*. In the terminology of irreversible thermodynamics, the potential differences for unit length were called *thermodynamic forces*.[18] This name was introduced by Maxwell.[19]

The term *force* has been used also to denote energy, i.e. *vis viva* for kinetic energy, *vis mortua* for potential energy,[20] and the term *conservation of forces* was used by Mayer (1842) and Helmholtz (1847) instead of *conservation of energy*.[21] Today it is called *energy*.

Even today we use the terms *electromotive force* and *magnetomotive force*: these are variables with the nature of an electric or magnetic voltage. There are two new terms: *electromotance* and *magnetomotance*, meaning (literally) tendency to move ("-motance") an electric charge. Electromotance and magnetomotance are semantically more accurate, but not commonly adopted.[22] It is curious to note that the electric potential was introduced by Ohm, in analogy to temperature in thermal conduction, and he called it *electroscopic force*.[23]

Table 2.3 shows the modern nomenclature for many *forces* used in the past.

## 2.8 Many Meanings of "Flux"

To consider the right association of physical variables with space and time elements, which is one of the goals of this book, we must avoid any ambiguous use of nomenclature. Unfortunately, the term *flux* is used throughout the literature with different meanings, and this fact can create some misunderstandings.

---

17 See p. 467.

18 de Groot and Mazur [49, p. 25].

19 Maxwell, [152, p. 227].

20 Lindsay [139, p. 20].

21 Helmholtz [90, pp. 12–75].

22 See the well-written article "Electromotive force" on Wikipedia. The term *electromotance* is used in the book by Lorrain et al. [143, pp. 367, 413, 431].

23 Whittaker [254, Vol. I, p. 90].

**Table 2.3** Modern equivalent of various quantities improperly called 'forces'

| *Old* | *New* |
|---|---|
| Living *force* (Leibniz) | Kinetic energy |
| Dead *force* (Leibniz) | Potential energy |
| Conservation of *force* (Mayer, Helmholtz) | Conservation of energy |
| Thermodynamic *force* (Maxwell) | (None) |
| Electroscopic *force* (Ohm) | Electric potential |
| Electromotive *force* (Volta) | (Electromotance) |
| Magnetomotive *force* | (Magnetomotance) |

Let us start by observing that the term *flow* refers to the passage of a physical quantity through a surface.[24] This is the use in, for example, *traffic flow*, *water flow*, *cash flow* and *air flow*. What about the term *flux*? The scientific literature gives the following meanings.

- Some authors use the term *flux* and *flow* as synonyms[25];
- Some authors use the term *flux* as a synonym for *flow rate*. Examples are *energy flux* (watt) and *luminous flux* (lumen).[26] A more appropriate name is *current*. So electric current $I$ is the electric charge flow for unit time (rate of charge flow), and nobody calls it flux. In a similar way, we can use the terms *energy current, heat current, entropy current, particle current, momentum current=force*.
- Many authors use the term *flux* as a synonym for *flow rate density* ≡ *current density*. Examples are *energy flux* (watt/m$^2$), *entropy flux, particle flux*.[27] A convenient name is *current density*, and the proper symbol is $\mathbf{J}$ with an index like $\mathbf{J}_s$, $\mathbf{J}_e$ to denote *entropy current density, energy current density* respectively. This is the current notation in irreversible thermodynamics.
- We will see that *magnetic flux* $\Phi$, *vortex flux* $W$ and *electric flux* $\Psi$ are physical variables 'associated' with surfaces and instants, i.e. they have nothing to do with a flow of something through a surface. We will use the term flux only with this meaning.

---

[24] As a curiosity, Thomson (= Lord Kelvin) called *flow* the line integral of velocity; see Lamb [119, p. 33].

[25] de Groot and Mazur [49, p. 17; p. 21], Born and Wolf [18, p. 796]. Prigogine [190, p. 2] says: '*For simplicity, we will also use the term flow instead of density of flow (or current or flux).*' Four different quantities with the same name: no comment!

[26] Fleury and Mathieu [71, Vol. II, p. 423], Paterson [172, p. 523], Lamb [119, p. 38], Landau and Lifshitz [125, p. 11].

[27] Callen [33, p. 286], de Groot and Mazur [49, p. 17], Bird et al. [16, p. 3], Misner et al. [161, p. 138], Lamarsh [118, p. 321], Batchelor [10, p. 56]. The Gray and Isaacs dictionary of physics [84] says "*neutron flux Syn. neutron flux density.*"(!)

- *Flux density*, as in *magnetic flux density* **B**, *electric flux density* **D** (better named *electric displacement*), *vortex flux density* **w** (better named *vorticity*), denotes a flux per unit area.

Faced with this chaotic terminology, we need to make a clear distinction between five terms:

1. The term *flow* denotes something passing through a surface during an *interval*. This is the case of mass, charge, momentum, heat, particles.
2. The term *current* denotes the *rate of flow* or *flow per unit time*.
3. The term *current density* denotes *the rate of flow for unit area*. So the electric current density **J** is the charge flow rate for unit area.
4. The term *flux* will be used in this book only to denote a global physical variable associated with a surface and a time instant.
5. The term *flux density* denotes the flux for unit area. This is the case of electric surface charge density $\sigma$.

Table 2.4, summarizes the terminology we use in this book.

**Table 2.4** Unambiguous terminology for *flux* and its related terms

| 'Associated' with intervals | 'Associated' with instants |
|---|---|
| **Flow** | **Flux** |
| **Current** = flow rate | **Flux density** |
| **Current density** = flow density rate | |

## 2.9 Many Kinds of Equality

In physics, as in mathematics, the = sign is universally used without distinguishing the meaning of the equality. So faced with the three 'equations'

$$\frac{\mathrm{d}f}{\mathrm{d}x} = \lim_{\Delta x \to 0} \frac{f(x+\Delta x) - f(x)}{\Delta x}, \qquad (a+b)(a-b) = a^2-b^2, \qquad 3x^2+2x-4 = 0,$$

we can easily see that the first is a *definition*, the second an *identity*, i.e. valid *regardless of* the values of the variables, and the third simply an *equation*, i.e. it is valid only for *some* values of the variables. This suggests that one should specify the = sign as follows:

$$\frac{\mathrm{d}f}{\mathrm{d}x} \stackrel{\text{def}}{=} \lim_{\Delta x \to 0} \frac{f(x+\Delta x) - f(x)}{\Delta x}, \qquad (a+b)(a-b) \equiv a^2 - b^2, \qquad 3x^2 + 2x - 4 = 0.$$

Even in mathematics it is convenient, for educational purposes, mainly for students, to make such distinctions. It is strange that the same mathematicians who are so careful with the terminology they use do not feel the need to make systematic use of these distinctions.

**Notation.** The = symbol was introduced for the first time in 1557 by the mathematician and physician Rober Recorde.[28] Later, in 1801, the symbol ≡ was introduced for identities.[29]

This analysis is even more appropriate in physics, where it is good to distinguish between a *constitutive* or *material* equation and an equation describing a *universal law*. The material laws specify the behaviour of a material or a medium, considering the vacuum as a limit case, and contains, in general, physical parameters and material constants. These are *particular laws*, of a phenomenological nature, representing the result of measurements that are valid within a definite range of values. The universal laws, in contrast, are valid in any situation, do not depend on the material and, hence, do not contain material parameters. Thus, faced with the equations

$$V = RI, \qquad \sigma = E\,\varepsilon, \qquad \partial_t \rho + \operatorname{div} \mathbf{J} = 0, \qquad \operatorname{grad} p = \mathbf{f},$$

we see that the first two are material laws, whereas the third expresses a conservation (of charge, of mass, of energy) and the fourth the equilibrium of a fluid (of whatever fluid). This suggests that one should write them as follows:

$$V \overset{\text{mat}}{=} RI, \qquad \sigma \overset{\text{mat}}{=} E\,\varepsilon, \qquad \partial_t \rho + \operatorname{div} \mathbf{J} \overset{\text{law}}{=} 0, \qquad \operatorname{grad} p \overset{\text{law}}{=} \mathbf{f}\,.$$

In summary, we will use the following notations:

$$\begin{array}{llll} \overset{\text{def}}{=} & \text{definition} & H \overset{\text{def}}{=} U + pV & \text{definition of enthalpy,} \\ \equiv & \text{identity} & a^2 - b^2 \equiv (a+b)(a-b) & \text{for all } a \text{ and } b, \\ = & \text{equation} & 3x^2 - 2x = 5 & \text{the variable } x \text{ unknown,} \\ \overset{\text{mat}}{=} & \text{material law} & V \overset{\text{mat}}{=} RI & \text{Ohm's law,} \\ \overset{\text{law}}{=} & \text{general law} & \partial_t \rho + \operatorname{div} \mathbf{J} \overset{\text{law}}{=} 0 & \text{conservation law.} \end{array} \tag{2.19}$$

We admit that this distinction cannot be easily maintained in the elaboration of the equations. Nevertheless, we think that, at least at the stage of introducing a theory, it can contribute to clarifying the meaning and the physical content of the equations. For this reason, even if faced with some uncertainties, in this book we will make such distinctions as much as possible.

---

[28] See the Web site *http://en.wikipedia.org/wiki/Equals_sign* and the books cited therein.

[29] See *http://en.wikipedia.org/wiki/Table_of_mathematical_symbols_by_introduction_date*.

**Notation.** The notation $\stackrel{\text{def}}{=}$ is not completely new; it is used by Schouten [204](p. 1), although more common are the symbols $\triangleq$ and $:=$. We prefer that symbol because it is more readable for non-mathematicians and is similar to the symbols $\stackrel{\text{mat}}{=}$ and $\stackrel{\text{law}}{=}$. The last two symbols seem to be new.

## 2.10 Irrotational and Solenoidal Vector Fields

The terminology of vector fields is strongly based on the differential formulation. Thus, a vector field is commonly said to be *irrotational* if the curl vanishes and *solenoidal* if the divergence vanishes. In this book, we want to stress the *finite* (i.e. not differential) formulation of physics because a premature appeal to the process of performing the limit hides some geometric features. For this reason it is better to give the following definitions:[30]

Definition. *A vector field is called* irrotational *in a space region if the line integral of the vector along any* reducible *closed line contained in the region vanishes; it is called* solenoidal *in a space region if the surface integral through any* reducible *closed surface contained in the region vanishes.*

What is not explicit in the corresponding differential formulation, i.e. $\nabla\times\mathbf{u} = 0$ and $\nabla\cdot\mathbf{u} = 0$, is that closed lines and closed surfaces must be *reducible* in the region. This specification is essential when we want to pass from a local form to a global one: the condition $\nabla\times\mathbf{u} = 0$ implies the existence of a scalar potential $\phi$, i.e. $\mathbf{u} = \nabla\phi$ only in a space region which is *simply connected with respect to lines*, i.e. if all its closed lines contained in the region are reducible. Moreover, the condition $\nabla\cdot\mathbf{u} = 0$ implies the existence of a vector potential $\mathbf{v}$, i.e. $\mathbf{u} = \nabla\times\mathbf{v}$ only in a space region which is *simply connected with respect to surfaces*, i.e. if all its closed surfaces contained in the region are reducible.

Another limitation of the traditional definition $\nabla\times\mathbf{u} = 0$ and $\nabla\cdot\mathbf{u} = 0$ is that the field must be continuous and must have continuous partial derivatives. This condition is not always satisfied. Thus, the electric field strength $\mathbf{E}$ of a metallic body, for instance, is discontinuous on its surface, and hence one cannot evaluate its curl at the points of the surface, and therefore one cannot say whether the field is irrotational. According to the definition just given, since the line integral through every closed line vanishes, the electric field, even if discontinuous, is irrotational.

[30] Recall that a closed line and a closed surface are called *reducibles* in a given region if they can be contracted to a point always remaining in the given region, or, which is equivalent, without crossing the boundary of the region.

## 2.11 Reversal of Motion Is Not Time Reversal

Almost all authors consider time reversal to be equivalent to the reversal of motion.[31] Few authors consider the two as different notions;[32] in particular, Post distinguishes between *active* and *passive* time reversals.[33]

By the expression *time reversal* is meant the change of time variable $t$ into $-t$, while by the expression *reversal of motion* is meant the process of reversing the time sequence of the natural motion of a system.

It is commonly assumed that *time reversal* is equivalent to *reversal of motion*. In contrast to this, we want to show that these are two distinct operations; the first is a purely mathematical operation, whereas the latter has a physical meaning.

Let us first consider a particle in motion. According to time reversal, $t \longrightarrow -t$, we have $\mathrm{d}t \longrightarrow -\mathrm{d}t$; hence, the velocity $\overline{\mathbf{v}}$ after the time reversal is given by

$$\mathbf{v}' = \frac{(\mathrm{d}\mathbf{r})'}{(\mathrm{d}t)'} = \frac{\mathrm{d}\mathbf{r}}{(-\mathrm{d}t)} = -\mathbf{v}\,. \tag{2.20}$$

This is a *purely mathematical* operation. In this operation, the infinitesimal displacement $\mathrm{d}\mathbf{r}$ is left unchanged. In contrast, if we consider the *physical* aspect of

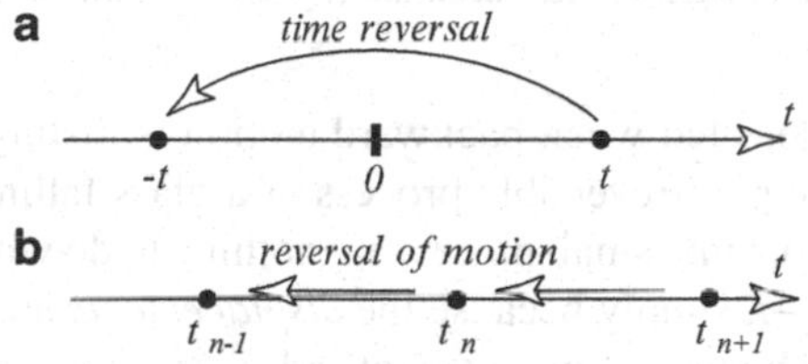


**Fig. 2.1** (**a**) The change of $t$ into $-t$ means the reflection of the time axis. (**b**) Reversal of order of time instants, i.e. reversal of motion

the reversal of motion, then we see that reversing the order of the time sequence causes the displacement $\mathrm{d}\mathbf{r}$ to change sign (Fig. 2.1).

In fact, the most impressive manifestation of the backward motion is the reversal of the *displacement*, i.e. $\mathrm{d}\mathbf{r} \longrightarrow -\mathrm{d}\mathbf{r}$. This becomes clear when we see a film running backwards in time and we compare the backward motion in the pictures with the forward motion of a clock we hold in our hands while watching the film. This means that the orientation of the time interval $\mathrm{d}t$ used to compare the backward motion with the forward motion has not changed. It is $(\mathrm{d}t)' = \mathrm{d}t$. It follows that

---

[31] Wigner [250, p. 325].

[32] Ballentine [8, p. 377].

[33] Post [186, p. 136], [185, p. 287], [187, p. 834].

$$\mathbf{v}' = \frac{\mathrm{d}\mathbf{r}'}{\mathrm{d}t'} = \frac{(-\,\mathrm{d}\mathbf{r})}{\mathrm{d}t} = -\mathbf{v}\,. \tag{2.21}$$

We see that both in the time reversal and in reversal of motion the velocity changes sign, and this leads us to believe that time reversal and reversal of motion are synonyms, which is wrong. In time reversal, we perform a purely mathematical substitution of $t$ into $-t$, which leaves the displacement $\mathrm{d}\mathbf{r}$ unchanged, whereas in reversal of motion, the displacement changes sign, i.e. $\mathrm{d}\mathbf{r} \longrightarrow -\,\mathrm{d}\mathbf{r}$, and the sign of the time interval is unchanged. This is important in this book because we must examine the behaviour of global variables under reversal of motion. Henceforth, *we will consider only reversal of motion as physically significant.*
Post wrote:[34]

> ... one should keep in mind that the active time reversal characteristics so obtained cannot be directly identified with the passive transformational characteristics of physical quantities under the formal operation of coordinate time reversal. In the active time reversal it is the object that is being changed, while the reference is unaffected, whereas in the passive time reversal it is the reference that is being changed while the object remains unaffected.

Ballentine wrote:[35]

> In the first place, the term "time reversal" is misleading, and the operation that is the subject of this section would be more accurately described as *motion reversal*. We shall continue to use the traditional but less accurate expression "time reversal", because it is so firmly entrenched.

Irreversibility is detected when backward motion is distinguishable from forward motion. The simple irreversible process of a glass falling down, impacting the ground and breaking into small pieces, has nothing to do with the mathematical substitution of $t$ into $-t$, simply because the *displacements* are inverted. We summarize the difference between time reversal and reversal of motion in Table 2.5.

**Table 2.5** Reversal of motion and time reversal

| **Reversal of motion** | **Time reversal** |
|---|---|
| or 'active time reversal' | or 'passive time reversal' |
| displacements reversed | displacements unchanged |
| $\mathrm{d}\mathbf{r} \longrightarrow -\,\mathrm{d}\mathbf{r}$ | $\mathrm{d}\mathbf{r} \longrightarrow \mathrm{d}\mathbf{r}$ |
| Time intervals unchanged | Time intervals reversed |
| $\mathrm{d}t \longrightarrow \mathrm{d}t$ | $\mathrm{d}t \longrightarrow -\,\mathrm{d}t$ |

Let us denote by $\mathcal{R}$ the operation of reversal of motion. Denoting by $\mathbf{u}$ a displacement we have the two defining relations

$$\mathcal{R}(\mathbf{u}) = -\mathbf{u} \qquad \mathcal{R}(t) = t\,. \tag{2.22}$$

[34] Post [186, p. 136].

[35] Ballentine [8, p. 377].

Denoting by **r** the position vector of a particle, **v** its velocity, **u** the displacement in a time interval, **F** the force acting on it, **J** the impulse of the force in a time interval and **p** its momentum, the following relations are valid:

$$\begin{aligned}
\mathbf{u} &= \int_{t^-}^{t^+} \mathbf{v}(t)\,\mathrm{d}t &&\longrightarrow & \mathcal{R}(\mathbf{u}) &= \int_{t^-}^{t^+} \mathcal{R}[\mathbf{v}(t)]\,\mathcal{R}(\,\mathrm{d}t) = \int_{t^-}^{t^+} (-\mathbf{v})\,\mathrm{d}t = -\mathbf{u},\\
\mathbf{r}(t) &= \int_{0}^{t} \mathbf{v}(t')\,\mathrm{d}t' &&\longrightarrow & \mathcal{R}(\mathbf{r}) &= \int_{t}^{0} \mathcal{R}[\mathbf{v}(t')]\,\mathcal{R}(\,\mathrm{d}t') = -\int_{0}^{t} -\mathbf{v}(t')\,\mathrm{d}t' = +\mathbf{r},\\
\mathbf{J} &\stackrel{\text{def}}{=} \int_{t^-}^{t^+} \mathbf{F}(t)\,\mathrm{d}t &&\longrightarrow & \mathcal{R}(\mathbf{J}) &= \int_{t^-}^{t^+} \mathcal{R}[\mathbf{F}(t)]\,\mathcal{R}(\,\mathrm{d}t) = \int_{t^-}^{t^+} \mathbf{F}(t)\,\mathrm{d}t = +\mathbf{J},\\
\mathbf{p}(t) &\stackrel{\text{def}}{=} \int_{0}^{t} \mathbf{F}(t')\,\mathrm{d}t' &&\longrightarrow & \mathcal{R}(\mathbf{p}) &= \int_{t}^{0} \mathcal{R}[\mathbf{F}(t')]\,\mathcal{R}(\,\mathrm{d}t') = -\int_{0}^{t} +\mathbf{F}(t')\,\mathrm{d}t' = -\mathbf{p}.
\end{aligned} \tag{2.23}$$

This states that the radius vector and the impulse do not change signs under reversal of motion, whereas the displacement and the momentum change signs.

# Chapter 3
# Space and Time Elements and Their Orientation

## 3.1 Space Elements

Points, lines, surfaces and volumes: these will be called *space elements* and will be denoted by their initial capital letter, i.e. **P** for points, **L** for lines, **S** for surfaces, **V** for volumes (Fig. 3.1). We have chosen *capital* letters to render the notation uniform with the common practice of denoting points by capital letters, such as $A, B, C, P, Q$, and we have chosen *boldface* type because we usually denote a point function by $f(t, \mathbf{x})$ or $f(t, \mathbf{P})$. Another reason to use boldface type lies in the fact that we find it natural to use the corresponding letters $L, S, V$, in plain text, to denote the extension of the corresponding space elements, i.e. *length, area* and *volume*.[1] In summary:

$$\begin{array}{r} \text{Point } \mathbf{P} \\ \text{Line } \mathbf{L} \longrightarrow L \text{ length} \\ \text{Surface } \mathbf{S} \longrightarrow S \text{ area (sometimes } A) \\ \text{Volume } \textit{(as region)}\ \mathbf{V} \longrightarrow V \text{ volume } \textit{(as magnitude)} \end{array} \tag{3.1}$$

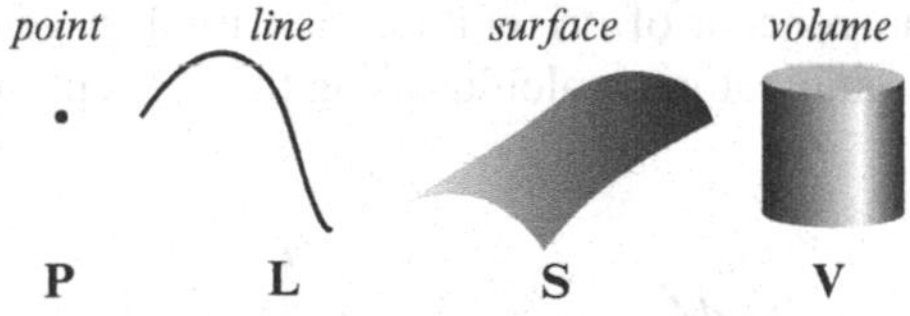


**Fig. 3.1** The four space elements: points, lines, surfaces, volumes

[1] Unfortunately, the term *volume* has two meanings: it denotes a space region and its magnitude. This ambiguity does not exist for lines and surfaces whose size, length and area respectively have a different name.

E. Tonti, *The Mathematical Structure of Classical and Relativistic Physics*, Modeling and Simulation in Science, Engineering and Technology,
DOI 10.1007/978-1-4614-7422-7_3, © Springer Science+Business Media New York 2013

## 3.2 Orientation of Space Elements

The pages of a book have a front and back, a box has an inside and an outside, a road can be travelled in one of its two directions, a wheel can spin clockwise or anticlockwise. All these facts are manifestations of a single concept: orientation. *The notion of orientation is, essentially, a notion of order*. All museums have a port of entry and a port of exit; it follows that the path of the visitors is oriented.

DEFINITION. *Given a set composed of two elements, one can say that this set is* oriented *when we have decided on an order of the two elements, which is the preceding element and which is the next or, put another way, which is first and which is second. Terms such as front/back, before/after, input/output and similar terms fix an order and therefore provide an orientation.*

This notion is important in physics because the sign of a physical variable associated with a space element is inverted when we invert the orientation of the space element. Thus, to describe the motion of a particle along a line, we must choose a positive sense along the line; when we speak of mass flow, we must discriminate between the two sides of the surface, and when we write the mass balance, we choose the outward normal to the volume, as shown in Fig. 3.2. These are some of the cases in which we make use of the notion of orientation. We will see that the oriented elements in Fig. 3.2 are particular cases of the notion of orientation of a space element. As we will see, there is a more systematic way of orienting a space element which also includes the points and which will lead to the establishment of the concepts of *inner* and *outer* orientation. This distinction will replace the common notions of Fig. 3.2 by a broader and more systematic classification scheme, as shown in Table 3.1 on p. 41. The important role of orientation in physics motivates a detailed analysis of the two types of orientation of space and time elements.

We stress that the concept of orientation, first used in geometry, can be expressed in terms of combinatorial calculus using the concept of permutation.

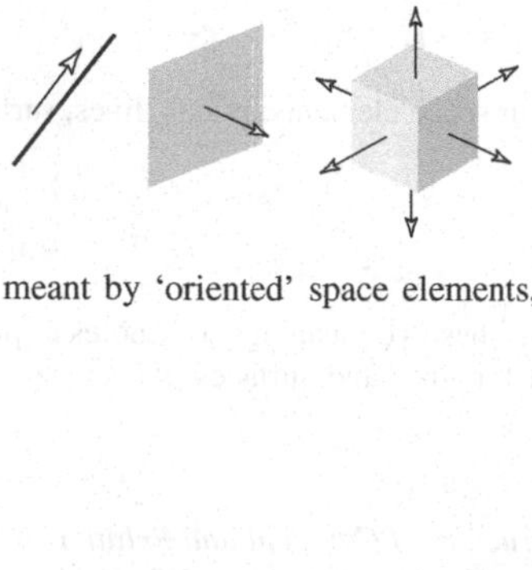

**Fig. 3.2** What is commonly meant by 'oriented' space elements, but there are other kinds of orientation

**Table 3.1** The eight oriented space elements

| | | Inner orientation | Outer orientation | | |
|---|---|---|---|---|---|
| $\overline{\mathbf{P}}$ | 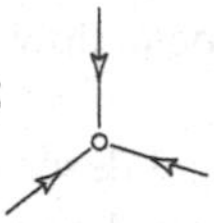 | Inner orientation of a point: a positive orientation is the one of sink | Outer orientation of a volume: a positive orientation is the one with outwards normals | 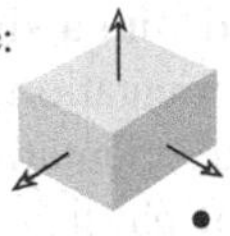 | $\widetilde{\mathbf{V}}$ |
| $\overline{\mathbf{L}}$ |  | Inner orientation of a line: this is the basic notion used to give a meaning to all other space elements | Outer orientation of a surface: this is the inner orientation of a line crossing the surface |  | $\widetilde{\mathbf{S}}$ |
| $\overline{\mathbf{S}}$ |  | Inner orientation of a surface: this is a compatible orientation of its edges, i.e. a direction along its boundary | Outer orientation of a line: this is the inner orientation of a surface crossing the line |  | $\widetilde{\mathbf{L}}$ |
| $\overline{\mathbf{V}}$ | 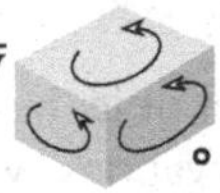 | Inner orientation of a volume: this is a compatible orientation of its faces. It is equivalent to the screw line | Outer orientation of a point: this is the inner orientation of the volume containing the point |  | $\widetilde{\mathbf{P}}$ |

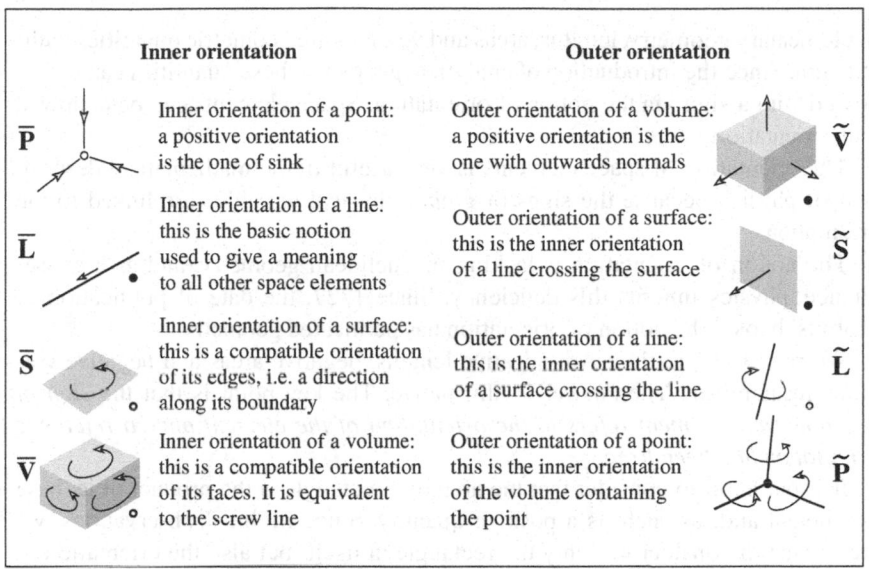

The filled balls indicate the most common orientations of lines, surfaces and volumes used in physics; the empty balls denote orientations seldom mentioned. Finally, the inner and outer orientations of the points are never mentioned!

## 3.3 Combinatorial Side of Orientation

In combinatorics, we study the *permutations* of $n$ objects, and we introduce the notion of *transposition* of two elements. Given a reference permutation of three elements, which we denote by (**ABC**) by making all the possible inversions, one obtains other permutations which can be divided into two classes: those resulting from the reference permutation with an even number of inversions and those obtained with an odd number of inversions. The first are called permutations of an *even* class, the second are called permutations of an *odd* class.

If we agree to use the + sign to indicate all the permutations of the even class, including the reference one, it is natural use the − sign to indicate those which belong to the odd class. Hence, if (**PQR**) is the reference permutation of three elements, we obtain the following two classes:

$$\begin{aligned} (\mathbf{PQR}) = (\mathbf{QRP}) = (\mathbf{RPQ}) \qquad & \text{even: } +\,, \\ (\mathbf{RQP}) = (\mathbf{QPR}) = (\mathbf{PRQ}) \qquad & \text{odd : } -\,. \end{aligned} \tag{3.2}$$

At this point, the notion of the **orientation** of a line, a surface and a volume comes into play.

## 3.4 Orientation in Geometry

In elementary geometry, lengths, areas and volumes are geometric quantities without sign. Since the introduction of analytical geometry these quantities can be endowed with a sign via the notion of orientation. Space elements can be endowed with orientation.

The orientation of space elements is very useful in the mathematical description of physics because the signs of *global* physical variables are linked to the orientation.

The notion of orientation is lacking in Euclidean geometry, and as a consequence, physics inherits this deficiency. Since 1827, the date of publication of Möbius' book,[2] the notion of orientation has permeated geometry.[3]

We raise the question: Are negative lengths, negative areas and negative volumes meaningful? The answer is affirmative. The key point is that the *sign of a geometrical element refers to the orientation of the element once a reference orientation has been fixed.*

It is our habit to consider that the area of a rectangle is the product of its base and height and, as such, is a positive quantity. Since Möbius' observations, we are invited to consider not only the rectangle in itself, but also the orientation of its edges. Thus we can say that the area of a rectangle is ± the product of its base and height. We are not accustomed to considering negative magnitudes of space elements despite the fact that mathematics provides formulae for measuring the extension of a space element endowed with sign. Hence, with reference to Fig. 3.3, the formulae

$$L_{OP} = \begin{vmatrix} 1 & 0 \\ 1 & x_P \end{vmatrix}, \qquad A_{OPQ} = \frac{1}{2!}\begin{vmatrix} 1 & 0 & 0 \\ 1 & x_P & y_P \\ 1 & x_Q & y_Q \end{vmatrix}, \qquad V_{OPQR} = \frac{1}{3!}\begin{vmatrix} 1 & 0 & 0 & 0 \\ 1 & x_P & y_P & z_P \\ 1 & x_Q & y_Q & z_Q \\ 1 & x_R & y_R & z_R \end{vmatrix} \tag{3.3}$$

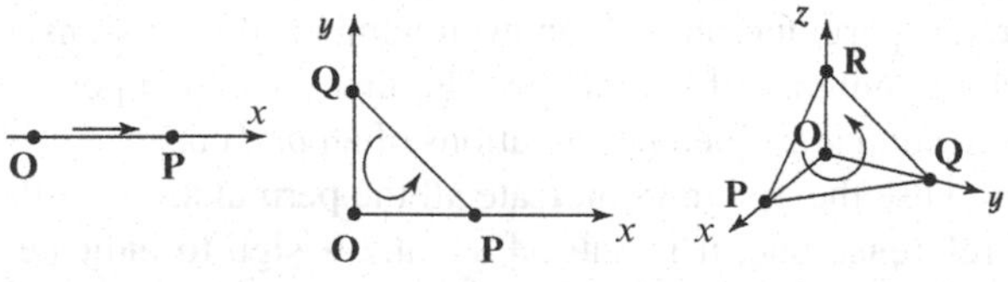


**Fig. 3.3** Simplexes of one, two and three dimensions. Recall that the term *simplex* denotes the most simple polygon, i.e. the triangle, and the most simple polyhedron, i.e. the tetrahedron

---

2 Möbius [162].

3 Klein [114, p. 16].

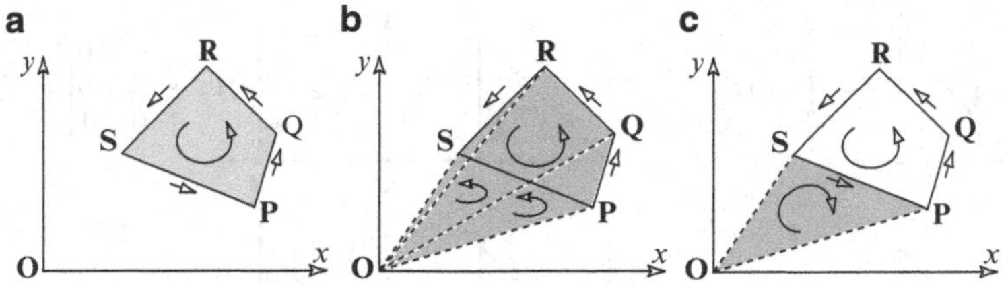


**Fig. 3.4** The oriented area of the polygon is the sum of the oriented areas of the triangles with a common vertex

give the oriented magnitude of a segment of the $x$-axis, a triangle in the $xy$-plane and a tetrahedron in $xyz$ space respectively.

NOTATION. In contrast to the traditional notation in the determinant, we set the column composed of ones as the first column instead of the last column: in this way, for a segment OP we obtain a length of OP $= x_P$ and the volume of the tetrahedron OPQR indicated in Fig. 3.3 is positive. The elements of the first row are the coordinates of the origin of the coordinate system.

In particular, the sign of the area of a triangle depends on the order in which its vertices are traversed. One advantage of considering areas with a sign is that if we want to evaluate the area of an oriented polygon, then we can decompose the polygon into triangles by taking a reference point.

For example, in Fig. 3.4, let us consider the triangles with a common vertex in the reference point **O** and with an oriented edge of the polygon as their sides. By summing the oriented areas of every triangle we obtain the area of the oriented polygon. This procedure can be used even if the polygon is non-convex and if the reference point is inside the polygon. Let us remark that we also use oriented areas in the integration of a function. As is known, the geometric interpretation of the definite integral of a function of one variable is the area enclosed under the representative curve. If we reverse the extremes of integration, the orientation of the time interval is reversed – $[a, b]$ into $[b, a]$ – and the sign of the definite integral is reversed. As Fig. 3.5 shows, the orientation of the area is also reversed; hence, this implies that the definite integral is associated with an oriented area, not simply ti its absolute value. If we give a line an orientation, the length of a line segment will be positive or negative; this depends on the order given to its vertices.

The great mathematician Felix Klein in his book *Elementary Mathematics from an Advanced Standpoint, Geometry* (which we recommend to our reader) states[4]:

> We say ordinarily that length or, as the case may be, the area or the volume, is equal to the *absolute value* of these several magnitudes, whereas, actually, our formulas furnish, over and above that, *a definite sign*, which depends upon the order in which points are taken. [...] Great advantages are thus gained over the ordinary elementary geometry which considers length and contents as absolute magnitudes.

[4] Klein [114, pp. 3–5].

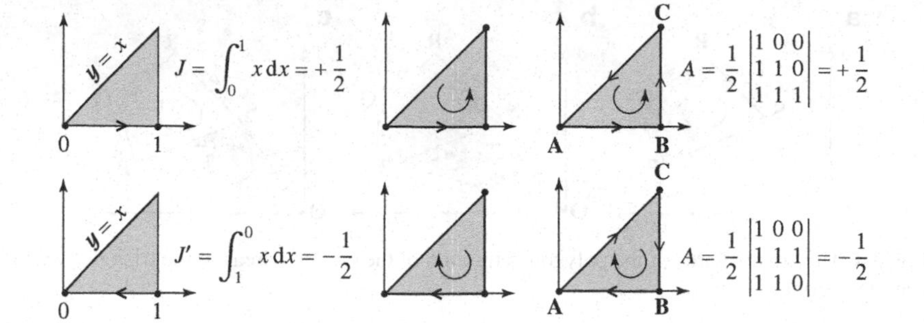


**Fig. 3.5** The definite integral of a function of one variable has a geometrical representation of the oriented area under the representative curve of the function

When we say that lengths, areas and volumes are *additive*, we implicitly admit that each of them can have a negative value. In fact, the additive property implies the existence of the null element and of the opposite of every element. Analytical geometry takes into account signs, whereas elementary geometry ignores them.

## 3.5 Two Kinds of Orientation: Inner and Outer

Physics books do not treat the notion of orientation in detail simply because books on geometry do the same! Even books on *algebraic topology*, which use the notion of orientation, do not usually distinguish between the two kinds of orientation: the *inner* and the *outer* orientation of a space element.[5] The orientation we presented in the preceding section can be properly called *inner* orientation.

Of course, the outer orientation of a space element can be reduced to an inner orientation by means of the screw rule. Nevertheless, we do not do this because this leads to identifying distinct geometric objects.

This is what happens when we replace "skew-symmetric tensors by vectors ...this may be justifiable on the ground of economy of expression, but in some way it hides the essential features," as Weyl stated.[6] In the present analysis the reduction of outer orientation to inner orientation and vice versa halves the number of distinct space elements and coalesces distinct geometrical objects.

What is meant by the 'inner' and 'outer' orientation of a space element? Let us consider a surface: given a curvilinear triangle and considering an order of its vertices we have fixed an inner orientation. The term *inner* refers to the fact that this orientation requires that only points of a surface be considered, i.e. we move and stay inside the surface. In contrast, an outer orientation requires us to consider both sides of the surface, i.e. to pass from one side to the other crossing

[5] This subject is extensively treated in the book by Schouten [204].

[6] Weyl [247, p. 46].

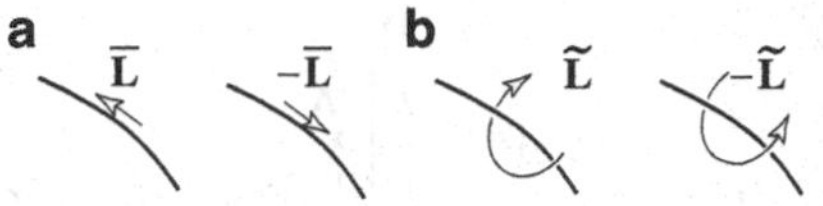


**Fig. 3.6** Orientation of a line: (**a**) inner orientation; (**b**) outer orientation

the surface, and this implies that we must go out from the surface. This justifies the term *outer* for this orientation.

## 3.5.1 Inner and Outer Orientation of Lines

A one-way street can be represented by a line with an *inner* orientation because there is a direction *along* the line, as seen in Fig. 3.6a. In what follows, a line endowed with an inner orientation will be denoted by putting a bar over the symbol, i.e. $\overline{\mathbf{L}}$.

The Earth's axis of rotation can be represented by a line with a sense of rotation *around* the axis, i.e. an outer orientation (Fig. 3.6b). The *outer* orientation of a line is useful for describing, for example, the rotatory polarization of a light beam. A line endowed with an outer orientation will be denoted by putting a *tilde* over the symbol, i.e. $\tilde{\mathbf{L}}$.

## 3.5.2 Inner and Outer Orientation of Surfaces

A surface has an *inner* orientation when a direction of its boundary has been selected (Fig. 3.7a). A surface endowed with inner orientation will be denoted by placing a bar over the symbol, i.e. $\overline{\mathbf{S}}$.

Example 1. When the boundary is traveled in the counterclockwise sense, the area enclosed is positive, the system expands giving work to the environment, as shown in Fig. 3.8 a. On the contrary when the boundary is traveled clockwise, the area enclosed is negative, we have a

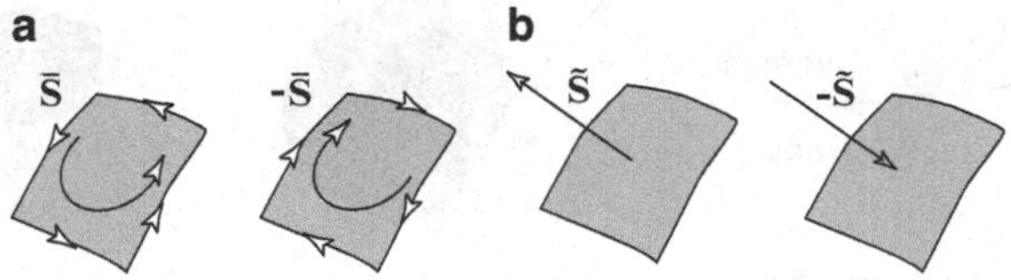


**Fig. 3.7** Orientation of a surface: (**a**) inner orientation; (**b**) outer orientation

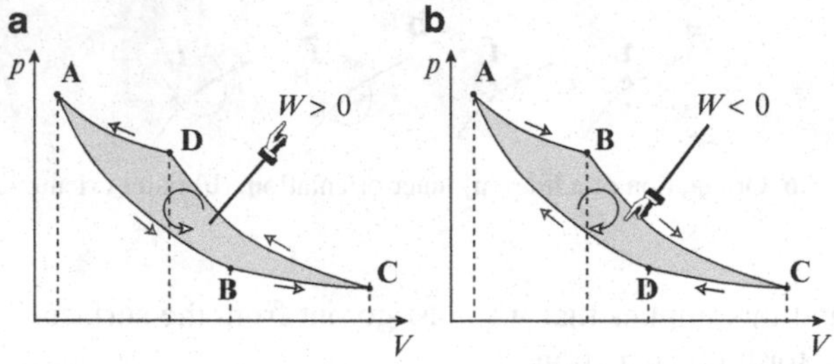


**Fig. 3.8** Pressure (p)–volume (V) diagram: the orientation of the area is linked with the sign of the work (W). (**a**) positive area implies a positive work; (**b**) a negative area implies a negative work

compression and the system absorbes work from the environment, as shown in Fig. 3.8 b. This means that the sign of the oriented area is linked with the sign of the work.

If we use the modern convention[7] of considering positive the work entering a system, as is considered positive the heat entering the system, we see that a positive area (oriented anticlockwise) implies positive work.

Therefore, if we invert the orientation of the area of the cycle, then the sign of the associated physical variable *work* is inverted. Hence the work involved in a thermodynamic cycle is an example of a physical variable whose sign depends on the inner orientation of the area which the cycle includes.

A surface has an *outer* orientation when one of the two sides is chosen as the front and the other as the back (Fig. 3.9, left). If we choose two colours, say white and black, we can decide that the front of the surface can be coloured white while the rear can be in black, as shown in Fig. 3.10. We can also mark the front with a + sign and the back with a − sign. Equivalently, we can draw an arrow crossing the surface from back to front.

The outer orientation of a surface is widely used in physics. Thus, the heat which crosses a surface is considered positive when it enters the surface, i.e. when it goes in the opposite direction of the arrow which crosses the surface. The same

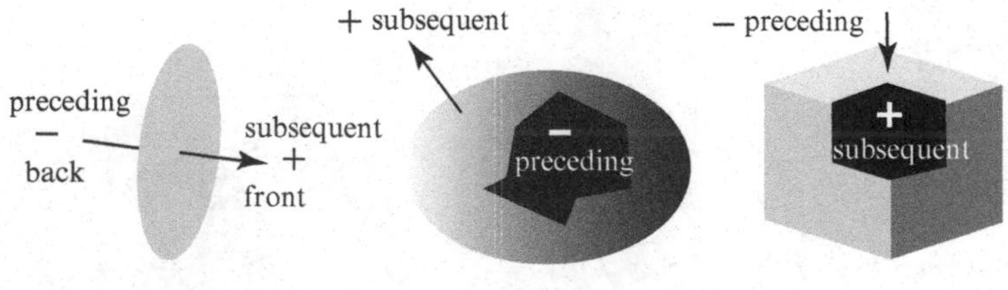


**Fig. 3.9** Orientation of a surface and a volume

[7] Clausius and his followers considered positive the work given *by* the system.

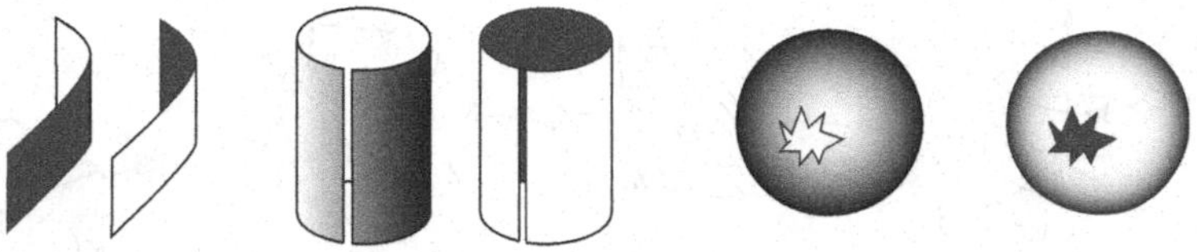

**Fig. 3.10** Outer orientation of a surface represented by two colours on two sides, *white* and *black*

convention is used in continuum mechanics: the surface force associated with a small piece of a flat surface is the one which the matter that lies on the front exerts on the matter that lies on the back.[8]

In contrast to these conventions, mass current, energy current, and charge current, which are associated with surfaces, are considered positive when they move in the same direction of the arrow, i.e. from back to front. A surface endowed with an outer orientation will be denoted by placing a *tilde* over the symbol, i.e. $\widetilde{\mathbf{S}}$.

### *3.5.3 Inner and Outer Orientation of Volumes*

With reference to Fig. 3.11a, a cube has an *inner* orientation when an inner orientation is chosen on one of its faces and, hence, is *coherently* propagated to all faces. 'Coherently' means that the orientation of the two faces induces an opposite orientation on the common edge. This is *Möbius' law of edges*.[9] If we consider two opposite faces of a cube, then we see that their orientations, if viewed from a single point, are opposite: if we rotate the two faces in the direction of their own orientation, then we obtain a screw, as shown in Fig. 3.12. Thus *the passage from a left-handed screw to a right-handed screw is equivalent to a change in the inner orientation of a volume.*

A volume endowed with an inner orientation will be denoted by placing a bar over the symbol, i.e. $\overline{\mathbf{V}}$.

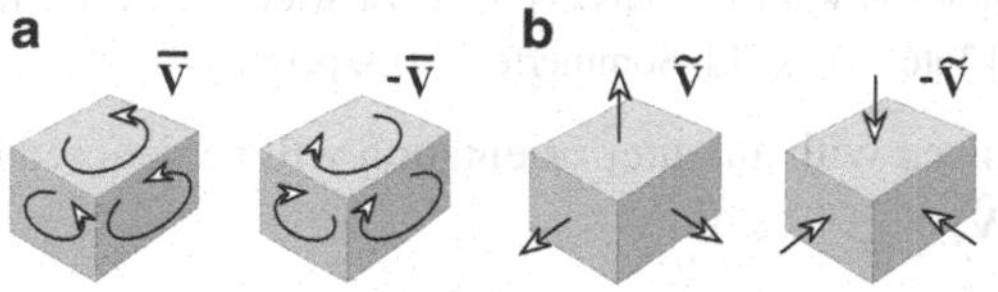


**Fig. 3.11** Orientation of a volume: (**a**) inner orientation; (**b**) outer orientation

[8] Fung [77, p. 58], Paterson [172, p. 96], Prager [188, p. 43].

[9] Klein [114, p. 17].

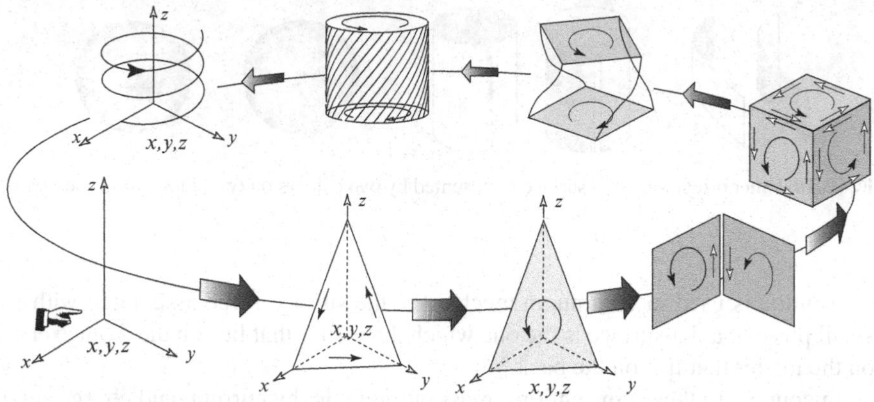


**Fig. 3.12** The inner orientation of a cube is equivalent to the choice of a right-handed or a left-handed helix

A volume is endowed with an *outer* orientation when the interior of the volume is considered as preceding and the exterior as following, or vice versa, i.e. when we consider positive the negative and interior the exterior or vice versa. Since to give an outer orientation to a surface we have selected one of its faces as preceding and considered an arrow going from the preceding side to the following one, it follows that we can do the same with a volume. Hence, if we have chosen the interior as the preceding part, then the arrow crossing its boundary must be directed outwards. In contrast, if we consider the exterior of the volume as the preceding part, the arrows must go inside the volume, as shown in Fig. 3.9. Recall that the divergence is the limit of the ratio outflow/volume.

**Remark.** In the old scientific literature, the positive orientation of volumes was the one with inward arrows. Later it was decided to invert this choice and take the outward normals as positive. In Maxwell's time, volumes were oriented with inward normals, and Maxwell introduced the notion of *convergence*, which was later changed to *divergence*: see [152, p. 231]. The old convention is used by Lamb [119, p. 44], Milne-Thomson [160, p. 54], Fer [66, Vol. II, p. 36]. This convention is still used in hydraulics (see De Marchi [52, Vol. I, p. 24]) and in geotechnics. The modern convention (outward normals) is the most widely used one today; see Fung [77, p. 66], Billington and Tate [15, p. 70], Sommerfeld [214, p. 60].

A volume endowed with an outer orientation will be denoted by placing a *tilde* over the symbol: $\tilde{\mathbf{V}}$.

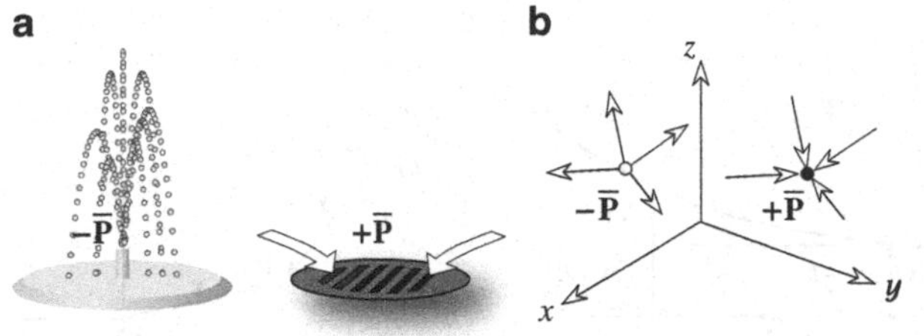


**Fig. 3.13** (**a**) A fountain is a source, whereas a manhole is a sink. (**b**) A point with an inner orientation means that it is conceived as a source or a sink

### *3.5.4 Inner Orientation of Points*

Is the orientation of a point meaningful? At first glance the answer would seem to be negative because a point has no spatial extension. However, we distinguish water emanating from a source from water disappearing into a hole in the ground: the first is a *source* whereas the second is a *sink* (Fig. 3.13a). When a point is conceived as a source or a sink, it will be endowed with *inner* orientation. This is used in electromagnetism to distinguish a positive from a negative charge. Let us give the following definition:

DEFINITION. *In three-dimensional space, the* **dual** *of a space element of dimension p is an element of dimension* $(3 - p)$.

In particular, the dual of a point is a volume, the dual of a line is a surface, the dual of a surface is a line and the dual of a volume is a point.

As we can see, the notions of inner and outer orientation are linked by the following rule: *the outer orientation of a space element is by definition the inner orientation of the dual space element.* Thus, the outer orientation of a line coincides with the inner orientation of a surface crossing the line and vice versa. This rule also gives a geometrical meaning to the inner and outer orientations of points.

The best way to give meaning to the inner orientation of a point is to define it as *the outer orientation of its dual space element, i.e. the volume*. Since a volume is endowed with an outer orientation when its boundary has an outgoing or ingoing normal, we will say that *a point is endowed with an inner orientation when the outcoming or incoming lines are considered positive or negative.*

REMARK. To illustrate this point, when a concept cannot be applied *directly* to an object, it may be possible to apply the notion *indirectly* to preserve some formal properties. For example, let us consider the integer powers of a number, such as $a^m$. According to its original definition, $a^4$ means $a^4 = a \times a \times a \times a$, and the power $a^0$ is meaningless. Later, we discover the rule

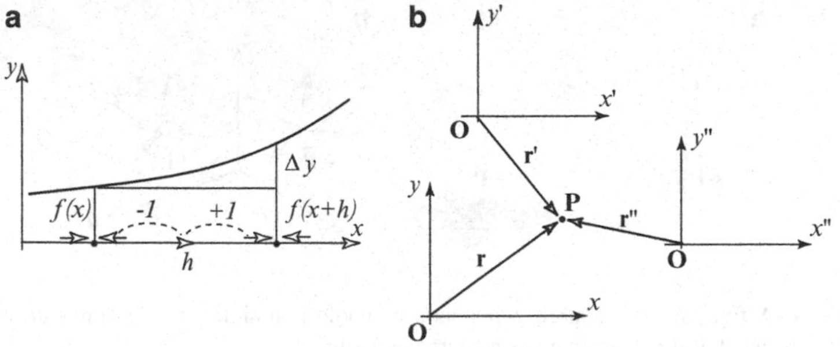


**Fig. 3.14** (**a**) The definition of the increment of a function implies that points have been oriented as sinks. (**b**) The radius vector is always oriented from the origin of a reference system towards a point, and hence the point is oriented as a sink

$a^m/a^n = a^{m-n}$ for $m > n$. We can make this rule valid also for $m = n$ because $a^m/a^m = a^0 = 1$. Thus the expression $a^0$, which has no *direct* meaning, can have an *indirect* meaning by setting $a^0 = 1$.

In particular, we will show that *points are implicitly oriented as sinks*, as shown in Fig. 3.14. This is never explicitly stated, but it can be inferred from the fact that the usual definition of the increment $\Delta f(x) \stackrel{\text{def}}{=} f(x+h) - f(x)$ can be reinterpreted as the following sum:

$$\Delta f = (+1)f(x+h) + (-1)f(x) . \tag{3.4}$$

In fact, if we consider that all line segments are oriented as the $x$-axis, we see that the plus sign in front of the term $f(x+h)$ and the minus sign in front of the term $f(x)$ can be interpreted as *incidence numbers*[10] between the oriented line segment and the oriented boundary points, as shown in Fig. 3.14a. Hence, *an increment deals implicitly with the notion of orientation of points.*

Another reason for considering points oriented as sinks is the choice of the radius vector **r** in geometry and physics which starts at the origin of a coordinate system and ends at the point, not vice versa, as shown in Fig. 3.14b. A point endowed with an inner orientation will be denoted by placing a bar over the symbol, i.e. $\overline{\mathbf{P}}$.

REMARK. There is a significant discrepancy in books on algebraic topology regarding the notion of the orientation of a point. To the author's knowledge, none of the books gives a physical or geometric meaning to the concept of orientation of a point. Many of them simply state that a point is oriented when we associate it with the number −1 or +1.[11] Some authors say

---

[10] See p. 186.

[11] Veblen and Whitehead [241, p. 24], Patterson [173, Sect. 40], Seifert and Threlfall [210, p. 42], Schouten [204, p. 55], Ciarlet and Lions [42, p. 123], Alexandrov [3, p. 4], Springer [218, p. 107].

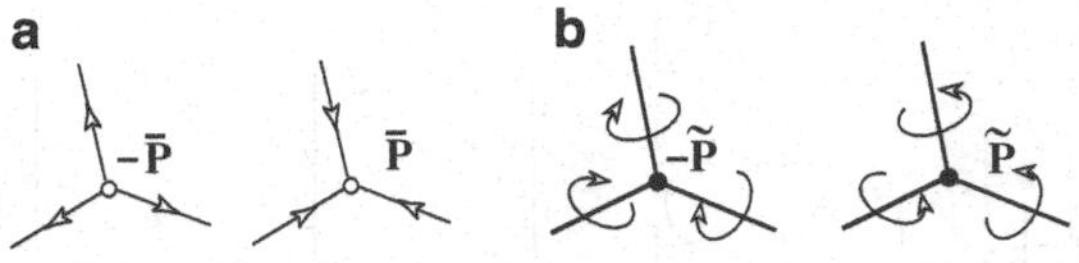


**Fig. 3.15** Orientation of a point: (**a**) inner orientation; (**b**) outer orientation

that 0-simplexes (points) 'admit only one orientation'.[12] One author says that 'a point has no orientation property'.[13]

The notion of source and sink, borrowed from electromagnetism and fluid dynamics, can be defined as an inner orientation of points because it allows us to maintain the notion of incidence number from a $(p+1)$-cell and a $p$-cell even when $p = 0$. We note also that in network theory, to apply Kirchhoff's law of currents, we must decide if the currents coming to a node must be computed as positive or negative, i.e. if nodes must be oriented as sinks or sources.

### *3.5.5 Outer Orientation of Points*

We can define the *outer* orientation of a point in a similar fashion: a point will be endowed with an outer orientation when an outer orientation around any line starting at the point has been selected, as shown in Fig. 3.15b. Of course, the outer orientation of every line must be coherent with those of the other lines. A point endowed with an outer orientation will be denoted by placing a *tilde* over the symbol: $\widetilde{\mathbf{P}}$. Another way to define the outer orientation of a point is to embed the point in a volume and to consider the inner orientation of the volume: such an inner orientation of the volume induces an outer orientation to the point, as shown in Fig. 3.16.

In conclusion, we can say that points can have two kinds of orientation, an inner and an outer one, and these two orientations can be easily visualized. The important thing in physics is that there are physical variables associated with points with an inner orientation, such as *electric potential* and *velocity potential*, whereas others are associated with points endowed with an outer orientation, such as *scalar magnetic potential* and *stream function* in fluid dynamics.[14] The author believes that this alleged lack of meaning about the concepts of inner and outer orientation of points is due to a lack of interest by scholars of algebraic topology with respect to the physical applications of the orientation of points.

---

[12] Hilton and Wylie [92, p. 54], Naber [166, p. 113].

[13] Whitney [253, pp. 4, 360].

[14] See pp. 287 and 363 respectively.

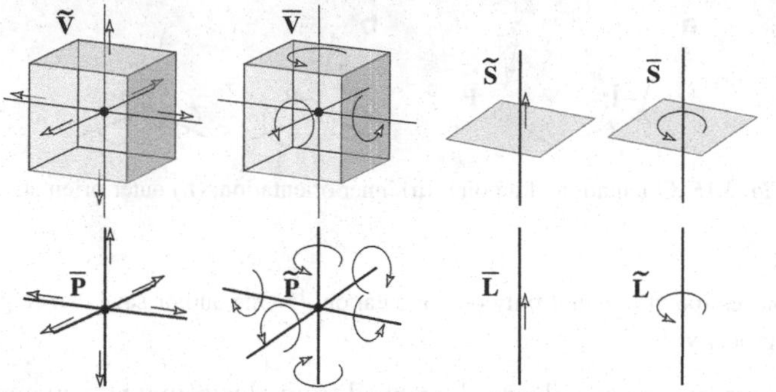

**Fig. 3.16** The inner orientation of a space element induces an outer orientation of the dual element and vice versa

$$\oint \mathbf{v} \cdot \mathbf{t} \, dL = 0 \qquad \oint \mathbf{q} \cdot \mathbf{n} \, dL = 0 \qquad \varphi = \int_{A}^{P} \mathbf{v} \cdot \mathbf{t} \, dL \qquad \psi = \int_{B}^{R} \mathbf{q} \cdot \mathbf{n} \, dL$$

irrotational          solenoidal          velocity potential          stream function

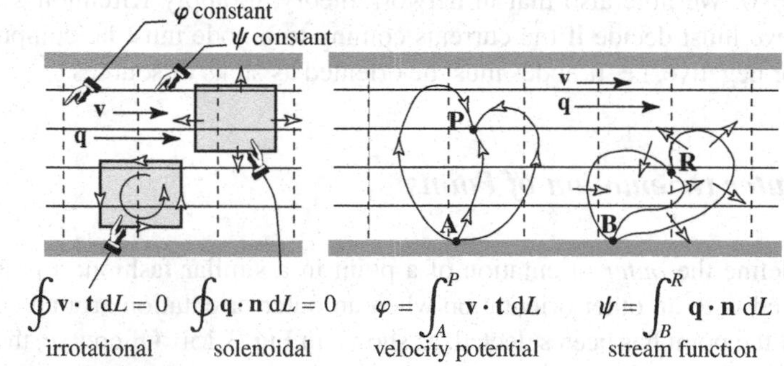

**Fig. 3.17** Fluid flow in plane motion

**EXAMPLE 2.** To show the need to endow points with an inner or outer orientation, let us consider the motion of an inviscid fluid in a rectangular channel. The motion of the fluid is plane, irrotational, isochoric and steady, as in Fig. 3.17. The irrotationality condition means that the line integral of the velocity along every closed line vanishes. The isochoric condition means that the flux across every closed line vanishes. Since the velocity field is irrotational, by choosing a fixed point **A** as the origin we can associate with any point **P** the line integral of velocity from **A** to **P**. To this end, we must consider every line connecting the two points endowed with an inner orientation, and the orientation is from **A** to **P**. Hence, all lines are incoming at **P**, and this point is oriented as a *sink* or positively inner oriented. Let us now consider the flux across a line. By choosing a fixed point **B** as the origin, we can associate with any point **R** the flux of the vector $\mathbf{q} = \rho \mathbf{v}$ across every line connecting **B** with **R**. Since the motion is isochoric, the flux crossing a closed line vanishes, and hence the flux across a line connecting **B** with **R** is the same for all lines. Since we need the normal to the line for the evaluation of the flux, we see from Fig. 3.17 that all lines coming at **R** induce a direction of rotation around **R**, i.e. an *outer* orientation of **R**.

## 3.6 Role of Space of Immersion

While the inner orientation of a space element does not depend on the dimension of the embedding space, the outer orientation does. Thus the outer orientation of a segment in three-dimensional space is an arrow around the segment, as in Fig. 3.18a. When the same segment is embedded in a two-dimensional space, its outer orientation is the direction of an arrow crossing it, as in Fig. 3.18b. When the same segment belongs to a one-dimensional line, its outer orientation is like a traction or a compression, as in Fig. 3.18c. In the theory of trusses, the bars subjected to traction are considered positive. Let us now consider a plane area. If it is immersed in three-dimensional space, its outer orientation is the direction of an arrow crossing the surface to the plane, as shown in Fig. 3.19a. When it is immersed in a surface, its outer orientation is a direction in which its boundary in the plane is crossed, as shown in Fig. 3.19b. The outer orientation of a point immersed in three-dimensional space is represented in Fig. 3.20a. When it is immersed in a surface, its outer orientation is a direction of rotation around the point, as shown in Fig. 3.20b. Lastly, when it is immersed in a line, an orientation is a crossing direction of the point, as in Fig. 3.20c. With reference to Fig. 3.21a we see that the outer orientation of the $x$-axis in a plane is the inner orientation of the $y$-axis and the outer orientation of the $y$-axis is the inner orientation of the $x$-axis. Figure 3.21b shows that the outer orientation of the $xy$-plane is the inner orientation of the $z$-axis and the outer orientation of the $z$-axis is the inner orientation of the $xy$-plane.

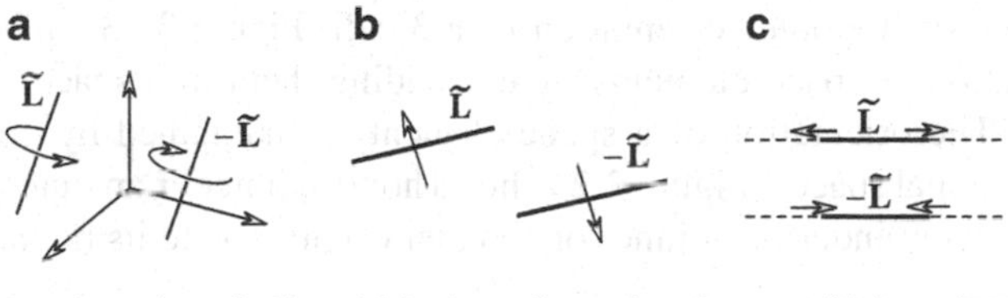


**Fig. 3.18** Outer orientation of a line: (**a**) in three-dimensional space; (**b**) in a plane; (**c**) on a line

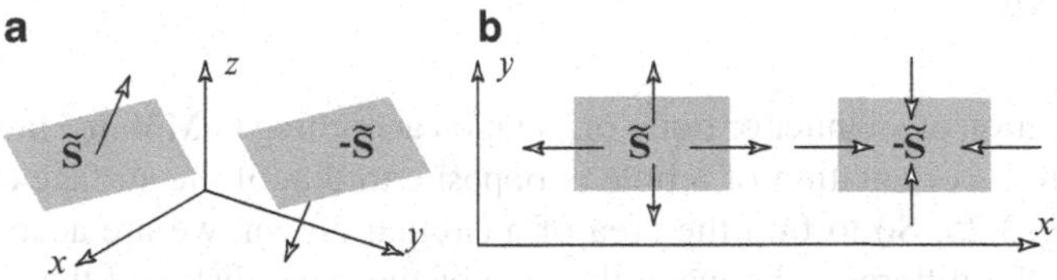


**Fig. 3.19** Outer orientation of a surface element: (**a**) in three-dimensional space; (**b**) in a plane

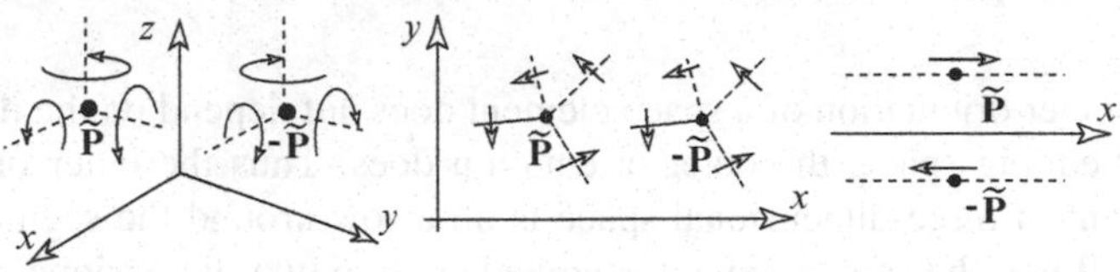


**Fig. 3.20** Outer orientation of a point: (**a**) in three-dimensional space; (**b**) in a plane; (**c**) on a line

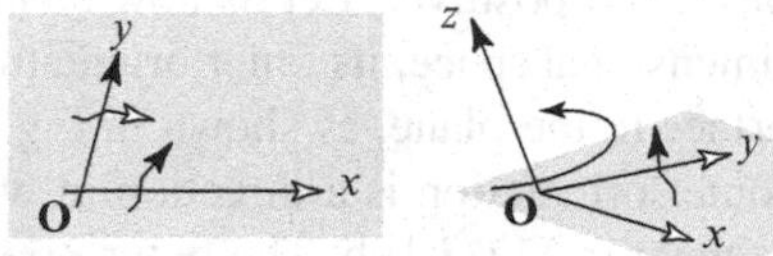


**Fig. 3.21** Relation between inner and outer orientations of coordinate axes: (**a**) in two dimensions: the outer orientation of one axis coincides with the inner orientation of the other and viceversa; (**b**) in three dimensions: the inner orientation of a coordinate plane coincides with the outer orientation of the third axis

### *3.6.1 Induced Orientation*

Figure 3.22 shows how the inner and outer orientations of a 3-cell is transformed into the corresponding orientations of a 2-cell and a 1-cell. In particular, the figure shows that the outer orientation of a 1-cell in a one-dimensional manifold is the equivalent of the outer orientation of a 3-cell. Figure 3.23 summarizes the outer orientation of space elements by embedding them into spaces of different dimensions. The orientation of a space element is maintained by projection on lower-dimensional spaces. Figure 3.24 shows how an inner or an outer orientation of a space element induces an inner or an outer orientation to its boundary.

### *3.6.2 Holes*

The propagation of an inner or outer orientation according to Möbius' law of edges implies that the orientation of a hole is opposite to that of the surface containing it, as in Fig. 3.25. So to find the area of a circular crown, we are accostumed to calculating the difference between the area of the large disk and the area of the small disk, the hole. Using the notion of oriented area we can say that the area of the krone is the *algebraic sum* of the areas of the two disks where the area of the small disk is negative whereas that of the large disk is positive. This can be done if we use an inner or an outer orientation of the disks.

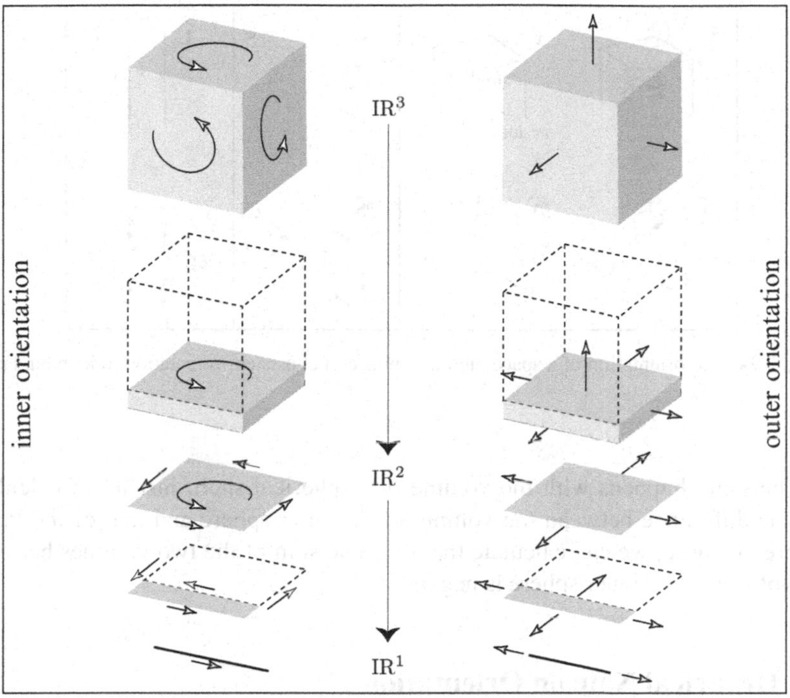


**Fig. 3.22** Picture showing how the inner and outer orientations of a cube in three dimensions are projected in two dimensions and in one dimension

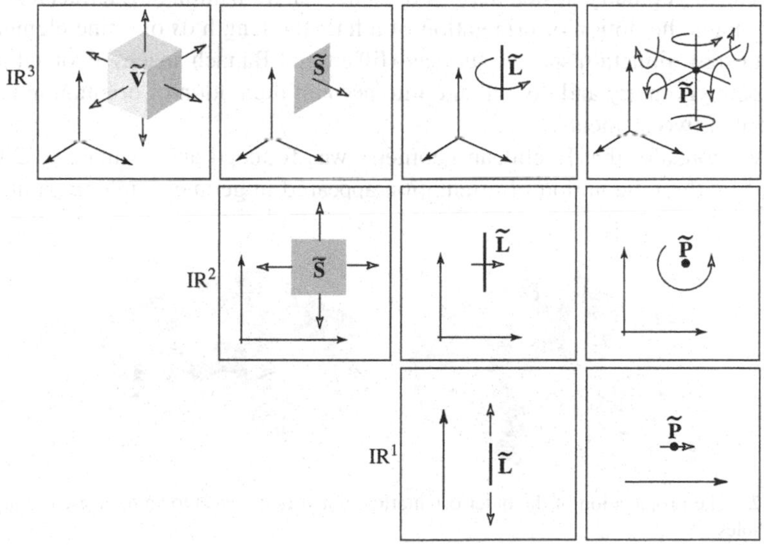


**Fig. 3.23** The outer orientation of a space element depends on the dimensions of the embedding space. The windows must be read in rows

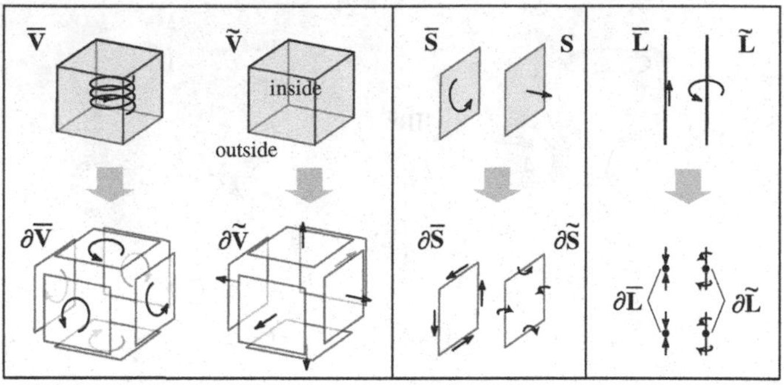


**Fig. 3.24** The orientation of a space element induces a consistent orientation to its boundary

The same happens with the volume of a spherical shell: instead of calculating the difference between the volume of the outer sphere and that of the inner sphere, the hole, we can calculate the algebraic sum of the two volumes because the volume of the inner sphere is negative.

## 3.7 Historical Note on Orientation

Euclidean geometry was developed without reference to the notion of orientation. Riemannian geometry, which derives from the generalization of Euclidean geometry, ignores the notion of orientation of a line: the length d$s$ of a line element is without sign. Function spaces, such as Hilbert and Banach spaces, evolved from Euclidean geometry and do not take into account the notion of orientation in the distance of two elements.

If we consider that Euclidean geometry was founded approximately 22 centuries ago, then the notion of orientation appeared in geometry in comparatively

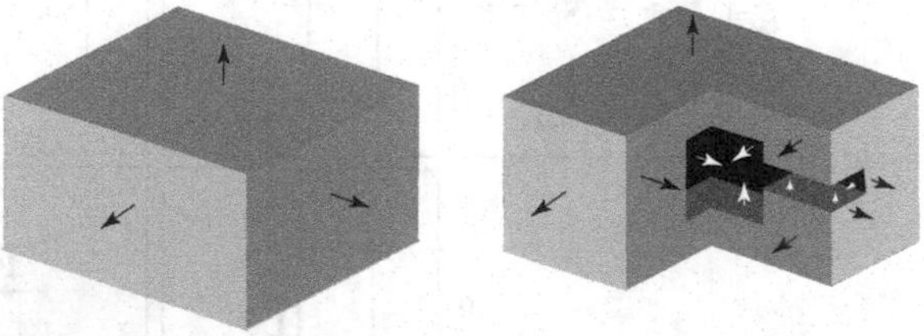

**Fig. 3.25** The propagation of the outer orientation of a volume leads to an opposite orientation of the holes

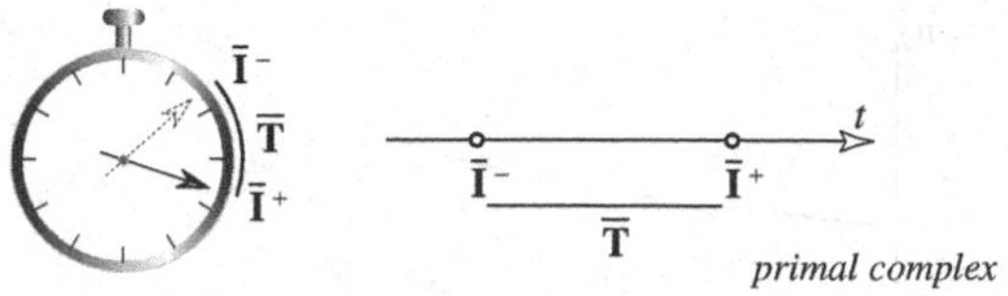


**Fig. 3.26** Two time elements: time instants and intervals

recent times. It was Möbius who, in 1827, with his book *Der baryzentrische Kalkül*, introduced the sign principle into geometry.[15]

Physics developed without considering the concept of orientation with explicit reference to the two types of orientation, inner and outer. We are indebted to Schouten for the systematic use of the two kinds of orientation.[16]

In what follows, we will consider oriented space and time elements systematically.

## 3.8 Time Elements

Global variables refer not only to space elements but to time elements as well. What is a *time element*? By this we mean a time instant and a time interval, which we denote by $I$ and $T$ respectively. Hence, for uniformity with space elements, we use capital letters for time instants, too. Moreover, we believe it is better to use boldface capital letters **I**, **T** because it seems natural to reserve the letter $T$ to denote the *duration* of the time interval **T**. To analyse time elements, a geometric representation of time is appropriate. An analogue clock (Fig. 3.26) allows us to relate time instants with the points of a circumference. The hand on the clock indicates the position of a point on the circumference; hence, a time instant and each angle of rotation of the hand correspond to a time interval. The natural sequence of time instants, which allows us to speak about *before* and *after*, is represented by the rotation of the hand of the analogue clock. In place of the circumference of the clock, we can use a straight line to indicate the sequence of time instants: this is the *time axis*. Matching time instants with the points of the time axis we obtain a geometric representation of time which is frequently used in physics to make a diagram to describe the time behaviour of a phenomenon. The correspondence between the time instants and the points of a line and the correspondence between the time intervals and the segments of the line enables us to apply geometric concepts to time elements.

---

15 Klein [114, p. 16].

16 Schouten [204].

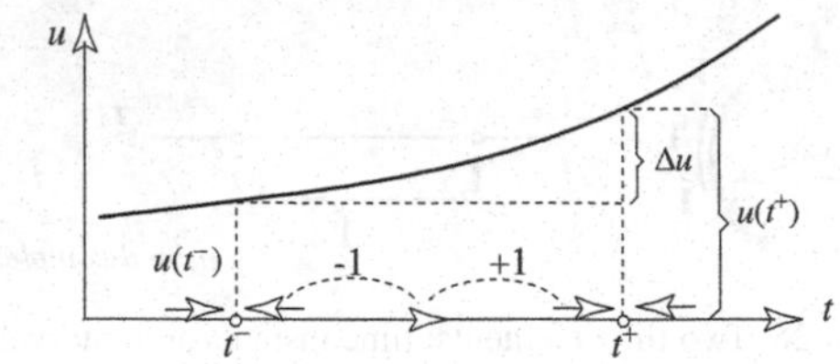


**Fig. 3.27** Incidence numbers in a time complex

### 3.8.1 Primal Time Elements and Their Orientation

If we subdivide the time axis into intervals, as shown in Fig. 3.26, we obtain a subdivision which gives rise to a *cell complex in time:* the cells are the time intervals and the boundaries of each cell are the time instants. This cell complex will be called *primal.* We denote the elements of a primal complex by placing a bar over the symbol, i.e. $\bar{\mathbf{I}}$ and $\bar{\mathbf{T}}$ respectively.

**Primal time intervals.** With reference to Fig. 3.27, the *inner* orientation of the primal time intervals is the same as the time axis: each time interval is oriented from the preceding time instant to the following one in a natural time sequence. The arrow along the time axis is a geometric representation of the progression of time. The orientation of a primal time interval can be visualized by placing an arrow over each primal interval.

**Primal time instants.** Are time instants orientable? At first glance this seems non-sensical because time instants have no extension: the situation is similar to that discussed for points.[17] Let us consider a function $f(t)$ and draw up its graph. The definition of the increment of the function is

$$\Delta_t f \stackrel{\text{def}}{=} (+1)\, f(t^+) + (-1)\, f(t^-)\,. \tag{3.5}$$

In Eq. 3.5 we draw the reader's attention to the coefficients +1 and −1, which can be interpreted as *incidence numbers* once we orient time instants as sinks, as shown in Fig. 3.27. Hence, the number (+1) means that the orientation of the time interval agrees with the orientation of the time instant $t^+$, whereas the number (−1) means that the orientation of the time interval disagrees with that of the time instant $t^-$.[18] These signs are shown in the bottom of the right part of Fig. 3.28.

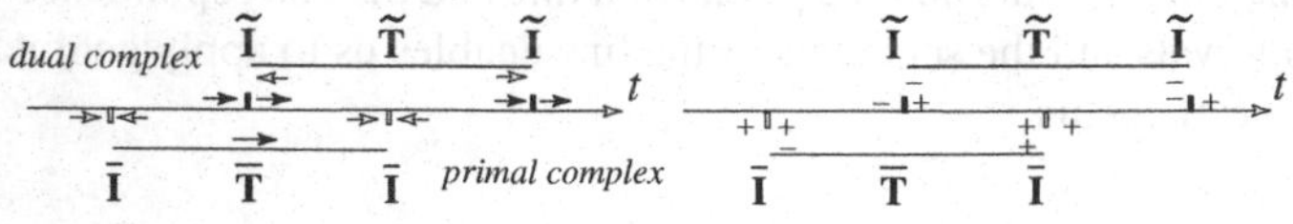


**Fig. 3.28** The two kinds of orientation of time elements using signs

[17] See p. 50.

[18] See the corresponding notion for oriented space segments and oriented points of Fig. 3.14.

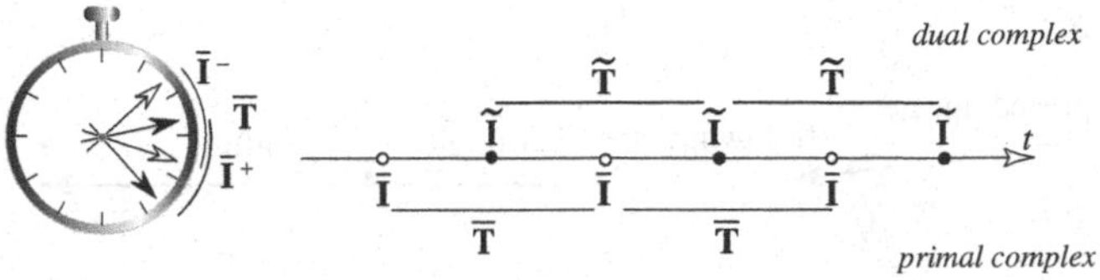


**Fig. 3.29** A primal and a dual cell complex on the time axis: four time elements

### 3.8.2 Dual Time Elements and Their Orientation

By means of the geometric representation of time, we introduce the notion of dual complex in time. The intermediate time instant of each interval (Fig. 3.29) will be called a *dual instant* and the time intervals between dual time instants will be called *dual intervals*. In this way, we have constructed a second cell complex along the time axis, which can be called a *dual* of the first: its elements will be denoted by placing a tilde over the symbol, i.e. $\tilde{\mathbf{I}}$ and $\tilde{\mathbf{T}}$.

REMARK. We are not required to choose the dual time instants in the *middle* time of the primal intervals: any time instant which belongs to a primal time interval is a possible candidate as a dual instant.

The reason for introducing the dual time elements will become clear when we examine the notion of acceleration of a particle. To define acceleration, let us consider three time instants $t_1, t_2, t_3$ with the same duration $\tau$ of the time interval between them, i.e. $\tau = t_3 - t_2 = t_2 - t_1$. If we denote by $x_1, x_2, x_3$ the three abscissae, then the mean velocity and the mean acceleration are given respectively by

$$v_{\rm m} = \frac{x_2 - x_1}{\tau}, \qquad v'_{\rm m} = \frac{x_3 - x_2}{\tau}, \qquad a_{\rm m} = \frac{v'_{\rm m} - v_{\rm m}}{\tau} = \frac{x_3 - 2x_2 + x_1}{\tau^2}. \tag{3.6}$$

From these formulae we see that while velocities change sign under reversal of motion, acceleration does not change sign. In fact, by inverting the time sequence $t_1, t_2, t_3$, we have the sequence $t_3, t_2, t_1$, and the mean acceleration remains the same despite the change $x_1 \leftrightarrow x_3$.

The fact that acceleration does not change its sign under reversal of motion implies that it cannot be associated with primal time intervals which are endowed with inner orientation; otherwise the *oddness principle*[19] would be violated. This leads us to consider intervals not endowed with an inner orientation. Considering that the mean acceleration makes use of the middle time instants of two consecutive intervals, we see that the time intervals with which it is associated are the dual ones.

---

[19] See p. 122.

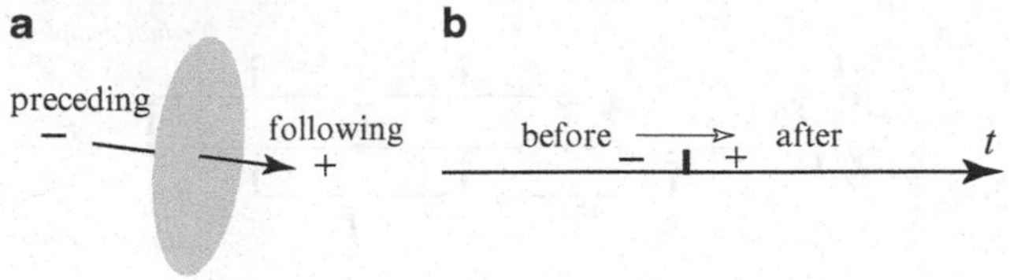


**Fig. 3.30** (**a**) Outer orientation of surface in space compared with (**b**) outer orientation of time interval

We have seen that in a space cell complex, the inner orientation of a *p*-dimensional cell of a primal complex induces an outer orientation on its dual cell. It is therefore natural to define as outer orientation of the dual time elements the one induced by the inner orientation of the primal time elements, as shown in Fig. 3.28.

**Dual time instants.** Let us examine the meaning of the arrow that crosses the dual instant $\tilde{\mathbf{I}}$. Referring to Fig. 3.30, let us compare the outer orientation of a surface embedded in space and the dual instant on the time axis. What does *outer orientation* mean? In the case of a surface, it means the choice of one of the two sides of the surface, like the preceding one and the other as the following one. It is natural to indicate the preceding part with a minus sign and the following part with a plus sign. It is also natural to use an arrow that goes from minus to plus. The arrow passes through the surface.

On the time axis, it is natural to consider the instants that precede and those that follow a given dual instant and consider these two sets in this order, i.e. to speak of *before* and *after* the instant. It is also natural to assign a minus sign to *before* and a plus sign to *after*. Following the rule used up to now we can draw an arrow from the minus sign to the plus sign, i.e. from *before* to *after*. Hence the arrow that crosses the dual instant simply means the order before/after or after/before.

Since the orientation of a dual instant is, by definition, the one induced by the inner orientation of the corresponding primal interval, it follows that a dual instant $\tilde{\mathbf{I}}$ acquires an outer orientation, which is represented by an arrow that 'crosses' the instant (Fig. 3.28). An alternative representation of the orientations is given in the right part of Fig. 3.28: instead of the arrows we use the + and – signs.

**Dual time intervals.** Let us raise the question of whether we can call *oriented* a time interval endowed with arrows directed outside or inside the interval. Recall that in the theory of bars one can consider positive the traction and negative the compression, as shown in Fig. 3.31. The notion of *orientation* arises whenever one entity has two opposite determinations. The outer orientation of dual intervals,




**Fig. 3.31** This is the convention used in books on the strength of materials

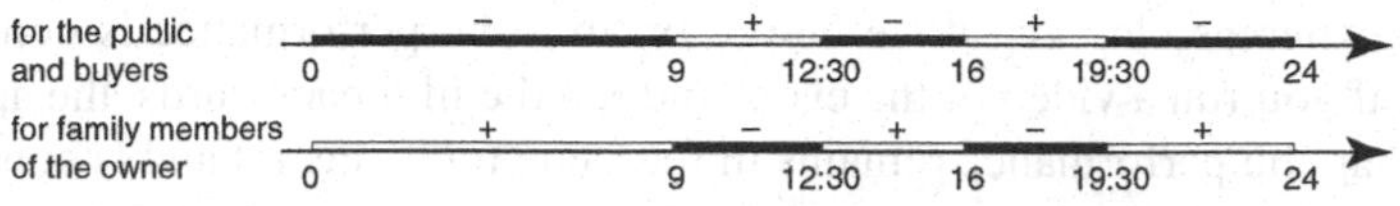


**Fig. 3.32** Time intervals endowed with outer orientation

induced by the inner orientation of primal time instants, is represented by arrows pointing inside the interval, like a bar under compression. We stated that the outer orientation of a volume induced by the inner orientation of a point (sink) was opposite to that currently used in physics, i.e. the orientation with outward normals. Hence, to agree with existing conventions, we take as the outer orientation of volumes the *opposite* of the orientation induced by primal points.

Therefore, we will do the same here for the outer orientation of intervals: whereas primal time instants induce an orientation on the dual time intervals with the arrows directed inwards, we will choose, according to the existing convention, to direct the arrows outwards, i.e. opposite to the orientation induced by the primal time instants, as in Fig. 3.28. This sign inversion is needed to agree with the formula which defines acceleration.

When time instants and time intervals are represented geometrically, as in Fig. 3.28, the arrows give a geometric meaning to the time orientation. In fact, in Eq. 3.6, we used the relation

$$a_{\mathrm{m}} = \frac{(+1)v'_{\mathrm{m}} + (-1)v_{\mathrm{m}}}{\tau}, \tag{3.7}$$

in which the coefficients +1 and −1 can be interpreted as incidence numbers between the time interval with outgoing arrows and the time instants with arrows crossing them, as shown in Fig. 3.28.

What is the physical meaning of the outer orientation of dual time intervals? We will show that this type of orientation is widely used in our daily lives. For example, consider the opening hours of a store. In the morning, the shop is open from 9 a.m. to 12:30 p.m. and in the afternoon from 4 to 7:30 p.m. Considering the matter from the point of view of the public and of shoppers these two intervals are positive, whereas from the point of view of the shop owner and his or her family members the complementary intervals are positive (Fig. 3.32).

To say that a time interval has an outer orientation means that we must consider as positive or negative the value of a physical variable associated with such an interval, disregarding the 'arrow of time'. The best example is that of the impulse of a force which does not invert its sign when we perform a reversal of motion.[20]

In other words, to say that an interval has an outer orientation means that any event that happens in the interval is considered positive, whereas when it happens outside of the interval it is considered negative. To take another example, clapping that is done during the execution of an orchestral performance is considered

[20] See p. 131.

negative, whereas clapping done before or after the performance is considered positive. If you run a video of the event and run the film backwards, the applause done during the performance remains in the same time interval and is considered negative even in this case.

The notion of orientation in physics is made difficult to grasp because we are accustomed to using specific terms to denote the two opposite determinations of an attribute. Table 3.2 gives examples of such terms.

**Table 3.2** Adjectives referring to motion: the + sign is conventional

| *Minus* (-) | *Plus* (+) | *Minus* (-) | *Plus* (+) |
|---|---|---|---|
| To receive | To give | Backwards | Forwards |
| Incoming | Outgoing | Preceding | Following |
| Arrival | Departure | Before | After |

Table 3.3 shows the two possible meanings of orientation of time instants.

**Table 3.3** Inner and outer orientation of time instants

| Inner orientation | | Outer orientation | |
|---|---|---|---|
| To receive<br>Incoming<br>Arrival | 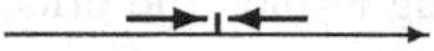 | Forward<br>Following<br>After | 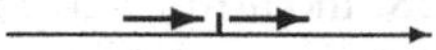 |
| | $\bar{I}$ | | $\tilde{I}$ |
| To give<br>Outcoming<br>Departure | 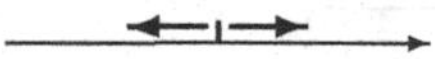 | Backward<br>Preceding<br>Before |  |

# Chapter 4
# Cell Complexes

## 4.1 Coordinate Systems and Cell Complexes

Before Descartes, there were two branches of science which developed separately for centuries: algebra on the one hand and geometry on the other. Descartes had the idea of associating a couple of numbers with each point of the plane: a simple idea, but one with an explosive content. With the birth of Cartesian coordinates, in fact, the wall that divided algebra and geometry was demolished. This fusion gave rise to analytic geometry.

The result is surprising: every elementary geometrical figure, such as a line, parabola, circle, ellipse, hyperbola, spiral of Archimedes, cardioid, cycloid, tractrix, catenary or lemniscate of Bernoulli, can be described by an equation which expresses the relationship between the coordinates of its points. Conversely, to each function (of one variable) we can associate a line. This correspondence is commonly used, for example, in seismography for recording earthquakes and in cardiology for recording heart activity.

The merger between geometry and algebra gave many mathematical operations a corresponding visual representation: thus, for example, the derivative of a function is related to the tangent to the curve which represents the function, and the integral of a function is related to the area enclosed by the curve. The great power of geometric representation is to capture with immediacy the main features of a function, for example, its maxima and minima, its stationary points, the regions of increase and decrease, the zeros and the asymptotes.

Physics greatly benefits from the correspondence between the functions which represent the behaviour of a phenomenon and their graphical representation: think of thermodynamic cycles, or the study of oscillations, or the huge number of diagrams that express the evolution in time of a physical variable.

Since physics describes phenomena arising in space, it makes extensive use of geometric concepts such as, for example, distances, lengths, angles, areas and volumes. Physical notions are translated into mathematical notions by the

E. Tonti, *The Mathematical Structure of Classical and Relativistic Physics*, Modeling and Simulation in Science, Engineering and Technology,
DOI 10.1007/978-1-4614-7422-7_4, 

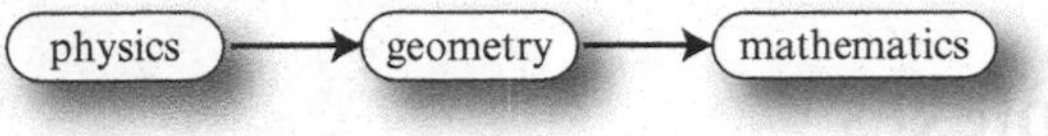


**Fig. 4.1** Geometry is the indispensable bridge between physics and mathematics

**Table 4.1** Summary of symbols and names used to denote space elements

| Symbol | Space element | Cell complex | Algebraic topology |
|---|---|---|---|
| **P** | Point | Vertex | 0-cell $\mathbf{e}^0$ |
| **L** | Line | Edge | 1-cell $\mathbf{e}^1$ |
| **S** | Surface | Face | 2-cell $\mathbf{e}^2$ |
| **V** | Volume | Cell | 3-cell $\mathbf{e}^3$ |

intermediary of geometry (see Fig. 4.1) via the use of coordinate systems. At the same time physical variables are reduced to the function of points in order to permit the differential formulation of physics. It follows that coordinate systems have emerged as an indispensable tool to deal with geometrical notions when using the differential formulation.

On the other hand, if we want to provide a mathematical description of physical fields using global variables, we must consider that global variables are set functions, not point functions, i.e. they are associated with lines, surfaces and volumes, not only with points. Hence, a coordinate system is not the most appropriate framework: the most natural framework is a *cell complex*.

A cell complex is a subdivision of a space region into small elements called *cells*. In numerical analysis cells are usually called *elements:* this is the term used in the *Finite Element Method (FEM)*. We prefer to use the term *cell* and consequently the term *cell complex* because this is the name given in algebraic topology, where a complete theory has been developed on this topic. *In the algebraic formulation of physics cell complexes play the same role that coordinate systems play in the differential formulation*. Cell complexes offer all the space elements needed for the algebraic formulation: points (vertices), lines (edges), surfaces (faces) and volumes (cells).

In algebraic topology, vertices are called zero-dimensional cells, or 0-cells for short; edges are called one-dimensional cells, or 1-cells for short; faces are called two-dimensional cells, or 2-cells for short; volumes are called three-dimensional cells, or 3-cells for short. These $p$-cells will be denoted by $\mathbf{e}^0, \mathbf{e}^1, \mathbf{e}^2, \mathbf{e}^3$ respectively. Table 4.1 summarizes this nomenclature.

In numerical analysis one speaks about *meshes*, *grids*, *nets* or *lattices* instead of cell complexes;[1] in geometry one speaks about *tessellation*.[2] The notion of

[1] Isaacson and Keller [99, p. 364].

[2] Coxeter [44].

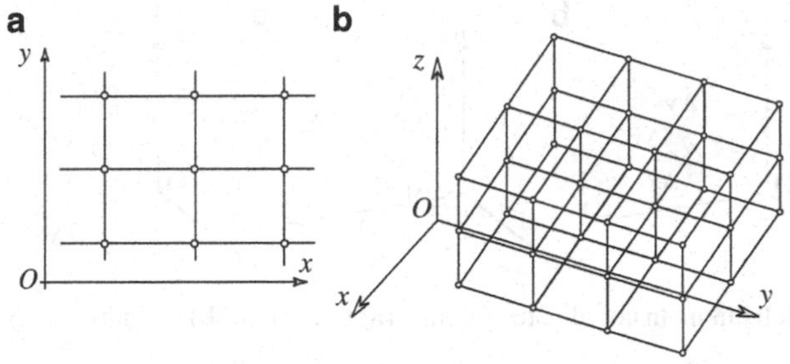


**Fig. 4.2** Cartesian cell complexes in two and three dimensions

*cell complex* is more general and contains these geometric structures as particular cases.

There are two possible alternatives for creating a cell complex:

1. Discretizing a space region by using the coordinate surfaces of a coordinate system, be it Cartesian, cylindrical, spherical or otherwise. A coordinate cell complex obtained in this way is useful for deducing the differential formulation of physical equations from the algebraic formulation;
2. Discretizing a space region by subdividing it into elements of an arbitrary shape. The simplest subdivisions are formed by tetrahedra (in space) and triangles (in a plane). Since these elements are the simplest polyhedra and polygons, respectively, they are called simplices, and a cell complex formed by these simplices is called a *simplicial complex*.

## *4.1.1 Coordinate Cell Complexes*

Figure 4.2 shows a two-dimensional and a three-dimensional cell complex obtained from a Cartesian coordinate system, while Fig. 4.3 shows cells derived from Cartesian, cylindrical and spherical coordinate systems.

The most natural cell complexes are those formed by the coordinate lines and surfaces of a coordinate system, such as Cartesian coordinate system (Fig. 4.2) cylindrical and spherical coordinate systems (Fig. 4.3). We will call them *coordinate cell complexes*.

The use of coordinate cell complexes is implicit in the customary derivation of *gradient*, *curl* and *divergence* in Cartesian and in curvilinear coordinate systems.

Thus, to evaluate the gradient of a scalar function, we must evaluate the function increments along the coordinate lines; to evaluate the curl of a vector field, we must perform the line integral of the vector along the boundary of a cell face; to evaluate the divergence of a vector field, we must evaluate the flow through the boundary of a cell.

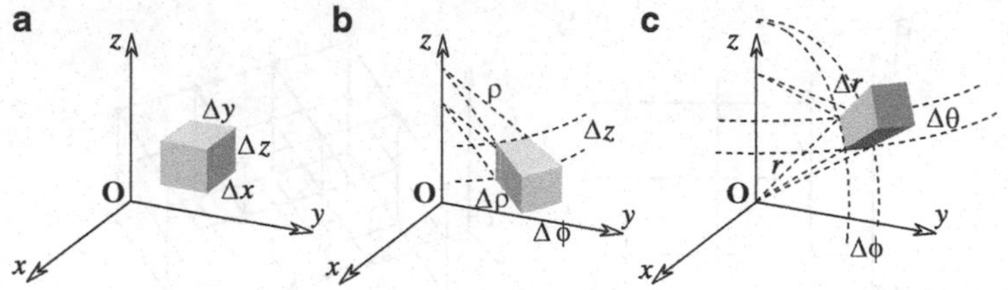


**Fig. 4.3** Cells in main coordinate systems: (**a**) Cartesian; (**b**) cylindrical; (**c**) spherical

### *4.1.2 Families of Cells*

Only in the case of cell complexes obtained from coordinate systems can we speak of a *family of lines* and a *family of surfaces*. In other words, the 1-cells which are relatives of the same kind of coordinate line constitute a *family* and the 2-cells which are relatives of the same kind of coordinate surface constitute a *family*.

To this end, let us consider a Cartesian cell complex in three-dimensional space (Fig. 4.2). We see that there are three families of lines, those parallel to the $x$-, $y$- and $z$-axes. There are also three families of surfaces, those parallel to the $xy$-, $yz$- and $xz$-planes. We see that there is only a single family of points and a single family of volumes.

Hence, for a coordinate cell complex, and only for this, in three-dimensional space we can place the number of families in front of the letters denoting the space elements and write **1P**, **3L**, **3S**, **1V**, as shown in Fig. 4.4.

Note that a reduction in the dimension of the ambient space produces a corresponding reduction in the number of families. In fact, as Fig. 4.5 shows, in two-dimensional space, there are two families for lines but one family for points and surfaces. Hence, we can write **1P**, **2L**, **1S**.

The same considerations hold for cell complexes obtained from other kinds of coordinate systems. To summarize, in three-dimensional space, the 1-cells can be grouped into 3 families: those along the coordinates $x, y, z$ in the Cartesian complex; $\phi, \rho, z$ in the cylindrical cell complex; and $\phi, r, \theta$ in the spherical cell complex. The 2-cells can also be grouped into three families: $xy, yz, zx$ in the Cartesian cell complex; $\phi\rho, \rho z, z\phi$ in the cylindrical cell complex; and $\phi r, r\theta, \theta\phi$ in the spherical cell complex. The 0-cells and the 3-cells have only 1 family.

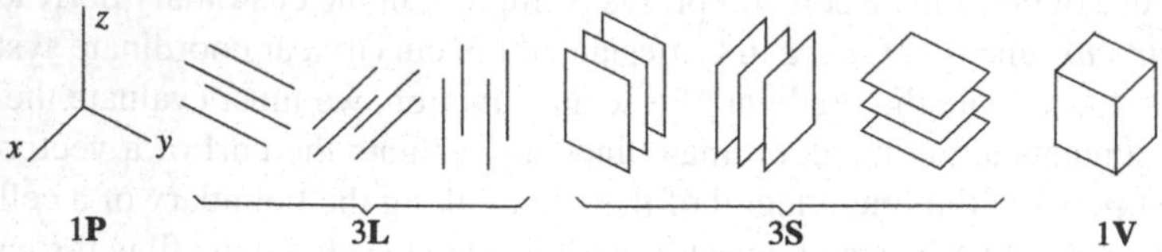


**Fig. 4.4** Families of points, lines, surfaces and volumes of a Cartesian coordinate system in three-dimensional space

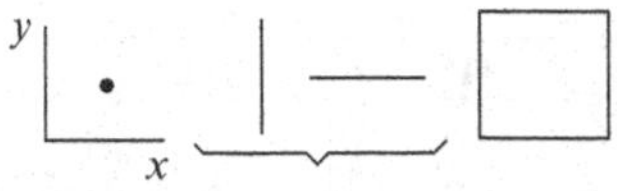


**Fig. 4.5** Families of points, lines and surfaces of a Cartesian coordinate system in two-dimensional space

**Table 4.2** Cells of the same dimension of a coordinate cell complex can be grouped into families

| | | | |
|---|---|---|---|
| One-dimensional region | One family of 0-cells<br>One family of 1-cells | 1P<br>1L | **P**<br>$\mathbf{L}_x$ |
| Two-dimensional region | One family of 0-cells<br>Two families of 1-cells<br>One family of 2-cells | 1P<br>2L<br>1S | **P**<br>$\mathbf{L}_x, \mathbf{L}_y$<br>$\mathbf{S}_{xy}$ |
| Three-dimensional region | One family of 0-cells<br>Three families of 1-cells<br>Three families of 2-cells<br>One family of 2-cells | 1P<br>3L<br>3S<br>1V | **P**<br>$\mathbf{L}_x, \mathbf{L}_y, \mathbf{L}_z$<br>$\mathbf{S}_{xy}, \mathbf{S}_{xz}, \mathbf{S}_{yz}$<br>$\mathbf{V}_{xyz}$ |
| Four-dimensional region | One family of 0-cells<br>Four families of 1-cells<br>Six families of 2-cells<br>Four families of 3-cells<br>One family of 4-cells | 1P<br>4L<br>6S<br>4V<br>1H | **P**<br>$\mathbf{L}_t, \mathbf{L}_x, \mathbf{L}_y, \mathbf{L}_z$<br>$\mathbf{S}_{tx}, \mathbf{S}_{ty}, \mathbf{S}_{tz}, \mathbf{S}_{xy}, \mathbf{S}_{xz}, \mathbf{S}_{yz}$<br>$\mathbf{V}_{txy}, \mathbf{V}_{txz}, \mathbf{V}_{tyz}, \mathbf{V}_{xyz}$<br>$\mathbf{H}_{txyz}$ |

Note that in a four-dimensional space, typically space-time, the number of families is **1P**, **4L**, **6S**, **4V**, **1H**, where the symbol **H** stands for *hypercell*. This can be viewed in Table 4.2 and in the last level of Fig. 4.6. In fact, if we add the coordinate $t$ to the three space coordinates, we have four families of lines, i.e. $t, x, y, z$; six families of surfaces, i.e. $tx, ty, tz, xy, xz, yz$; four families of three-dimensional cells, i.e. $txy, txz, tyz, xyz$, and, lastly, one family of points (called **events**) and hypervolumes, i.e. $txyz$.

### 4.1.3 Simplicial Cell Complexes

As stated previously, even though squares and cubes are easy to draw, they are not the simplest polygons and polyhedra: triangles and tetrahedra are the simplest objects in two- and three-dimensional spaces respectively. For this reason triangles and tetrahedra are called *simplices* of the corresponding space, and a cell complex

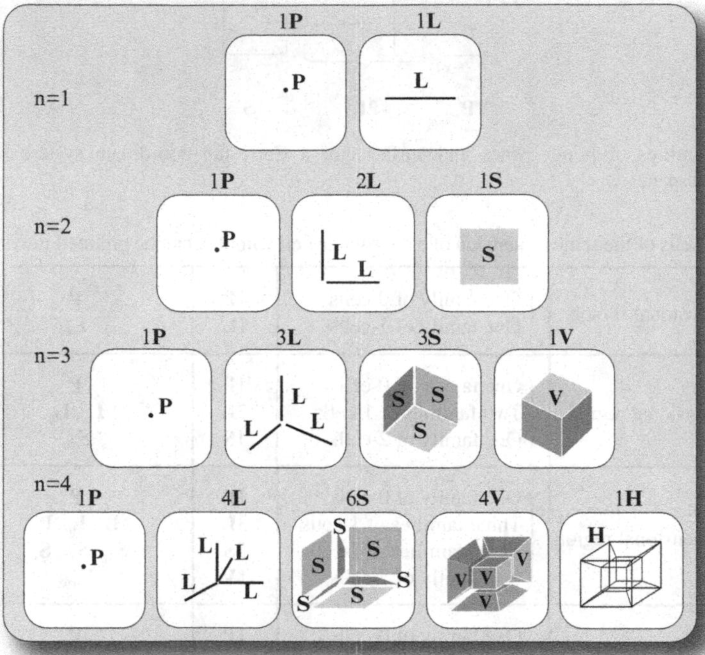


**Fig. 4.6** Pascal's triangle gives the number of families of elements in spaces of various dimensions

formed by *simplices* is called a *simplicial complex* (Fig. 4.7). They are the most useful complexes for dealing with numerical solutions of physical problems in complex geometries.

Simplicial complexes[3] are mainly used in *algebraic topology*, one of the two branches into which topology is divided, the other being *point set topology*. Many properties of cell complexes have been developed in algebraic topology, among them the notions of *orientation*, *duality* and *incidence numbers*. A cell complex is said to be $n$-dimensional, and will be denoted by $\mathbf{K}^n$, if the highest-dimensional cells have dimension $n$. For example, a lattice of points (nodes), such as that used in the finite difference method, is a 0-dimensional cell complex, $\mathbf{K}^0$. A *planar graph*[4] formed by *vertices* and *edges* is a one-dimensional cell complex, $\mathbf{K}^1$.

For simplicial cell complexes we cannot establish families as we did for Cartesian cell complexes. Moreover, we must assign a label to every $p$-dimensional simplex: this is analogous to the assignment of the coordinates to the points by means

[3] Wallace [244, p. 168]; Alexandrov [2].

[4] Lefschetz [132, p. 93]; Lefschetz [133, p. 47]; Bourgin [24, p. 17]; Singer and Thorpe [211, p. 101]; Hilton and Wylie [92, p. 64].

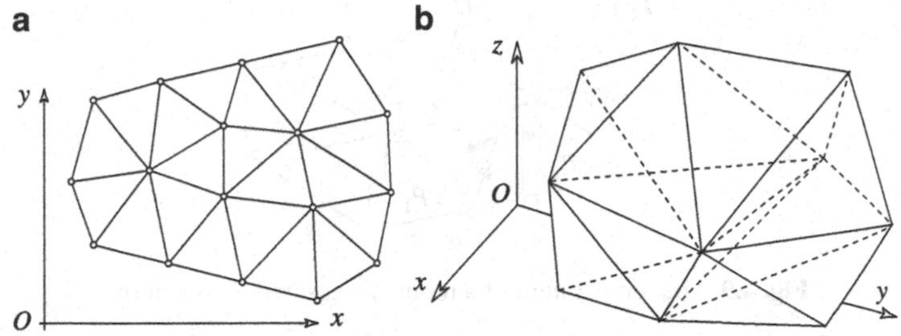


**Fig. 4.7** A simplicial complex: (**a**) in two dimensions; (**b**) in three dimensions

of a coordinate system. The criterion for assigning a label to each $p$-dimensional, and such assignment is automatically performed by the mesh generators used in computational physics (Fig. 4.8).

## 4.2 Dual Cell Complexes

In the differential formulation, physical laws are expressed by differential operators. Let us consider, as an example, the Laplace operator $\nabla^2$. An equation such as Poisson's equation

$$-k\nabla^2 u(\mathbf{P}) = \rho(\mathbf{P}) \tag{4.1}$$

links the source density $\rho$ evaluated at a point $\mathbf{P}$ with the Laplacian function $\nabla^2 u(\mathbf{P})$ evaluated at the same point. Even though this relation appears to be a pointwise relation, it is not really pointwise. In fact, the operations involved in the calculation of the Laplacian requires us to consider a neighbourhood of the point $\mathbf{P}$ to perform the partial derivatives. *This means that a differential equation implies a neighbourhood of every point and not only the point itself.* This neighbourhood acts as an auxiliary region of the point, and its extension is undefined. When one uses a discrete formulation, e.g. the finite difference method, this extended auxiliary region can be identified with the 'stencil' used in the forward and

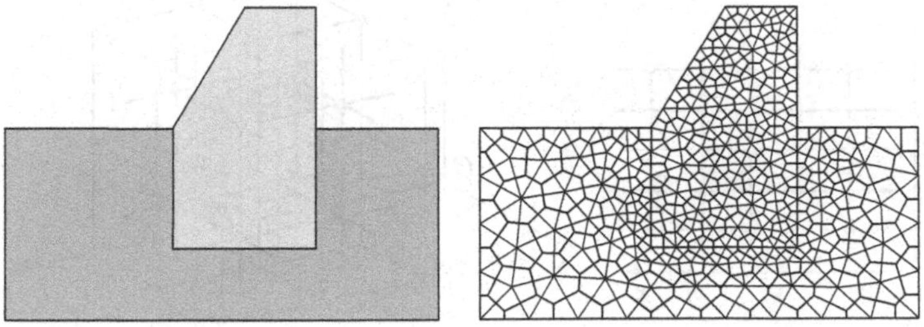

**Fig. 4.8** Space region containing different materials discretized by a simplicial complex and its dual

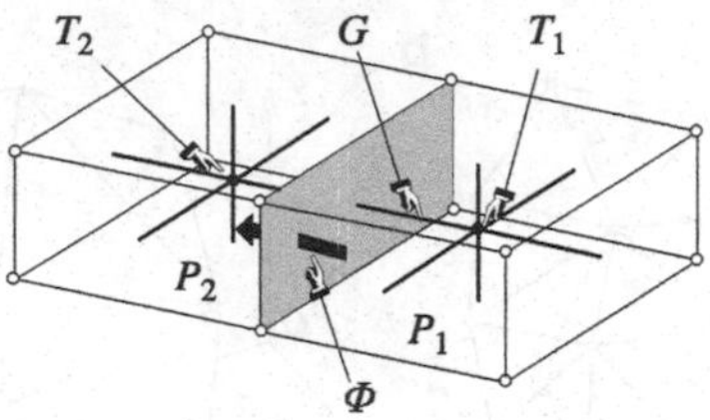


**Fig. 4.9** The temperature of a room is measured at its centre

backward differences. As a general principle, *the algebraic formulation requires a surrounding auxiliary region around the nodes of a cell complex.*

These simple physical considerations suggest considering an auxiliary region around the cell complex for every one of its nodes. All these auxiliary regions constitute another cell complex, called a *dual* cell complex of the complex.

**Example 1.** Let us consider two adjacent rooms (Fig 4.9). Let us consider at a time instant the heat generation rate $P_1$ and $P_2$ (in watt) within each room and the heat current $\Phi$ (W) through the wall separating the rooms. Let $T_1$ and $T_2$ be the temperatures of the two rooms at the same time instant. What point can be chosen to measure the temperature of each room? An obvious choice for measuring the temperature is the centre of each room, as shown in Fig. 4.9. When such a choice is made, the difference between the temperatures of the two rooms is associated with the line connecting the two centres. Now, if we consider more than two rooms, as in Fig. 4.10b, then we refer the temperatures to the centres of each cell and the temperature differences to the lines which connect these centres. The centres of the cells and the lines connecting them can be conceived as vertices and edges, respectively, of another cell complex, which is *dual* of the first complex.

**Example 2.** Let us consider the deflection of a horizontal plate subjected to a vertical load, and let us subdivide it into squares, as shown in Fig. 4.11a. These squares will be considered cells of a cell complex. If we consider the vertical load acting on each cell, then it is natural to consider the centre of the cell as being representative of the vertical displacement of every cell. In turn,

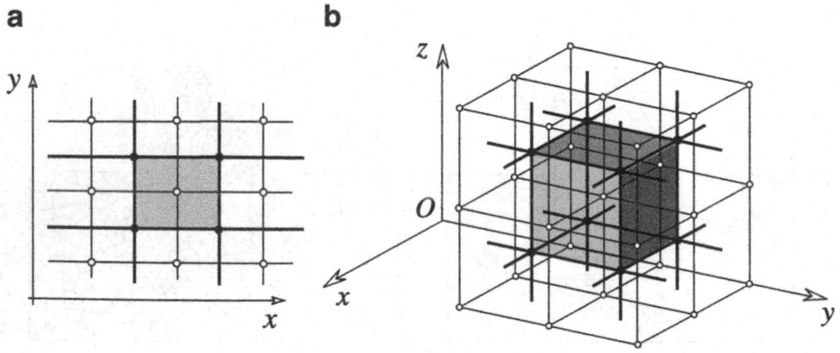


**Fig. 4.10** Cartesian cell complexes in two and three dimensions: the dual complex is shown in *thick lines*, the primal complex in *thin lines*

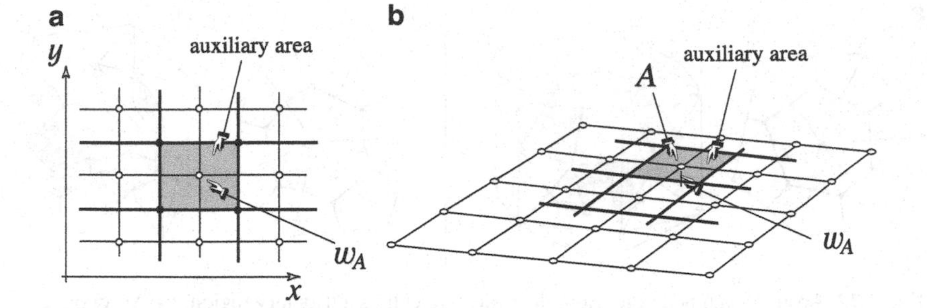


**Fig. 4.11** Cell complex and its dual for plate under normal load

the cells can be thought of as auxiliary regions of every central node. Also in this case, we have obtained another cell complex which is dual to the first complex.

These two examples suggest that by starting with a cell complex called a *primal*, we can construct another complex called a *dual* cell complex, or its *dual* for short. We will denote the *primal* cell complex by $\overline{\mathbf{K}}$ and the *dual* cell complex by $\widetilde{\mathbf{K}}$.

It is a matter of convenience to consider the first complex as primal and the second as dual, or vice versa. Experience tell us that it is convenient to consider as dual that complex whose cells contain the sources of the field. Hence, in thermal conduction, the rooms are taken to be a dual complex because each of them can have a heat-generation rate and in the bending of plates the squares of the first subdivision on which the load is considered are taken to be dual.

The 1-cells of the dual (dual edges) are the lines connecting the dual 0-cells (dual nodes) contained in two adjacent $n$-cells of the primal complex. The 2-cells of the dual are the faces bounded by the 1-cells of the dual: in $\mathbb{E}^3$ they are intersected by the 1-cells of the primal complex. The 3-cells of the dual contain the 0-cells of the primal complex. Then for every $p$-cell of the primal complex there is a corresponding $(n-p)$-cell of the dual. The correspondence lies in the fact that

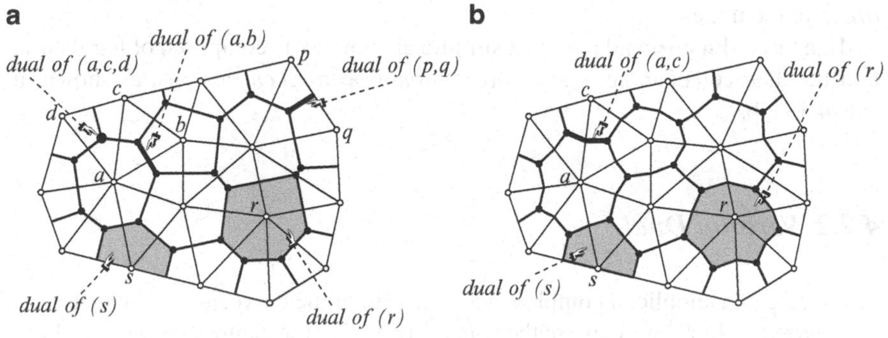


**Fig. 4.12** Dual of a cell complex: (**a**) Voronoi dual; (**b**) barycentric dual

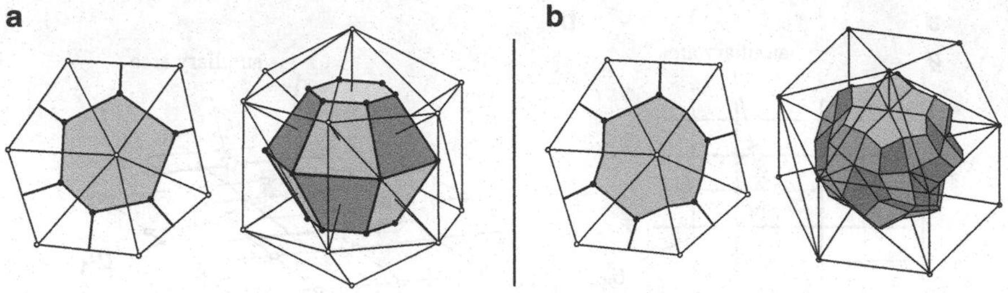


**Fig. 4.13** Around each node one can select the dual cell as a tributary region: (**a**) Voronoi dual; (**b**) barycentric dual

a $p$-cell of the primal contains or crosses or is contained in an $(n-p)$-cell of the dual, as shown in Figs. 4.19 and 4.30 (p. 86). This one-to-one mapping between $p$-cells in the primal and $(n-p)$-cells in the dual motivates the term *dual* given to the complex. Moreover, this one-to-one mapping allows us to assign the same label to a $p$-cell and the dual $(n\text{-}p)$-cell.

We will denote the primal cell complex by thin lines and the dual one by thick lines in the figures.

### *4.2.1 Duals of Simplicial Complexes*

Let us consider a simplicial complex in a two-dimensional space, as shown in Fig. 4.12. In principle, there are many possible ways to construct a dual complex. The simplest way is by using *Voronoi* cells (Fig. 4.12a) and *barycentric* cells (Fig. 4.12b), which are commonly used in computational physics. Voronoi cells are polygons whose sides are the axes of the edges of a primal complex, as shown in Fig. 4.12a, whereas barycentric cells are polygons obtained by connecting the barycentre of every triangle with the midpoint of the edges of the triangles, as shown in Fig. 4.12b. These polygons are the natural *areas of influence* or *tributary areas* of the nodes.

In a three-dimensional region, a simplicial complex is composed of tetrahedra, and the dual cells can be, once more, *Voronoi* or *barycentric* cells, as shown in Fig. 4.13a, b.

### *4.2.2 Voronoi Dual*

In the case of a simplicial complex in $\mathbb{E}^2$, instead of the barycentre one can use the *circumcentre*. In $\mathbb{E}^3$ we can use the *spherocentre*, i.e. the centre of a sphere whose

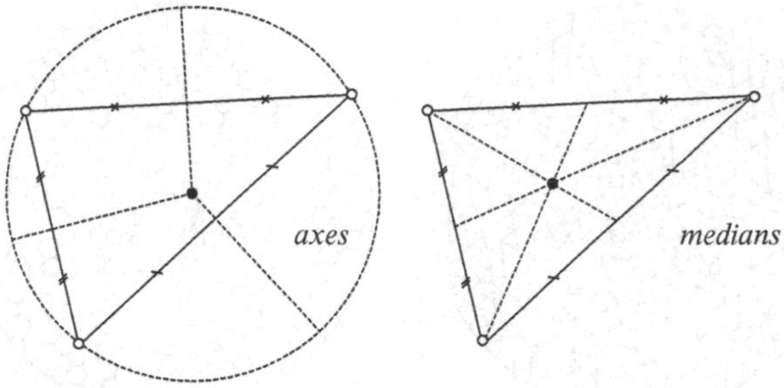


**Fig. 4.14** Circumcentre (*left*) and barycentre (*right*) of a triangle

surface passes through the four vertices of a tetrahedron. The choice of the circumcentre has a disadvantage: it is possible that in a simplicial complex the simplices will have circumcentres which lie outside the simplex. In a plane, this happens when there are triangles with an angle greater than 90°. The circumcentres that lie outside the corresponding triangles generate numerical errors. The circumcentres which lie outside the corresponding triangles generate numerical errors. A triangulation such that the circumcentre of each triangle lies inside the triangle is called a Delaunay triangulation.[5]

In numerical analysis, where metrical concepts are essential, for simplicial complexes the use of circumcentres and spherocentres is preferred because they possess the useful property that the line connecting the circumcentres of two adjacent cells is orthogonal to their common face, as shown in Fig. 4.14a.

With reference to Fig. 4.12a in a two-dimensional simplicial complex, the dual edges (1-cells) which start from a common dual node (0-cell) bound a polygon. This polygon possesses the property that its interior points have a distance from the common node which is less than the distance from any other node of the mesh. This gives rise to a new complex called a *Voronoi* complex.[6] When a Voronoi complex is made on a Delaunay complex, we obtain the well-known Voronoi–Delaunay complex (Fig. 4.12a).[7]

Simplicial complexes are widely used in numerical solutions. The dual is often of the barycentric type, as shown in Figs. 4.15 and 4.16.

**Remark.** We stated that the notion of a cell complex in a discrete setting corresponds to the notion of a coordinate system in a differential setting. We can add that a Voronoi–Delaunay couple of cell complexes in an algebraic setting corresponds to an *orthogonal* coordinate system in a differential setting.

---

[5] Cavendish et al. [37]; Frey and Cavendish [76].

[6] Also called a Dirichlet complex. See Frey and Cavendish [76].

[7] Cavendish et al. [37].

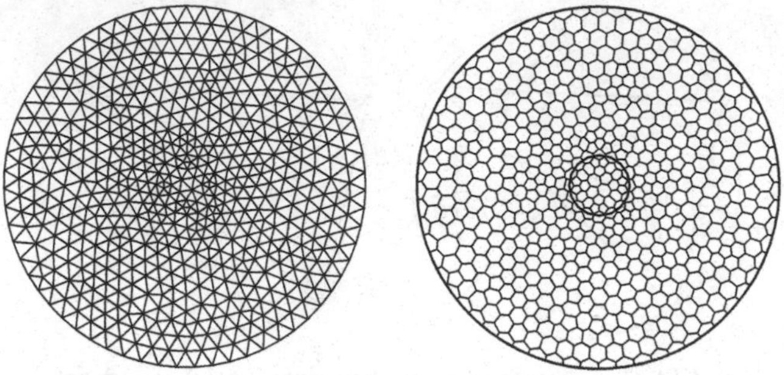

Fig. 4.15 Region around an electric wire decomposed using a simplicial complex (*left*) and its barycentric dual (*right*)

## 4.2.3 Barycentric Dual

In the barycentric subdivision of a plane cell complex, the dual of a 1-cell is composed of two line segments connecting the midpoint of the 1-cell with the two barycentres of the adjacent cells, as shown in Fig. 4.12b. In other words, the dual 1-cell is not a straight line.

The same happens for the dual of a 1-cell in three dimensions: this dual cell is formed by many faces, as shown in Fig. 4.17b.[8]

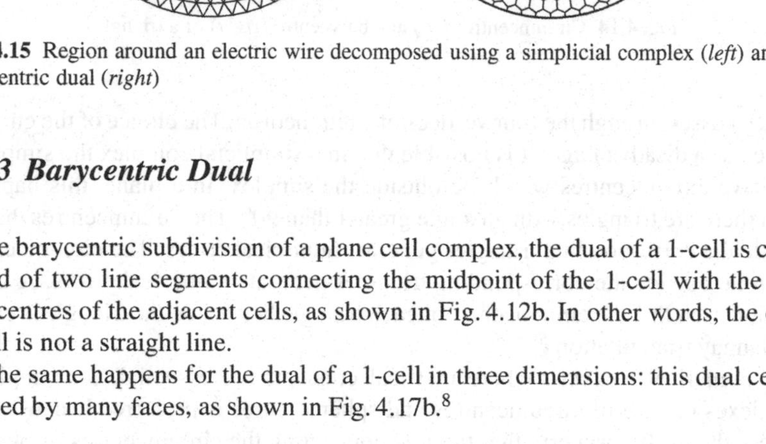

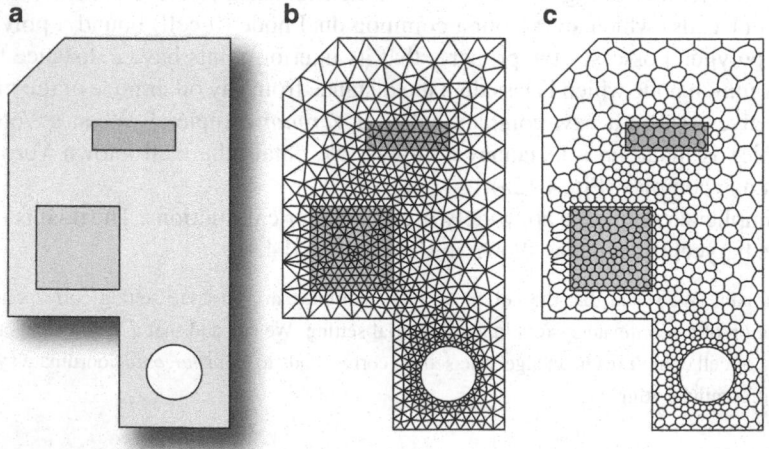

Fig. 4.16 (**a**) Domain composed of different materials. (**b**) Simplicial complex. (**c**) Its barycentric dual

[8] Munkres [165, p. 378]; Dubrovin et al. [58, Sect. 7].

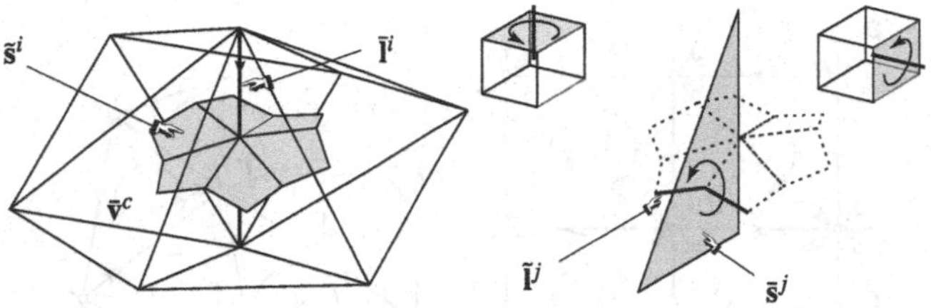


**Fig. 4.17** The dual of a 1-cell in three dimensions is a face composed of many flat small faces

## 4.3 Inner and Outer Oriented Cell Complexes

In a planar graph, the graph is said to be *directed* when all edges have been endowed with an arbitrary orientation, as shown in Fig. 4.18b. In contrast, nodes and loops are all endowed with a compatible orientation, as shown in Fig. 4.18a, c.

Hence, given a cell complex, one can assign an inner orientation to all its $p$-cells as follows:

- First, we can consider *all* 0-simplices as sinks or sources: sinks are the common choice.
- Then we must give an inner orientation to all $n$-simplices. To do this, we choose one of the two possible orientations of one $n$-simplex and we *propagate* the same orientation to all $n$-simplices. The propagation criterion is provided by the Möbius law of edges:[9] *two adjacent $n$-cells in $\mathbb{E}^n$ have compatible orientations when they induce opposite orientations on their common face.*

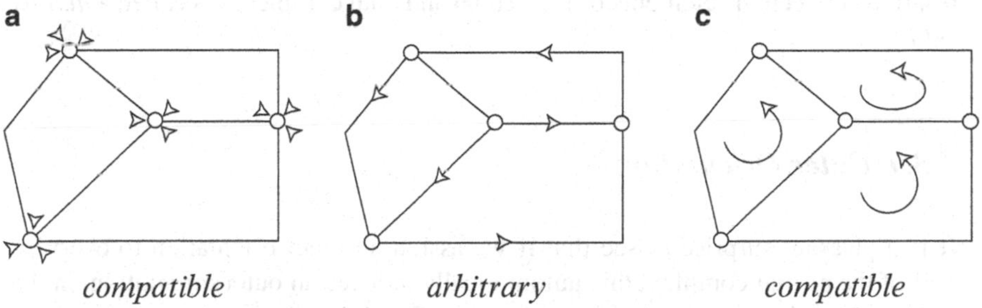


**Fig. 4.18** Oriented graph, also called a *directed* graph. (**a**) The 0-cells (nodes) all have the same orientation: e.g. sinks. (**b**) The 1-cells (edges or branches) have arbitrary orientations. (**c**) The 2-cells (loops) have a compatible orientation

[9] Klein [114, p. 17].

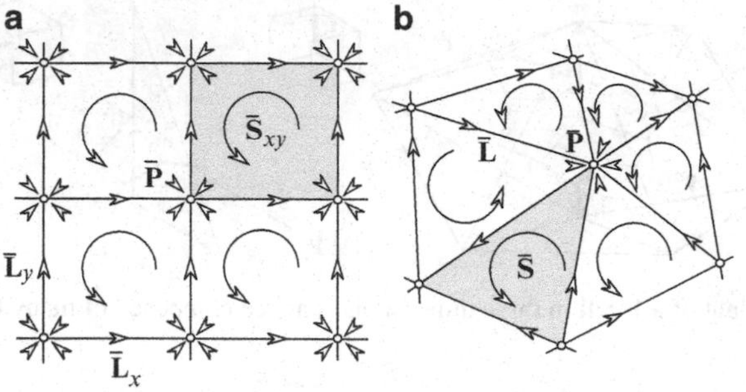


**Fig. 4.19** Oriented cell complexes. (**a**) Oriented Cartesian complex. (**b**) Oriented simplicial complex

- For cells of dimension $p \neq 0$ and $p \neq n$ it is not possible to give, in general, a compatible orientation,[10] and this must be done arbitrarily for every $p$-cell. In computational physics, it is customary to orient 1-cells from the node with the smaller label to the node with the greater label.

In conclusion, we can say that a cell complex has an *inner* orientation when *all* the $p$-dimensional cells, with $0 \leq p \leq n$, have been oriented (Fig 4.19).[11]

Since physical variables refer to space elements endowed with an orientation, it will be useful to assign an orientation to all $p$-cells of a cell complex (Fig. 4.20). A Cartesian cell complex with all $p$-cells endowed with an inner orientation is shown in Fig. 4.21. This figure has an interesting feature: every vertex of the primal cell complex lies in the centre of a dual cell; each edge of the primal complex crosses a face of the dual complex; each face of the primal is crossed by an edge of the dual, and each cell of the primal has a vertex of the dual as its centre. This relation between the elements of the primal and dual complexes is called a *duality relation*.

## *4.3.1 Outer Orientation*

It is a pleasant surprise to see that if we assign an inner orientation to every $p$-cell of the primal complex, this automatically induces an outer orientation on the dual cells of dimension $(n - p)$, as shown in Fig. 4.22. As a consequence of this beautiful marriage between the two kinds of cell complex and the two kinds of orientation, it follows that the same symbol, say $\bar{\mathbf{L}}$, can be used to denote two things: a line element endowed with an inner orientation and an edge of the primal

---

[10] Franz [74, p. 31].

[11] Hocking and Young [94, p. 223].

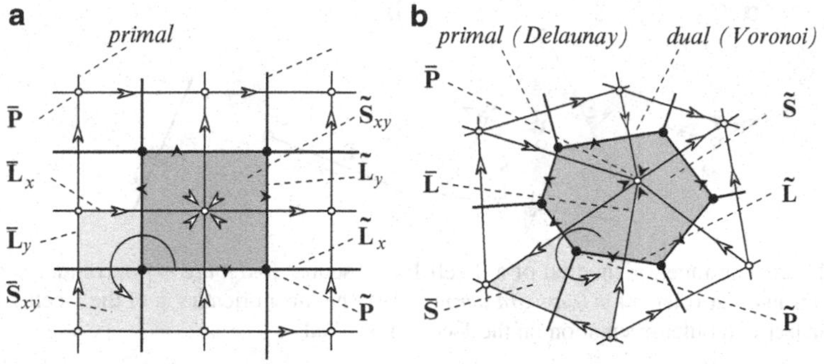


**Fig. 4.20** Oriented cell complexes and their dual: (**a**) oriented Cartesian complex; (**b**) oriented simplicial complex and its Voronoi dual

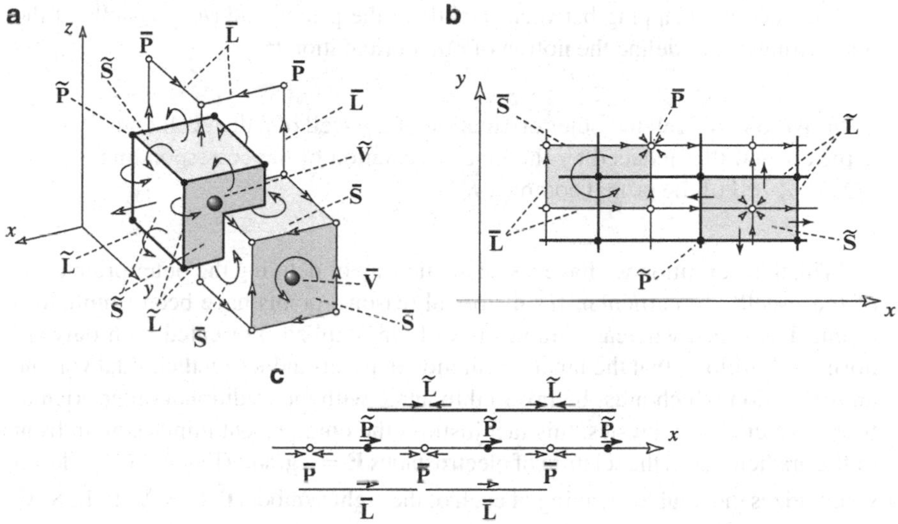


**Fig. 4.21** Primal and dual cell complexes and their orientation: (**a**) in three dimensions; (**b**) in two dimensions; (**c**) in one dimension.

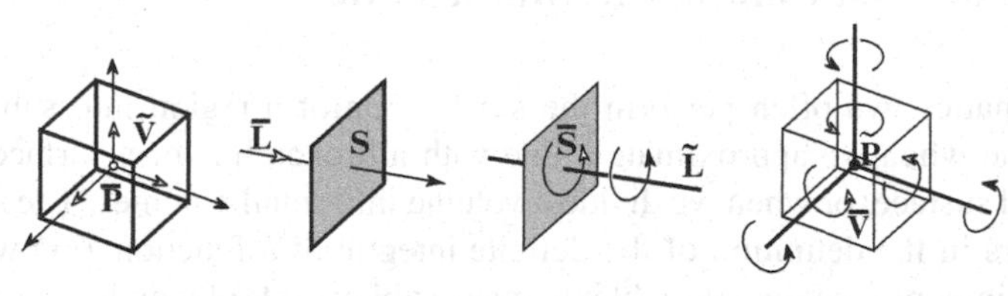

**Fig. 4.22** The inner orientation of an element of the primal complex induces an outer orientation on the corresponding element of the dual

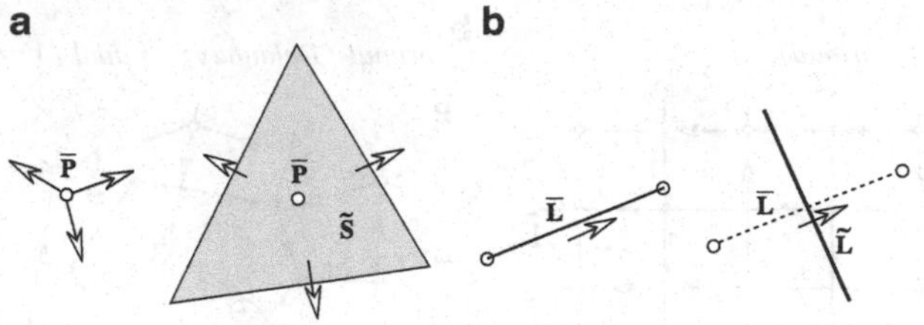


**Fig. 4.23** (**a**) The inner orientation of a 0-cell $\overline{\mathbf{P}}$ as a source (*outgoing arrows*) induces on the dual cell $\tilde{\mathbf{S}}$ an outer orientation (*outward arrows*). (**b**) The inner orientation of the 1-cell $\overline{\mathbf{L}}$ of the primal induces an outer orientation on the 1-cell of the dual

complex. Similarly, the symbol $\tilde{\mathbf{S}}$ denotes a surface element endowed with an outer orientation and a face of the dual complex, and so forth.

The bijective mapping between $p$-cells of the primal and $(n-p)$-cells of the dual permits us to define the notion of outer orientation.[12]

DEFINITION. We call the outer orientation of a $p$-cell of a dual cell complex $\tilde{K}$ and that induced by the inner orientation of the corresponding $(n-p)$-cell of the primal complex $\overline{K}$.

With this definition we have a systematic way of defining the outer orientation of the $p$-cells. In particular, for historical reasons, points have been (implicitly) oriented as sinks, whereas volumes have been (explicity) oriented with outward normals. It follows that the inner orientation of points induce on their dual volume an orientation which must be inverted to agree with the traditional outer orientation of volumes. In physics, this fact justifies the omnipresent minus sign in front of the gradient, as in the relation of electrostatics $\mathbf{E} = -\text{grad}\phi$ (Fig. 4.23). Table 4.3 summarizes the double meaning of each of the eight symbols $\overline{\mathbf{P}}, \overline{\mathbf{L}}, \overline{\mathbf{S}}, \overline{\mathbf{V}}, \tilde{\mathbf{P}}, \tilde{\mathbf{L}}, \tilde{\mathbf{S}}, \tilde{\mathbf{V}}$.

## 4.4 Role of Dual Complex in Mathematics

In mathematics, we often perform the subdivision of a region into subregions: this is done when we approximate a line with a broken line or a surface with a polyhedral surface, or when we divide a volume into small volumes. The simplest case occurs in the definition of the definite integral of a function $f(x)$ when we divide the integration interval $[a, b]$ into many subintervals of length $\delta_i = x_k - x_{k-1}$, as shown in Fig. 4.24a. Later we take a value $\xi_i$ inside each subinterval and define the definite integral as the limit of the sum

[12] Veblen and Whitehead [241, p. 55].

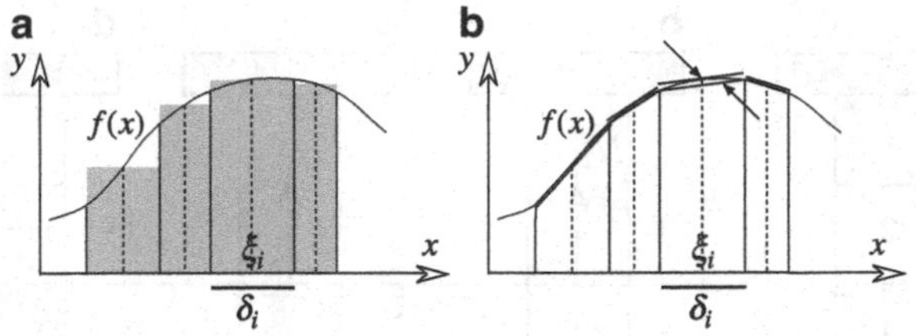


**Fig. 4.24** A primal and a dual cell complex useful for performing numerical integration and numerical derivation

**Table 4.3** Summary of symbols used to denote space elements endowed with inner or outer orientation

| | | |
|---|---|---|
| Point with **inner** orientation | $\overline{\mathbf{P}}$ | Vertex of **primal** complex |
| Line with **inner** orientation | $\overline{\mathbf{L}}$ | Edge of **primal** complex |
| Surface with **inner** orientation | $\overline{\mathbf{S}}$ | Face of **primal** complex |
| Volume with **inner** orientation | $\overline{\mathbf{V}}$ | Cell of **primal** complex |
| Point with **outer** orientation | $\widetilde{\mathbf{P}}$ | Vertex of **dual** complex |
| Line with **outer** orientation | $\widetilde{\mathbf{L}}$ | Edge of **dual** complex |
| Surface with **outer** orientation | $\widetilde{\mathbf{S}}$ | Face of **dual** complex |
| Volume with **outer** orientation | $\widetilde{\mathbf{V}}$ | Cell of **dual** complex |

$$J = \int_a^b f(x)\,\mathrm{d}x \stackrel{\text{def}}{=} \lim \sum_i f(\xi_i)\,\delta_i\,. \tag{4.2}$$

The position of the point $\xi_i$ within the subinterval $\delta_i$ is not important; the midpoint is a convenient choice for a numerical approximation. The resulting subdivision can be called *dual*. Figure 4.24b shows that the incremental ratio of a function relative to a subinterval gives a good evaluation of the derivative of the function at the midpoint of the subinterval. This is an approximate version of the mean value theorem.

## 4.5 Role of Dual Complex in Physics

Let us consider a simple problem, such as the traction of a rod with uniform cross section suspended at one end and subjected to its own weight, as shown in Fig. 4.25a. Suppose also that the material is not homogeneous, so that the elastic modulus is variable along the axis of the stalactite. The fundamental problem

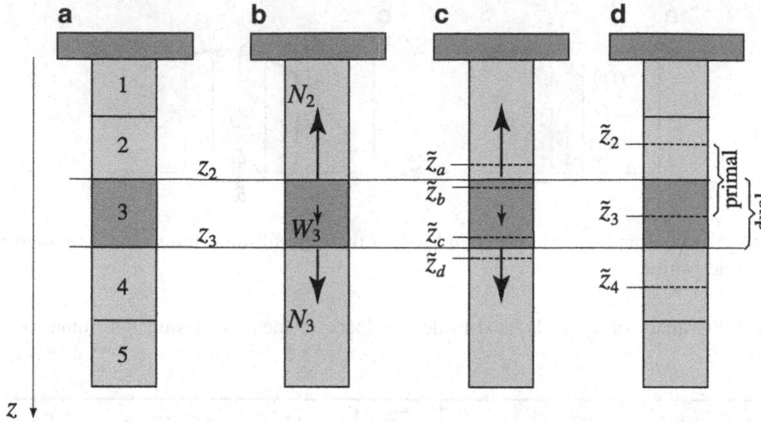


**Fig. 4.25** The reason for the introduction of the dual cell complex shown on a stalactite

is the following: *given the distribution of the weights along the z-axis of the rod, find the displacements u(z) of every normal section.*

To solve this problem, we start with the discrete case, i.e. we divide the whole rod into blocks (shaded area) and we impose equilibrium on every block. Considering, for example, the two faces of block 3, the equation of equilibrium is

$$+N_3 - N_2 + W_3 = 0, \tag{4.3}$$

where $W_3$ denotes the weight of the block, as shown in Fig. 4.25b. The axial forces $N_2$ and $N_3$ are linked to the elastic deformation by Hooke's law $N = EA\varepsilon$, where $\varepsilon$ is the linear strain defined as $\Delta L/L$ and $A$ is the area of the section.

We must now examine which $L$ must be considered. A first idea is to consider the elongation of the same block on which we have imposed equilibrium. But this is improper because it will give $N_2 = E_2 A_2 \varepsilon$ and $N_3 = E_3 A_3 \varepsilon$, so that the equilibrium equation would be reduced to $\varepsilon (E_3 A_3 - E_2 A_2) + W_3 = 0$, which is wrong, as can be easily seen when the rod is homogeneous, i.e. $E_2 = E_3$, since it has a uniform section, i.e. $A_2 = A_3$, because this would imply $W_3 = 0$. Hence it must be $\varepsilon_2 \neq \varepsilon_3$. To evaluate $\varepsilon_2$, we must consider a small region that contains a face at $z_2$. To do this, we find it natural to introduce two sections at $\tilde{z}_a \leq z_2 \leq \tilde{z}_b$. For the same reason we must introduce two sections such that $\tilde{z}_c \leq z_3 \leq \tilde{z}_d$, as shown in Fig. 4.25c. Hence, denoting by $u$ the displacement of a section from its initial position, we have

$$\varepsilon_2 = \frac{\Delta L_2}{L_2} = \frac{u_b - u_a}{\tilde{z}_b - \tilde{z}_a} \qquad \varepsilon_3 = \frac{\Delta L_3}{L_3} = \frac{u_d - u_c}{\tilde{z}_d - \tilde{z}_c} . \tag{4.4}$$

Inserting these relations into Eq. 4.3 we obtain the equation

$$+E_3 A_3 \frac{u_d - u_c}{\tilde{z}_d - \tilde{z}_c} - E_2 A_2 \frac{u_b - u_a}{\tilde{z}_b - \tilde{z}_a} + W_3 = 0. \tag{4.5}$$

This obligates us to introduce two sections for every face of the primal blocks. It follows that the number of displacements ($u_k$) involved is twice the number of blocks on which we can write the equilibrium; hence, we do not have enough equations to find all the unknown displacements.

How can we solve the problem? The idea is to increase the lengths $L_2 = \tilde{z}_b - \tilde{z}_a$ and $L_3 = \tilde{z}_d - \tilde{z}_c$ in such a way to make the two sections inside the same block, i.e. $\tilde{z}_b$ and $\tilde{z}_c$, coincide. The simplest way to do this is by considering the sections at the midpoint of every block, as shown in Fig. 4.25d. In this way, we have one section $\tilde{z}$ for every block, and the number of displacements becomes equal to the number of blocks increased by 2. The two equations which are needed are provided by the boundary conditions, which is typical of the differential formulation. Equation 4.5 becomes

$$+E_3 A_3 \frac{u_3 - u_2}{\tilde{z}_3 - \tilde{z}_2} - E_2 A_2 \frac{u_4 - u_3}{\tilde{z}_4 - \tilde{z}_3} + W_3 = 0, \tag{4.6}$$

and the number of unknowns is equal to the number of equilibrium equations. In doing this we are naturally prompted to use a dual cell complex. The choice of the midpoint is not imperative; what we need is to make $\tilde{z}_b$ coincident with $\tilde{z}_c$ for every face of the blocks.

## 4.6 Global Variables and Computational Physics

The discovery that global physical variables are associated with space elements endowed with inner or outer orientation, coupled with the use of two cell complexes, the primal and dual, allows us to write physical equations directly in algebraic form.

In fact, the traditional mathematical treatment of physics, based on the differential formulation of its physical laws, easily leads to differential equations. This is the case for the equations of Poisson, Maxwell, Fourier, Navier–Stokes, etc. But the solution of these partial differential equations in the cases of practical interest in physics and engineering, involves the use of numerical techniques which are generally not part of physics books. In these books the solution is found only for the simplest cases, leaving to computational physics the task of solving more complex problems numerically. Computational physics uses different methods to *discretize* differential equations, such as FEM, BEM, FDM, FVM etc.

In contrast, the algebraic formulation which we will present in this book, which is obtained by using global variables and taking into account their association with the space and time elements, *directly* provides systems of algebraic equations needed for the numerical solution.

Hence, an analysis of the mathematical structure of physical theories, made to explain the reasons for the analogies in physics as well as to provide a rational classification of variables and equations of any physical theory, also allows for a numerical solution of any specific problem because it *directly* provides the algebraic equations. This opens the way for a new method of calculation based on geometry rather than on mathematics, because it is not necessary to make the discretization of differential equations.

It is a fact that all discretization methods are based on purely mathematical approaches, i.e. without relation to the physical phenomenon described by the equation. This is the case of the entire family of *weighted residuals methods*: these include the methods of least squares, collocation, sub-domains, Galerkin, moments.

Since the Galerkin method is based on energy principles, it appears more "physical" compared to other methods. However, the criterion to approximate the unknown solution of a physical problem by a linear combination of shape functions that are chosen in an arbitrary way, is a purely mathematical one. Furthermore, in order to find the coefficients of the linear combination, the criterion to require that the residual of the differential equation is orthogonal to the every shape functions, is also a purely mathematical criterion. The same Ritz method, which requires the minimization of the energy functional, is faithful to the physical principle of minimum potential energy. Nevertheless, it introduces an arbitrary mathematical expedient, to express the unknown solution of the problem using a linear combination of an arbitrary set of basis functions.

The path to be taken in this book is free from this arbitrariness because it directly gives us the set of discrete equations, all the while remaining closely attached to the physical reality, thanks to the association of the variables with space and time oriented elements.

### 4.6.1 Finite Difference Method Reinterpreted

Historically, the first procedure to find the numerical solution to partial differential equations is the method of finite differences. Its starting point is the discretization of the differential equation by discretizing the partial derivatives: this is the case for the Laplace's equation $\nabla^2 T(x,y) = 0$. The backdrop is a regular lattice in Cartesian coordinates, as shown in Fig. 4.26a. A computational "stencil" for the discrete approximation of $\nabla^2 T$ is

$$\nabla^2 T(x,y) \approx \frac{T_{i-1,j} - 2T_{i,j} + T_{i+1,j}}{h^2} + \frac{T_{i,j-1} - 2T_{i,j} + T_{i,j+1}}{k^2} . \tag{4.7}$$

*A new discretization method.* We propose to show that the same result can be obtained by a more detailed process, i.e. by separately discretizing those

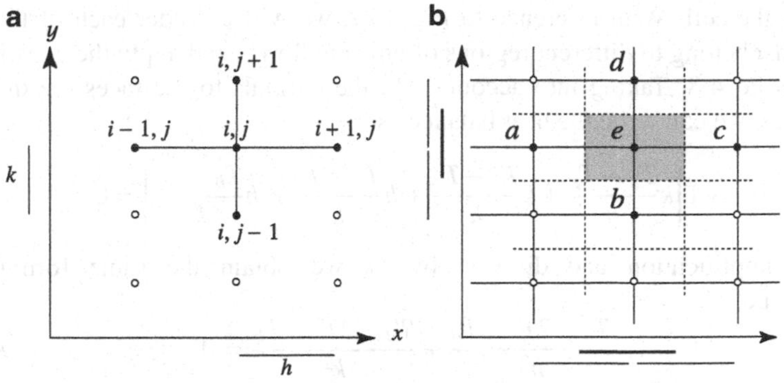


**Fig. 4.26** The reinterpretation of the finite difference method

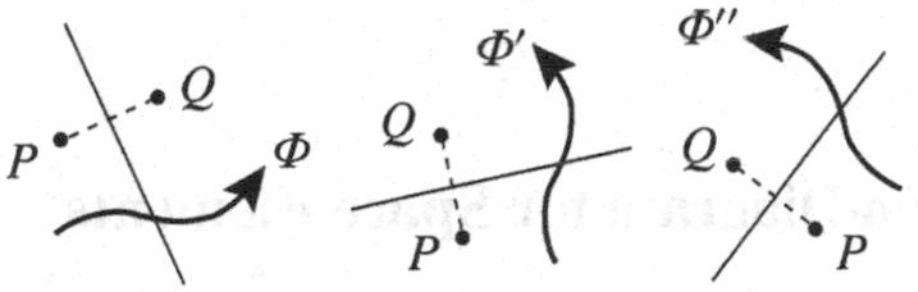


**Fig. 4.27** In a region of uniform heat, the heat current through a flat surface, for any surface orientation, is proportional to the temperature gradient in the direction normal to the surface

elementary equations that, by composition, give rise to the Laplace's equation. The advantage of this new discretizing procedure is that it can be applied to cells of any shape, not necessarily cubic cells. *This allows us to extend the finite difference method to unstructured meshes*, for example to cell complexes with triangular cells (in 2D problems) and with tetrahedral cells (for 3D problems).

To this end, we extend the lattice of Fig. 4.26a to a cell complex and its dual, as shown in Fig. 4.26b. The grey area denotes a dual cell. In this complex, we use a single label to denote the vertices of the primal complex, as is commonly done by mesh generators.

Considering the constitutive relation of heat conduction, Fourier's law asserts that in regions of *uniform* flow, the heat current for unit area is proportional to the temperature difference measured in the direction *normal* to the surface, and inversely proportional to the distance $d$ between the two points $P$ and $Q$ that lie on a straight line normal to the surface, as shown in Fig. 4.27, i.e.

$$\frac{\Phi}{A} = -\lambda \left( \frac{T_Q - T_P}{d} \right) . \tag{4.8}$$

To write the heat balance on the dual cell (note that it must be a dual cell because of its outer orientation), we must consider the heat current across each face of the cell. With reference to Fig. 4.27, we will consider each of the four faces that belong to different regions of uniform flow[13] and apply the constitutive relation Eq. 4.8. Taking into account that the normals to the faces are directed outwards, we can write the heat balance as

$$-\lambda\left[k\frac{T_a - T_e}{h} + k\frac{T_c - T_e}{h} + h\frac{T_d - T_e}{k} + h\frac{T_b - T_e}{k}\right] = 0\,. \tag{4.9}$$

After simplification and division by $hk$ we obtain the same formula of Eq. 4.7, i.e.

$$\frac{T_a - 2T_e + T_c}{h^2} + \frac{T_d - 2T_e + T_b}{k^2} = 0\,. \tag{4.10}$$

The same discretization process can be applied to any cell complex, in particular those with cells formed by triangles (in two-dimensional problems) or by tetrahedra (in three-dimensional problems). This is the *cell method.*[14]

## 4.7 Classification Diagram for Space Elements

With reference to Fig. 4.28, the duality relation between space elements can be emphasized by inverting the order of the dual space elements in the right column. In this way, dual elements appear on the same level. Doing this we obtain a diagram which classifies the space elements and summarizes all the properties we have emphasized until now, i.e. primal and dual complex, inner and outer orientation and the duality relation.

Note that the sum of the dimensions of the elements on the same level is three (i.e. the dimension of space) such that the elements on the same level are complementary (i.e. dual) to one another. Hence the dual of a point with an inner orientation ($p = 0$) is a volume with an outer orientation ($p' = 3 - p = 3$); the dual of a line with an inner orientation ($p = 1$) is a surface with an outer orientation ($p' = 3 - p = 2$); and so forth.

For space elements in a plane, $n = 2$, and in a line, $n = 1$, the pairs of dual space elements are represented in Fig. 4.29. 0.8Dual elements in two- and one-dimensional space If we consider a cell complex and its dual, an $(n - p)$-cell of the dual complex is dual to the $p$-cell of the primal complex and vice versa; hence, it is natural to consider the corresponding cells as dual (Fig. 4.30).

[13] A flow is uniform in a region when the velocity of all particles is invariant under translation.

[14] The direct algebraic formulation of physical laws is the starting point of the cell method; see Tonti [230–232, 234] and the papers cited on the Web site *discretephysics.dicar.units.it*.

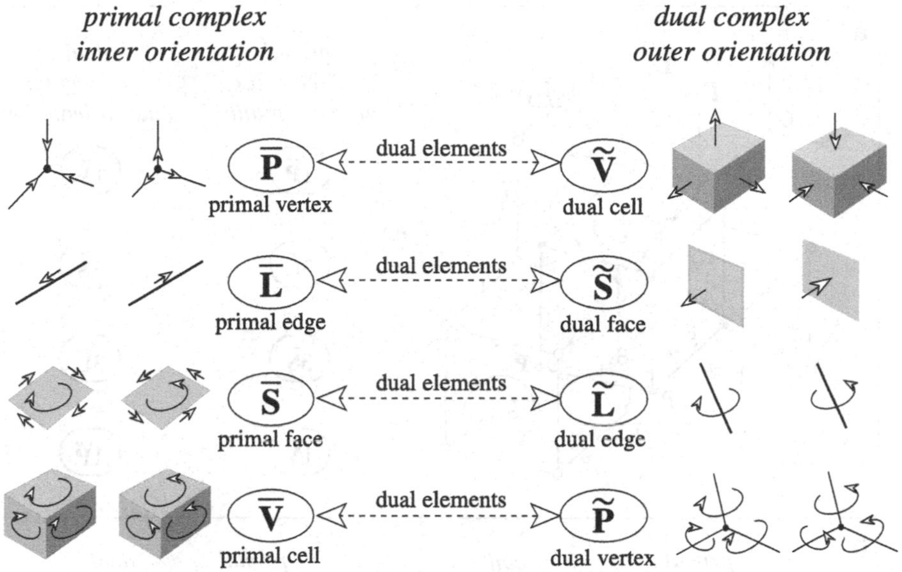


**Fig. 4.28** The two kinds of space elements: there are eight in total

## 4.8 Classification of Time Elements

In Chap. 3 on p. 57 we introduced a cell complex and its dual on the time axis. Now the four time elements $\bar{\mathbf{I}}, \bar{\mathbf{T}}, \tilde{\mathbf{I}}, \tilde{\mathbf{T}}$ can be placed in a diagram in a similar way as was done for space elements. Whereas for space elements the left column contains elements with an inner orientation, for time elements it will be useful to consider two classifications, as shown in Table 4.4.

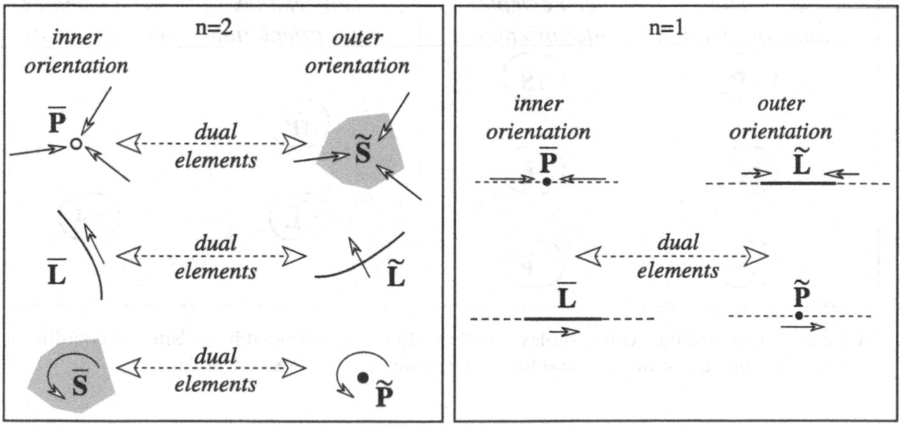


**Fig. 4.29** Dual elements in two- and one-dimensional space

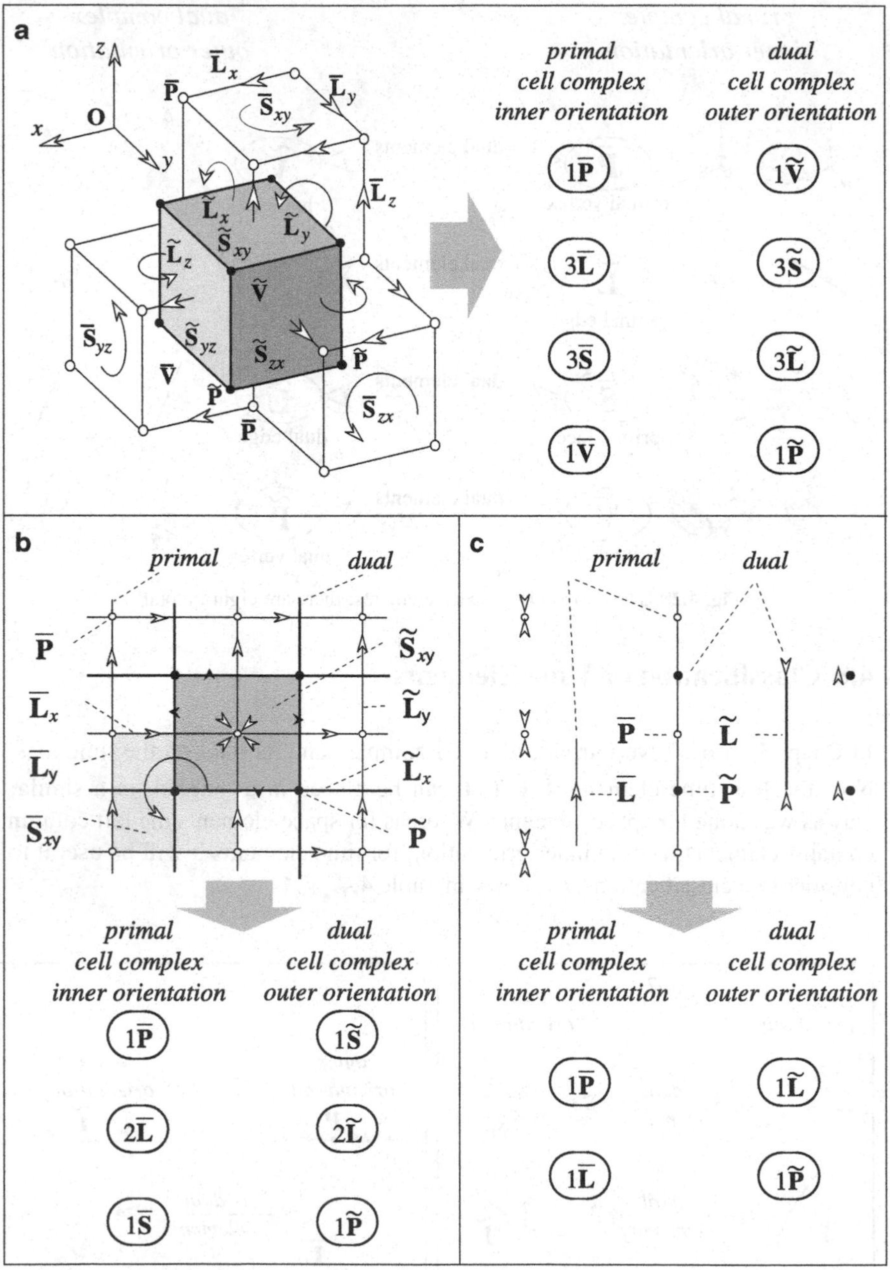


**Fig. 4.30** Primal and dual cell complexes in three dimensions formed by a Cartesian coordinate system. The same diagrams are valid for a coordinate cell complex and its dual

**Table 4.4** Two possible classification diagrams of time elements

| *Inner orientation* | *Outer orientation* |
|---|---|
| $\bar{\mathbf{I}}$ | $\tilde{\mathbf{T}}$ |
| $\bar{\mathbf{T}}$ | $\tilde{\mathbf{I}}$ |
| Mechanical theories | |

| *Outer orientation* | *Inner orientation* |
|---|---|
| $\tilde{\mathbf{I}}$ | $\bar{\mathbf{T}}$ |
| $\tilde{\mathbf{T}}$ | $\bar{\mathbf{I}}$ |
| Field theories | |

The arrangement of the diagram on the left is useful in mechanical theories where forces are the sources of the phenomenon: this is the case of particle mechanics, analytical mechanics, solid mechanics and fluid mechanics. The arrangement of the diagram on the right is useful in field theories where the source variables are charge, mass and heat. Such are electromagnetism, gravitation, diffusion, thermal conduction and acoustics in fluid and non-equilibrium thermodynamics.

As with space elements, the order of elements in the right column is also opposite to the order of the time elements of the left column. The bars over the letters $\bar{\mathbf{I}}$ and $\bar{\mathbf{T}}$ denote two things: that time elements belong to a primal time complex and that they are endowed with an inner orientation. Similarly, the tilde over the letters $\tilde{\mathbf{I}}$ and $\tilde{\mathbf{T}}$ denote two things: that time elements belong to a dual time complex and that they are endowed with an outer orientation.

In computational physics, the cell complex drawn by a mesh generator is usually taken as the dual one.

# Chapter 5
# Analysis of Physical Variables

> *... It is only through the progress of science in recent times that we have become acquainted with such a large number of physical quantities that a classification of them is desirable.*
>
> J.C. Maxwell, *Proc. London Math. Soc.*, Vol.III, (1871), p. 224

## 5.1 Role of Mathematics in Physics

Why is mathematics used in physics? Because of the existence of *physical quantities*. In fact, mathematics is used in physics thanks to the fact that there are quantitative attributes that suggest the introduction of physical quantities, and these lead to numbers. Physical variables contain more information than the simple numerical value. Of course, the numerical value is accompanied by the physical dimensions of the variable. But there is more: each of them is associated with a particular space element and a particular time element, each endowed with an inner or an outer orientation. Mathematics deliberately ignores the geometric or physical content of physical quantities. It is precisely this feature which makes mathematics so broadly applicable: it can be applied whenever numbers are used, without regard to the things to which they refer. Poincaré wrote:[1]:

> Mathematicians study not objects, but relations between objects; the replacement of these objects by others is therefore indifferent to them, provided the relations do not change. The matter is for them unimportant, the form alone interests them.

Mathematics deliberately ignores the physical objects with which numbers are associated. In fact, if we consider two containers, each being 30 $m^3$, then the whole

[1] *Science and Hypothesis*, p. 44.

E. Tonti, *The Mathematical Structure of Classical and Relativistic Physics*,
Modeling and Simulation in Science, Engineering and Technology,
DOI 10.1007/978-1-4614-7422-7_5, 

capacity is $2 \times 30 = 60\,\text{m}^3$. Two cables, each conducting 30 A to the same device, are equivalent to a cable conducting $2\times 30 = 60$ A. Mathematics simply considers the operation $2 \times 30 = 60$ and ignores whether the numbers refer to containers or cables.

Mathematics deliberately ignores both the physical dimensions and the units of measure of physical variables. While for mathematics 5 is different from 500, for physics 5 m is equal to 500 cm and 1 kcal is equal to 4,186 J. The numerical values can be different even if the physical quantity is the same because the units of measure make the difference.

For these reasons, if we want to perform an investigation on the mathematical structure of physical theories, the starting point must be an analysis of physical quantities. Many geometric features of physical variables are hidden because we usually describe physical theories using differential calculus. In fact, differential calculus requires the use of field functions, most of which are obtained from *global variables* by extracting their densities.[2] Field functions, which are required by the notion of the derivative, hide the association of physical variables with space and time elements other than points and instants. In fact, it is just the association with space and time elements, endowed with an inner or an outer orientation, that makes it possible to discover the intrinsic nature of physical variables, which leads to a new classification of them.

## 5.2 Material and Spatial Descriptions

To introduce these two kinds of description,[3] let us consider two simple examples.

**Example 1.** Consider a person looking at an anthill. At first he has a broad view and sees a lot of ants in disorderly motion. At a given moment his attention is attracted by a group of ants carrying a large leaf; he follows the group until it disappears from his eyes. At this moment the observer goes back to a broad view of the anthill. He sees that there are regions in which the ants are less active and others where they are highly industrious. Continuing to watch, he is attracted by an ant which drags a small leaf towards a hole in the ground, and he follows the ant up to the moment in which it disappears inside the hole, and so on.

**Example 2.** The same may happen to a person watching city traffic from a window. He sees two columns of cars that stop at a traffic light and notices a busy stream of people walking on the pavement. Suddenly, his glance is attracted by a limousine of the diplomatic corps with little flags on the front lights. He follows the car with curiosity until it enters the embassy gate. At this point, he goes back to a broad view again watching the cars that have divided now into two different lanes: one continuing straight ahead and the other turning right. He notices that in a

[2] For the definition of global variables see p. 106.

[3] The term *description* is equivalent to the terms *viewpoint* and *approach*.

certain area there is a higher concentration of cars, whereas in another the traffic is free flowing. Then, an eccentrically dressed woman catches his attention: he watches her stopping at a shop window, then setting off again, then stopping to greet a friend of hers and finally stopping at a traffic light waiting for the signal to cross the street. Crossing the street, she disappears from sight. He returns to an overall view of the city traffic, and so on.

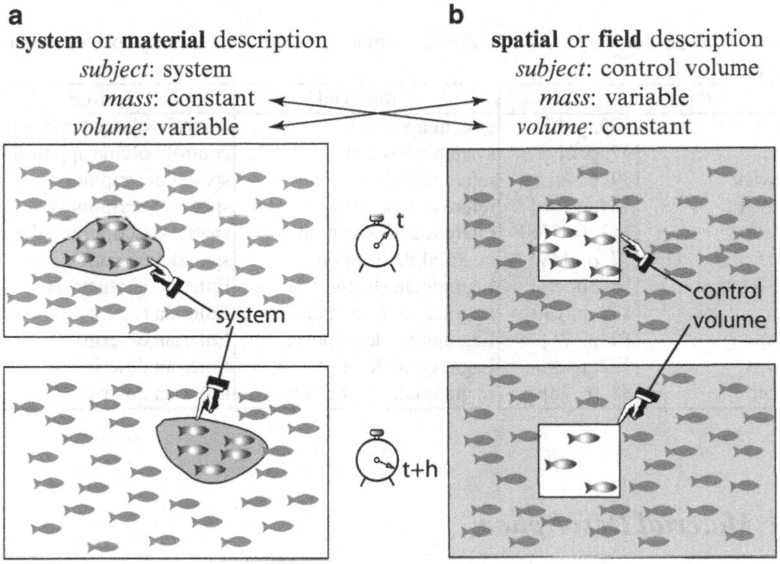


**Fig. 5.1** Material and spatial descriptions. (**a**) system description; (**b**) spatial description

Starting from these examples we can see that the passage from an indefinite vision of many objects to a vision concentrated on a particular object is a common practice in our daily lives. In a broad vision, we look at a space region as a whole without focusing on a particular object: in physics, this is a *field description*. In contrast, in a concentrated vision, we follow the behaviour of a single object and ignore what is happening in the whole space region: in physics, this is a *material description*. Figure 5.1 shows the difference between the two descriptions. In summary we have a

- *Material* or *system* or *referential* or *Lagrangian* description and a
- *Spatial* or *field* or *Eulerian* description.

When we describe physical phenomena, we commonly use one of two complementary descriptions.[4] Material descriptions are typical in the mechanics of rigid

[4] A clear distinction between the *material* and *spatial* descriptions can be found in Shames [209, p. 132], where the term *field* is substituted by the term *control volume*. The two descriptions are also clearly analysed by Hughes [97, p. 2]. In fluid dynamics the *material* description is also called a *system* (Shames [209, p. 78]) or *referential* (Chadwick, [39, p. 54]) or *Lagrangian* description (Hunter, [98, p. 23]). The field description is also called an *Eulerian* descrip-

bodies, analytical mechanics, mechanics of deformable solids and thermodynamics. Spatial descriptions are typical of field theories, such as electromagnetism, gravitation, heat conduction, diffusion and irreversible thermodynamics.

Since in a material description the volume of a system is variable (think of thermodynamics), we will denote it by the calligraphic letter $\mathcal{V}$.[5] In what follows, we will deal mainly with spatial descriptions (Table 5.1).

**Table 5.1** The terminology used by different authors about the two descriptions of continuum mechanics

| *in this book* | | **material** | **spatial** |
|---|---|---|---|
| Shames | [209, p. 13] | system approach | control volume approach |
| Hughes | [97, p. 2] | system approach | control volume approach |
| Chadwick | [39, p. 54] | referential description | spatial description |
| Truesdell | [239, p. 96] | referential description | spatial description |
| Malvern | [147, p. 138] | referential description | spatial description |
| Fung | [77, p. 119] | material description | spatial description |
| Milne-Thomson | [160, p. 81] | Lagrangian (historical) | Eulerian (statistical) |
| Prager | [188, p. 189] | Lagrangian (...of a driver) | Eulerian (...of a policeman) |
| Batchelor | [10, p. 71] | Lagrangian description | Eulerian description |
| Paterson | [172, p. 38] | Lagrangian description | Eulerian description |
| Granger | [82, p. 28] | Lagrangian description | Eulerian description |

## 5.2.1 Material Description

When the object of our observation is a *physical system* which evolves over time, we adopt a *material description*. Only in this case a system can be broken down into *bodies*, and bodies into *particles*.[6]

The quantitative attributes of a system are expressed by physical quantities which refer to the system as a whole. Thus we speak, for example, about the potential and kinetic energies *of a system*, the temperature *of a body*, the electric charge *of a capacitor* and the momentum *of a particle*. All of these variables are *system variables* which change over time. This is the traditional description of mechanics of rigid bodies and equilibrium thermodynamics. Therefore, we consider, for example, *mechanical* systems, *thermodynamic* systems and *optical* systems. In a material description, physical variables are 'global' quantities in space which change over time, i.e. all variations are time variations. Hence, in the differential

tion (Hunter, [98, p. 27]) or *spatial* description (Chadwick, [39, p. 54], Eringen and Suhubi [64, p. 11]).

[5] We will use this notation in Chap. 13.

[6] In his engaging book Hunter wrote: *Particles are mythical mathematical entities which in no sense are to be confused with atoms or molecules* [98, p. 23].

formulation physical laws are written as *ordinary* differential equations with time derivatives only, as in particle mechanics, mechanics of rigid bodies and analytical mechanics.

In a *material* description (Fig. 5.1a), we follow the motion of a system, e.g. a fluid body, and the global physical variables are associated with systems ($\mathscr{S}$), bodies ($\mathscr{B}$) and particles ($\mathscr{P}$).

### 5.2.2 Spatial Description

We speak of a *spatial description* when the object of our observation is a *space region* and we want to know what happens inside it. Only in a spatial description can we divide the space region into subregions, called *control volumes*,[7] which are formed by *points*.

In a spatial description (Fig. 5.1b), global physical variables are associated with points, lines, surfaces and volumes. So we speak, for example, of mass contained *in a volume*, flux *through a surface*, voltage *along a line*, temperature *at a point*, electric charge contained *in a volume* and mass flowing *through a surface*.

To express physical notions in terms of the differential formulation, we need to use densities and rates by determining the limit to obtain point functions, the so-called *field variables*. In this fashion, we can obtain derivatives, and as a consequence, physical laws are expressed by *partial* differential equations.

In differential formulations we consider an infinitesimal Cartesian cell of sides $\mathrm{d}x$, $\mathrm{d}y$, $\mathrm{d}z$: this cell is usually conceived of as a small control volume around an arbitrary point **P**, but it can also be conceived of as a member of a cell complex which fills the whole region.

In a spatial description, every system variable is split into two variables: content and flow, as shown in Table 5.2. For example, let us consider the notion of mass: usually it is an attribute of a body, and we simply speak about the mass of a body. In a spatial description, the reference structure is not a body but a control volume. In this case, we are led to introduce two variables: the mass contained in the control volume at an instant and the mass leaving the control volume (flow) during a time interval. This splitting up can be carried out on many physical variables (extensive variables in thermodynamics). Such a split is found in all *balance equations* (Fig. 5.2). A balance equation is a statement which involves four aspects of an *extensive* quantity:

- The amount *produced* in a space region during a time interval,
- The amount *contained* in a space region at a given time instant,
- The amount *stored* in a space region during a time interval,
- The amount *flowing out* from the space region during the same time interval.

---

[7] Shames [209, p. 78].

A balance can be expressed as follows:

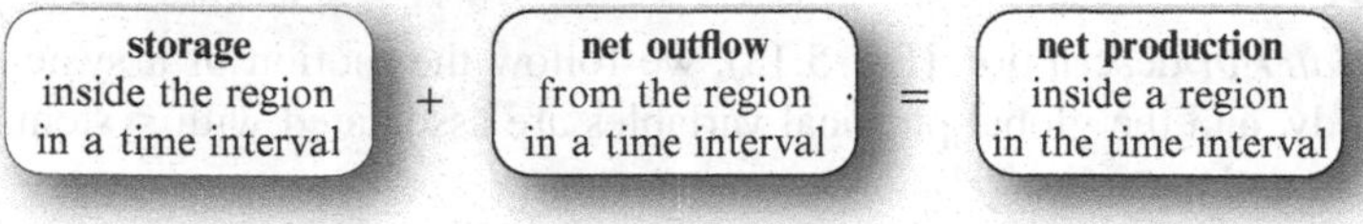


**Fig. 5.2** Balance equation in spatial formulation

**Table 5.2** Passing from a *material* to a *spatial* description global physical variables are split into *content* and *flow*

| *Material description* | | *Spatial description* | | | |
|---|---|---|---|---|---|
| Mass | $M$ | Mass content | $M^{\mathrm{c}}$ | Mass flow | $M^{\mathrm{f}}$ |
| Energy | $E$ | Energy content | $E^{\mathrm{c}}$ | Energy flow | $E^{\mathrm{f}}$ |
| Momentum | $\mathbf{P}$ | Momentum content | $\mathbf{P}^{\mathrm{c}}$ | Momentum flow | $\mathbf{P}^{\mathrm{f}}$ |
| Ang. momentum | $\mathbf{L}$ | Ang. mom. content | $\mathbf{L}^{\mathrm{c}}$ | Ang. mom. flow | $\mathbf{L}^{\mathrm{f}}$ |
| Charge | $Q$ | Charge content | $Q^{\mathrm{c}}$ | Charge flow | $Q^{\mathrm{f}}$ |
| Particle number | $N$ | Particle content | $N^{\mathrm{c}}$ | Particle flow | $N^{\mathrm{f}}$ |
| Probability | $P$ | Probability content | $P^{\mathrm{c}}$ | Probability flow | $P^{\mathrm{f}}$ |
| Entropy | $S$ | Entropy content | $S^{\mathrm{c}}$ | Entropy flow | $S^{\mathrm{f}}$ |

In short, we can say that *a body is composed of particles, a control volume is composed of points.*

### 5.2.3 Material and Spatial Descriptions: An Overview

In physics, as in everyday life, material and spatial descriptions are continuously interchanged. Let us prove the existence of these two descriptions in physical theories.

1. In **classical mechanics**, one considers discrete mechanical systems formed by particles or rigid bodies: this is a *material description*.
2. In the **mechanics of deformable solids**, under the hypothesis of small displacements, a natural description is a *spatial* description. The main kinematic variable, the displacement vector $\mathbf{u}(t, x, y, z)$, depends on time and space coordinates.
3. In **fluid dynamics**, one may use a *material* description or a *spatial* description. Material descriptions are useful for dealing with the convection of vorticity, with the actual path followed by fluid particles and with the idea of

virtual masses.[8] The lift and drag of an airfoil are examples of global variables which refer to a system.

In spatial descriptions, typical variables are *velocity*, *pressure* and *temperature*. These variables are conceived of as field functions, i.e. functions of points and instants. Hence, in a spatial description, the motion of a fluid can be described as in a field theory, such as in electromagnetism.

4. In **thermodynamics**, we use *material* descriptions: the object of investigation is a thermodynamic *system* of which we consider, for example, the internal energy, enthalpy, entropy, phase changes and heat exchanged with the surroundings. Classical thermodynamics uses quasi-static transformations of a *system*, such as expansion, compression, cooling and heating. All of these transformations involve time. In fact, we speak about initial and final states of a *system*, and this implies a time evolution.

   A *closed system* of thermodynamics corresponds to material description, while an *open system* of thermodynamics corresponds to a control volume description and, hence, to a *spatial* description.[9]
5. **Irreversible thermodynamics** is developed using a *field* description. The electrochemical potential, the degree of advancement of a chemical reaction, and current densities of energy, entropy and mass depend on space and time coordinates.[10]
6. In **thermal conduction**, we consider, for example, energy density, entropy density, energy current density, temperature, temperature gradient as functions of space and time coordinates, and hence we use a *field* description.
7. In **electromagnetism**, when we consider a conductor, we speak about the charge, potential and capacitance of the conductor, and hence we use a *material* description. In contrast, when we consider the electromagnetic field in the region around the conductor, we use a *spatial* description.
8. In **analytical mechanics**, Lagrange's and Hamilton's functions refer to a mechanical system; hence, we use a *material* description.
9. **Statistical mechanics** uses a *spatial* description in the phase space. The variables are functions of the phase coordinates $q^k$ and $p_k$.
10. In **quantum mechanics**, the Schrödinger picture is a *spatial* description because the probability amplitude $\Psi(t, \mathbf{x})$ is a function of time and space

[8] Temple [223, p. 11].

[9] Shames [209, p. 133].

[10] de Groot and Mazur [49, p. 3].

coordinates. In contrast, the Heisenberg picture is a *material* description because the dynamic variables (matrices) depends only on time: $q^k(t)$, $p_k(t)$. The wave-particle duality is essentially a field-system duality.

## 5.3 Physical Quantities

Physical quantities can be classified according to at least three different criteria:

1. The *physical theory* to which they belong, for example mechanics, electromagnetism or thermodynamics;
2. Their *mathematical nature*, for example scalar, vector, tensor, axial scalar or axial vector;
3. Whether they are *constants*, *parameters* or *variables*.

Since the first two classifications are self-evident or well known, we will focus our attention on the third criterion, which has received less treatment.[11]

### *5.3.1 Physical Constants*

Some physical quantities have a constant value. To this class belong, for example, the universal constants such as Planck's constant $h$, the speed of light in vacuum $c$, the gas constant $R$, Boltzmann's constant $k_B$, Avogadro's number $N_A$ and the gravitational constant $G$.

### *5.3.2 Physical Parameters*

A *parameter* is a physical quantity, characteristic of a material, system or process, which is constant in the context in which it appears but can vary under certain circumstances.

Broadly speaking, physical constants and physical parameters are those physical quantities whose values are collected in tables. Some parameters characterize the medium in which a process takes place; some are critical values of variables which discriminate between two regimes of a process; some characterize the interaction between two phenomena; and so forth. Physical parameters constitute a larger class of physical quantities which include *system parameters* (e.g. inductance of a coil, diopter of a lens, stiffness constant $k$ of a rod), *material parameters* (e.g. decay constant, thermal conductivity of a material), *process parameters* (e.g. Reynolds number, Nusselt number), *modules* (e.g. Young modulus), *coefficients*

[11] See p. 493 for the meanings of symbols.

**Table 5.3** A classification of physical quantities

| | | |
|---|---|---|
| Physical **quantities** | **Constants** | $h, c, R, k_B, N_A, G, e, \hbar \ldots$ |
| | **Parameters** | System parameters $(R, C, L, M, \ldots)$ |
| | | Material parameters $(\lambda, E, \nu, \mu, \epsilon, \sigma, \ldots)$ |
| | | Coupling coefficients $(\alpha_l, \ldots)$ |
| | | Process parameters $(Re, Nu, Fr, \eta, \ldots)$ |
| | **Variables** | Space global variables and system variables $(M, Q, U, V, S, \mathbf{r}, \mathbf{v}, \mathbf{P}, M^f, Q^f, M^c, T, \Phi, \Psi, \phi, \ldots)$ |
| | | Time global variables $(M, Q, A_H, S, \mathbf{u}, \mathbf{P}, M^f, Q^f, M^c, \Phi, \Psi, \varphi, \ldots)$ |
| | | Field variables $(\rho, p, \mathbf{E}, \mathbf{D}, \mathbf{H}, \mathbf{B}, \mathbf{J}, \omega, \mathbf{q}, \mathbf{g}, \mathbf{k} \ldots)$ |

(e.g. coupling coefficient), *factors* (e.g. transmission factor) and *ratios* (e.g. Poisson's ratio).

Some terms are synonymous:

1. *Physical constants*, which include the *fundamental* or *universal* constants;
2. *System* parameters, also called *lumped* parameters;
3. *Material* parameters, also called *distributed* parameters;
4. *Coupling* parameters, also called *interaction* parameters.

Parameters are generally constant, but many of them suffer slight variations as a function of certain variables such as temperature. This is the case, for example, of specific heat, which depends on temperature, mainly near absolute zero, and of the refractive index, which depends on the wavelength of light.[12]

Table 5.3 shows a general classification of physical quantities. Table 5.4 gives examples of physical parameters of the various classes.

### 5.3.3 Physical Variables

Physical variables[13] are all physical quantities that are neither physical constants nor parameters. A physical variable can be defined in one of the following ways:

[12] In mathematics, the distinction between *constants, parameters* and *variables* was introduced by Leibniz; see Bourbaki [22, p. 244].

[13] All the physical variables considered in this book are collected in the *List of Physical Variables* at the end of the book; see p. 493.

**Table 5.4** Physical parameters

| | | | |
|---|---|---|---|
| System or lumped parameters | | | |
| Capacity | *Of a capacitor* | Inductance | *Of a coil* |
| Resistance | *Of a wire* | Stiffness | *Of a spring* |
| Mass | *Of a star* | Magnetic moment | *Of a particle* |
| Electric dipole | *Of a molecule* | Luminosity | *Of a star* |
| Moment of inertia | *Of a body* | Mean life | *Of an atomic system* |
| Heat capacity | *Of a body* | Power | *Of a lens* |
| Period of revolution | *Of a planet* | Spin | *Of a nucleon* |
| Gyromagnetic ratio | *Of an atomic system* | Strangeness | *Of a particle* |
| Proton number | *Of a nucleus* | Voltage | *Of a battery* |
| Reverberation time | *Of a room* | Emissivity | *Of a radiation body* |
| Damping coefficient | *Of an oscillating system* | Transport coefficient | *Of an ion* |
| Charge | *Of an electron* | Mass | *Of a proton* |
| Material or distributed parameters | | | |
| Permittivity | *Of a medium* | Resistivity | *Of a medium* |
| Elastic modulus | *Of a material* | Surface tension | *Of a liquid* |
| Specific weight | *Of a substance* | Conductance | *Of a medium* |
| Refractive index | *Of a medium* | Diffusion coefficient | *Of a medium* |
| Fusion temperature | *Of a material* | Transmission factor | *Of a substance* |
| Speed of light | *In a medium* | Poisson's ratio | *Of a material* |
| Curie temperature | *Of a ferrom. material* | Shear modulus | *Of a material* |
| Thermoelectric coeff. | *Of a material* | Friction coefficient | *Of two materials* |
| Coupling or interaction parameters | | | |
| Thermoelectric coefficient | *Of a material* | Friction coefficient | *Of two materials* |
| Piezoelectric coefficient | *Of a material* | Peltier coefficient | *Of two metals* |
| Process parameters | | | |
| Reynolds number | *Of a fluid flow* | Mach number | *Of a fluid flow* |
| Nusselt number | *In heat transfer* | Efficiency | *Of a thermodyn. cycle* |

- *Directly* by its measurement process, e.g. length (metre), time interval (chronometer), electric charge (electrometer), force (dynamometer), temperature (thermometer), mass (inertial balance) and a few others;
- *Indirectly* i.e. in terms of other quantities already defined, e.g. work (force × displacement), electric field vector (force/charge), density (charge/volume), pressure (force/area), velocity (displacement/duration).

Physical variables can be classified according to at least two different criteria:

1. The *role* they play in a theory: we will call them *configuration*, *source* and *energy* variables (see the following section);

2. Their *global* or *local* nature. For example, mass, charge and force are global variables, while mass density, charge density and pressure are local variables.

## 5.4 Configuration, Source and Energy Variables

The first criterion, i.e. the *role* which a physical variable plays in a theory, is not rooted in the physical literature. It is somewhat similar to the classification of people into functional classes according to the role they play in society: workers, employees, self-employed workers, managers, and so forth.

In mechanics, variables can be divided into three classes:

1. *Static* and *dynamic variables*, such as, for example, force, impulse, moment of force, torque, momentum, angular momentum and angular impulse.
2. *Geometric* and *kinematic variables*, such as, for example, Lagrangian coordinates, angle of position, angle of rotation, Eulerian angles, position vector, arc length, curvature, velocity, acceleration, angular velocity, angular acceleration, period and frequency.
3. *Energy variables*, which include, for example, work, power, kinetic and potential energy, Lagrangian, Hamiltonian, dissipation function and action.

Static and dynamic variables are those variables which describe the equilibrium configuration and motion respectively. Hence they are used to describe the 'source' of a deformation or of motion. For this reason we can call them *source variables*. Geometric and kinematic variables describe the configuration of a system and can be called *configuration variables*.

It is remarkable that material parameters and physical constants appear only in equations which link source variables with configuration variables, and these are the *constitutive equations*. This is the case, for example, with *mass*, which links momentum and velocity; the *moment of inertia*, which links angular momentum and angular velocity; *stiffness*, which links force and elongation; and the *damping coefficient* which links force and velocity.

Moreover, it is remarkable that the energy variables are obtained from the product of a source variable and a configuration variable without the intervention of physical constants or material parameters, as shown in Table 5.5. Later, when we need to express the energy variables in terms of the configuration variables alone or of the source variables alone, we must resort to the constitutive equations.

This grouping of physical variables into three classes, *source*, *configuration* and *energy* variables, is not commonly done in field theories, such as, for example, electromagnetism, irreversible thermodynamics and thermal conduction. *We propose to extend such a classification to all theories of the macrocosm.*

**Table 5.5** Energy variables of mechanics

| | | | |
|---|---|---|---|
| Kinetic energy | $T(\mathbf{p}) \overset{\text{def}}{=} \int_0^{\mathbf{p}} \mathbf{v}(\mathbf{p}) \cdot \mathbf{dp}$ | Kinetic co-energy | $T^*(\mathbf{v}) \overset{\text{def}}{=} \int_0^{\mathbf{v}} \mathbf{p}(\mathbf{v}) \cdot \mathbf{dv}$ |
| Work | $W(t^-, t^+) \overset{\text{def}}{=} \int_{t^-}^{t^+} \mathbf{F}(t, \mathbf{r}, \mathbf{v}) \cdot \mathrm{d}_t\mathbf{r} = \int_{t^-}^{t^+} \mathbf{F}(t, \mathbf{r}, \mathbf{v}) \cdot \mathbf{v}(t)\, \mathrm{d}t = \int_{t^-}^{t^+} P(t)\, \mathrm{d}t$ | | |
| Potential energy | $V(\mathbf{r}) \overset{\text{def}}{=} \int_{\mathbf{r}}^{\mathbf{r}_0} \mathbf{F}(\mathbf{r}) \cdot \mathrm{d}_s\mathbf{r}$ | Power | $P(t) \overset{\text{def}}{=} \lim_{t^+ \to t^-} \frac{W(t^+, t^-)}{t^+ - t^-}$ |
| Lagrangian | $L \overset{\text{def}}{=} T^* - V$ | Hamiltonian | $H \overset{\text{def}}{=} T + V$ |
| Hamiltonian action | $A_H(t^-, t^+) \overset{\text{def}}{=} \int_{t^-}^{t^+} L(t)\, \mathrm{d}t$ | Hamilton's principal function | $S(t) \overset{\text{def}}{=} \int_0^t L(t')\, \mathrm{d}t'$ |

### 5.4.1 Source Variables

Every physical field has its sources. So an electric charge is the source of an electric field, an electric current is the source of a magnetic field, mass is the source of gravitational fields, force is the source of displacement, heat production is the source of a thermal field and so forth.

The source of an electromagnetic field is an electric charge, both at rest or in motion, i.e. electric charge content $Q^c$ and electric charge flow $Q^f$. Hence also the current $I$, which is the electric charge flowing in a time interval divided by its duration, is a variable which describes the source. The current density $\mathbf{J}$ is also used to describe a source. In short: it is natural to consider as source variables all those variables obtained from the main source variable by forming the density, the rate and the space derivatives.

DEFINITION. *We call* source variables *all variables which describe the source of a field, the cause of a phenomenon and all variables linked to them by the operations of sum, difference, division by a length, an area, a volume or an interval; by a limit process and, hence, by time and space derivatives; line, surface, volume and time integrals. These relations must not contain physical constants.*

What follows is a list of source variables for the main physical theories:

**Mechanics of continua:** All *static* and *dynamic variables*, i.e. volume force, impulse of volume force, surface force, momentum, impulse of surface force, pressure, stress vector, stress tensor, stress deviator, stress function, Airy's function, bending moment, surface torque, impulse of surface torque, couple stress tensor, traction; momentum content; momentum density, momentum flow, momentum current=force, momentum current density; angular momentum content, angular momentum density, angular momentum flow, angular momentum current = torque, angular momentum current density;

**Electromagnetism:** Electric charge, electric charge content, electric charge density, electric charge flow, electric current, electric current density, electric flux, electric displacement, magnetic voltage, magnetic voltage impulse, magnetomotive force, magnetomotive force impulse, magnetic field strength, magnetic scalar potential;

**Gravitation:** Gravitational mass, mass content, mass density, mass flow, mass current, mass current density;

**Thermodynamics:** Entropy, entropy content, entropy density, entropy flow, entropy current, entropy current density, entropy production, entropy source.

## 5.4.2 Configuration Variables

At the same time, all physical fields have variables which describe their *configuration*, typically the field potentials. This is the case with, for example, the electric potential, gravitational potential, magnetic vector potential, position vector for particle mechanics, displacement for deformation of solids, velocity for fluid dynamics, and temperature which describes the thermal configuration of a body.

DEFINITION. *We call* configuration variables *all variables which describe the configuration of a physical system, in particular the potential of fields. Belonging to this class are all variables linked to a potential by the operations of sum, difference, division by a length, an area, a volume or a time interval; by a limit process and, hence, by time derivatives or space derivatives; or by integrals on lines, surfaces, volumes and time intervals. These relations must not contain physical constants.*

An important point is the following: *the same variable may play a different role in different physical theories*. This is similar to a person who is a member of various clubs, societies and companies: his role may differ depending on the organization. So he could be an *employee* of a company, the *treasurer* of a sports club or the *president* of a charity.

A physical variable which plays different roles is the mass. In particle mechanics and in rigid body mechanics, mass is a *physical parameter*, such as the mass of a proton, the mass of a star or the mass of a body. In contrast, in fluids, mass is a source variable because one can consider the *mass production* and then the mass source $\sigma_{\mathrm{m}}$.

What follows is a list of configuration variables for the main physical theories.

**Mechanics of continua:** All *geometric* and *kinematic variables*, i.e. displacement, relative displacement, displacement gradient, Burgers vector, dislocation density tensor, extension, strain, strain deviator, strain rate, strain rate deviator, bulk strain, velocity line integral, velocity potential, angular velocity, angular acceleration, vortex strength, vorticity vector;

**Electromagnetism:** Electric potential, electric field strength, electric potential impulse, voltage, voltage impulse, voltage, voltage impulse; magnetic flux, magnetic vector potential, magnetic induction, gauge function;

**Gravitation:** Acceleration of gravity, gravitational potential, gravitational potential difference, metric tensor, gravitational tensor, linear connexion, Riemann curvature tensor, Ricci tensor;

**Thermodynamics:** Temperature, temperature difference, temperature gradient, indefinite time integral of temperature (thermacy), thermodynamic forces.

### *5.4.3 Energy Variables*

DEFINITION. *We call* energy variables *all variables obtained by multiplying a configuration variable by a source variable and all variables linked to them, by the operations of sum, difference, division by a length, an area, a volume or an interval; by time or space derivatives; or by integrals on lines, surfaces, volumes and time intervals.*

What follows is a list of energy variables:

**Energy variables**: Work, heat, power (energy current), energy current density, energy content, energy density, energy flow, kinetic energy, kinetic co-energy, potential energy, internal energy light exposure, light exposure rate, Lagrangian, Hamiltonian, action, Lagrangian density, Hamiltonian density, electric energy, electric energy density, magnetic energy, magnetic energy density, Poynting vector, enthalpy, Helmholtz free energy, Gibbs free energy; Rayleigh dissipation function; absorbed dose, absorbed dose rate (Fig. 5.3).

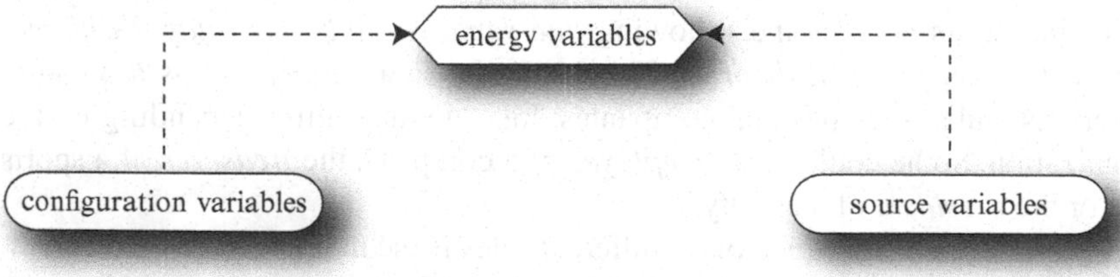


**Fig. 5.3** Classification of variables of a physical theory

**REMARK.** The terms *configuration variables* and *source variables* were introduced by the author[14] and correspond to those previously introduced by Penfield and Haus[15] with the names *geometric* and *force* variables. We prefer the term *configuration* variables because variables like velocity, angular velocity, vorticity, temperature, temperature gradients, potentials and magnetic induction are hardly conceived of as *geometric*. The term *configuration* is borrowed from analytical mechanics and means the configuration of a mechanical system. Similarly, we prefer the term *source* variables because it denotes variables like electric charges, electric currents, gravitational masses, magnetic field strength and heat current, which are not usually considered *forces*. Moreover, Hallen[16] uses the term force with a meaning opposite to Penfield's use of the same term. A comparison between different nomenclatures is given in Table 5.6.

**Table 5.6** The three kinds of variables of all physical theories: configuration, source and energy variables

| Present | Configuration | Source | Energy |
|---|---|---|---|
| Hallen | Force | Source | Mechanical |
| Penfield | Geometric | Force | |

## 5.5 Fundamental Problem of a Physical Theory

In every physical theory of the macrocosm,[17] there is one or more variables which describe the *configuration* of the system (e.g. the field potential) and onc or more variables which describe the *source* of the phenomenon (or of the field). The problem of finding the configuration of a system (or of a field) once the sources are assigned will be called the *fundamental problem* of the theory. What follows is a list of the fundamental problems of some physical theories:

**Mechanics**: *given the forces acting on a system (or a body or particle), find the configuration of the system at every instant.*

**Electrostatics**: *given the space distribution of electric charges, find the electric potential at every point.*

**Magnetostatics**: *given the space distribution of electric currents, find the magnetic vector potential at every point.*

**Gravitostatics**: *given the space distribution of masses, find the gravitational potential at every point.*

---

[14] Tonti [227, 228].

[15] Penfield and Haus [175, p. 155].

[16] Hallen [87, p. 1].

[17] With the exception of reversible thermodynamics because it is the science of energy.

**Dynamics of deformable solids**: *given the forces acting on a solid, find the displacements of all points at every instant.*
**Fluid dynamics**: *given the forces acting on a fluid, find the velocity at all points and at every instant.*
**Thermal conduction**: *given the space distribution and intensities of heat sources, find the temperature at all points at every instant.*

The fundamental problem of a physical field can be stated, in more detail, as follows:

- *Assign the space and time distributions of the field sources in a region:*
  - *Assign the shape and dimensions of the region.*
  - *Assign the nature of the materials which fill the region.*
  - *Assign the boundary conditions (which summarize the action of the external sources on the field region).*
  - *Assign the initial conditions (for time variable fields).*
- *Find the configuration of the field at every instant.*

In the second part of the book, where we present the individual physical theories in more detail, we will also state their fundamental problem.

## 5.6 Set Functions

Among physical quantities there are some which refer to a system as a whole, whereas others refer to points. In the first class we have the charge *of a capacitor*, the kinetic energy *of an airplane* and the weight *of a body*, whereas in the second class we have the temperature *at a point*, the density of air *at a point* and the pressure *at a point*. Mathematics uses the term *set functions* and *point functions* respectively.

Given a set $\mathscr{A}$ and a vector space $\mathsf{V}$, a *set function*, also called a *domain function*, is a map which associates an element of the vector space $\mathsf{V}$ with every subset of $\mathscr{A}$. The minimum requirement, always satisfied in physics, is that the set function be additive. This means that if $\mathscr{A}_1$ and $\mathscr{A}_2$ are two *disjoint* subsets of $\mathscr{A}$, if $v_1$ and $v_2$ are the elements of the vector space $\mathsf{V}$ corresponding to the two subsets, the to the subset $\mathscr{A}_1 \cup \mathscr{A}_2$ it associates the element $v_1 + v_2$.

Examples of set functions include *the length* of a rope, *the area* of a geographical region, *the volume* of a body, *the number* of passengers on a bus at a given instant, *the momentum* of a car at a given instant or *the heat* needed to raise the temperature of a boiler by 20°.[18]

Henri Lebesgue wrote: "les grandeurs de la physique directement mesurables apparissent d'ailleurs toujours comme des fonctions de domaine; ... il peut s'agir

[18] Lebesgue [131, p. 150].

de domaines sur la droite, c'est a dire d'intervals, de domains plans ou de domains a plus de trois dimensions ...".[19]

## 5.7 Global Variables and Field Variables

The differential formulation of physical fields requires the use of *field variables* because they make possible the use of partial derivatives. Those physical variables which refer to extended space elements are *global variables*; they include line integrals, fluxes and contents.

Most field variables arise as space densities of global variables. The introduction of densities and rates is prompted by the desire to obtain physical variables which are independent of the extension of the space and time domains. In turn, global variables are usually obtained by the integration of field variables on lines, surfaces and volumes. For this reason they are commonly called *integral variables*.

There is a subclass of field variables which are not space densities of global variables. Therefore, whereas pressure $p(\mathbf{P})$, which is force/area, and electric charge density $\rho(\mathbf{P})$, which is charge/volume, are field variables of the density kind, in contrast, temperature $T(\mathbf{P})$ and electric potential $\phi(\mathbf{P})$ are field variables, without being densities of another global variable. Table 5.7 shows the two kinds of field variables.

The process of forming field variables from global variables can be divided into two steps:

- We obtain *mean densities* and *mean rates* by dividing the global variable by the extension of the space or time element to which the global variable is referred.
- We perform the *limit* process by making the extension of the space or time element tend towards zero. In this way, we obtain the pointwise densities and rates, i.e. field variables.

Note that while the first step gives mean values and can be done without particular care, the second step requires the existence of a limit. As we showed previously, the limit can be discontinuous through the separation surface between two media; hence, the field variables obtained cannot be differentiable. The notions of scalar, vector and tensor *fields* imply the notion of field variables. Field variables are completely independent of geometric attributes. While field variables are necessary ingredients for the *differential* formulation of physical laws, global variables are the natural ingredients for an *algebraic* or *finite* or *discrete* formulation of physical laws.

---

[19] "Physical variables which are directly measurable always appear as domain functions; ... this may be a line domain (i.e. an interval), a plane domain or a domain of more than three dimensions ..." Lebesgue [130, p. 20].

**Table 5.7** The two kinds of field variables

| *Field variables which are not densities of global variables* | | *Field variables which are densities of global variables* | | |
|---|---|---|---|---|
| Temperature | $T[\bar{\mathbf{P}}]$ | Internal energy density | $u[\tilde{\mathbf{V}}]$ | $U[\tilde{\mathbf{V}}]$ |
| Electric potential | $\phi[\bar{\mathbf{P}}]$ | Mass density | $\rho[\tilde{\mathbf{V}}]$ | $M^{c}[\tilde{\mathbf{V}}]$ |
| Gravitational potential | $U_{g}[\bar{\mathbf{P}}]$ | Momentum density | $\mathbf{p}[\tilde{\mathbf{V}}]$ | $\mathbf{P}[\tilde{\mathbf{V}}]$ |
| Phase function | $\phi[\bar{\mathbf{P}}]$ | Entropy source strength | $\sigma_{s}[\tilde{\mathbf{V}}]$ | $S^{p}[\tilde{\mathbf{V}}]$ |
| Position vector | $\mathbf{r}[\bar{\mathbf{P}}]$ | Electric current density | $\mathbf{J}[\tilde{\mathbf{S}}]$ | $I[\tilde{\mathbf{S}}]$ |
| Total displacement | $\mathbf{u}[\bar{\mathbf{P}}]$ | Pressure | $p[\tilde{\mathbf{S}}]$ | $F[\tilde{\mathbf{S}}]$ |
| Velocity potential | $\phi[\bar{\mathbf{P}}]$ | Magnetic flux density | $\mathbf{B}[\bar{\mathbf{S}}]$ | $\Phi[\bar{\mathbf{S}}]$ |
| Gauge function | $\chi[\bar{\mathbf{P}}]$ | Mass current density | $\mathbf{q}[\tilde{\mathbf{S}}]$ | $M^{f}[\tilde{\mathbf{S}}]$ |
| Stream function | $\psi[\tilde{\mathbf{P}}]$ | Electric field strength | $\mathbf{E}[\bar{\mathbf{L}}]$ | $V[\bar{\mathbf{L}}]$ |
| Airy's stress function | $\phi[\tilde{\mathbf{P}}]$ | Magnetic field strength | $\mathbf{H}[\tilde{\mathbf{L}}]$ | $F[\tilde{\mathbf{L}}]$ |
| Scalar magnetic potential | $\phi_{m}[\tilde{\mathbf{P}}]$ | Acceleration of gravity | $\mathbf{g}[\bar{\mathbf{L}}]$ | $\Delta_{s}U_{g}[\bar{\mathbf{L}}]$ |
| *(perhaps a few others)* | | *(definitely many others)* | | |

The formation of field variables such as the density of the variable domain implies an assumption of regularity, namely continuity and differentiability, whereas these conditions are not required for global variables. It follows that the range of application of the differential formulation to physics is restricted to regions of regularity, i.e. without material discontinuities and concentrated sources.

## 5.8 Global Variables

For three centuries the mathematical description of physics has been based on the differential formulation, that is, total and partial derivatives occur that act on field variables.

Let us pose the following question: *since the mathematical formulation of physics rests on physical variables, do they appear directly as functions of points and instants?* In other words: *is the differential formulation the most natural form for a mathematical description of physics?*

Let us try to answer this question. Few physical variables arise directly as functions of points and instants, whereas most of them arise in association with a volume, a surface, a line or a time interval. Most physical variables used in the differential formulation become functions of points and instants because they are obtained from variables associated with extended space elements and with time intervals when the corresponding densities and rates are calculated.

To give some examples, the mass of a body is associated with a body (in a material description) or with a volume (in a spatial description), the internal surface force in continuum mechanics is associated with a surface, and voltage is associated with a line. Internal energy, entropy, potential energy and kinetic energy are associated with a system (in a material description) or with a volume (in a spatial description), and hence they are not functions of a point, but they are functions of a time instant; the number of moles of a substance refers to a volume.

The displacement of a particle, impulse of a force, work and heat are associated with a time interval. The energy emitted by a hot body is associated with its bounding surface and with a time interval. The radiation emitted by a specimen of radioactive material depends on the volume of the specimen and on the time interval considered.

The differential formulation uses mass density and not directly mass, pressure and not directly the force normal to a plane surface, strain and not extension, concentration and not the mole number, heat current density and not heat, and so forth.

These considerations suggest giving a proper status, and hence a proper name, to all physical variables which are not densities or rates.

Since most measured physical variables are *global*, in order to obtain the corresponding field functions we must evaluate their densities, such as lines, surfaces and volume densities. Hence, by starting from a global physical variable, we create a field function: the global variable, when needed, is reconstructed by a process of integration.

It is important to remark that *the formation of densities of global variables hides their geometric and physical contents*. Let us make a simple consideration: all field functions refer to points. Hence, when we perform the limit process, starting from a global variable, to obtain a density, i.e. a field function, we lose information about the association of the variable with space and time elements.[20] At first glance this seems unimportant, but it is not. In fact, *it is precisely the link between global physical variables and space and time elements that will allow us to obtain a new classification for physical variables.*

Some readers may think that we are suggesting the abandonment of field functions and the differential formulation of physics; far from it! *We are simply stating that global physical variables contain some geometric information that field variables do not contain: consequently, we suggest developing the algebraic formulation* before *the differential formulation and not vice versa.*

---

20 See p. 113.

### 5.8.1 Extension of the Notion of Density

The most common notion of density is that obtained from a quantity associated with volume: mass density, charge density, energy density, density of entropy, density of momentum, probability density.

Less used is the concept of surface density, such as that of the population of a territory, which is the ratio between the number of inhabitants and the area of the territory. In electromagnetism, one has the surface density of charge (C/m$^2$). In the theory of liquid membranes, one considers the energy per unit area.

The global variables associated with a volume can be written as the product of the volume density and the extension of the volume, and if this variable is different from point to point, then we must write the integral on the volume of the product $\rho\,dV$. A global variable relative to a surface will be written as the integral of the product $\sigma\,dS$. Lastly, in the case of linear density, we may write the global variable as an integral of the product $\lambda\,dL$:

$$Q[\mathbf{V}] = \int_{\mathbf{V}} \rho\,dV, \qquad Q[\mathbf{S}] = \int_{\mathbf{S}} \sigma\,dS, \qquad Q[\mathbf{L}] = \int_{\mathbf{L}} \lambda\,dL\,. \tag{5.1}$$

The three quantities $Q[\mathbf{V}]$, $Q[\mathbf{S}]$, $Q[\mathbf{L}]$ are usually called integral quantities and as such are automatically global quantities. In thermodynamics, variables associated with volumes are called *extensive*. If the term extensive is conceived in a broader sense as equivalent to *additive* not only on volumes but also on surfaces and lines, then we can say that the integrals are extensive magnitudes. Quoting Tolman:[21]

> If we regard the different kinds of quantities that are used by physicists, we find that they fall with reasonable lack of ambiguity into two general classes, those having a certain additive nature so that a given quantity can be regarded as being the sum of a number of smaller quantities of the same kind and those which have no such additive properties. We say that quantities of the first class have extensive magnitude, and that quantities not having an additive nature have intensive magnitude.

We remark that in thermodynamics the additivity is limited to volumes, whereas in this authoritative citation additivity is associated not only with volumes, but also with surfaces, lines and time intervals.

The quantities $\rho$, $\sigma$, $\lambda$ are respectively volume, surface and line densities.[22] When we compute the flow of a vector through a surface or the line integral of a vector along a line, we write

$$\begin{aligned} \Phi[\mathbf{S}] &= \int_{\mathbf{S}} \mathbf{q}\cdot d\mathbf{S} = \int_{\mathbf{S}} (\mathbf{q}\cdot\mathbf{n})\,dS = \int_{\mathbf{S}} q_n\,dS, \\ \Gamma[\mathbf{L}] &= \int_{\mathbf{L}} \mathbf{v}\cdot d\mathbf{L} = \int_{\mathbf{L}} (\mathbf{v}\cdot\mathbf{t})\,dL = \int_{\mathbf{L}} v_t\,dL\,. \end{aligned} \tag{5.2}$$

---

[21] Tolman [225, p. 239].

[22] The French literature speaks about densité *volumique, surfacique, linéique*.

The quantities in parentheses, which can be written as $q_n$ and $v_t$, are respectively the surface density and the line density. Considering as a special case that of uniform vector fields, as in the motion of a perfect fluid in a pipe, and choosing the unit vectors **n** and **t** in the direction of the field lines, the terms in brackets are reduced to the modules of the respective vectors.

These modules are respectively the surface and the line density of the global quantities $\Phi[\mathbf{S}]$ and $\Gamma[\mathbf{L}]$. It becomes natural to extend the term density to the respective vectors **q** and **v**. Hence the vectors used to calculate a flow or a line integral may be called *densities* in a generalized sense.

According to this extension, the mass current density vector **q** can be seen as a surface density; hence, it will be associated with surfaces. Similarly, in fluid dynamics and in a spatial description, the velocity vector **v** is the line density of its line integral and will be associated with lines.

### 5.8.2 Global Variables in Space

The fact that a physical variable can be described by a mathematical entity does not authorize us to perform all possible mathematical operations on it: *an operation which is mathematically meaningful can be physically meaningless.* In mathematics, one can add two numbers, but in physics one cannot add two physical quantities if they are not of the same kind and do not refer to the same unit of measure.

This happens because a physical quantity belongs to a definite *species*, and its value is referred to a definite *unit of measure*. Moreover, even if two physical quantities are of the same species and are referred to the same unit of measure, their sum can be meaningless: thus, the sum of two temperatures $T_1$ and $T_2$ or of two time instants $t_1$ and $t_2$ is devoid of physical interest.[23]

In mathematics one can always calculate an indefinite integral on an integrable function of one variable. In contrast, *in physics one cannot perform mathematical operations at will.* If we perform a time integration on a physical variable, it can be meaningless to perform subsequent time integrations. So the indefinite time integral of the acceleration $\mathbf{a}(t)$ gives the velocity $\mathbf{v}(t)$; the indefinite time integral of the velocity gives the position vector $\mathbf{r}(t)$, but the time integral of the position vector is devoid of physical interest. It is meaningful to perform the indefinite time integral of a force $\mathbf{F}(t)$ to obtain the momentum $\mathbf{p}(t)$, but the indefinite time integral of momentum is devoid of physical interest. The ratio of a displacement to the time interval with which it is associated is meaningful, but the product of the displacement for the time interval is devoid of physical interest.

[23] Lebesgue [131, p. 130].

We stated previously that many physical variables are densities of other variables: this suggests the need to introduce a specific name for those variables which are not densities of other variables.

DEFINITION. *We call a* global variables in space *any variable which is not a line, surface or volume density of another variable.*

These include position vector, displacement, momentum, force, mass, mass flow, electric charge, voltage, magnetic flux, electric flux, heat, work, vortex flux, entropy, entropy flow and entropy content. From this definition it follows that there are global variables which are associated with points, such as temperature, electric potential, chemical potential and velocity potential.

**Integral Variables.** Recall that *integral variables* are those arising by integration on lines, surfaces and volumes of field variables. In the differential formulation we are led to introduce integral variables because we start from field variables. Of course, integral variables are global variables, but there are global variables which are not integral variables, and these are variables associated with points (Fig. 5.4).

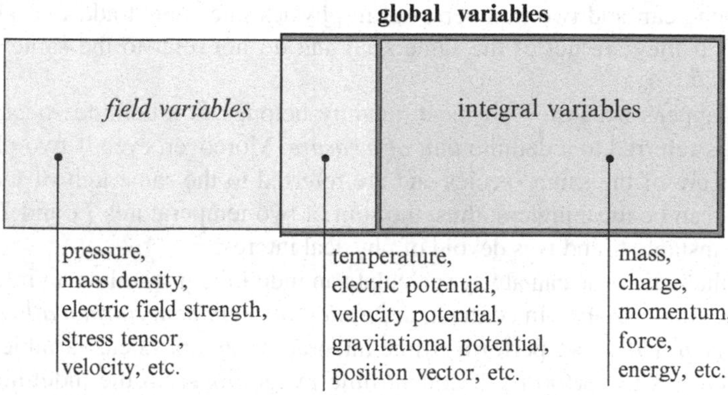


**Fig. 5.4** Relation between global, integral and field variables. A longer list of field variables which are also global variables is given in Table 5.7

We use the term global instead of integral because it also includes variables associated with points which are not densities or rates and, therefore, they are not obtained from space or time integration of field variables. In Table 5.7 (p. 106) the variables in the left column are global variables that are not integral variables.

In electromagnetism, the magnetic flux $\Phi$ is usually expressed as the surface integral of the magnetic flux density $\mathbf{B}$; hence, it is considered an integral variable. What is curious is that from an experimental viewpoint we find it easier to measure the magnetic flux and to *deduce* the magnetic flux density, i.e. we follow the opposite procedure to that used in the differential formulation.

If we consider the typical *intensive* variables of thermodynamics, i.e. temperature $T$, pressure $p$ and mass density $\rho$, we see that only temperature is not a density, and hence it is a global variable. This shows that the class of *global* variables is larger than the class of *extensive* variables.

In many cases, global variables are those variables we measure, mainly because probes have a spatial extension. For example, it is easier to measure mass and charge than their densities; it is easier to measure voltage than electric field strength $\mathbf{E}$; it is easier to measure a magnetic flux $\Phi$, e.g. with a flip-flop coil, than the magnetic flux density $\mathbf{B}$. There are also global variables which we do not measure directly: these include, for example, entropy, work, energy and action.[24]

**Intensive, Extensive, Through and Across Variables.** Other classifications of physical variables exist, even if their field of interest is sometime limited to a single theory. Table 5.8 shows these four kinds of variables.

Thus, in *thermodynamics*, variables are classified as *intensive* and *extensive*. This classification is practically limited to thermodynamics.

In *systems theory*, there is another classification, limited to energy flows, which distinguishes between *through* and *across* variables, as shown in Table 5.9. *Across* variables are associated with lines endowed with an inner orientation, whereas *through* variables are associated with surfaces with an outer orientation. Typical variables of this kind are the voltage along a line and the current through a surface.[25]

**Global Variables and Computational Physics.** We have just stated that global variables, which refer to space elements endowed with extension, i.e. lines, surfaces and volumes, are *set functions*, whereas field functions are *point functions*. We cannot perform derivatives on set functions; we can only perform algebraic operations. For this reason global variables are the most natural variables used in computational physics, where only algebraic operations are possible. This approach is new in computational physics. In fact, it is customary to start with differential equations and then perform a discretization process on them to obtain algebraic equations.[26]

---

[24] Many physicists believe that quantities which are not measurable should not be used in physics. This statement is wrong, as expressed by many illustrious physicists. See Appendix C.

[25] MacFarlane [146, p. 17].

[26] The direct algebraic formulation of physical laws is the starting point of the *cell method*: see Tonti [230–232, 234] and the papers quoted on the Web site *discretephysics.dicar.units.it*.

**Table 5.8** Connection between various classifications used in physics and in engineering

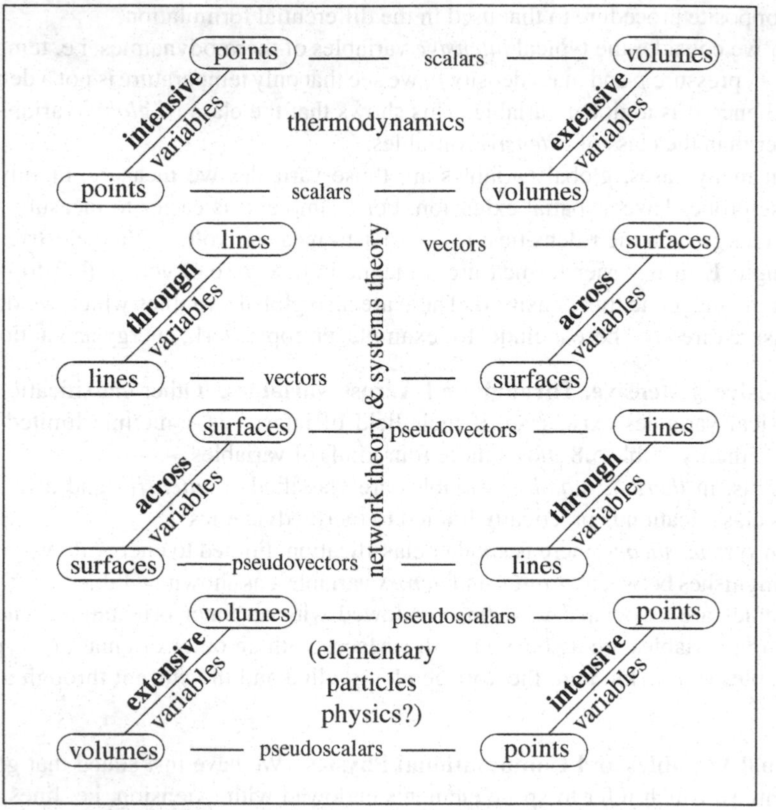


This table can be understood in reference to Fig. 8.9 at page 236

### 5.8.3 Continuity of Global Variables in Space

An analysis of global variables shows the following important property: *global variables are continuous through the interface of two different media.* Hence, by crossing a shock wave in fluids, the mass flow, the energy flow and the momentum flow are continuous.[27] Examples of this property are shown in Fig. 5.5. Whereas global variables are continuous, their *variations* can be discontinuous through the separation surface between two media. Even field variables, which are densities and rates, are generally discontinuous. Thus, in the rebound of a ball against a wall, the position vector (global variable in space and in time) is continuous, whereas the velocity (its rate) is discontinuous; mass content (global variable in

[27] Landau and Lifshitz [125, Sect. 81].

**Table 5.9** Conjugate physical variables: the product *intensive* × *extensive* gives an energy; the product *across* × *through* gives a power

| *Type* | | *Variables* | Symbol | *Space element* |
|---|---|---|---|---|
| | | *Intensive* × *extensive* = *energy* | | |
| | ♡ | Pressure | $-p$ | Point |
| | ♠ | Temperature | $T$ | Point |
| *Intensive* variables | △ | Chemical potential | $\mu$ | Point |
| | ▷ | Electric potential | $\phi$ | Point |
| | ◁ | Gravitational potential | $U_g$ | Point |
| | | *Across* × *through* = *power* | | |
| | ◇ | Voltage | $E$ | Line |
| *Across* variables | ♣ | Velocity | $\mathbf{v}$ | Line |
| | ◇ | Angular velocity | $\breve{\omega}$ | Line |
| | ◇ | Electric current | $I$ | Surface |
| *Through* variables | ♣ | Force | $\mathbf{F}$ | Surface |
| | ◇ | Torque | $\breve{\tau}$ | Surface |
| | | *Intensive* × *extensive* =*energy* | | |
| | ♡ | Volume | $V$ | Volume |
| | ♠ | Entropy | $S$ | Volume |
| *Extensive* variables | △ | Mole number | $n$ | Volume |
| | ▷ | Electric charge | $Q$ | Volume |
| | ◁ | Gravitational mass | $M$ | Volume |

space) is continuous, whereas mass density is discontinuous; temperature (global variable in space) is continuous, whereas the temperature gradient is discontinuous; electric potential (global variable in space) is continuous, whereas the electric field strength is discontinuous.

### 5.8.4 Space Association

We have said that global variables in space are associated with space elements. What do we mean by *association*? To explain this, let us consider as an example the notions of flux and flow: whenever we talk about flux or flow, e.g. magnetic flux $\Phi$, energy flow $E^{\mathrm{f}}$, entropy flow $S^{\mathrm{f}}$ or mass flow $M^{\mathrm{f}}$, we are referring to a surface. If we wish to highlight this attribute of a variable, we will say that the flux is *associated* with the surface.

Also, variables which are obtained by a line integral of a vector, such as the voltage $E$, the magnetomotive force $F_{\mathrm{m}}$, the line integral of the velocity $\Gamma$, the

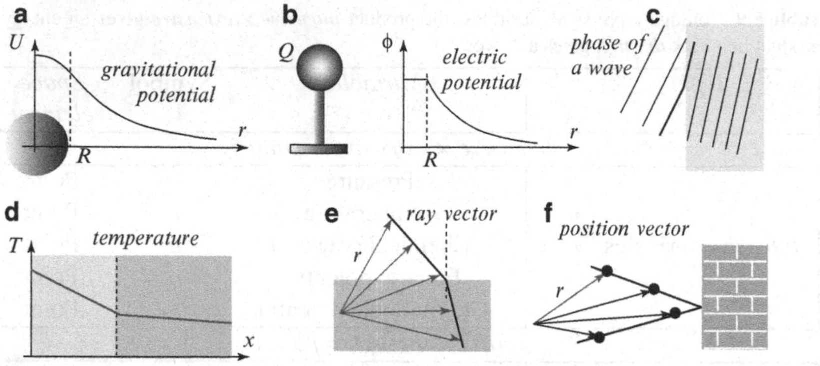


**Fig. 5.5** The global variables associated with points are continuous through the separation surface of two media. (**a**) The gravitational potential is continuous passing from the exterior to the interior of a massive sphere; (**b**) the electrical potential is continuous passing from the exterior to the interior of a metallic sphere; (**c**) the phase of a plane wave is continuous through the separation surface between two transparent media; (**d**) temperature is continuous through the separation surface of different materials; (**e**) the ray vector in optics is continuous in refraction; (**f**) the position vector is continuous in the rebound of a particle

line integral of a force $W$, since they involve a line in their definition, are said to be *associated* with a line.

Similarly, the concept of the *content* of a variable, which is typical of spatial descriptions, e.g. the mass content $M^c$, the energy content $E^c$, the entropy content $S^c$, the momentum content $\mathbf{P}^c$, refers to a volume. We express this fact by saying that these variables are *associated* with a volume.

There are also variables which are referred to points *and that are not densities of other variables*: these include temperature $T$, electric potential $\phi$ and total displacement $\mathbf{u}$ in continuum mechanics. We say that these variables are *associated* with points.

To highlight this association, next to the physical variable we place its corresponding space element enclosed within square brackets. Table 5.10 provides examples of this nomenclature. This association with the space element highlights a characteristic aspect of the variable, a sort of DNA.

**Table 5.10** Examples of associations of physical variables with space elements

| | | |
|---|---|---|
| $M^c$[**V**] mass content | $\mathbf{F}^v$ [**V**] volume force | $V$[**V**] potential energy |
| $Q$[**S**] heat | $\mathbf{T}$ [**S**] internal surface force | $\Phi$[**S**] magnetic flux |
| $E$[**L**] voltage | $e$[**L**] extension | $\Gamma$[**L**] velocity line integral |
| $T$[**P**] temperature | $\mathbf{u}$ [**P**] total displacement | $\phi$[**P**] electric potential |

As we will show in the second part of the book, which deals with single physical theories, some fluxes are associated with surfaces endowed with an inner orientation, whereas other fluxes are associated with surfaces endowed with an outer orientation. For example, we will show that the magnetic flux $\Phi$ and the vortex flux $W$ are associated with surfaces $\overline{\mathbf{S}}$ endowed with inner orientations, while the electric flux $\Psi$, the mass flow $M^{\mathrm{f}}$, energy flow $E^{\mathrm{f}}$, charge flow $Q^{\mathrm{f}}$, entropy flow $S^{\mathrm{f}}$, momentum flow $\mathbf{P}^{\mathrm{f}}$ and heat $Q$ are associated with surfaces $\widetilde{\mathbf{S}}$ endowed with an outer orientation.

Similarly, the voltage $E$, the line integral of the velocity $\Gamma$ and the work $W$ of a force in a force field are associated with lines $\overline{\mathbf{L}}$ endowed with an inner orientation, whereas the magnetomotive force $F_{\mathrm{m}}$ is associated with lines $\widetilde{\mathbf{L}}$ with an outer orientation.

Almost all contents – e.g. mass content $M^{\mathrm{c}}$, energy content $E^{\mathrm{c}}$, charge content $Q^{\mathrm{c}}$, entropy content $S^{\mathrm{c}}$, momentum content $\mathbf{P}^{\mathrm{c}}$ and angular momentum content $\check{\mathbf{L}}^{\mathrm{c}}$ – are associated with volumes $\widetilde{\mathbf{V}}$ with an outer orientation: only the hypothetical magnetic charge $\check{G}$ is associated with volumes $\overline{\mathbf{V}}$ endowed with an inner orientation. With regard to the variables associated with points, some of them, such as

**Table 5.11** Global variables in space

| | *Inner orientation* | *Outer orientation* |
|---|---|---|
| **P** | $\phi[\overline{\mathbf{P}}], U[\overline{\mathbf{P}}], \phi[\overline{\mathbf{P}}], \mathbf{u}[\overline{\mathbf{P}}]$ | $\check{\phi}_{\mathrm{m}}[\widetilde{\mathbf{P}}], \check{\psi}[\widetilde{\mathbf{P}}]$ |
| **L** | $E[\overline{\mathbf{L}}], \Gamma[\overline{\mathbf{L}}], W[\overline{\mathbf{L}}]$ | $F_{\mathrm{m}}[\widetilde{\mathbf{L}}]$ |
| **S** | $\Phi[\overline{\mathbf{S}}], W[\overline{\mathbf{S}}]$ | $M^{\mathrm{f}}[\widetilde{\mathbf{S}}], E^{\mathrm{f}}[\widetilde{\mathbf{S}}], Q^{\mathrm{f}}[\widetilde{\mathbf{S}}], S^{\mathrm{f}}[\widetilde{\mathbf{S}}], \mathbf{P}^{\mathrm{f}}[\widetilde{\mathbf{S}}], Q[\widetilde{\mathbf{S}}]$ |
| **V** | $\check{G}[\overline{\mathbf{V}}]$ | $M^{\mathrm{c}}[\widetilde{\mathbf{V}}], E^{\mathrm{c}}[\widetilde{\mathbf{V}}], Q^{\mathrm{c}}[\widetilde{\mathbf{V}}], S^{\mathrm{c}}[\widetilde{\mathbf{V}}], \mathbf{P}^{\mathrm{c}}[\widetilde{\mathbf{V}}], \check{\mathbf{L}}^{\mathrm{c}}[\widetilde{\mathbf{V}}]$ |

the electric potential $\phi$, the gravitational potential $U_{\mathrm{g}}$, the velocity potential $\phi$ and the total displacement $\mathbf{u}$ are associated with points $\overline{\mathbf{P}}$ endowed with an inner orientation, while others, such as, for example, the magnetic scalar potential $\check{\phi}_m$ and the stream function $\check{\psi}$, are associated with points $\widetilde{\mathbf{P}}$ endowed with an outer orientation. Hence we enrich the element enclosed in square brackets by also indicating its type of orientation. Table 5.11 shows the variables we have mentioned.

We pose the following question: is the marriage between physics and geometry so strict that the notion of inner and outer orientation is really necessary? Our answer is a definite yes.

To prove this, let us consider a polygon which overlaps on itself and which has a zero area,[28] as shown in Fig. 5.6b. If we consider a circuit with the same shape embedded in a *uniform* magnetic field, as in Fig. 5.6b, the magnetic flux through the circuit is zero. This can be tested by switching off the magnetic field

[28] Klein [114, p. 9].

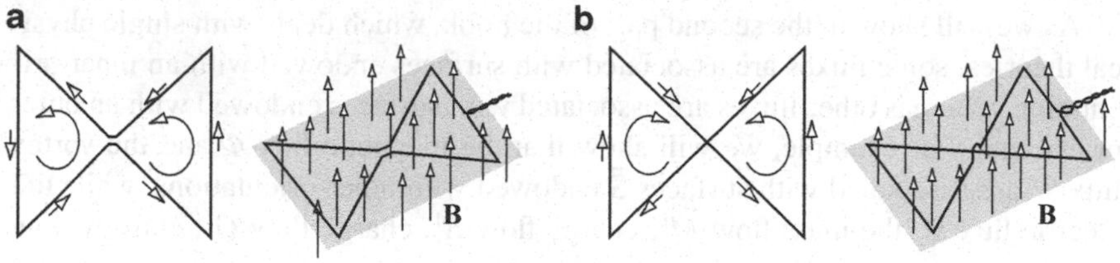


**Fig. 5.6** (**a**) A polygon with non vanishing area; (**b**) a polygon with zero area

and measuring the voltage impulse at the terminals. Does this association of global physical variables with oriented space elements suggest the introduction of a new classification of global physical variables? The answer is affirmative, as we will show in the book. We can state the following principle:

ASSOCIATION PRINCIPLE. *In physical theories, in a* spatial *description, global physical variables have a natural association with one of the four space elements endowed with an inner or outer orientation, i.e.* $\overline{\mathbf{P}}$, $\overline{\mathbf{L}}$, $\overline{\mathbf{S}}$, $\overline{\mathbf{V}}$, $\widetilde{\mathbf{P}}$, $\widetilde{\mathbf{L}}$, $\widetilde{\mathbf{S}}$, $\widetilde{\mathbf{V}}$, *whereas both, in the material and in the spatial description, the global physical variables are associated with one of the two time elements endowed with an inner or outer orientation, i.e.* $\overline{\mathbf{I}}$, $\overline{\mathbf{T}}$, $\widetilde{\mathbf{I}}$, $\widetilde{\mathbf{T}}$.

How do we decide if a given physical variable is associated with a space element endowed with an inner or outer orientation? The analysis of many physical theories has witnessed the progressive emergence of the following empirical rule:

EMPIRICAL RULE. *Configuration variables are associated with space elements endowed with an inner orientation, whereas source variables and energy variables are associated with space elements endowed with an outer orientation.*

The author is unable to give a justification of this empirical rule which reveals an unexpected marriage between physics and geometry. Table 5.12 provides an intuitive insight into this association for four physical theories.

In physics, the distinction between the two kinds of orientation of a space element is not considered seriously. Nobody speaks of points endowed with inner or outer orientations, of lines with outer orientations, of surfaces and volumes endowed with inner orientations. In contrast, the distinction between inner and outer orientation plays a pivotal role in the present analysis because it corresponds to

**Table 5.12** Analogies between four physical fields

| | | |
|---|---|---|
| temperature $T$ refers to vertices of the primal cell complex<br>temperature difference $G$ refers to edges of the primal cell complex | **thermal field**<br>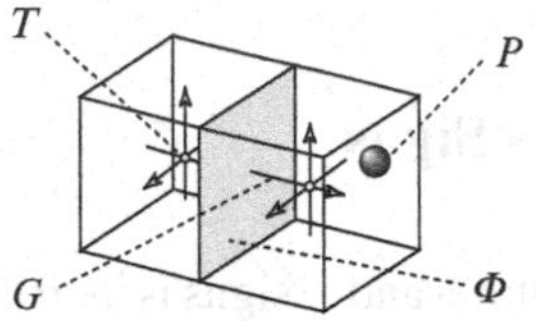 | heat generation rate $P$ refers to cells of the dual cell complex<br>heatcurrent $\Phi$ refers to faces of the dual cell complex |
| electric potential $\phi$ refers to vertices of the primal cell complex<br>voltage $V$ refers to edges of the primal cell complex | **electric field**<br>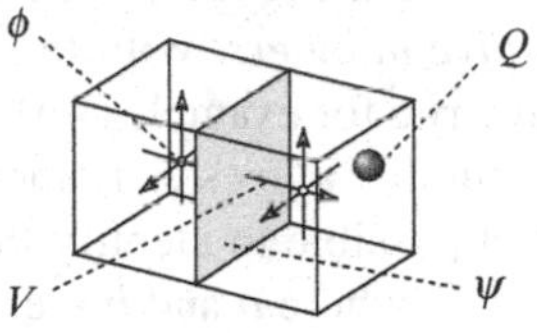 | electric charge $Q$ refers to cells of the dual cell complex<br>electric flux $\psi$ refers to faces of the dual cell complex |
| total displacement **u** refers to vertices of the primal cell complex<br>relative displacement **h** refers to edges of the primal cell complex | **elastic field**<br>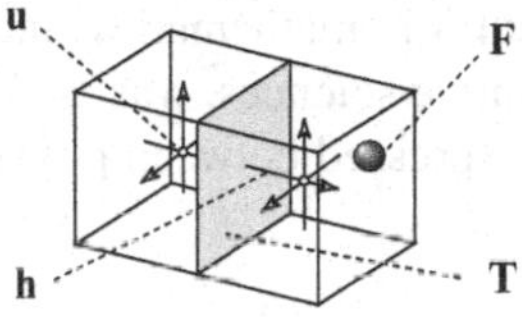 | volume force $\mathbf{F}^{\mathrm{v}}$ refers to cells of the dual cell complex<br>surface force **T** refers to faces of the dual cell complex |
| magnetic flux $\Phi$ refers to faces of the primal cell complex<br>hypotetical magnetic charge $G$ refers to cells of the primal cell complex | **magnetic field**<br>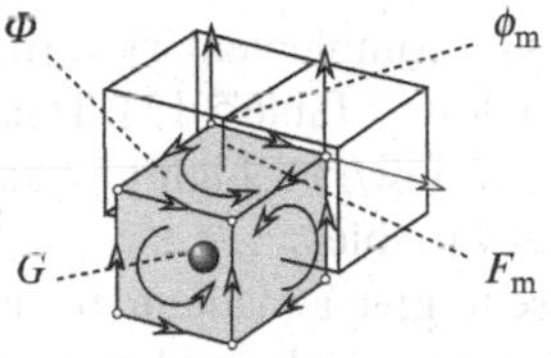 | scalar magnetic potential $\phi_{\mathrm{m}}$ refers to vertices of the dual cell complex<br>magnetom. force $F$ refers to edges of the dual cell complex |

the distinction between configuration and source variables which are mapped respectively onto the left and right parts of the classification diagram of physical variables.[29] In turn, this distinction allows us to highlight the collocation of the constitutive relations in the same classification diagram as a link between vari-

[29] See Chap. 8.

ables associated with space elements endowed with an inner orientation and space elements endowed with an outer orientation.

## 5.9 Genesis of + and – Signs

The first time we encounter the + and – signs is in *arithmetic*, in the two operations on numbers: addition and subtraction. In primary school we are taught to figure out the difference between two numbers in which the first number is greater than the number following it, e.g. $7-3$. Then, in high school, we learn an important notion whereby we consider the signs attached to the numbers and write $(+7)+(-3)$. In this way we introduce the *relative numbers*. With this introduction, we can alter the order of the summands and write for example $(-3)+(+7)$. Hence, considering the minus sign attached to the number allows us immediately to change the order of the summands. This makes it possible to identify the numbers with the letters and write the algebraic sum $a+b$, where $a$ and $b$ are two letters which can have positive or negative values.

The result is the birth of *literal calculus*, which allows us to merge the two operations of addition and subtraction in a single operation called *algebraic sum*. The introduction of literal calculus in mathematics, and therefore its use in geometry, physics and other quantitative sciences, was a wonderful invention because it allowed relationships to be expressed between physical variables regardless of the specific numerical values.[30]

## 5.10 Sign of Physical Variables

In everyday life, we often deal with complementary attributes such as, for example, beautiful/ugly, good/bad and up/down (Table 5.13). These include some attributes which, being quantitative, such as slow/fast, high/low, small/large and cold/warm, can evolve and become physical variables.

Since physical variables arise to give a mathematical description to the quantitative attributes of a physical system, the plus and minus signs in front of them can be used to distinguish between two opposite states of the same attribute. Typical is the use of the + and – signs to denote the two kinds of electric charge, *vitreous* and *resinous*, or the temperature in Celsius, above and below the freezing point of water. Since global physical variables are associated with space and time elements endowed with an inner or outer orientation, once we select a preferred orientation of the space or time element, it is possible and appropriate to invert the sign of the variable when we invert the orientation of the associated element (Table 5.14).

---

[30] Bashmakova and Smimova [9].

**Table 5.13** Examples of opposite attributes: the + sign is conventional

| *Minus* (−) | *Plus* (+) | *Minus* (−) | *Plus* (+) | *Minus* (−) | *Plus* (+) |
|---|---|---|---|---|---|
| Left | Right | Sink | Source | Male | Female |
| Down | Up | Empty | Filled | Off | On |
| Longitude west | Longitude east | Resinous | Vitreous | Low | High |
| Outside | Inside | No | Yes | Repulsion | Attraction |
| Destruction | Production | Bad | Good | Small | Large |
| Latitude south | Latitude north | Compression | Expansion | Cold | Warm |
| Left handed | Right handed | Weak | Strong | Black | White |

**Table 5.14** Paradigm to find the association of a physical variable with its space and time element. Filled circles denote space global variables, empty circles time global variables

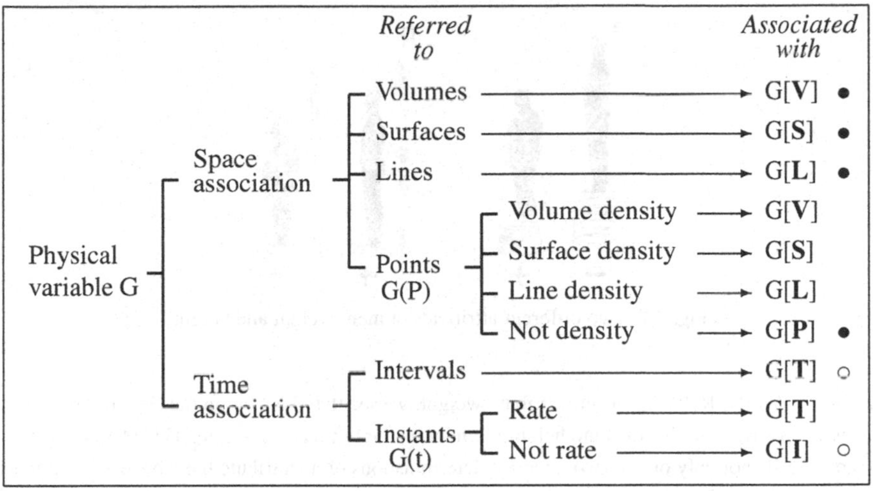


This is what we commonly do with many, *but not all*, physical quantities. So we can decide to consider positive the work given *to* a system, the force acting on a bar when it acts as a *traction*, or a rotation when it is *anticlockwise*. This implies that the work done by a system or a compressive force on a bar or a clockwise rotation must to be considered negative.

For many physical variables we are not accustomed to using the sign: this is the case of an area, a volume or a mass. The idea of negative areas, negative volumes and negative masses seems absurd!

Thus, it is common to state that a negative mass does not exist. Why? Because in speaking of negative mass we think that the sign is necessarily linked to the attribute of attraction and repulsion. Since masses never repel each other, contrary to electric charges, we conclude that a negative mass does not exist. This conclusion presupposes that the term 'negative' refers to the properties of attraction and repulsion. But we will show shortly that there exist other attributes of the mass and the positive or negative state when related to one of these attributes can be fully meaningful.

The fact is that when a system has more than one attribute, we can assign the + sign or the adjective *positive*, with reference to *one* of these attributes: if we make reference to another attribute of the system, then we can associate the − sign or the adjective *negative* with the system. Hence, if we do not specify the referenced attribute, then the sign remains ambiguous.

**Example 3**. To show that the same object can be assigned a plus or minus sign according to different attributes, let us consider two men, as shown in Fig. 5.7. We may refer to the attributes

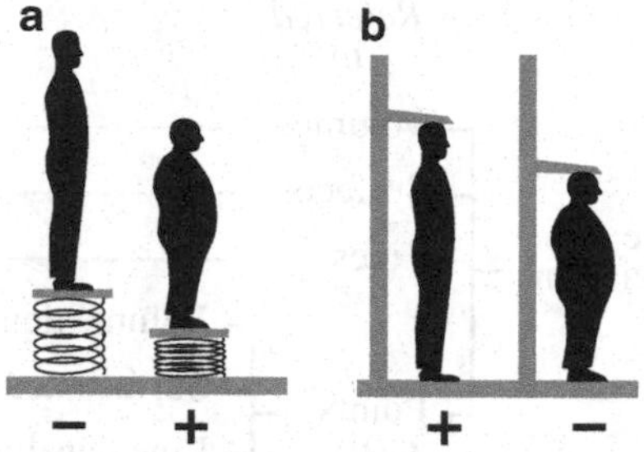


**Fig. 5.7** Two different attributes of men: weight and height

*weight* or *height*. Referring to the attribute weight, we see that the man on the right has the sign +, while referring to the attribute height the man on the left has the + sign. This shows that the sign depends not only on the two opposite determinations of an attribute but, above all, requires that the attribute to which it refers must be expressly specified.

**Example 4**. Let us consider a space region, say a building, oriented with outward normals: this means that the interior of the building will be considered as the *active* region. A given number of persons inside the building is then considered positive. If we change the outward normals into inward normals, the interior of the building becomes the *passive* region. It follows that the same number of persons becomes negative. To take a concrete case, let us consider the embassy of a country $B$ in nation $A$, as shown in Fig. 5.8. An embassy can be considered a hole in the host nation. Let us suppose that two citizens of nation $A$, seeking political asylum, take refuge inside the embassy of nation $B$. From the standpoint of nation $B$ they are considered to be +2, while from the standpoint of nation $A$ they are considered to be −2. The sign changes depending on the viewpoint or, put another way, the sign changes depending on which reference system is chosen, a system or its complement. In this case, the attribute considered is the people's nationality.

**Example 5**. If some woman's ring falls *into* a manhole, she considers the ring lost, and as such considers it negative capital. This is because we regard as lost that which is *inside* a region to which we do not have access, in this case the manhole.

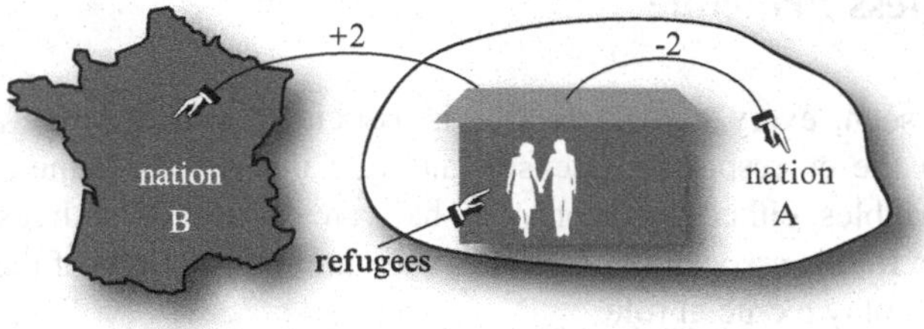


**Fig. 5.8** The sign depends on the viewpoint: two refugees inside the embassy of nation *B* inside nation *A*

**Example 6.** Let us consider a road connecting two cities, P and Q. If someone told you "*I saw a car travelling at a speed of* −20 *km/h*," our first thought is probably that the car was moving ... backwards! But the justification of the sign can be different: the person may have implicitly taken a positive direction, say from P to Q, as the reference, and since the car moves from Q to P at a speed of 20 km/h, he correctly states that its velocity was −20 km/h. The minus sign is the result of the choice of a positive direction and of a comparison of the direction of motion of the car with the direction chosen as positive.

This is a general property of signs: when we consider an attribute with two complementary states, we may select one of the two states as the *active* one and the other as the *passive* one. Other adjectives which can be used depending on circumstances are, for example, favourable/unfavourable, and opportune/inopportune. It is random to consider the *active*, *favourable* and *opportune* as positive attributes.

Every time we detect a state coinciding with the active one, we can assign the plus sign to it; otherwise the minus sign is assigned to it. Stated in this fashion, it is obvious that by inverting the active state with the passive one, the sign of the physical quantity is inverted. This property will be called the *oddness principle*.

In everyday life, we almost always use the absolute value of a magnitude such as, for example, the length of a road, the height of a building, the area of an apartment or the extension of agricultural land. In some cases, however, the absolute value is not enough. Thus, if we tell a sailor to change the direction of his ship by 90°, we must add the information *to the right* or *to the left*. If we order the operator of a crane to move a load vertically, it is not enough to indicate how many metres; we must say whether the movement is *up* or *down*. In all these cases, the content of information required is greater than a simple number. To specify one of the two aspects of a measure, we can use the + or − signs.

## 5.11 Oddness Principle

As we have seen, every global variable is associated with a space and a time element, and the corresponding density and rate inherits the same association. Hence all variables, either global or local, have an association with a space and a time element. In this association, the inner and outer orientations of the space and time elements play a crucial role.

While it is fairly easy to determine which is the element of space or time to which a physical variable is associated, it is less easy to determine the corresponding type of orientation of the space or time element. To this end, based on experimental evidence for many global variables, we introduce the following general principle.

ODDNESS PRINCIPLE. *Every* global *physical variable referring to an oriented space or a time element reverses its sign when the orientation of the space or the time element is reversed.*

1. The line integral of a force along a line endowed with an inner orientation changes sign when we invert the orientation of the line: hence $W^*[-\overline{\mathbf{L}}] = -W^*[\overline{\mathbf{L}}]$.[31]
2. The sign of the magnetic flux is reversed when we reverse the inner orientation of the surface to which it is referred because the sign of the magnetic flux is determined by the direction of the current which flows in a search coil; hence, $\Phi[-\overline{\mathbf{S}}] = -\Phi[\overline{\mathbf{S}}]$.
3. In continuum mechanics, one introduces the internal surface force $\mathbf{T}$ acting on a piece of plane surface defined as the resultant force that the matter contained on the positive side of the surface exerts on the matter of the negative side through the surface element. Reversing the normal to the surface element, the two sides change roles, and then the force changes its sign. It is customary to introduce the stress vector $\mathbf{t}$ as the ratio of the surface force $\mathbf{T}$ to the area of a plane element. This is one of the few cases where the oddness principle is expressly stated in the literature and is written in the form $\mathbf{T}(-\mathbf{n}) = -\mathbf{T}(\mathbf{n})$ and we write it in the form $\mathbf{T}[-\widetilde{\mathbf{S}}] = -\mathbf{T}[\widetilde{\mathbf{S}}]$.[32]
4. The flow through a surface endowed with an outer orientation changes sign if we reverse the outer orientation of the surface; hence, $Q^{\mathrm{f}}[-\widetilde{\mathbf{S}}] = -Q^{\mathrm{f}}[\widetilde{\mathbf{S}}]$.
5. The electric flux collected on one face of a surface is opposite to the electric flux on the opposite face; hence, $\Psi[-\widetilde{\mathbf{S}}] = -\Psi[\widetilde{\mathbf{S}}]$.

[31] We use the star on the letter $W$ because the line integral in a force field is the virtual work.

[32] This relation is an immediate consequence of the principle of action and reaction: see Milne-Thomson [160, p. 630], Chadwick [39, p. 85]. In contrast, other authors deduce this property from the equation of motion: see Billington and Tate [15, p. 67], Jaunzemis [106, p. 209].

6. The displacement of a point changes into its opposite if we reverse the inner orientation of the time interval to which it is referred; hence, $\mathbf{u}[-\overline{\mathrm{T}}] = -\mathbf{u}[\overline{\mathrm{T}}]$.
7. The radius vector of a point **P**, usually defined as the vector connecting a fixed origin **O** with the point **P**, changes sign when points are oriented as sources instead of sinks: $\mathbf{r}[-\overline{\mathbf{P}}] = -\mathbf{r}[\overline{\mathbf{P}}]$. Thus, in planetary motion, because the force of the Sun is attractive, it is conceivable to consider the radius vector of a planet as the vector from the planet to the Sun, rather than the opposite.

The validity of this principle is needed in order for the form of the equations expressing physical laws to be independent of the *orientation* of the space and time elements. It is analogous to the principle according to which the equations expressing physical laws must be independent of the *coordinate system* chosen, and to the principle of relativity according to which the form of the equations expressing physical laws must be independent of the *reference system* chosen.

For the majority of global physical variables the principle is evident, but in a few cases it is not, as in the case of the mass, particle number and energy. Table 5.11 (p. 124) presents the main global variables of physics, showing the oddness principle with respect to the inner or outer orientation of the element of space and time to which they are associated. All line integrals change sign with the inversion of the orientation of the line; all surface integrals change sign with the inversion of the orientation of the surface. Since flow refers to the motion of something through a surface (e.g. matter, charge, energy), it changes sign under a reversal of motion. A potential, defined as the line integral of a vector, changes sign when the inner or outer orientation of the point is inverted. In fact, the line integral of a vector from a fixed point to an arbitrary point implies that the point is considered a sink, whereas that from an arbitrary point towards a fixed point implies that the arbitrary point is considered a source.

As was stated about mass, it is easily seen that every content changes sign under an inversion of the outer orientation of volumes.[33]

When dealing with the mathematical description of physical theories it is better to start with global variables rather than field variables for at least two main reasons:

1. They clearly show the space and time elements to which they refer.
2. They are often directly measurable.

The fact of making a mathematical description of physics starting with global variables instead of field variables is a radical departure from the traditional description of physics, which uses the differential formulation from the very beginning.

[33] For mass and particle number, see p. 118; for the impulse of a force see p. 243.

## Oddness principle [ODD]

The product of the two space and time global variables, in the same strip, is an action

| | | | | | | |
|---|---|---|---|---|---|---|
| 1 | Electromotive force impulse | $\mathcal{E}[\tilde{\mathbf{T}},\bar{\mathbf{L}}]$ | $\mathbf{E}$ | $\mathcal{E}[\tilde{\mathbf{T}},-\bar{\mathbf{L}}]=-\mathcal{E}[\tilde{\mathbf{T}},\bar{\mathbf{L}}]$ | $\mathcal{E}[-\tilde{\mathbf{T}},\bar{\mathbf{L}}]=-\mathcal{E}[\tilde{\mathbf{T}},\bar{\mathbf{L}}]$ | [ELE3] |
| 2 | Electric flux | $\Psi[\bar{\mathbf{I}},\tilde{\mathbf{S}}]$ | $\mathbf{D}$ | $\Psi[\bar{\mathbf{I}},-\tilde{\mathbf{S}}]=-\Psi[\bar{\mathbf{I}},\tilde{\mathbf{S}}]$ | $\Psi[-\bar{\mathbf{I}},\tilde{\mathbf{S}}]=-\Psi[\bar{\mathbf{I}},\tilde{\mathbf{S}}]$ | [ELE3] |
| 3 | Electric potential impulse | $\varphi[\tilde{\mathbf{T}},\bar{\mathbf{P}}]$ | $\phi$ | $\varphi[\tilde{\mathbf{T}},-\bar{\mathbf{P}}]=-\varphi[\tilde{\mathbf{T}},\bar{\mathbf{P}}]$ | $\varphi[-\tilde{\mathbf{T}},\bar{\mathbf{P}}]=-\varphi[\tilde{\mathbf{T}},\bar{\mathbf{P}}]$ | [ELE3] |
| 4 | Electric charge content | $Q^c[\bar{\mathbf{I}},\tilde{\mathbf{V}}]$ | $\rho$ | $Q^c[\bar{\mathbf{I}},-\tilde{\mathbf{V}}]=-Q^c[\bar{\mathbf{I}},\tilde{\mathbf{V}}]$ | $Q^c[-\bar{\mathbf{I}},\tilde{\mathbf{V}}]=-Q^c[\bar{\mathbf{I}},\tilde{\mathbf{V}}]$ | [ELE3] |
| 5 | Line integral of **A** | $a[\tilde{\mathbf{I}},\bar{\mathbf{L}}]$ | $\mathbf{A}$ | $a[\tilde{\mathbf{I}},-\bar{\mathbf{L}}]=-a[\tilde{\mathbf{I}},\bar{\mathbf{L}}]$ | $a[-\tilde{\mathbf{I}},\bar{\mathbf{L}}]=-a[\tilde{\mathbf{I}},\bar{\mathbf{L}}]$ | [ELE3] |
| 6 | Charge flow | $Q^f[\bar{\mathbf{T}},\tilde{\mathbf{S}}]$ | $\mathbf{J}$ | $Q^f[\bar{\mathbf{T}},-\tilde{\mathbf{S}}]=-Q^f[\bar{\mathbf{T}},\tilde{\mathbf{S}}]$ | $Q^f[-\bar{\mathbf{T}},\tilde{\mathbf{S}}]=-Q^f[\bar{\mathbf{T}},\tilde{\mathbf{S}}]$ | [ELE3] |
| 7 | Magnetic flux | $\Phi[\tilde{\mathbf{I}},\bar{\mathbf{S}}]$ | $\check{\mathbf{B}}$ | $\Phi[\ ,-\bar{\mathbf{S}}]=-\Phi[\tilde{\mathbf{I}},\bar{\mathbf{S}}]$ | $\Phi[-\tilde{\mathbf{I}},\bar{\mathbf{S}}]=-\Phi[\tilde{\mathbf{I}},\bar{\mathbf{S}}]$ | [ELE3] |
| 8 | Magnetomotive force impulse | $\mathcal{F}_m[\bar{\mathbf{T}},\tilde{\mathbf{L}}]$ | $\check{\mathbf{H}}$ | $\mathcal{F}_m[\bar{\mathbf{T}},-\tilde{\mathbf{L}}]=-\mathcal{F}_m[\bar{\mathbf{T}},\tilde{\mathbf{L}}]$ | $\mathcal{F}_m[-\bar{\mathbf{T}},\tilde{\mathbf{L}}]=-\mathcal{F}_m[\bar{\mathbf{T}},\tilde{\mathbf{L}}]$ | [ELE3] |
| 9 | Total displacement | $\mathbf{u}[\bar{\mathbf{I}},\bar{\mathbf{P}}]$ | $\mathbf{u}$ | $\mathbf{u}[\bar{\mathbf{I}},-\bar{\mathbf{P}}]=-\mathbf{u}[\bar{\mathbf{I}},\bar{\mathbf{P}}]$ | $\mathbf{u}[-\bar{\mathbf{I}},\bar{\mathbf{P}}]=-\mathbf{u}[\bar{\mathbf{I}},\bar{\mathbf{P}}]$ | [SOL9] |
| 10 | Impulse of volume forces | $\mathbf{J}^v[\tilde{\mathbf{T}},\tilde{\mathbf{V}}]$ | $\mathbf{f}^v$ | $\mathbf{J}^v[\tilde{\mathbf{T}},-\tilde{\mathbf{V}}]=-\mathbf{J}^v[\tilde{\mathbf{T}},\tilde{\mathbf{V}}]$ | $\mathbf{J}^v[-\tilde{\mathbf{T}},\tilde{\mathbf{V}}]=-\mathbf{J}^v[\tilde{\mathbf{T}},\tilde{\mathbf{V}}]$ | [SOL9] |
| 11 | Displacement | $\boldsymbol{\eta}[\bar{\mathbf{T}},\bar{\mathbf{P}}]$ | $\boldsymbol{\eta}$ | $\boldsymbol{\eta}[\bar{\mathbf{T}},-\bar{\mathbf{P}}]=-\boldsymbol{\eta}[\bar{\mathbf{T}},\bar{\mathbf{P}}]$ | $\boldsymbol{\eta}[-\bar{\mathbf{T}},\bar{\mathbf{P}}]=-\boldsymbol{\eta}[\bar{\mathbf{T}},\bar{\mathbf{P}}]$ | [SOL12] |
| 12 | Momentum content | $\mathbf{P}^c[\tilde{\mathbf{I}},\tilde{\mathbf{V}}]$ | $\mathbf{p}$ | $\mathbf{P}^c[\tilde{\mathbf{I}},-\tilde{\mathbf{V}}]=-\mathbf{P}^c[\tilde{\mathbf{I}},\tilde{\mathbf{V}}]$ | $\mathbf{P}^c[-\tilde{\mathbf{I}},\tilde{\mathbf{V}}]=-\mathbf{P}^c[\tilde{\mathbf{I}},\tilde{\mathbf{V}}]$ | [SOL12] |
| 13 | Velocity circulation | $\Gamma[\tilde{\mathbf{I}},\bar{\mathbf{L}}]$ | $\mathbf{v}$ | $\Gamma[\tilde{\mathbf{I}},-\bar{\mathbf{L}}]=-\Gamma[\tilde{\mathbf{I}},\bar{\mathbf{L}}]$ | $\Gamma[-\tilde{\mathbf{I}},\bar{\mathbf{L}}]=-\Gamma[\tilde{\mathbf{I}},\bar{\mathbf{L}}]$ | [FLU4] |
| 14 | Mass flow | $M^f[\bar{\mathbf{T}},\tilde{\mathbf{S}}]$ | $\mathbf{q}$ | $M^f[\bar{\mathbf{T}},-\tilde{\mathbf{S}}]=-M^f[\bar{\mathbf{T}},\tilde{\mathbf{S}}]$ | $M^f[-\bar{\mathbf{T}},\tilde{\mathbf{S}}]=-M^f[\bar{\mathbf{T}},\tilde{\mathbf{S}}]$ | [FLU4] |
| 15 | Vortex flux | $W[\tilde{\mathbf{I}},\bar{\mathbf{S}}]$ | $\check{\mathbf{w}}$ | $W[\tilde{\mathbf{I}},-\bar{\mathbf{S}}]=-W[\tilde{\mathbf{I}},\bar{\mathbf{S}}]$ | $W[-\tilde{\mathbf{I}},\bar{\mathbf{S}}]=-W[\tilde{\mathbf{I}},\bar{\mathbf{S}}]$ | [FLU4] |
| 16 | Impulse of line integral of stream vector | $\beta[\bar{\mathbf{T}},\tilde{\mathbf{L}}]$ | $\check{\boldsymbol{\psi}}$ | $\beta[\bar{\mathbf{T}},-\tilde{\mathbf{L}}]=-\beta[\bar{\mathbf{T}},\tilde{\mathbf{L}}]$ | $\beta[-\bar{\mathbf{T}},\tilde{\mathbf{L}}]=-\beta[\bar{\mathbf{T}},\tilde{\mathbf{L}}]$ | [FLU4] |
| 17 | Impulse of temperature | $\mathcal{T}[\tilde{\mathbf{T}},\bar{\mathbf{P}}]$ | $T$ | $\mathcal{T}[\tilde{\mathbf{T}},-\bar{\mathbf{P}}]=-\mathcal{T}[\tilde{\mathbf{T}},\bar{\mathbf{P}}]$ | $\mathcal{T}[-\tilde{\mathbf{T}},\bar{\mathbf{P}}]=-\mathcal{T}[\tilde{\mathbf{T}},\bar{\mathbf{P}}]$ | [TCO3] |
| 18 | Entropy content | $S^c[\bar{\mathbf{I}},\tilde{\mathbf{V}}]$ | $S$ | $S^c[\bar{\mathbf{I}},-\tilde{\mathbf{V}}]=-S^c[\bar{\mathbf{I}},\tilde{\mathbf{V}}]$ | $S^c[-\bar{\mathbf{I}},\tilde{\mathbf{V}}]=-S^c[\bar{\mathbf{I}},\tilde{\mathbf{V}}]$ | [TCO3] |

### *5.11.1 Global Variables in Time*

By a careful examination of the time dependence of physical variables we are led to introduce the notion of *association* of a physical variable with oriented time elements, i.e. with one of the four elements $\bar{\mathbf{I}}, \bar{\mathbf{T}}, \tilde{\mathbf{I}}, \tilde{\mathbf{T}}$. To do this, we start by observing that physical variables are referred to an instant or an interval.

Many physical variables *refer to* instants, i.e. they are *functions of* instants: this is the case with, for example, temperature, velocity, pressure, mass density, momentum, electric potential, electric field vector, electric current, force, stress, strain and stream function. This is what we usually mean when we write $T(t), \mathbf{v}(t), p(t), \rho(t), \mathbf{p}(t)$, and so forth.

Other variables refer to intervals. This is the case, for example, with the displacement of a particle, work given to a system, heat exchanged between two systems, an electric charge stored in a battery, the impulse given to a body, entropy production, and flows of energy, mass, or entropy.

This suggests that global physical variables may be divided into two classes: those associated with instants and those associated with intervals. This is indeed the case, but we must make an important remark: *among physical variables that refer to instants, there are some which, nevertheless, require a time interval for their definition and for their measure.* This happens with those variables which are *rates* of variables referred to intervals.

**Example 7.** As a first example, let us consider the position vector $\mathbf{r}$ of a particle and the velocity $\mathbf{v}$ of the particle; both are functions of time instants. Despite this, to define (and measure) the velocity, we need a small time interval, whereas to measure the position vector, we do not need a time interval. In other words, velocity is attributed to an instant after calculating a rate, i.e. displacement/(time interval) and a limit, wheres the position vector is attributed to an instant directly. This shows that, although both variables refer to instants, the velocity needs a time interval for its definition. We will say that velocity is *associated* with an interval of time, like the displacement from which it derives.

**Example 8.** As a second example, let us consider the mass content $M^c$ of a given volume at a given instant and the power $P$ emitted by an energy generator at a given instant. Both make reference to instants. Despite this, to define (and measure) the power, we need a small time interval to measure the work given by the generator, whereas the measure of the mass content does not involve a time interval. Power is attributed to an instant after calculating a rate (work/time interval) and a limit, whereas mass content is attributed to an instant directly. This shows that, although both variables refer to instants, power needs a time interval for its definition. We will say that power is *associated* with an interval of time, like the work from which it derives.

This distinction suggests the need to introduce the following definition.

DEFINITION. *We call* a global variable in time *any variable which is not the rate of another variable.*

## 5.12 Time Association

A global variable in time which, by its own definition, involves an interval or an instant is said to be *associated* with an interval or with an instant respectively. Moreover, the rate of a global variable associated with an interval is said to *inherit* the association with an interval. Therefore, since velocity is the rate of displacement and the latter is associated with an interval, velocity *inherits* the association with an interval. In an analogous way, since power is the rate of work and the latter is associated with an interval, we say that power *inherits* the association with intervals. The same can be said of an electric current: since it is the rate of the electric charge flow, it *inherits* the association with intervals.

The distinction between the notion of *association* with a time element on the one hand and the notions of *reference* to a time element or of a *function* of a time element on the other plays a pivotal role in the present classification of physical variables.

In contrast, there are variables referring to instants which are not rates of other variables: this is the case with the position vector **r**, the total displacement **u** of a point in a continuum from a reference configuration, momentum **P**, magnetic flux $\Phi$, mass content $M^c$, entropy content $S^c$, electric charge content $Q^c$, velocity line integral $\Gamma$, velocity potential $\phi$, vorticity **w**, vortex flux $W$ and a few others. These variables not only refer to instants but are also *associated* with instants.[34]

Hence, we make a sharp distinction in the use of the terms *referred to* and *function of*, on the one hand, and *associated with* on the other:

- Variables *referred* to an interval are automatically *associated* with that interval.
- Among variables *referred* to an instant we must distinguish between the following:
  - Those which are rates of other variables *associated* with intervals are also *associated* with intervals.
  - Those which are not rates of other variables are *associated* with instants.

In a sense, the association takes into account the "genealogy" of the variable and its measurement process.

---

[34] The fact that velocity line integral is associated with time instants while velocity is associated with time intervals is a consequence of a peculiar ambiguity of velocity in fluid dynamics, as we will explain in Sect. 12.2, p. 356.

**Energy.** Objections may be raised concerning the association of energy and temperature to time intervals. To justify this, let us remark that kinetic and potential energies can be composed to give the Lagrangian, and this function is integrated on a time interval to give the action. Hence, the Lagrangian is an action rate. This implies that all energies are rates of the action; hence, they inherit the association with the interval, as does the action.

**Temperature.** With regard to temperature, let us remark that its integration in time is justified by the following consideration: in a gas, the mean kinetic energy $\overline{T}$ per molecule is proportional to the absolute temperature $T$ via a universal constant, Boltzmann's constant $k_B$:

$$\overline{T} = \frac{3}{2} k_B T . \tag{5.3}$$

On account of this relation, if the time integration of energy is meaningful, then the time integration of temperature also becomes meaningful. As a matter of fact, some authors have considered the indefinite time integral of the thermodynamic temperature, calling it *thermacy*.[35]

**Volume Dilatation.** Volume dilatation is the increase in volume per unit volume and is measured from a reference configuration. Hence, it refers to an instant, like the displacement of points from a reference position.[36]

**Velocity.** In continuum mechanics, velocity in a spatial (= Eulerian) description has a double connotation: it is used to compute a line integral along a line; hence, it is the line density of the line integral and is associated with dual time instants. It is also used to perform the velocity gradient; hence, it is associated with points and time intervals.[37]

As was done for space elements, we will highlight the association of a physical variable with a time element by writing the physical variable followed by the time element inside square brackets, as shown in the following examples.

1. Variables *associated* with time instants: position vector $\mathbf{r}\,[\mathbf{I}]$; strain $\varepsilon\,[\mathbf{I}]$; magnetic flux $\Phi[\mathbf{I}]$; momentum $\mathbf{p}[\mathbf{I}]$.
2. Variables *associated* with time intervals are of two kinds:
   (a) Those which are directly associated with time intervals because they are global variables over time: displacement $\mathbf{u}\,[\mathbf{T}]$; work $W[\mathbf{T}]$; heat $Q[\mathbf{T}]$; electric charge stored $Q^{s}[\mathbf{T}]$; impulse of a force $\mathbf{J}\,[\mathbf{T}]$.
   (b) Those which inherit an association with time intervals because they are rates of global variables over time: force $\mathbf{F}\,[\mathbf{T}]$; velocity $\mathbf{v}\,[\mathbf{T}]$; power $P[\mathbf{T}]$; electric current $I[\mathbf{T}]$.

---

35 See p. 389.

36 See the analysis of the displacement, p. 331.

37 See p. 356.

A useful method to find the association of physical variables with time elements is offered by the following criterion:

CRITERION 1: EQUILIBRIUM. A physical variable whose measurement presupposes that an equilibrium has been reached is associated with an interval.

This is the case with

| | |
|---|---|
| *Temperature* | Thermal equilibrium |
| *Force* | Mechanical equilibrium |
| *Electric potential* | Electric equilibrium |
| *Velocity* | Dynamic equilibrium |
| *Electric current* | Steady state |

Therefore, to measure a *force*, say with a dynamometer, we must wait for the apparatus to reach mechanical equilibrium.[38] To measure the *velocity*, say of a car with a speedometer, we need to reach the dynamic equilibrium; in fact, during a rapid acceleration of the car, the pointer on the speedometer moves up to the moment when the motion is uniform. To measure an *electric current*, say with a galvanometer, we must wait for the magnetic needle to reach equilibrium.

The equilibrium criterion applied to temperature $T$ and to electric potential $\phi$ shows us that they are *associated* with time intervals even though they *refer* to time instants. In fact, to measure the *temperature* of a body, we must wait for a time interval in order to reach thermal equilibrium between the thermometer bulb and the body. Similarly, to measure the electric potential of a body, we need a (very short) time interval to reach the electric equilibrium between the probe of the voltmeter and the body. Hence, the notation $T$[T] and $\phi$[T].

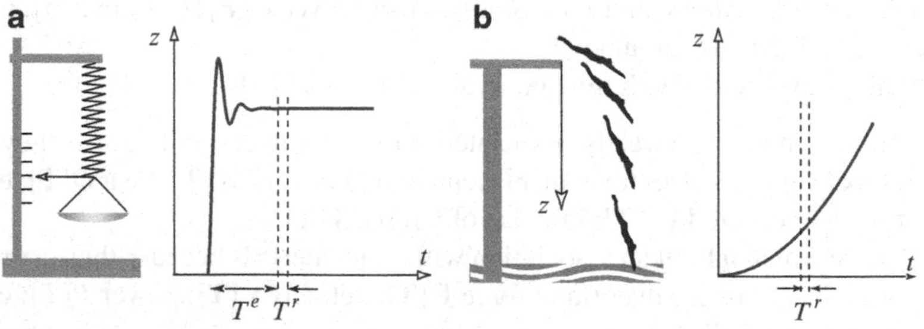


**Fig. 5.9** **(a)** The time needed to reach equilibrium $T^{\mathrm{e}}$ is distinct from the registration time $T^{\mathrm{r}}$. **(b)** The registration time $T^{\mathrm{r}}$ for the position of a particle must be very short

[38] Redlich [191, p. 588].

The time interval needed to reach equilibrium must not be confused with that required for the registration of a signal. Thus, to register the position of a particle by a photographic film, we need a short time interval to capture a number of photons to impress the film. This does not suggest that the position is associated with a time interval. In fact, the shorter the exposure time, the more precise the determination of the particle position, as shown in Fig. 5.9b. To make this point clear, let us consider the registration of the position of the pointer of a dynamometer to measure the weight of a body, as shown in Fig. 5.9a. When we put the body on the dynamometer, it starts to oscillate, and we must wait for the pointer to stop oscillating. This is the interval $T^{\mathrm{e}}$ needed to reach equilibrium. At this time we need a time $T^{\mathrm{r}}$ of registration by photography to expose the photographic film. Both time intervals $T^{\mathrm{e}}$ and $T^{\mathrm{r}}$ can be reduced using a damped dynamometer and fast photographic film; nevertheless, they remain distinct.

Another useful criterion to associate variables with time elements is the following:

CRITERION 2: To decide whether a physical variable associated with instants is global or not, we can see whether or not the product of the variable for a time interval has a physical meaning.

Thus, the products of velocity **v** and force **F** for a time interval have a physical meaning, and they give rise to displacement and impulse respectively. It follows that velocity and force are time rates; hence, **v** [**T**] and **F** [**T**]. In contrast, the product of momentum and a time interval is physically meaningless. It follows that momentum is not a rate; hence, **p** [**I**].

The products of the kinetic co-energy $T^*$ or the potential energy $V$ and a time interval have physical meaning, and they give rise to the kinetic and potential part of the Hamiltonian action

$$A_{\mathrm{H}} = \int_{t^-}^{t^+} (T^* - V)\,\mathrm{d}t\,. \qquad (5.4)$$

It follows that the Hamiltonian action is a global variable in time and the kinetic and potential energies are its rates. We can write

$$A_{\mathrm{H}}[\mathbf{T}] \quad T^*[\mathbf{T}] \quad V[\mathbf{T}]\,. \qquad (5.5)$$

All kinds of energy must have the same behaviour of kinetic and potential energies; otherwise, they would not be summable. Thus, internal energy $U$, enthalpy $H$, Helmholtz free energy $F$, Gibbs free energy $G$, the Lagrangian $L$ and the Hamiltonian $H$ are time rates, and so they inherit an association with [**T**]:

$$U[\mathbf{T}] \quad H[\mathbf{T}] \quad F[\mathbf{T}] \quad G[\mathbf{T}] \quad L[\mathbf{T}]\,. \qquad (5.6)$$

### 5.12.1 Primal or Dual Time Elements?

In the preceding chapter we stated the fundamental distinction between a *time reversal* and a *reversal of motion*.[39] To decide between primal and dual time elements, we use the oddness principle applied to time elements, as shown in Table 5.15. Thus, the displacement of a particle makes sense when it is associated

**Table 5.15** Procedure to find time association

| | *Time element* | *Changes sign?* | |
|---|---|---|---|
| A given variable is associated with: | Instant | No: $\longrightarrow \bar{\mathbf{I}}$ | Yes: $\longrightarrow \tilde{\mathbf{I}}$ |
| | Interval | No: $\longrightarrow \tilde{\mathbf{T}}$ | Yes: $\longrightarrow \bar{\mathbf{T}}$ |

with a time interval; the same is true for the work on a system. Moreover, when a reversal of motion is performed, both the displacement and the work change signs: the work given *to* a system is changed into the work given *by* the system. The oddness principle implies that the time interval is the one endowed with an inner orientation. Hence, for the displacement $\eta$ and the work $W$ we can write

$$\eta\,[-\bar{\mathbf{T}}] = -\eta\,[\bar{\mathbf{T}}] \quad \textit{and} \quad W\,[-\bar{\mathbf{T}}] = -W[\bar{\mathbf{T}}]\,. \tag{5.7}$$

All flows are associated with primal time intervals. Thus, to measure a mass flow $M^{\mathrm{f}}$ from a tap, we can collect water in a container for a time interval; to measure an electric charge flow $Q^{\mathrm{f}}$, say with an electrolytic cell, we need a time interval. Moreover, a reversal of motion changes the sign of flows. Hence, flows are associated with time intervals endowed with an inner orientation. This is the case with work (= energy flow), mass flow, charge flow, entropy flow and heat:

$$W[\bar{\mathbf{T}}] = E^{\mathrm{f}}[\bar{\mathbf{T}}], \quad M^{\mathrm{f}}[\bar{\mathbf{T}}], \quad Q^{\mathrm{f}}[\bar{\mathbf{T}}], \quad S^{\mathrm{f}}[\bar{\mathbf{T}}], \quad Q[\bar{\mathbf{T}}]\,. \tag{5.8}$$

The association with a time interval leads to the introduction of the corresponding rates, i.e. velocity $\mathbf{v}$, power $P$, mass current $\Phi$, energy current $I_{\mathrm{e}}$, electric current $I$, entropy current $I_{\mathrm{s}}$ and heat current $\Phi$ respectively. These variables inherit the association with time intervals. Hence, we can write

$$P[\bar{\mathbf{T}}] \quad I_{\mathrm{m}}[\bar{\mathbf{T}}] \quad I_{\mathrm{e}}[\bar{\mathbf{T}}] \quad I[\bar{\mathbf{T}}] \quad I_{\mathrm{s}}[\bar{\mathbf{T}}] \quad \Phi[\bar{\mathbf{T}}]\,. \tag{5.9}$$

Since mass content $M^{\mathrm{c}}$, charge content $Q^{\mathrm{c}}$ and entropy content $S^{\mathrm{c}}$ are invariant under a reversal of motion, it follows that they are associated with primal time instants, i.e. $M^{\mathrm{c}} \equiv M[\bar{\mathbf{I}}, \tilde{\mathbf{V}}]$, $Q^{\mathrm{c}} \equiv Q[\bar{\mathbf{I}}, \tilde{\mathbf{V}}]$, $S^{\mathrm{c}} \equiv S\,[\bar{\mathbf{I}}, \tilde{\mathbf{V}}]$.

---

39 See p. 35.

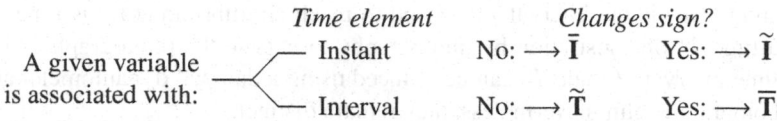

**Remark.** The present classification of global physical variables and densities or rates has nothing to do with the traditional distinction of variables into two classes, *fundamental* and *derived*. The last classification is based on the choice of a few physical variables (in the SI system of units they are seven) to express all other variables as products of the fundamental ones.

Of course, the *definite* time integral of a function is associated with a time interval and the *indefinite* time integral is associated with a time instant. What is not obvious is deciding whether the instant or the interval is primal or dual. To decide this, we use the following criterion:

> Criterion 3: If a function is invariant under a reversal of motion, its definite time integral is associated with the dual time intervals $\tilde{\mathbf{T}}$ and its indefinite time integral is associated with the dual time instants $\tilde{\mathbf{I}}$. In contrast, if the function changes sign under a reversal of motion, its definite time integral is associated with the primal time intervals $\bar{\mathbf{T}}$ and its indefinite time integral is associated with the primal time instants $\bar{\mathbf{I}}$.

The best way to test these rules is to apply them to the main physical variables for which a time integration is significant.

**Radius Vector, Displacement and Velocity.** The position vector, which gives the position of a particle, does not change sign under a reversal of motion; hence, $\mathbf{r}[\bar{\mathbf{I}}]$. Since the displacement $\mathbf{u}[\bar{\mathbf{T}}]$ changes sign under a reversal of motion, and since the velocity $\mathbf{v}$ is the time rate of the displacement, it follows that velocity changes sign under a reversal of motion $\mathbf{v} \longrightarrow -\mathbf{v}$, i.e. it inherits the association with $\bar{\mathbf{T}}$. We will write $\mathbf{v}[\bar{\mathbf{T}}]$.

**Force, Impulse and Momentum.** To analyse the impulse $\mathbf{J}$ of a force, we must know if $\mathbf{J}$ changes its sign after a reversal of motion. Let us consider the rebound of a ball against a wall, as shown in Fig. 5.10.

The ball gives an impulse $\mathbf{J}$ to the wall. If we take a movie of the rebound and then run the movie backwards in time, the velocities and the momentum are reversed but the ball still gives the same impulse to the wall. This means that, while the velocity and the momentum change sign under a reversal of the motion, the impulse does not change sign. Since the impulse is associated with a time interval, in accordance with the oddness principle, it cannot refer to an inner orientation; otherwise, it would change sign. Then, by exclusion, it should refer to a time interval endowed with an *outer* orientation: $\mathbf{J}[\tilde{\mathbf{T}}]$.

What happens when we change the outer orientation of the time interval? Coherently we must require that the impulse changes sign. How is this possible? Well, we remark that the impulse that the ball gives to the wall is the opposite of the impulse that the ball receives from the wall on account of the action and reaction principle. Hence, to satisfy the oddness principle we must change the impulse *given to* the wall with the impulse *received* from the wall. We see that *the action and reaction principle becomes the key to give a physical meaning to the outer orientation of time intervals.*

This is analogous to refugees in an embassy: their number is positive or negative depending on the reference country, that of the embassy or that of the nation that houses the embassy: this is an inversion of an *outer* orientation of volumes.

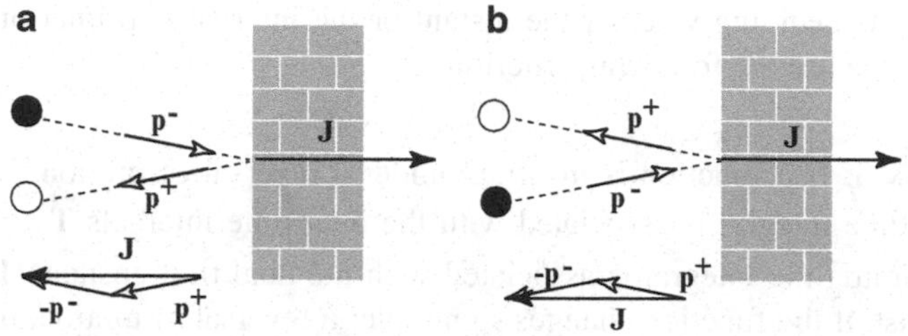


**Fig. 5.10** The rebounding of a ball against a wall shows that the impulse does not change sign under a reversal of motion. (**a**) With forward motion the impulse given to the ball points to the left. (**b**) With backward motion the impulse also points to the left

Since force is the rate of impulse, it does not change sign under a reversal of motion, i.e. $\mathbf{F}[\tilde{\mathbf{T}}]$. To enforce this association of force with time intervals with an outer orientation, we remark that the gravitational force, which is attractive, remains attractive even when viewed in a movie running backwards in time.

It is easy to prove that a force, whether its nature is gravitational, electric, magnetic, elastic, nuclear, always has the same behaviour under a reversal of motion. Indeed, if we start from the observation that the equilibrium of a body remains unchanged for a reversal of motion, it follows that two forces of different natures which are in equilibrium must have the same behaviour for a reversal of motion. Thus, Millikan's experience shows that an electric force which attracts an oil drop upwards is balanced by the gravitational force which attracts the oil drop downwards. Since the gravitational force does not change sign for a reversal of motion, the electric force does not changes its sign either. Hence, whatever their physical nature, the forces are all associated with a dual time interval.

This indicates that the analysis we are conducting highlights the intrinsic characteristics of physical variables, i.e. their association with space and time elements, which does not seem to be considered in the literature. This fact differs from dimensional analysis, which assigns dimensions to each variable by taking into account the choice of the fundamental variables; for this reason, it is not intrinsic.

The momentum **p** of a particle is the *indefinite* time integral of the force,[40]

[40] See Chap. 9 for a more detailed presentation.

$$\mathbf{p}(t) \stackrel{\text{def}}{=} \int_0^t \mathbf{F}(t')\,\mathrm{d}t' \qquad \longrightarrow \qquad \mathbf{p}[\widetilde{\mathbf{I}}], \tag{5.10}$$

and must be associated with dual time instants $\widetilde{\mathbf{I}}$ in order to satisfy the relation between impulse and momentum:[41]

$$\mathbf{J}(t^-, t^+) = \mathbf{p}(t^+) - \mathbf{p}(t^-) \qquad \longrightarrow \qquad \mathbf{J}[\widetilde{\mathbf{T}}]\,. \tag{5.11}$$

**Work and Power.** Since power satisfies the relation

$$P(t) \stackrel{\text{def}}{=} \mathbf{F}\cdot\mathbf{v} \longrightarrow \mathscr{R}\,P(t) = \mathscr{R}\,\mathbf{F}(t)\cdot\mathscr{R}\,\mathbf{v}(t) = [+\mathbf{F}(t)]\cdot[-\mathbf{v}(t)] = -P(t), \tag{5.12}$$

it changes sign under a reversal of motion: the power absorbed by a system becomes power released by the system. We can say that power is *time odd*. It follows that work is also time odd under a reversal of motion, i.e. $W[\overline{\mathbf{T}}]$, as we just stated in Eq. 5.7.

**Heat and Temperature.** Since the measurement of temperature requires thermal equilibrium, according to Criterion 1 it is associated with time intervals, i.e. $T[\mathbf{T}]$. A reversal of motion has no reason to change the temperature of a body; hence, according to the oddness principle, it cannot be associated with primal intervals; hence, it is associated with dual intervals, i.e. $T[\widetilde{\mathbf{T}}]$.

A further argument is that the absolute temperature is the average kinetic energy per molecule according to the formula $\overline{T} = 3/2k_BT$: since energy is associated with dual time intervals, temperature is also associated with dual time intervals; hence, $T[\widetilde{\mathbf{T}}]$.

### 5.12.2 Global Variables in Space and Time

DEFINITION. *We call a* global variable in space and time *any variable which is global both in space and in time.*

Let us give some examples of variables which are global in space and time (Table 5.16).

- *Electric charge content* $Q^c$, i.e. the charge contained at an instant in a volume, is global in space because it is not a density of another variable, and it is global

[41] The global variables and equations of particle dynamics are displayed in the diagram in Table 9.1 at p. 242.

**Table 5.16** Global variables in time associated with time intervals which are increments of other global variables in time associated with instants

| | | | |
|---|---|---|---|
| Displacement | $\mathbf{u}(t^-, t^+) =$ | $+\mathbf{r}(t^+) - \mathbf{r}(t^-)$ | Position vector |
| Mass flow | $M^{\mathrm{f}}(t^-, t^+) =$ | $+M^{\mathrm{c}}(t^+) - M^{\mathrm{c}}(t^-)$ | Mass content |
| Charge flow | $Q^{\mathrm{f}}(t^-, t^+) =$ | $+Q^{\mathrm{c}}(t^+) - Q^{\mathrm{c}}(t^-)$ | Charge content |
| Work | $W(t^-, t^+) =$ | $+T(t^+) - T(t^-)$ | Kinetic energy |
| Work+heat | $W(t^-, t^+) + Q(t^-, t^+) =$ | $+U(t^+) - U(t^-)$ | Internal energy |
| Impulse | $\mathbf{J}(t^-, t^+) =$ | $+\mathbf{p}(t^+) - \mathbf{p}(t^-)$ | Momentum |
| Action | $A_H(t^-, t^+) =$ | $+S(t^+) - S(t^-)$ | Hamilton's principal function |

**Table 5.17** The two kinds of field variables as functions of time

| *Field variables which are not rates of other variables* associated with instants $\bar{\mathbf{I}}$ or $\tilde{\mathbf{I}}$ | | | *Field variables which are rates of other variables* associated with intervals $\bar{\mathbf{T}}$ or $\tilde{\mathbf{T}}$ | | |
|---|---|---|---|---|---|
| Mass density | $\rho(t)$ | $\rho[\bar{\mathbf{I}}]$ | Velocity ♠ | $\mathbf{v}(t)$ | $\mathbf{v}[\bar{\mathbf{T}}]$ |
| Surface charge density | $\sigma(t)$ | $\sigma[\bar{\mathbf{I}}]$ | Electric current density | $\mathbf{J}(t)$ | $\mathbf{J}[\bar{\mathbf{T}}]$ |
| Probability density | $\rho(t)$ | $\rho[\bar{\mathbf{I}}]$ | Particle current density | $\mathbf{J}_{\mathrm{p}}(t)$ | $\mathbf{J}_{\mathrm{p}}[\bar{\mathbf{T}}]$ |
| Particle density | $n(t)$ | $n[\bar{\mathbf{I}}]$ | Magnetic scalar potential | $\breve{\phi}_{\mathrm{m}}(t)$ | $\breve{\phi}_{\mathrm{m}}[\bar{\mathbf{T}}]$ |
| Burgers vector | $\breve{\mathbf{b}}(t)$ | $\breve{\mathbf{b}}[\bar{\mathbf{I}}]$ | Mass current density | $\mathbf{q}(t)$ | $\mathbf{J}_{\mathrm{m}}[\bar{\mathbf{T}}]$ |
| Concentration | $c(t)$ | $c[\bar{\mathbf{I}}]$ | Angular velocity | $\breve{\omega}(t)$ | $\breve{\omega}[\bar{\mathbf{T}}]$ |
| Total displacement | $\mathbf{u}(t)$ | $\mathbf{u}[\bar{\mathbf{I}}]$ | Magnetic field strength | $\breve{\mathbf{H}}(t)$ | $\breve{\mathbf{H}}[\bar{\mathbf{T}}]$ |
| Eikonal | $S(t)$ | $S[\bar{\mathbf{I}}]$ | Volume dilatation | $\Theta(t)$ | $\Theta[\bar{\mathbf{T}}]$ |
| Charge density | $\rho(t)$ | $\rho[\bar{\mathbf{I}}]$ | Volume dilatation rate | $\theta(t)$ | $\theta[\bar{\mathbf{T}}]$ |
| Position vector | $\mathbf{r}(t)$ | $\mathbf{r}[\bar{\mathbf{I}}]$ | Entropy source strength | $\sigma_{\mathrm{s}}(t)$ | $\sigma_{\mathrm{s}}[\bar{\mathbf{T}}]$ |
| Electric displacement | $\mathbf{D}(t)$ | $\mathbf{D}[\bar{\mathbf{I}}]$ | Heat source strength | $\sigma_{\mathrm{q}}(t)$ | $\sigma_{\mathrm{q}}[\bar{\mathbf{T}}]$ |
| Entropy density | $s(t)$ | $s[\bar{\mathbf{I}}]$ | Energy current density | $\mathbf{J}_{\mathrm{e}}(t)$ | $\mathbf{J}_{\mathrm{e}}[\bar{\mathbf{T}}]$ |
| Strain tensor | $\boldsymbol{\varepsilon}(t)$ | $\boldsymbol{\varepsilon}[\bar{\mathbf{I}}]$ | Entropy current density | $\mathbf{J}_{\mathrm{s}}(t)$ | $\mathbf{J}_{\mathrm{s}}[\bar{\mathbf{T}}]$ |
| Phase | $\phi(t)$ | $\phi[\bar{\mathbf{I}}]$ | | | |
| Magnetic vector potential | $\mathbf{A}(t)$ | $\mathbf{A}[\tilde{\mathbf{I}}]$ | Acceleration of gravity | $\mathbf{g}(t)$ | $\mathbf{g}[\tilde{\mathbf{T}}]$ |
| Momentum density | $\mathbf{p}(t)$ | $\mathbf{p}[\tilde{\mathbf{I}}]$ | Heat current density | $\mathbf{q}(t)$ | $\mathbf{q}[\tilde{\mathbf{T}}]$ |
| Vorticity | $\breve{\mathbf{w}}(t)$ | $\breve{\mathbf{w}}[\tilde{\mathbf{I}}]$ | Temperature | $T(t)$ | $T[\tilde{\mathbf{T}}]$ |
| Angular momentum density | $\breve{\mathbf{l}}(t)$ | $\breve{\mathbf{l}}[\tilde{\mathbf{I}}]$ | Internal energy density | $u(t)$ | $u[\tilde{\mathbf{T}}]$ |
| Magnetic flux density | $\breve{\mathbf{B}}(t)$ | $\breve{\mathbf{B}}[\tilde{\mathbf{I}}]$ | Symmetric stress tensor | $\sigma(t)$ | $\sigma[\tilde{\mathbf{T}}]$ |
| Velocity potential | $\phi(t)$ | $\phi[\tilde{\mathbf{I}}]$ | Electric potential | $\phi(t)$ | $\phi[\tilde{\mathbf{T}}]$ |
| Velocity ♠ | $\mathbf{v}(t)$ | $\mathbf{v}[\tilde{\mathbf{I}}]$ | Angular acceleration | $\breve{\alpha}(t)$ | $\breve{\alpha}[\tilde{\mathbf{T}}]$ |
| | | | Chemical potential | $\mu(t)$ | $\mu[\tilde{\mathbf{T}}]$ |
| | | | Force | $\mathbf{F}(t)$ | $\mathbf{F}[\tilde{\mathbf{T}}]$ |
| | | | Electric field strength | $\mathbf{E}(t)$ | $\mathbf{E}[\tilde{\mathbf{T}}]$ |

The notation $\breve{a}$ denotes a pseudoscalar, whereas $\breve{\mathbf{a}}$ denotes a pseudovector; see p. 145. For the double connotation of velocity see p. 355

in time because it is not a rate of another variable. The same goes for mass content $M^{\mathrm{c}}$, entropy content $S^{\mathrm{c}}$ and energy content $E^{\mathrm{c}}$, which are global in space and time.

- *Electric charge flow* $Q^{\mathrm{f}}$, i.e. the charge which flowed during a time interval through a surface, is global in space and time. The same goes for mass flow $M^{\mathrm{f}}$, energy flow $E^{\mathrm{f}}$, entropy flow $S^{\mathrm{f}}$ and momentum flow $\mathbf{P}^{\mathrm{f}}$.
- *Position vector* $\mathbf{r}$ is a global variable in space and time because it is neither a space density nor a time rate.
- *Displacement* $\mathbf{u}$ during a time interval of a point of a body, in a given time interval, is global in space and time.
- *Magnetic flux* $\Phi$ at an instant through a surface is a global variable in space and time; in fact, it is neither a density nor a rate.
- The *impulse of a volume force* $\mathbf{J}^{\mathrm{v}}$ is global in space and global in time.

## 5.13 Field Variables: Inherited Association

Once we realize that global variables are directly related to space and time elements, it is natural to consider their line, surface and volume densities and their rates as indirectly related to the corresponding space and time elements. This remark leads us to introduce the notion of *inherited association* as follows.

> DEFINITION. *We say that densities and rates* inherit *an association with the space and time elements to which the corresponding global variables refer.*

Hence

- Mass density $\rho$ *inherits* an association with volumes because the mass content $M^{\mathrm{c}}$ is associated with volumes $\widetilde{\mathbf{V}}$ and time instants $\bar{\mathbf{I}}$. Hence,

$$\textit{from} \quad M^{\mathrm{c}}[\bar{\mathbf{I}},\widetilde{\mathbf{V}}] \quad \textit{it follows that} \quad \rho[\bar{\mathbf{I}},\widetilde{\mathbf{V}}]. \tag{5.13}$$

- Electric field strength $\mathbf{E}$ *inherits* from electric voltage $E$ an association with time intervals endowed with an outer orientation and with lines endowed with an inner orientation. Hence,

$$\textit{from} \quad E[\widetilde{\mathbf{T}},\bar{\mathbf{L}}] \quad \textit{it follows that} \quad \mathbf{E}[\widetilde{\mathbf{T}},\bar{\mathbf{L}}]. \tag{5.14}$$

- Heat current density $\mathbf{q}$ *inherits* an association with time intervals and surfaces because it is the rate of heat and the surface density of the heat $Q$. Hence,

$$\textit{from} \quad Q[\bar{\mathbf{T}},\widetilde{\mathbf{S}}] \quad \textit{it follows that} \quad \mathbf{q}\,[\bar{\mathbf{T}},\widetilde{\mathbf{S}}]. \tag{5.15}$$

- Force, which is the time rate of impulse, *inherits* an association with dual time intervals; hence, $\mathbf{F}[\widetilde{\mathbf{T}}]$.

The inherited association with the space-time elements enables the preservation of the noteworthy information, which is commonly lost.

The space global variables which are associated with points, such as, for example, temperature, electric potential, chemical potential, displacement and velocity potential, are automatically field variables. Nevertheless, most field variables arise from those global variables which are associated with space elements endowed with extension, i.e. lines, surfaces and volumes, by performing a two-step process:

- Divide the global variable by the extension of the space element, i.e. length, area, volume, to obtain a *mean* density.
- Perform a limit process on this mean density to obtain a *density* which is a point function.

Hence, field variables are characterized by a double notation:

1. *Traditional notation:* the point **P** is placed inside *round* brackets; e.g. heat current density is denoted $\mathbf{q}(t, \mathbf{P})$.
2. *Additional notation:* the inherited association with the temporal and the spatial elements is highlighted by placing the corresponding elements within *square* brackets, for example the heat current density will be referred to as $\mathbf{q}\,[\overline{\mathbf{T}}, \widetilde{\mathbf{S}}]$.

**Notation.** Usually to denote a function which depends on time and space variables, it is common to use the notations $f(\mathbf{x}, t)$, i.e. the space variable may precede the time variable or vice versa. In this book the time variable always precedes the space one. To justify this choice, we note the following:

1. Firstly, the fundamental equations of thermal conduction and the wave equation are usually written in the form
$$\rho c \frac{\partial T}{\partial t} - \lambda \nabla^2 T = 0 \qquad \frac{1}{v^2}\frac{\partial^2 \phi}{\partial t^2} - \nabla^2 \phi = 0\,. \tag{5.16}$$
In both cases, the position of the time derivatives *precedes* the position of the space derivatives. This suggests that we should consider the time variable as the first variable, i.e. $f(t, \mathbf{x})$.
2. Secondly, when passing from a material description to a spatial description, we must add space variables, i.e. from $f(t)$ to $f(t, x, y, z)$, and this leads to placing space variables *after* time variables.
3. Thirdly, when space variables depend on time, the total derivative is commonly written
$$\frac{df(t, \mathbf{x})}{dt} = \frac{\partial f}{\partial t} + \frac{\partial f}{\partial x}\frac{\mathrm{d}x}{\mathrm{d}t} + \frac{\partial f}{\partial y}\frac{\mathrm{d}y}{\mathrm{d}t} + \frac{\partial f}{\partial z}\frac{\mathrm{d}z}{\mathrm{d}t}, \tag{5.17}$$
i.e. once more, the time variable *precedes* the space variables.
4. A fourth reason is that in the theory of relativity, the main invariant $\mathrm{d}s^2$ can be written in one of two ways:
$$\mathrm{d}s^2 = c^2\,\mathrm{d}t^2 - (\mathrm{d}x^2 + \mathrm{d}y^2 + \mathrm{d}z^2); \qquad \mathrm{d}s^2 = (\mathrm{d}x^2 + \mathrm{d}y^2 + \mathrm{d}z^2) - c^2\,\mathrm{d}t^2\,. \tag{5.18}$$

The first choice is preferable because it allows us to avoid the introduction of the imaginary unit in relativity; see Misner et al. [161, p. 51].

We will adopt this notation and write expressions such as $[\mathbf{I},\mathbf{P}]$, $[\mathbf{T},\mathbf{V}]$ instead of $[\mathbf{P},\mathbf{I}]$, $[\mathbf{V},\mathbf{T}]$.

In Table 5.18 we have assembled a list of global physical variables with the corresponding field functions. A comparison of the fourth and fifth columns in this table reveals that the information content of field variables, using the space and time elements inside the square brackets, is greater than the information content inside the round brackets. In fact, when we perform the limit process, all variables become functions of $t$ and $\mathbf{P}$, regardless of their origin. In Tables 5.17 and 5.18, we have specially marked velocity because it has a double connotation in continuum mechanics.[42]

## 5.14 How to Find the Space and Time Association

We give here the operative rules to find the space and time elements with which a physical variable is associated.

**Space Association.** Given a physical variable $Q$, you have to decide, at first, whether it is a global variable in space or not. Recall that a variable is called *global in space* if it is not a line, surface or volume density of another variable.

- If the variable $Q$ is *global in space*, then we use the definition of the variable and take into account the measurement process (if it is a measurable quantity). We must see if the variable makes reference to a volume, surface, line or point. Following this detailed definition and the possible measurement process we can find the correct space element. In these cases we will put, provisionally, the space element in square brackets as follows: $Q[\mathbf{V}]$, $Q[\mathbf{S}]$, $Q[\mathbf{L}]$, $Q[\mathbf{P}]$ respectively. Now we must determine the orientation of the space element. To this end, we pose the following question: does the sign of the variable change when we reverse the *inner* orientation of the space element? If the variable *changes* sign, then the variable is associated with an element endowed with an inner orientation, and we will write $Q[\overline{\mathbf{V}}]$, $Q[\overline{\mathbf{S}}]$, $Q[\overline{\mathbf{L}}]$, $Q[\overline{\mathbf{P}}]$, depending on the space element with which the variable $Q$ is associated. In contrast, if the sign *does not* change, then the variable is associated with an element endowed with an *outer* orientation, and we will write $Q[\tilde{\mathbf{V}}]$, $Q[\tilde{\mathbf{S}}]$, $Q[\tilde{\mathbf{L}}]$, $Q[\tilde{\mathbf{P}}]$, depending on the space element with which the variable $Q$ is associated.
- If the variable $Q$ is a **density**, then it inherits the association with the same space element of the global variable of which it is a density. Hence, also in this case, we will write $Q[\overline{\mathbf{V}}]$, $Q[\overline{\mathbf{S}}]$, $Q[\overline{\mathbf{L}}]$ or $Q[\tilde{\mathbf{V}}]$, $Q[\tilde{\mathbf{S}}]$, $Q[\tilde{\mathbf{L}}]$ respectively.

[42] As will be explained in Chap. 12, at p. 356.

**Table 5.18** Global variables and corresponding field functions

| *Global variable* | | *Corresponding field function* | | |
|---|---|---|---|---|
| *Associated with volumes* | | | | |
| Mass content | $M^{c}[\bar{\mathbf{I}}, \tilde{\mathbf{V}}]$ | *Mass density* | $\rho[\bar{\mathbf{I}}, \tilde{\mathbf{V}}]$ | $\rho(t, \mathbf{P})$ |
| Entropy content | $S^{c}[\bar{\mathbf{I}}, \tilde{\mathbf{V}}]$ | *Entropy density* | $s[\bar{\mathbf{I}}, \tilde{\mathbf{V}}]$ | $s(t, \mathbf{P})$ |
| Entropy production | $S^{p}[\bar{\mathbf{T}}, \tilde{\mathbf{V}}]$ | *Entropy source strength* | $\sigma_{s}[\bar{\mathbf{T}}, \tilde{\mathbf{V}}]$ | $\sigma_{s}(t, \mathbf{P})$ |
| Impulse of volume force | $\mathbf{J}^{v}[\tilde{\mathbf{T}}, \tilde{\mathbf{V}}]$ | *Volume force* | $\mathbf{F}^{v}[\tilde{\mathbf{T}}, \tilde{\mathbf{V}}]$ | $\mathbf{F}(t, \mathbf{P})$ |
| *Associated with surfaces* | | | | |
| Magnetic flux | $\Phi[\tilde{\mathbf{I}}, \bar{\mathbf{S}}]$ | *Magnetic flux density* | $\mathbf{B}[\tilde{\mathbf{I}}, \bar{\mathbf{S}}]$ | $\mathbf{B}(t, \mathbf{P})$ |
| Electric flux | $\Psi[\bar{\mathbf{I}}, \tilde{\mathbf{S}}]$ | *Electric displacement* | $\mathbf{D}[\bar{\mathbf{I}}, \tilde{\mathbf{S}}]$ | $\mathbf{D}(t, \mathbf{P})$ |
| Vortex flux | $W[\tilde{\mathbf{I}}, \bar{\mathbf{S}}]$ | *Vorticity vector* | $\mathbf{w}[\tilde{\mathbf{I}}, \bar{\mathbf{S}}]$ | $\mathbf{w}(t, \mathbf{P})$ |
| Mass flow | $M^{f}[\bar{\mathbf{T}}, \tilde{\mathbf{S}}]$ | *Mass current density* | $\mathbf{q}[\bar{\mathbf{T}}, \tilde{\mathbf{S}}]$ | $\mathbf{q}(t, \mathbf{P})$ |
| Electric charge flow | $Q^{f}[\bar{\mathbf{T}}, \tilde{\mathbf{S}}]$ | *Electric current density* | $\mathbf{J}[\bar{\mathbf{T}}, \tilde{\mathbf{S}}]$ | $\mathbf{J}(t, \mathbf{P})$ |
| Entropy flow | $S^{f}[\bar{\mathbf{T}}, \tilde{\mathbf{S}}]$ | *Entropy current density* | $\mathbf{J}^{s}[\bar{\mathbf{T}}, \tilde{\mathbf{S}}]$ | $\mathbf{J}^{s}(t, \mathbf{P})$ |
| Heat | $Q[\bar{\mathbf{T}}, \tilde{\mathbf{S}}]$ | *Heat current density* | $\mathbf{q}[\bar{\mathbf{T}}, \tilde{\mathbf{S}}]$ | $\mathbf{q}(t, \mathbf{P})$ |
| Impulse of surface force | $\mathbf{J}^{s}[\tilde{\mathbf{T}}, \tilde{\mathbf{S}}]$ | *Internal surface force* | $\mathbf{T}[\tilde{\mathbf{T}}, \tilde{\mathbf{S}}]$ | $\mathbf{T}(t, \mathbf{P})$ |
| | | *Stress tensor* | $\sigma[\tilde{\mathbf{T}}, \tilde{\mathbf{S}}]$ | $\sigma(t, \mathbf{P})$ |
| | | *Pressure* | $p[\tilde{\mathbf{T}}, \tilde{\mathbf{S}}]$ | $p(t, \mathbf{P})$ |
| *Associated with lines* | | | | |
| Relative displacement | $\mathbf{h}[\bar{\mathbf{I}}, \bar{\mathbf{L}}]$ | *Strain tensor* | $\boldsymbol{\varepsilon}[\bar{\mathbf{I}}, \bar{\mathbf{L}}]$ | $\boldsymbol{\varepsilon}(t, \mathbf{P})$ |
| Velocity line integral | $\Gamma[\tilde{\mathbf{I}}, \bar{\mathbf{L}}]$ | *Velocity* ♠ | $\mathbf{v}[\tilde{\mathbf{I}}, \bar{\mathbf{L}}]$ | $\mathbf{v}(t, \mathbf{P})$ |
| Relative velocity | $\mathbf{e}[\bar{\mathbf{T}}, \bar{\mathbf{L}}]$ | *Strain rate tensor* | $\mathbf{d}[\bar{\mathbf{T}}, \bar{\mathbf{L}}]$ | $\mathbf{d}(t, \mathbf{P})$ |
| Magnetomotive force impulse | $\mathscr{E}[\tilde{\mathbf{T}}, \bar{\mathbf{L}}]$ | *Electric field strength* | $\mathbf{E}[\tilde{\mathbf{T}}, \bar{\mathbf{L}}]$ | $\mathbf{E}(t, \mathbf{P})$ |
| Voltage impulse | $\mathscr{E}[\bar{\mathbf{T}}, \tilde{\mathbf{L}}]$ | *Magnetic field strength* | $\mathbf{H}[\bar{\mathbf{T}}, \tilde{\mathbf{L}}]$ | $\mathbf{H}(t, \mathbf{P})$ |
| *Associated with points* | | | | |
| Position vector | $\mathbf{r}[\bar{\mathbf{I}}, \bar{\mathbf{P}}]$ | *Radius vector* | $\mathbf{r}[\bar{\mathbf{I}}, \bar{\mathbf{P}}]$ | $\mathbf{r}(t, \mathbf{P})$ |
| Velocity potential | $\phi[\tilde{\mathbf{I}}, \bar{\mathbf{P}}]$ | *Velocity potential* | $\phi[\tilde{\mathbf{I}}, \bar{\mathbf{P}}]$ | $\phi(t, \mathbf{P})$ |
| Thermacy | $\mathscr{T}[\tilde{\mathbf{I}}, \bar{\mathbf{P}}]$ | *Temperature* | $T[\tilde{\mathbf{T}}, \bar{\mathbf{P}}]$ | $T(t, \mathbf{P})$ |
| Displacement | $\mathbf{u}[\bar{\mathbf{T}}, \bar{\mathbf{P}}]$ | *Velocity* ♠ | $\mathbf{v}[\bar{\mathbf{T}}, \bar{\mathbf{P}}]$ | $\mathbf{v}(t, \mathbf{P})$ |
| Electric potential impulse | $\varphi[\tilde{\mathbf{T}}, \bar{\mathbf{P}}]$ | *Electric potential* | $\phi[\tilde{\mathbf{T}}, \bar{\mathbf{P}}]$ | $\phi(t, \mathbf{P})$ |
| Scalar magnetic potential impulse | $\varphi_{m}[\bar{\mathbf{T}}, \tilde{\mathbf{P}}]$ | *Scalar magnetic potential* | $\phi_{m}[\bar{\mathbf{T}}, \tilde{\mathbf{P}}]$ | $\phi_{m}(t, \mathbf{P})$ |

**Time Association.** Given a physical variable $Q$, we must decide, firstly, whether it is *global in time* or not. Recall that a variable is said to be *global in time* if it is not the rate of another variable.

- If the variable is *global in time* using the definition of the variable and taking into account the way we measure it (if is a measurable quantity), then we must see whether it makes reference to a time interval or to a time instant. In these cases, we will put, provisionally, the time element in square brackets as follows: $Q[\mathbf{T}]$, $Q[\mathbf{I}]$ respectively. Now we must determine the orientation of the time element. To do this, we ask ourselves: does the sign of the variable change when we perform a reversal of motion? If the sign changes, then the variable $Q$ is associated either with primal time intervals (inner orientation), hence $Q[\overline{\mathbf{T}}]$, or with dual time instants (outer orientation), hence $Q[\widetilde{\mathbf{I}}]$, depending on which time element it is associated with. In contrast, if the sign does not change, then the variable $Q$ is associated either with primal time instants (inner orientation), hence $Q[\overline{\mathbf{I}}]$, or with dual time intervals (outer orientation), hence $Q[\widetilde{\mathbf{T}}]$, depending on which time element it is associated with.
- If the variable $Q$ is a *rate*, then it inherits the association with the same time element of the global variable of which it is the rate. Hence, we will write $Q[\overline{\mathbf{T}}]$ or $Q[\widetilde{\mathbf{T}}]$ respectively.

Table 9.6 (p. 265) shows the time association of many energy variables. For a description see Chaps. 9, 10, and 13, in particular p. 253.

## 5.15 Physical Variables Can Be Grouped into Families

Physical variables, which are linked to each other by the process of forming a rate or a line, surface or volume density, will be considered as belonging to the same *family.* In every family there is a variable which is the "head" of the family: this is a global variable in space and time. In every family there is a *field variable*, i.e. a variable which is obtained from the head variable by computing the density and the rate, thereby eliminating the dependence on space and time extensions.

1. Mass and density belong to the same family: mass is the head and density is the field variable:

$$\textbf{mass } m \longrightarrow \textit{mass density } \rho.$$

2. Displacement and velocity belong to the same family: velocity is the rate of displacement, hence the displacement is the head of the family:

$$\textbf{displacement } \mathbf{u} \longrightarrow \textit{displacement rate} = \textit{velocity } \mathbf{v}.$$

3. The family of surface forces is as follows:

   **impulse of surface force $\mathbf{J}^s$** $\longrightarrow$ surface impulse rate = surface force $\mathbf{T}$

   surface impulse density $\longrightarrow$ *surface force density = stress vector* $\mathbf{t}$.

4. The electric current density $\mathbf{J}$, the electric current $I$ and the electric charge flow $Q^f$ belong to the same family: the charge flow is the head and the electric current density is the field variable:

   **electric charge flow $Q^f$** $\longrightarrow$ electric charge flow rate = current $I$

   electric charge flow density $\longrightarrow$ *electric current density* $\mathbf{J}$.

5. The family of voltages is

   **voltage impulse $\mathscr{E}$** $\longrightarrow$ voltage $E$

   electric impulse strength $\longrightarrow$ *electric field strength* $\mathbf{E}$.

6. Electric potential is a function of a point and an instant. Since it is physically meaningful to perform a time integration on it, we obtain the electric potential impulse:

   **electric potential impulse $\varphi$** $\longrightarrow$ *electric potential* $\phi$.

7. Work, heat, energy flow, energy flow rate (= power transmitted) and energy flow rate density belong to the same family. Which is the head? Since work and heat are two forms of energy flow, the head of the family is energy flow: Then we have

   **energy flow** $\longrightarrow$ energy flow rate = energy current=power $P$

   energy flow density $\longrightarrow$ *energy current density* (*e.g. Poynting vector* $\mathbf{S}$)

   These examples show that there are four kinds of families: one with four members, two with two members and one with a single member, as shown in Fig. 5.11.

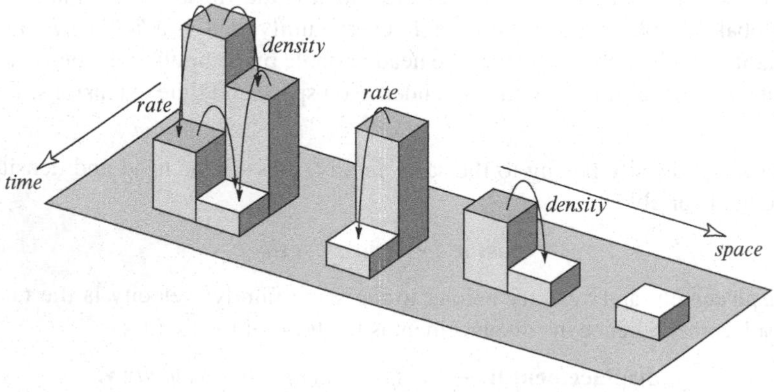


**Fig. 5.11** The four families of physical variables: some are composed of four members, some of two members and some of a single member

The identification of the head of the family is not always immediate. Thus, one may think that electric potential and temperature are the heads of the corresponding families, while they are rates. The corresponding heads are the electric potential impulse and the time integral of temperature respectively. Lastly, the temperature family is[43]

$$\textbf{thermacy}\ \mathcal{T} \longrightarrow \text{thermodynamic } \textit{temperature}\ T\,.$$

## 5.16 Identification Criteria

A general criterion for identifying the space and time elements associated with a physical variable is offered by the adjectival modifiers preceding or following the name of the variable, or by the name of the variable. We list here the more usual ones.

1. **Produced, stored, released, generated, dissipated, source, supply** are typical names of variables associated with volumes endowed with an outer orientation and time intervals endowed with an inner or outer orientation. Examples include *entropy production* $S^{\mathrm{p}}[\overline{\mathbf{T}}, \widetilde{\mathbf{V}}]$, *entropy source strength* $\sigma_{\mathrm{s}}[\overline{\mathbf{T}}, \widetilde{\mathbf{V}}]$ and *momentum production*, i.e. the *impulse* (of a force) $\mathbf{J}\,[\widetilde{\mathbf{T}}, \widetilde{\mathbf{V}}]$.
2. **Content, amount, volume density, specific** are typical names of variables associated with volumes endowed with an outer orientation and time instants endowed with an inner or outer orientation. Examples include *mass content* $M^{\mathrm{c}}[\overline{\mathbf{I}}, \widetilde{\mathbf{V}}]$ and *momentum content* $\mathbf{P}^{\mathrm{c}}[\widetilde{\mathbf{I}}, \widetilde{\mathbf{V}}]$.
3. **Absorbed, emitted, transmitted, flow, current** are typical of variables associated with surfaces endowed with an outer orientation and time intervals endowed with an inner or outer orientation. Examples include *entropy flow* $S^{\mathrm{f}}[\mathbf{T}, \tilde{\mathbf{S}}]$ and *mass flow* $M^{\mathrm{f}}[\overline{\mathbf{T}}, \tilde{\mathbf{S}}]$.
4. **Flux**[44] is typical of variables associated with surfaces endowed with an inner or outer orientation and time instants endowed with an inner or outer orientation. Examples include *magnetic flux* $\Phi[\widetilde{\mathbf{I}}, \overline{\mathbf{S}}]$, *electric flux* $\Psi[\overline{\mathbf{I}}, \widetilde{\mathbf{S}}]$ and *vortex flux* $W[\widetilde{\mathbf{I}}, \overline{\mathbf{S}}]$.
5. **Potential difference, voltage, strength, thermodynamic force** are associated with lines endowed with an inner or outer orientation and time intervals endowed with an inner or outer orientation. Examples include *voltage* $E[\widetilde{\mathbf{T}}, \overline{\mathbf{L}}]$, *magnetomotive force* $F_{\mathrm{m}}[\overline{\mathbf{T}}, \widetilde{\mathbf{L}}]$ and *temperature difference* $G[\widetilde{\mathbf{T}}, \overline{\mathbf{L}}]$.
6. **Circulation** denotes a physical variable associated with lines with an inner or outer orientation and time instants endowed with an inner or outer orientation.

---

[43] See p. 387.

[44] Since the term *flux* is used with many different meanings, see p. 30 for the restricted meaning we apply to this term.

Examples include *velocity line integral* $\Gamma[\widetilde{\mathbf{I}},\overline{\mathbf{L}}]$, *optical path length* $s[\overline{\mathbf{I}},\overline{\mathbf{L}}]$, *spatial phase difference* $\Delta_s\phi[\widetilde{\mathbf{I}},\overline{\mathbf{L}}]$ and *line integral of the magnetic vector potential* $a[\widetilde{\mathbf{I}},\overline{\mathbf{L}}]$.

7. **Potential** is associated with points endowed with an inner or outer orientation and with time intervals endowed with an inner or outer orientation. Examples include *electric potential* $\phi[\widetilde{\mathbf{T}},\overline{\mathbf{P}}]$, *gravitational potential* $U_g[\widetilde{\mathbf{T}},\overline{\mathbf{P}}]$, *magnetic scalar potential* $\check{\phi}_m[\overline{\mathbf{T}},\widetilde{\mathbf{P}}]$ and *thermal potential* ≡ *temperature* $T[\widetilde{\mathbf{T}},\overline{\mathbf{P}}]$. *Kinetic potential* $\phi[\widetilde{\mathbf{I}},\overline{\mathbf{P}}]$ refers to time instants.
8. **Function** is associated with points endowed with an inner or outer orientation and time instants endowed with an inner or outer orientation. Examples include *gauge function* $\chi[\widetilde{\mathbf{I}},\overline{\mathbf{P}}]$, *Hamilton's principal function* $S[\widetilde{\mathbf{I}},\overline{\mathbf{P}}]$,[45] *eikonal function* $S[\overline{\mathbf{I}},\overline{\mathbf{P}}]$, *phase function* $\phi[\widetilde{\mathbf{I}},\overline{\mathbf{P}}]$, *velocity potential* $\phi[\widetilde{\mathbf{I}},\overline{\mathbf{P}}]$, *stream function* $\psi[\overline{\mathbf{T}},\widetilde{\mathbf{P}}]$ and *Airy's function* $\phi[\widetilde{\mathbf{I}},\widetilde{\mathbf{P}}]$.

These notions allow us to construct a classification diagram of physical variables.

Global variables associated with volumes are additive on volumes; they are commonly called *extensive* variables in thermodynamics. The class of *intensive* variables has no precise definition. The usual presentation of intensive variables is *those which are not additive, like, for example, pressure, temperature and density.*[46] It includes variables associated with points, like temperature $T[\widetilde{\mathbf{T}},\overline{\mathbf{P}}]$ and electric potential $\phi[\widetilde{\mathbf{T}},\overline{\mathbf{P}}]$; others associated with volumes, like mass density $\rho[\widetilde{\mathbf{I}},\widetilde{\mathbf{V}}]$ and internal energy density $u[\widetilde{\mathbf{T}},\widetilde{\mathbf{V}}]$; and still others associated with surfaces, like pressure $p[\widetilde{\mathbf{T}},\widetilde{\mathbf{S}}]$. The absence of a definition of intensive variables[47] may explain why they are of no use outside of thermodynamics.

From the preceding discussion we can see that *many functions of points and instants (field variables) arise from the need to compute ratios in order to make possible the limit process.* Physical variables associated with lines, surfaces and volumes give rise, after the limit process on their mean densities, to point functions. This implies a loss of information. Thus, in thermostatics, temperature, pressure and density are considered *intensive* variables. When dealing with intensive variables it is not noted that they may be associated with different space elements: thus, temperature and electric potential are associated with points, pressure is associated with surfaces and density is associated with volumes (Table 5.18). Why should we lose this information?

---

[45] This is the indefinite time integral of the Lagrangian function along a path: see Lur'é, [144, vol. II, p.705], Landau and Lifshitz [123, p. 138].

[46] Lewis and Randal [137].

[47] Redlich [191].

### *5.16.1 Possible Ambiguities*

The association of physical variables with space and time elements must be made with care because we are led to associate many variables with points and instants. It is important to note that *this association pertains to global variables*, and only indirectly to densities.

**Remark.** In integrals, the space element, which is written at the bottom of the integral symbol, is marked with an overline, or a tilde, according to whether the element is endowed with an inner or outer orientation. This practice is never applied in physics or mathematics simply because, in dealing with multiple integrals, there is not the custom of distinguishing between inner and outer orientations.

1. We may think that pressure is *associated* with points and instants. This is incorrect because pressure is not a global variable; rather, it is the density of a surface force which, in turn, is the time rate of the surface impulse. The impulse of the normal force acting on a surface $\widetilde{\mathbf{S}}$ with normal $\mathbf{n}$ is

$$\mathbf{J}^{\mathrm{s}} = \int_{\widetilde{\mathbf{T}}} \int_{\widetilde{\mathbf{S}}} -p(t, \mathbf{P})\, \mathbf{n}\; \mathrm{d}S\; \mathrm{d}t \qquad \textit{surface impulse.} \tag{5.19}$$

Hence, pressure, which is the density rate of the surface impulse, *inherits* an association with surfaces and intervals. The intervals are those endowed with an outer orientation. To stress this relation, we will write

$$\mathbf{J}^{\mathrm{s}}[\widetilde{\mathbf{T}}, \widetilde{\mathbf{S}}] \longrightarrow \mathbf{T}[\widetilde{\mathbf{T}}, \widetilde{\mathbf{S}}] \longrightarrow p\,[\widetilde{\mathbf{T}}, \widetilde{\mathbf{S}}]. \tag{5.20}$$

All three of these variables *belong to the same family*, the family of surface impulses. Note that surface impulse is the head of the family.

**Remark.** Some authors state that pressure coincides with energy density, and as consequence, one can think that pressure is associated with volumes. This is not correct. In a material description, pressure is equal to the increase in energy for unit volume, according to the thermodynamic relation $\mathrm{d}U = -p\; \mathrm{d}\mathcal{V}$. This presupposes a variation in the volume, possible in a material description, while in a spatial description, the volume is fixed (control volume). Pressure has the same dimensions of an energy density, but this is no reason to confuse the two variables. In addition, the moment of a couple has the same dimensions of a work, but they are two distinct variables. One argument to contrast this identification is that energy is defined up to an arbitrary constant; hence, its density also depends on an arbitrary constant. In contrast, pressure is a measurable quantity. The identification of pressure with energy density is possible and convenient only in cases where there is no arbitrary constant, as is the case with magnetic energy density, which vanishes when the magnetic field is zero.

2. Similarly we may think that mass density $\rho$ is associated with points. This is incorrect because mass density is a mass content $M^{\rm c}$ for unit volume and mass content is associated with volumes and time instants. Since

$$M^{\rm c} = \int_{\widetilde{\mathbf{V}}} \rho\, \mathrm{d}V \qquad \textit{mass content,} \tag{5.21}$$

we will write

$$M^{\rm c}[\overline{\mathbf{I}}, \widetilde{\mathbf{V}}] \longrightarrow \rho\,[\overline{\mathbf{I}}, \widetilde{\mathbf{V}}] \tag{5.22}$$

and say that these two variables belong *to the same family*.

3. Let us consider the electric field **E**. It appears natural to consider it as being *associated with* points and instants; this is incorrect. In fact, the vector **E** is not a global variable: its line integral is the voltage $V$, which in turn is the time rate of the *voltage impulse*

$$\mathscr{V} = \int_{\widetilde{\mathbf{T}}} \int_{\overline{\mathbf{L}}} \mathbf{E}(t, \mathbf{P}) \cdot \mathbf{t}\, \mathrm{d}L\, \mathrm{d}t \qquad \textit{voltage impulse.} \tag{5.23}$$

We will write

$$\mathscr{V}[\widetilde{\mathbf{T}}, \overline{\mathbf{L}}] \longrightarrow V[\widetilde{\mathbf{T}}, \overline{\mathbf{L}}] \longrightarrow \mathbf{E}[\widetilde{\mathbf{T}}, \overline{\mathbf{L}}], \tag{5.24}$$

and we consider the three variables of *the same family*, the family of the voltage impulse, which is the head of the family.

4. Let us consider the electric current density **J**; it appears natural to consider it as being *associated with* points and instants. This is incorrect. In fact, the vector **J** is not a global variable; it is the surface density of the electric current $I$, which in turn is the time rate of the *charge flow*

$$Q^{\rm f} = \int_{\overline{\mathbf{T}}} \int_{\widetilde{\mathbf{S}}} \mathbf{J}(t, \mathbf{P}) \cdot \mathbf{n}\, \mathrm{d}S\, \mathrm{d}t \qquad \textit{charge flow.} \tag{5.25}$$

We will write

$$Q^{\rm f}[\overline{\mathbf{T}}, \widetilde{\mathbf{S}}] \longrightarrow I[\overline{\mathbf{T}}, \widetilde{\mathbf{S}}] \longrightarrow \mathbf{J}\,[\overline{\mathbf{T}}, \widetilde{\mathbf{S}}], \tag{5.26}$$

and we consider that the three variables *belong to the same family*, the family of charge flow, which is the head. With these notations we see that Ohm's law, which is a constitutive law

$$V(t) = R\, I(t), \tag{5.27}$$

can be written as

$$\textit{voltage} \qquad E[\widetilde{\mathbf{T}}, \overline{\mathbf{L}}] = R\, I[\overline{\mathbf{T}}, \widetilde{\mathbf{S}}] \qquad \textit{current} \tag{5.28}$$

or, using space and time global variables,

$$\textit{voltage impulse} \qquad \mathscr{E}[\widetilde{\mathbf{T}}, \overline{\mathbf{L}}] = R\, Q^{\rm f}[\overline{\mathbf{T}}, \widetilde{\mathbf{S}}] \qquad \textit{charge flow.} \tag{5.29}$$

This shows that the time integral of the electric voltage is meaningful. Even more, we see that Ohm's law links a variable associated with a time interval with an inner orientation with one associated with a time interval with an outer orientation. As we will discuss later (p. 164), this shows that Ohm's law describes an *irreversible* phenomenon.

## 5.17 Differential Formulation and Orientation

A differential formulation does not explicitly employ the concept of orientation of the space element involved. To show this, we note that when we deduce integral variables from field variables, the integration along lines, over surfaces and in volumes does not specify the kind of orientation of the space element. Therefore, we usually write line, surface and volume integrals with the notation

$$\int_L \quad \int_S \quad \int_V \quad \textit{instead of} \quad \int_{\overline{\mathbf{L}}} \quad \int_{\overline{\mathbf{S}}} \quad \int_{\overline{\mathbf{V}}} \quad \textit{and} \quad \int_{\tilde{\mathbf{L}}} \quad \int_{\tilde{\mathbf{S}}} \quad \int_{\tilde{\mathbf{V}}} \tag{5.30}$$

for inner and outer orientations respectively.

The fact that the magnetic flux $\Phi$ and the vortex flux $W$ are associated with a surface endowed with an *inner* orientation whereas the electric flux $\Psi$ and the electric current $I$ are associated with a surface endowed with an *outer* orientation does not appear in the formulas. Moreover, when the surfaces are endowed with an inner orientation, to introduce the unit normal $\mathbf{n}$, we must have recourse to the screw rule, and this implies that the normal is a pseudovector, a fact that is not commonly noted. In addition, the unit vector tangent to a line endowed with an outer orientation is a pseudovector.

**Notation.** A *pseudovector* is a vector which by definition requires a direction of translation to be associated with a direction of rotation. A typical example is the vector of angular velocity of a cylinder around its axis, which is defined as a vector parallel to the axis of rotation whose modulus is the angle of rotation for unit time and whose direction is associated with the direction of rotation *by making use of the rule of the right-handed screw*. The transformation law of pseudovectors is similar to the transformation law of (polar) vectors except for the sign of the determinant of the *transition matrix*[48] from two sets of base vectors. Similarly, a *pseudoscalar* is a scalar whose sign depends on the screw chosen. Following the notation used in the French literature, which puts a curved arrow over pseudovectors,[49] we will denote pseudovectors and pseudoscalars by placing a check sign above the symbol: $\check{\mathbf{n}}$ and $\check{\mathbf{t}}$ respectively.[50] This is the case with the scalar magnetic potential $\check{\phi}_{\mathrm{m}}$.

---

48 See Appendix B.

49 See for example Fleury and Mathieu [71, vol. 6], Brillouin [30], Fournet [73], Jouguet [110].

50 This symbol is used by Corson [43, p. 8].

It follows that a more appropriate notation for integrals is

$$\underbrace{\Phi=\int_{\mathbf{\tilde{S}}}\check{\mathbf{B}}\cdot\check{\mathbf{n}}\,\mathrm{d}S}_{magnetic\ flux},\quad \underbrace{W=\int_{\mathbf{\tilde{S}}}\check{\mathbf{w}}\cdot\check{\mathbf{n}}\,\mathrm{d}S}_{vortex\ flux},\quad \underbrace{\Psi=\int_{\tilde{\mathbf{S}}}\mathbf{D}\cdot\mathbf{n}\,\mathrm{d}S}_{electric\ flux},\quad \underbrace{I=\int_{\tilde{\mathbf{S}}}\mathbf{J}\cdot\mathbf{n}\,\mathrm{d}S}_{electric\ current}, \tag{5.31}$$

$$\underbrace{E=\int_{\mathbf{L}}\mathbf{E}\cdot\mathbf{t}\,\mathrm{d}L}_{electric\ voltage},\quad \underbrace{F=\int_{\tilde{\mathbf{L}}}\check{\mathbf{H}}\cdot\check{\mathbf{t}}\,\mathrm{d}L}_{magnetic\ voltage},\quad \underbrace{\check{G}=\int_{\mathbf{V}}\check{\rho}_{\mathrm{m}}\,\mathrm{d}V}_{magnetic\ charge},\quad \underbrace{Q=\int_{\tilde{\mathbf{V}}}\rho\,\mathrm{d}V}_{electric\ charge}. \tag{5.32}$$

To show that the differential formulation fails to take into serious consideration orientation, let us consider the line integral along a line of a pseudovector, such as magnetic field strength $\check{\mathbf{H}}$. Such a line integral is called *magnetomotive force*.[51] We know that the line integral of the pseudovector $\check{\mathbf{H}}$ along a closed line is equal to the current which crosses any surface which has the line as its boundary (Ampère's law). Since the current is a true scalar, we can deduce that the line integral is also a true scalar. On the other hand, we know that the line integral of the pseudovector $\check{\mathbf{H}}$ from a fixed point **A** to an arbitrary point **P** gives the scalar magnetic potential at **P**, and this is a pseudoscalar $\check{\phi}_{\mathrm{m}}$.[52]

Hence, we have found that the line integral of a pseudovector is sometimes a scalar and sometimes a pseudoscalar! How can we overcome this contradiction?

We can do so as follows: for Ampère's law we must use a line endowed with an *outer* orientation, hence the unit tangent vector must be a pseudovector $\check{\mathbf{t}}$; whereas for the calculus of the scalar magnetic potential we must use a line endowed with an *inner* orientation and, hence, a unit tangent vector which is a true vector **t**. The two line integrals are

$$F=\int_{\tilde{\mathbf{L}}}\check{\mathbf{H}}\cdot\check{\mathbf{t}}\,\mathrm{d}L, \qquad \check{\phi}_{\mathrm{m}}=\int_{\mathbf{L}}\check{\mathbf{H}}\cdot\mathbf{t}\,\mathrm{d}L. \tag{5.33}$$

Thus, the magnetomotive force is a scalar (as it must be taken into account that it is equal to a current) and must be evaluated along a line endowed with an outer orientation. In contrast, the scalar magnetic potential must be evaluated along a line endowed with an inner orientation; hence, it is a pseudoscalar.[53]

The same observation can be made with regard to an integration in time where we are not accustomed to distinguishing the integration on a primal or a dual time interval (Table 5.19):

---

[51] See p. 287.

[52] Jouguet [110, vol. II; p. 31].

[53] Fournet [73, p. 49].

**Table 5.19** Pseudovectors and pseudoscalars

Pseudovectors

| | | | |
|---|---|---|---|
| Angular velocity | $\check{\omega}$ | Angular momentum | $\check{\mathbf{L}}$ |
| Angular acceleration | $\check{\alpha}$ | Magnetization | $\check{\mathbf{M}}$ |
| Electric vector potential | $\check{\mathbf{F}}$ | Torque | $\check{\tau}$ |
| Magnetic flux density | $\check{\mathbf{B}}$ | Magnetic field strength | $\check{\mathbf{H}}$ |
| Vorticity | $\check{\mathbf{w}}$ | Stream vector | $\check{\psi}$ |
| Burger's vector | $\check{\mathbf{b}}$ | Magnetic moment | $\check{\mu}$ |

Pseudoscalars

| | | | |
|---|---|---|---|
| Scalar magnetic potential | $\check{\phi}_{\mathrm{m}}$ | Stream function | $\check{\phi}$ |
| Magnetic charge | $\check{G}$ | Airy's function | $\check{\phi}$ |
| Magnetic charge density | $\check{\rho}_{\mathrm{m}}$ | | |

$$\underbrace{\mathbf{u}=\int_{\mathrm{T}}\mathbf{v}\,\mathrm{d}t,}_{displacement}\qquad \underbrace{\mathbf{J}=\int_{\tilde{\mathrm{T}}}\mathbf{F}\,\mathrm{d}t}_{impulse\ of\ a\ force}\ ,\qquad \underbrace{Q=\int_{\mathrm{T}}I\,\mathrm{d}t,}_{electric\ charge}\qquad \underbrace{A_H=\int_{\tilde{\mathrm{T}}}L\,\mathrm{d}t}_{Hamiltonian\ action}\ . \tag{5.34}$$

The pseudoscalar nature of Airy's function is linked to its definition; see [212].

## 5.18 What Vector Calculus Ignores

Not all operations on vectors, although meaningful from a mathematical point of view, make sense from a physical point of view. Thus:

- It makes sense to sum the forces acting on different particles which form a body, but it does not make sense to sum their velocities.
- In a fluid flow it make sense to compute the line integral of the velocity along a line and to compute the flux of the mass current density vector through a surface. It makes no sense to compute the line integral of the mass current density vector along a line.
- It makes sense to calculate the line integral of a force along a line in a force field, but it makes no sense to calculate the flux of force through a surface.

Vector calculus does not take into account the *physical meaning* of vectors, although it would be appropriate to take account, at least, some geometrical features such as the distinction between polar vectors and pseudovectors and the distinction between line vectors and surface vectors, as introduced by Maxwell.[54]

[54] See the quotation of Maxwell on p. 12.

### 5.18.1 Energy as a Potential of Constitutive Equations

In accordance with the definition of the potential energy of a system[55] in a given configuration, we must evaluate the work of the external forces to bring the system from a reference configuration to the given configuration. This requires the use of the relation between the forces and the configuration variables, i.e. the *constitutive equations*.

Let us start by considering the simple case of an elastic spring which works both in traction and in compression. If we denote by $\mathbf{u}$ the total displacement vector from its reference configuration, the restoring force is always opposite to it, i.e. $\mathbf{F} = -k\,\mathbf{u}$. The work is given to the elastic spring by an external force $\mathbf{F}^{\text{ext}} = -\mathbf{F}$ and is $\int k\,\mathbf{u} \cdot \mathrm{d}\mathbf{u} = \frac{1}{2} k\, u^2$.

The *reversible* constitutive equations of physical theories have a common property: they are usually linear and their operator, usually a matrix, is symmetric. In the rare cases where they are non-linear, the derivative of the operator[56] is symmetric. This implies some interesting mathematical properties:

- The equations express the stationarity of a functional, usually the potential energy, but in some cases the kinetic energy, as shown in Fig. 5.12. In general, the stationary value is a minimum.
- The stationary property, combined with the property of adjointness of the operators,[57] gives rise to a reciprocity principle and to a variational principle for the fundamental equation.[58]

Since the potential and kinetic energies, as well as their space densities, arise by integration of the constitutive equations which link the source with the configuration variables, it follows that they play the role of potentials of the constitutive equations, as shown in Fig. 5.13. If we introduce the variables

$$\underbrace{V(\mathbf{u}) \stackrel{\text{def}}{=} -\int_0^{\eta} F_k(\mathbf{u})\, \mathrm{d}u^k}_{\text{elastic potential energy}}, \qquad \underbrace{T(\mathbf{p}) \stackrel{\text{def}}{=} \int_0^{p} v^k(\mathbf{p})\, \mathrm{d}p_k}_{\text{kinetic energy}}, \qquad \underbrace{T^*(\mathbf{v}) \stackrel{\text{def}}{=} \int_0^{v} p_k(\mathbf{v})\, \mathrm{d}v^k}_{\text{kinetic co-energy}},$$

$$\underbrace{u_{\mathrm{e}} \stackrel{\text{def}}{=} \int_0^{\epsilon} \sum_{h,k} \sigma_{hk}(\boldsymbol{\epsilon})\, \mathrm{d}\epsilon^{hk}}_{\text{elastic energy density}}, \qquad \underbrace{u_{\mathrm{e}} \stackrel{\text{def}}{=} \int_0^{E} D^k(\mathbf{E})\, \mathrm{d}E_k}_{\text{electric energy density}}, \qquad \underbrace{u_{\mathrm{m}} \stackrel{\text{def}}{=} \int_0^{H} B_k(\mathbf{H})\, \mathrm{d}H^k}_{\text{magnetic energy density}}, \tag{5.35}$$

then we can deduce the constitutive equations

---

55 See Chap. 9, p. 255.
56 See p. 429.
57 See Chap. 15.
58 Tonti [228].

$$\begin{aligned} F_k &= -\frac{\partial V(\mathbf{u})}{\partial u^k}, & v^k &= \frac{\partial T(\mathbf{p})}{\partial p_k}, & p_k &= \frac{\partial T^*(\mathbf{v})}{\partial v^k}, \\ \sigma_{hk} &= \frac{\partial u_\mathrm{e}(\epsilon)}{\partial \epsilon^{hk}}, & D^k &= \frac{\partial u_\mathrm{e}(\mathbf{E})}{\partial E_k}, & B_k &= \frac{\partial u_\mathrm{m}(\mathbf{H})}{\partial H^k}. \end{aligned} \tag{5.36}$$

which are displayed in Table 5.12. The *elastic energy density* is also called the *strain energy function.*[59]

## 5.19 Conjugated Variables

In physics and mathematics, it is common to speak about, for example, *conjugated* variables, *conjugated* functions and *conjugated* quantities.

In physics and engineering, the term *conjugate* refers to one of the three quantities *action*, *energy* or *power*. More precisely:

- Analytical mechanics and quantum mechanics call *canonically conjugate* those pairs of variables whose product gives an *action* (energy × time).
- Thermodynamics calls *conjugate variables* those pairs of variables whose product gives an *energy*. These conjugated variables are always linked by a *reversible* constitutive relation.
- System theory and network theory call *conjugate variables* those pairs of variables whose product gives *power* (work/time). These conjugated variables are always linked by an *irreversible* constitutive relation, as shown in Table 5.20 (p. 151).

With reference to Fig. 5.14, we observe that the product of two variables which belong to the same level has the dimensions of an energy; the product of two variables on the secondary diagonal has the dimension of power; the product of two variables on the main diagonal has the dimension of action.

It is interesting to note that the product of two *global* variables which lie on the same level has the dimensions of an action. Thus, while the scalar product of a force to a displacement has the dimensions of an energy, the scalar product of an impulse (time global variable of the force) and a displacement has the dimensions of an action. In fact, since $[\mathbf{u}] = \mathsf{L}$ and $[\mathbf{J}] = \mathsf{MLT}^{-1}$, it follows that $[\mathbf{J}\cdot\mathbf{u}] = \mathsf{ML}^2\mathsf{T}^{-1}$.

In electromagnetism, the product of the impulse of the electromotive force $[\mathcal{E}] = \mathsf{ML}^2\mathsf{T}^{-2}\mathsf{I}^{-1}$ and the electric flux $[\Psi] = \mathsf{TI}$ gives $[\mathcal{E}\Psi] = \mathsf{ML}^2\mathsf{T}^{-1}$. Hence, *while the field functions which lie on the same level of a diagram are conjugated with respect to energy, the corresponding global variables are conjugated with respect to action.*

[59] Fung [77, pp. 285, 347].

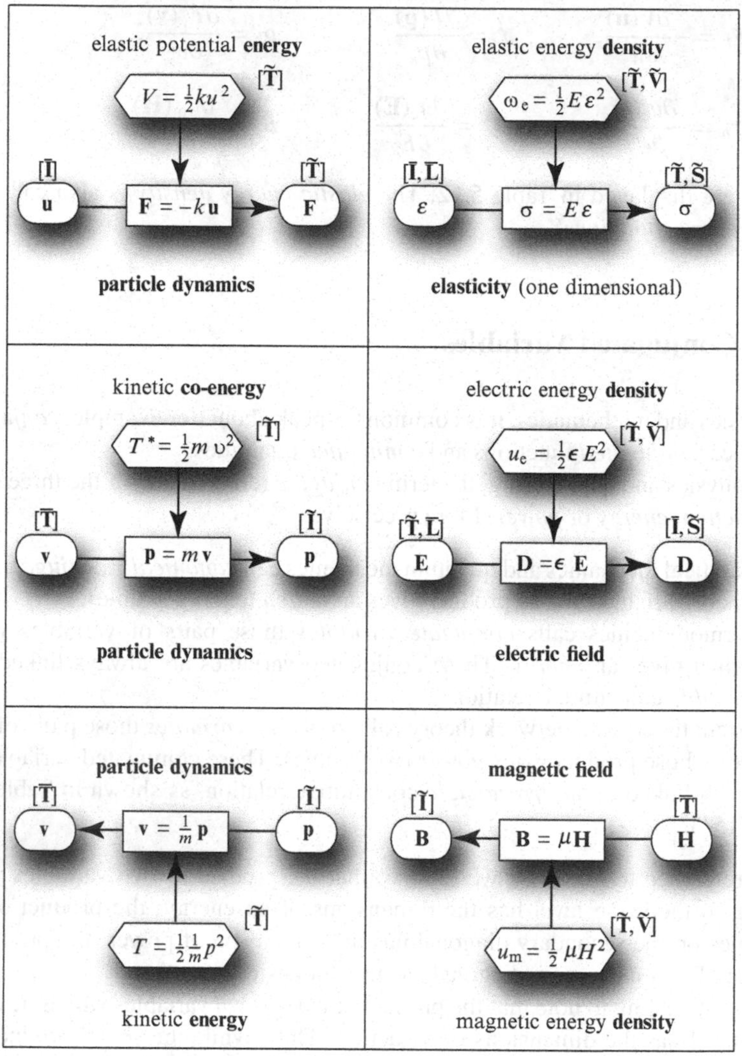


**Fig. 5.12** Energy and energy densities as potentials of reversible constitutive equations

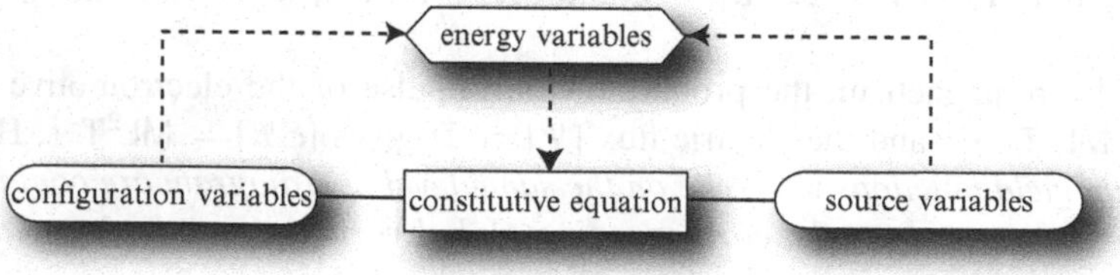


**Fig. 5.13** The elementary unit of the diagram

**Table 5.20** Conjugate variables with respect to power are linked by irreversible constitutive equations

| Class | Variable | Constitutive equation | Variable | Class | Code |
|---|---|---|---|---|---|
| $[\bar{\mathbf{T}}, \mathscr{P}]$ | $\mathbf{v}$ | $\mathbf{F} = -h\,\mathbf{v}$ | $\mathbf{F}$ | $[\tilde{\mathbf{T}}, \mathscr{P}]$ | [PAR1] |
| $[\bar{\mathbf{T}}, \mathscr{R}]$ | $\check{\omega}$ | $\check{\mathbf{T}} = -h\,\check{\omega}$ | $\check{\mathbf{T}}$ | $[\tilde{\mathbf{T}}, \mathscr{R}]$ | [ANA2] |
| $[\tilde{\mathbf{T}}, \bar{\mathbf{L}}]$ | $\mathbf{E}$ | $\mathbf{J} = \sigma\,\mathbf{E}$ | $\mathbf{J}$ | $[\bar{\mathbf{T}}, \tilde{\mathbf{S}}]$ | [ELE3] |
| $[\bar{\mathbf{T}}, \bar{\mathbf{L}}]$ | $Q$ | $\Delta p = -\mathscr{R}\,Q$ | $p$ | $[\tilde{\mathbf{T}}, \tilde{\mathbf{P}}]$ | [FLU9] |
| $[\bar{\mathbf{T}}, \bar{\mathbf{L}}]$ | $\gamma$ | $\tau = -\mu\,\gamma$ | $\tau$ | $[\tilde{\mathbf{T}}, \tilde{\mathbf{P}}]$ | [FLU9] |
| $[\tilde{\mathbf{T}}, \bar{\mathbf{L}}]$ | $\mathbf{g}$ | $\mathbf{J}_s = -\dfrac{\lambda}{T}\,\mathbf{g}$ | $\mathbf{J}_s$ | $[\bar{\mathbf{T}}, \tilde{\mathbf{S}}]$ | [TCO3] |

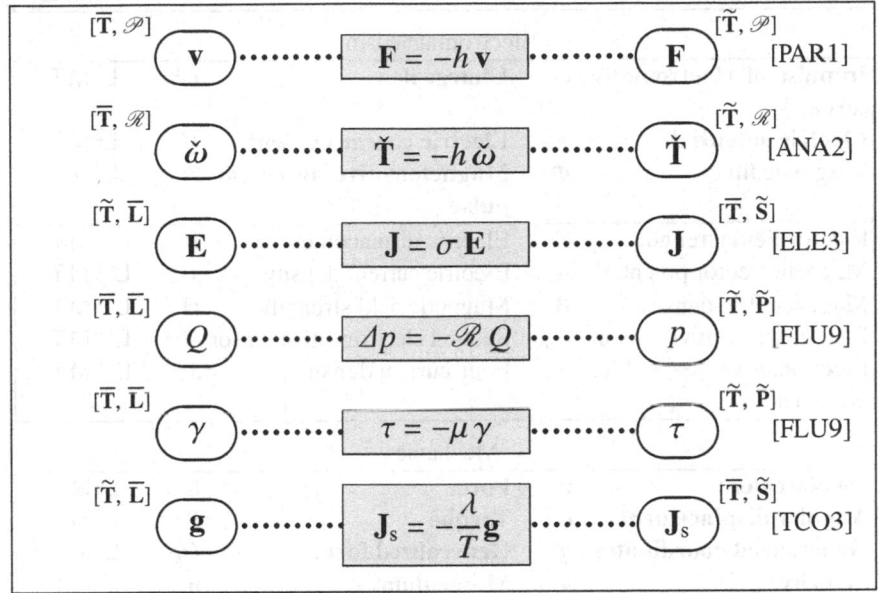

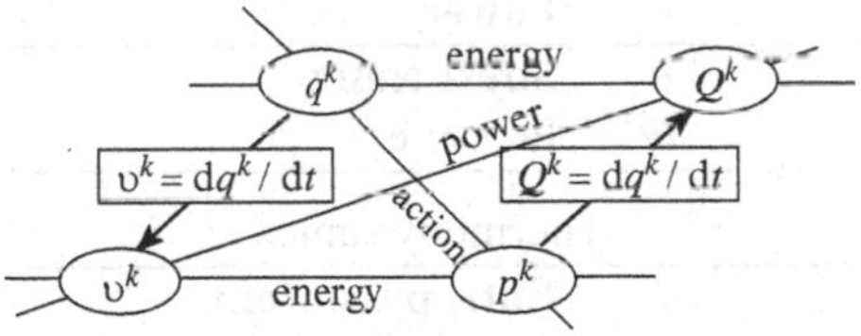


**Fig. 5.14** The three kinds of conjugation, with respect to power, energy and action

Conjugate variables have the same tensorial order and opposite variance: if one is covariant, then its conjugate is contravariant and vice versa. This assures that their product is an invariant, as an action, an energy and a power must be. Table 5.21 displays the main couples of conjugate variables of physical theories.

$$\textit{analytical mechanics} \qquad \underbrace{V = \int Q_k \,\mathrm{d}q^k}_{\textit{potential energy}}, \quad \underbrace{T = \int \upsilon^k \,\mathrm{d}p_k}_{\textit{kinetic energy}}, \quad \underbrace{P = Q_k \upsilon^k}_{\textit{power}}, \quad \underbrace{A_\mathrm{L} = \int p_k \,\mathrm{d}q^k}_{\textit{Lagrange action}} . \tag{5.37}$$

**Table 5.21** Conjugate variables with respect to energy (and energy density)

| *Configuration variable* | | *Source variable* | | Product |
|---|---|---|---|---|
| | | Electromagnetism | | |
| **Impulse of electromotive force** | $\mathcal{E}$ | **Charge flow** | $Q^{f}$ | $L^2 M T^{-2}$ |
| **Electric potential** | $\phi$ | **Electric charge content** | $Q^{c}$ | $L^2 M T^{-2}$ |
| **Magnetic flux** | $\Phi$ | **Magnetomotive force impulse** | $\mathcal{F}$ | $L^2 M T^{-2}$ |
| Electric field strength | $\mathbf{E}$ | Electric displacement | $\mathbf{D}$ | $L^{-1} M T^{-2}$ |
| Magnetic vector potential | $\mathbf{A}$ | Electric current density | $\mathbf{J}$ | $L^{-1} M T^{-2}$ |
| Magnetic flux density | $\mathbf{B}$ | Magnetic field strength | $\check{\mathbf{H}}$ | $L^{-1} M T^{-2}$ |
| First electromotive tensor | $F_{\mu\nu}$ | Second electromotive tensor | $G^{\mu\nu}$ | $L^{-1} M T^{-2}$ |
| Electromotive four-potential | $A_{\alpha}$ | Four-current density | $J^{\alpha}$ | $L^{-1} M T^{-2}$ |
| | | Mechanics | | |
| **Displacement** | $\mathbf{u}$ | **Force** | $\mathbf{F}$ | $L^2 M T^{-2}$ |
| **Angular displacement** | $\alpha$ | **Torque** | $\check{\tau}$ | $L^2 M T^{-2}$ |
| **Generalized coordinates** | $q^k$ | **Generalized forces** | $Q_k$ | $L^2 M T^{-2}$ |
| **Velocity** | $\mathbf{v}$ | **Momentum** | $\mathbf{p}$ | $L^2 M T^{-2}$ |
| **Angular velocity** | $\check{\omega}$ | **Angular momentum** | $\check{\mathbf{L}}$ | $L^2 M T^{-2}$ |
| **Time** | $t$ | **Power** | $P$ | $L^2 M T^{-2}$ |
| Strain tensor | $\epsilon$ | Stress tensor | $\sigma$ | $L^{-1} M T^{-2}$ |
| Volume dilatation | $\Theta$ | Pressure | $p$ | $L^{-1} M T^{-2}$ |
| | | Thermodynamics | | |
| **Temperature** | $T$ | **Entropy content** | $S^{c}$ | $L^2 M T^{-2}$ |
| **Temperature integral** | $\mathcal{T}$ | **Entropy production** | $S^{p}$ | $L^2 M T^{-2}$ |
| **Chemical potential** | $\mu$ | **Mole number** | $n$ | $L^2 M T^{-2}$ |
| | | Gravitation | | |
| **Gravitational potential** | $U_g$ | **Mass** | $M$ | $L^2 M T^{-2}$ |
| Acceleration of gravity | $\mathbf{g}$ | Gravitational flux density | $\mathbf{h}$ | $L^{-1} M T^{-2}$ |

Global space variables are in boldface

## 5.20 Phase, Angular Frequency and Wave Vector

The phase $\varphi$ of a wave is a relativistic invariant function.[60] For this reason it is natural to associate it with time instants and space points, i.e. $\varphi[\mathbf{I}, \mathbf{P}]$. To discriminate between primal and dual instants, let us recall the relation $\varphi = \omega t - \mathbf{k} \cdot \mathbf{r}$. The

[60] Möller [163, p. 6], Rosser [194, p. 17], Jackson [102, Sect. 11.9].

wave vector $\mathbf{k}$ has a modulus given by $k = 2\pi/\lambda$, where $\lambda$ is the wavelength and is orthogonal to the surfaces of equal phase and directed towards the increasing values of the phase. In summary, $\mathbf{k} = -\nabla\varphi = (2\pi/\lambda)\,\mathbf{n}$. Now a reversal of motion implies $\mathbf{n} \rightarrow -\mathbf{n}$: think of the water waves generated by a falling stone when they are filmed and then projected back. Hence the wave vector changes sign under a reversal of motion, and this means that it is associated with dual instants. In conclusion, $\mathbf{k}[\widetilde{\mathbf{I}}]$. Since its measurement presupposes the measure of a length (the wavelength), it is associated with lines. Hence, $\mathbf{k}[\widetilde{\mathbf{I}}, \overline{\mathbf{L}}]$. This agrees with its being a gradient of the phase. It follows that the phase, its mother variable, has the following association: $\varphi[\widetilde{\mathbf{I}}, \overline{\mathbf{P}}]$. The angular frequency $\omega = 2\pi\nu$ is the time derivative of the phase and, hence, has the association $\omega[\widetilde{\mathbf{T}}, \overline{\mathbf{P}}]$. In summary:

$$\begin{array}{ccc} \varphi = \omega t - \mathbf{k}\cdot\mathbf{r} & \omega = \partial_t\varphi & \mathbf{k} = -\nabla\varphi \\ \varphi[\widetilde{\mathbf{I}}, \overline{\mathbf{P}}], & \omega[\widetilde{\mathbf{T}}, \overline{\mathbf{P}}], & \mathbf{k}[\widetilde{\mathbf{I}}, \overline{\mathbf{L}}]\,. \end{array} \tag{5.38}$$

These associations are in harmony with the relations $\mathbf{p} = \hbar\,\mathbf{k}$ and $E = \hbar\,\omega$, which denote conservative relations because $\mathbf{p}$ inverts its sign under a reversal of motion, whereas $E$ does not.

# Chapter 6
# Analysis of Physical Equations

## 6.1 Introduction

Since physical variables are introduced to describe the quantitative attributes of a system and of a field, the equations that link them describe the behaviour of a phenomenon.

Before we start analysing the structures of equations used in physics, let us summarize some of the commonest names applied to physical equations:

| | | |
|---|---|---|
| *Constitutive equation* | *Material equation* | *Circuital equation* |
| *Phenomenological equation* | *Topological equation* | *Continuity equation* |
| *Interaction equation* | *Coupling equation* | *Defining equation* |
| *Balance equation* | *Conservation equation* | *Equation of state* |
| *Field equation* | *Canonical equation* | *Equation of constraint* |
| *Subsidiary equation* | *Auxiliary equation* | *Equation of variation* |
| *Compatibility equation* | *Wave equation* | *Equation of condition* |
| *Fundamental equation* | *Equation of motion* | *Equation of evolution* |

Since these terms developed in different periods of history and in different physical theories, it is natural that many of them are equivalent. In addition, the same equation can be interpreted as belonging to one class or another, depending on which aspect we want to highlight. We can subdivide the physical equations into five main classes:

1. **Defining equations** define a new variable in terms of variables already known. A defining equation can be *explicit*, like the one which introduces the Poynting vector in electromagnetism, $\mathbf{S} \stackrel{\text{def}}{=} \mathbf{E}\times\check{\mathbf{H}}$, or *implicit*, like the one that introduces the stream function $\psi$ in fluid dynamics, $q_x \stackrel{\text{def}}{=} \partial_y\psi;\, q_y \stackrel{\text{def}}{=} -\partial_x\psi$.

E. Tonti, *The Mathematical Structure of Classical and Relativistic Physics*, Modeling and Simulation in Science, Engineering and Technology, DOI 10.1007/978-1-4614-7422-7_6, 

2. **Topological equations** link the value of a variable which is associated with a space element, such as a line, a surface or a volume, with another variable associated with its boundary. As such they have a *global* character: when they are applied to the neighbourhood of a point, they become local.[1] They *do not depend* on the nature of the medium; hence, they do not contain material parameters. This class includes *balance equations*, *continuity equations*, *conservation equations*, *circuital equations* and some *equations of behaviour* (see below). The topological equations, when applied to regions across two different media, give rise to *jump conditions*.
3. **Equations of behaviour** or **equations of condition** specify a particular behaviour of a phenomenon or a particular class of transformations. They do not contain physical constants; they specify whole classes of transformations, motions or materials, by expressing the condition to which the physical variables must obey, so that a given condition is fulfilled. Examples include *incompressible* fluids ($\nabla \cdot \mathbf{v} = 0$), *stationary* motion ($\partial_t \mathbf{v} = 0$), *isothermal* transformations ($T = T_0$) and *irrotational* motion ($\nabla \times \mathbf{v} = 0$).
4. **Constitutive equations** or **material equations** or **equations of state** link the configuration variables with the source variables of the same theory. They are characterized by the presence of physical parameters which *depend* on the medium, with vacuum being considered a particular medium. Constitutive equations have a *local* character, i.e. they link the values of variables near every point. They are valid in a particular range of values, may be linear or non-linear, may describe a local or non-local law, may describe a hereditary link, and so forth. They have an empirical origin and are inferred from experiment and, hence, are *phenomenological* equations. A constitutive equation can be *reversible*, like Hooke's law, or *irreversible*, like Ohm's law.
5. **Interaction equations** or **coupling equations** link the configuration variables of one physical theory with the source variables of another theory. They are characterized by the presence of physical parameters which *depend* on the medium, with vacuum being considered a particular medium. Interaction equations have a *local* character, i.e. they link the values of variables near every point. An interaction equation can be *reversible*, as with piezoelectric effects, or *irreversible*, as with thermoelasticity. In addition, these equations have an empirical origin and are inferred from experiment and, hence, are phenomenological equations.

The topological equations become *metric* equations after the introduction of the densities of global variables; this is a consequence of the fact that we divide global variables for the extension of the corresponding space elements, and extension

[1] We remember that those global variables which are associated with an extended space element, such as a line, a surface or a volume, coincide with *integral* variables, i.e. are obtained by integration of the field functions. But there are also global variables associated with points, and these cannot be integral variables; see pp. 106 and 110.

**Table 6.1** Various kinds of basic equations

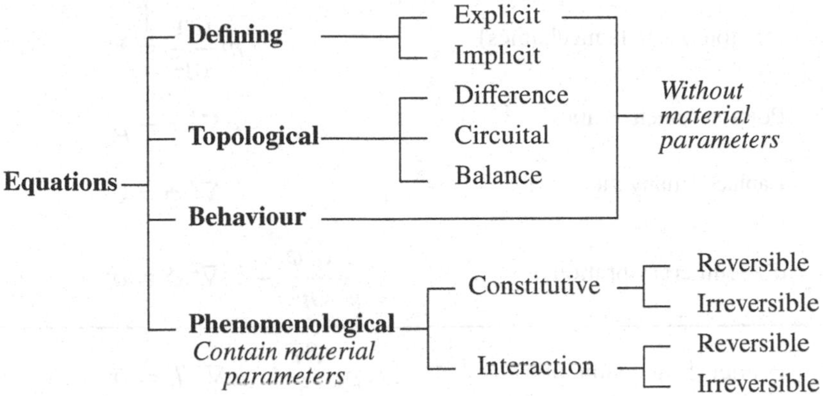


is a metric notion. The densities, however, are indispensable for the differential formulation. It is important to remark that the first three classes (defining, topological and compatibility equations) do not depend on the material media, in contrast to the two last classes (constitutive and interaction equations), which contain the physical parameters of the media. Table 6.1 shows the present classification.

When we compose equations of these four classes, we obtain a *fundamental equation*, such as *equation of motion*, *equation of evolution*, *wave equation*, this is the case of the equations of Newton (for particle mechanics); Laplace and Poisson (arising in many physical fields); Navier, Airy, Lagrange–Sophie Germain (for solid mechanics); Euler and Navier–Stokes (for inviscid and viscous fluid mechanics respectively); Fourier (for thermal conduction); Fick (for diffusion); d'Alembert (for waves); Klein–Gordon (for relativistic quantum mechanics); Einstein (for relativistic gravitation). Table 6.2 displays these fundamental equations.

This classification should be seen as an initial attempt to organize terms which are of widespread use in the literature.

To introduce the subject, let us give some examples. For each of them we will show how to decompose the fundamental equation into its primitive elements, variables and equations.

Example 1. **Damped oscillations.** The fundamental equation of one-dimensional damped oscillations is

$$m\,\ddot{x}(t) = F^{imp}(t) - h\dot{x}(t) - kx(t)\ . \tag{6.1}$$

It is composed of the following physical quantities:

**Table 6.2** Fundamental equations of the main physical theories

| | |
|---|---|
| Newton (particle mechanics) | $m\dfrac{\mathrm{d}^2\mathbf{r}}{\mathrm{d}t^2} = \mathbf{F}$ |
| Poisson (electrostatics) | $-\epsilon\,\nabla^2\,\phi = \rho$ |
| Laplace (many theories) | $\nabla^2\,\phi = 0$ |
| d'Alembert (vibrations) | $\dfrac{1}{v^2}\dfrac{\partial^2\phi}{\partial t^2} - k^2\nabla^2\,\phi = \sigma$ |
| Fourier (heat conduction) | $\rho c\,\dfrac{\partial T}{\partial t} - k\,\nabla^2\,T = \sigma_q$ |
| Fick (diffusion) | $\dfrac{\partial c}{\partial t} - D\,\nabla^2\,c = \sigma_c$ |
| Maxwell<br>Non-homogeneous wave equations | $\begin{cases} \dfrac{1}{c^2}\dfrac{\partial^2\phi}{\partial t^2} - \nabla^2\,\phi = \dfrac{1}{\epsilon_0}\rho \\ \dfrac{1}{c^2}\dfrac{\partial^2\mathbf{A}}{\partial t^2} - \nabla^2\,\mathbf{A} = \mu_0\mathbf{J} \end{cases}$ |
| Navier (elasticity) | $\rho\,\dfrac{\partial^2\mathbf{u}}{\partial t^2} - \mu\,\nabla^2\mathbf{u} - (\lambda+\mu)\,\nabla\,(\nabla\cdot\mathbf{u}) = \mathbf{f}$ |
| Navier–Stokes (fluid dynamics) | $\rho\left[\dfrac{\partial\mathbf{v}}{\partial t} + \nabla\left(\dfrac{v^2}{2}\right) + \left(\nabla\times\mathbf{v}\right)\times\mathbf{v}\right] - (\lambda+\mu)\nabla\left(\nabla\cdot\mathbf{v}\right) - \mu\,\nabla^2\mathbf{v} = \mathbf{f} - \nabla p$ |
| Schrödinger (quantum mechanics) | $i\hbar\dfrac{\partial\psi}{\partial t} + \left(\dfrac{\hbar^2}{2m}\right)\nabla^2\,\psi = eV\psi$ |
| Klein–Gordon (relativistic quantum mechanics) | $\dfrac{1}{c^2}\dfrac{\partial^2\psi}{\partial t^2} - \nabla^2\,\psi - \left(\dfrac{m_0 c}{\hbar}\right)^2\psi = 0$ |
| Einstein (relativistic gravitation) | $R_{\mu\nu} - \dfrac{1}{2}Rg_{\mu\nu} = \left(\dfrac{8\pi G}{c^4}\right)T_{\mu\nu}$ |

$$\left\{\begin{array}{lll} x(t) & \text{Elongation} & \textit{Configuration variable (geometric)} \\ \dot{x}(t) & \text{Velocity} & \textit{Configuration variable (kinematic)} \\ \ddot{x}(t) & \text{Acceleration} & \textit{Configuration variable (kinematic)} \\ m & \text{Mass} & \textit{System parameter} \\ h & \text{Damping coefficient} & \textit{System parameter} \\ k & \text{Stiffness} & \textit{System parameter} \\ F^{imp}(t) & \text{Impressed force} & \textit{Source variable} \end{array}\right. \tag{6.2}$$

It is generated by assembling the following equations:[2]

$$\left|\begin{array}{lll} v(t) \stackrel{\text{def}}{=} \dot{x}(t) & \textit{Definition of velocity} \\ a(t) \stackrel{\text{def}}{=} \ddot{x}(t) & \textit{Definition of acceleration} \\ p(t) \stackrel{\text{def}}{=} \int_{t_0}^{t} F(t')\,\mathrm{d}t' & \textit{Definition of momentum, with } v(t_0) = 0 \\ p(t) \stackrel{\text{mat}}{=} m\,v(t) & \textit{Constitutive equation} \\ F_{\text{e}}(t) \stackrel{\text{mat}}{=} -k\,x(t) & \textit{Constitutive equation} \\ F_{\text{d}}(t) \stackrel{\text{mat}}{=} -h\,\dot{x}(t) & \textit{Constitutive equation} \end{array}\right. \tag{6.3}$$

**Example 2. D'Alembert's equation** for one-dimensional longitudinal waves in a bar:

$$\frac{1}{c^2}\frac{\partial^2 u(t,x)}{\partial t^2} - \frac{\partial^2 u(t,x)}{\partial x^2} = 0\,. \tag{6.4}$$

It is composed of the following quantities:

$$\left|\begin{array}{lll} u(t,x) & \text{(Total) displacement} & \textit{Geometric variable} \\ \varepsilon(t,x) & \text{Strain} & \textit{Geometric variable} \\ v(t,x) & \text{Velocity} & \textit{Kinematic variable} \\ p(t,x) & \text{Momentum / length} & \textit{Source variable} \\ N(t,x) & \text{Traction} & \textit{Source variable} \\ A & \text{Cross-sectional area} & \textit{System parameter} \\ E & \text{Elastic modulus} & \textit{Material parameter} \\ \rho & \text{Mass density} & \textit{Material parameter} \\ c & \text{Sound velocity} & \textit{Material parameter} \end{array}\right. \tag{6.5}$$

The fundamental equation of d'Alembert is the result of the assembly of the following equations:

$$\left|\begin{array}{rl} \partial_t p(t,x) \stackrel{\text{law}}{=} \partial_x N(t,x) & \textit{Momentum balance} \\ v(t,x) \stackrel{\text{def}}{=} \partial_t u(t,x) & \textit{Definition of velocity} \\ \varepsilon(t,x) \stackrel{\text{def}}{=} \partial_x u(t,x) & \textit{Definition of strain} \\ p(t,x) \stackrel{\text{mat}}{=} A\,\rho\,v(t,x) & \textit{Constitutive equation} \\ N(t,x) \stackrel{\text{mat}}{=} A\,E\,\varepsilon(t,x) & \textit{Constitutive equation} \\ c \stackrel{\text{def}}{=} \sqrt{\frac{E}{\rho}} & \textit{Definition of sound velocity} \end{array}\right. \tag{6.6}$$

[2] In Chap. 9 we will show that the relation $\mathbf{p} = m\,\mathbf{v}$ is not the definition of momentum but a constitutive law, whereas momentum is, by definition, the indefinite time integral of force.

## 6.2 Phenomenological Equations

The term *phenomenological* means a 'description of observed phenomena, without being prejudicially affected by what is believed to be an understanding of the mechanism causing those phenomena'.[3]

Or, equivalently: 'Theories expressed in terms of the field concept are called *phenomenological*, because they represent the immediate phenomena of experience, not attempting to explain them in terms of corpuscles or other inferred quantities.'[4]

When we declare that a law is phenomenological, we make reference to the childhood of the law; later, in possession of a model of the phenomenon, the law can be interpreted and becomes, as it were, an adult. Typically, phenomenological equations are the constitutive and interaction ones. Examples are the gas law, before the kinetic model of Maxwell, and Balmer's series in atomic spectra, before the planetary model of Bohr's atom.

## 6.3 Constitutive Equations

Constitutive equations, also called *material equations*, describe the behaviour of a material, substance or medium. Examples include the equations of electric conduction (Ohm's law), diffusion (Fick's law), the elastic behaviour (Hooke's law) of a solid and the equation of the state of a gas.

Constitutive equations share some common features:

1. They link a configuration variable with a source variable; hence, they link a physical variable associated with a space element endowed with an inner orientation with another physical variable associated with a space element endowed with an outer orientation. Moreover, the corresponding space elements are dual: that corresponding to a line is a surface; that corresponding to a point is a volume. The main properties are as follows:
2. They are local relations;
3. They contain material and system parameters;
4. They contain metric notions, such as lengths, areas, volumes and orthogonality.

All constitutive equations introduce physical parameters; in this sense they are also defining equations for the *physical parameters*. Thus, for example, for an elastic bar under traction, the stress is proportional to the strain, i.e. $\sigma \propto \varepsilon$. This is the statement of a constitutive law, which is valid for a class of materials, i.e. linear elastic materials. When we change this proportionality into an equation, by writing $\sigma = E\,\varepsilon$, we *define* the elastic modulus $E$ (Table 6.3).

[3] Post [185, p. 291].

[4] Truesdell and Toupin [237, p. 227].

**Table 6.3** Constitutive equations in the case of affine behaviour of a 'potential'. $A$ is the area of a plane section and $d$ is the distance between two points **P** and **Q**

Irreversible

| | | |
|---|---|---|
| **Heat conduction** (Fourier's law) | $\Phi \overset{\text{mat}}{=} -\lambda A \dfrac{T_Q - T_P}{d}$ | $\Phi$ Heat current<br>$\lambda$ Thermal conductivity<br>$T$ Temperature |
| **Electric conduction** (Ohm's law) | $I \overset{\text{mat}}{=} -\sigma A \dfrac{\phi_Q - \phi_P}{d}$ | $I$ Electric current<br>$\sigma$ Electric conductivity<br>$\phi$ Electric potential |
| **Diffusion** (Fick's law) | $\Phi \overset{\text{mat}}{=} -D A \dfrac{c_Q - c_P}{d}$ | $\Phi$ Mass current<br>$D$ Diffusion coefficient<br>$c$ Concentration |
| **Fluids in porous media** (Darcy's law) | $\Phi \overset{\text{mat}}{=} -\dfrac{k}{\mu} A \dfrac{p_Q - p_P}{d}$ | $\Phi$ Mass current<br>$k$ Permeability<br>$\mu$ Viscosity<br>$p$ Pressure |
| **Viscosity** (Newton's law) | $T \overset{\text{mat}}{=} \mu A \dfrac{v_Q - v_P}{d}$ | $T$ Shear force<br>$\mu$ Viscosity<br>$v$ Velocity |

Reversible

| | | |
|---|---|---|
| **Electrostatics** (no name) | $\Psi \overset{\text{mat}}{=} \varepsilon A \dfrac{\phi_Q - \phi_P}{d}$ | $\Psi$ Surface charge<br>$\epsilon$ Permittivity<br>$\phi$ Electric potential |
| **Elasticity** (Hooke's law) | $N \overset{\text{mat}}{=} E A \dfrac{u_Q - u_P}{d}$ | $N$ Traction<br>$E$ Elastic modulus<br>$u$ Longitudinal displacement |

### *6.3.1 Reversible and Irreversible Constitutive Equations*

There are two kinds of constitutive equations: those expressing a reversible law and those expressing an irreversible law. The criterion for distinguishing between these two classes of constitutive equations is based on the behaviour of a reversal of motion of the related physical variables: 'Consider the equations that describe time-dependent physical processes; if these equations are invariant with regard to the algebraic sign of the time, the process is called a reversible process; otherwise it is called an irreversible process.'[5]

Hence

- *Reversible* constitutive equations link a variable associated with a time element with another associated with its dual time element, i.e. $\bar{\mathbf{I}} \leftrightarrow \tilde{\mathbf{T}}$ or $\tilde{\mathbf{I}} \leftrightarrow \bar{\mathbf{T}}$.

[5] Demirel [54, p. 6].

- *Irreversible* constitutive equations link a variable associated with a time interval with another associated with its dual time interval, i.e. $\tilde{\mathbf{T}} \leftrightarrow \bar{\mathbf{T}}$.

To graphically distinguish the irreversible link from the reversible link, we will shade the box grey and substitute the line connecting the boxes of the two variables with a sequence of small bullets, as shown in Figs. 6.3 and 6.4.

The irreversible links are those which link a variable associated with $\bar{\mathbf{T}}$ with another associated with $\tilde{\mathbf{T}}$. In fact, a reversal of motion changes the sign of those associated with $\bar{\mathbf{T}}$, whereas it maintains the sign of those associated with $\tilde{\mathbf{T}}$. Table 6.4 (p. 166) shows many constitutive relations of these two kinds. Let us start by considering two typical constitutive laws.

**Hooke's law.** Let us analyse the prototype of the constitutive equation:[6] the stretching of an elastic rod (left part of Fig. 6.1). Let $L$ be the length of the specimen, $A$ its cross section and $F$ the axial force acting on it. Hooke's law can be expressed as

$$s \stackrel{\text{def}}{=} (L' - L) \qquad \longrightarrow \qquad F \stackrel{\text{mat}}{=} k\, s\,. \tag{6.7}$$

The constant $k$ is called the *stiffness* of the rod. This relation can be described by the right part of Fig. 6.1.

Since $F[\tilde{\mathbf{T}}]$ and $s[\bar{\mathbf{I}}]$, it follows that the reversal of motion does not change the sign, either of $F$ or of $s$. Hence, the relation describes a *reversible* link, as is well known. If we examine several samples of the same material, we find that each has its own stiffness. Great progress was made when in 1705 Jacob Bernoulli discovered that the constant $k$ is proportional to the cross-sectional area $A$ of the specimen and inversely proportional to its length $L$.[7] At this point Hooke's law becomes

$$F \stackrel{\text{mat}}{=} \left(E\frac{A}{L}\right) s, \tag{6.8}$$

where $E$, called the *elastic modulus*, is a constant depending only on the material, not on the geometry of the specimen. The passage from the description of the deformation of a structure as a *whole* to a point description made possible the separation between size and shape (geometry) on the one hand and the material (physics) on the other.[8]

---

[6] This was the first constitutive equation in history; see Truesdell [237, p. 702].

[7] Benvenuto [12, Ch. 8, Sect. 1].

[8] This topic is well described in the book by Gordon and Wagley [81, Ch. 2].

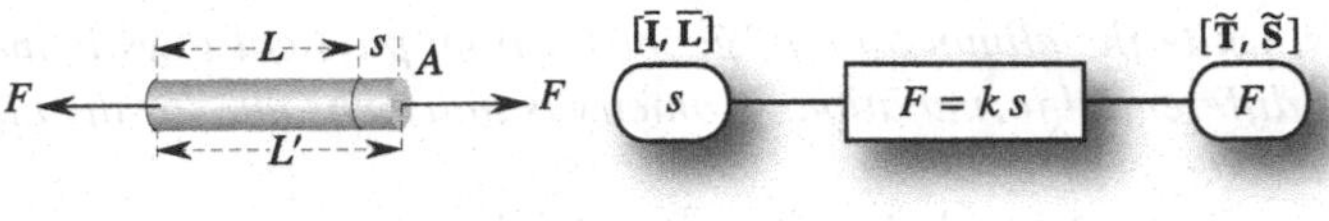


**Fig. 6.1** Hooke's law for a bar: a reversible link

The peculiar form of this relation suggests the need to introduce two quantities: the surface density of the force called *stress* and the line density of the stretching called *strain*, given respectively by

$$\sigma \stackrel{\text{def}}{=} \frac{F}{A} \qquad\qquad \varepsilon \stackrel{\text{def}}{=} \frac{s}{L}\,. \tag{6.9}$$

Notice that the idea of introducing these two quantities is suggested by the constitutive law because it assumes the very simple form

$$\sigma \stackrel{\text{mat}}{=} E\,\varepsilon\,. \tag{6.10}$$

Written in this form, Hooke's law no longer depends on the geometric parame-

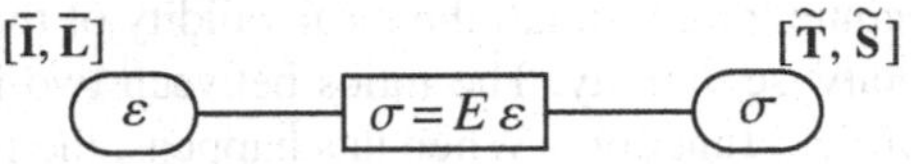


**Fig. 6.2** Hooke's law for simple traction: a reversible link

ters of the specimen but only on the nature of the material. This relation can be represented in the elementary diagram of Fig. 6.2.

| contains the geometry | does not contain the geometry | |
|---|---|---|
| $F \stackrel{\text{mat}}{=} k\,s$ | $\sigma \stackrel{\text{mat}}{=} E\,\varepsilon$ | (6.11) |
| $k$: system parameter | $E$: material constant | |

The simplicity of relation 6.10 is obtained by eliminating the geometry from the original Eq. 6.8. With this process the constitutive equation leads to the introduction of line and surface densities of the physical quantities, which are associated with lines and surfaces respectively.

In this process, it is assumed that the specimen is *homogeneous*, that it is subjected to a *uniform* strain and *uniform* stress, so that the quotient $s/L$ has the same value for every part of the specimen, i.e. if $l$ is the distance between two cross sections of the specimen, then we must have $\Delta_s l/l = s/L$ . In the same way, if $a$ is the area of a piece of the cross section and $f$ the force acting on it, then we must have $f/a = F/A$ . Hence, we can say that the proportionality of the force to the area and of the stretching to the length enables the introduction of densities and then the reduction of a physical law to a form that is independent of geometric notions.

However, *while the elimination of geometry from physical laws is indispensable for the differential formulation, geometry is indispensable for the numerical formulation!*

Since space elements, with the exception of points, have an extension, we *need* lengths, areas and volumes; usually they are obtained by integration, which is the natural inverse of the process of differentiation. When we are interested in the numerical solution of a field problem, the differential formulation can be avoided to return later to an algebraic formulation: maintaining physical laws in their algebraic formulation and using global variables is enough.

Let us clarify this fundamental point. As stated previously, constitutive equations are tested under the hypothesis of a *uniform* field and *homogeneous* material. Since fields, in general, are not uniform, we need to consider *small* regions in which the field can be considered approximately uniform. How *small* should this be? This depends on the tolerance required by the solution. Since the differential formulation requires exactness, i.e. ignores tolerance, we need to consider *infinitesimal* regions as uniform. Then to obey the differential formulation, we must introduce field functions and express constitutive laws as relations between them.

It is at this stage that scalar-valued functions, vector-valued functions and tensor-valued functions are introduced. When the differences between two points and two instants become infinitesimal, the local validity of the constitutive equations becomes a pointwise validity. The ratios between two increments become partial derivatives of field functions. When this happens, the field equations must also be expressed by field functions and the equations become differential in nature. In this way, balance equations lead to divergence, circuital equations lead to curls, spatial differences give rise to gradients and time differences give rise to time derivatives.

**Ohm's law.** Given a wire, if we apply the voltage $E$ between its two ends, a current $I$ flows in the conductor. Ohm's law states that $V = RI$, where $R$ is the *resistance* of the wire. This law can be depicted as in Fig. 6.3. Introducing the

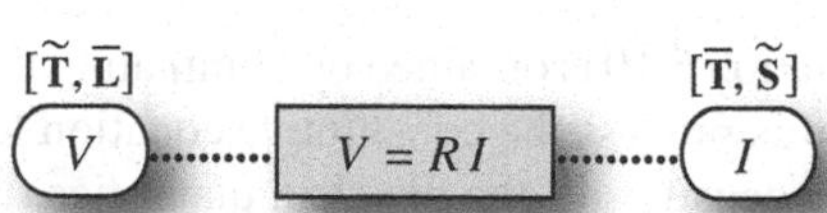


**Fig. 6.3** Ohm's law for a wire: an irreversible process

variables

$$E \stackrel{\text{def}}{=} \frac{V}{L}, \quad J \stackrel{\text{def}}{=} \frac{I}{A}, \quad \sigma \stackrel{\text{def}}{=} \frac{1}{\rho} \tag{6.12}$$

we can write Ohm's equation as follows:

$$V \stackrel{\text{mat}}{=} RI \qquad\qquad \mathbf{J} \stackrel{\text{mat}}{=} \sigma \mathbf{E} \tag{6.13}$$

$R$: system parameter contains the geometry, $\qquad$ $\sigma$: material constant does not contain the geometry,

which is described in the elemental diagram in Fig. 6.4. It is important to note that

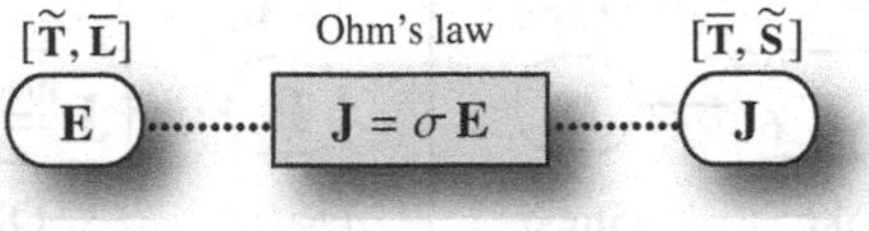


**Fig. 6.4** Ohm's law for a wire: an irreversible phenomenon

the two variables $V, I$ and $\mathbf{E}, \mathbf{J}$, connected by a constitutive relation, are both associated with time intervals, but one is associated with the primal interval and the other with the dual interval. In fact, $V[\tilde{\mathbf{T}}, \bar{\mathbf{L}}]$ and $I[\bar{\mathbf{T}}, \tilde{\mathbf{S}}]$. This implies that the reversal of motion changes the sign of $I$ but does not change the sign of $V$. It follows that the relations $V = RI$ and $\mathbf{J} = \sigma \mathbf{E}$ are not invariant under a reversal of motion, and this lack of invariance indicates that the relation describes an *irreversible* process, as is well known.

These two examples show that constitutive equations can be organized in a diagram formed by two (rounded) boxes containing two related physical variables and a (rectangular) box containing a linking equation.

### *6.3.2 Interaction Equations*

An *interaction equation*, also called a *coupling equation*, links the variables of two distinct phenomena or fields.

Interaction equations are, in general, phenomenological in nature. Some of them are reversible, and then describe conservative links, whereas others are irreversible, and then describe dissipative links. We have listed some of them in Table 6.5.

To reveal the reversible or irreversible nature of interaction equations, it is enough to consider the behaviour of the two variables linked by an equation and verify whether or not both variables change sign under a reversal of motion. Thus, in the piezoelectric effect neither variable, $\sigma[\bar{\mathbf{I}}, \tilde{\mathbf{S}}]$ or $p[\tilde{\mathbf{T}}, \tilde{\mathbf{S}}]$, changes sign under a reversal of motion; hence, the piezoelectric effect is *reversible*. In contrast, in the magneto-optic effect, the first variable, $\alpha[\bar{\mathbf{I}}, \tilde{\mathbf{L}}]$, does not change sign but the second variable, $F_{\text{m}}[\bar{\mathbf{T}}, \tilde{\mathbf{L}}]$, changes sign under a reversal of motion; hence, the magneto-optic effect is *irreversible*.

**Table 6.4** Constitutive equations

| **Reversible links** | | | **Irreversible links** | | |
|---|---|---|---|---|---|
| *Configuration variables* | | *Source variables* | *Configuration variables* | | *Source variables* |
| $[\bar{\mathbf{I}},\bar{\mathbf{L}}]$<br>$\varepsilon$<br>Strain<br>*Time even* | **Elasticity**<br>$\sigma \overset{\text{mat}}{=} E\,\varepsilon$<br>Hooke | $[\tilde{\mathbf{T}},\tilde{\mathbf{S}}]$<br>$\sigma$<br>Stress<br>*Time even* | $[\tilde{\mathbf{T}},\bar{\mathbf{L}}]$<br>$\mathbf{E}$<br>Electric field strength<br>*Time even* | **Electric conduction**<br>$\mathbf{J} \overset{\text{mat}}{=} \sigma\,\mathbf{E}$<br>Ohm | $[\bar{\mathbf{T}},\tilde{\mathbf{S}}]$<br>$\mathbf{J}$<br>Electric current density<br>*Time odd* |
| $[\tilde{\mathbf{T}},\bar{\mathbf{L}}]$<br>$\mathbf{E}$<br>Electric field strength<br>*Time even* | **Electrostatics**<br>$\mathbf{D} \overset{\text{mat}}{=} \epsilon\,\mathbf{E}$ | $[\bar{\mathbf{I}},\tilde{\mathbf{S}}]$<br>$\mathbf{D}$<br>Electric displacement<br>*Time even* | $[\tilde{\mathbf{T}},\bar{\mathbf{L}}]$<br>$\mathbf{g}$<br>Chemical potential gradient<br>*Time even* | **Diffusion**<br>$\mathbf{J}_{\mathrm{m}} \overset{\text{mat}}{=} -D\,\mathbf{g}$<br>Fick | $[\bar{\mathbf{T}},\tilde{\mathbf{S}}]$<br>$\mathbf{J}_{\mathrm{m}}$<br>Mass current density<br>*Time odd* |
| $[\tilde{\mathbf{I}},\bar{\mathbf{S}}]$<br>$\check{\mathbf{B}}$<br>Magnetic flux density<br>*Time odd* | **Magnetism**<br>$\check{\mathbf{H}} \overset{\text{mat}}{=} \frac{1}{\mu}\,\check{\mathbf{B}}$ | $[\bar{\mathbf{T}},\tilde{\mathbf{L}}]$<br>$\check{\mathbf{H}}$<br>Magnetic field strength<br>*Time odd* | $[\bar{\mathbf{T}},\bar{\mathbf{L}}]$<br>$\gamma$<br>Shear strain rate<br>*Time odd* | **Fluid dynamics**<br>$\tau \overset{\text{mat}}{=} \mu\,\gamma$<br>Newton | $[\tilde{\mathbf{T}},\tilde{\mathbf{S}}]$<br>$\tau$<br>Shear stress<br>*Time even* |
| $\bar{\mathbf{T}}$<br>$\mathbf{v}$<br>Velocity<br>*Time odd* | **Particle dynamics**<br>$\mathbf{p} \overset{\text{mat}}{=} m\,\mathbf{v}$ | $\tilde{\mathbf{I}}$<br>$\mathbf{p}$<br>Momentum<br>*Time odd* | $[\tilde{\mathbf{T}},\bar{\mathbf{L}}]$<br>$\mathbf{i}$<br>Piezometric gradient<br>*Time even* | **Percolation**<br>$\mathbf{q} \overset{\text{mat}}{=} -k\,\mathbf{i}$<br>Darcy | $[\bar{\mathbf{T}},\tilde{\mathbf{S}}]$<br>$\mathbf{q}$<br>Mass current density vector<br>*Time odd* |
| $[\tilde{\mathbf{I}},\bar{\mathbf{L}}]$<br>$\mathbf{v}$<br>Velocity<br>*Time odd* | **Fluid dynamics**<br>$\mathbf{q} \overset{\text{mat}}{=} \rho\,\mathbf{v}$ | $[\bar{\mathbf{T}},\tilde{\mathbf{S}}]$<br>$\mathbf{q}$<br>Mass current density vector<br>*Time odd* | $[\tilde{\mathbf{T}},\bar{\mathbf{L}}]$<br>$\mathbf{g}$<br>Temperature gradient<br>*Time even* | **Thermal conduction**<br>$\mathbf{q} \overset{\text{mat}}{=} -\lambda\,\mathbf{g}$<br>Fourier | $[\bar{\mathbf{T}},\tilde{\mathbf{S}}]$<br>$\mathbf{q}$<br>Heat current density<br>*Time odd* |

**Table 6.5** Some interaction equations

| *Effect* | *Equation* | *First term* | *Second term* | *Citation* |
|---|---|---|---|---|
| **Reversible** | | | | |
| Piezoelectric (*Curie*) | $\boxed{\sigma = d\,p}$ $d$ = piezoelectric modulus | $\sigma[\bar{\mathbf{I}}, \tilde{\mathbf{S}}]$ Surface charge density | $p[\tilde{\mathbf{T}}, \tilde{\mathbf{S}}]$ Pressure | Fleury-Mathieu [71, v. 6, Sect. 4.10] |
| Thermoelectric (*Seebeck*) | $\boxed{\Delta_s\phi = S\,\Delta_s T}$ Seebeck coefficient | $\phi[\tilde{\mathbf{T}}, \bar{\mathbf{P}}]$ Electric potential | $T[\tilde{\mathbf{T}}, \bar{\mathbf{P}}]$ Temperature | Perucca [176, p. 534] |
| **Irreversible** | | | | |
| Thermomechanical (*Duhamel–Neumann*) | $\boxed{\varepsilon = \alpha_l\,\Delta_t T}$ $\alpha_l$ = linear expansion coefficient | $\varepsilon[\bar{\mathbf{I}}, \bar{\mathbf{L}}]$ Strain | $(\Delta_t T)[\bar{\mathbf{T}}, \bar{\mathbf{P}}]$ Temperature | Fung [77, Sect. 14.1] |
| Photoelectric (*Einstein*) | $\boxed{E = \hbar\,\omega}$ $\hbar = h/2\pi$ $h$ = Planck constant | $E[\tilde{\mathbf{T}}, \mathscr{S}]$ Energy | $\omega[\bar{\mathbf{T}}]$ Angular frequency | de Broglie [46, p. 99] |
| Magneto-optic (*Faraday*) | $\boxed{\alpha = \rho\,F_{\mathrm{m}}}$ $\rho$ = Verdet constant | $\alpha[\bar{\mathbf{I}}, \tilde{\mathbf{L}}]$ Angle of rotation | $F_{\mathrm{m}}[\bar{\mathbf{T}}, \tilde{\mathbf{L}}]$ Magnetic voltage | Bruhat [31, p. 559] |
| Magnetoelectric (*Hall effect*) | $\boxed{J_h = \sigma_h^k(\mathbf{H})E_k}$ Conductivity tensor | $\mathbf{J}[\bar{\mathbf{T}}, \tilde{\mathbf{S}}]$ Current density | $\mathbf{E}[\tilde{\mathbf{T}}, \bar{\mathbf{L}}]$ Electric vector | Landau–Lifshitz [124, Sect. 21] |

With some hesitation we insert into the class of interaction equations the following expressions:

$$E = \hbar\,\omega, \qquad \mathbf{p} = \hbar\,\mathbf{k}, \qquad E = c^2\,m, \qquad S = k \ln P, \tag{6.14}$$

which refer to the wave–particle duality (Einstein and de Broglie), mass–energy conversion (Einstein) and the entropy-probability relation (Boltzmann) because these expressions contain physical constants. The hesitation (of the author) arises from the fact that they do not link different physical theories but different descriptions of the same physical theory.

Despite this we can see that all four equations describe a reversible relation, as can be observed comparing the behaviour of the variables of each relation under a reversal of motion.[9]

[9] See the list of physical variables on p. 485.

## 6.4 Topological Equations

We will call *topological equations* those equations which express a relationship between a variable associated with a space element and a variable associated with its boundary. Denoting by **M** a space element, if $A[\mathbf{M}]$ is a physical variable associated with it and $B[\partial\mathbf{M}]$ is another variable associated with its boundary, then a topological equation will have the form

$$B[\partial\mathbf{M}] = \pm A[\mathbf{M}] \,. \tag{6.15}$$

Figure 6.5 shows the geometric ingredients for topological equations. Topological equations share some common features:

- They are valid for any shape and extension of the space elements involved: this is why they can be called *topological equations*;
- They are valid both on a large scale as well as on a small scale, i.e. they are *global equations*;
- They are valid for any medium contained in a region, i.e. they do not contain material or system parameters; hence, they are valid even across material discontinuities. This is why they are used to find jump conditions in the interfaces between two media.

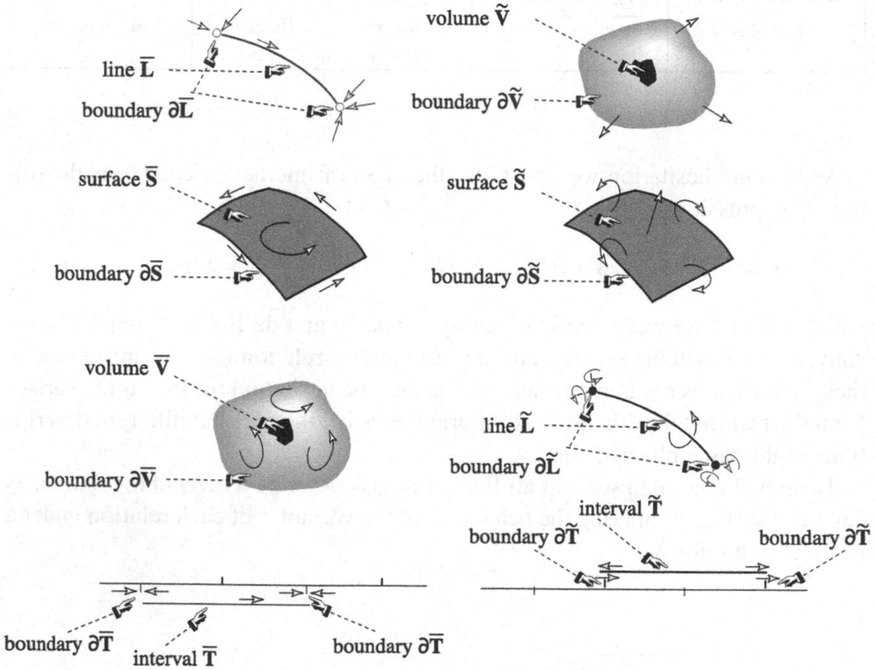


**Fig. 6.5** Space and time elements and their boundaries

Topological equations are the pillars of all physical fields. When topological equations are expressed in a differential setting, they are often mixed with metric notions; hence, they lose their purely topological nature. To show this, let us consider the process of forming the gradient of a function at a point. We must first select a direction from the point, hence consider another point in that direction, then compute the difference of the function between the two points, then compute the ratio of this difference to the distance of the two points and, lastly, compute the limit. This limit is the directional derivative. We must now search for the direction in which this limit is at a maximum and introduce a vector whose modulus is the maximum limit we have just found and whose direction is this privileged direction: this is the gradient of the function. Hence the gradient of a function starts from the difference in the function between two points, a topological notion, but later also requires metric notions, like those of direction and the distance between two points.

Balance and circuital equations start with topological notions, like a volume and its boundary, a surface and its boundary line. To reduce them to the differential formulation, we must introduce metric notions – lengths, areas, volumes, angles, scalar and vector products. Hence they become metrical equations.[10] Since topological equations are valid on both large and small scales, *they are not the cause of the recourse to the differential formulation.*

In a differential formulation, topological equations are expressed by first-order differential operators such as the gradient, curl, divergence without the intervention of physical constants. The connection between these differential operators and the topological relation Eq. 6.15 becomes clear when we switch from differential formulation to the integral formulation using the theorems of Leibniz, Stokes and Gauss. We will not make use of these theorems in our book because we take the global ($\simeq$ integral) variables as starting points and deduce the differential formulation from them instead of the converse.

The following are topological equations mixed with metric notions:

$$\begin{array}{ccc} \mathbf{v} = \nabla\phi & \mathbf{E} = -\nabla\phi & \mathbf{g} = \nabla T \\ \textit{fluid dynamics} & \textit{electrostatics} & \textit{heat conduction} \\[2ex] \check{\mathbf{B}} = \nabla\times\mathbf{A} & \check{\mathbf{w}} = \nabla\times\mathbf{v} & \nabla\times\mathbf{v} = 0 \\ \textit{electromagnetism} & \textit{fluid dynamics} & \textit{irrotational motion} \\[2ex] \nabla\cdot\mathbf{v} = 0 & \nabla\cdot\mathbf{D} = \rho & \nabla\cdot\mathbf{q} = \sigma \\ \textit{fluid dynamics} & \textit{electromagnetism} & \textit{heat conduction} \end{array} \tag{6.16}$$

[10] Topological equations can be described by algebraic topology using *cochains*, also called *discrete forms*, and the *coboundary* operator. In the differential setting, they can be described by scalar- or vector-valued exterior differential forms and by an exterior differential.

If in Eq. 6.15 we consider **M** not only a space element but also a space-time element, we can also consider as topological equations, mixed with metric notions, those which contain time derivatives, such as

$$\underset{\textit{electromagnetism}}{\mathbf{E} = -\nabla\phi - \partial_t\mathbf{A},} \qquad \underset{\textit{electromagnetism}}{\nabla\times\mathbf{E} + \partial_t\check{\mathbf{B}} = 0,} \qquad \underset{\textit{fluid dynamics}}{\partial_t\rho + \nabla\cdot\mathbf{q} = 0\,.} \tag{6.17}$$

In some cases, topological equations express a physical law (e.g. equations of continuity), and in some cases they express the behaviour of a process (e.g. irrotational motion), while in still others they define a new physical variable (e.g. temperature gradient).

It should be noted that the presence of physical constants in an equation implies that it is not a purely topological equation. Hence the following equations represent a mix of constitutive and topological equations:

$$\mathbf{q} = -\lambda\nabla T, \qquad \nabla\times\left(\frac{1}{\mu}\check{\mathbf{B}}\right) = \mathbf{J}, \qquad \nabla\cdot(\rho\mathbf{v}) = 0. \tag{6.18}$$

To analyse topological equations, we will consider in detail the various kinds of space elements involved in Eq. 6.15.

### 6.4.1 Balance Equations

Let us consider a physical variable $A[\mathbf{V}]$ associated with a volume, endowed with an inner or outer orientation, and let us consider also a physical variable $B[\partial\mathbf{V}]$ associated with the boundary of the volume. The relation

$$B[\partial\mathbf{V}] = \pm A[\mathbf{V}] \qquad \textit{topological equation} \tag{6.19}$$

In a balance equation the production of something in a region during a time interval is equal to the sum of what comes out of the region and of what is stored in the region in the same time interval. Denoting by $Q$ a physical variable associated with volumes (i.e. an extensive variable) we can write

$$Q^{produced} = \varDelta_t Q^{content} + Q^{outgoing}\,. \tag{6.20}$$

In particular, when there is no production, the balance reduces to a conservation and can be written in one of the two forms

$$\varDelta_t Q^{content} + Q^{outgoing} = 0 \qquad\qquad \varDelta_t Q^{content} = Q^{ingoing}, \tag{6.21}$$

which reduce to the form

$$\partial_t\rho + \operatorname{div}\mathbf{q} = 0 \qquad\qquad \varDelta_t U[\widetilde{\mathbf{T}},\widetilde{\mathbf{V}}] = W[\overline{\mathbf{T}},\partial\widetilde{\mathbf{S}}] + Q[\overline{\mathbf{T}},\partial\widetilde{\mathbf{S}}], \tag{6.22}$$

typical of the mass and charge conservation in the differential formulation and of the first principle of thermodynamics.

**Example 3**. In fluid dynamics, the statement that the mass stored inside a *volume* in a time interval is equal to the mass flow *entering* the *boundary* of the volume in the same time interval is a *continuity equation*, which is a topological equation. We can write

$$M^{\text{stored}}[\overline{\mathbf{T}},\widetilde{\mathbf{V}}] \stackrel{\text{def}}{=} M^{\text{c}}[\overline{\mathbf{I}}^{+},\widetilde{\mathbf{V}}] - M^{\text{c}}[\overline{\mathbf{I}}^{-},\widetilde{\mathbf{V}}] \qquad \text{and} \qquad M^{\text{out}}[\overline{\mathbf{T}},\partial\widetilde{\mathbf{V}}] \equiv -M^{\text{in}}[\overline{\mathbf{T}},\widetilde{\mathbf{V}}], \tag{6.23}$$

where $M^{\text{out}}[\overline{\mathbf{T}},\partial\widetilde{\mathbf{V}}]$ is the mass that *flowed out* through the boundary of the volume in the time interval $\overline{\mathbf{T}}$. We can write the continuity equation as

$$M^{\text{c}}[\overline{\mathbf{I}}^{+},\widetilde{\mathbf{V}}] - M^{\text{c}}[\overline{\mathbf{I}}^{-},\widetilde{\mathbf{V}}] + M^{\text{out}}[\overline{\mathbf{T}},\partial\widetilde{\mathbf{V}}] \stackrel{\text{law}}{=} 0\,. \tag{6.24}$$

To obtain the differential formulation, we must introduce field functions, and this obligates us to use metric notions, as is evident from the following equations:

$$\frac{\mathrm{d}}{\mathrm{d}t}\int_{\widetilde{\mathbf{V}}} \rho\,\mathrm{d}V + \int_{\partial\widetilde{\mathbf{V}}} \mathbf{q}\cdot\mathbf{n}\,\mathrm{d}S \stackrel{\text{law}}{=} 0 \qquad \longrightarrow \qquad \partial_t\rho + \nabla\cdot\mathbf{q} \stackrel{\text{law}}{=} 0, \tag{6.25}$$

where $\mathbf{q}$ is the mass current density vector $\mathbf{q} \stackrel{\text{def}}{=} \rho\,\mathbf{v}$.

**Example 4**. A fundamental law of electromagnetism, the law of electrostatic induction,[11] states that if we consider a metallic shell which contains an electric charge $Q^{\text{c}}$, then on the exterior part of the shell exactly a charge $Q^{\text{c}}$ is induced. This surface charge is called *electric flux* and it is denoted by the symbol $\Psi$. The surface charge so induced does not depend on the nature of the metal or on the shape of the shell. This makes it possible to associate the surface charge to any closed surface which encloses a charge $Q^{\text{c}}$ and, hence, to write

$$\Psi[\partial\mathbf{V}] \stackrel{\text{law}}{=} Q^{\text{c}}[\mathbf{V}]\,. \tag{6.26}$$

In this case, the charge $Q^{\text{c}}$ and the flux $\Psi$ can both be measured,[12] hence the equality expresses a law, not the definition of one of the two physical variables.

### 6.4.2 Circuital Equations

Let us consider a physical variable $A[\mathbf{S}]$ associated with a surface, endowed with an inner or outer orientation, and a physical variable $B[\partial\mathbf{S}]$ associated with its boundary. The relation

$$B[\partial\mathbf{S}] = \pm A[\mathbf{S}] \qquad \textit{topological equation} \tag{6.27}$$

is a topological equation.

---

[11] Due to Faraday [65], but commonly called *Gauss's law*, see Schelkunoff [201, p. 23].

[12] See Chap. 10 for the experimental apparatus.

**Example 5.** We may start with a physical variable $B$ associated with the boundary of a surface and use Eq. 6.27 to *define* the variable $A$ associated with the surface. This happens when we start from the velocity line integral $\Gamma$ along the boundary of a surface to define the vortex flux $W$ on that surface. Hence,

$$\textit{putting} \qquad \Gamma[\mathbf{L}] \stackrel{\text{def}}{=} \int_{\mathbf{L}} \mathbf{v}\cdot\mathbf{t}\,\mathrm{d}L \quad \longrightarrow \quad W[\mathbf{S}] \stackrel{\text{def}}{=} \Gamma[\partial\mathbf{S}]\,. \tag{6.28}$$

This relation is the origin of the vorticity vector $\mathbf{w}$. The corresponding differential formulation will be $\check{\mathbf{w}} = \nabla \times \mathbf{v}$.

**Example 6.** In contrast to the preceding example, sometimes we start with a physical variable $A$ associated with the surface to define a variable $B$ associated with the boundary, and we use Eq. 6.27 to *define* the variable. This is the case of the electric current $I$ through a surface $\mathbf{S}$ by which we define a variable $F$ associated with its boundary. Hence,

$$I[\mathbf{S}] \stackrel{\text{def}}{=} \int_{\mathbf{S}} \mathbf{J}\cdot\mathbf{n}\,\mathrm{d}S \quad \longrightarrow \quad F[\partial\mathbf{S}] \stackrel{\text{def}}{=} I[\mathbf{S}]\,. \tag{6.29}$$

The variable $F$ so defined is called a *magnetomotive force*. Traditionally, we first define the vector $\check{\mathbf{H}}$ and then the variable $F$ as the integral of $\check{\mathbf{H}}$ along a line. Operating in this fashion, Eq. 6.29 becomes a *law*, Ampère's law, which is a topological law.[13]

For the differential formulation we are obligated to introduce field functions, and this implies the use of metric notions, as is evident from the following equations:

$$\int_{\partial\tilde{\mathbf{S}}} \check{\mathbf{H}}\cdot\check{\mathbf{t}}\,\mathrm{d}L \stackrel{\text{law}}{=} \int_{\tilde{\mathbf{S}}} \mathbf{J}\cdot\mathbf{n}\,\mathrm{d}S \quad \longrightarrow \quad \nabla\times\check{\mathbf{H}} \stackrel{\text{law}}{=} \mathbf{J}\,. \tag{6.30}$$

**Example 7.** In the electrostatic field, the integral of the *electric field strength* $\mathbf{E}$ along a line is called the *voltage*. The electrostatic field has the property that its integral along *any* closed line vanishes, i.e.

$$V \stackrel{\text{def}}{=} \int_{\mathbf{L}} \mathbf{E}\cdot\mathbf{t}\,\mathrm{d}L \quad \longrightarrow \quad V[\partial\mathbf{S}] \stackrel{\text{law}}{=} 0\,. \tag{6.31}$$

The last statement is a law. When the integral of a vector field along all *reducible* closed lines vanishes, we say that the field is *irrotational*. This property, the vanishing of the integral along all closed lines, reducible or not, implies that the electrostatic field is not only irrotational but also *conservative*.

**Example 8.** In the motion of a fluid, we use a vector $\mathbf{q}$ called a *mass current density vector* (kg/s/m$^2$). When the fluid is *incompressible*, like water, or the motion is *isochoric*, like air in an air-conditioning system, i.e. $\partial_t\rho = 0$, the mass current $\Phi$ across a closed surface vanishes:

$$\textit{putting} \qquad \Phi[\mathbf{S}] = \int_{\mathbf{S}} \mathbf{q}\cdot\mathbf{n}\,\mathrm{d}S \quad \longrightarrow \quad \Phi[\partial\mathbf{S}] = 0\,. \tag{6.32}$$

This equation is not a law, but the statement of a particular behaviour of the fluid motion. The vector field $\mathbf{q}(\mathbf{P})$ is said to be *solenoidal*, i.e. in the differential formulation $\nabla\cdot\mathbf{q} = 0$.

---

[13] Pauli [174, pp. 20–21].

**Example 9.** Thermal conduction makes use of a vector **q** describing the *heat current density* (W/m$^2$). In a steady thermal field, if through the boundary of the volume **V** there is an outgoing heat current $\Phi$, this means that the heat-generation rate $P$ (in watts) inside the volume is

$$P[\tilde{\mathbf{V}}] \stackrel{\text{def}}{=} \Phi[\partial\tilde{\mathbf{V}}] \qquad \longrightarrow \qquad \Phi = \int_{\tilde{\mathbf{S}}} \mathbf{q}\cdot\mathbf{n}\,\mathrm{d}S\ . \tag{6.33}$$

This relation *defines* the heat-generation rate $P$.

### 6.4.3 Space Differences

Let us consider a line **L**, endowed with an inner or outer orientation, and its boundary formed by the points **P** and **Q**: we will denote the set of the two points by $\partial$**L**. Let us consider a physical variable $A$[**L**] associated with the line and a physical variable $B$[**P**] associated with points.

Introducing the notation $B[\partial\,\mathbf{L}] \stackrel{\text{def}}{=} B(\mathbf{Q})$ - $B(\mathbf{P})$ we can consider the relation

$$B[\partial\,\mathbf{L}] = A[\mathbf{L}] \qquad \textit{topological equation}\ . \tag{6.34}$$

In some cases, this relation is used to define the variable $A$ when the variable $B$ is known; in other cases, it serves to define the variable $B$ when the variable $A$ is known.

**Example 10.** In electrostatics, the statement that the voltage $E$ along a *line* is the difference between the electric potentials of its end points is a topological relation. This equation defines the electric potential $\phi$:

$$V \stackrel{\text{def}}{=} \phi(\overline{\mathbf{P}}) - \phi(\overline{\mathbf{Q}}) \qquad \longrightarrow \qquad V[\overline{\mathbf{L}}]\ . \tag{6.35}$$

To introduce field functions, we are obligated to use metric notions, as is evident from the following forms:

$$\int_{\overline{\mathbf{L}}} \mathbf{E}\cdot\mathbf{t}\ \mathrm{d}L \stackrel{\text{def}}{=} \phi^- - \phi^+ \qquad \longrightarrow \qquad \mathbf{E} = -\nabla\phi\ . \tag{6.36}$$

**Example 11.** In the thermal field, a temperature difference $G$ along a line connecting two points $\overline{\mathbf{P}}$ and $\overline{\mathbf{Q}}$ is defined as the difference between the temperature at the second point and that at the first point:

$$G \stackrel{\text{def}}{=} T(\overline{\mathbf{Q}}) - T(\overline{\mathbf{P}}) \qquad \longrightarrow \qquad G[\overline{\mathbf{L}}]\ . \tag{6.37}$$

This leads to the introduction of a vector **g** such that its integral along the line **L** is equal to the temperature difference $G$; hence,

$$G = \int_{\mathbf{L}} \mathbf{g}\cdot\mathbf{t}\,\mathrm{d}L \qquad \longrightarrow \qquad \mathbf{g} = \nabla T\ . \tag{6.38}$$

The vector **g** thus created is the temperature gradient.

**Example 12.** In the motion of a fluid, it may happen that the velocity integral along all *reducible* closed lines vanishes; in this case, the velocity field is said to be *irrotational*. By choosing a fixed point as origin, any other point in the field can be associated with the line integral from the origin to the point since this does not depend on the line joining the two points.

In this way, the line integral from the origin to the point is a scalar called the *kinetic potential* and is denoted by $\phi$. After doing this, the velocity integral associated with a line joining two points $\mathbf{P}$ and $\mathbf{Q}$ can be written as the difference in the kinetic potential of two points, namely

$$\Gamma = \phi(\mathbf{Q}) - \phi(\mathbf{P}) \qquad \longrightarrow \qquad \Gamma[\mathbf{L}] = \phi[\partial\mathbf{L}] \,. \tag{6.39}$$

In this case, the equation defines the potential from the integral associated with the line.

## 6.5 Invariance of Equations Under Inversion of Orientation

It is natural to request that *all physical equations maintain their invariance in form when we invert the orientation of the* space *element involved*. For what concerns the invariance under inversion of orientation of time elements, we must distinguish topological equations from phenomenological equations.[14] Some phenomenological equations describe a reversible process, whereas other equations describe an irreversible process:[15] *the equations that describe a reversible process are those which are invariant in form when we change the orientation of the time element involved.*

**Invariance of Topological Equations.** In some cases, this is evident, in other cases no. Let us consider the case of *circuital equations* (which are topological equations), where the invariance is evident. Ampère's circuital law states that the magnetomotive force $F_{\mathrm{m}}$ along the boundary of a surface is equal to the intensity of current $I$ through the surface. By writing the law in global form, i.e. $F_{\mathrm{m}}[\partial\tilde{\mathbf{S}}] = I[\tilde{\mathbf{S}}]$, we see that by changing the direction of the normal to the surface, i.e. $\tilde{\mathbf{S}} \to -\tilde{\mathbf{S}}$ and consequently $\partial\tilde{\mathbf{S}} \to -\partial\tilde{\mathbf{S}}$, the form invariance is ensured by the fact that the variables change their sign if we change the outer orientation, i.e. $I[-\tilde{\mathbf{S}}] = -I[\tilde{\mathbf{S}}]$ and $F_{\mathrm{m}}[-\partial\tilde{\mathbf{S}}] = -F_{\mathrm{m}}[\partial\tilde{\mathbf{S}}]$. In this way, the equation $F_{\mathrm{m}}[\partial\tilde{\mathbf{S}}] = I[\tilde{\mathbf{S}}]$ is transformed into the equation $-F_{\mathrm{m}}[\partial\tilde{\mathbf{S}}] = -I[\tilde{\mathbf{S}}]$, which respects the form invariance.

The property of invariance in form is *not* evident for those physical variables which are associated with volumes and points. This is due to the fact that *we are not accustomed to considering volumes and points endowed with a sign*. In fact, while elementary geometry considers lengths, areas and volumes as always positive variables, analytical geometry, by introducing the concept of orientation of a space element, gives a sign to these geometric quantities. This is an opportunity

[14] See the next chapter.

[15] See page 161.

which is little used in geometry and, as a consequence, is rarely used in physics. Since we have shown that the global physical variables are associated with space elements, we can take the opportunity to give an orientation to space elements by endowing their extensions wih a sign, i.e. speaking about negative lengths, areas and volumes. It seems natural to extend this possibility to the global variables. So when we accept that the volume of a region endowed with an outer orientation can be either positive or negative, it seems natural to consider the mass contained in the volume as having a sign as well. We mean that the sign of the mass can be adapted to the sign of the volume whenever such a property can be useful for the theory. This is not an obligation but a possibility offered to us.

The advantage of the systematic introduction of the sign of physical variables in relation to the orientation of the associated space element is that it lends itself beautifully to highlighting the invariance of physical equations with respect to the change in orientation.

The change in sign may be justified by two complementary points of view. For some variables the change in sign is simply a consequence of the definition of the variable, and it has the result of maintaining the form invariance of the equations expressing the laws of physics in which the variable is involved. For other variables the sign reversal is not obvious, and it *must be imposed* in order to maintain the invariance in the form of the equations expressing the laws of physics in which the variable is involved.

Let us give two examples of *balance equations* (which are topological equations): the balance of mass and the balance of energy.

**Example 13.** Let us consider the mass balance inside a volume and a time interval. We can write

$$M^{c}(t^{+}) - M^{c}(t^{-}) + M^{f}(t^{-}, t^{+}) = 0, \tag{6.40}$$

where we follow the convention that the mass flow $M^{f}$ is positive when outgoing.

We now ask: what is the behaviour under a reversal of motion of the physical variables $M^{c}$ and $M^{f}$ involved in the balance? Let us denote by $\overline{M}^{c}$ and $\overline{M}^{f}$ the two variables after the reversal of motion. The sign of the mass content at each instant is the same, but the order of the two time instants is inverted; the sign of the mass flow is inverted because the mass is now ingoing. It follows that

$$\overline{M}^{c}(t^{+}) \to M^{c}(t^{-}) \qquad \overline{M}^{c}(t^{-}) \to M^{c}(t^{+}) \qquad \overline{M}^{f}(t^{+}, t^{-}) \to -M^{f}(t^{-}, t^{+})\,. \tag{6.41}$$

It follows that, by inverting the order of the two time instants, Eq. 6.40 becomes

$$M^{c}(t^{-}) - M^{c}(t^{+}) - M^{f}(t^{-}, t^{+}) = 0. \tag{6.42}$$

This equation coincides with Eq. 6.40 multiplied by $-1$; hence, the relation Eq. 6.40 is invariant in form under a reversal of motion.

We now ask: what is the behaviour of the physical variables $M^{c}$ and $M^{f}$ involved in the balance if we change the outer orientation of volumes? We follow two conventions: the mass is positive where it belongs to a region and the mass flow is positive when it leaves that region. Let us denote by $\overline{M}^{c}$ and $\overline{M}^{f}$ the two variables after inversion of the outer orientation.

To change the outer space orientation of a volume means changing the point of view referring the physical variables to the complementary of the volume instead of to the volume. Since the mass content $M^c$ within a region is outside the complementary region, it is considered negative for the complement. Moreover, the mass flow out of the region (i.e. positive) enters the complementary region and is therefore considered negative with respect to the latter. In conclusion,

$$\overline{M}^c \rightarrow -M^c \qquad \textit{and} \qquad \overline{M}^f \rightarrow -M^f . \tag{6.43}$$

Equation 6.40 becomes

$$[-M^c(t^+)] - [-M^c(t^-)] + [-M^f(t^-, t^+)] = 0, \tag{6.44}$$

which, again, is equal to Eq. 6.40 multiplied by $-1$. It follows that the inversion of the sign of the mass content leads to the form invariance of the balance Eq. 6.40.

**Example 14.** In thermodynamics the first principle, expressing the energy balance, can be written in the form

$$U(t^+) - U(t^-) = Q(t^-, t^+) + W(t^-, t^+). \tag{6.45}$$

In this case the heat and the work are considered positive when entering the system. If we take the outside of the system as our reference, the heat and the work are outgoing and hence become negative: $\overline{Q} = -Q$ and $\overline{W} = -W$. From the viewpoint of the exterior of the system, the internal energy of the system is considered negative, hence $\overline{U} = -U$. The result is that the first principle of thermodynamics, a balance law, is invariant under inversion of the outer orientation of volumes, like mass balance and the refugees in an embassy.

## 6.6 Defining Equations

Once we have defined some physical variable directly by its measuring process, e.g. force and temperature, we introduce new variables by means of *defining equations*. New variables can be introduced by an explicit or implicit definition, as we will see in the next section. The typical operations are as follows:

- The operation of computing the *ratio* of two variables. This operation is largely used to form densities and rates. This is the case of mass density (mass/volume), pressure (normal force/area), strain (increase in length/length), velocity (displacement/duration) and power (work/duration).
- The operation of computing the *product* of two variables, such as work which is force times displacement; the moment of a force, which is a force times a distance; the moment of inertia of a particle with respect to an axis, which is mass times the square of the distance from the axis; or the electric dipole moment, which is charge times distance.

- The operation of computing the *sum* or the *difference* of two variables. This is the case of the Hamiltonian, which is the sum of the kinetic and potential energies, and the Lagrangian, which is the difference between the kinetic co-energy and potential energy. A typical case are equations which define the space and time increments of a variable. This is the case with displacement, which is the difference between two radius vectors; voltage, which is the difference between the electric potential at two space points; and temperature increase, which is the difference between temperatures at two time instants.
- The operation of computing the *derivative* or the *integral* of a variable.

Some topological equations can be used as equations of definition. This is the case with the first principle of thermodynamics, which can be used to define heat;[16] it is also the case with the process of defining a space difference which is a topological equation.

## 6.7 Equations of Behaviour

This class contains the equations that prescribe a particular behaviour of a process, of a transformation, of a material, of a flow, etc.

- The transformations of a gas: *adiabatic, isothermal, isobaric, isochoric*;
- The equations which prescribe the behaviour of a body or of a material: *rigid body, incompressible material, perfect fluid*;
- The equations which prescribe the behaviour of scalar or a vector field: *uniform (in space), static, stationary, irrotational, solenoidal, constant (in time);*
- The equations which prescribe a particular fluid flow: *isochoric, irrotational, stationary, barotropic.*

Some authors[17] consider equations which characterize a perfect fluid and a rigid body, for example, as constitutive equations. The notion of rigid is an ideal condition, which is only approximated by a body; it seems more appropriate to say that the body *behaves* like a rigid body. For this reason we prefer to consider the latter equations as equations of behaviour. Moreover these equations are not specific to a material, i.e. they do not contain material parameters.

**Remark.** Although obvious, it is better to stress that in science all physical variables must have a definition. More specifically, a physical variables must have an *operative definition* if it is measurable or a definition which comes from a *defining equation*. We observe that among

[16] Born [19]; Guggenheim [86, p. 41].

[17] Truesdell and Toupin [237, p. 233].

**Table 6.6** Some defining equations

**Ratios**

| | | | |
|---|---|---|---|
| Pressure | $p \overset{\text{def}}{=} \dfrac{F_n}{A}$ | Elastic modulus | $E \overset{\text{def}}{=} \dfrac{\sigma}{\varepsilon}$ |
| Resistance | $R \overset{\text{def}}{=} \dfrac{V}{I}$ | Electric current | $I \overset{\text{def}}{=} \dfrac{Q^f}{T}$ |
| Mass density | $\rho \overset{\text{def}}{=} \dfrac{m}{V}$ | Frequency | $\nu \overset{\text{def}}{=} \dfrac{1}{T}$ |
| Kinematic viscosity | $\nu \overset{\text{def}}{=} \dfrac{\mu}{\rho}$ | Wave number | $k \overset{\text{def}}{=} \dfrac{2\pi}{\lambda}$ |
| Reynolds number | $Re \overset{\text{def}}{=} \dfrac{\rho u_\infty l}{\mu}$ | Boltzmann's constant | $k_B \overset{\text{def}}{=} \dfrac{R}{N_A}$ |
| Surface tension | $\sigma \overset{\text{def}}{=} \dfrac{E}{A}$ | Gravitational mass | $m_g \overset{\text{def}}{=} \dfrac{w}{g}$ |

**Products**

| | | | |
|---|---|---|---|
| Angular momentum | $\check{\mathbf{L}}_A \overset{\text{def}}{=} \mathbf{r} \times \mathbf{p}$ | Moment of a force | $\check{\mathbf{M}}_A \overset{\text{def}}{=} \mathbf{r} \times \mathbf{F}$ |
| Poynting vector | $\mathbf{S} \overset{\text{def}}{=} \mathbf{E} \times \check{\mathbf{H}}$ | Moment of inertia | $I_a \overset{\text{def}}{=} mr^2$ |

**Algebraic sums**

| | | | |
|---|---|---|---|
| Lagrangian | $L \overset{\text{def}}{=} K^* - V$ | Hamiltonian | $H \overset{\text{def}}{=} K + V$ |
| Heat | $Q \overset{\text{def}}{=} \Delta_t U - W$ | Enthalpy | $H \overset{\text{def}}{=} U + pV$ |
| Free energy | $F \overset{\text{def}}{=} U - TS$ | Free enthalpy | $G \overset{\text{def}}{=} H - TS$ |

**Derivatives**

| | | | |
|---|---|---|---|
| Velocity | $\mathbf{v} \overset{\text{def}}{=} \dfrac{d\mathbf{r}}{dt}$ | Acceleration | $\mathbf{a} \overset{\text{def}}{=} \dfrac{d\mathbf{v}}{dt}$ |
| Chemical potential | $\mu \overset{\text{def}}{=} \dfrac{\partial G}{\partial n}$ | Angular velocity | $\omega \overset{\text{def}}{=} \dfrac{d\theta}{dt}$ |

**Integrals**

| | | | |
|---|---|---|---|
| Kinetic energy | $K(\mathbf{p}) \overset{\text{def}}{=} \int_0^{\mathbf{p}} \mathbf{v}(\mathbf{p}) \cdot d\mathbf{p}$ | Work | $W[L] \overset{\text{def}}{=} \int_L \mathbf{F}(\mathbf{r}) \cdot d\mathbf{r}$ |
| Gravitational potential | $U_g(\mathbf{r}) \overset{\text{def}}{=} \int_L \mathbf{g}(\mathbf{r}) \cdot d\mathbf{r}$ | Velocity line integral | $\Gamma[L] \overset{\text{def}}{=} \int_L \mathbf{v}(\mathbf{r}) \cdot d\mathbf{r}$ |
| Electric potential | $\phi(\mathbf{r}) \overset{\text{def}}{=} \int_r^{\infty} \mathbf{E}(\mathbf{r}) \cdot d\mathbf{r}$ | Magnetic flux | $\Phi[\mathbf{S}] \overset{\text{def}}{=} \int_S \check{\mathbf{B}}(\mathbf{r}) \cdot d\check{\mathbf{S}}$ |
| Momentum | $\mathbf{p}(t) \overset{\text{def}}{=} \int_0^t \mathbf{F}(t')\, dt'$ | Impulse | $\mathbf{J}[T] \overset{\text{def}}{=} \int_0^T \mathbf{F}(t)\, dt$ |

**Table 6.7** Examples of equations of behaviour

| | | | |
|---|---|---|---|
| *Adiabatic* | $S = constant$ | *Barotropic* | $p = p(\rho)$ |
| *Isobaric* | $p = constant$ | *Stationary* | $\partial_t \phi = 0$ |
| *Incompressible* | $\rho = constant$ | *Inviscid* | $\mu = 0$ |
| *Isothermal* | $T = constant$ | *Uniform* | $\nabla \phi = 0$ |
| *Isochoric* | $V = constant$ | *Irrotational* | $\nabla \times \mathbf{v} = 0$ |
| *Rigid* | $\vert \mathbf{r}_h - \mathbf{r}_k \vert = constant$ | *Solenoidal* | $\nabla \cdot \mathbf{v} = 0$ |

physical variables only a few are directly measurable; most of them arise from a definition in terms of other variables, as shown in Table 6.6. Many physicists state that in physics one must use only measurable physical quantities. This is a bias, as can be seen in the opinions of eminent physicists quoted in Appendix C, on p. 477 (Table 6.7).

# Chapter 7
# Algebraic Topology

## 7.1 Why Algebraic Topology?

Algebraic topology is closer to the discrete nature of physical variables, is simpler than differential calculus, has an unexpected power to unify the mathematical description of physics and leads directly to computational physics.

Since its foundation, which occurred more than three centuries ago, differential calculus has formed the basis of the mathematical description of physics. In recent decades a new mathematical tool has become very popular, the theory of exterior differential forms, which is an enrichment of infinitesimal calculus.

**Exterior Differential Forms.** This mathematical formalism has the great advantage of taking into account the geometric aspect which calculus ignores. The basic idea is to associate appropriate *exterior differential forms* to the field functions, thereby performing on them the operation of the *exterior differential*. This operation, which, starting from a form of $p$ degrees, gives rise to a form of $(p + 1)$ degrees, expresses in a wonderfully unitary way the traditional operators gradient, curl and divergence. The theory of exterior differential forms has the merit of highlighting the geometric background of physical variables and providing a description which is independent of the coordinate system used. Nevertheless, this formalism makes use of the field functions and not of global variables, and for this reason it must use the notion of derivative. This formalism is widely used in electromagnetism.

**Geometric Content of Physical Variables.** Our aim is to show that it is possible to obtain a geometric description of physical theories which stems directly from the operative definition of the variables, in particular from their measurement procedure, at least for those quantities which are measurable.

We are so accustomed to using the differential formulation that we believe that the description of physics by means of field functions and by coordinate systems is the most convenient description for physics. However, the nature of physical

E. Tonti, *The Mathematical Structure of Classical and Relativistic Physics*, Modeling and Simulation in Science, Engineering and Technology, DOI 10.1007/978-1-4614-7422-7_7, 

quantities reveals that we mostly create and measure physical quantities referred to space elements which have an extension such as the flow through a surface, the mass contained in a volume, the energy of a system, the electromotive force along a line, the work of a force along a line, the surface charge, and so forth, i.e. physical variables which are not simply functions of the point. To arrive at a differential description, we need to introduce the densities of these physical quantities. We are accustomed to using the functions of the field and then to deduce the corresponding global quantities, by integrating them on domains with one, two and three dimensions. This has led to the introduction of the term *integral variables*. But in physics we often measure integral variables more than field functions!

**Why Do We Make Use of Infinitesimal Volumes and Infinitesimal Surfaces?** The most important equation in physics is the balance equation because all the fundamental equations of physical theories have a balance equation as the starting point. We compute the balance of mass, energy, electric charge, momentum, angular momentum, entropy and number of particles. A balance equation links a physical variable associated with a volume with a variable associated with its boundary. For example, in electromagnetism, there are two circuital laws and two balance laws, to which is added the balance of the electric charge (charge conservation). The conservation of charge is a physical law independent of the Maxwell equations. The sources of the electromagnetic field, i.e. charges and currents, must satisfy the conservation of charge in advance. The result is that the charge conservation becomes a compatibility condition for the equations of Mawell.

In fluid dynamics, the fundamental equations of Navier–Stokes are formed by a momentum balance and a mass balance, and so on.

Why do we apply the circuital and balance laws to *infinitesimal* volumes and *infinitesimal* circuits by introducing the densities, thereby obtaining differential equations which contain the gradient, curl and divergence? Why do we introduce densities and rates of these global physical variables? The answer is simple: we are accustomed, from three centuries, to using the differential formulation. Differential formalism is deeply rooted in our culture, and the only alternative to it seems to be that of an integral formulation of physical laws. But the same integral formulation is obtained from the differential one, and it is difficult to describe the physics using integral relations.

The purpose of this book is to show that it is possible to avoid both the differential and the integral formulation using a formalism which is close to experimental measurements and is local in nature, such as the differential formalism, without carrying out the limit process. Moreover, it has the great advantage that it can be used directly for a numerical formulation.

**Algebraic Topology Instead of Differential Calculus and Differential Forms.** The question therefore arises: *If we do not want to use the differential formulation, is there a branch of mathematics which develops notions corresponding to those of the differential formulation but is based on global variables instead of field functions?* The answer is yes; that branch is *algebraic topology*.

Algebraic topology divides the region of space within which we operate into many cells, thereby building a *cell complex* on which are introduced very simple notions like *chains* and *cochains*. These terms sound unusual to the ears of physicists and engineers, but they describe notions with which we are already familiar. These notions allow us to express physical laws directly in global form, i.e. without passing through a field's functions, bypassing the differential forulation and, in particular, without going through the exterior differential forms.

While the differential formulation requires that field functions satisfy differentiability conditions which are foreign to physics, the corresponding notion of discrete form does not need this restriction and as such is closer to the nature of global variables.

**Physics Is Based on Measurements, and These Imply Approximation.** To avoid the differential formulation as our starting point, we need to completely revise our attitude. In our culture, formed by three centuries of differential formulation of physical laws, we find the differential formulation so familiar that we are led to think that it is natural for physics. However, we know very well that only in a few elementary cases, with space regions of simple geometric shapes and under particular boundary conditions, can one obtain a solution in closed form: hence the "exact solution" promised by differential formulations is almost never achieved in practice. Moreover, the great scientific and technological advancements achieved in our day by numerical solutions to physical problems, which do not admit a solution in closed form, suggest that this progress has arisen mainly because we have found a way to obtain *approximate* solutions to our problems.

To our culture, modelled on mathematical analysis, the term *approximate* sounds flawed. Nevertheless, *the goal of a numerical simulation is agreement with experimental measurements*. Reducing the error of an approximate solution does not mean making the error *as small as we like*, as a limit process requires; rather, it means making the error smaller than a preassigned tolerance. We are well aware that all measurements are affected by a tolerance because every measuring instrument belongs to a given class of precision.

In measurements, an infinite precision, in the sense of a limit process of mathematics, is not attainable. The same positioning of the measuring probe in a field implies a tolerance. The notion of precision in a measuring apparatus plays the same role as the notion of tolerance in manufacturing and as the notion of error in numerical analysis. In conclusion, one cannot deny the satisfaction of knowing the "exact" solution of a physical problem when such a solution is available. What we deny is the *need* to refer to an idealized exact solution when this is not available in order to compare a numerical result with experimental facts.

**The Same Field Can Be Described by Different Formalisms.** The formalism with which we can describe a physical theory is comparable to the clothes we wear; some suit well, others have been shortened over the years, others are cheap and comfortable, while still others are elegant and expensive. Just as a given outfit must be appropriate for a given occasion, in a similar way a formalism must be adapted both to the problem that must be treated and to the level of the intended reader. For example, electromagnetism can be described using the formalism of vectors, quaternions, Clifford numbers, tensors or exterior differential forms. This can be done in a three-dimensional space or in a four-dimensional space. Optics can be described with the formalisms of geometrical optics, of ondulatory optics or of quantum optics, three different "clothes" for three different occasions. Another example is offered by quantum mechanics, for which there exist different descriptions, the matrix description of Heisenberg, the wave description of Schrödinger and the algebraic description of Dirac.

## 7.2 Topology and Algebraic Topology

Topology can be defined as the science which studies the properties of geometric figures which are preserved under continuous deformations without tearing or overlaps. Since a deformation can be viewed as a transformation, the transformations considered by topology are continuous, invertible and with continuous inverses; for such transformations Poincaré introduced the term *homeomorphism*. Thus, briefly put, topology studies the properties of geometric figures which are invariant under homeomorphisms.

> The subject of topology deals with those properties of geometric figures which are unchanged by topological mappings, that is, by mappings which are bijective (i.e. one-to-one correspondences) and bicontinuous (i.e. continuous, with continuous inverses). Those properties which remain unchanged under topological mappings are called the topological properties of the figures. Two figures which can be mapped topologically onto each other are said to be homeomorphic.[1]

Topology can be divided into two parts: one is the *analytic* topology, also called *point set* topology or *general* topology; the other is *algebraic* topology, which initially was called *combinatorial* topology. The second part studies the topological properties of manifolds by means of algebraic methods.[2]

*One important class of manifolds is the class of differentiable manifolds. This differentiable structure allows calculus to be performed on manifolds.* Algebraic topology, in turn, is divided into two parts: *homotopy* and *homology* (see Table 7.1). Homology theory studies the topological properties of spaces and manifolds by means of algebraic methods applied to cell complexes. As a representative of the algebraic method of homology theory one can take Euler's famous formula for polyhedra, which states that regardless of the polyhedron, the following rule is valid: faces-edges + vertices = 2.

---

[1] See Seifert and Threlfall [210, p. 1]

[2] Franz [74, p. 1], Hocking and Young [94, p. 218].

**Table 7.1** Partition of topology

$$\begin{cases} \textbf{Analytic} \text{ topology (= point set topology = general topology)} \\ \textbf{Algebraic} \text{ topology (= combinatorial topology)} \begin{cases} \textbf{homology} \leftarrow \textit{here} \\ \textbf{homotopy} \end{cases} \end{cases}$$

## 7.3 Role of Cell Complexes

The algebraic formulation of physics, which rests on the fact that global physical variables are associated with space and time elements, is described in a natural way if we introduce cell complexes. To do this, we must define some notions and some operations to be performed on the cells of the primal and dual complexes.

### *7.3.1 Faces of a $p$-Cell and Boundary*

The faces of a cube are the six squares which bound the cube. Since a cube is a three-dimensional element of space, its faces are two-dimensional elements. Algebraic topology extends the term *face* of a $p$-cell to every $(p-1)$-cell which bounds the $p$-cell. Thus the *faces* of a 1-cell are the two terminal nodes, i.e. two 0-cells; the *faces* of a 2-cell are the edges (1-cells) that bound the 2-cell. Stated in general, the *faces* of a $p$-cell are those $(p-1)$-cells which are adjacent to the $p$-cell (think of the faces of a cube): the set of all faces of a $p$-cell is called the *boundary* of the $p$-cell.[3]

### *7.3.2 Cofaces of a $p$-Cell and Coboundary*

Let us consider two rooms separated by a wall; the wall is the face common to both rooms, and the two rooms can be called the **cofaces** of the wall.[4]

More generally, in a cell complex, given a $p$-cell, one can consider the $(p+1)$-cells that have the given $p$-cell as a common face: these $(p+1)$-cells are called the *cofaces* of the $p$-cell. Figure 7.1 shows the cofaces of $p$-cells in a Cartesian cell complex. The set of cofaces of a $p$-cell is called the *coboundary* of the $p$-cell.

While the notion of *face* has meaning for a single cell, the notion of *coface* has meaning only when the cell is considered as a member of a cell complex.

---

[3] See Schwarz [206, p. 109].

[4] The prefix 'co-' must be understood as an abbreviation of *complementary*, as in the terms *co-energy* and *co-tree* used in network theory.

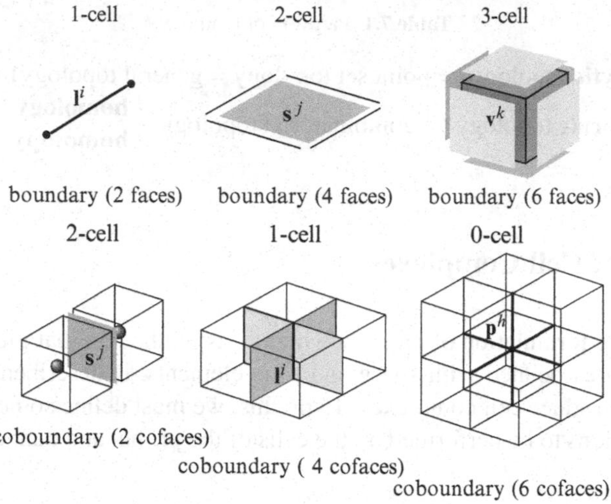


**Fig. 7.1** Faces and cofaces of 0-, 1-, 2-, 3-cells respectively in a Cartesian complex

### *7.3.3 Incidence Numbers and Incidence Matrices*

In the theory of oriented graphs used in planar network theory, one considers oriented *nodes, branches* and *meshes*. In this setting, we introduce the notion of *incidence numbers* between nodes and branches and between branches and meshes. Incidence numbers have values 0, $-1$ and $+1$, and they tell us whether two $p$-cells are mutually incident (number) and whether their mutual orientations are compatible or not (sign). *Incidence numbers* can be collected to form *incidence matrices*. The notion of incidence number can be extended both to a primal complex endowed with an inner orientation and to a dual complex.

We can number the $p$-cells of a cell complex according to any criterion we wish. After having numbered, i.e. labelled, the $p$-cells of a primal complex according to a convenient criterion, and since all $(n-p)$-cells of the dual complex correspond to a $p$-cell of the primal, it is natural to assign to every $(n-p)$-cell of the dual complex the same number (label) of the $p$-cell of the primal complex. Moreover, it follows that the number of $p$-cells of the primal complex, which we will denote by $N_{\mathrm{p}}$, is also the number of $(n-p)$-cells of the dual. This is shown in Fig. 7.1, where the numbers 2, 4 and 6 denote the number of faces of a primal cell and the numbers of the cofaces of a dual cell.

We are now in a position to define the *incidence number* of a $p$-cell $\mathbf{e}_{\mathrm{p}}^h$ with a $(p-1)$-cell $\mathbf{e}_{\mathrm{p-1}}^k$: this is the relative integer $q_{hk} = [\mathbf{e}_{\mathrm{p}}^h : \mathbf{e}_{\mathrm{p-1}}^k]$ whose values are[5]

[5] The lower index denotes the dimension of the cell, and we will write it in roman font, while the upper index denotes the label of the cell, and we will write it in italics. The letter 'e' to denote a cell is used by Massey.

**Table 7.2** Incidence numbers of a planar cell complex

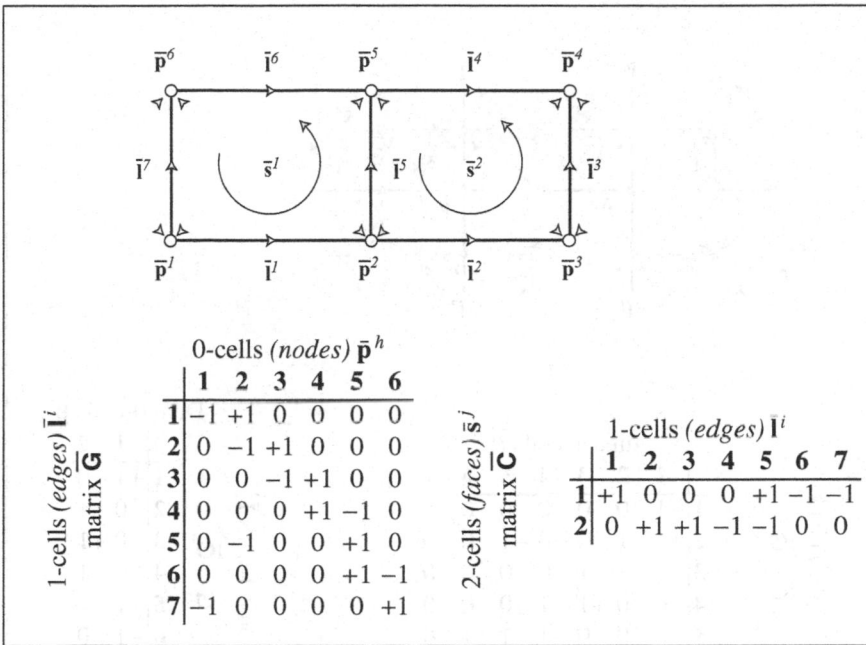

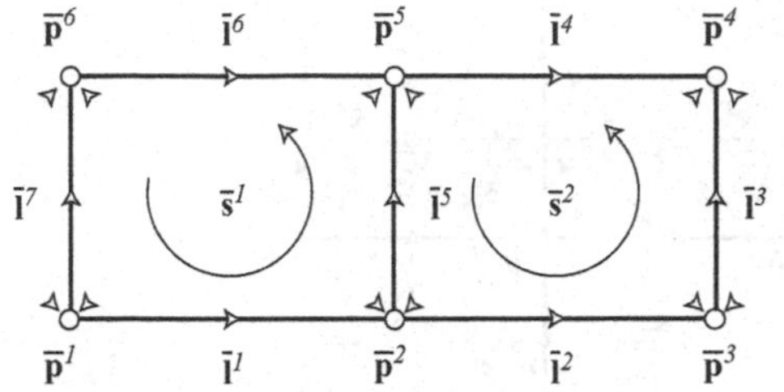


1-cells *(edges)* $\bar{\mathbf{l}}^i$, matrix $\overline{\mathbf{G}}$; 0-cells *(nodes)* $\bar{\mathbf{p}}^h$

| | 1 | 2 | 3 | 4 | 5 | 6 |
|---|---|---|---|---|---|---|
| **1** | −1 | +1 | 0 | 0 | 0 | 0 |
| **2** | 0 | −1 | +1 | 0 | 0 | 0 |
| **3** | 0 | 0 | −1 | +1 | 0 | 0 |
| **4** | 0 | 0 | 0 | +1 | −1 | 0 |
| **5** | 0 | −1 | 0 | 0 | +1 | 0 |
| **6** | 0 | 0 | 0 | 0 | +1 | −1 |
| **7** | −1 | 0 | 0 | 0 | 0 | +1 |

2-cells *(faces)* $\bar{\mathbf{s}}^j$, matrix $\overline{\mathbf{C}}$; 1-cells *(edges)* $\bar{\mathbf{l}}^i$

| | 1 | 2 | 3 | 4 | 5 | 6 | 7 |
|---|---|---|---|---|---|---|---|
| **1** | +1 | 0 | 0 | 0 | +1 | −1 | −1 |
| **2** | 0 | +1 | +1 | −1 | −1 | 0 | 0 |

- +1 if $\mathbf{e}^k_{\text{p}-1}$ is a face of $\mathbf{e}^h_{\text{p}}$ and the orientations of $\mathbf{e}^k_{\text{p}-1}$ and $\mathbf{e}^h_{\text{p}}$ are compatible;
- −1 if $\mathbf{e}^k_{\text{p}-1}$ is a face of $\mathbf{e}^h_{\text{p}}$ and the orientations of $\mathbf{e}^k_{\text{p}-1}$ and $\mathbf{e}^h_{\text{p}}$ are not compatible;
- 0 if $\mathbf{e}^k_{\text{p}-1}$ is not a face of $\mathbf{e}^h_{\text{p}}$.

In the notation $q_{hk}$, the first index refers to the cell of greater dimension.[6]

Tables 7.2 and 7.3 show the important fact that *when the outer orientation of a dual complex (dual 2-cells denoted by shaded areas) is induced by the inner orientation of a primal one, then the incidence number between a p-cell and a (p–1)-cell of the primal cell complex coincides with the incidence number between the corresponding dual cells.*

Thus, with reference to Table 7.2 since the incidence number between the primal edge $\bar{\mathbf{l}}^6$ and the primal vertex $\bar{\mathbf{p}}^5$ is +1, the incidence number between the dual edge $\tilde{\mathbf{l}}^6$ and the dual face $\tilde{\mathbf{s}}^5$ is also +1 as shown in Table 7.3.

The incidence number between the primal edge $\bar{\mathbf{l}}^1$ and primal point $\bar{\mathbf{p}}^2$ is +1; the incidence numbers between the corresponding dual elements, which are respectively the dual line $\tilde{\mathbf{l}}^1$ and the dual surface $\tilde{\mathbf{s}}^2$, have the same incidence number, i.e. +1.

This property is the consequence of the following three circumstances:

6 Alexandrov [4, p. 275], Franz [74, p. 30], Patterson [173, p. 103], Hocking and Young [94, p. 223], Lefschetz [133, p. 49]. The reader should be aware that some authors use the opposite convention for the order of indices.

**Table 7.3** Incidence numbers of dual of planar cell complex of Fig. 7.2

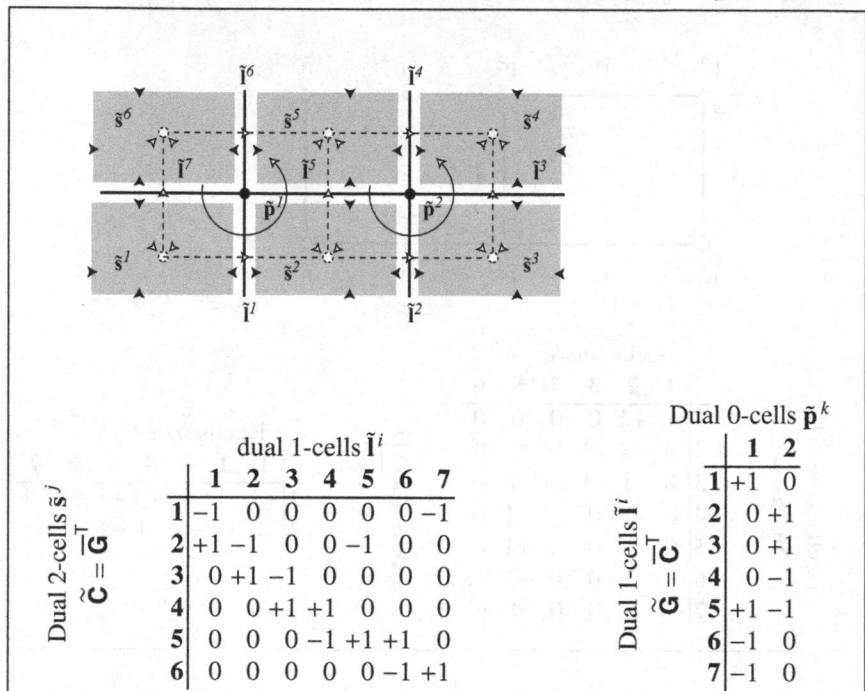

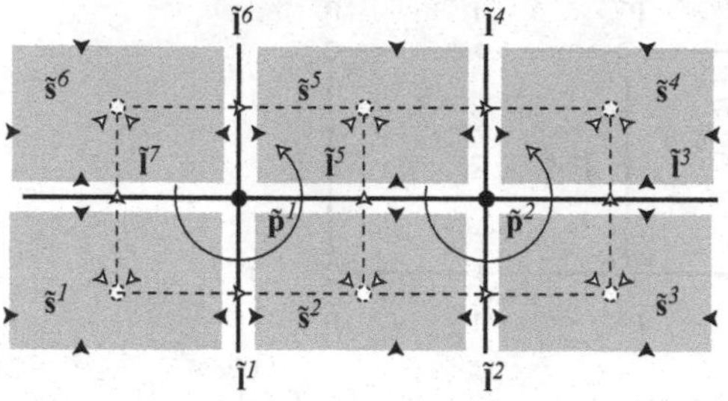


Dual 2-cells $\tilde{\mathbf{s}}^j$, $\tilde{\mathbf{C}} = \overline{\mathbf{G}}^{\mathsf{T}}$; dual 1-cells $\tilde{\mathbf{l}}^i$

| | **1** | **2** | **3** | **4** | **5** | **6** | **7** |
|---|---|---|---|---|---|---|---|
| **1** | −1 | 0 | 0 | 0 | 0 | 0 | −1 |
| **2** | +1 | −1 | 0 | 0 | −1 | 0 | 0 |
| **3** | 0 | +1 | −1 | 0 | 0 | 0 | 0 |
| **4** | 0 | 0 | +1 | +1 | 0 | 0 | 0 |
| **5** | 0 | 0 | 0 | −1 | +1 | +1 | 0 |
| **6** | 0 | 0 | 0 | 0 | 0 | −1 | +1 |

Dual 1-cells $\tilde{\mathbf{l}}^i$, $\tilde{\mathbf{G}} = \overline{\mathbf{C}}^{\mathsf{T}}$; Dual 0-cells $\tilde{\mathbf{p}}^k$

| | **1** | **2** |
|---|---|---|
| **1** | +1 | 0 |
| **2** | 0 | +1 |
| **3** | 0 | +1 |
| **4** | 0 | −1 |
| **5** | +1 | −1 |
| **6** | −1 | 0 |
| **7** | −1 | 0 |

1. Every $(n - p)$-cell of the dual complex intersects, belongs to or contains the corresponding $p$-cell of the primal.
2. The labels of the $(n - p)$-cell of the dual complex were chosen to be equal to the labels of the corresponding primal $p$-cells.
3. The outer orientation of a dual $(n - p)$-cell is the one induced by the inner orientation of the corresponding primal $p$-cell.

In three-dimensional space there are three incidence matrices:

1. Matrix **G** of the incidence numbers between edges and nodes,
2. Matrix **C** of the incidence numbers between faces and edges,
3. Matrix **D** of the incidence numbers between cells and faces.

In two-dimensional space, we have only **G** and **C**, as can be seen in Tables 7.2 and 7.3.

As we will see, these matrices are the discrete version of the differential operators 'grad', 'curl', 'div', and this fact justifies the letters **G**, **C**, **D** chosen to denote them.

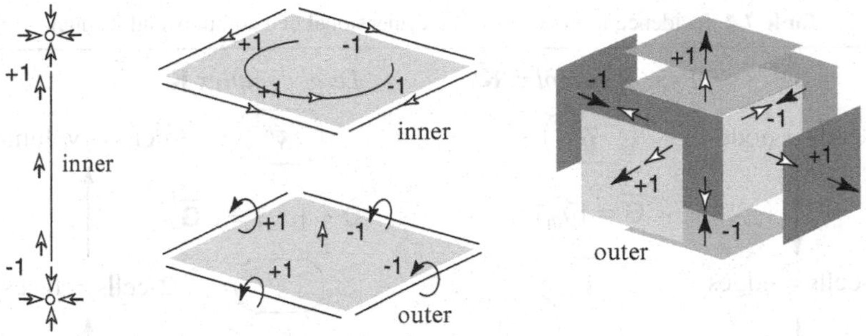


**Fig. 7.2** Incidence numbers between a cell and its faces

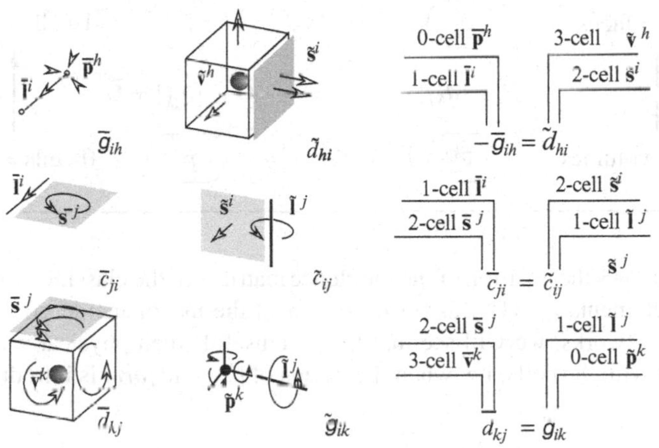


**Fig. 7.3** Incidence numbers between cells of a cell complex and its dual

Figures 7.2 and 7.3 show the incidence numbers of a 3-cell complex and its dual. As stated previously (p. 78), for historical reasons points are oriented as *sinks* while volumes are oriented with *outward* normals.

Hence, the orientation of volumes is not the orientation induced by the inner orientation of points but its opposite. It follows that $\tilde{\mathbf{D}} = -\bar{\mathbf{G}}^{\mathsf{T}}$. In summary, we have

$$
\begin{aligned}
&\bar{\mathbf{G}} \overset{\text{def}}{=} [\bar{g}_{ih}], \quad \tilde{\mathbf{D}} \overset{\text{def}}{=} [\tilde{d}_{hi}], \quad \tilde{d}_{hi} \overset{\text{def}}{=} [\tilde{\mathbf{v}}^h : \tilde{\mathbf{s}}^i] = -[\bar{\mathbf{l}}^i : \bar{\mathbf{p}}^h] \overset{\text{def}}{=} -\bar{g}_{ih}, \quad \tilde{\mathbf{D}} = -\bar{\mathbf{G}}^{\mathsf{T}}, \\
&\bar{\mathbf{C}} \overset{\text{def}}{=} [\bar{c}_{ji}], \quad \tilde{\mathbf{C}} \overset{\text{def}}{=} [\tilde{c}_{ij}], \quad \tilde{c}_{ij} \overset{\text{def}}{=} [\tilde{\mathbf{s}}^i : \tilde{\mathbf{l}}^j] = [\bar{\mathbf{s}}^j : \bar{\mathbf{l}}^i] \overset{\text{def}}{=} \bar{c}_{ji}, \quad \tilde{\mathbf{C}} = \bar{\mathbf{C}}^{\mathsf{T}}, \\
&\bar{\mathbf{D}} \overset{\text{def}}{=} [\bar{d}_{kj}], \quad \tilde{\mathbf{G}} \overset{\text{def}}{=} [\tilde{g}_{jk}], \quad \tilde{g}_{jk} \overset{\text{def}}{=} [\tilde{\mathbf{l}}^j : \tilde{\mathbf{p}}^k] = [\bar{\mathbf{v}}^k : \bar{\mathbf{s}}^j] \overset{\text{def}}{=} \bar{d}_{kj}, \quad \tilde{\mathbf{G}} = \bar{\mathbf{D}}^{\mathsf{T}}.
\end{aligned}
\tag{7.1}
$$

**Table 7.4** Incidence numbers of three-dimensional cell complex and its dual

| | *Primal complex* $\overline{\mathbf{K}}$ | *Dual complex* $\widetilde{\mathbf{K}}$ | |
|---|---|---|---|
| 0-cells = nodes | $\overline{\mathbf{p}}^h$ | $\tilde{\mathbf{v}}^h$ | 3-cells = volumes |
| ↓ | $\overline{\mathbf{G}} = [\bar{g}_{ih}]$ | $\widetilde{\mathbf{D}} = [\tilde{d}_{hi}] = -\overline{\mathbf{G}}^{\mathsf{T}}$ | ↑ |
| 1-cells = edges | $\overline{\mathbf{l}}^i$ | $\tilde{\mathbf{s}}^i$ | 2-cells = faces |
| ↓ | $\overline{\mathbf{C}} = [\bar{c}_{ji}]$ | $\widetilde{\mathbf{C}} = [\tilde{c}_{ij}] = \overline{\mathbf{C}}^{\mathsf{T}}$ | ↑ |
| 2-cells = faces | $\overline{\mathbf{s}}^j$ | $\tilde{\mathbf{l}}^j$ | 1-cells = edges |
| ↓ | $\overline{\mathbf{D}} = [\bar{d}_{kj}]$ | $\widetilde{\mathbf{G}} = [\tilde{g}_{jk}] = \overline{\mathbf{D}}^{\mathsf{T}}$ | ↑ |
| 3-cells = volumes | $\overline{\mathbf{v}}^k$ | $\tilde{\mathbf{p}}^k$ | 0-cells = nodes |

Table 7.4 shows the position of the incidence matrices in the classification diagram of space elements (p. 84). Taking into account the use of incidence matrices in electrical networks, we will see that they are useful when physical variables are associated with the cells, i.e. when the notion of *discrete form* is introduced.

## 7.4 The Notion of a Chain

Let us consider the two-dimensional complex of Fig. 7.4a. Let us select arbitrarily a reference orientation of its 1-cells, e.g.[7]

$$\mathbf{e}_1^1 = (\mathbf{p}^1, \mathbf{p}^2), \quad \mathbf{e}_1^2 = (\mathbf{p}^1, \mathbf{p}^3), \quad \mathbf{e}_1^3 = (\mathbf{p}^3, \mathbf{p}^2), \quad \mathbf{e}_1^4 = (\mathbf{p}^2, \mathbf{p}^4)\,. \tag{7.2}$$

With reference to Fig. 7.4b–d, with every oriented 1-cell $\mathbf{e}_1^h$ we can associate a *multiplicity* or a *weight*, i.e. a relative integer $n_h$. We have pictorially represented the multiplicity drawing many copies of the 1-cells. The multiplicity $n_h$ associated with the cell $\mathbf{e}_1^h$ is denoted by putting the number $n_h$ before the corresponding cell as follows: $n_h\,\mathbf{e}_1^h$. Each couple formed by a 1-cell $\mathbf{e}_1^h$ and the number +1 as multiplicity is called an *elementary 1-chain*. A collection of oriented 1-cells, each taken with a certain multiplicity, is called a *1-chain* and is written as[8]

[7] We follow Flanders [70, p. 63] using the boldface to denote cells and chains.

[8] Hilton and Wylie [92, p. 56], Alexandrov [4, p. 285], Franz [74, p. 31], Hocking and Young [94, p. 297].

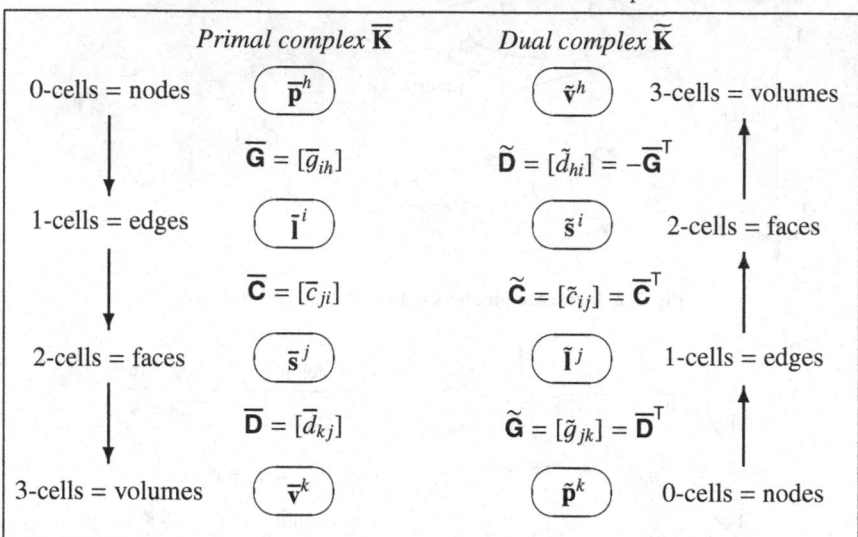

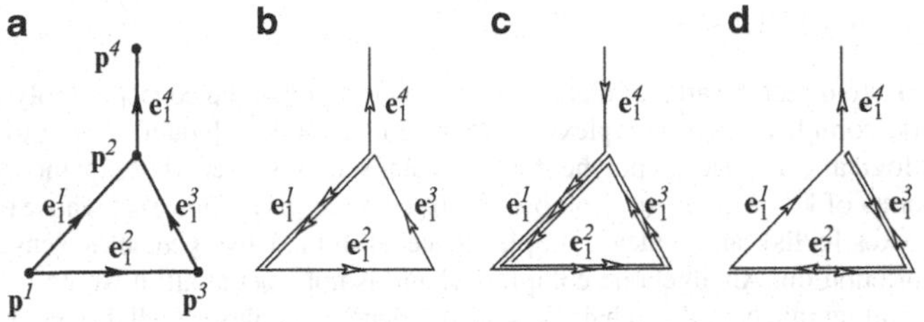


**Fig. 7.4** One-dimensional chains: **(a)**, **(b)**, **(c)**, **(d)** are different 1-chains on the same complex

$$\mathbf{c}_1 = +3\mathbf{e}_1^1 + 2\mathbf{e}_1^2 + 2\mathbf{e}_1^3 - 1\mathbf{e}_1^4 \qquad \text{or, in general,} \quad \mathbf{c}_1 = \sum_h^{\mathrm{f}} n_h\, \mathbf{e}_1^h. \tag{7.3}$$

One can also write chains of Fig. 7.4b–d as algebraic vectors as follows:[9]

$$(-2, +1, +1, +1), \qquad (+3, +2, +2, -1), \qquad (+1, 0, 0, +1)\,. \tag{7.4}$$

The notion of a 1-chain can be extended to a $p$-chain as follows:[10]

DEFINITION. *Given a cell complex* **K**, *endowed with an inner or outer orientation, i.e.* $\overline{\mathbf{K}}$ *or* $\widetilde{\mathbf{K}}$, *let us consider a collection of* $p$*-dimensional cells* $\mathbf{e}_\mathrm{p}^h$, *each being assigned a relative integer* $n_h$, *called its multiplicity. It is required that if the orientation of the cell is changed into its opposite, then the integer* $n_h$ *changes sign, i.e.*

$$n_h(-\mathbf{e}_\mathrm{p}^h) = (-n_h)\,(\mathbf{e}_\mathrm{p}^h) \qquad \text{oddness condition.} \tag{7.5}$$

*The collection thus obtained is called a* $p$-dimensional chain *with integer coefficients and will be denoted by* $\mathbf{c}_\mathrm{p}$. *A* $p$*-chain can be represented by the* formal *sum*

$$\mathbf{c}_\mathrm{p} = n_1\,\mathbf{e}_\mathrm{p}^1 + n_2\,\mathbf{e}_\mathrm{p}^2 \ldots n_N\,\mathbf{e}_\mathrm{p}^N = \sum_h^{\mathrm{f}} n_h\,\mathbf{e}_\mathrm{p}^h\,. \tag{7.6}$$

*Each couple formed by a p-cell* $\mathbf{e}_\mathrm{p}^h$ *and the number* $+1$ *as multiplicity is called an* elementary chain. *The integers* $n_h$ *are called* coefficients *of the chain and denote the multiplicity (the weight) with which a cell enters the chain. A chain can also be described by a column vector* $[n_1, n_2, \cdots n_{N_p}]^\mathrm{T}$.

---

[9] Seifert and Threlfall [210, p. 61].

[10] Seifert and Threlfall [210, p. 61], Franz [74, p. 31], Hilton and Wylie [92, p. 56], Hocking and Young [94, p. 225], Paterson [173, p. 117], Alexandrov [2, p. 18], Alexandrov [4, p. 285], Wallace [244, p. 105].

Quoting Alexandroff:[11]

> I would like to direct the attention of the reader to the fact that the concepts "polyhedron", "geometric complex" [cells complex] and "algebraic complex" [chain] belong to entirely different logical categories: a polyhedron is a point set, thus a set whose elements are ordinary points of $\mathbb{R}^n$; a geometric complex [cells complex] is a (finite) set whose elements are simplexes [cells], and, indeed, simplexes [cells] in the naive geometric sense, that is without orientation. An algebraic complex [chain] is not a set at all: it would be false to say that an algebraic complex [chain] is a set of oriented simplexes [cells] since the essential thing about an algebraic complex [chain] is that the simplexes [cells] which appear in it are provided with coefficients and, therefore in general, are to be counted with a certain multiplicity. This distinction between the three concepts, which often appear side by side, reflects the essential difference between the set-theoretic and the algebraic approaches to topology.

While $p$-cells are point sets, i.e. topological entities, elementary $p$-chains are algebraic entities; whereas point sets cannot be added, elementary $p$-chains can be.[12]

**Remark.** Let us give an example of a formal sum:

$$\begin{aligned} \mathbf{a} &= 2 \text{ apples} + 3 \text{ peaches}, \\ \mathbf{b} &= 5 \text{ apples} + 1 \text{ peach} + 4 \text{ bananas}, \\ \mathbf{a+b} &= (5{+}2) \text{ apples} + (3{+}1) \text{ peaches} + 4 \text{ bananas}. \end{aligned} \tag{7.7}$$

Chains of the same dimension can been added:[13]

$$\mathbf{c}_\mathrm{p} = \sum_h^\mathrm{f} n_h \, \mathbf{e}_\mathrm{p}^h, \qquad \mathbf{d}_\mathrm{p} = \sum_h^\mathrm{f} m_h \, \mathbf{e}_\mathrm{p}^h, \qquad \mathbf{c}_\mathrm{p} + \mathbf{d}_\mathrm{p} = \sum_h^\mathrm{f} (n_h + m_h) \, \mathbf{e}_\mathrm{p}^h; \tag{7.8}$$

hence, $p$-chains form an additive group.[14] In particular, a $p$-chain all of whose coefficients are vanishing is called a *null* $p$-chain and will be denoted by $\Theta_\mathrm{p}$.

**Remark.** It is possible to extend the notion of a $p$-chain as a function which to every oriented $p$-cell assigns an element of a additive group.[15] The handicap of this extension is that when the coefficients belong to an arbitrary additive group, the intuitive geometric content is lost. If we want to maintain a geometric content, as indeed we do, then the group must be that of integers, i.e. $\mathbf{Z}$.[16]

---

[11] Alexandroff [2, p. 19]. The terms in square brackets were inserted by the present author. For the oddness condition see Alexandroff [4, Chap. VII, Sect. 5.1], Hocking and Young [94, p. 297].

[12] Alexandrov [2, p. 20].

[13] Lefschetz [134, p. 19].

[14] Hilton and Wylie [92, p. 57].

[15] Hocking and Young [94, p. 225].

[16] Hilton and Wylie [92, p. 58], Bourgin [24, p. 22], Wallace [245, p. 6].

The introduction of a cell complex in the space allows us to select some $p$-dimensional region formed by a set of $p$-cells. The algebraic entities which describe them are $p$-chains.

### 7.4.1 Boundary of a Line, Surface and Volume

The boundary of a line is the set formed by its two ending points: if the line is closed, like a ring, then it has no boundary. Similarly, the boundary of a surface is a closed line: if the surface is closed, like the shell of an egg, then it has no boundary. Hence the boundary of a surface is a *closed* line and the boundary of a volume is a *closed* surface. In every case, the boundary of a manifold is a closed manifold with one dimension less than the dimension of the manifold. This primitive topological property plays an important role in the applications of algebraic topology to physical fields; thus, it deserves particular consideration.

### 7.4.2 Boundary of a Chain

The notion of the boundary of a manifold of dimension $p$ can be extended to a $p$-chain. Let us denote by

$$\mathbf{c}_{\mathrm{p}} = \sum_h^{\mathrm{f}} n_h\, \mathbf{e}_{\mathrm{p}}^h \quad \textit{and} \quad \mathbf{c}_{\mathrm{p-1}} = \sum_k^{\mathrm{f}} m_k\, \mathbf{e}_{\mathrm{p-1}}^k \tag{7.9}$$

two chains of dimensions $p$ and $(p-1)$ respectively. With reference to Fig. 7.5, let us consider the integer number $n_h$ associated with a cell $\mathbf{e}_{\mathrm{p}}^h$ and transfer this integer to each bounding cell $\mathbf{e}_{\mathrm{p-1}}^k$ of one dimension less, with the plus or minus sign depending on whether the mutual incidence number is $+1$ or $-1$. When this is done, every bounding cell receives the values from all its cofaces: we sum all these values, obtaining an integer value which we associate with the bounding cell $\mathbf{e}_{\mathrm{p-1}}^k$. In this way, starting from a $p$-chain we arrive at a $(p-1)$-chain.

In general we can define the *boundary of a* $p$*-chain* $\mathbf{c}_{\mathrm{p}}$ *as the* $(p–1)$*-chain* $\mathbf{c}_{\mathrm{p-1}}$ *whose coefficients* $m_h$ *are the sum of the coefficients* $n_h$ *multiplied by the incidence numbers of the* $p$*-cells* $\mathbf{e}_{\mathrm{p}}^h$ *and the* $(p–1)$*-cells* $\mathbf{e}_{\mathrm{p-1}}^k$*, i.e.*

$$m_k \stackrel{\mathrm{def}}{=} \sum_h n_h \underbrace{[\mathbf{e}_{\mathrm{p}}^h : \mathbf{e}_{\mathrm{p-1}}^k]}_{\textit{incidence number}} \;. \tag{7.10}$$

This $(p-1)$-chain is written as

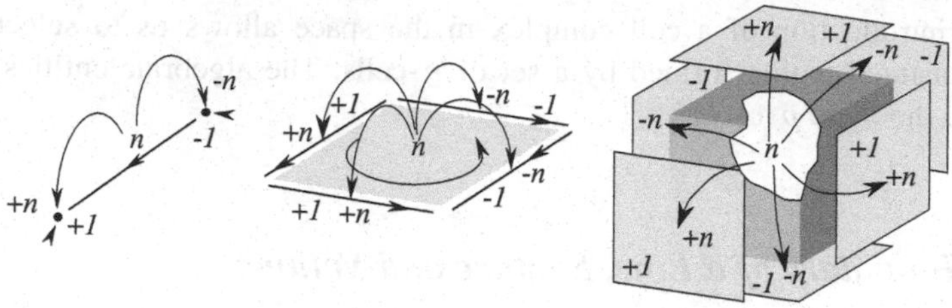


**Fig. 7.5** Pictorial view of boundary process

$$\mathbf{c}_{p-1} = \partial\, \mathbf{c}_p \,. \tag{7.11}$$

The process of forming the boundary of a $p$-chain is called a *boundary process*, and the operator $\partial$ is called a *boundary operator*. Figure 7.5 gives a pictorial view of the process of transferring the coefficients $n_h$ of a $p$-chain from the $p$-cells to the $(p-1)$-cells which form the boundary of the $p$-chain.[17] We see that a boundary operator is a linear mapping with integer coefficients between the group of $p$-chains and that of $(p-1)$-chains, i.e. $\partial : \mathscr{C}_p \rightarrow \mathscr{C}_{p-1}$.

A closed line is a line without a boundary; its algebraic description is a closed 1-chain, i.e. a chain such that $\partial\, \mathbf{c}_1 = \boldsymbol{\Theta}_0$, where $\boldsymbol{\Theta}_0$ denotes the null 0-chain. This is called a 1-*cycle*. Similarly, a closed surface is a surface without a boundary. Its algebraic description is a closed 2-chain, i.e. a 2-chain such that $\partial\, \mathbf{c}_2 = \boldsymbol{\Theta}_1$, where $\boldsymbol{\Theta}_1$ denotes the null 1-chain. This is called a 2-*cycle*.

### *7.4.3 The Boundary of a Boundary Is a Null Chain*

With reference to Fig. 7.6, we can easily see that performing the boundary process twice in sequence on an elementary $p$-chain we obtain a null $(p-2)$-chain. The left part of Fig. 7.6 shows the boundary process starting with an elementary 2-chain, in this case a 2-cell with assigned multiplicity $n$. The integer $n$ is transferred to the four edges with the plus or minus sign according to the mutual orientation of the face with each edge. When we repeat the boundary process starting from the elementary 1-chains of the boundary, the multiplicity $\pm n$ of the elementary 1-chain is transferred to each of its bounding nodes with the same or opposite sign, depending on the mutual incidence edges/nodes. If for every node we add the integers thus obtained, then we obtain the number 0. Hence, since each 0-cell of the complex has a multiplicity 0, we have obtained a null 0-chain.

In the bottom part of Fig. 7.7 is shown a boundary process starting with an elementary 3-chain, in this case a 3-cell with assigned multiplicity 1. The integer

[17] This pictorial representation by means of fountains is introduced in this book to grasp immediately the boundary process (and later also the coboundary process).

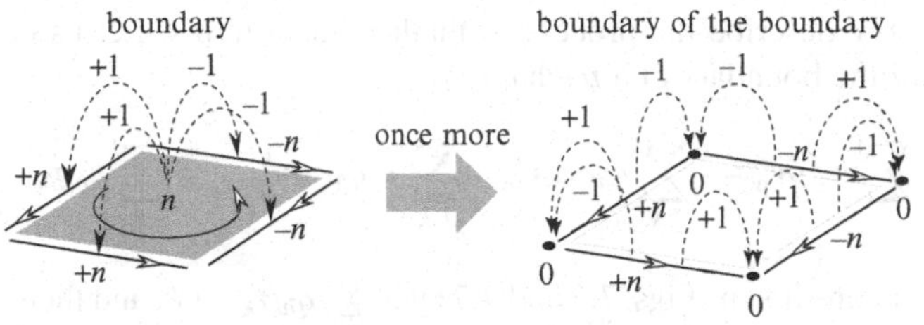


**Fig. 7.6** The boundary of a boundary is a null chain. This pictorial view illustrating a process, similar to fountains, which transfers the coefficient $n$ of the 2-cell in the chain to the 1-cells that form the edge, and from these, in turn, to the 0-cells that form the edge thereof

1 is transferred to the six faces with a plus or minus sign according to the mutual orientation of the cell with each face. Here we have considered cells endowed with an outer orientation. When we repeat the boundary process (right part of Fig. 7.7), starting from the elementary 2-chains of the boundary, the multiplicity 1 of the elementary 2-chain is transferred to each of its bounding edges with the same or opposite sign, according to the mutual incidence faces/edges. If for every edge we sum the integers thus arrived at, we obtain the number 0. Hence we have obtained a null 1-chain.

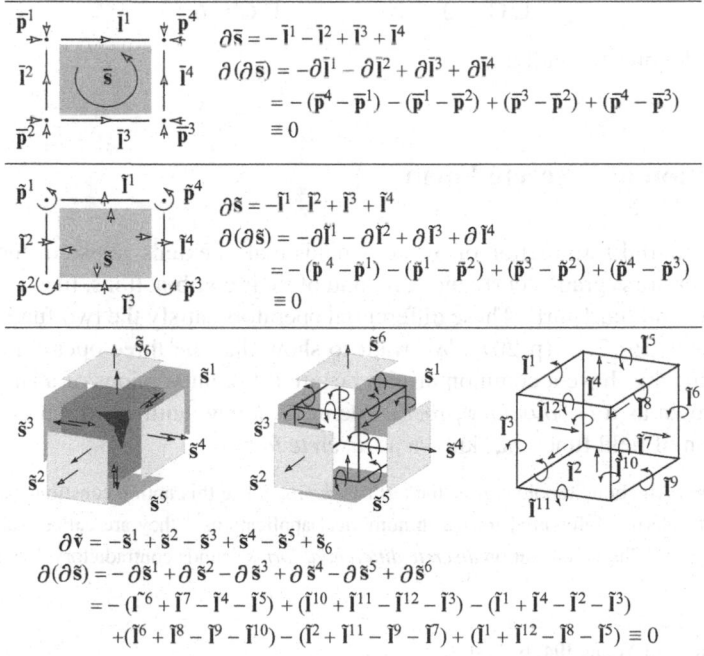

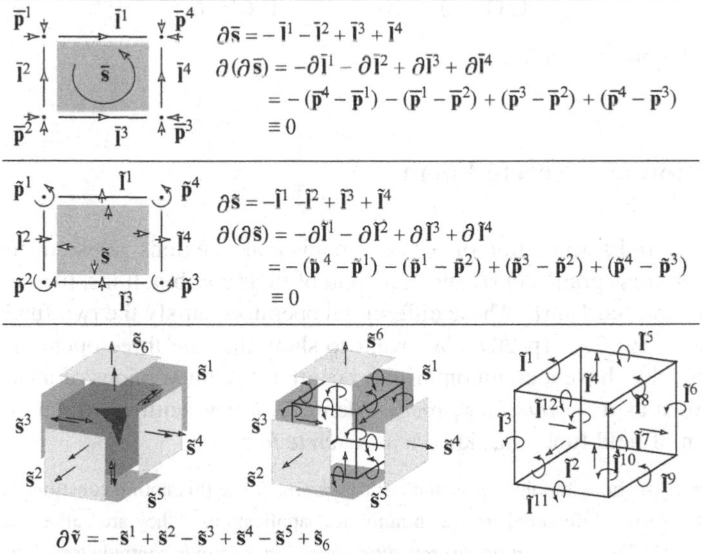

$$\partial \bar{s} = -\bar{l}^1 - \bar{l}^2 + \bar{l}^3 + \bar{l}^4$$
$$\partial(\partial \bar{s}) = -\partial\bar{l}^1 - \partial\bar{l}^2 + \partial\bar{l}^3 + \partial\bar{l}^4 = -(\bar{p}^4 - \bar{p}^1) - (\bar{p}^1 - \bar{p}^2) + (\bar{p}^3 - \bar{p}^2) + (\bar{p}^4 - \bar{p}^3) \equiv 0$$

$$\partial \tilde{s} = -\tilde{l}^1 - \tilde{l}^2 + \tilde{l}^3 + \tilde{l}^4$$
$$\partial(\partial \tilde{s}) = -\partial\tilde{l}^1 - \partial\tilde{l}^2 + \partial\tilde{l}^3 + \partial\tilde{l}^4 = -(\tilde{p}^4 - \tilde{p}^1) - (\tilde{p}^1 - \tilde{p}^2) + (\tilde{p}^3 - \tilde{p}^2) + (\tilde{p}^4 - \tilde{p}^3) \equiv 0$$

$$\partial \tilde{v} = -\tilde{s}^1 + \tilde{s}^2 - \tilde{s}^3 + \tilde{s}^4 - \tilde{s}^5 + \tilde{s}_6$$
$$\partial(\partial \tilde{s}) = -\partial\tilde{s}^1 + \partial\tilde{s}^2 - \partial\tilde{s}^3 + \partial\tilde{s}^4 - \partial\tilde{s}^5 + \partial\tilde{s}^6$$
$$= -(\tilde{l}^6 + \tilde{l}^7 - \tilde{l}^4 - \tilde{l}^5) + (\tilde{l}^{10} + \tilde{l}^{11} - \tilde{l}^{12} - \tilde{l}^3) - (\tilde{l}^1 + \tilde{l}^4 - \tilde{l}^2 - \tilde{l}^3)$$
$$+(\tilde{l}^6 + \tilde{l}^8 - \tilde{l}^9 - \tilde{l}^{10}) - (\tilde{l}^2 + \tilde{l}^{11} - \tilde{l}^9 - \tilde{l}^7) + (\tilde{l}^1 + \tilde{l}^{12} - \tilde{l}^8 - \tilde{l}^5) \equiv 0$$

**Fig. 7.7** Computing the boundary of a boundary we obtain systematically zero

Let us now describe the process in mathematical terms. Let us consider the boundary of the boundary of a $p$-chain:

$$\partial\partial \mathbf{c}_{\mathrm{p}} = \sum_{h}^{\mathrm{f}} n_h \, \partial \mathbf{e}_{\mathrm{p}}^h = \partial \sum_{h,k}^{\mathrm{f}} n_h q_{hk} \, \mathbf{e}_{(\mathrm{p}-1)}^k \sum_{h,k}^{\mathrm{f}} n_h q_{hk} \, \partial \mathbf{e}_{(\mathrm{p}-1)}^k = \sum_{h,k,j}^{\mathrm{f}} n_h q_{hk} r_{kj} \, \mathbf{e}_{(\mathrm{p}-2)}^j . \tag{7.12}$$

Now it has been shown (Figs. 7.6 and 7.7) that $\sum_k q_{hk} r_{kj} \equiv 0$, and then

$$\partial\,(\partial \mathbf{c}_{\mathrm{p}}) \equiv \Theta_{\mathrm{p}-2}, \tag{7.13}$$

where $\Theta_{\mathrm{p}-2}$ is the null $(p-2)$-chain. This relation can be read as follows: *the boundary of the boundary of a chain vanishes.*

The identity 7.13 may be expressed by the relations[18]

$$\sum_k [\mathbf{e}_{\mathrm{p}+1}^i : \mathbf{e}_{\mathrm{p}}^k]\,[\mathbf{e}_{\mathrm{p}}^k : \mathbf{e}_{\mathrm{p}-1}^h] \equiv 0\,. \tag{7.14}$$

Referring to Fig. 7.3 and recalling Eq. 7.1, we may write

$$\sum_i [\mathbf{s}^j : \mathbf{l}^i][\mathbf{l}^i : \mathbf{p}^h] = \sum_i c_{ji}\, g_{ih} \equiv 0, \qquad \sum_j [\mathbf{v}^k : \mathbf{s}^j][\mathbf{s}^j : \mathbf{l}^i] = \sum_j d_{kj}\, c_{ji} \equiv 0 \tag{7.15}$$

or

$$\mathbf{C}\mathbf{G} \equiv \mathbf{0}, \qquad \mathbf{D}\mathbf{C} \equiv \mathbf{0}\,, \tag{7.16}$$

where $\mathbf{0}$ denotes the null matrix.

## 7.5 Notion of Discrete Form

The differential formulation of physical fields makes extensive use of the differential operators 'grad', 'curl' and 'div' and of their combinations, like 'div grad', 'grad div' and 'curl curl'. These differential operators satisfy the two fundamental identities of Eq. 7.42 (p. 207). We want to show that the three operators 'grad', 'curl' and 'div' have a common origin, i.e. are different versions of a single process, known as a *coboundary process*, defined on new entities which generalize the notion of field functions, known as *discrete forms*.

**Remark.** Algebraic topology uses the term *cochains*. Since this notion constitutes a discrete version of exterior differential forms, in numerical applications[19] they are called *discrete differential forms.* The juxtaposition *discrete differential forms* sounds contradictory, like *beautiful*

[18] Hocking and Young [94, p. 224].

[19] Desbrun et al. [55, 56], Castillo et al. [35, 36], Stern et al. [219]

*ugly flower* or *good bad ice cream*. For this reason we prefer to use the simplest locution, *discrete forms*, as some authors do.[20]

The starting point for this unified view is, once more, the fundamental remark that global physical variables have a natural association with space elements. If we build a cell complex in a space region in which a field is defined, we have at our disposal many space elements, say many nodes, edges, faces and cells. Therefore, if we consider a physical variable $\Phi$ associated, for example, with surfaces, i.e. $\Phi[\mathbf{S}]$, then to every face $\mathbf{s}_i$ of the cell complex we can assign the amount $\phi_i \stackrel{\text{def}}{=} \Phi[\mathbf{s}_i]$ of the physical variable. In doing this we have constructed something that generalizes a field function. In fact, while a field function, say $u = f(\mathbf{P})$, associates a value of a physical variable $u$ with every point of the region, here we associate the value of a physical variable $\Phi$ with every face of the complex. Whereas $f(\mathbf{P})$ is a *point* function, $\Phi[\mathbf{S}]$ is a *set* function. This function in algebraic topology is called a *two-dimensional cochain* or, *2-cochain* for short. In numerical methods, it is called a *discrete 2-form*.[21] Franz [74, p. 144] uses the term *complementary* instead of *dual* to avoid confusion with the algebraic duality concept (e.g. dual modulus, dual mapping, dual basis).

Thus, the potential of a vector field, evaluated at all 0-cells, gives rise to a discrete 0-form; the line integral of a vector, evaluated on the 1-cells of a complex, gives rise to a discrete 1-form; the flux evaluated on all 2-cells gives rise to a discrete 2-form and the contents, e.g. mass content, evaluated on all 3-cells, gives rise to a discrete 3-form.

**Notation.** Chains are usually denoted by lower indices, discrete forms by upper indices. See p. 299 in [94], p. 56 in [92], p. 31 in [61]. Dubrovin [58, p. 3] uses $\sum^f$ where $f$ means *formal*.

**Definition.** *Given an oriented cell complex* $\mathbf{K}$*, endowed with an inner or outer orientation, hence* $\overline{\mathbf{K}}$ *or* $\widetilde{\mathbf{K}}$*, and an additive and commutative group* $\mathcal{G}$ *(e.g. a scalar, vector, matrix), a discrete p-form,* $c^{\mathrm{p}}$*, is a linear function on p-chains with integer coefficients with values in the group* $\mathcal{G}$.

This means that a discrete $p$-form associates an element $g \in \mathcal{G}$ with every chain $\mathbf{c}_{\mathrm{p}}$ in such a way that

$$\overbrace{c^{\mathrm{p}}(\mathbf{c}_{\mathrm{p}} + \mathbf{c}'_{\mathrm{p}}) = c^{\mathrm{p}}(\mathbf{c}_{\mathrm{p}}) + c^{\mathrm{p}}(\mathbf{c}'_{\mathrm{p}})}^{\textit{additive}}, \qquad \overbrace{c^{\mathrm{p}}(n\,\mathbf{c}_{\mathrm{p}}) = n\,c^{\mathrm{p}}(\mathbf{c}_{\mathrm{p}})}^{\textit{homogeneous}} .$$

[20] Hirani [93], Frauendiener [75].

[21] The prefix 'co-' means *conjugate* (p. 461 in [253]), which is synonymous with *dual*. In modern mathematics, the passage to the dual is traditionally denoted by the prefix 'co-', e.g. *vector* and *covector*.

In particular, denoting by $g_h$ the element associated with the elementary chain $\mathbf{e}_{\mathrm{p}}^h$, a discrete form, on account of its linearity, is completely defined when all the values $g_h$ are given. As for the function of one or more variables $u = f(x_1, x_2)$, a linear function can be written in the form $u = a_1x_1 + a_2x_2$, so instead of the notation $c^{\mathrm{p}}(\mathbf{c}_{\mathrm{p}})$, it is convenient to use the notation

$$(c^{\mathrm{p}}, \mathbf{c}_{\mathrm{p}}) \qquad \text{or} \qquad \langle c^{\mathrm{p}}, \mathbf{c}_{\mathrm{p}} \rangle \,. \tag{7.17}$$

We prefer the second notation, the one with angle brackets, because it complies with the one used in the theory of spaces put in duality, as shown in Chap. 15.

In particular, the group $\mathcal{G}$ may be a vector space on real or complex numbers, or it can be an algebra, like matrix and Clifford algebras.

A discrete $p$-form, $c^{\mathrm{p}}$, can also be represented by a row vector whose elements are the values $g_k$ of the discrete form on the elementary chains

$$c^{\mathrm{p}} = [g_1, g_2, \cdots g_N] \,. \tag{7.18}$$

**Notation.** There are many notations for cochains in books on algebraic topology ($c^p$ denotes a cochain and $d_p$ a chain):

| | |
|---|---|
| $c^{\mathrm{p}}(d_{\mathrm{p}})$ | Dubrovin et al. [58, p. 32]), Hocking and Young [94, p. 300] |
| $(c^{\mathrm{p}}, d_p)$ | Hilton and Wylie [92, p. 67] |
| $c^{\mathrm{p}} \cdot d_p$ | Hocking and Young [94, p. 300] |
| $(d_p, c^{\mathrm{p}})$ | Franz [74, p. 92] |
| $(d_{\mathrm{p}})\, c^p$ | Hilton and Wylie [92, p. 68] |

### 7.5.1 Value of a Discrete Form on a Chain

It is useful to introduce the notion of the *value of the discrete form* $c^{\mathrm{p}} = [g_1, g_2, \cdots g_N]$ *on the chain* $\mathbf{c}_p = [n_1, n_2, \cdots n_{N_p}]^{\mathrm{T}}$ *as the element* $g \in \mathcal{G}$ *given by the sum of the values of the discrete form on single cells, each multiplied by the coefficient of the cell in the chain. In symbols:*

$$g = \sum_{h=1}^{N} g_h\, n_h \equiv [\, g_1\; g_2\; \cdots\; g_N \,] \begin{bmatrix} n_1 \\ n_2 \\ \dots \\ n_N \end{bmatrix} , \tag{7.19}$$

which can also be written as

$$g = \langle c^{\mathrm{p}}, \mathbf{c}_{\mathrm{p}} \rangle = \left\langle c^{\mathrm{p}}, \sum_h^{\mathrm{f}} n_h \mathbf{e}_{\mathrm{p}}^h \right\rangle = \sum_h n_h \langle c^{\mathrm{p}}, \mathbf{e}_{\mathrm{p}}^h \rangle = \sum_h n_h\, g_h \,. \tag{7.20}$$

This property fits like a hand in a glove with the additivity of global variables. Thus the line integral of a vector along a line is equal to the sum of the line integrals along the parts of the line; the flux associated with a surface is the sum of the fluxes associated with the parts of the surface; the mass contained in a volume is the sum of the masses of the parts of the volume; and so forth. In particular, from the homogeneous property it follows that[22]

$$\langle c^{\mathrm{p}}, -\mathbf{e}_{\mathrm{p}}^{k} \rangle = -\langle c^{\mathrm{p}}, \mathbf{e}_{\mathrm{p}}^{k} \rangle = -g_k \qquad \text{oddness principle.} \tag{7.21}$$

This means that the inversion of the orientation of a $p$-cell (this is the meaning of $-\mathbf{e}_{\mathrm{p}}^{k}$) implies a change in sign of the value of the cochain associated with the cell. This is the *oddness principle* of discrete forms of degree $p$, which corresponds to the oddness principle of chains, shown by Eq. 7.5.

The cells and their labels used in algebraic topology correspond to the points and their coordinates which are commonly used in physics. The role of field functions in a differential setting is played by discrete forms in a discrete setting. In passing from a discrete to a differential setting, the discrete $p$-forms on the primal complex become even differential $p$-forms,[23] whereas the discrete $p$-forms on the dual complex become odd differential forms, and for this reason they will be called *odd discrete* $p$-forms.[24] The elements of the group $\mathcal{G}$ are vectors instead of scalars, and the corresponding differential form is a vector valued differential form.

**An Analogy.** When we walk into a supermarket to purchase goods, such as fruit, we find a number on each type of fruit and a price.

The analogy lies in the fact that the fruits correspond to the cells of a complex, the shopping list of the day corresponds to a chain and the set of prices corresponds to a discrete form.

Let us denote the fruits by $\mathbf{e}^1, \mathbf{e}^2, \cdots \mathbf{e}^n$.

The prices are the elements of a 'discrete form' $p = [p_1, p_2, \cdots p_n]$ which we can represent by a row vector. We can also write $p_h \stackrel{\text{def}}{=} p(\mathbf{e}^h)$ in the same spirit of the notation $y_h = f(x_h)$ for a function of one variable.

Our shopping list of the day specifies the number of kilos of each fruit that we want to buy. These can be collected in a vector $\mathbf{s} = [n_1, n_2, \cdots n_n]^{\mathrm{T}}$ which we represent as a column vector ($\mathbf{s}$ = 'shopping').

Referring to Table 7.5, we see that the number of kilos for every fruit on the shopping list can be viewed as coefficients (weights) of a 'chain' $\mathbf{s}$ and *the total cost C of our shopping can be viewed as the value of the discrete form of prices on the chain of the shopping list.*

---

[22] Hocking and Young [94, p. 298].

[23] Franz [74, p. 45], Hilton and Wylie [92, p. 72].

[24] Burke [32, p. 183].

**Table 7.5** Shopping in a supermarket provides an analogue of chains and cochains

| Fruits | Apples | Oranges | Bananas | Pears | Peaches |
|---|---|---|---|---|---|
| 'Cells' | $\mathbf{e}^1$ | $\mathbf{e}^2$ | $\mathbf{e}^3$ | $\mathbf{e}^4$ | $\mathbf{e}^5$ |
| Prices | $p_1$ | $p_2$ | $p_3$ | $p_4$ | $p_5$ |

Cochain of prices: $p = [p_1, p_2, p_3, p_4, p_5]$

*Day's shopping list*
2 kg apples $n_1 = 2$
3 kg oranges $n_2 = 3$
6 kg pears $n_4 = 6$

'Chain' of day's shopping list

$$\mathbf{s} = 2\mathbf{e}^1 + 3\mathbf{e}^2 + 0\,\mathbf{e}^3 + 6\mathbf{e}^4 + 0\,\mathbf{e}^5 = \begin{bmatrix} 2 \\ 3 \\ 0 \\ 6 \\ 0 \end{bmatrix}$$

Total cost: $C = \sum_{h=1}^{5} n_h\, p_h$

## 7.5.2 Coboundary of a Discrete p-Form

The *coboundary process* on a discrete $p$-form is a process which generates a discrete $(p+1)$-form. It is remarkable that this process plays a key role in physics because balance, circuital equations and the equations forming a difference can be expressed by the coboundary process performed on discrete forms of degree 3, 2 and 1 respectively. This process is analogous, in an algebraic setting, to exterior differentiation on exterior differential forms and leads to typical operators such as 'grad', 'curl' and 'div', along with time derivatives.[25]

In this way we have introduced the coboundary process in a geometric language. This is much more intuitive than the corresponding analytical definition of exterior differential on exterior differentiation forms.

## 7.5.3 Coboundary of a Discrete 0-Form

Let us refer to the first row of Table 7.6, and let us start by considering a physical variable $\Phi$ associated with the points (0-cells) of a primal complex (in number of $N_0$). This fact gives rise to a discrete 0-form, $\phi^0 = [\phi_1, \phi_2, \cdots \phi_{N_0}]$. Let us consider the point $\mathbf{p}^h$, and let $\phi_h$ be the value associated with it (Table 7.7). The process we are describing is performed in two steps:

1. *Step 1.* Transfer the value $\phi_h$ to every coface of the point $\mathbf{p}^h$ (i.e. to every edge $\mathbf{l}^i$ incident with the point $\mathbf{p}^h$) with a plus or minus sign according to whether the mutual incidence is compatible or incompatible. Thus, considering points

[25] Hilton and Wylie [92, p. 72].

**Table 7.6** The two steps of the coboundary process performed on primal cell complex

**Inner orientation**

*First step*
*transfer to the cofaces*

*Second step*
*collect from the faces*

*from points to lines*

$\gamma_\alpha = \phi_i - \phi_h$

*From lines to surfaces*

$\psi_\mu = -\gamma_\alpha + \gamma_\beta + \gamma_\gamma - \gamma_\delta$

*From surfaces to volumes*

$w_h = +\psi_\alpha - \psi_\beta + \psi_\gamma - \psi_\delta + \psi_\epsilon - \psi_\eta$

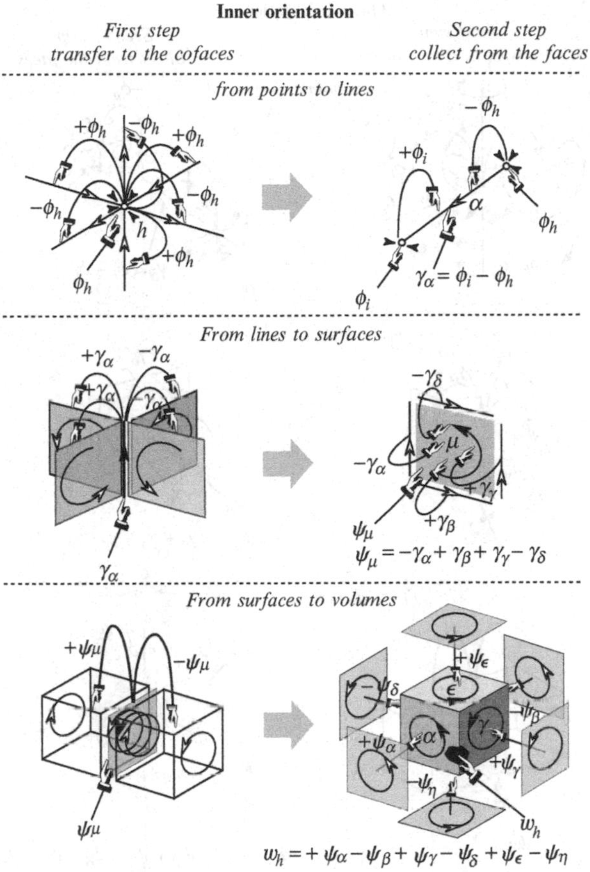

as sinks, the edge which arrives at the point has a compatible orientation; hence, the incidence number is +1. We associate the value $+\phi_h$ with this edge. This process is repeated for all cofaces of $\mathbf{p}^h$, i.e. for the entire coboundary of $\mathbf{p}^h$. The same transfer to the coboundary is repeated for all points of the complex.

2. *Step 2.* For every edge $\mathbf{l}^i$ of the complex (in number of $N_1$) add the values coming from its boundary. In this way, we obtain the quantity $\gamma_i = +\phi_i - \phi_h$. Thus, we have formed a discrete 1-form, $\gamma^1 = [\gamma_1, \gamma_2, \cdots, \gamma_{N_1}]$.

Hence, starting with a discrete 0-form, $\phi^0$, we have formed a discrete 1-form, $\gamma^1$:

$$\phi^0 = [\phi_1, \phi_2, \cdots \phi_{N_0}] \quad \longrightarrow \quad \gamma^1 = [\gamma_1, \gamma_2, \cdots, \gamma_{N_1}] . \tag{7.22}$$

**Table 7.7** The two steps of the coboundary process performed on dual cell complex

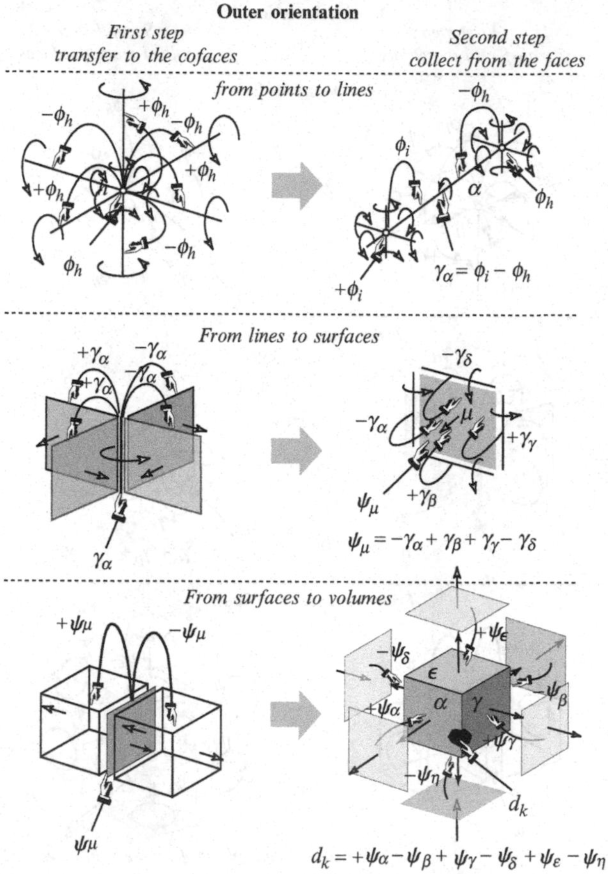


This is the coboundary process performed on a discrete 0-form and denoted by

$$\gamma^1 = \delta\,\phi^0 \,. \tag{7.23}$$

## 7.5.4 Coboundary of a Discrete 1-Form

Let us start by considering a physical variable $\Gamma$ associated with the edges (1-cells) of a primal complex. This gives rise to a discrete 1-form, $\Gamma^1 = [\Gamma_1, \Gamma_2, \cdots \Gamma_{N_1}]$. Let us consider the edge $\mathbf{l}^i$, and let $\Gamma_i$ be the value associated with it. Always with reference to Table 7.6, the process we are describing is performed in two steps as follows:

1. *Step 1.* We transfer the value $\Gamma_i$ to every coface of the edge $\mathbf{l}^i$, with the plus or minus sign according to whether the mutual incidence is compatible or incompatible. Hence this process involves the whole coboundary of $\mathbf{l}^i$. The same transfer to the coboundary is repeated for all edges of the complex.
2. *Step 2.* For every coface $\mathbf{s}^\mu$ of the complex (in number of $N_2$) add the values coming from its boundary. In this way, we obtain the quantity $\psi_\mu = -\Gamma_i + \Gamma_j + \Gamma_\gamma - \Gamma_\delta$,[26] i.e. we have formed a discrete 2-form, $\psi^2 = [\psi_1, \psi_2, \cdots, \psi_{N_2}]$.

Hence, starting with a discrete 1-form, $\Gamma^1$, we have formed a discrete 2-form, $\psi^2$:

$$\Gamma^1 = [\Gamma_1, \Gamma_2, \cdots \Gamma_{N_1}] \qquad \longrightarrow \qquad \psi^2 = [\psi_1, \psi_2, \cdots, \psi_{N_2}] \,. \tag{7.24}$$

This is the coboundary process performed on a discrete 1-form and denoted by

$$\psi^2 = \delta\, \Gamma^1 \,. \tag{7.25}$$

### 7.5.5 Coboundary of a Discrete 2-Form

Let us now describe the coboundary process on a discrete 2-form. Let us start by considering a physical variable $\Psi$ associated with the faces (2-cells) of a primal complex. This gives rise to a discrete 2-form, $\Psi^2 = [\Psi_1, \Psi_2, \cdots \Psi_{N_2}]$. Let us consider the face $\mathbf{s}^i$, and let $\Psi_i$ be the value associated with it. With reference to Table 7.6, the process we are describing is performed in two steps:

1. *Step 1.* Transfer the value $\Psi_i$ to every coface (i.e. every 3-cell incident with the face $\mathbf{s}^i$) with a plus or minus sign according to whether the mutual incidence is compatible or incompatible. Hence this process involves the entire coboundary of $\mathbf{s}^i$. The same transfer to the coboundary is repeated for all faces of the complex.
2. *Step 2.* For every volume $\mathbf{v}^k$ of the complex (in number of $N_3$) add the values coming from its boundary. In this way, we obtain the quantity $\rho_k = +\Psi_i - \Psi_j + \Psi_\gamma - \Psi_\delta + \Psi_\epsilon - \Psi_\eta$.[27] Thus, we have formed a discrete 3-form, $\rho^3 = [\rho_1, \rho_2, \cdots, \rho_{N_3}]$.

Hence, starting with a discrete 2-form, $\Psi^2$ we have formed a discrete 3-form, $\rho^3$:

$$\Psi^2 = [\Psi_1, \Psi_2, \cdots \Psi_{N_2}] \qquad \longrightarrow \qquad \rho^3 = [\rho_1, \rho_2, \cdots, \rho_{N_3}] \,. \tag{7.26}$$

This is the coboundary process performed on a discrete 2-form and denoted by

$$\rho^3 = \delta\, \Psi^2 \,. \tag{7.27}$$

[26] The signs in this sum depend on the mutual orientation of the face and its edges.

[27] The signs in this sum depend on the mutual orientation of the cell with its faces.

The coboundary process we have described using a primal complex can be performed in the same way on a dual complex; in fact, what we need are the incidence numbers, and these exist for an oriented complex which is endowed with an inner or with outer orientation.

### 7.5.6 General Definition of Coboundary Process

DEFINITION. *Given a discrete p-form,* $c^{\mathrm{p}}$, *we may obtain a discrete* $(p+1)$*-form,* $c^{\mathrm{p}+1}$, *by the following process. We associate with every* $(p+1)$*-cell* $\mathbf{e}^k_{\mathrm{p}+1}$ *the sum of the products of the quantities* $g_h$ *associated with its faces, each multiplied by the corresponding incidence number. In this way, with every* $(p+1)$*-cell we associate the quantity*

$$f_k \stackrel{\text{def}}{=} \sum_h \underbrace{[\mathbf{e}^k_{\mathrm{p}+1} : \mathbf{e}^h_{\mathrm{p}}]}_{\textit{incidence number}} \; g_h \,. \tag{7.28}$$

*We can write*

$$g_h = \langle c^{\mathrm{p}}, \mathbf{e}^h_{\mathrm{p}} \rangle \qquad f_k = \langle c^{\mathrm{p}+1}, \mathbf{e}^k_{\mathrm{p}+1} \rangle \,. \tag{7.29}$$

*The discrete form thus obtained is denoted by*

$$c^{\mathrm{p}+1} = \delta\, c^{\mathrm{p}} . \tag{7.30}$$

*This process is called the* coboundary process *and the operator* $\delta$ *which describes it is called the* coboundary operator.

Whereas the boundary process on a chain lowers the dimension of the chain by one unit, the coboundary process on a discrete form raises the degree of the form by one unit. This fact is easily verified by observing that in Eq. 7.10 the first index of the incidence matrix is summed over, whereas in Eq. 7.28 the second index of the incidence matrix is summed over.

### 7.5.7 Discrete Version of Stokes' Theorem

We now show the relation between the boundary operator $\partial$ on chains and the coboundary operator $\delta$ on discrete forms (= cochains). This relation is a simple

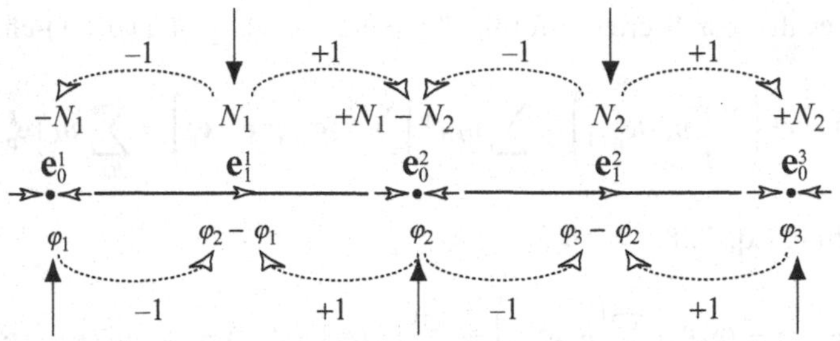


**Fig. 7.8** The simplest, one-dimensional, cell complex allows one to capture the algebraic root of the generalized Stokes theorem

and wonderful mathematical identity which serves as the foundation of three classical identities – Gauss's theorem, Stokes's theorem and Leibniz's theorem.

**Introductory Consideration.** Let us consider the simplest example of a one-dimensional cell complex composed of two 1-cells $\mathbf{e}_1^1, \mathbf{e}_1^2$ and, as a consequence, of three 0-cells $\mathbf{e}_0^1, \mathbf{e}_0^2, \mathbf{e}_0^3$, as shown in Fig. 7.8.

Let us consider a 1-chain $\mathbf{c}_1 = N_1 \mathbf{e}_1^1 + N_2 \mathbf{e}_1^2$ and a 0-cochain $c^0 = [\varphi_1, \varphi_2, \varphi_3]$. Let us form the boundary of the 1-chain $\mathbf{c}_1$,

$$\partial\, \mathbf{c}_1 = (-N_1)\, \mathbf{e}_0^1 + (+N_1 - N_2)\, \mathbf{e}_0^2 + (+N_2)\, \mathbf{e}_0^3, \tag{7.31}$$

and the coboundary of the 0-cochain $c^0$,

$$\delta\, c^0 = [(\varphi_2 - \varphi_1), (\varphi_3 - \varphi_2)]\ . \tag{7.32}$$

Let us find the *value* (here a real number) of the 0-cochain $c^0$ on the boundary of the 1-chain $\mathbf{c}_1$, i.e. the 0-chain $\partial\, \mathbf{c}_1$,

$$\langle c^0, \partial\, \mathbf{c}_1 \rangle \stackrel{\text{def}}{=} \varphi_1\, (-N_1) + \varphi_2\, (+N_1 - N_2) + \varphi_3\, (+N_2), \tag{7.33}$$

and the *value* of the 1-cochain $\delta\, c^0$ on the 1-chain $\mathbf{c}_1$,

$$\begin{aligned} \langle \delta\, c^0, \mathbf{c}_1 \rangle &\stackrel{\text{def}}{=} (\varphi_2 - \varphi_1)\, N_1 + (\varphi_3 - \varphi_2)\, N_2 \\ &\equiv \varphi_1\, (-N_1) + \varphi_2\, (+N_1 - N_2) + \varphi_3\, (+N_2) \\ &= \langle c^0, \partial\, \mathbf{c}_1 \rangle\ . \end{aligned} \tag{7.34}$$

Hence,

$$\langle c^0, \partial\, \mathbf{c}_1 \rangle \equiv \langle \delta\, c^0, \mathbf{c}_1 \rangle \tag{7.35}$$

Stated in words: *the value of a 0-cochain on the boundary of a 1-chain is equal to the value of the coboundary of the 0-cochain on the 1-chain.*

After this introductory consideration, let us prove the same property for a $p$-cochain on an arbitrary cell complex.

Let us evaluate a discrete $p$-form, $c^p$, on the boundary of a $(p+1)$-chain $\mathbf{c}_{p+1}$:

$$c^{\mathrm{p}}(\partial \mathbf{c}_{p+1}) = c^{\mathrm{p}}\left(\sum_k^{\mathrm{f}} n_k\, \partial \mathbf{e}_{\mathrm{p+1}}^k\right) = \sum_k n_k\, c^{\mathrm{p}}\left(\sum_h^{\mathrm{f}} [\mathbf{e}_{\mathrm{p+1}}^k : \mathbf{e}_{\mathrm{p}}^h]\, \mathbf{e}_{\mathrm{p}}^h\right) = \sum_{h,k} n_k\, [\mathbf{e}_{\mathrm{p+1}}^k : \mathbf{e}_{\mathrm{p}}^h]\, g_h \tag{7.36}$$

and, recalling Eq. 7.28,

$$(\delta c^{\mathrm{p}})(\mathbf{c}_{p+1}) = (\delta c^{\mathrm{p}})\left(\sum_k^{\mathrm{f}} n_k\, \mathbf{e}_{\mathrm{p+1}}^k\right) = \sum_k n_k\, (\delta c^{\mathrm{p}})(\mathbf{e}_{\mathrm{p+1}}^k) = \sum_{h,k} n_k\, [\mathbf{e}_{\mathrm{p+1}}^k : \mathbf{e}_{\mathrm{p}}^h]\, g_h\ . \tag{7.37}$$

Hence,

$$c^{\mathrm{p}}(\partial \mathbf{c}_{\mathrm{p+1}}) \equiv (\delta c^{\mathrm{p}})(\mathbf{c}_{\mathrm{p+1}})\ . \tag{7.38}$$

It is much better to rewrite this identity in the equivalent form[28]

$$\langle c^{\mathrm{p}}, \partial \mathbf{c}_{\mathrm{p+1}} \rangle \equiv \langle \delta c^{\mathrm{p}}, \mathbf{c}_{\mathrm{p+1}} \rangle\ . \tag{7.39}$$

**Theorem.** *The value of a discrete p-form on the p-dimensional boundary of a (p+1)-dimensional chain is equal to the value of the coboundary of the discrete p-form on a (p+1)-dimensional chain.*[29]

This theorem is the algebraic form of Stokes' theorem and is called a *generalized* or *combinatorial* form of Stokes' theorem.[30] This shows that Stokes' theorem is an immediate consequence of the definition of the coboundary process. This also shows that Stokes' theorem *is a purely topological relation*[31] and that the various continuity and differentiability conditions usually required in its proof depend on the fact that one uses field functions and derivatives. This is a recurrent situation in physics: *many differentiability requirements do not belong to physical laws but are required by the differential apparatus used in their description.*

This equality is the discrete form[32] of the generalized Stokes theorem, which includes Gauss's theorem, the proper Stokes theorem and the theorem of Leibniz:

[28] This notation is used by Teixeira and Chew [222] and by Desbrun et al. [56].

[29] Hocking and Young [94, p. 301], Dubrovin et al. [58, p. 35], Franz [74, p. 46], Auchmann and Kurz [7].

[30] Franz [74, p. 43], Hocking and Young [94, p. 301].

[31] Synge and Schild [221, p. 269].

[32] Franz [74, p. 45].

$$\begin{cases} \int_{\mathbf{V}} \nabla \cdot \mathbf{u} \, dV = \int_{\partial \mathbf{V}} \mathbf{u} \cdot d\mathbf{S} & \textit{(Gauss)}, \\ \int_{\mathbf{S}} (\nabla \times \mathbf{v}) \cdot d\mathbf{S} = \int_{\partial \mathbf{S}} \mathbf{v} \cdot d\mathbf{L} & \textit{(Stokes)}, \\ \int_{\mathbf{L}} \nabla \phi \cdot d\mathbf{L} = \phi(\partial \mathbf{L}) \equiv \phi(\mathbf{B}) - \phi(\mathbf{A}) & \textit{(Leibnitz)}. \end{cases} \tag{7.40}$$

Note that the generalized Stokes theorem highlights the non-metric nature of the theorem[33] and also applies regardless of the assumptions of differentiability of the vectors $\mathbf{u}, \mathbf{v}$ or of the function $\phi$.

Conversely, Eq. 7.39 can be used to define the coboundary operator.

### *7.5.8 The Coboundary of the Coboundary Vanishes*

The notion of a coboundary operator which raises the degree of a discrete form by one unit corresponds to the notion of an exterior differential which raises the degree of an exterior differential form by one unit.[34]

From identity 7.38 and recalling Eq. 7.13, we obtain the important identity

$$\delta(\delta c^{\mathrm{p}})(\mathbf{c}_{\mathrm{p+2}}) = (\delta c^{\mathrm{p}})(\partial\, \mathbf{c}_{\mathrm{p+2}}) = c^{\mathrm{p}}(\partial\, \partial\, \mathbf{c}_{\mathrm{p+2}}) = c^{\mathrm{p}}(\Theta_{\mathrm{p}}) = 0. \tag{7.41}$$

This means that when the coboundary process is performed twice in sequence, it gives rise to the null element of the group $\mathscr{G}$. This is the algebraic root of many differential identities.

As we will show, these relations are the discrete counterpart of the differential identities:

$$\text{curl}\ (\text{grad} f) \equiv 0, \qquad \text{div}\,(\text{curl}\ \mathbf{v}) \equiv 0\,. \tag{7.42}$$

**Remark.** Physical laws in their original form, as inferred from experiments, naturally involve *global variables* associated with spatial entities and *not* field functions. The formation of densities and rates and the passage to a limit to form field functions deprive physical variables of much of their physical content.

---

[33] Synge and Schild [221, p. 269].

[34] Hilton and Wylie [92, p. 72].

### 7.5.9 Chains or Discrete Forms in Physics?

Some authors on algebraic topology consider chains with real coefficients instead of integer coefficients, i.e. consider the group **R** instead of **Z**. In physics one is tempted to consider these real coefficients as the values of a physical variable.[35] For example, in thermal conduction, the temperatures at the nodes of a cell complex are taken as coefficients of a 0-chain, so

$$\mathbf{T} = T_1\mathbf{e}^1 + T_2\mathbf{e}^2 + \cdots + T_N\mathbf{e}^N = \sum_k^{\mathrm{f}} T_k\mathbf{e}^k \,. \tag{7.43}$$

In this way you do not have available the coboundary process, a fundamental process in the description of physics as it summarizes the three operators gradient, curl and divergence. The coboundary process is defined on the discrete forms and not on the chains.

Moreover, what happens when we want to apply algebraic topology to continuum mechanics? In this case, the physical variable associated with the nodes is not a scalar but a vector, the displacement **u**, and the 0-chains would look like this:

$$\mathbf{u} = \mathbf{u}_1\mathbf{e}^1 + \mathbf{u}_2\mathbf{e}^2 + \cdots + \mathbf{u}_N\mathbf{e}^N = \sum_k^{\mathrm{f}} \mathbf{u}_k\mathbf{e}^k, \tag{7.44}$$

which is a strange object. It is much more natural to consider the coefficients of a chain as integers giving the weights of the single cells and to make use of discrete forms to describe the distribution of the physical variables on the cells.

The boundary process on chains *distributes* the weight of a $p$-cell to the $(p-1)$-cells forming its faces, whereas the coboundary process *collects* the values associated with the $p$-cells forming the faces of a $(p+1)$-cell. This process is the equivalent of the integration of the field variables on lines, surfaces and volumes.

For the description of physics it is natural to separate geometric entities, such as points, lines, surfaces and volumes, from the many possible physical variables associated with them. The additive nature of space elements disappears when the coefficients of a chain are considered as values of a physical variable.

With reference to the use of chains or discrete forms in physics, Post stated: 'Epistemologically it is better if the cohomology of fields is given a primal physical role than the homology of chains over which these fields are being integrated.'[36]

Some authors justify the choice of chains instead of discrete forms due to the fact that in finite-dimensional cell complexes one may establish a one-to-one mapping between $p$-chains and discrete $p$-forms with the same coefficient group $\mathscr{G}$.

---

[35] This was for years the choice of the present author.

[36] Post [182, p. 517].

Hilton and Wylie wrote: 'This does not justify us in concealing the distinction between these two notions, for the correspondence depends on a choice of basis of $\mathscr{C}_{\rm p}$. The situation in vector space theory is entirely analogous. Although a finite-dimensional vector space is isomorphic to its dual, the elements of the dual are linear functions and it is essential that the two spaces be not confused.'[37]

Since a $p$-dimensional region can be covered by a $p$-chain with integer coefficients, it follows that discrete forms are a natural tool for describing set functions.

Henry Lebesgue wrote: "Si pourtant, on parle peu de ces fontions [les fonctions d'ensemble], c'est que les mathématiciens n'ont pas encore créé l'Algèbre et l'Analyse des fonctions de domaine."[38] The book by Lebesgue was written in 1928, and the notion of discrete form was introduced in subsequent years; we see that the notion of discrete form and the corresponding theory, the theory of cohomology, are the tools Lebesgue was talking about.

## 7.6 Coboundary Process in Two Dimensions

With reference to Fig. 7.9, let us consider, as a problem type, a two-dimensional cell complex formed by four squares. The cell complex has 9 nodes, 12 edges and 4 bidimensional cells, i.e. 9 **P**, 12 **L**, 4 **S**. The nodes, edges and faces are labelled according to an arbitrary rule. The corresponding dual complex is shown in Fig. 7.9. On account of the one-to-one mapping of the edges of the primal complex and those of the dual complex, and of the one-to-one mapping of the cells of the dual complex with the nodes of the primal nodes, it is spontaneous to assign to every element of the dual complex the same label as the corresponding element of the primal complex. In this way, for example, the cell number 9 of the dual complex corresponds to the node number 9 of the primal complex.

### *7.6.1 Coboundary Process on Primal Complex*

At this point our purpose is to describe the coboundary process. Let us assign a value of a scalar variable $\Phi$ with every node; in doing this, we have formed a discrete 0-form, $[\varphi_1, \varphi_2, \cdots, \varphi_9]$, on the primal cell complex. We now show how to perform the coboundary process on this discrete 0-form to obtain a discrete 1-form. Referring to Fig. 7.10, this process is divided into two steps. The first step is to transfer the values of $\varphi_k$ from every node to all edges which are incident to the

---

37 Hilton and Wylie [92, p. 67].

38 Lebesgue [130, p. 293]. *Thus, one talks less of set functions because mathematicians have not invented algebra and analysis of set functions.*

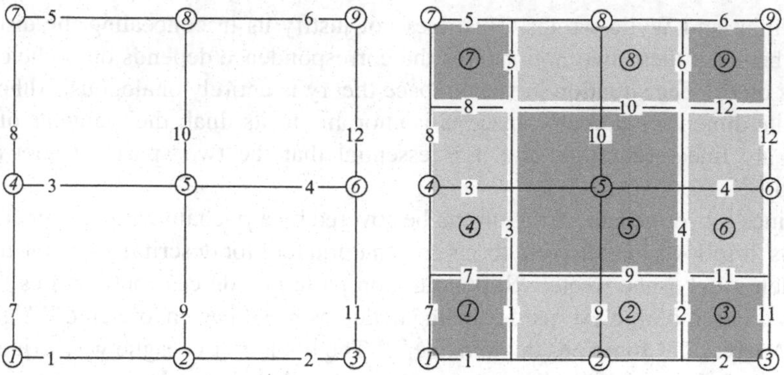

**Fig. 7.9** (*Left*) A cell complex; (*right*) its dual

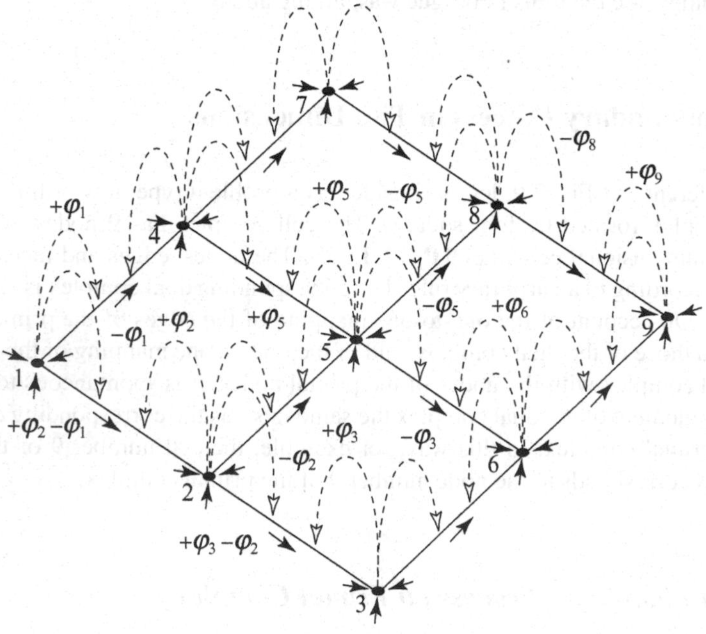

**Fig. 7.10** Coboundary of a discrete 0-form in a two-dimensional complex

node, with a plus or minus sign according to the mutual incidence nodes/edges. The second step is to sum, for every edge, the values coming from its bounding nodes. If we denote by $U$ the global variable associated with the edges of the primal complex, for edge number 2 (Fig. 7.9), then we obtain, say, $U_2 = \varphi_3 - \varphi_2$. In general,

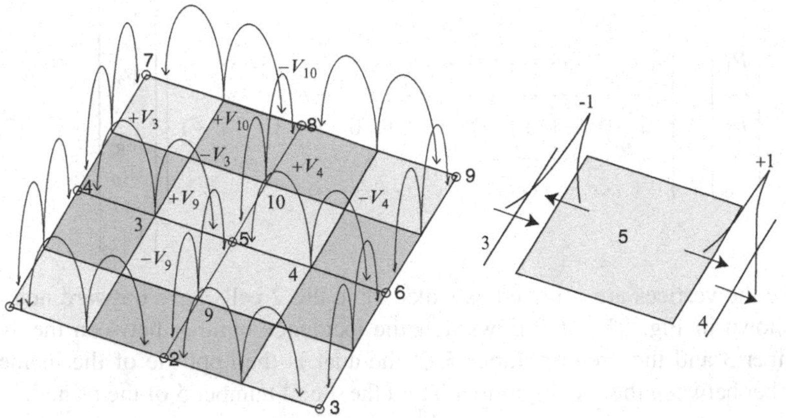


**Fig. 7.11** Coboundary process from a discrete 1-form to a discrete 2-form on dual complex

$$\begin{bmatrix} U_1 \\ U_2 \\ \cdots \\ U_{12} \end{bmatrix} = \begin{bmatrix} -1 & +1 & 0 & \cdots \\ 0 & -1 & +1 & \cdots \\ \cdots & \cdots & \cdots & \cdots \\ 0 & 0 & -1 & +1 \end{bmatrix} \begin{bmatrix} \varphi_1 \\ \varphi_2 \\ \cdots \\ \varphi_9 \end{bmatrix} \qquad or \qquad \mathbf{U} = \mathbf{G}\boldsymbol{\Phi}\,. \tag{7.45}$$

Hence, we have obtained a discrete 1-form, $[U_1, U_2, \cdots, U_{12}]$, starting from the discrete 0-form, $[\varphi_1, \varphi_2, \cdots, \varphi_9]$, by the coboundary process. Denoting by $\Phi^0$ the discrete 0-form and by $U^1$ the discrete 1-form, using the language of algebraic topology we can write $U^1 = \delta\,\Phi^0$, where $\delta$ denotes the coboundary operator.

## 7.6.2 Coboundary Process on Dual Complex

Let us recall that, as far as physics is concerned, a discrete $p$-form arises when we assign a value of a physical variable with every elementary $p$-chain. Hence $p$ is referred to the dimension of the cell, $p = 0, 1, 2, 3$. Referring to Fig. 7.9, let us associate with every 1-cell of the dual complex a value of a physical variable $V$; in doing this, we form a discrete 1-form, $[V_1, V_2, \cdots, V_n]$, on the dual. Since the orientation of the dual cells is the outer one, it follows that the $p$-form is odd (or pseudo $p$-form). Performing the coboundary process on this discrete 1-form we obtain a discrete 2-form, $[P_1, P_2, \cdots, P_m]$. The two-step process can be described as follows: first we transfer the value $V_\alpha$ associated with the 1-cell number $\alpha$ to the two 2-cells adjacent to it; hence, for every 2-cell, say $k$, we sum all values coming from its boundary, obtaining a value $P_k$. Thus, for example, we have $P_5 = +V_4 + V_{10} - V_3 - V_9$.

$$\begin{bmatrix} P_1 \\ \cdots \\ P_5 \\ \cdots \\ P_9 \end{bmatrix} = \begin{bmatrix} \cdots & \cdots & \cdots & \cdots & \cdots & \cdots & \cdots & \cdots & \cdots & \cdots & \cdots & \cdots \\ \cdots & \cdots & \cdots & \cdots & \cdots & \cdots & \cdots & \cdots & \cdots & \cdots & \cdots & \cdots \\ 0 & 0 & -1 & +1 & 0 & 0 & 0 & 0 & 0 & 0 & -1 & +1 \\ \cdots & \cdots & \cdots & \cdots & \cdots & \cdots & \cdots & \cdots & \cdots & \cdots & \cdots & \cdots \\ \cdots & \cdots & \cdots & \cdots & \cdots & \cdots & \cdots & \cdots & \cdots & \cdots & \cdots & \cdots \end{bmatrix} \begin{bmatrix} V_1 \\ \cdots \\ V_3 \\ \cdots \\ V_9 \\ V_{10} \\ \cdots \\ V_{12} \end{bmatrix}. \qquad (7.46)$$

Since the vertices are oriented as sinks while the 2-cells have outward normals, as shown in Fig. 7.11, it follows that the incidence number between the 1-cell number 3 and the 2-cell number 5 of the dual is the opposite of the incidence number between the 1-cell number 3 and the 0-cell number 5 of the primal.

Hence, $\tilde{d}_{k\alpha} = -\overline{g}_{\alpha k}$, i.e. $\tilde{\mathbf{D}} = -\overline{\mathbf{G}}^{\mathrm{T}}$. Denoting by $P^2$ the discrete 2-form and by $V^1$ the discrete 1-form we can write, in the language of algebraic topology, $P^2 = \delta V^1$, where $\delta$ denotes the coboundary operator.

## 7.7 The Wonderful Role of the Coboundary Process in Physics

A physical law is a link between the variables that describe a phenomenon. The field equations, when expressed in terms of the differential formulation, involve space and time derivatives; thus, Faraday's law, curl $\mathbf{E} = -\partial\mathbf{B}/\partial t$, links the curl of the vector $\mathbf{E}$ with the partial derivative of the vector $\mathbf{B}$. This implies that, to write the field equations, we must perform a process on each variable. So we perform the curl on $\mathbf{E}$ and the time derivative on $\mathbf{B}$. The processes which we perform on field variables include

1. The space or time derivative,
2. The gradient,
3. The curl,
4. The divergence.

If, instead of the differential formulation, we resort to an algebraic formulation using global variables and discrete forms, then these four processes become a single process, one of forming the coboundary of discrete forms. This unification is the algebraic analogue of the process of performing the exterior differential on exterior differential forms of various degrees.

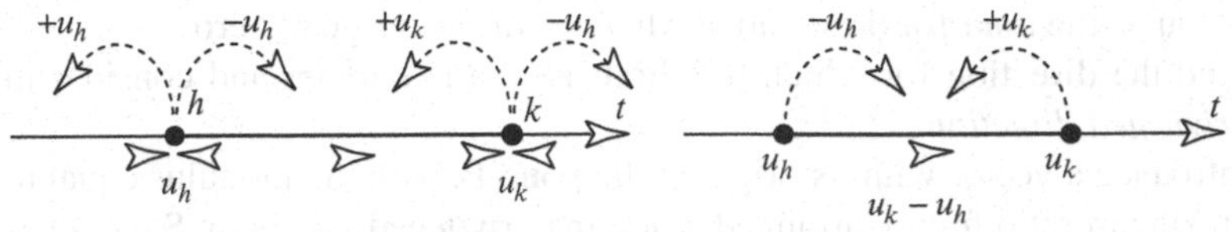


**Fig. 7.12** The coboundary process in one dimension, here on the time axis, is the discrete analogue of a derivative

### *7.7.1 Algebraic Analogue of Derivative*

We recall that the derivative of a function is a process which can be divided into three steps: (a) compute an increment; (b) compute the ratio of this increment for the increment of the variable; (c) perform the limit process for the increment of the variable going to zero. In a discrete setting, the fundamental process is that of computing the increment. To this end, let us consider a one-dimensional cell complex, similar to one we can build on the time axis and a function $u(t)$. With reference to Fig. 7.12, let us caryy out this two-step process:

1. Transfer the value $u_h$ of the function $u(t)$ at the instant $t_h$ to the two intervals which have the given instant as common faces (these intervals are the *cofaces* of the instant) with a plus or minus sign in accordance with the corresponding incidence number.
2. For every 1-cell take the sum of the values coming from its bounding instants (the *faces*).

Therefore, starting with the discrete 0-form, $[u_1, u_2, \ldots u_{N_0}]$, one obtains a discrete 1-form, $[v_1, v_2, \ldots v_{N_1}]$, where $N_0$ is the number of 0-cells (nodes) and $N_1$ is the number of 1-cells (edges). This two-step process is the *coboundary* of the discrete 0-form in one dimension and is the analogue of the differential of a function of one variable. Since the differential can be decomposed into the product of the derivative and the differential of the independent variable, we can say that the coboundary process performed on a discrete 0-form in one dimension is analogous to the derivative.

### *7.7.2 Algebraic Analogue of Gradient*

We recall that the gradient of a *scalar-valued* function $\phi(\mathbf{P})$ is a *vector-valued* function $\mathbf{u}(\mathbf{P})$, obtained performing a process which can be divided into five steps:

1. Considering a point $\mathbf{P}$ of a field, evaluate the *increments* of the function $\phi$ along various oriented directions going out from the point.
2. Evaluate the *ratios* of these increments to the corresponding *distances*.

3. Evaluate the *limit* for these ratios when the distances go to zero.
4. Find the direction for which this limit is at a maximum and consider this a *privileged direction*.
5. Introduce a vector with its origin at the point **P**, with the modulus equal to the maximum ratio found, arranged along the privileged direction. Such a vector-valued function **u**(**P**) is defined as the *gradient* of the scalar-valued function $\phi$ at the point considered and will be denoted by **u**(**P**) = grad $\phi(\mathbf{P})$.

What is the corresponding notion in an algebraic setting? Introducing a cell complex the function $\Phi(\mathbf{x})$ assigns the value $\phi_h$ to each node $\mathbf{p}^h$. This gives rise to a discrete 0-form, $\Phi^0 = [\phi_1, \phi_2, \ldots \phi_{N_0}]$. We evaluate the increments of the function along the edges $\mathbf{l}^i$ of the complex; this is done by adding the function's values corresponding to extreme points of each segment, with the plus or minus sign according to the mutual incidence number between nodes and edges.

Let $\gamma_i$ be the increment of the function along the edge $\mathbf{l}^i$; the set of all $\gamma_i$ gives rise to a discrete 1-form, $\gamma^1 = [\gamma_1, \gamma_2, \ldots \gamma_{N_1}]$, which is the coboundary of the discrete 0-form, $\phi^0 = [\phi_1, \phi_2, \ldots \phi_{N_0}]$:

$$
\begin{array}{ccc}
\phi(x, y, z) & \longrightarrow & \mathbf{u}(x, y, z) = \nabla\phi(x, y, z), \\
\phi^0 = [\phi_1, \phi_2, \ldots \phi_{N_0}] & \longrightarrow & \gamma^1 = \delta\phi^0
\end{array}
\tag{7.47}
$$

Thus, in an algebraic context, an operation similar to the formation of the gradient of a scalar field is the formation of a *coboundary* of a discrete 0-form starting from a discrete 1-form.

### 7.7.3 Algebraic Analogue of the 'Curl'

Recall that the curl of a *vector-valued* function **u**(**P**) is another vector-valued function **v**(**P**) obtained by performing a process that can be divided into five steps:

1. Considering a point **P** and taking a small plane surface centred at the point, evaluate the line *integral* of the vector along the boundary of the surface.
2. Evaluate the *ratios* between such circulation and the *area* of the surface.
3. Perform the *limit* of this *ratio* when the surface shrinks to the point **P**.
4. Varying the direction of the plane surface, search for the direction for which this limit is at a *maximum* and consider this a *privileged direction*.
5. Introduce a vector with its origin at the point **P**, with modulus equal to the maximum ratio found, situated along the privileged direction and oriented in such a way that the orientation of the closed line (the one used to evaluate the circulation) and the vector form a clockwise screw. This vector is the *curl* of the given vector field at the point considered and will be denoted by **v**(**P**) = curl **u**(**P**).

What is the corresponding notion in an algebraic setting? Let us evaluate the line integral along each edge $\mathbf{l}_i$ of a cell complex, and let us denote by $\gamma_i$ the value of this line integral; the set of all these values gives rise to a discrete 1-form, $\gamma^1 = [\gamma_1, \gamma_2, \ldots \gamma_{N_1}]$. At this point, for each 1-cell $\mathbf{l}_i$ we *transfer* the value $\gamma_i$ thus evaluated to all cofaces of the 1-cell, each multiplied by the mutual incidence number between the 1-cell and the coface. As a second step, for every 2-cell $s_j$ we *add* all values thus transferred. Denoting by $\psi_j$ the sum thus obtained, we can construct a discrete 2-form, $\psi^2 = [\psi_1, \psi_2, \ldots \psi_{N_2}]$. This discrete 2-form is the *coboundary* of the discrete 1-form $\gamma^1$, and we write $\psi^2 = \delta\gamma^1$. Hence, we have the following equivalence:

$$\begin{array}{ccc} \mathbf{u}(x,y,z) & \longrightarrow & \mathbf{v}(x,y,z) = \nabla \times \mathbf{u}(x,y,z) \\ \gamma^1 = [\gamma_1, \gamma_2, \ldots \gamma_{N_1}] & \longrightarrow & \psi^2 = \delta\gamma^1 . \end{array} \tag{7.48}$$

Hence, in an algebraic setting, the analogue of curl of a vector field is coboundary of the discrete 1-form of circulations that gives rise to a discrete 2-form.

### 7.7.4 Algebraic Analogue of Divergence

Recall that the divergence of a vector-valued function $\mathbf{v}(\mathbf{P})$ is a scalar-valued function $\psi(\mathbf{P})$ obtained by performing a process which can be divided into three steps:

1. Considering a point $\mathbf{P}$ and taking a small *space region* centred at the point, consider the *flux* of the vector across the boundary of such a region.
2. Evaluate the *ratio* between this flux and the *volume* of the small region.
3. Compute the *limit* of this ratio when the region contracts to the point $\mathbf{P}$.

In this way, we obtain a scalar-valued function $\psi(\mathbf{P})$, called the *divergence* of the given vector-valued function, which will be denoted by $\psi(\mathbf{P}) = \operatorname{div} \mathbf{v}(\mathbf{P})$.

What is the corresponding notion in an algebraic setting? Let us denote by $\psi_j$ the flux of the vector through every 2-cell $\mathbf{s}_j$ of a cell complex. In this fashion one obtains a discrete 2-form of fluxes $\psi^2 = [\psi_1, \psi_2, \ldots \psi_{N_2}]$. The first step is to *transfer* the flux associated with every 2-cell to its cofaces, each flux being multiplied by the incidence number between the 2-cell and the coface. The second step is to *add* the values of the fluxes thus transferred for every 3-cell $\mathbf{v}_h$. In this way, we obtain a quantity $d_h$ for every cell $\mathbf{v}_h$. This gives rise to a discrete 3-form, $d^3 = [d_1, d_2, \ldots d_{N_3}]$:

$$\begin{array}{ccc} \mathbf{v}(x,y,z) & \longrightarrow & w(x,y,z) = \nabla \cdot \mathbf{v}(x,y,z), \\ \psi^2 = [\psi_1, \psi_2, \ldots \psi_{N_2}] & \longrightarrow & d^3 = \delta\psi^2 . \end{array} \tag{7.49}$$

Hence, in an algebraic setting, the analogue of a divergence of a vector field, which is described by the discrete 2-form of fluxes, is a discrete 3-form formed by the *coboundary* of the discrete 2-form. In summary:

- A *gradient* arises from scalar quantities associated with *points*.
- A *curl* arises from scalar quantities associated with *lines*.
- A *divergence* arises from scalar quantities associated with *surfaces*.

## 7.8 Examples of Coboundary Process in Physics

**Example 1.** Let us consider a cell complex in a region in which there is an electric field; an electric voltage is associated with every 1-cell of the complex. Since all 1-cells have a number as a label, one can describe the distribution of the voltages with an array:

$$E^1 \stackrel{\text{def}}{=} [e_1, e_2, \dots e_{N_1}] \qquad e_k \text{ are voltages,} \tag{7.50}$$

i.e. with an *algebraic* vector whose components are the voltages of the single 1-cells. This association is a map from the 1-cells to the real numbers which are the values of the voltage. Therefore, instead of writing the voltages as $e_k$, one can write $E[\mathbf{e}_1^k]$. We use square brackets instead of round ones because we have a *set function*:

$$e_k \stackrel{\text{def}}{=} E[\mathbf{e}_1^k] \qquad \text{analogous to} \qquad y_k = f(x_k)\ . \tag{7.51}$$

This notation is analogous to that used in the theory of functions where $y_k$ is the value of the function $f$ in $x_k$. One says that $e_k$ is the *value* of the map $E$ on the 1-cell $\mathbf{e}_1^k$. Since voltages are additive on lines, this map is easily extended from cells to chains. One can write

$$E[\mathbf{c}_1] = E\left[\sum_k^{\mathrm{f}} n_k\, \mathbf{e}_1^k\right] = \sum_k n_k\, E[\mathbf{e}_1^k] = \sum_k n_k\, e_k\ . \tag{7.52}$$

Hence, one can say that the map $E$ assigns a real number to every 1-chain. This map is a discrete 1-form.

**Example 2.** Let us consider a cell complex in a region in which there is a magnetic field; with every 2-cell of the complex is associated a magnetic flux. Since all 2-cells $\mathbf{e}_2^h$ have a number as a label, we can describe the distribution of the magnetic fluxes with an algebraic vector:

$$\Phi^2 \stackrel{\text{def}}{=} [\Phi_1, \Phi_2, \dots \Phi_{N_2}] \qquad \Phi_i \text{ are magnetic fluxes.} \tag{7.53}$$

This association is a map from the 2-cells of the complex to the real numbers which are the values of the magnetic flux. If we denote this map by $\Phi$, then we can write $\Phi_i = \Phi[\mathbf{e}_2^h]$ and say that $\Phi_i$ is the *value* of $\Phi$ on the cell $\mathbf{e}_2^h$. The value of the map $\Phi$ on the 2-chain $\mathbf{c}_2$ is given by

$$\Phi[\mathbf{c}_2] = \Phi\left[\sum_h^{\mathrm{f}} n_h \mathbf{e}_2^h\right] = \sum_h n_h \Phi[\mathbf{e}_2^h] = \sum_h n_h \Phi_h\ . \tag{7.54}$$

This map is a discrete 2-form.

In the following subsections, we will give two examples of the coboundary process which we want to emphasize because they play a major role, respectively, in electromagnetism and continuum mechanics.

### 7.8.1 Gauss's Law of Electrostatics

Let us consider an electric field in a space region $\Omega$. Since electric charges and electric fluxes are source variables, they must be associated with space elements endowed with an outer orientation.

Gauss's law of electrostatics asserts that the electric flux $\Psi$ associated with the boundary $\partial\widetilde{\mathbf{V}}$ of a volume $\widetilde{\mathbf{V}}$ is equal to the electric charge contained in the volume. This can be expressed in a formula as follows:

$$\Psi[\partial\widetilde{\mathbf{V}}] \overset{\text{law}}{=} Q[\widetilde{\mathbf{V}}] \,. \tag{7.55}$$

Let us cover the whole region with a cell complex and its dual. Let $\tilde{\mathbf{v}}^k$ be the $k$th dual 3-cell and $\tilde{\mathbf{s}}^i$ the $i$th dual 2-cell. Let us remark that, on account of the one-to-one correspondence between the $p$-cells of the primal complex and the $(p–1)$-cells of the dual complex, the number $\tilde{N}_3$ of the dual 3-cells is equal to the number $N_0$ of the primal 0-cells; the number $\tilde{N}_2$ of the dual 2-cells is equal to the number $N_1$ of the primal 1-cells. If we introduce the variables

$$\Psi_i \overset{\text{def}}{=} \Psi[\tilde{\mathbf{s}}^i] \qquad\qquad Q_k \overset{\text{def}}{=} Q[\tilde{\mathbf{v}}^k], \tag{7.56}$$

then we can describe the space distribution of the electric charges and electric fluxes by a discrete 2-form $\Psi^2$ and a discrete 3-form $Q^3$ respectively, where

$$\begin{cases} \Psi^2 \overset{\text{def}}{=} [\Psi_1, \Psi_2, \dots \Psi_{N_1}] & \longrightarrow \quad \text{discrete 2-form of electric fluxes} \\ Q^3 \overset{\text{def}}{=} [Q_1, Q_2, \dots Q_{N_0}] & \longrightarrow \quad \text{discrete 3-form of electric charges}\,. \end{cases} \tag{7.57}$$

Since

$$\Psi[\partial\tilde{\mathbf{v}}^k] = \sum_i [\tilde{\mathbf{v}}^k : \tilde{\mathbf{s}}^i]\, \Psi[\tilde{\mathbf{s}}^i] = \sum_i \tilde{d}_{ki}\, \Psi_i, \tag{7.58}$$

Gauss's law 7.55 can be written in the form

$$\sum_i \tilde{d}_{ki}\, \Psi_i = Q_k \,. \tag{7.59}$$

Using the coboundary operator in compact form we obtain

$$\delta\Psi^2 = Q^3 \,. \tag{7.60}$$

This is equivalent to the notation of the differential forms

$$\mathrm{d}\Psi^{(2)} = Q^{(3)} \tag{7.61}$$

**Table 7.8** Various forms of Gauss's law of electrostatics

| | |
|---|---|
| *Global notation:* | $\Psi[\partial\tilde{\mathbf{V}}] = Q[\tilde{\mathbf{V}}]$ |
| *Integral notation:* | $\int_{\partial\tilde{\mathbf{V}}} \mathbf{D}\cdot\mathbf{n}\,\mathrm{d}S = \int_{\tilde{\mathbf{V}}} \rho\,\mathrm{d}V$ |
| *Algebraic notation:* | $\sum_i \tilde{d}_{ki}\,\Psi_i = Q_k$ |
| *Algebraic topology: coboundary operator:* | $\delta\Psi^2 = Q^3$ |
| *Differential forms: exterior differential:* | $\mathrm{d}\Psi^{(2)} = Q^{(3)}$ |
| *Differential notation: divergence:* | $\nabla\cdot\mathbf{D} = \rho$ |

and to the traditional differential formulation

$$\nabla\cdot\mathbf{D} = \rho\,. \tag{7.62}$$

Table 7.8 shows different ways of expressing Gauss's law of electrostatics.

### 7.8.2 Equilibrium Law of Continuum Mechanics

Let us consider the equilibrium of an element of volume subjected to its weight and to internal surface forces. Let us denote by $\mathbf{F}[\tilde{\mathbf{V}}]$ the resultant of the body forces acting on a volume $\tilde{\mathbf{V}}$ of the continuum and by $\mathbf{F}[\partial\tilde{\mathbf{V}}]$ the resultant of the internal surface forces acting on the boundary of the volume. The equilibrium condition can be expressed as

$$\mathbf{T}[\partial\tilde{\mathbf{V}}] + \mathbf{F}[\tilde{\mathbf{V}}] = \mathbf{0} \qquad (global). \tag{7.63}$$

Let us introduce a cell complex and its dual. Since forces are associated with an outer orientation, we must impose the balance of forces on a dual 3-cell $\tilde{\mathbf{v}}^k$, and the equilibrium Eq. 7.63 becomes

$$\mathbf{F}[\partial\tilde{\mathbf{v}}^k] + \mathbf{F}[\tilde{\mathbf{v}}^k] = \mathbf{0}\,. \tag{7.64}$$

Setting

$$\mathbf{F}_i \stackrel{\text{def}}{=} \mathbf{F}[\tilde{\mathbf{s}}^i] \qquad \mathbf{F}_k \stackrel{\text{def}}{=} \mathbf{F}[\tilde{\mathbf{v}}^k], \tag{7.65}$$

and since

$$\mathbf{F}[\partial\tilde{\mathbf{v}}^k] = \sum_i \tilde{d}_{hi}\,\mathbf{F}_i, \tag{7.66}$$

**Table 7.9** Various forms of equilibrium equation in continuum mechanics

| | |
|---|---|
| (1) *Global formulation*: | $\mathbf{T}[\partial\tilde{\mathbf{V}}] + \mathbf{F}[\tilde{\mathbf{V}}] = 0$ |
| (2) *Integral formulation*: | $\int_{\partial\tilde{\mathbf{V}}} \boldsymbol{\tau}\cdot\mathbf{t}\,\mathrm{d}S + \int_{\tilde{\mathbf{V}}} \mathbf{f}\,\mathrm{d}V = 0$ |
| (3) *Algebraic formulation* (global): | $\sum_i \tilde{d}_{ki}\,\mathbf{T}_i + \mathbf{F}_k = 0$ |
| (4) *Algebraic topology: coboundary operator*: | $\delta T^2 + F^3 = 0$ |
| (5) *Differential forms: exterior differential*: | $D\psi^{(2)} + f^{(3)} = 0$ |
| (6) *Differential formulation: divergence*: | $\nabla_i\,\sigma^i_k + f_k = 0$ |
| | $(\mathrm{Div}\,\boldsymbol{\sigma} + \mathbf{f} = 0)$ |

the equilibrium Eq. 7.64 can be written as

$$\sum_i \tilde{d}_{hi}\,\mathbf{T}_i + \mathbf{F}_k = 0 \qquad (\textit{global on a cell } \tilde{\mathbf{v}}^k), \tag{7.67}$$

where ($i = 1, 2, \ldots N_2;\ k = 1, 2, \ldots N_3$). Setting

$$T^2 \stackrel{\text{def}}{=} [\mathbf{T}_1, \mathbf{T}_1, \cdots \mathbf{T}_{N_2}] \qquad F^3 \stackrel{\text{def}}{=} [\mathbf{F}_1, \mathbf{F}_2, \cdots \mathbf{F}_{N_3}] \tag{7.68}$$

we can write

$$\delta T^2 + F^3 = 0\,. \tag{7.69}$$

This is the discrete analogue of the equation

$$\nabla_i\,\sigma^i_k + f_k = 0 \qquad (\textit{differential}) \tag{7.70}$$

used in the differential setting. We remark that in the algebraic Eq. 7.67, the indices are the labels of faces and cells of the dual complex respectively,[39] whereas in the differential formulation 7.70, $k, i$ are coordinate indices.

Table 7.9 shows different forms of the equilibrium equation in statics of continua.

As we have seen, the four operations – derivative, gradient, curl and divergence – are particular embodiments of a single process, the coboundary process. If we pause for a few moments to think that these four operations are ubiquitous in physics, we are struck by the unifying role played by the coboundary process in the topological equations which describe fundamental physical laws. We are also

[39] Remember that $\tilde{N}_2 = N_1$ and $\tilde{N}_3 = N_0$.

struck by the great unifying power of algebraic topology in the description of physical fields. We are familiar with the great role played by differential calculus in the description of field laws; now we see that an even greater role is played by an algebraic-topological process, i.e. the *coboundary process*.

The given examples are part of a general result emerging from a comparative analysis of physical theories, the *coboundary principle*, which states:

COBOUNDARY PRINCIPLE: *in every physical theory there are elementary physical laws which assert that a global physical variable associated with a p-dimensional manifold* $\boldsymbol{\Omega}$ *is equal to a global physical variable associated with its* $(p-1)$*-dimensional boundary* $\partial\boldsymbol{\Omega}$.

# Chapter 8
# Birth of Classification Diagrams

## 8.1 Classification Diagram of Physical Variables

As was shown in Sect. 4.7 (p. 84), the four space elements can be organized in a diagram. The diagram is composed of two columns: on the left we have the four space elements endowed with an inner orientation, whereas on the right we have the four elements endowed with an outer orientation. The order of the space elements in the right column is inverted with respect to the order of the elements in the left column. This inversion is a natural consequence of considering a cell complex and its dual.

On account of the association of global physical variables with oriented space elements, the same classification diagram can be used to classify global physical variables. In particular, since the configuration variables of all theories are associated with space elements endowed with an inner orientation, as stated in the association principle (p. 116), they find their place in the left column of the diagram. By the same principle, source variables, which are referred to the space elements endowed with an outer orientation, find their place in the right column.

Let us give an example using arbitrary letters such as $A, B$ to represent space global variables which describe the configuration and $C, D$ to represent the space global variables which describe the sources. Hence, for a generic scalar physical field such as electrostatics, stationary thermal conduction and stationary perfect fluid motion, the association with space global physical variables and oriented space elements can be expressed as follows:

$A[\overline{\mathbf{P}}]$: a global variable associated with points of a primal complex,
$B[\overline{\mathbf{L}}]$: a global variable associated with lines of a primal complex,
$C[\widetilde{\mathbf{S}}]$: a global variable associated with surfaces of a dual complex,
$D[\widetilde{\mathbf{V}}]$: a global variable associated with volumes of a dual complex.

E. Tonti, *The Mathematical Structure of Classical and Relativistic Physics*, Modeling and Simulation in Science, Engineering and Technology,
DOI 10.1007/978-1-4614-7422-7_8, © Springer Science+Business Media New York 2013

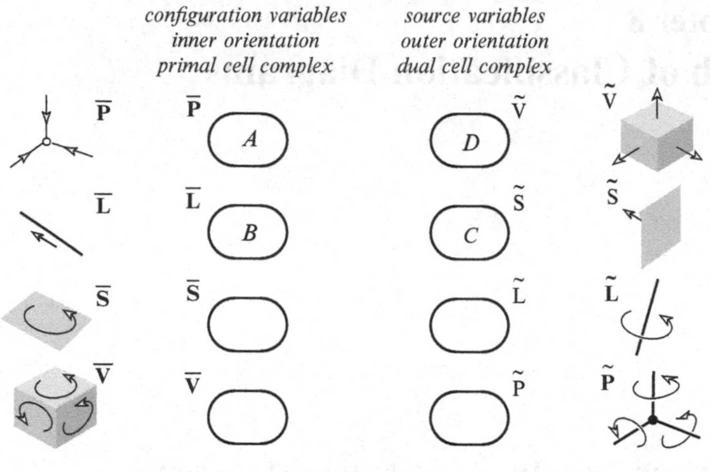


**Fig. 8.1** Classification of space global variables

Hence, shifting the symbols of the oriented space elements outside the boxes and inserting the corresponding space global variables $A, B, C, D$ we obtain the classification diagram shown in Fig. 8.1.

Let us consider the densities of these space global variables. As a result of the position of the global variables in the classification diagram which we have choosen for this example, we can say that since $A$ is associated with points, it coincides with the corresponding density, i.e. it is both a global and a field variable at the same time. The density of the scalar variable $B$, associated with lines, is a vectorial variable $\mathbf{b}$ whose integral along a line gives the scalar variable $B$. The density of the scalar variable $C$, associated with surfaces, is another vectorial variable $\mathbf{c}$ whose flux across a surface gives the scalar variable $C$. The density of the scalar variable $D$, associated with volumes, is a scalar $d$ whose volume integral gives the scalar variable $D$. We see that $\mathbf{b}$ is a *line vector* and $\mathbf{c}$ is a *surface vector*, according to Maxwell's nomenclature. In summary:

$$\begin{array}{l|cc|cc}
 & \text{Configuration variables} & & \text{Source variables} & \\
\text{Global variables} & A[\overline{\mathbf{P}}] & B[\overline{\mathbf{L}}] & C[\tilde{\mathbf{S}}] & D[\tilde{\mathbf{V}}] \\
 & \downarrow & \downarrow & \downarrow & \downarrow \\
\text{Field variables} & A[\overline{\mathbf{P}}] & \mathbf{b}[\overline{\mathbf{L}}] & \mathbf{c}[\tilde{\mathbf{S}}] & d[\tilde{\mathbf{V}}]
\end{array} \tag{8.1}$$

The corresponding classification diagram for the densities is shown in Fig. 8.2. In this diagram the numbers 1, 3, 3, 1 in each column denote the number of families of cells of the same dimension (p. 66). We see that the number of families of cells coincides with the number of components of the local density variable. In other words, in the box marked $3\overline{\mathbf{L}}$ we have a vector $\mathbf{b}$ with three components; in the box marked $3\tilde{\mathbf{S}}$ we have a vector $\mathbf{c}$ with three components; in the box marked $1\overline{\mathbf{P}}$

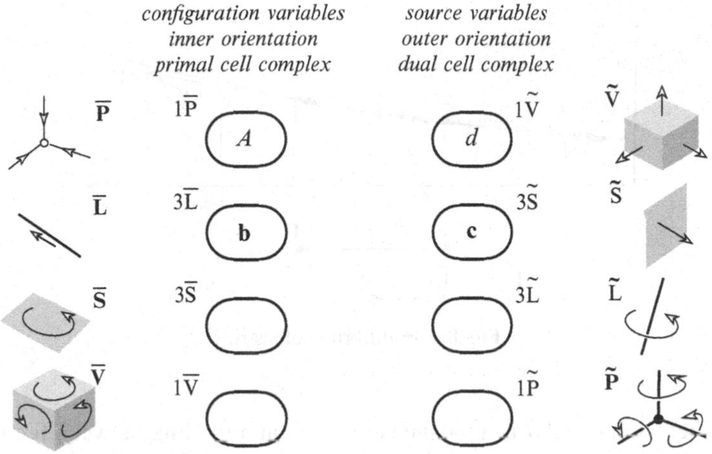


**Fig. 8.2** Corresponding classification of field variables

we have a scalar $d$, i.e. one component. Note that a vector has three components, equal to the number of vectors of a local basis, which in turn is equal to the number of coordinate lines of a coordinate system. Hence the number 3 in the symbols $3\overline{\mathbf{L}}$ and $3\tilde{\mathbf{L}}$ denotes two things: the number of components of the vector and the number of coordinate lines passing through the point. In a similar way, the number 3 in the symbol $3\overline{\mathbf{S}}$ and $3\tilde{\mathbf{S}}$ denotes two things: the number of components of the vector and the number of the coordinate surfaces passing through the point.

Since the differential formulation makes implicit use of the cell complex formed by a coordinate system, for which the coordinate lines and coordinate surfaces can be divided into families, it is reasonable to include the numbers 1, 3, 3, 1 in diagrams dealing with field functions.

The diagram of Fig. 8.2 can be completed by adding the relations between the variables, as shown in Fig. 8.3. The link between $A$ and $\mathbf{b}$ is a defining equation,

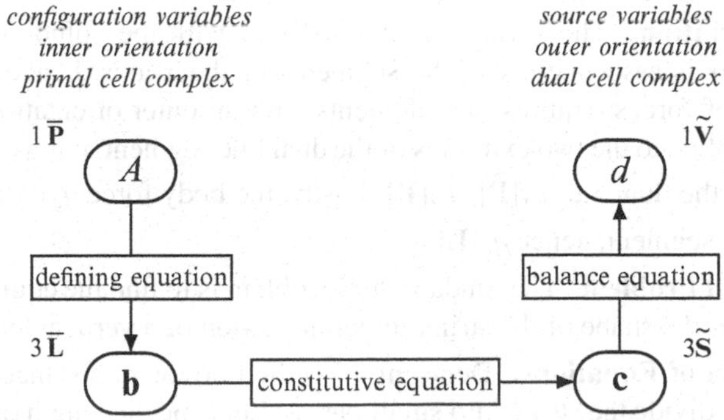


**Fig. 8.3** Classification of field variables enriched with corresponding relations

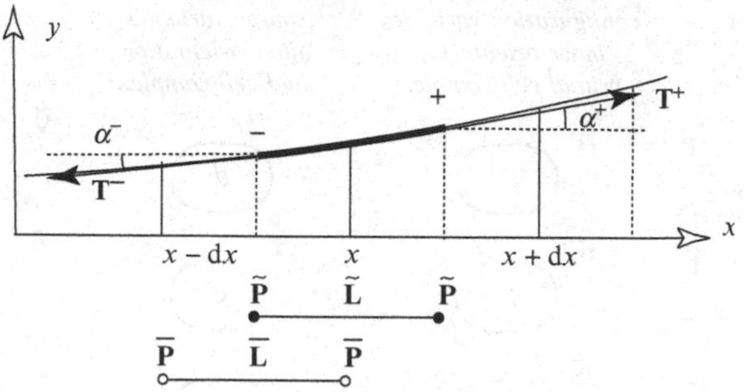


**Fig. 8.4** Equilibrium of a string

the link between $c$ and $d$ is a balance equation, and the link between **b** and **c** is a constitutive equation.

In this table we have represented only the upper part of the whole diagram because the other boxes are empty.

## 8.2 Statics of Strings

In this section we give an example of classification of physical variables and equations of the statics of a string and show how to build a corresponding classification diagram.

Let us consider the equilibrium of a tense string of length $L$, slightly deformed from a straight horizontal configuration, as shown in Fig. 8.4. We describe the most common case of a vertical load.

**Configuration and Source Variables.** The configuration variables are the vertical displacement $y$, the incremental displacement $\Delta y$ and its line density $\alpha = \Delta y/\Delta x$. The source variables are the vertical load $Q_y$, its line density $q_y$, the traction $T$ and its components $T_x, T_y$.

**Space Association.** The variable $y$ is associated with the points of the $x$-axis, hence $y[\overline{\mathbf{P}}]$; $\alpha$ is associated with the segments of the $x$-axis, hence $\alpha[\overline{\mathbf{L}}]$. Since the balance of forces requires line elements with an outer orientation, and since tension is applied to the two extremes of the dual line segment, it is associated with the points of the dual, i.e. $T_x[\widetilde{\mathbf{P}}]$, $T_y[\widetilde{\mathbf{P}}]$. Lastly, the body force $q_y\,\mathrm{d}x$ is associated with the dual segment, hence $q_y[\widetilde{\mathbf{L}}]$.

**Fundamental Problem.** The fundamental problem is to find the equilibrium configuration, i.e. the shape of the string under the action of a vertical load.

**Classification of Equations.** Denoting by $a$ the horizontal distance between its extremes, we divide the string into small pieces. Let $l_k$ be the length of the generic piece. The points of the subdivision will be denoted by $x_0 = 0, x_1, x_2, \dots x_n = a$. This is a primal subdivision.

Since the slope $\alpha$ is small, we can make the position $\tan(\alpha) \approx \alpha$. This is equivalent to the approximation $\mathrm{d}s \approx \mathrm{d}x$. The equilibrium is expressed by the *equilibrium equations*

$$-T_x^- + T_x^+ \overset{\text{law}}{=} 0, \qquad -T_y^- + T_y^+ + q_y\,\mathrm{d}x \overset{\text{law}}{=} 0\,. \tag{8.2}$$

The first equation states that the horizontal component of the tension is constant, hence $T_x^+ = T_x^- = T_x$.

The slopes $\alpha^-$ and $\alpha^+$ must be calculated at the extremities of the dual interval using the following *defining* equations:

$$\alpha^+ \overset{\text{def}}{=} \frac{y(x+\mathrm{d}x) - y(x)}{\mathrm{d}x} \qquad \alpha^- \overset{\text{def}}{=} \frac{y(x) - y(x-\mathrm{d}x)}{\mathrm{d}x}\,. \tag{8.3}$$

The relation between $T_y$ and $\alpha$ is[1]

$$T_y = T_x\,\alpha\,. \tag{8.4}$$

Combining Eqs. 8.2–8.4 we obtain the *fundamental* equation

$$T_x\left[\frac{y(x+\mathrm{d}x) - y(x)}{\mathrm{d}x} - \frac{y(x) - y(x-\mathrm{d}x)}{\mathrm{d}x}\right] + q_y\,\mathrm{d}x = 0\,, \tag{8.5}$$

which can be written in the usual form

$$T_x\frac{\mathrm{d}^2 y(x)}{\mathrm{d}x^2} + q_y = 0\,. \tag{8.6}$$

In the differential formulation, the three equations which compose the fundamental equation are

$$\alpha \overset{\text{def}}{=} \frac{\mathrm{d}y(x)}{\mathrm{d}x}, \qquad T_y \overset{\text{mat}}{=} T_x\,\alpha, \qquad \frac{\mathrm{d}T_y(x)}{\mathrm{d}x} + q_y \overset{\text{law}}{=} 0\,. \tag{8.7}$$

The variables and the equations can be collocated in the diagram as shown in Fig. 8.5.

[1] The nature of this equation stems from the hypothesis that the string is perfectly flexible and inextensible; the latter is similar to the condition of stiffness, which is a property of the material.

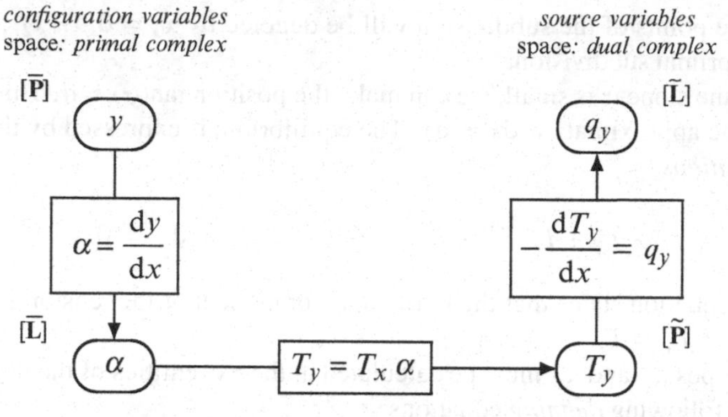


**Fig. 8.5** Differential formulation of statics of strings

## 8.3 A Time Diagram

Up to now we have shown how to arrange physical variables in a diagram which classifies space elements. We now show how to obtain a diagram which includes time elements. To this end, we start by considering particle dynamics. The physical variables of particle dynamics which are global in time are the *radius vector* $\mathbf{r}$, *displacement* $\mathbf{u}$, *momentum* $\mathbf{p}$ and *impulse* $\mathbf{J}$. They are global in time because none of them needs to be integrated in time (p. 126).

Since displacement and force are associated with intervals, we can introduce the corresponding rates, i.e. *velocity* $\mathbf{v}$ and *force* $\mathbf{F}$ respectively. These four global variables and the corresponding rates are summarized as follows:

| | Configuration variables | | Source variables | |
|---|---|---|---|---|
| Global variables | $\mathbf{r}[\bar{\mathbf{I}}]$ | $\mathbf{u}[\bar{\mathbf{T}}]$ | $\mathbf{p}[\tilde{\mathbf{I}}]$ | $\mathbf{J}[\tilde{\mathbf{T}}]$ |
| | $\downarrow$ | $\downarrow$ | $\downarrow$ | $\downarrow$ |
| Field variables | $\mathbf{r}[\bar{\mathbf{I}}]$ | $\mathbf{v}[\bar{\mathbf{T}}]$ | $\mathbf{p}[\tilde{\mathbf{I}}]$ | $\mathbf{F}[\tilde{\mathbf{T}}]$ |

(8.8)

Figure 8.6 shows a classification diagram for particle dynamics for the field variables and includes the most common constitutive equations and the relations between the variables on the left side of the diagram (configuration variables) and on the right side of the diagram (source variables). In this case there is no space association because the particle $\mathscr{P}$ is the subject of this theory. In the next section, the reader will see how to obtain a classification diagram which includes the association of variables with both space and time elements.

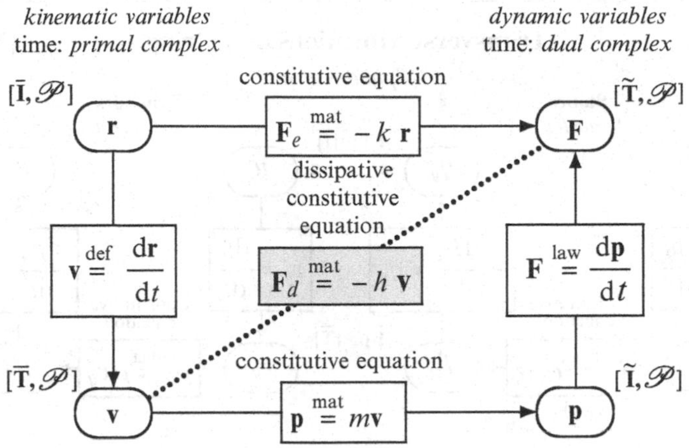


**Fig. 8.6** Classification diagram for particle dynamics

## 8.4 How to Combine Space and Time

As we stated in the introduction (Chap. 1), each space element can be coupled in two possible ways with a time element. These two combinations of space and time elements give rise to two kinds of diagram: one for mechanical theories, where the sources are forces, and another for field theories, where the sources are of different nature, as shown in Fig. 3.3 (p. 7). Table 1.3 (p. 7) shows how to combine space and time aspects in a single diagram. In the next section we will give an example of a diagram of this type.

### *8.4.1 Transversal Vibrations of Strings*

Returning to a taut string, we consider its motion by analysing its transversal vibrations. The corresponding diagram is [VIB]. The difference in passing from statics to dynamics is that the sum of the forces is not zero but equal to the time derivative of the momentum. Momentum is associated with dual instants, i.e. $\mathbf{p}[\tilde{\mathbf{I}}]$, whereas velocity is associated with primal time intervals, i.e. $\mathbf{v}[\overline{\mathbf{T}}]$. The relation between momentum and velocity is a constitutive equation. From diagram [VIB] we see that statics belongs to a vertical plane in the diagrams, while dynamics belongs to a horizontal plane. This fact makes the diagram tridimensional.

## Transverse vibrations of a string [VIB]

*Space + time*

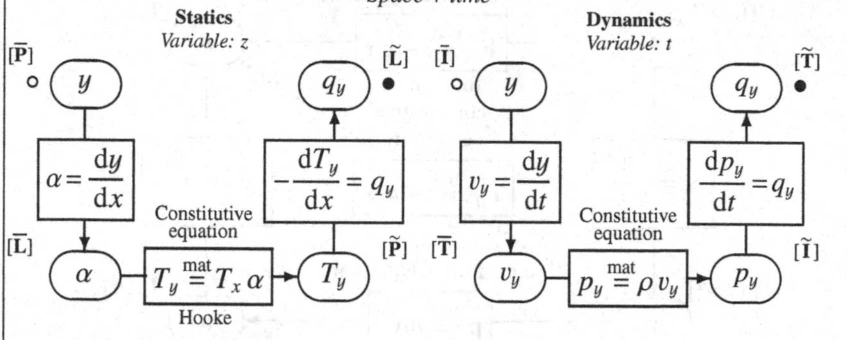


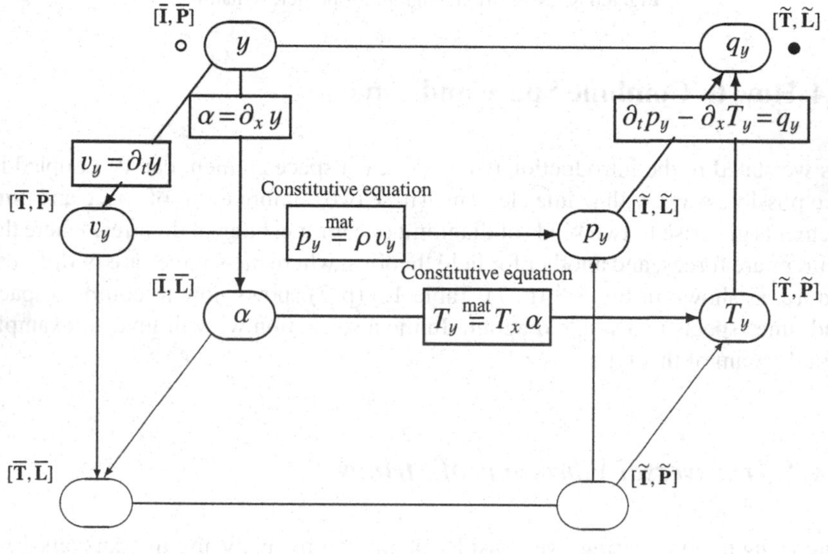


| | | | |
|---|---|---|---|
| $y(t,x)$ | *Transversal displacement* | $q_y(t,x)$ | *Transversal force / length* |
| $\alpha(t,x)$ | *Slope* | $T_x$ | *Horizontal component of traction* |
| $v_y(t,x)$ | *Transversal velocity* | $T_y(t,x)$ | *Transversal component of traction* |
| $\rho$ | *Linear density* | $p_y(t,x)$ | *Transversal momentum / length* |

Fundamental equation:

$$\rho\,\frac{\partial^2 y}{\partial t^2} - T\,\frac{\partial^2 y}{\partial x^2} = q_y$$

$$\partial_t(\rho\partial_t y) - \partial_x(T\partial_x y) = q_y$$

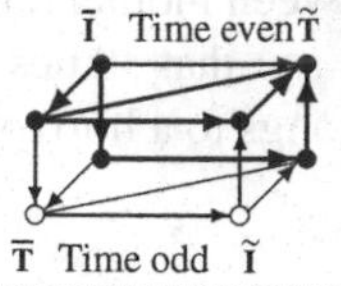


Ref: Crawford F.S., Berkeley Physics Course: Waves, Mc Graw Hill, 1971.

VIB-7; http://discretephysics.dicar.units.it

## 8.5 The Structure of the Diagrams

Each diagram is composed of many elementary units formed by two rounded boxes on the same level, one containing a configuration variable (on the left) and the other containing a corresponding source variable (on the right). The two boxes may be connected by a rectangular box containing a constitutive equation.

The number of vertical boxes in a diagram is equal to the number of independent variables plus one. Thus, in dealing with a one-dimensional field, like the longitudinal traction or compression of a rod, the diagram reduces to two levels.

Each of these diagrams gathers together the physical variables and the equations connecting them, according to a criterion of a geometric nature. Each diagram arises from the following features:

1. *Global* physical variables are associated with space and time elements.
2. This association is contained in the definition of each physical variable and is reflected in its measurement process (when it is measurable).
3. The *densities* of space global variables and the *rates* of time global variables inherit the association with space and time elements of the corresponding space or time global variable.
4. Space and time elements must be *oriented*. There are two kinds of orientation, *inner* and *outer*.
5. Each global physical variable changes sign when we invert the orientation of the corresponding space or time element (oddness principle). This inversion becomes a powerful criterion for deciding which is the orientation of the space element with which a given global variable is associated.
6. The physical variables of each theory can be classified into three broad classes, *configuration*, *source* and *energy* variables.
7. *Configuration* variables are associated with space elements endowed with an inner orientation, whereas *source* and *energy* variables require an outer orientation.
8. This association of physical variables with oriented space elements suggests the opportunity to divide the working region into cells. In this way we construct a cell complex. From this we can build up a dual cell complex.
9. If the cells of a primal complex are endowed with an inner orientation, then those of the dual complex are automatically endowed with an outer orientation.
10. The association of global physical variables with space and time elements leads to a consideration of the distribution of these variables on the various cells of a cell complex and its dual.
11. It follows that *configuration* variables are associated with the cells of the *primal* complex, whereas *source* variables are associated with the cells of the *dual* complex.
12. The basic equations of each physical theory include *defining* equations, *topological* equations and *constitutive* equations.

13. *Topological equations* link configuration variables with configuration variables and source variables with source variables; hence, they are contained in vertical links. When topological equations are reduced to differential equations, they are combined with metrical notions; in electromagnetism a net separation between purely topological and metrical notions is possible. There are four kinds of topological equations: equations forming spatial differences, equations forming time differences, balance equations and circuital equations.
14. In a differential setting, *topological equations* are expressed by the operators 'grad', 'curl' and 'div' and the derivatives $d/dx$, $d/dt$ or $\partial/\partial t$.
15. *Constitutive equations* link configuration variables with source variables; hence, they are contained in horizontal links.

The fact that in every theory there are adjoint differential operators, such as the operators 'div' and '-grad', is taken for granted. In contrast, in an algebraic setting, when using balance and circuital relations in a finite form, these relations appear as a consequence of the geometrical duality that pervades every field of physics.

## 8.6 What the Diagram Shows

The diagram demonstrates several properties. One of these is that physical parameters are contained only in constitutive laws. Another is that the formal differential operators '-grad' and 'div' are formally adjoint (Chap. 15) and give rise to adjoint operators when the corresponding (homogeneous) boundary conditions are assigned (Chap. 15). Moreover, the formal differential operator 'curl' is self-adjoint.

It is not a problem if some boxes of the diagram remain empty: this means that in the given theory there are boxes for possible physical variables which are not introduced because, in general, they are not interesting. Let us take the following example: we know that an electric charge can be neither produced nor destroyed, i.e. it is conserved. We can introduce a physical variable *electric charge production*, but this would be identically zero. Therefore, we have no interest in introducing it.

### *8.6.1 Global Variables Versus Field Variables*

We can pose the following question: why use global variables instead of the corresponding field variables, which are more familiar? The reason is that field variables, being point functions, do not make evident their relations with space elements endowed with extension (lines, surfaces and volumes). In contrast, global variables allow us to find the correct association with the corresponding space elements.

The differential formulation of physics hides a lot of geometric features that, in contrast, are obvious if you avoid premature recourse to the concept of limit. Thus, the limit of a mean density reduces the density to a function of the point and of the instant. This is exactly what the differential formulation requires in order to perform space and time derivatives. In contrast, the global variables from which a density originates can be associated with a volume (e.g. mass, charge, entropy), a surface (e.g. surface force, flux) or a line (e.g. voltage, line integral).

### 8.6.2 Reversible and Irreversible Links

Constitutive equations are shown in horizontal links connecting a configuration variable with a source variable. When a link connects the left with the right column in the front or in the rear, a constitutive equation describes a *reversible* phenomenon. In contrast, when the link connects the left column in the front with the right column in the rear (and there are no other connections), a constitutive equation describes an *irreversible* phenomenon.

The distinction between these two classes of constitutive equations is shown in Fig. 8.7. Reversible relations connect two variables whose product is an energy or an energy density; irreversible relations connect two variables whose product is a power. Reversible relations give rise to variational formulations (Chap. 15), whereas irreversible relations give rise to maximum or minimum principles such as the *principle of minimum heat* and the principle of *maximum entropy production*.

### 8.6.3 Topological Relations

Vertical links do not contain physical constants, and for this reason they do not depend on the nature of the medium. Once the sources of a field are assigned, the corresponding configuration must be obtained via the constitutive laws, and for this reason they depend on the material.

Concerning the operators which appear in vertical columns, we have the following cases:

1. When a diagram uses **global variables**, these links are formed by the *coboundary operator*.
2. When a diagram uses **exterior differential forms**, these links are formed by the *exterior differential*.
3. When a diagram uses **field functions**, these links are formed by *differential operators*, grad, curl, div. In the last case also metrical notions are involved.

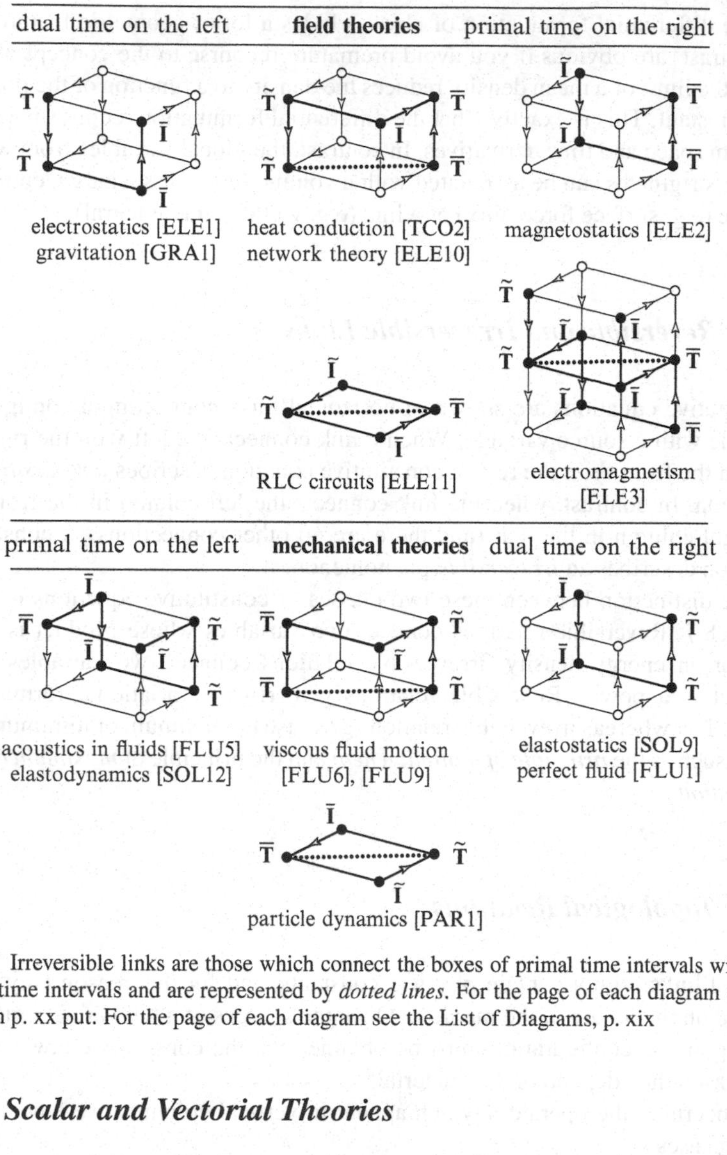


**Fig. 8.7** Irreversible links are those which connect the boxes of primal time intervals with that of dual time intervals and are represented by *dotted lines*. For the page of each diagram see the index on p. xx put: For the page of each diagram see the List of Diagrams, p. xix

### *8.6.4 Scalar and Vectorial Theories*

We will call *scalar* a physical theory when all its *space global* variables are scalars. This is the case with electromagnetism; in fact, electric and magnetic fluxes, electric charge content and charge flows, and electromotive and magnetomotive forces are all scalars. Other scalar theories are the (classical) gravitational field, heat conduction and diffusion.

We will call *vectorial* a physical theory when all its global variables are vectors. This is the case of mechanics of deformable solids, as shown in diagram [SOL12]. In fact displacement, relative displacement, force and momentum are vectors.

Fluid dynamics is a mixed theory because it is formed by a scalar theory (velocity potential, velocity line integral, mass content, mass flow), as shown in diagram [FLU3], and a vectorial theory (velocity, relative velocity, force, momentum), as shown in diagram [FLU7]. In fact, its main equations are the scalar equation of mass conservation and the vectorial equation of momentum balance.

### 8.6.5 Exterior Differential Forms

Configuration variables are described by even forms and source variables by odd forms. This is a consequence of the fact that configuration variables refer to space elements endowed with an inner orientation, whereas source variables refer to space elements endowed with an outer orientation (Chap. 14).

Moreover, there is a strict link between the degree of a differential form and the dimension of the space element: a $p$-form refers to a $p$-dimensional element.

The coboundary operator on discrete forms becomes the exterior differential on differential forms. Then we have the following correspondence:

| *Left side of diagram* | *Right side of diagram* |
|---:|:---|
| Configuration variables | Source variables |
| Inner orientation | Outer orientation |
| Even discrete forms | Odd discrete forms |
| Even differential forms | Odd differential forms |

(8.9)

### 8.6.6 Material Parameters

The material constants of each physical theory are contained in constitutive equations (also called material equations) and, as such, are located in the horizontal links of the diagrams, as Table 8.1 shows. The inclined links in the diagrams of this table must be horizontal links when time is added to the diagram, as shown in diagrams [PAR1], [ELE3], [TCO2] and [FLU9].

### 8.6.7 Configuration Variables are Material Dependent

Usually the space and time distribution of the source variables of a field is assigned while the resultant configuration of the field depends on the material medium in which the field is embedded.

**Table 8.1** Material parameters are contained in constitutive equations

| Electrostatics [ELE1] | | Magnetostatics [ELE2] | |
|---|---|---|---|
| Configuration | Source | Configuration | Source |
| $\phi$ | $\rho$ | $\mathbf{A}$ | $\mathbf{J}$ |
| $\mathbf{E} \longrightarrow \epsilon \longrightarrow \mathbf{D}$ | | $\mathbf{B} \longrightarrow \mu \longrightarrow \mathbf{H}$ | |
| Medium dependent | Medium independent | Medium dependent | Medium independent |
| Heat conduction [TCO1] | | Particle mechanics [PAR1] | |
| Configuration | Source | Configuration | Source |
| $T$ | $\sigma_\mathrm{u}$ | $\mathbf{r} \longrightarrow k \longrightarrow \mathbf{F}$ | |
| $\mathbf{g} \cdots\cdots \lambda \cdots\cdots \mathbf{q}$ | | $\mathbf{v} \longrightarrow m \longrightarrow \mathbf{p}$ ($h$) | |
| Medium dependent | Medium independent | Medium dependent | Medium independent |
| Mechanics of fluids [FLU6] | | Mechanics deformable solids [SOL9] | |
| Configuration | Source | Configuration | Source |
| $v_h$ | $f_h$ | $u_h$ | $f_h$ |
| $L_{hk} \cdots \mu, \dfrac{\lambda}{\rho} \cdots \sigma_{hk}$ | | $\varepsilon_{hk} \longrightarrow E, \nu \longrightarrow \sigma_{hk}$ | |
| Medium dependent | Medium independent | Medium dependent | Medium independent |
| Diffusion | | Viscous fluid motion [FLU9] | |
| Configuration | Source | Configuration | Source |
| $\mu$ | $\sigma$ | $v$ | $G$ |
| $\mathbf{g} \cdots\cdots D \cdots\cdots \mathbf{J}$ | | $\gamma \cdots\cdots \mu \cdots\cdots \tau$ | |
| Medium dependent | Medium independent | Medium dependent | Medium independent |

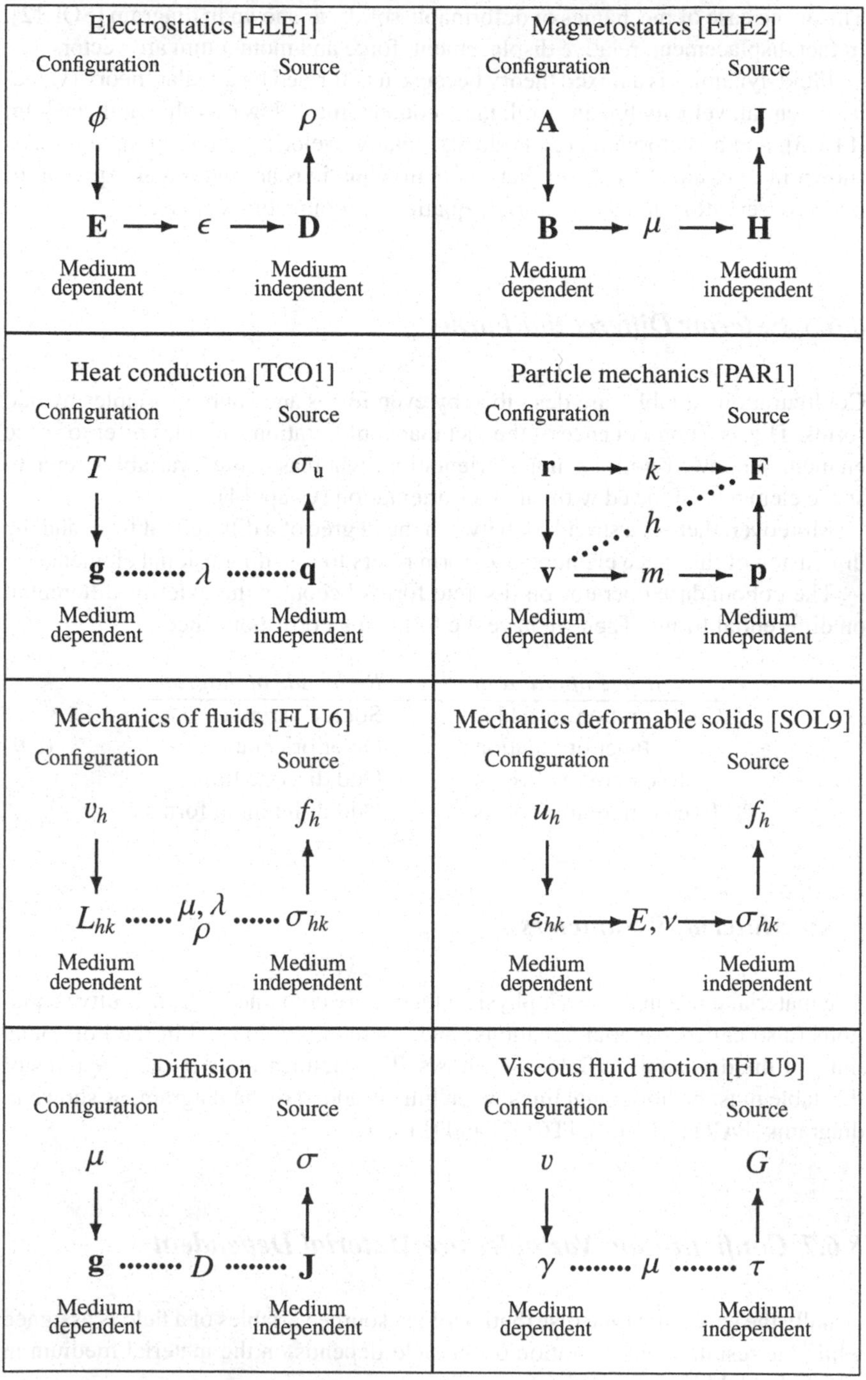

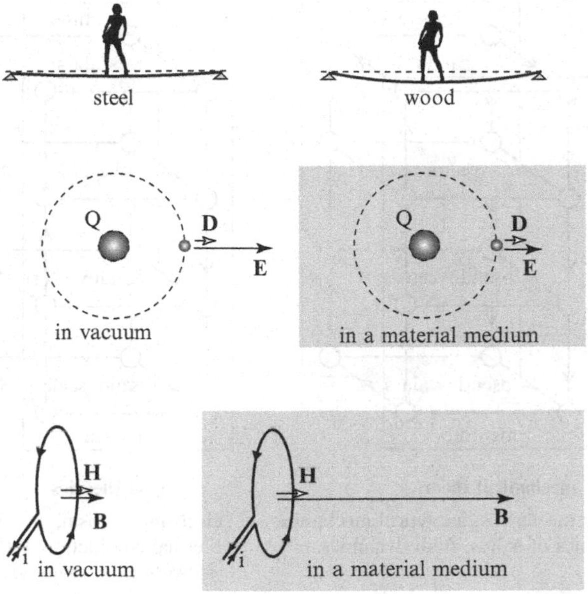


**Fig. 8.8** The same source gives rise to different configurations depending on the material medium

This can be easily grasped from Fig. 8.8. In the upper part is shown that the same weight (source of deformation) acting on an elastic beam supported at both ends produces different deformed configurations depending on the material of the beam.

The middle part of Fig. 8.8 shows the electric field created by a charge $Q$ (source of the field). We can see that the electric displacement **D** (a source variable) at a point is the same for any material medium. This follows from Gauss's law because the electric flux $\Psi$ on a spherical surface located in the centre of the charge is equal to the charge $Q$. In contrast, according to Coulomb's law, the electric field strength **E** (a configuration variable) at the same point depends on the material medium, and in particular attains its maximum value in vacuum.

In the bottom part of Fig. 8.8 we consider the magnetic field of a loop carrying a current $i$ (source variable). The magnetic field strength **H** (a source variable) in the centre of the loop does not depend on the medium, as we can see from Ampere's law of currents; in contrast, the magnetic flux density **B** (a configuration variable) depends on the material. The proof of this fact is given in Chap. 10.

## *8.6.8 Tensorial Nature of Field Functions*

One of the beautiful features of a classification diagram is that it respects the tensorial nature of the variables inside every box. If the configuration variable in

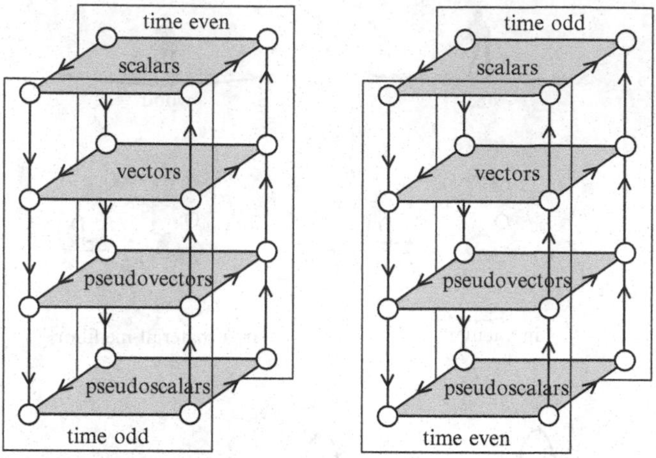


**mechanical theories**
particle mechanics, analytical mechanics
mechanics of solids, fluid dynamics

**field theories**
electromagnetism, gravitation
thermal conduction, diffusion
irreversible thermodynamics

**Fig. 8.9** Behaviour of field functions under space and time transformations

the left columns at a given level is, say, covariant, then the corresponding source variable at the same level is contravariant, and vice versa. As a consequence, the scalar product of the two variables which lie on the same level is invariant, as shown in the following equations:

$$\overbrace{\overline{a}_k = a_h \frac{\partial x^h}{\partial \overline{x}^k}}^{covariance}, \qquad \overbrace{\overline{b}^k = b^h \frac{\partial \overline{x}^k}{\partial x^h}}^{contravariance}, \qquad \overbrace{a_h b^h = \overline{a}_k \overline{b}^k}^{invariance} \ . \tag{8.10}$$

This invariance is reflected in the fact that the energy and its density are invariant with respect to a change in the coordinate system.

With reference to Fig. 8.9, the field functions of the diagram which belong to the same level have the same tensorial nature. Counting from the top level, it can be seen that the functions of the first level are *scalars* (sometimes called *true* scalars), those of the second level are *vectors* (rarely called *polar* vectors), those of the third level are *pseudovectors*, also called *axial vectors*, and those of the fourth level are *pseudoscalars*, also called *axial scalars*.

Figure 8.10 shows that the number of components of the field variables increases up to the centre of the diagram and then decreases.

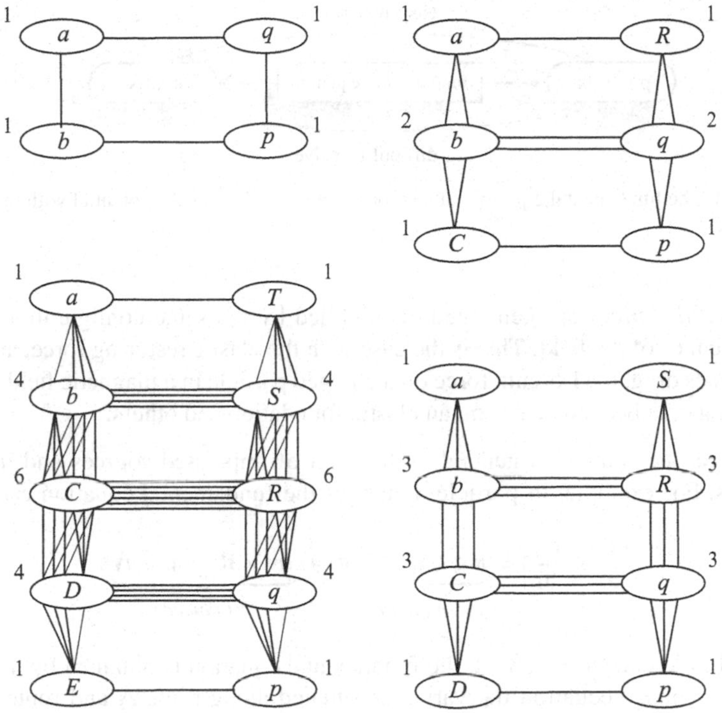


**Fig. 8.10** Possible links between variables of a physical theory for one , two-, three- and four-dimensional cases

## *8.6.9 Composing the Equations*

Each classification diagram contains variables in round boxes and arrows connecting them, as in Fig. 8.12. The main configuration variable, i.e. the one from which all configuration variables can be deduced, is usually the potential of the field, and for this reason we will decide to use for it the term *potential*.

The main source variable, i.e. the one to which all other source variables are linked, will be called simply the *source*.

The equation that links the potential with the source will be called *fundamental equation*. The list of fundamental equations of physical theories can be found on p. 158. The fundamental equation describes mathematically *fundamental problem* of the theory (Fig. 8.11). To show how to compose the equation of a field (topological and constitutive) to obtain the fundamental equation, we must distinguish between two kinds of source:

- *Impressed sources* can be assigned at will, for example heating devices in a room or electric generators in an electric circuit.

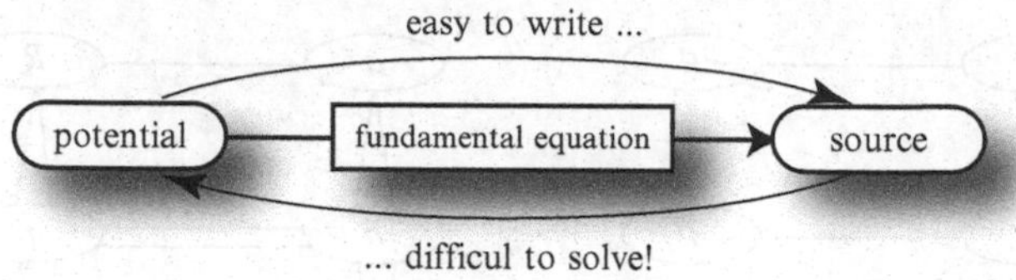


**Fig. 8.11** The fundamental equation of a theory is that which links the potential with the source of the field

- *Induced sources* are generated or modified by the same configuration of the system or of the field. This is the case with the elastic restoring force, air drag, viscous drag, the Lorentz force on a charged particle in a magnetic field, forces acting on a beam that lies on an elastic foundation and others.

Hence, the source, in general, is the sum of impressed sources and induced sources. For example, in particle dynamics the fundamental equation can be as follows:

$$\frac{\mathrm{d}}{\mathrm{d}t}\left[m\frac{\mathrm{d}\mathbf{r}}{\mathrm{d}t}\right] = \underbrace{m\mathbf{g} + \mathbf{f}_0 sin(\omega t)}_{impressed} + \underbrace{q\mathbf{v}\times\mathbf{B} - k\mathbf{r} - h\mathbf{v}}_{induced} . \tag{8.11}$$

With reference to Fig. 8.12, the fundamental equation is obtained by inserting into the balance equation the variables situated along route A and route B. On the right side of the balance equation is the total source $\sigma$, i.e. the sum of induced sources $\sigma^{ind}$ and of the impressed sources $\sigma^{imp}$, i.e. those situated along the dashed routes.

**Example 1**. In the diagram of particle dynamics, see diagram [PAR1], we compose the three equations $\mathbf{F} = \mathrm{d}\mathbf{p}/\mathrm{d}t$, $\mathbf{p} = m\,\mathbf{v}$ and $\mathbf{v} = \mathrm{d}\mathbf{r}/\mathrm{d}t$ to obtain the left part of the fundamental equation and the two equations $\mathbf{F}_e = -k\,\mathbf{r}$ and $\mathbf{F}_d = -h\,\mathrm{d}\mathbf{r}/\mathrm{d}t$ of the induced forces, to which we add the impressed force $\mathbf{F}_{\mathrm{imp}}$, to obtain the right term of the fundamental equation.

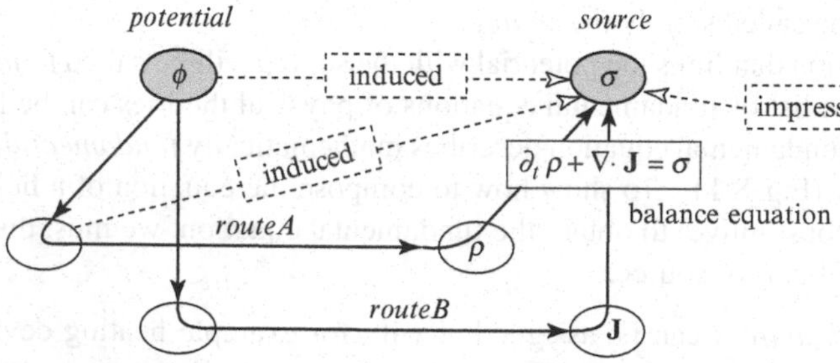


**Fig. 8.12** Paths which must be followed to compose fundamental equations

# Part II
# Analysis of Physical Theories

# Chapter 9
# Particle Dynamics

## 9.1 Fundamental Problem

Particle mechanics deals only with the time variable, and for this reason the term *global* means here *global in time*.

The *fundamental problem* of particle dynamics can be stated as follows:

- Given a particle,
- Given a time interval,
- Given the constraints,
- Given the initial position and the initial velocity of the particle,
- Given the force acting on the particle during the time interval,
- Find the position of the particle at every instant of the interval.

We present the main physical variables of particle dynamics in a new way, stressing their operative definition, which for some variables differs from those commonly given. So momentum is usually *defined* as the product of mass times velocity, and force is *defined* as the time derivative of momentum. In contrast to this approach, we will consider force as a primitive variable and *define* momentum as the indefinite time integral of force. The relation momentum/velocity is considered a constitutive equation. Moreover, we will distinguish kinetic energy from kinetic *co-energy*, showing that kinetic energy enters into the Hamiltonian whereas kinetic co-energy enters into the Lagrangian.

We will start by exploring the association between the physical variables of mechanics and the elements of the time axis. We will refer to the two cell complexes on the time axis.[1]

[1] See Table 9.2, p. 247.

E. Tonti, *The Mathematical Structure of Classical and Relativistic Physics*, Modeling and Simulation in Science, Engineering and Technology,
DOI 10.1007/978-1-4614-7422-7_9, 

**Table 9.1** Main variables of particle dynamics

| Kind of variable | Variable | | Global variable | |
|---|---|---|---|---|
| *Configuration* | $\mathbf{r}$ | Radius vector | $\mathbf{r}$ | *(Same)* |
| | $\mathbf{v}$ | Velocity | $\mathbf{u}$ | Displacement |
| *Source* | $\mathbf{F}$ | Force | $\mathbf{J}$ | Impulse |
| | $\mathbf{p}$ | Momentum | $\mathbf{p}$ | *(Same)* |
| *Energy* | $P$ | Power | $W$ | Work |
| | $T$ | Kinetic energy | $A_{\mathrm{H}}$ | Action |
| | $V$ | Potential energy | $A_{\mathrm{H}}$ | Action |

## 9.2 Source Variables

The source variables of particle dynamics are the static and dynamic variables, i.e. force, impulse and momentum.

### 9.2.1 Force

Force is a primitive physical variable which is measured easily in statics. The measuring device known as a *dynamometer* is essentially a spring that undergoes elastic deformation, lengthening and shortening under the action of force. For small forces a torsion balance can be used. All measurements of a force require that equilibrium be reached; ***hence, a time interval is involved in the measurement of force.***[2] It follows that force is associated with time intervals: $\mathbf{F}[\mathbf{T}]$. Since a reversal of motion does not change the dynamometer indication, it follows that the time interval is endowed with an outer orientation: $\mathbf{F}[\tilde{\mathbf{T}}]$.

Unlike in statics, force is not easily measurable in dynamics. For example, the force acting on an aircraft moving in the air cannot be measured directly in motion, so it is measured in a wind tunnel by having a model of the aircraft at rest and the air in motion. In dynamics, the primitive physical variable is the impulse.

### 9.2.2 Impulse

The impulse communicated to a particle during a time interval is the *definite* time integral of the force acting on the particle during that time interval, i.e.

[2] See p. 128.

$$\boxed{\mathbf{J}(t^-, t^+) \stackrel{\text{def}}{=} \int_{t^-}^{t^+} \mathbf{F}(t)\,\mathrm{d}t} \qquad \text{impulse.} \tag{9.1}$$

**Time Association.** It follows that an impulse is associated with a time interval. As we saw when considering a rebound,[3] it does not change sign under a reversal of motion; hence, it is associated with dual time intervals, i.e. $\mathbf{J}[\widetilde{\mathbf{T}}]$. From the definition given by Eq. 9.1 it follows that the *mean value* of the force during a time interval is the time rate of the impulse:

$$\langle \mathbf{F}\,[\widetilde{\mathbf{T}}]\rangle = \frac{\mathbf{J}\,[\widetilde{\mathbf{T}}]}{T} \qquad \mathbf{F}[\widetilde{\mathbf{T}}] = \lim_{T \to 0} \frac{\mathbf{J}\,[\widetilde{\mathbf{T}}]}{T}\,. \tag{9.2}$$

Hence, an impulse is global in time and force is its time rate. This fact confirms, with dynamical considerations, that force is associated with dual time intervals simply because it inherits this association from the impulse.

In Chap. 5 (p. 122) we laid out the oddness principle. We ask now: for this principle to be valid with respect to impulses, inverting the outer orientation of the dual time intervals, the impulse must change sign, i.e. $\mathbf{J}[-\widetilde{\mathbf{T}}] = -\mathbf{J}[\widetilde{\mathbf{T}}]$. Is this true?

We know that the impulse transmitted to a wall by a ball bouncing on it is opposite to the impulse which the wall gives the ball. Hence, to find the physical meaning of the inversion of the outer orientation of a time interval, we can say that *we must invert the role from active to passive to the body on which the impulse is impressed.* With this meaning in mind the oddness principle is satisfied.

### 9.2.3 Momentum

The momentum of a particle in motion must be defined as the *indefinite* time integral of the force acting on the particle, starting from rest, up to the instant $t$, i.e.

$$\boxed{\mathbf{p}(t) \stackrel{\text{def}}{=} \int_{(rest)}^{t} \mathbf{F}(t')\,\mathrm{d}t' \qquad \longrightarrow \qquad \mathbf{F}(t) = \frac{\mathrm{d}\mathbf{p}(t)}{\mathrm{d}t}}\,. \tag{9.3}$$

The notion of momentum, defined as an indefinite time integral of force, is suitable for describing the flight of an object, be it the flight of, for example, a javelin, ball or discus, as Fig. 9.1 shows. The conservation of momentum of an isolated system is a consequence of the action and reaction principle.

**Time Association.** A stone falls under the action of gravity because it receives an impulse downward and its momentum increases. Under a reversal of

---

[3] See p. 131.

**Fig. 9.1** The momentum imparted on a body is the time integral of force, starting from rest

motion the impulse of gravity is always downwards, but the stone goes upwards and its momentum decreases. Going upwards the final momentum is smaller than the initial one. The impulse must be the difference between the final momentum and the initial one, both in progressive motion and retrograde motion; hence,

$$\mathbf{J} = \mathbf{p}^+ - \mathbf{p}^- \quad \textit{and} \quad \overline{\mathbf{J}} = \overline{\mathbf{p}}^+ - \overline{\mathbf{p}}^- . \tag{9.4}$$

Since the impulse has not changed its sign, i.e. $\overline{\mathbf{J}} = \mathbf{J}$, it follows that the final momentum in retrograde motion has the same value as the initial momentum in progressive motion, but with the opposite sign. The same can be said of the initial momentum in retrograde motion. Hence,

$$\overline{\mathbf{p}}^+ = -\mathbf{p}^- \quad \textit{and} \quad \overline{\mathbf{p}}^- = -\mathbf{p}^+. \tag{9.5}$$

This shows that momentum changes sign with reversal of motion, i.e. $\mathbf{p}[\widetilde{\mathbf{I}}]$.

### 9.2.4 Force from Momentum or Vice Versa?

It is commonly stated that force can be *defined* as the time derivative of momentum. This presupposes that momentum is known before force. This contradicts the fact that the notion of force arises in statics, where the notion of momentum does not arise. We measure the force between two charged bodies (Coulomb's law) and between two masses (Newton's law) with a torsion balance, i.e. with a static procedure. We measure the viscous force acting on a body moving in a viscous fluid (Stokes' law); we measure the buoyancy force acting on a body in a fluid (Archimedes' principle); we measure the force of the air drag on a body in wind tunnels. In all these cases we measure the *force*, not the momentum! Why, then, should we define force as the time derivative of momentum if force is measured directly? Hermann Weyl wrote:

> Indeed Newton recognized that the force is composed additively (according to the parallelogram law of vector addition) of individual forces exerted upon $k$ by each of the bodies $k_1, k_2, \ldots$, and that this occurs in such a manner that, for example, the force exerted by $k_1$ on $k$ at a certain moment depends solely on the condition of these two bodies (location and velocity) at that instant. This is the real meaning of the decomposition of the one force into several component forces. Looking at these facts one cannot escape the conclusion that the definition "force = time-derivative of momentum" does not reflect the nature of force adequately but that the real state of affairs is the other way round: force is the expression of an independent power that connects the bodies according to their inner nature and their relative position and motion, and that power causes the change of momentum with time. Thus the living metaphysical interpretation conforms to the theoretical construction. Through the basic mechanical law of motion, physics is given the task of exploring the forces operating among bodies in their dependence on position, motion, and inner condition. The latter will enter the laws of force by way of numbers characteristic of the inner state of the reacting bodies, like the electrical charge in the case of Coulomb's law of electrostatic attraction and repulsion. Thus the concept of force becomes a source of new measurable physical characteristics of matter.[4]

Jammer wrote:

> Since Newton clearly distinguishes between definitions and axioms (or laws of motion), it is obvious that the second law of motion was not intended by Newton as a definition of force, although it is sometimes interpreted as such by modern writers on the foundations of mechanics. Nor was it meant to be merely the statement of a method of measuring forces. Force, for Newton, was a concept given a priori, intuitively, and ultimately in analogy to human muscular force. Definition IV, therefore, is not to be interpreted as a nominal definition, but as summarizing the characteristic property of forces to determine accelerations. [5]

**Remark.** We cannot accept the relation $\mathbf{p} = m\mathbf{v}$ as the *definition* of momentum for (at least) three reasons:

1. A definition cannot contain an unknown parameter (here the inertial mass);
2. If this were a definition, it would not change when passing from classical to relativistic mechanics, contrary to what actually happens;
3. The position of this relation in the diagram (p. 268) is the same as all other constitutive relations of physical theories, i.e. a horizontal link between configuration and source variables.

Can we accept the *definition* of voltage as the product of the resistance for current intensity, i.e. $V \stackrel{\text{def}}{=} RI$? Can we accept the *definition* of stress as the product of elastic modulus for strain, i.e. $\sigma \stackrel{\text{def}}{=} E\varepsilon$? Surely not! Because the physical constants $R$ and $E$ are evaluated from these two relations, which come from experiments and are valid only for linear materials. So how can we possibly consider the relation $\mathbf{p} = m\mathbf{v}$ as the *definition* of momentum if it contains a parameter of the particle?

To raise the intuitive notion of mass (quantity of matter or a measure of inertia) to the rank of a physical quantity, like the intuitive notions of electrical resistance ($R$) and of material stiffness ($E$), we must use a constitutive equation. This presupposes that the two physical variables that are linked by the equations should already be defined.

---

[4] Weyl [248, p. 148].

[5] Jammer [104, p. 124].

That being said, let us show how a ballistic pendulum can be used to measure the momentum of a particle. The box of the ballistic pendulum has a large mass compared with the mass of bullets.[6] The device is calibrated by shooting simultaneously $1, 2, 3, \ldots n$ *identical* bullets *with the same velocity* and measuring the amplitude of the recoil for each $n$, as shown in Fig. 9.2. The physical variable thus measured will be called the *impulse*. This is similar to what we do to define force with a dynamometer. The scale thus constructed has an arbitrary unit which depends on the mass of one of the bullets, its velocity and the characteristics of the pendulum.

A similar procedure is used to measure electric charge by transforming an electroscope into an electrometer: we insert $1, 2, 3, \ldots n$ *identical* charges into a Faraday sink placed on top of an electroscope. Also, in this case the calibration depends on the charge considered and on the characteristic of the electroscope.

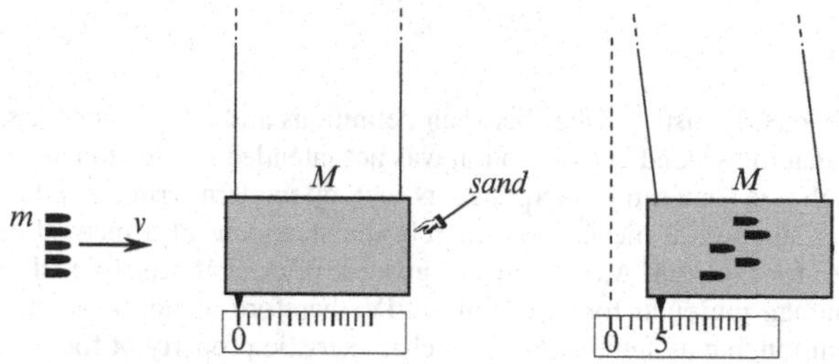


**Fig. 9.2** Calibration of an impulsometer. We shoot $n$ bullets in a box containing sand; the bullets are captured and stopped. The maximum displacement is the measure of the impulse received by the box, and this is equal to the momentum of the bullet before impact

From the definitions given we see that *the momentum of a particle is the impulse released by the particle when it is stopped*:

$$\mathbf{p} \stackrel{\text{def}}{=} \mathbf{J}\,(\textit{released stopping the particle})\,. \tag{9.6}$$

It follows that the impulse released by a particle when its velocity is reduced is the difference between the momenta of the particle at the two instants bounding the time interval. Since we find it useful to consider the impulse *supplied to* a particle instead of that *released by* the particle, we must write

$$\mathbf{J}(t^-, t^+) = \mathbf{p}(t^+) - \mathbf{p}(t^-) \qquad \text{with } \mathbf{J}\,[\widetilde{\mathrm{T}}] \text{ and } \mathbf{p}\,[\widetilde{\mathrm{I}}], \tag{9.7}$$

which is the classical relation used in impulsive motion.

[6] Since, up to now, we have not introduced the measurement of mass, the latter is considered in its intuitive notion as an amount of matter.

## 9.3 Configuration Variables

Configuration variables are geometric and kinematic variables, and knowledge of them provides the position and state of motion of a particle at every instant.

### 9.3.1 Radius Vector

In choosing a reference system, one may consider the position vector of a particle as the vector connecting the origin **O** of the reference system with the particle position **P**. The introduction of the radius vector enables us to write

$$\mathbf{u}\,(t^-, t^+) \overset{\text{def}}{=} \mathbf{r}\,(t^+) - \mathbf{r}\,(t^-) \qquad \text{with } \mathbf{u}\,[\overline{\mathbf{T}}] \text{ and } \mathbf{r}\,[\overline{\mathbf{I}}]. \qquad (9.8)$$

The radius vector is a global variable associated with instants; hence, we write $\mathbf{r}[\mathbf{I}]$. Since the radius vector does not change sign under a reversal of motion, it is associated with primal instants, i.e. $\mathbf{r}[\overline{\mathbf{I}}]$.

**Table 9.2** Association of mechanical variables with time elements

| | Hamilton's principal function | | Hamiltonian action | |
|---|---|---|---|---|
| | $S[\tilde{\mathbf{I}}^-]$ | ———— | $A_H[\tilde{\mathbf{T}}]$ ———— | $S[\tilde{\mathbf{I}}^+]$ |
| | | | $T[\tilde{\mathbf{T}}], V[\tilde{\mathbf{T}}], L[\tilde{\mathbf{T}}], H[\tilde{\mathbf{T}}]$ | |
| | | Kinetic energy, potential energy, Lagrangian, Hamiltonian | | |
| | $\mathbf{v}(\tilde{\mathbf{I}}^-)$ | ———— | $(\Delta_t \mathbf{v})[\tilde{\mathbf{T}}]$ ———— | $\mathbf{v}(\tilde{\mathbf{I}}^+)$ |
| | | | $\mathbf{a}[\tilde{\mathbf{T}}]$ Acceleration | |
| | $\mathbf{p}[\tilde{\mathbf{I}}^-]$ | ———— | $\mathbf{J}[\tilde{\mathbf{T}}]$ Impulse —— | $\mathbf{p}[\tilde{\mathbf{I}}^+]$ |
| | **Momentum** | | $\mathbf{F}[\tilde{\mathbf{T}}]$ Force | **Momentum** |
| *Dual time complex* | $\tilde{\mathbf{I}}^-$ | ———— | $\tilde{\mathbf{T}}$ ———— | $\tilde{\mathbf{I}}^+$   $t$ |
| | $\overline{\mathbf{I}}^-$ ———— | $\overline{\mathbf{T}}$ | ———— $\overline{\mathbf{I}}^+$ | *Primal time complex* |
| | $\mathbf{r}[\overline{\mathbf{I}}^-]$ ———— | $\mathbf{u}[\overline{\mathbf{T}}]$ Displacement —— $\mathbf{r}[\overline{\mathbf{I}}^+]$ | | |
| | **Radius vector** | $\mathbf{v}[\overline{\mathbf{T}}]$ Velocity | **Radius vector** | |
| | $T(\overline{\mathbf{I}}^-)$ ———— | $W[\overline{\mathbf{T}}]$ Work ———— $T(\overline{\mathbf{I}}^+)$ | | |
| | | $P[\overline{\mathbf{T}}]$ Power | | |

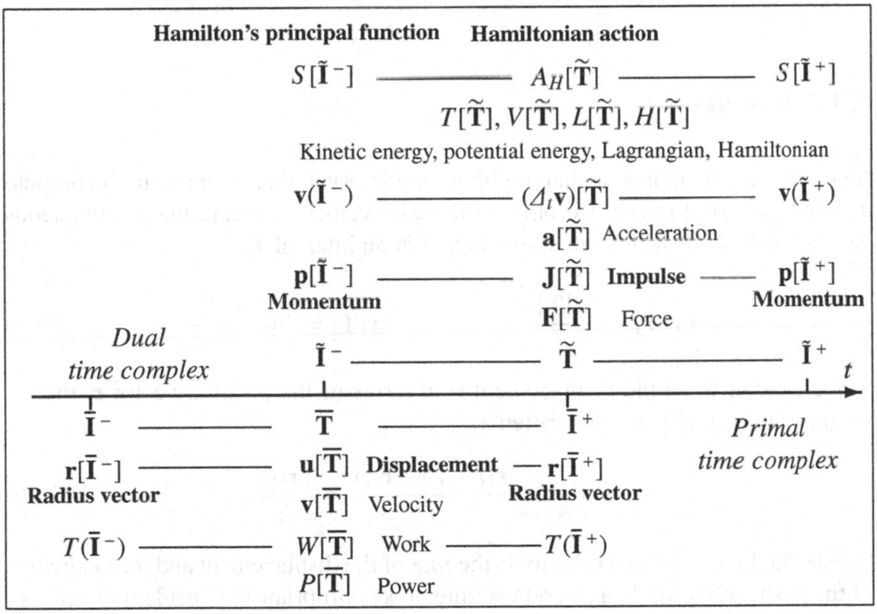

The main reason for introducing a radius vector is the possibility of expressing the displacement as a difference between two radius vectors. In this sense, the radius vector is a kind of time potential of displacement: like every potential, it may be changed by adding an arbitrary vector, and this is equivalent to a change in the origin of the reference system.

Equation 9.7 shows that momentum plays the same role with respect to impulses as radius vector plays with respect to displacement, as shown by Eq. 9.8. This remarkable analogy is shown in Table 9.4. Momentum is associated with dual instants and is a kind of time potential of an impulse. Consequently, an arbitrary vector can be added to it, and this is the same as considering a reference system in motion with respect to the first system.

### 9.3.2 Displacement

The displacement of a particle in a given time interval is a vector connecting the initial position of the particle with its final position. As we have already said, a reversal of motion changes a displacement into its opposite. This means that displacement is associated with time intervals endowed with an inner orientation, i.e. $\mathbf{u}[\overline{\mathbf{T}}]$.

### 9.3.3 Velocity

Since displacement is associated with a time interval, it is meaningful to compute its time rate. In this way we obtain the *mean* velocity $\langle \mathbf{v} \rangle$ and the instantaneous velocity $\mathbf{v}$, both of which are associated with an interval $\overline{\mathbf{T}}$:

$$\langle \mathbf{v} \rangle [\overline{\mathbf{T}}] \stackrel{\text{def}}{=} \frac{\mathbf{u}[\overline{\mathbf{T}}]}{T} \qquad \mathbf{v}[\overline{\mathbf{T}}] = \lim_{T \longrightarrow 0} \frac{\mathbf{u}[\overline{\mathbf{T}}]}{T} . \tag{9.9}$$

Since we can write the displacement $\mathbf{u}$ in terms of the position vector $\mathbf{r}$, the instantaneous velocity can be written as

$$\mathbf{v}(t) = \lim_{T \longrightarrow 0} \frac{\mathbf{r}(t+T) - \mathbf{r}(t)}{T} = \frac{\mathrm{d}\mathbf{r}(t)}{\mathrm{d}t} . \tag{9.10}$$

Hence, the instantaneous velocity is the *rate* of the displacement and the *derivative* of the position vector. In a discrete setting, it is appropriate to consider the velocity as referred to the middle instant of the interval $\overline{\mathbf{T}}$, i.e. to the dual time instant $\widetilde{\mathbf{I}}$. To show this let us consider a uniformly accelerated motion whose formulae are

$$\mathbf{v}(t) = \mathbf{v}(0) + \mathbf{a}\,t, \qquad \mathbf{r}(t) = \mathbf{r}(0) + \mathbf{v}(0)\,t + \frac{1}{2}\mathbf{a}\,t^2 \tag{9.11}$$

and consider the mean velocity in an interval $[t_1, t_2]$. We have

$$\langle \mathbf{v} \rangle = \frac{\mathbf{r}(t_2) - \mathbf{r}(t_1)}{t_2 - t_1} = \frac{\mathbf{v}(0)\,(t_2 - t_1) + \frac{1}{2}\mathbf{a}\,(t_2^2 - t_1^2)}{t_2 - t_1} = \mathbf{v}(0) + \mathbf{a}\left(\frac{t_2 + t_1}{2}\right), \tag{9.12}$$

which is the velocity at the middle instant of the interval. Since every regular motion in every short interval can be approximated by a uniformly accelerated motion, it follows that velocity is *approximately* associated with the middle instant of the interval. In a differential setting, we are obligated to compute the limit of the velocity when $t^{n-1} \rightarrow t^n$ to obtain the instantaneous velocity at $t^n$; in doing so, we lose the information that velocity is associated with dual time instants.

We point out the important fact that the *instantaneous* speed cannot in any way be measured: it is the result of our idealization because it makes use of the limit process. In a discrete setting, we consider only *small* intervals; hence, we have no reason to compute the limit and will use only the mean velocity.

Under a reversal of motion the velocity changes sign; this follows from the definition in Eq. 9.9 recalling that $T$ is the duration of the interval $\overline{\mathbf{T}}$, i.e. it is the absolute value of the extension of $\overline{\mathbf{T}}$ and, as such, is always positive. Hence,[7]

$$\textit{from} \qquad \mathbf{u}[-\overline{\mathbf{T}}] = -\mathbf{u}[\overline{\mathbf{T}}] \qquad \textit{it follows} \qquad \mathbf{v}[-\overline{\mathbf{T}}] = -\mathbf{v}[\overline{\mathbf{T}}]\,. \tag{9.13}$$

### 9.3.4 Acceleration

Mean acceleration $\langle \mathbf{a} \rangle$ is defined as the time rate of the velocity variation, and the instantaneous acceleration is the corresponding limit

$$\langle \mathbf{a} \rangle\,(t^-, t^+) \stackrel{\text{def}}{=} \frac{\mathbf{v}\,(t^+) - \mathbf{v}\,(t^-)}{t^+ - t^-}, \qquad \mathbf{a}\left(\frac{t^- + t^+}{2}\right) = \lim_{(t^+ - t^-) \longrightarrow 0} \frac{\mathbf{v}\,(t^+) - \mathbf{v}\,(t^-)}{t^+ - t^-}\,. \tag{9.14}$$

We must stress that acceleration has no proper place in a classification diagram because it is the result of the combination of the first time derivative of the right column with the first time derivative of the left column, via the constitutive equation $\mathbf{p} = m\,\mathbf{v}$, where $m$ is a system parameter. To emphasize this fact, we remark that in relativity, where the relation $\mathbf{v} \longrightarrow \mathbf{p}$ is non-linear, the acceleration is devoid of interest.

Acceleration is associated with dual time intervals, $\mathbf{a}[\widetilde{\mathbf{T}}]$, and must be evaluated at primal time instants, $\mathbf{a}(\overline{\mathbf{I}})$. It follows that acceleration *is invariant under a*

[7] The fact that velocity changes sign under a reversal of motion is usually deduced in an incorrect way, as discussed on p. 35.

*reversal of motion* (p. 59). In fact, in the gravitational field, a stone going downwards or upwards (time reversed motion) always has its acceleration directed downwards.

## 9.4 Constitutive Laws

We are now in a position to explore the constitutive laws of particle dynamics.

### *9.4.1 Momentum–Velocity Relation*

Having given an operative definition of momentum, we ask now what the relation v between momentum and velocity. To this end, let us consider our calibrated ballistic pendulum. (p. 246). We can shoot the same bullet with various velocities and measure the corresponding momenta.[8] The data thus obtained can be mapped in a diagram $v - p$, as shown in Fig. 9.3a. Measurements show that momentum is

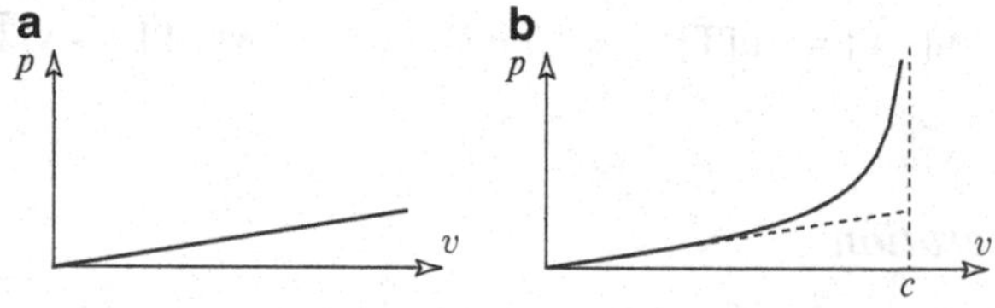


**Fig. 9.3** (**a**) Diagram $p - v$ in classical mechanics. (**b**) Corresponding diagram in relativistic mechanics

proportional to velocity and has the same direction, i.e. $\mathbf{p} \propto \mathbf{v}$. This leads us to introduce a parameter $m$ and to write

$$\mathbf{p}(t) \overset{\text{mat}}{=} m\,\mathbf{v}(t). \tag{9.15}$$

The parameter thus introduced as a proportionality constant between two variables is called (inertial) *mass*.

Mass is a system parameter like the capacitance of a condenser, the inductance of a coil, the thermal capacity of a body or the stiffness of a spring. To reinforce this presentation, we observe that in relativity the momentum–velocity relation must be substituted by

---

[8] Recall that calibration of the pendulum was obtained by shooting multiples of a given mass with the same velocity.

$$\mathbf{p}(t) \stackrel{\text{mat}}{=} m_0 \frac{\mathbf{v}(t)}{\sqrt{1 - \left(\frac{v(t)}{c}\right)^2}}, \tag{9.16}$$

as shown in Fig. 9.3b, and this cannot be reconciled with the interpretation of the momentum–velocity relation as a defining equation. The interpretation of relation 9.15 as a constitutive equation was given by Van Dantzig,[9] who said:

> Historically Newton's law did not state that the force is the product of mass and acceleration, but that it is the fluxion of the "impetus" (kinetic momentum), which is the product of mass and velocity. Moreover, the relation between momentum and velocity $\mathbf{p} = m\mathbf{v}$ is a linking equation and implies metric.

Equation 9.15 is a constitutive equation[10] which permits the definition of the physical constant $m$.

**Remark.** It is not contradictory to state that the same equation expresses a *law* and *defines* a physical constant. A constitutive law states that two physical quantities, already defined, are proportional. The proportionality constant, whose value must be inferred from experiments, is defined by the same equation. So when we state Hooke's law $F = k\,s$ for linear elastic materials, we express a law, but at the same time we define the stiffness $k$ of the elastic force. What is contradictory is to state that Eq. 9.15 is, *at the same time*, a constitutive equation and a defining equation![11]

**Remark.** It is curious to recall that Newton and Euler defined mass as the product of volume times mass density,[12] thus considering density as a basic variable. This is in contrast with the actual measurement of density, which is obtained by measuring mass and volume.

Ernst Mach wrote:

> With regard to the concept of "mass", it is to be observed that the formulation of Newton, which defines mass to be the quantity of matter of a body as measured by the product of its volume and density, is unfortunate. As we can only define density as the mass of unit of volume, the circle is manifest.[13]

Relation 9.15 can be written in a discrete setting as

$$\mathbf{p}[\tilde{\mathbf{I}}] \stackrel{\text{mat}}{=} m\,\mathbf{v}[\overline{\mathbf{T}}] \qquad \textit{reversible}\,. \tag{9.17}$$

In the velocity–momentum relation, a reversal of motion implies that both variables change sign; hence, in reversed motion the equation maintains the same form it has in forward motion. This implies that the relation describes a *reversible* process.

---

[9] Van Dantzig [240, p. 78].

[10] Preumont [189, p. 2].

[11] As in Williams [251, p. 138].

[12] Jammer [105, p. 66].

[13] Mach [145, p. 194].

### 9.4.2 Force–Radius Vector Relation

Force can depend on the position of a particle, on its velocity and on other field variables. In the simple case of an elastic restoring force, we have the constitutive relation

$$\mathbf{F}[\widetilde{\mathbf{T}}] \stackrel{\text{mat}}{=} -k\,\mathbf{r}[\bar{\mathbf{I}}] \qquad \textit{reversible}, \tag{9.18}$$

where $k$ is the *stiffness* of the elastic force. Every possible link between force and position is a constitutive relation. In the relation force position, under a reversal of motion both variables are unchanged, and hence this relation also describes a *reversible* process.

### 9.4.3 Force–Velocity Relation

Let us consider a force–velocity relation, like that of viscous force:

$$\mathbf{F}[\widetilde{\mathbf{T}}] \stackrel{\text{mat}}{=} -h\,\mathbf{v}[\overline{\mathbf{T}}] \qquad \textit{irreversible}. \tag{9.19}$$

Under a reversal of motion velocity changes sign while force does not. This implies that in reverse motion force increases with velocity: since this does not happen in nature, we conclude that this relation describes an *irreversible* process. Other irreversible laws are Newton's law of viscous fluids, Ohm's law of electric conduction, Fourier's law of heat conduction, Fick's law's of diffusion and Darcy's law for water motion.

In conclusion, we can summarize the various definitions which can be found in the literature about force and momentum in the following table:

| *Erroneous definition* | *Correct definition* |
|---|---|
| $\mathbf{F} \stackrel{\text{def}}{=} \dfrac{d\mathbf{p}}{dt}$ | $\mathbf{p}(t) \stackrel{\text{def}}{=} \displaystyle\int_{rest}^{t} \mathbf{F}(t')\,dt'$ |
| $\mathbf{p} \stackrel{\text{def}}{=} m\,\mathbf{v}$ | $\mathbf{p} \stackrel{\text{mat}}{=} m\,\mathbf{v}$ |

(9.20)

### 9.4.4 Classification of Forces

The notion of force precedes that of work, which in turn precedes that of energy. We can distinguish between forces which depend only on time and on position, $\mathbf{F}(t, \mathbf{P})$, giving rise to force fields, and forces which depend also on velocity $\mathbf{F}(t, \mathbf{P}, \mathbf{v})$. Among force fields there is a (large) subset of *conservative* forces (see

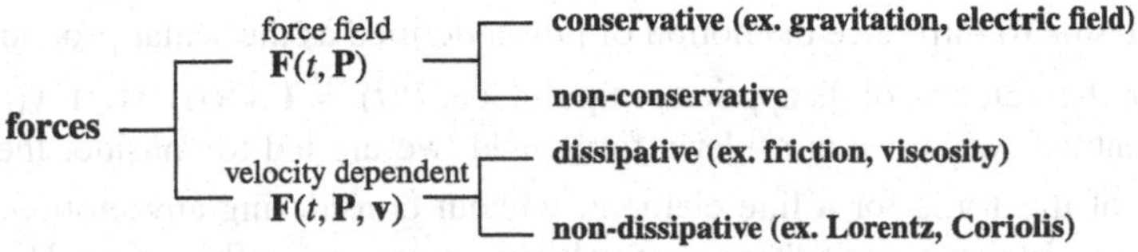


**Fig. 9.4** Classification of forces

later). Among velocity-dependent forces there is a (large) subset of *dissipative* forces. Some forces oppose the motion of a body or particle, i.e. those for which the power is negative, i.e. $\mathbf{F} \cdot \mathbf{v} < 0$. This implies a dissipation of the total energy (kinetic + potential). Such are forces of friction, those which cause air drag, in particular viscous forces. A notable exception is the Lorentz force acting on a moving charge in a magnetic field which is orthogonal to the velocity and, hence, does not dissipate energy: from $\mathbf{F}_L = q\mathbf{v} \times \check{\mathbf{B}}$ it follows that power is $P = \mathbf{F}_L \cdot \mathbf{v} = 0$. Another exception is the Coriolis apparent force in a rotating reference system: from $\mathbf{F}_C = -2m\check{\omega} \times \mathbf{v}_r$ it follows that $P = \mathbf{F}_C \cdot \mathbf{v}_r = 0$. Figure 9.4 summarizes this classification.

## 9.5 Energy Variables

Energy variables are the products of a source variable times a configuration variable. All of them originate from the notion of the *work* of a force. First of all, recall what is meant by energy: 'the energy of a system may be defined as the capacity it has of doing work, and is measured by the quantity of work it can do.'[14]

Every kind of energy is typically associated with a system $\mathscr{S}$; this is the case with kinetic energy $T[\mathscr{S}]$, potential energy $V[\mathscr{S}]$, internal energy $U[\mathscr{S}]$, entalpy $H[\mathscr{S}]$, Helmholtz free energy $F[\mathscr{S}]$, Gibbs free energy $G[\mathscr{S}]$ and electromagnetic energy $E[\mathscr{S}]$. In a spatial description each of these kinds of energy gives rise to a *content* and a *flow*.

### *9.5.1 Power, Work, Kinetic Energy*

When speaking of work it is better to distinguish between the work of a *single force* applied to a body in motion, in particular a single particle, and the work of a force in a *force field*. So when we push a shopping cart or pull a trolley, we exercise a single force which varies with time, place and speed. In an analogous way, a car on the road is pushed by a single force and is opposed by the air drag.

---

[14] Maxwell [154, p. 90].

This leads us to introduce the notion of power defined as the scalar product of the force for the velocity of its application point, i.e. $P(t) \stackrel{\text{def}}{=} \mathbf{F}(t, \mathbf{r}(t), \mathbf{v}(t)) \cdot \mathbf{v}(t)$.

In contrast, when we consider a force field, we are led to consider the scalar product of the force for a line element, without considering any motion: $w^* \stackrel{\text{def}}{=} \mathbf{F}(t, \mathbf{P}) \cdot \mathrm{d}\mathbf{L}$. In these cases, there is no velocity or, consequently, power. This scalar product must be called *virtual work*. It is common to denote an infinitesimal line element by $\delta\mathbf{r}$, even if a better notation would be $\mathrm{d}\mathbf{L}$. The symbol $\delta\mathbf{r}$, called the *virtual displacement*, arises from the need to distinguish it from the *effective displacement* $\mathrm{d}\mathbf{r}$.

**Single Force.** We will consider first the case of a single force applied to a particle in motion. From the power we can define two variables: the work as the *definite* time integral of power and the kinetic energy as the *indefinite* time integral of power, i.e.

$$\underbrace{W(t^-, t^+) \stackrel{\text{def}}{=} \int_{t^-}^{t^+} P(t)\,\mathrm{d}t}_{\textit{definite time integral}}^{\text{Work}} \qquad \underbrace{T(t) \stackrel{\text{def}}{=} \int_{(rest)}^{t} P(t')\,\mathrm{d}t'}_{\textit{indefinite time integral}}^{\text{Kinetic energy}} \tag{9.21}$$

These integrals can be evaluated only when the motion is known. From these definition it follows that

$$T(t^+) - T(t^-) = W(t^-, t^+), \qquad \text{from which} \qquad \frac{\mathrm{d}T(t)}{\mathrm{d}t} = P(t)\,. \tag{9.22}$$

Let us remark that the definition of kinetic energy as an indefinite time integral of power coincides with the classical definition, as the following relation shows:

$$T(t) \stackrel{\text{def}}{=} \int_{(rest)}^{t} P(t')\,\mathrm{d}t' \equiv \int_{(rest)}^{t} \mathbf{v} \cdot \mathbf{F}\,\mathrm{d}t \equiv \int_{0}^{\mathbf{p}} \mathbf{v}(\mathbf{p}) \cdot \mathrm{d}\mathbf{p} = T(\mathbf{p})\,. \tag{9.23}$$

Moreover, the definition of work as a definite time integral of power is equivalent to the traditional definition

$$W(t^-, t^+) \stackrel{\text{def}}{=} \int_{t^-}^{t^+} P(t)\,\mathrm{d}t \equiv \int_{t^-}^{t^+} \mathbf{F} \cdot \mathbf{v}\,\mathrm{d}t \equiv \int_{\mathbf{r}_1}^{\mathbf{r}_2} \mathbf{F}(\mathbf{r}) \cdot \mathrm{d}\mathbf{r}\,. \tag{9.24}$$

While work is an *energy flow*, power is an *energy flow rate* or *energy current*.

We stress the fact that force is a primitive quantity from which one can construct two other physical variables, the work and the impulse, as shown in Fig. 9.5.

**Force Field.** Let us consider a field of forces. This means that to all points of a space region is referred a force $\mathbf{F}(t, \mathbf{P})$. We will limit our consideration to a static field of forces, i.e. those of the kind $\mathbf{F}(\mathbf{P})$ which are the most widely used in physics. Such a field arises when we consider a particle of mass $m$ in a

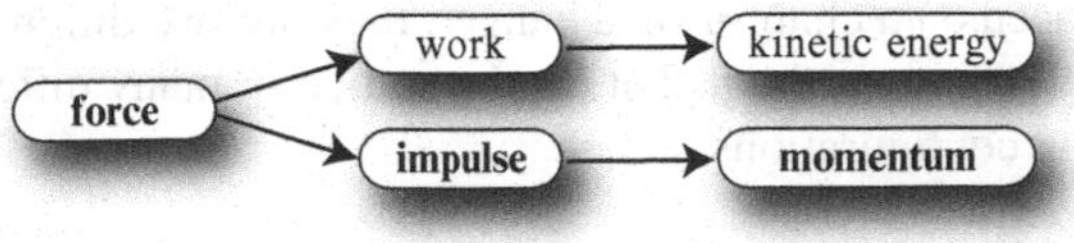


**Fig. 9.5** Force generates work and impulse

gravitational field, such as a stone in the gravitational field of the Earth or a planet in the solar system. Another case is that of a particle with a small charge $q$ in an electric field.

In a field of forces one can consider two kinds of work: the (effective) work performed on a particle in motion and the virtual work along a line

$$W(t^-, t^+) \stackrel{\text{def}}{=} \int_{t^-}^{t^+} \mathbf{F} \cdot \mathbf{v}\, dt \equiv \int_{\mathbf{r}_1}^{\mathbf{r}_2} \mathbf{F} \cdot d\mathbf{r}, \qquad W^*[\mathbf{L}] \stackrel{\text{def}}{=} \int_{\mathbf{L}} \mathbf{F} \cdot d\mathbf{L} \equiv \int_{\mathbf{L}} \mathbf{F} \cdot \delta\mathbf{r}\,. \tag{9.25}$$

Virtual work will give rise to the notion of potential energy, as we will show subsequently.

## 9.5.2 Potential Energy

While kinetic energy is a physical variable dealing with the motion of bodies, potential energy is used in all physical theories. It is a concept so important that it deserves a thorough examination and an explicit definition. In general, books on physics introduce this notion starting from particular cases, usually springs and gravity, without providing a general definition.

The adjective *potential* means that it has nothing to do with motion: potential energy is an energy of position. Usually books say that potential energy is work evaluated operating *slowly, with zero acceleration and zero kinetic energy*.[15] This is a subterfuge because one does not distinguish *effective* work, which presupposes a displacement, from *virtual* work, which is merely thought and does not involve time.

**Configuration.** Before introducing a definition of potential energy, we must analyse the notion of a system's *configuration* because the potential energy of a system $\mathscr{S}$ is a physical variable associated with the configuration of the system. The configuration of a system is *the geometric description of the simultaneous position of all the particles forming the system.* For a detailed definition of the concept of potential energy of a system it is better to distinguish between the term *internal configuration* and the *external position* of the system in a force field. Thus, a jacket

[15] See, for example, Lorrain et al. [143, p. 102].

which can be placed on a chair or on a hanger: they are two different positions; in addition, for each position, the jacket can be settled in many different ways, and these are different configurations.

DEFINITION. *The potential energy of a system in a given configuration is the energy stored in the system, which is equal to the work which the forces acting from* outside *on the system must perform to bring the system from a reference configuration to an actual configuration, under the hypothesis that such work is independent of intermediate configurations. In evaluating this work, velocity-dependent forces and dissipative forces (such as friction) should not be taken into account.*

Since each force acting from outside contrasts with a force exerted by a system on the outside, the work that enters the system is opposite to that done by the system; it follows that *the potential energy of a system in a given configuration is the work that the forces of the system must perform to bring the system from an* actual *configuration to a* reference *configuration.*

**Virtual Work.** In contrast with the notion of work of a single force applied to a particle in motion, let us consider a static *field* of forces $\mathbf{F}(\mathbf{P})$. This is the case with the force $\mathbf{F} = q\,\mathbf{E}(\mathbf{P})$ acting on an electric particle in an electrostatic field or the force $\mathbf{F} = m\,\mathbf{g}(\mathbf{P})$ acting on a massive particle in a gravitational field. These are two defining equations: one defines $\mathbf{E}$ and the other defines $\mathbf{g}$. Let us consider a line $\overline{\mathbf{L}}$ with an inner orientation and the force at a given instant. The line integral

$$\boxed{W^* \stackrel{\text{def}}{=} \int_{\overline{\mathbf{L}}} \mathbf{F}(\mathbf{r}) \cdot d\mathbf{L}} \qquad \text{virtual work} \tag{9.26}$$

recalls the notion of work, but it is very different. First of all, we do *not* have a single force applied to a *particle* but rather a *field* of forces, i.e. a force applied to *every point* of a region. Moreover, we do *not* have a particle which moves; we are simply considering a line element $d\mathbf{L}$ usually denoted in mechanics by $\delta\mathbf{r}$. Even more than this, the time is fixed. Also, in the case where the field of forces is variable in time, as happens in a variable electric field with $\mathbf{F}(t, \mathbf{P}) = q\,\mathbf{E}(t, \mathbf{P})$, the time instant is fixed. The analogy with work suggests that $\delta\mathbf{r}$ should be called a *virtual displacement* and the scalar product $\mathbf{F} \cdot \delta\mathbf{r}$ as *infinitesimal virtual work*. We have $w^*[\widetilde{\mathbf{T}}, \overline{\mathbf{L}}]$.

We will call *virtual work* the line integral of the force along a line in a force field. It may be objected that this name is already used in mechanics and is not explicitly considered in physics. The fact is that in mechanics one considers only the infinitesimal amount $w^* \stackrel{\text{def}}{=} \mathbf{F} \cdot \delta\mathbf{r}$, which is called *virtual work* instead of *infinitesimal* virtual work, which would be more appropriate.[16] The notion of virtual work as a line integral permits a simple definition of *potential energy*, as we will see in the next section.

In electromagnetism the electromotive force is defined as the line integral of the electric field $\mathbf{E}(t, \mathbf{P})$ at a fixed instant. Hence the electromotive force is the virtual work for unit charge: $E = W^*/q$.[17]

Referring to a force field we see that force is associated with lines, such as the vectors $\mathbf{E}$ and $\mathbf{g}$. Hence, $\mathbf{F}[\widetilde{\mathbf{T}}, \overline{\mathbf{L}}]$. Note that in physics, there are three types of forces: volume forces, surface forces and line forces. As a result, there are three types of work: volume work, surface work and line work, the last being a type of virtual work.

### 9.5.3 Potential Energy of a Particle in a Force Field

When a particle is at rest at a point $\mathbf{P}$ in an electric or in a gravitational field, it experiences a force. Taking into account that the force depends on the particle (its charge, its mass), we can refer the force directly to the point, writing $\mathbf{F}(\mathbf{P})$. In this way, we have a force field. Let us choose two points $\mathbf{A}$ and $\mathbf{B}$ of the region of definition and evaluate the line integral

$$W^*(\mathbf{A}, \mathbf{B}, \mathbf{L}) \stackrel{\text{def}}{=} \int_{\mathbf{A}}^{\mathbf{B}} \mathbf{F}(\mathbf{P}) \cdot d\mathbf{L} \,. \tag{9.27}$$

In general, the integral will depend on the line $\mathbf{L}$ connecting the two points.

**Conservative Forces.** The electrostatic and gravitational fields possess the property that this line integral does not depend on the line connecting the two points; in this case we say that the force field is *conservative*. In such a field let us choose a position $\mathbf{O}$ of the particle as reference position. In general, $\mathbf{O}$ is chosen where the field vanishes. For every point $\mathbf{P}$ of the field let us evaluate the line integral of the vector $\mathbf{F}(\mathbf{P})$ along any line $\mathbf{L}$ connecting $\mathbf{P}$ with $\mathbf{O}$, i.e.[18]

$$\boxed{V(\mathbf{P}) \stackrel{\text{def}}{=} \int_{\mathbf{P}}^{\mathbf{O}} \mathbf{F}(\mathbf{P}) \cdot \mathbf{t} \, dL} \quad \text{potential energy.} \tag{9.28}$$

Note that the line integral is taken from the actual point $\mathbf{P}$ to the fixed point $\mathbf{O}$, not vice versa. The function $V(\mathbf{P})$ is called the *potential energy* of the particle at

[16] Among the very few authors who use the term *infinitesimal virtual work* we mention Lanczos [120, p. 80].

[17] See page 267.

[18] Williams [251, p. 187]; Luré [144, Vol. I; p. 188].

the position **P**. One usually sets $V(\mathbf{O}) = 0$. Since the reference configuration is arbitrary, it follows that the potential energy is defined up to an arbitrary additive constant.

**Time Association.** Since Eq. 9.28 does not involve time variations (differences or derivatives) from $W^*[\widetilde{\mathbf{T}}]$, it follows that $V[\widetilde{\mathbf{T}}, \overline{\mathbf{P}}]$. The relation $\mathbf{F} = -\nabla V$ shows that **F** and $V$ must have the same time association; hence, the potential energy is, like force, associated with dual intervals.

The virtual work from a first point **A** to a second point **B** along a line connecting them is[19]

$$W^*(\mathbf{AB}) = \int_{\mathbf{A}}^{\mathbf{B}} \mathbf{F}(\mathbf{P}) \cdot \mathbf{t}\, dL = \int_{\mathbf{A}}^{\mathbf{O}} \mathbf{F}(\mathbf{P}) \cdot \mathbf{t}\, dL - \int_{\mathbf{B}}^{\mathbf{O}} \mathbf{F}(\mathbf{P}) \cdot \mathbf{t}\, dL = V(\mathbf{A}) - V(\mathbf{B}). \tag{9.29}$$

In a gravitational field the potential energy of a particle $V(\mathbf{P})$ depends on the mass and on the position of the particle. Since the force is proportional to the mass, the potential energy is also proportional to the mass. Hence, we can introduce the *gravitational potential* $U_g(\mathbf{P}) \stackrel{\text{def}}{=} V(\mathbf{P})/m$. In an electrostatic field, in the same way, one can introduce the electric potential (Table 2.2, p. 29).

It is important to remark that potential energy exists also when a system is subject to a mix of conservative and dissipative forces because only conservative forces are involved.

EXAMPLE 1. Let us consider two aeroplanes of equal weight moving side by side with the same velocity but coming from two distinct airports. They have the same potential energy, but the work done to bring them in that position is very different. This fact underlines that the potential energy must be calculated by means of the virtual work and not by means of the effective work.

**Energy Conservation.** Let us consider a moving particle in a static force field. The total force acting on the particle $\mathbf{F}^{\text{total}}$ can be decomposed into the force of the field $\mathbf{F}^{\text{field}}$ and the force depending on time and velocity $\mathbf{F}^{\text{others}}$:

$$\int_{t^-}^{t^+} \mathbf{F}^{\text{tot}} \cdot \mathbf{v}\, dt = \int_{\mathbf{A}}^{\mathbf{B}} \mathbf{F}^{\text{field}} \cdot d\mathbf{r} + \int_{t^-}^{t^+} \mathbf{F}^{\text{others}} \cdot \mathbf{v}\, dt\,. \tag{9.30}$$

The first term on the right side of this equation coincides with the virtual work calculated along the trajectory, which in turn is the difference between the potential energies at the two points **A** and **B** in which the particle passes at the two instants $t^-$ and $t^+$:

$$T(t^+) - T(t^-) = V(\mathbf{A}) - V(\mathbf{B}) + \int_{t^-}^{t^+} \mathbf{F}^{\text{others}} \cdot \mathbf{v}\, dt\,. \tag{9.31}$$

If there are no other forces, say velocity-dependent forces or friction forces, then we obtain the principle of energy conservation:

[19] Levi-Civita and Amaldi [135, Vol. II; p. 353].

$$T(t^+) + V(\mathbf{B}) = T(t^-) + V(\mathbf{A}) . \tag{9.32}$$

We remark that while kinetic energy depends on time, potential energy depends on a point.

### 9.5.4 Potential Energy of a System

The potential energy due to the internal configuration will be called the *internal potential energy* and will be denoted by $V^i$; the potential energy due to its position in a force field will be called the *external potential energy* and denoted by $V^e$. The separation of the internal from the external potential energy of a system is fundamental. Think of a pound of dynamite or a pound of butter located at the same height in the gravitational field: they have the same external potential energy while their internal potential energies are a bit different.

Figure 9.6 shows a system formed by two masses and a spring in different configurations and different positions. We can take configuration (a) as the reference. In configuration (b), the spring is compressed and the internal potential energy is increased. In configuration (c), the system is raised on the table with an uncompressed spring, and only the external potential energy is increased. In configuration (d) the system is raised on the table and the spring is compressed, and we see an increase in the external and the internal potential energies.

The potential energy is an attribute of a physical *system*, $V[\mathscr{S}]$; in a discrete system, such as a rigid body or a collection of rigid bodies, it depends on a finite number of generalized coordinates which specify the position+configuration of the system, i.e. $V[q^1, q^2, \ldots, q^n]$.

In particular, the potential energy of a particle $\mathscr{P}$ in a field depends only on the position of the particle in the field, hence it depends on the point $V[\overline{\mathbf{P}}]$. In a continuous system or in a field, we can consider the potential energy enclosed in each volume $\widetilde{\mathbf{V}}$; hence, $V[\widetilde{\mathbf{V}}]$. This is the case of the energy of deformation of a deformable solid or of the electric energy in an electric field.

Since the total energy of a system is split into kinetic and potential energy, $E[\mathscr{S}] = T[\mathscr{S}] + V[\mathscr{S}]$, when we move to a spatial description, both the kinetic and potential energies are split into content and flow:

$$E^c[\widetilde{\mathbf{V}}] = T^c[\widetilde{\mathbf{V}}] + V^c[\widetilde{\mathbf{V}}] \qquad E^f[\widetilde{\mathbf{S}}] = T^f[\widetilde{\mathbf{S}}] + V^f[\widetilde{\mathbf{S}}] . \tag{9.33}$$

The terms *kinetic energy flow* and *potential energy flow* are not commonly used, yet if we consider the water of a river, as well as forming a mass flow, it is a flow of kinetic energy and a flow of momentum, as is evident from the fact that bumping

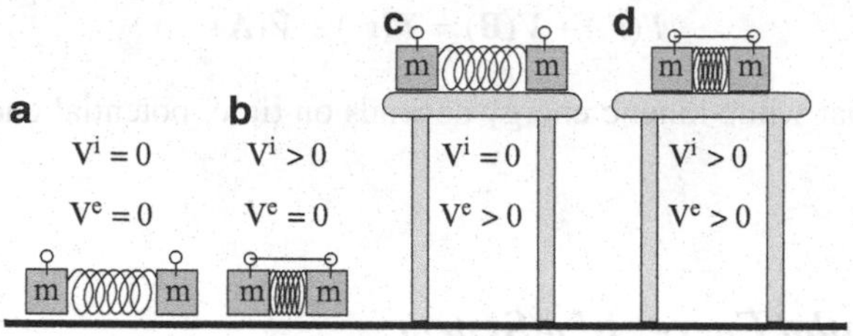


**Fig. 9.6** System formed by two masses and a compressible spring

up against the blades of a water mill puts them in motion. Moreover, if the river is sloping, there is also a flow of potential energy, as is clearly seen in the water of a waterfall.

One might be tempted to say that the term *energy flow* is nothing more than work and heat; but we must be careful. In a material description, energy flow refers to the boundary of a *system*, and this is synonymous with work and heat. But in a spatial description, energy flow refers simply to the motion of the system through the boundary of a fixed *control volume*, as Fig. 5.1b (p. 91) shows. In this case, the energy flow has nothing to do with work or heat because it exists even if the system (here the set of fish) moves in uniform and rectilinear motion. See also Fig. 9.7.

Let us consider some examples.

**Example 9.** An aeroplane of weight $w$ flying at a height $h$ has a gravitational potential energy of $V = w\,h$. The effective work needed to bring the aeroplane to that position is much greater than the potential energy acquired by the aeroplane. To evaluate the potential energy of the aeroplane, we must ignore the air drag and the kinetic energy: this is equivalent to the evaluation of the work of the gravitational force alone, the only conservative force.

**Example 10.** In a similar way, to calculate the potential energy of an ideal gas enclosed within a cylinder with insulated walls and compressed by a piston, we must neglect the friction of the piston with the inner wall of the cylinder. Moreover, if the compression is fast, we will notice the formation of vortices and then the transformation of a fraction of the work into kinetic energy at first and heat in the sequel. Hence the work needed to compress the gas is greater than

**Fig. 9.7** When a bird escapes from its cage, there is not only a mass flow, but also a flow of potential energy (internal and external)

the potential energy acquired. For these reasons thermodynamics has created the ideal notion of *reversible transformation* as a sequence of equilibrium states, i.e. a transformation that is 'infinitely' slow and without dissipation. In this way, the force acting on the piston becomes conservative.

EXAMPLE **11**. To lift a suitcase from the floor to a chair, we must perform more work than the gravitational potential energy gained by the suitcase, i.e. $V = wh$. In fact, innitially we must apply a force greater than the weight $w$, and hence we must lift the bag slightly above the chair and, lastly, reduce the force in order to set the bag on the chair.

EXAMPLE **12**. Dragging a case on a horizontal plane, we perform much work in overcoming the forces of friction, but the work done by the conservative force, gravity, is zero, and therefore the potential energy does not change. This applies to the motion of a car, a train or any other body.

**Space association.** Let us look at the association of potential energy with space elements. As the definition claims, potential energy is a property of a *system*, $V[\mathscr{S}]$, in particular of a *particle*, $V[\mathscr{P}]$. If the system has a finite number of degrees of freedom, then the potential energy is a function of these generalized coordinates, $V(q^1, q^2, \dots q^n)$. In particular, the potential energy *of a particle* in a field, say in the electric or in the gravitational field, is associated with the point in which the particle lies; hence, *for a particle* $V[\tilde{\mathbf{T}}, \mathscr{P}] \longrightarrow V[\tilde{\mathbf{T}}, \overline{\mathbf{P}}]$.

When we pass from a material description to a spatial one, the potential energy of a system splits into the potential energy content $V^c$ inside a control volume, and hence is associated with volumes, and the potential energy flow $V^f$ associated with surfaces $V[\tilde{\mathbf{T}}, \mathscr{S}] \longrightarrow V^c[\tilde{\mathbf{T}}, \tilde{\mathbf{V}}], V^f[\tilde{\mathbf{T}}, \tilde{\mathbf{S}}]$.

What kind of orientation of volumes is the potential energy associated with? Since the potential energy content $V^c$ is obviously invariant under space reflection, it follows that it cannot be associated with an inner orientation of a volume;[20] hence, it is associated with volumes endowed with an outer orientation, i.e. $V[\tilde{\mathbf{V}}]$. The potential energy flow $V^f$, i.e. work and heat, is associated with surfaces endowed with an outer orientation: $W[\tilde{\mathbf{S}}], Q[\tilde{\mathbf{S}}]$.

**The Minus Sign in the Potential Energy.** We now explain why, in a field, $\mathbf{F}$ is *minus* the gradient of $V$. In a field, like the electric and the gravitational fields, we can consider the potential energy of a test particle in a given point of the field. The external force needed to equilibrate the force of the field is opposite to the field force: $\mathbf{F}^{\text{ext}} = -\mathbf{F}$. The work needed *from outside* to bring the test particle from a reference point to the given point is an integral of the external force taken along a line from the reference point to the given point, i.e.

$$V(\mathbf{P}) \stackrel{\text{def}}{=} \int_{\mathbf{O}}^{\mathbf{P}} \mathbf{F}^{\text{ext}} \cdot d\mathbf{L}, \qquad \text{hence} \qquad \mathbf{F}^{\text{ext}} = \nabla V(\mathbf{P}) \,. \tag{9.34}$$

From the last relation it follows that $\mathbf{F} = -\mathbf{F}^{\text{ext}} = -\nabla V(\mathbf{P})$.

[20] Recall the *oddness principle* (p. 122).

### 9.5.5 Kinetic Energy and Kinetic Co-energy

Although commonly ignored, there are in fact two kinetic energies: *kinetic energy* $T$ and *kinetic co-energy* $T^*$ defined respectively as follows:

$$\boxed{\begin{array}{cc} \text{Kinetic energy} & \text{Kinetic co-energy} \\ T(\mathbf{p}) \overset{\text{def}}{=} \int_0^p \mathbf{v}(\mathbf{p}) \cdot \mathbf{dp} & T^*(\mathbf{v}) \overset{\text{def}}{=} \int_0^v \mathbf{p}(\mathbf{v}) \cdot \mathbf{dv} \end{array}} \tag{9.35}$$

The first is a function of momentum, the second of velocity.

**Remark.** At first sight, the fact of considering kinetic energy $T$ as a function of momentum $\mathbf{p}$ instead of velocity $\mathbf{v}$ contrasts with the common practice in mechanics. Nevertheless, we must observe that the differential under the integral is in $\mathbf{p}$ and it dictates the independent variable of the primitive. Once we take into account the classical relation $\mathbf{p} = m\mathbf{v}$, we can express kinetic energy as a function of velocity, as usual.

The advantage of the definitions given by Eq. 9.35 is that they are independent of the constitutive relation momentum–velocity. From these definitions it follows that

$$v^k = \frac{\partial T(\mathbf{p})}{\partial p_k}, \qquad p_k = \frac{\partial T^*(\mathbf{v})}{\partial v^k}. \tag{9.36}$$

Kinetic energy and its dual are linked by the relation

$$T^*(\mathbf{v}) \equiv \mathbf{p}(\mathbf{v}) \cdot \mathbf{v} - T(\mathbf{p}(\mathbf{v})), \tag{9.37}$$

i.e. by a duality transform, also called the Legendre transform. In classical mechanics, the constitutive relation momentum–velocity can be written in the form $\mathbf{p} = m\,\mathbf{v}$ or its inverse $\mathbf{v} = \mathbf{p}/m$. Inserting these relations into Eq. 9.35 we obtain

$$T(\mathbf{p}) = \frac{1}{2m}\,p^2 \qquad T^*(\mathbf{v}) \overset{\text{def}}{=} \frac{1}{2} m\,v^2. \tag{9.38}$$

In relativistic mechanics, the constitutive relation momentum–velocity can be written as

$$\mathbf{p} \overset{\text{mat}}{=} m_0\,\mathbf{v} \frac{1}{\sqrt{1 - \left(\frac{v}{c}\right)^2}}, \qquad \mathbf{v} \overset{\text{mat}}{=} \frac{\mathbf{p}}{m_0} \frac{1}{\sqrt{1 + \left(\frac{p}{m_0 c}\right)^2}}. \tag{9.39}$$

Inserting these relations into Eq. 9.35 we obtain

$$T(\mathbf{p}) = m_0c^2\left[\sqrt{1+\left(\frac{p}{m_0c}\right)^2}-1\right], \qquad T^*(\mathbf{v}) = m_0c^2\left[1-\sqrt{1-\left(\frac{v}{c}\right)^2}\right]. \quad (9.40)$$

If we consider two vector spaces, a three-dimensional space for velocities and another three-dimensional space for momenta, then the momentum–velocity relation can be viewed as describing a vector field in the velocity space while the inverse velocity–momentum relation can be viewed as defining a vector field in the momentum space. Since the two relations are the inverse of each other, we may call one vector field the inverse of the other. In this view, the constitutive relations 9.15 and 9.16 describe central vector fields and, hence, conservative vector fields. If one field admits a potential, then its inverse also admits a potential, and the two potentials are linked by the Legendre transform. It follows that *kinetic energy is the potential of velocities in the momentum space, while kinetic co-energy is the potential of momenta in the velocity space.*

In classical mechanics, since the relation $\mathbf{p} = m\,\mathbf{v}$ is linear, it follows that $T = p^2/2m = 1/2mv^2 = T^*$, i.e. kinetic co-energy is *numerically* equal to kinetic energy and has the same physical dimension. It is for this reason that kinetic co-energy is usually not considered (let alone defined!) in classical mechanics.

In relativistic mechanics, the momentum–velocity relation is non-linear, and then the equality of $T$ and $T^*$ is not valid, as shown in Fig. 9.8 (*right*). As potential energy is the potential of force, kinetic energy must therefore be viewed as the potential of velocity in the momentum space: both are evaluated as line integrals. We remark that energy conservation and the Hamiltonian function involve kinetic energy $T(\mathbf{p})$, while the Lagrangian involves kinetic co-energy $T^*(\mathbf{v})$.

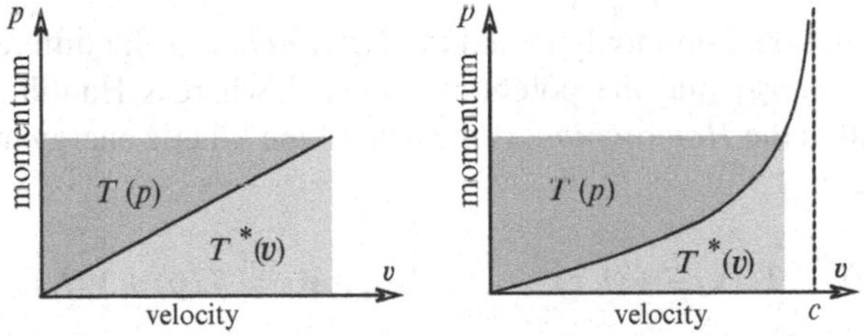


**Fig. 9.8** Kinetic energy and kinetic co-energy in classical (*left*) and in relativistic (*right*) cases

Historical Remark. The term *co-energy*, an abbreviation of *complementary energy*, is in common use in system theory, a branch of theoretical engineering (MacFarlane [146, p. 23]). Luré [144, Vol. II, p. 499] called *associated* kinetic energy our $T(\mathbf{p})$ and kinetic energy our $T^*(v)$, which he denoted by $\widetilde{T}(v)$. Penfield and Haus [175, p. 162] clarified the role of kinetic co-energy in a variational formulation. Sommerfeld [215, p. 266] denoted kinetic co-energy by $T$ and, following Helmholtz, called it *kinetic potential*. Kinetic co-energy was used by Preumont [189, p. 2]. A detailed treatment of kinetic co-energy can be found in Williams [251, p. 183], where it is denoted by $T^*(\mathbf{v})$.

Table 9.3 summarizes the terminology used for kinetic energy and kinetic co-energy.

**Table 9.3** Notions of kinetic energy and kinetic co-energy in the literature

| This book | $T(p) \overset{\text{def}}{=} \int_0^p \mathbf{v}(\mathbf{p}) \cdot d\mathbf{p}$ **Kinetic energy** | Page | $T^*(v) \overset{\text{def}}{=} \int_0^v \mathbf{p}(\mathbf{v}) \cdot d\mathbf{v}$ **Kinetic co-energy** | Page |
|---|---|---|---|---|
| MacFarlane | **Kinetic energy** $T(p)$ | 23 | Kinetic coenergy $T^*(v)$ | 23 |
| Williams | **Kinetic energy** $T(p)$ 'kinetic energy function' | 183 | Kinetic coenergy $T^*(v)$ 'kinetic coenergy function' | 183 |
| Preumont | **Kinetic energy** $T(p)$ | 2 | Kinetic coenergy $T^*(v)$ | 4 |
| Penfield, Haus | **Kinetic energy** $W(p)$ | 162 | Kinetic coenergy $\widetilde{W}(v)$ | 162 |
| Luré | Associated kinetic energy $T'$; $\widetilde{T}(p)$ | 138; 499 | **Kinetic energy** $T(v)$ | 138; 501 |
| Sommerfeld electrodynamics | **Kinetic energy** $T(v)$ | 264 | Kinetic potential $K(v)$ *(named by Helmholtz)* | 266 |

## 9.5.6 Lagrangian and Hamiltonian

Lagrange's function, commonly called the *Lagrangian*, is the difference between the kinetic *co-energy* and the potential energy,[21] whereas Hamilton's function, commonly called the *Hamiltonian*, is the sum of the kinetic energy and the potential energy, i.e.

$$L(t, \mathbf{r}, \mathbf{v}) \overset{\text{def}}{=} T^*(\mathbf{v}) - V(t, \mathbf{r}), \qquad H(t, \mathbf{r}, \mathbf{p}) \overset{\text{def}}{=} T(\mathbf{p}) + V(t, \mathbf{r}) . \tag{9.41}$$

Since the kinetic energy $T(\mathbf{p})$ is the potential of the constitutive equation $\mathbf{p} = m\mathbf{v}$ and $V(t, \mathbf{r})$ is the potential of the constitutive equation $\mathbf{F} = \mathbf{F}(t, \mathbf{r})$, it follows that the Hamiltonian can be conceived of as the overall potential of constitutive equations. Then assigning the Hamiltonian is equivalent to assigning the source of motion.

21 Williams [251, p. 214].

Hamilton's action $A_H$ is the *definite* time integral of the Lagrangian, whereas Hamilton's principal function $S$ is the *indefinite* time integral of the Lagrangian.[22] It follows that

$$A_H(t^-,t^+) = S(t^+) - S(t^-) \qquad S[\tilde{\mathbf{I}}] \longrightarrow A_H[\tilde{\mathbf{T}}] \,. \tag{9.42}$$

Since the kinetic energy and its dual are quadratic functions of momenta and velocities respectively, they do not change sign under a reversal of motion. It follows that

$$A_H \stackrel{\text{def}}{=} \int_{t^-}^{t^+} L(t)\,\mathrm{d}t \longrightarrow \mathscr{R} A_H = \int_{t^+}^{t^-} \mathscr{R} L(t)\, \mathscr{R}(\mathrm{d}t) = \int_{t^-}^{t^+} L(t)\,\mathrm{d}t = A_H \,. \tag{9.43}$$

Hence, the Hamiltonian action is associated with a physical system and time interval endowed with an outer orientation, i.e. $A_H[\tilde{\mathbf{T}}, \mathscr{S}]$. It follows that Hamilton's principal function $S$ is associated with dual time instants, i.e. $S[\tilde{\mathbf{I}}, \mathscr{S}]$, as shown in Table 2.1, p. 28.

In a *spatial description*, potential energy is associated with volumes endowed with an outer orientation and with intervals endowed with an outer orientation: $V[\tilde{\mathbf{T}}, \tilde{\mathbf{V}}]$ (Tables 9.5 and 9.6).

**Table 9.4** Classification diagram for particle dynamics

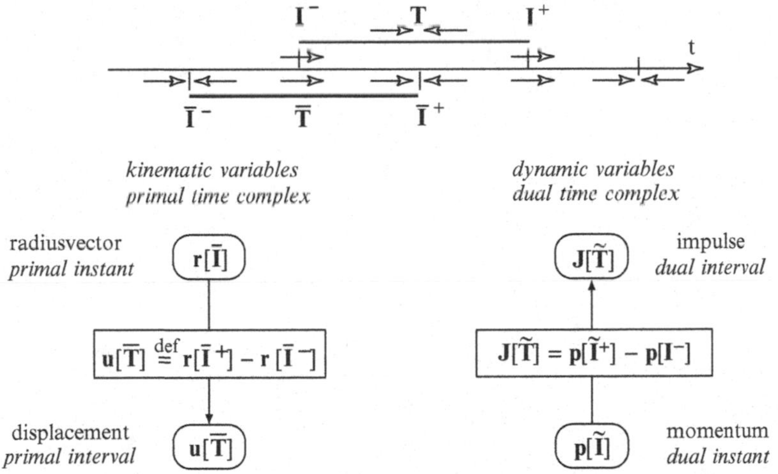


[22] See p. 100. Nowadays we must use the letter $W$ (IUPAP 1987), as do also Brillouin [30, p. 168]; Yourgrau and Mandelstam [257, p. 52]. The inconvenience with this is that $W$ also denotes work. To avoid this possible confusion, we use the letter $S$ used by Landau and Lifshitz [123, p. 138]; Sommerfeld [213, Sect. 44]; Goldstein [80, p. 274]; Luré [144, p. 705].

**Table 9.5** From field variables to global variables of particle mechanics

| Configuration | Configuration | Source | Source |
|---|---|---|---|
| $[\bar{I}]$ | $[\bar{I}]$ | $[\tilde{T}]$ | $[\tilde{T}]$ |
| **r** | **r** | **F** | **J** |
| $[\bar{T}]$ | $[\bar{T}]$ | $[\tilde{I}]$ | $[\tilde{I}]$ |
| **u** | **v** | **p** | **p** |

| Global variables | Field variables | Global variables |
|---|---|---|

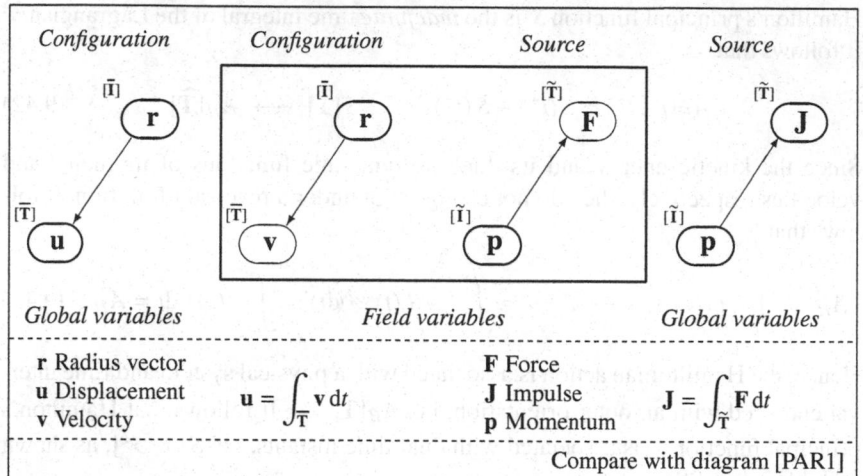

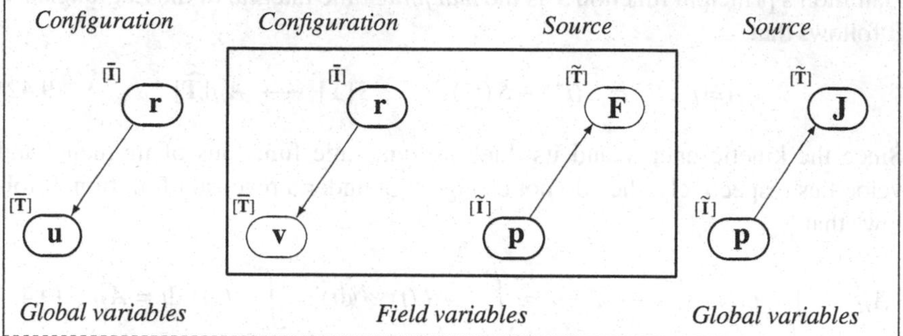

**r** Radius vector
**u** Displacement     $\mathbf{u} = \int_{\bar{T}} \mathbf{v}\, dt$     **F** Force
**v** Velocity                                             **J** Impulse     $\mathbf{J} = \int_{\tilde{T}} \mathbf{F}\, dt$
                                                           **p** Momentum

Compare with diagram [PAR1]

**Table 9.6** A paradigm for the time association

## A paradigm for the time association

Starting point: under reversal of motion
the displacement changes sign while the impulse do not.

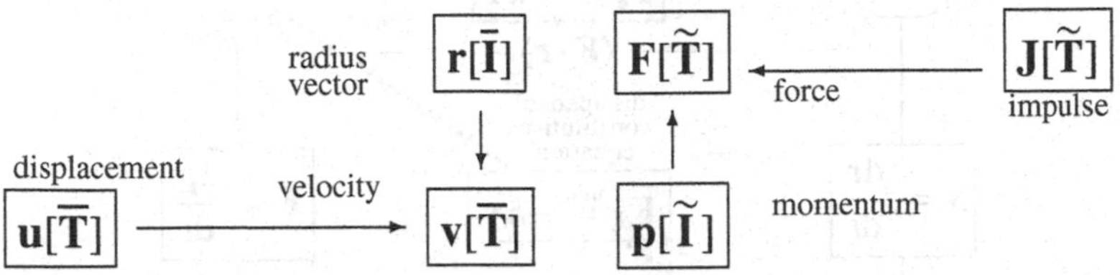


**work**

$$W \overset{\text{def}}{=} \int_{t^-}^{t^+} \mathbf{F}\cdot\mathbf{v}\,\mathrm{d}t \longrightarrow \boxed{W[\bar{\mathrm{T}}]}$$

**virtual work**

$$W^* \overset{\text{def}}{=} \int_{\mathbf{A}}^{\mathbf{B}} \mathbf{F}\cdot\mathbf{t}\,\mathrm{d}L \longrightarrow \boxed{W^*[\tilde{\mathrm{T}}]}$$

**kinetic energy**

$$T \overset{\text{def}}{=} \int_0^{\mathbf{p}} \mathbf{v}\cdot\mathrm{d}\mathbf{p} \longrightarrow \boxed{T[\tilde{\mathrm{T}}]}$$

**kinetic co-energy**

$$T^* \overset{\text{def}}{=} \int_0^{\mathbf{v}} \mathbf{p}\cdot\mathrm{d}\mathbf{v} \longrightarrow \boxed{T^*[\tilde{\mathrm{T}}]}$$

**potential energy**

$$V(\mathbf{P}) \overset{\text{def}}{=} \int_{\mathbf{P}}^{\mathbf{O}} \mathbf{F}\cdot\mathrm{d}\mathbf{L} \longrightarrow \boxed{V[\tilde{\mathrm{T}}]}$$

**Hamiltonian action**

$$A_H \overset{\text{def}}{=} \int_{t^-}^{t^+} L\,\mathrm{d}t \longrightarrow \boxed{A_H[\tilde{\mathrm{T}}]}$$

Hamiltonian $H \overset{\text{def}}{=} T + V$

$L \overset{\text{def}}{=} T^* - V$ Lagrangian

**electrostatics**

$$\mathbf{E}(\mathbf{P}) \overset{\text{def}}{=} \frac{\mathbf{F}(\mathbf{P})}{q} \longrightarrow \boxed{\mathbf{E}[\tilde{\mathrm{T}}]}$$

$$\phi(\mathbf{P}) \overset{\text{def}}{=} \frac{V(\mathbf{P})}{q} \longrightarrow \boxed{\phi[\tilde{\mathrm{T}}]}$$

**gravitation**

$$\mathbf{g}(\mathbf{P}) \overset{\text{def}}{=} \frac{\mathbf{F}(\mathbf{P})}{m} \longrightarrow \boxed{\mathbf{g}[\tilde{\mathrm{T}}]}$$

$$U_{\mathrm{g}}(\mathbf{P}) \overset{\text{def}}{=} \frac{V(\mathbf{P})}{m} \longrightarrow \boxed{U_{\mathrm{g}}[\tilde{\mathrm{T}}]}$$

The following relations confirm the preceding time associations

$$\mathbf{F} = -\nabla V \qquad \mathbf{J}(t^-, t^+) = \mathbf{p}(t^+) - \mathbf{p}(t^-)$$

$$\mathbf{g} = -\nabla U_{\mathrm{g}} \qquad W(t^-, t^+) = T(t^+) - T(t^-) \qquad w^* = \mathbf{F}\cdot\delta\mathbf{r}$$

$$\mathbf{E} = -\nabla\phi \qquad T + V = E$$

$$\phi(\mathbf{P}) = \int_{\mathbf{P}}^{\infty} \mathbf{E}\cdot\mathrm{d}\mathbf{L} \qquad U_{\mathrm{g}}(\mathbf{P}) = \int_{\mathbf{P}}^{\infty} \mathbf{g}\cdot\mathrm{d}\mathbf{L}$$

**Particle dynamics**                                                                    **[PAR1]**

*kinematic variables*                                          *dynamic variables*
time: *primal complex*                                         time: *dual complex*

$[\bar{\mathbf{I}}, \mathscr{P}]$     r     constitutive equation
$$F_e \overset{mat}{=} -k\,\mathbf{r}$$     F     $[\tilde{\mathbf{T}}, \mathscr{P}]$
$\mathbf{F}_{imp}$
$\langle \mathbf{F} \cdot \mathbf{r} \rangle$

$$\mathbf{v} = \frac{d\mathbf{r}}{dt}$$

dissipative
constitutive
equation
$$F_d \overset{mat}{=} -h\,\mathbf{v}$$

$$\mathbf{F} = \frac{d\mathbf{p}}{dt}$$

$[\bar{\mathbf{T}}, \mathscr{P}]$     v     constitutive equation
$$\mathbf{p} \overset{mat}{=} m\mathbf{v}$$     p     $[\tilde{\mathbf{I}}, \mathscr{P}]$
$\langle \mathbf{p} \cdot \mathbf{v} \rangle$

---

$[\bar{\mathbf{I}}, \mathscr{P}]$     r     constitutive equation
$$F_e \overset{mat}{=} -k\,\mathbf{r}$$     F     $[\tilde{\mathbf{T}}, \mathscr{P}]$
$\mathbf{F}_{imp}$

dissipative
constitutive equation
$$F_d \overset{mat}{=} -h\,\mathbf{v}$$

$$\mathbf{v} \overset{def}{=} d\mathbf{r}/dt$$     $\mathbf{F} = d\mathbf{p}/dt$

$[\bar{\mathbf{T}}, \mathscr{P}]$     constitutive equation
$$\mathbf{p} \overset{mat}{=} m\mathbf{v}$$
v     p     $[\tilde{\mathbf{I}}, \mathscr{P}]$

---

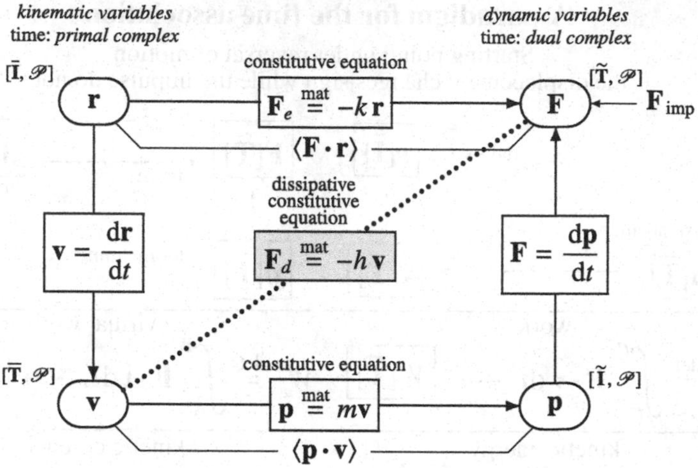

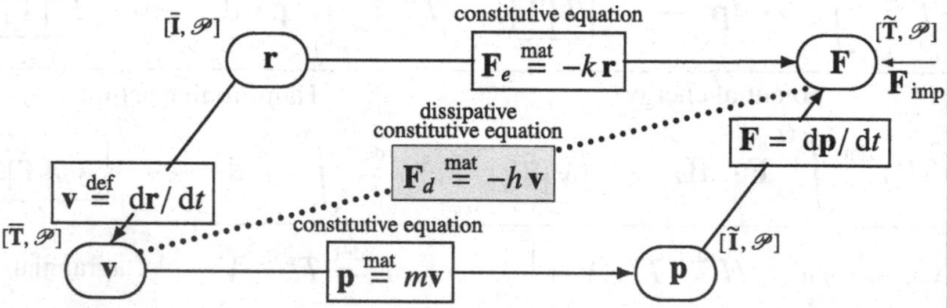

fundamental equation     $$\frac{d}{dt}\left( m\frac{d\mathbf{u}}{dt} \right) = -k\mathbf{u} - h\frac{d\mathbf{u}}{dt} + \mathbf{F}_{imp}$$

*dual complex*   $\tilde{\mathbf{I}}$      $\tilde{\mathbf{T}}$      $\tilde{\mathbf{I}}$                    $\tilde{\mathbf{I}}$      $\tilde{\mathbf{T}}$      $\tilde{\mathbf{I}}$
$t$                                    $t$
$\bar{\mathbf{I}}$      $\bar{\mathbf{T}}$      $\bar{\mathbf{I}}$   *primal complex*     $\bar{\mathbf{I}}$      $\bar{\mathbf{T}}$      $\bar{\mathbf{I}}$

---

**r**  *radius vector*                                          **F**     *force*
**v**  *velocity*                   $\bar{\mathbf{I}}$ time even $\tilde{\mathbf{T}}$     $\mathbf{F}_e$  *elastic restoring force*
**k**  *stiffness*                                              $\mathbf{F}_d$  *viscous force*
**h**  *dumping constant*           $\bar{\mathbf{T}}$ time odd $\tilde{\mathbf{I}}$     $\mathbf{F}_{imp}$  *impressed force*
**m**  *mass*                                                   **p**     *momentum*

PAR1-7; http://discretephysics.dicar.units.it

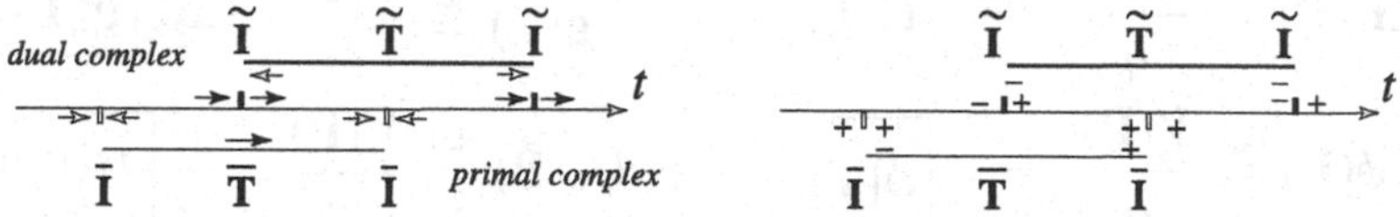

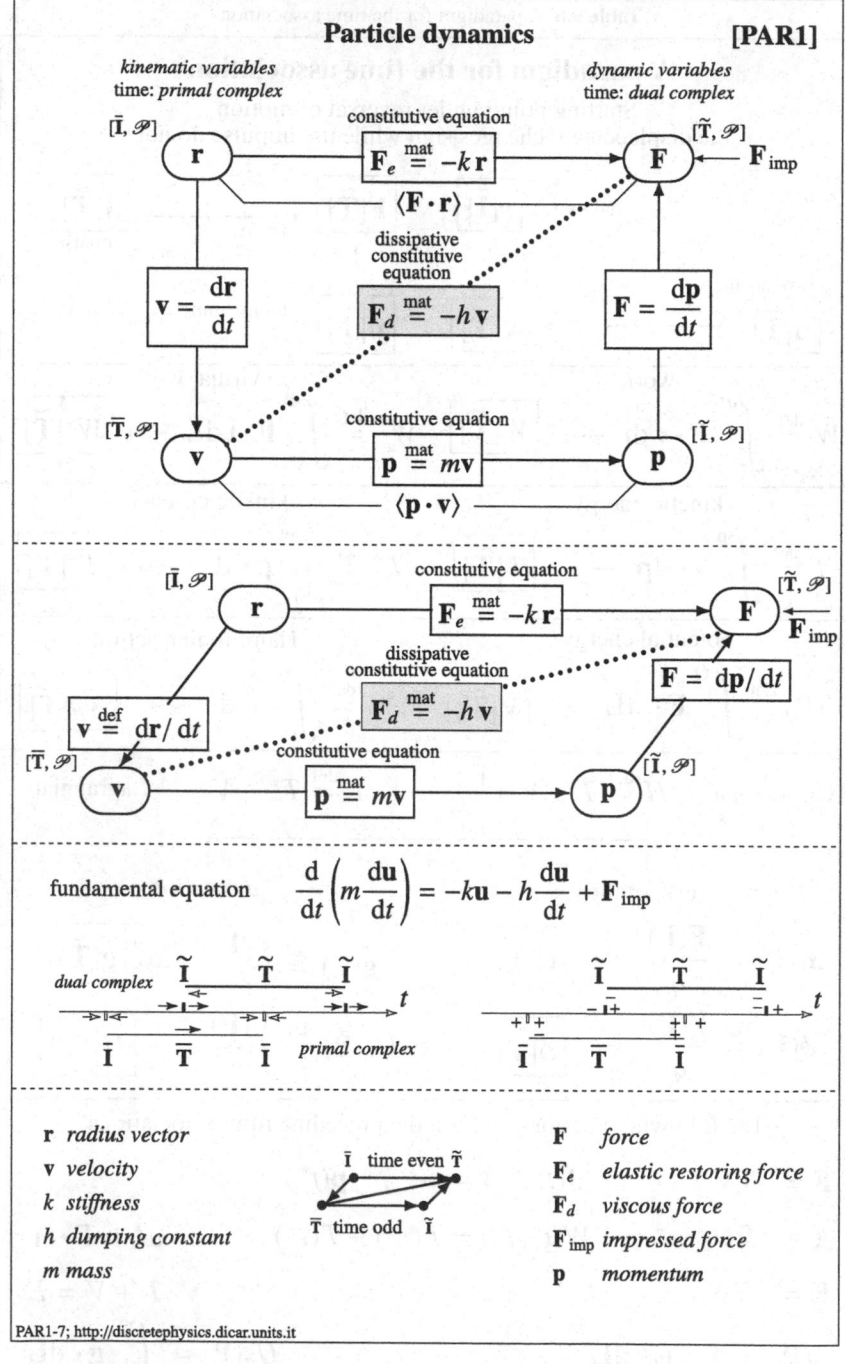

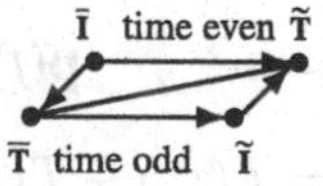

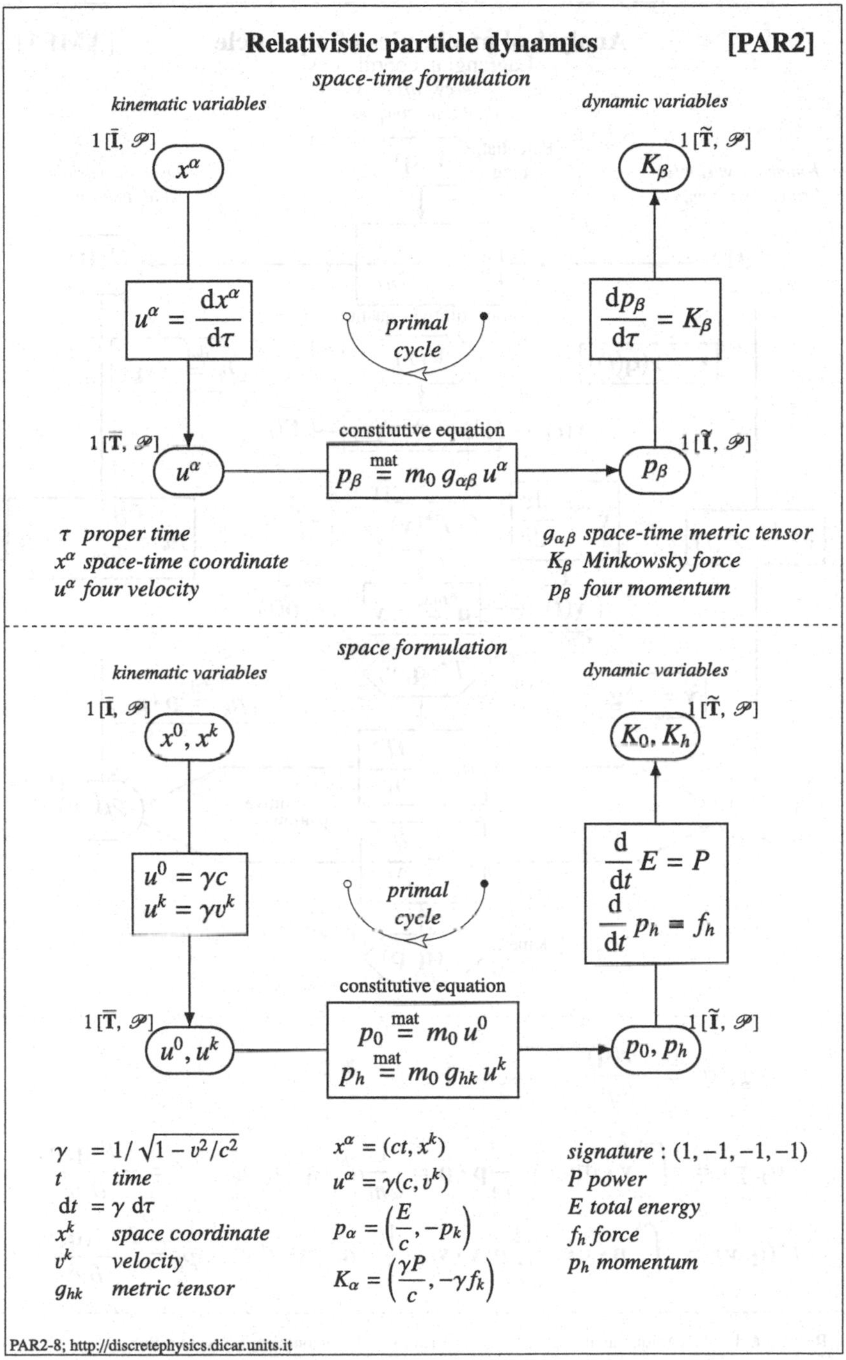
Relativistic particle dynamics
[PAR2]
space-time formulation
kinematic variables
dynamic variables
1 [$\bar{\mathbf{I}}$, $\mathscr{P}$]
$x^\alpha$
1 [$\tilde{\mathbf{T}}$, $\mathscr{P}$]
$K_\beta$
$u^\alpha = \frac{\mathrm{d}x^\alpha}{\mathrm{d}\tau}$
primal cycle
$\frac{\mathrm{d}p_\beta}{\mathrm{d}\tau} = K_\beta$
1 [$\bar{\mathbf{T}}$, $\mathscr{P}$]
$u^\alpha$
constitutive equation
$p_\beta \overset{\text{mat}}{=} m_0\, g_{\alpha\beta}\, u^\alpha$
1 [$\tilde{\mathbf{I}}$, $\mathscr{P}$]
$p_\beta$
$\tau$ proper time
$x^\alpha$ space-time coordinate
$u^\alpha$ four velocity
$g_{\alpha\beta}$ space-time metric tensor
$K_\beta$ Minkowsky force
$p_\beta$ four momentum
space formulation
kinematic variables
dynamic variables
1 [$\bar{\mathbf{I}}$, $\mathscr{P}$]
$x^0, x^k$
1 [$\tilde{\mathbf{T}}$, $\mathscr{P}$]
$K_0, K_h$
$u^0 = \gamma c$
$u^k = \gamma v^k$
primal cycle
$\frac{\mathrm{d}}{\mathrm{d}t} E = P$
$\frac{\mathrm{d}}{\mathrm{d}t} p_h = f_h$
1 [$\bar{\mathbf{T}}$, $\mathscr{P}$]
$u^0, u^k$
constitutive equation
$p_0 \overset{\text{mat}}{=} m_0\, u^0$
$p_h \overset{\text{mat}}{=} m_0\, g_{hk}\, u^k$
1 [$\tilde{\mathbf{I}}$, $\mathscr{P}$]
$p_0, p_h$
$\gamma = 1/\sqrt{1 - v^2/c^2}$
$t$ time
$\mathrm{d}t = \gamma\, \mathrm{d}\tau$
$x^k$ space coordinate
$v^k$ velocity
$g_{hk}$ metric tensor
$x^\alpha = (ct, x^k)$
$u^\alpha = \gamma(c, v^k)$
$p_\alpha = \left(\frac{E}{c}, -p_k\right)$
$K_\alpha = \left(\frac{\gamma P}{c}, -\gamma f_k\right)$
signature : (1, −1, −1, −1)
$P$ power
$E$ total energy
$f_h$ force
$p_h$ momentum
PAR2-8; http://discretephysics.dicar.units.it

## Analytical mechanics of a particle [AME1]

Lagrangian coordinates

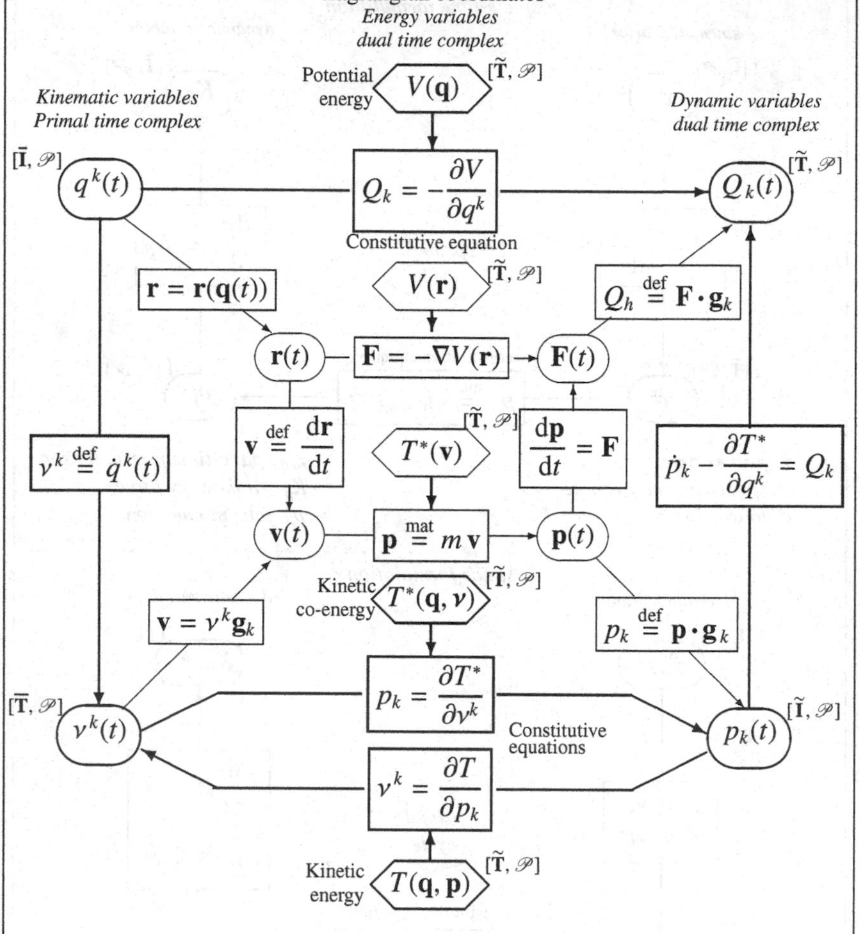


$$\mathbf{g}_k(\mathbf{q}) \stackrel{\text{def}}{=} \frac{\partial \mathbf{r}(\mathbf{q})}{\partial q^k} \qquad \text{Base vectors}$$

$$T(\mathbf{q},\mathbf{p}) \stackrel{\text{def}}{=} \int_0^{\mathbf{p}} \mathbf{v}\cdot \mathrm{d}\mathbf{p} = \frac{1}{2m}\mathbf{p}\cdot\mathbf{p} = \frac{1}{2m}a^{hk}(\mathbf{q})\, p_h\, p_k \qquad \nu^k = \frac{\partial T(\mathbf{q},\mathbf{p})}{\partial p_k}$$

$$T^*(\mathbf{q},\nu) \stackrel{\text{def}}{=} \int_0^{\mathbf{v}} \mathbf{p}\cdot \mathrm{d}\mathbf{v} = \frac{1}{2}m\,\mathbf{v}\cdot\mathbf{v} = \frac{1}{2}m\, a_{hk}(\mathbf{q})\,\nu^h\,\nu^k \qquad p_k = \frac{\partial T^*(\mathbf{q},\nu)}{\partial \nu^k}$$

Ref: Luré, L.: Mécanique analytique. Vols. I, II, Librairie Universitaire, Louvain (1968).

## Analytical mechanics [AME2]

Lagrange's and Hamilton's equations for scleronomic systems

| *Kinematic variables* | *Energy variables* | *Dynamic variables* |
| --- | --- | --- |
| *Primal time complex* | *Dual time complex* | *Dual time complex* |

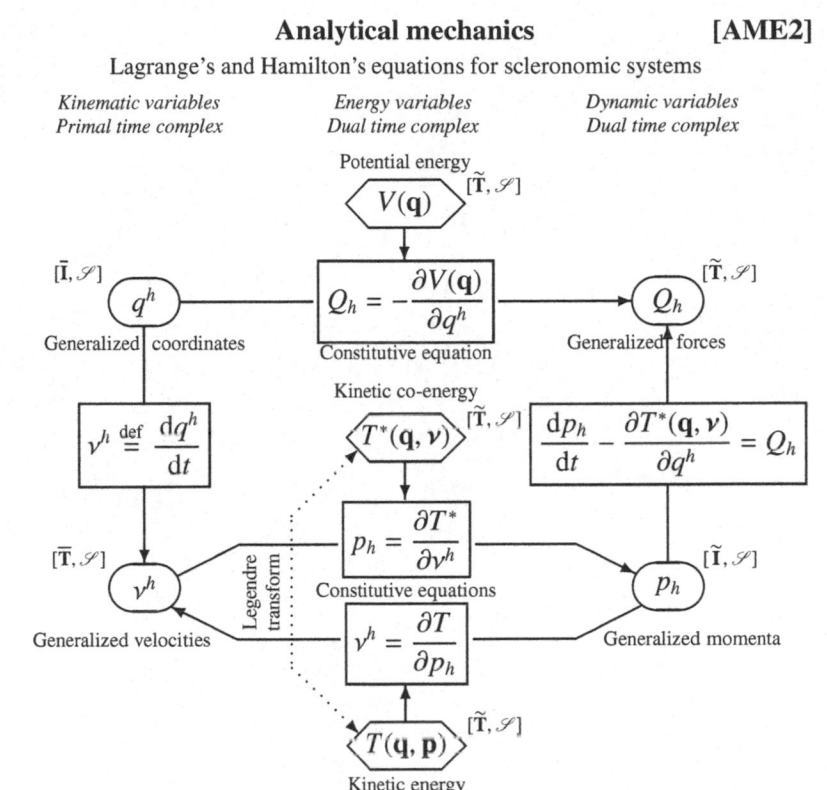

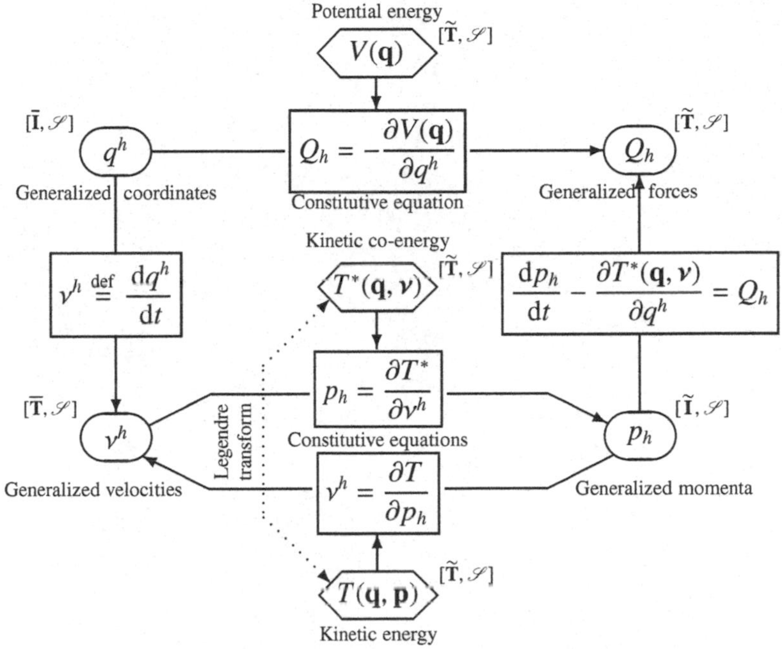


$$T^*(\mathbf{q}, \boldsymbol{v}) \stackrel{\text{def}}{=} \mathbf{p}(\mathbf{q}, \boldsymbol{v}) \cdot \boldsymbol{v} - T(\mathbf{q}, \mathbf{p}(\mathbf{q}, \boldsymbol{v})) \quad \longrightarrow \quad \frac{\partial T^*(\mathbf{q}, \boldsymbol{v})}{\partial q^h} = -\frac{\partial T(\mathbf{q}, \mathbf{p})}{\partial q^h}$$

$$\frac{\mathrm{d}p_h}{\mathrm{d}t} = \frac{\partial}{\partial q^h}\Big[T^*(\mathbf{q}, \mathbf{p}) - V(\mathbf{q})\Big] = -\frac{\partial}{\partial q^h}\Big[T(\mathbf{q}, \mathbf{p}) + V(\mathbf{q})\Big]$$

Lagrangian

$$L(\mathbf{q}, \boldsymbol{v}) \stackrel{\text{def}}{=} T^*(\mathbf{q}, \boldsymbol{v}) - V(\mathbf{q})$$

Hamiltonian

$$H(\mathbf{q}, \mathbf{p}) \stackrel{\text{def}}{=} T(\mathbf{q}, \mathbf{p}) + V(\mathbf{q})$$

Lagrange's equations

$$\frac{\mathrm{d}}{\mathrm{d}t}\frac{\partial L(\mathbf{q}, \boldsymbol{v})}{\partial v^h} - \frac{\partial L(\mathbf{q}, \boldsymbol{v})}{\partial q^h} = 0$$

Hamilton's canonical equations

$$\dot{q}^h = +\frac{\partial H(\mathbf{q}, \mathbf{p})}{\partial p_h}$$

$$\dot{p}_h = -\frac{\partial H(\mathbf{q}, \mathbf{p})}{\partial q^h}$$

Ref: Luré, L.: Mécanique analytique. Vols. I, II. Librairie Universitaire, Louvain (1968).
His $T'$ (Vol. I, p. 138) is the same as $\tilde{T}$ (Vol. II, p. 499) and is our $T$. His $T$ (Vol. II, p. 498) is our $T^*$.

AME2-7; http://discretephysics.dicar.units.it

# Chapter 10
# Electromagnetism

## 10.1 Fundamental Problem

The *fundamental problem* of electromagnetism can be stated as follows:[1]

- Given a space region and a time interval,
- Given the nature of the materials that fill the region,
- Given the boundary conditions,
- Given the initial values of the configuration variables,
- Given the space and time distribution of charges and currents,
- Find the configuration of the field at every point and at every later instant.

## 10.2 From Field Variables to Global Variables

The usual treatment of electromagnetism is done in differential terms by means of field functions.[2] In this section, our goal will be to present the global variables of electromagnetism. Although it is possible (and convenient) to bypass field variables by considering global variables from the start, we have decided to deduce the global variables from the field variables to help readers familiar with field functions.[3]

[1] This chapter is an expanded version of the author's paper [229]. A corresponding finite formulation of electromagnetism, useful for numerical purposes, can be found in the papers [230, 233, 234].

[2] This chapter presupposes a reading of Chaps. 1–9. Recall that in this book the 'check' symbol denotes the 'pseudo' nature of scalars and vectors (pseudoscalars and pseudovectors) as stated on p. 145.

[3] This unusual introduction to electromagnetic variables follows the presentation of Langevin [127], Sommerfeld [215], [178].

E. Tonti, *The Mathematical Structure of Classical and Relativistic Physics*, Modeling and Simulation in Science, Engineering and Technology, DOI 10.1007/978-1-4614-7422-7_10, © Springer Science+Business Media New York 2013

In accordance with the standard nomenclature,[4] we will use the following symbols and names for the field functions:

$$\begin{array}{ll} \mathbf{E} \text{ electric field strength} & \mathbf{D} \text{ electric displacement} \\ \check{\mathbf{B}} \text{ magnetic flux density} & \check{\mathbf{H}} \text{ magnetic field strength} \\ \rho \text{ volume charge density} & \mathbf{J} \text{ electric current density} \end{array} \tag{10.1}$$

Maxwell's field equations are

$$\text{homogeneous} \begin{cases} \operatorname{div} \check{\mathbf{B}} = 0 \\ \operatorname{curl} \mathbf{E} + \partial_t \check{\mathbf{B}} = 0 \end{cases}, \qquad \text{inhomogeneous} \begin{cases} \operatorname{div} \mathbf{D} = \rho \\ \operatorname{curl} \check{\mathbf{H}} - \partial_t \mathbf{D} = \mathbf{J} \end{cases} \tag{10.2}$$

to which we must add the three *constitutive equations*[5]

$$\mathbf{D} \overset{\text{mat}}{=} \epsilon\, \mathbf{E}, \qquad \check{\mathbf{H}} \overset{\text{mat}}{=} \frac{1}{\mu} \check{\mathbf{B}}, \qquad \mathbf{J} \overset{\text{mat}}{=} \sigma \mathbf{E} \tag{10.3}$$

which are valid for a material that is linear, homogeneous, isotropic and non-hereditary. These six variables can be divided, initially, into two classes, configuration and source variables. The sources of the electromagnetic field are the electric charges at rest and in motion, i.e. charges and currents. In a differential treatment, these *source variables* are the volume density of charge $\rho$ and the current density $\mathbf{J}$. Since the variables $\mathbf{D}$ and $\check{\mathbf{H}}$ are related to them by differential relations which do not contain physical parameters (Chap. 5), they are also source variables. The four variables, $\rho, \mathbf{J}, \mathbf{D}, \check{\mathbf{H}}$, are the protagonists of the second group of Maxwell's equations.

The two vectors $\mathbf{E}$ and $\check{\mathbf{B}}$, which are the protagonists of the first group of Maxwell's equations, define the configuration of the electromagnetic field, i.e. they are *configuration variables*.

To perform the classification in space and time, we need to find for each field variable the corresponding global variable. To this end, we will apply the rules explained in Chap. 5, i.e. we will ask ourselves if it is possible to compute a line, space or volume integral of the field variable we are analysing. If the answer is affirmative, then, after integration, we have found a space global variable. Then we will ask ourselves whether it is possible to perform a time integral of this space global variable. In this case, after integration, we have obtained a time global variable. In subsequent sections we will show how to obtain global variables directly from physical measurements, i.e. without using field variables as the starting point.

---

[4] See International Union of Pure and Applied Physics [101].

[5] It seems improper to call *material laws* the relations $\mathbf{D} = \epsilon\mathbf{E}$ and $\mathbf{B} = \mu\mathbf{H}$ even in vacuo. But this is equivalent to *constitutive law*: see Born and Wolf [18, p. 2].

Let us consider the vector **E**, called the *electric field strength* (see IUPAP). Note that it makes sense to compute the line integral of this quantity and that the variable we obtain, which is global in space, is called the *electromotive force* and will be denoted by $E$.[6] Does it make sense to compute the time integral of $E$? Yes, and this leads us to introduce the *electromotive force impulse*, which will be denoted by $\mathcal{E}$. Hence, this variable is global in space and time, and the vector **E** is its corresponding time rate and line density.

Let us consider the vector $\check{\mathbf{B}}$, called the *magnetic flux density* (see IUPAP). Let us observe that it makes sense to compute the surface integral of this quantity, and the variable we obtain, which is global in space, is called the *magnetic flux* and will be denoted by $\Phi$. Does it make sense to compute the time integral of $\Phi$? No, because the magnetic flux is already the time integral of the electromotive force (weber = volt × second). Hence, the variable $\Phi$ is global in space and time, and the vector $\check{\mathbf{B}}$ is its corresponding surface density.

Let us consider the vector $\check{\mathbf{H}}$, called the *magnetic field strength* (see IUPAP). Let us observe that it makes sense to compute the line integral of this quantity, and the variable we obtain, which is global in space, is called the *magnetomotive force*. It will be denoted by $F_{\mathrm{m}}$. Does it make sense to compute the time integral of $F_{\mathrm{m}}$? Yes, and this leads us to the *magnetomotive force impulse*, which will be denoted (by us) by $\mathcal{F}_{\mathrm{m}}$. Hence, this variable is global in space and time, and the vector $\check{\mathbf{H}}$ is its corresponding time rate and line density.

Let us consider the vector **D**, called the *electric displacement* (see IUPAP). Let us observe that it makes sense to compute the surface integral of this quantity, and the variable we obtain, which will be global in space, is called the *electric flux* and will be denoted by $\Psi$. Does it make sense to perform the time integral of $\Psi$? No, since $\Psi$ is an electric charge. Hence, this variable is global in space and time and **D** is only its corresponding surface density.

Let us consider the *electric charge density* $\rho$. Let us observe that it makes sense to compute the volume integral of this quantity, and the variable we obtain, which is global in space, is called *electric charge content* and will be denoted (by us) by $Q^{\mathrm{c}}$. Does it make sense to compute the time integral of $Q^{\mathrm{c}}$? No. Hence, this variable is global in space and time, and $\rho$ is only its corresponding volume density.

Lastly, let us consider the vector **J**, called *electric current density* (see IUPAP). We observe that it makes sense to perform its surface integral this quantity and the variable obtained, which is global in space, is called em electric current $I$. Does it make sense to perform the time integral of $I$? The answer is affirmative and leads us to the *electric charge flow*. It will be denoted (by us) by $Q^{\mathrm{f}}$. Hence, this variable is global in space and time and the vector **J** is its corresponding time rate and surface density.

---

[6] See p. 30.

Thus, we have obtained the six variables $\mathscr{E}$, $\Phi$, $\mathscr{F}_{\mathrm{m}}$, $\Psi$, $Q^{\mathrm{c}}$, $Q^{\mathrm{f}}$ which are global in space and time. It is interesting, and will be useful, to note that each of the mentioned space and space-time global variables has a corresponding measuring device. The tensorial nature of all six of these variables is that of (true) scalars. Table 10.1 summarizes the variables considered up to now. The second row of the table specifies the time and space elements, i.e. the classification.

**Table 10.1** Field variables of electromagnetism and corresponding global variables

| Classification | Configuration variable *Inner space orientation* $[\tilde{\mathbf{T}}, \bar{\mathbf{L}}]$ | $[\tilde{\mathbf{I}}, \bar{\mathbf{S}}]$ | Source variable *Outer space orientation* $[\bar{\mathbf{T}}, \tilde{\mathbf{L}}]$ | $[\bar{\mathbf{I}}, \tilde{\mathbf{S}}]$ | $[\bar{\mathbf{T}}, \tilde{\mathbf{S}}]$ | $[\bar{\mathbf{I}}, \tilde{\mathbf{V}}]$ |
|---|---|---|---|---|---|---|
| Space-time global variable | $\mathscr{E}$ Electro-motive force impulse | $\Phi$ Magnetic flux | $\mathscr{F}_{\mathrm{m}}$ Magneto-motive force impulse | $\Psi$ Electric flux | $Q^{\mathrm{f}}$ Electric charge flow | $Q^{\mathrm{c}}$ Electric charge content |
| Space global variable | $E$ Electro-motive force | | $F_{\mathrm{m}}$ Magneto-motive force | | $I$ Electric current | |
| Field variable | $\mathbf{E}$ Electric field strength | $\check{\mathbf{B}}$ Magnetic flux density | $\check{\mathbf{H}}$ Magnetic field strength | $\mathbf{D}$ Electric displacement vector | $\mathbf{J}$ Electric current density | $\rho$ Electric charge density |

## 10.3 Source Variables: Space and Time Classification

In the preceding section we deduced the global variables of electromagnetism from field variables. This is not the best way to find global variables. A more ‘physical’ approach rests on the *operational definition* of the variables. To do this, we will define each variable according to its measurement process.

This approach clearly shows the association of the physical variables with space and time elements endowed with inner or outer orientation.

The source variables are the *electric charge*, *electric flux*, *electric displacement*, *electric charge flow*, *electric current*, *electric current density* and *magnetomotive force*.

### *10.3.1 Electric Charge Content*

The main variable of electromagnetism is the *electric charge* $Q$. Since it is not a density or a rate of another variable, it is a global variable in space and time. It is a true scalar despite the fact that a few authors consider it a pseudoscalar.[7] Based on a system description, it is associated with a particle, a body or a system, whereas based on a spatial description it is split into two variables, i.e. *charge content*, $Q^{\mathrm{c}}$, and *charge flow*, $Q^{\mathrm{f}}$. We know that an electric charge cannot be produced; hence, there is no corresponding variable denoting charge production. A characteristic feature of an electric charge is that it is a relativistic invariant. In fact, the value of an electric charge contained in a given system measured from different inertial reference systems is the same.[8]

**Space Association.** An electric charge can be distributed inside a volume, and this gives rise to the electric charge density $\rho$ as the charge per unit volume, hence $Q^{\mathrm{c}}[\widetilde{\mathbf{V}}]$.

**Time Association.** The charge content $Q^{\mathrm{c}}$ is the amount of charge contained inside a volume at a given instant, hence $Q^{\mathrm{c}}[\mathbf{I}]$. Since it is assumed that an electric charge does not change sign under a reversal of motion, the time instant cannot be the dual, and hence it must be the primal, i.e. $Q^{\mathrm{c}}[\bar{\mathbf{I}}]$. In summary, we have the following space and time association: $Q^{\mathrm{c}}[\bar{\mathbf{I}}, \widetilde{\mathbf{V}}] \longrightarrow \rho[\bar{\mathbf{I}}, \widetilde{\mathbf{V}}]$.

### *10.3.2 Electric Flux*

If we place a plane metal surface of arbitrary shape, e.g. a metallic disk, in an electric field, then two electric charges of opposite sign will be collected by induction on both faces of the surface. This *charge-displacing property* is a consequence of the *force-exerting property*.[9] But while the forces of attraction/repulsion between two charges depend on the medium (recall Coulomb's law), the amount of charge collected on the faces does not depend on the medium. This agrees with the nature of source variables, as stated previously.

**Remark.** Be careful not to fall into the error of believing that force is always a source variable. In fact, the sources of the electromagnetic field are the charges at rest or in motion, while forces are configuration variables. The dependency on the medium underlines this fact. In contrast, in mechanical theories, forces are sources. Despite this different role (source and configuration), we will see that forces are always associated with dual intervals.

If we choose one of the two faces as preferred, the charge collected on it is called the *electric flux* and is denoted by the letter $\Psi$.

---

[7] This is the case with Post [183], Truesdell and Toupin [237, p. 682] and a few others. Nobody measuring a charge with an electrometer need refer to the right-handed or left-handed screw! Apart from this disagreement, the present author is a great admirer of Post.

[8] Jackson [102, Sect. 11.9].

[9] Rojansky [192, p. 230].

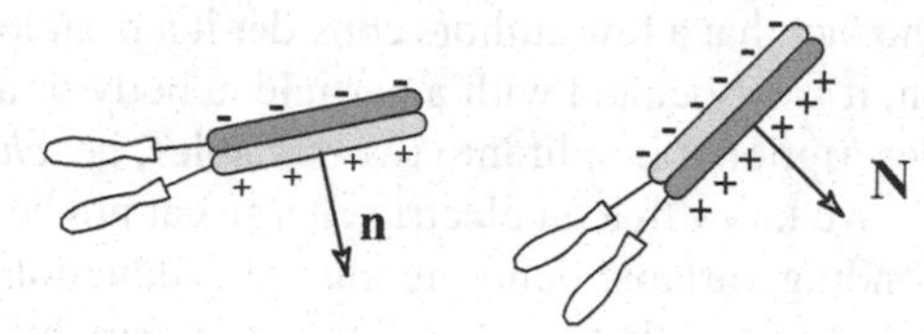


**Fig. 10.1** Measure of electric flux on element of surface with outer orientation. *N* is the normal, which shows the space direction for which the charge collected on the surface is maximum. This fact will be useful for introducing the electric displacement

**Measurement Process.** To measure the electric flux, it is convenient to use a probe consisting of two small metal discs in contact, each of which is endowed with an insulating handle.[10] When they are placed in a point of the electric field, two charges of opposite sign are collected on the two discs for electrostatic induction.[11] The amount of the surface charge depends on where you place the discs and on its space direction, as shown in Fig. 10.1. By placing the two plates in contact, as shown in Fig. 10.1, and then removing them, the induced charges remain trapped on the two plates and can be measured with an electrometer. Hence, we have that the electric flux is a function of the position **P** of the plate, the area of the surface, the space direction of the plate and the time instant, i.e. $\Psi(t, \mathbf{P}, \mathbf{n}, A)$.

**Space Association.** The following experimental observations are essential for classification and are usually overlooked in books on physics:

1. *The charge collected on a metal surface does not depend on the material with which the metal plate is made.*[12] This fact enables the electric flux to be directly associated with the geometric surface, i.e. $\Psi[\mathbf{S}]$.
2. *The electric flux does not depend on the medium in which the metal surface is immersed.* This fact can be seen by repeating the measurement after placing oil instead of air in the region.[13] This implies that the electric flux is a source variable. We remark that, in contrast, the electric field strength **E** depends on the medium.[14]

Since in the definition of the electric flux we have set a face as preferred, the direction in which the flux will go through the surface is automatically set, from the preferred face to the opposite one, which means that an *outer* orientation is given. For this reason we will write $\Psi[\widetilde{\mathbf{S}}]$. This can be marked by means of a

[10] Fouillé [72, p. 71], Fleury and Mathieu [71, p. 61], Maxwell [153, p. 47], Rojansky [192, p. 230], Schelkunoff [201, p. 25], Jefimenko [107, pp. 80, 225], Pohl [178, p. 66], Hehl and Obukhov [89, p. 133].

[11] See the superb and detailed description by Schelkunoff [201, p. 25].

[12] Langevin [126, p. 501].

[13] Fleury and Mathieu [71, p. 85]. It is understood that the charge on the plates is unchanged.

[14] See p. 234.

normal **n**. It is evident that if we invert the normal, then the preferred face changes into its opposite and the same happens for the sign of the electric charge. Hence,

$$\boxed{\Psi(-\mathbf{n}) = -\Psi(\mathbf{n})} \qquad \textit{oddness principle of } \Psi. \tag{10.4}$$

This property is common to all flows and flow rates, i.e. currents, such as the current of heat, mass, energy, entropy, probability and momentum (surface force), and it is not commonly stressed, with the exception of the internal surface force in the mechanics of continua, where it is written $\mathbf{T}(-\mathbf{n}) = -\mathbf{T}(\mathbf{n})$.

This is a further confirmation that source variables are associated with space elements endowed with an outer orientation.

**Time Association.** Since an electric flux is an electric charge, it is associated with primal time instants, i.e. $\Psi[\mathbf{I}]$. Let us determine the orientation of the time element. The sign of the electric charge collected on the two plates does not change if we perform a reversal of motion; hence, the electric flux is invariant under a reversal of motion. This means that the instant is primal, i.e. $\Psi[\bar{\mathbf{I}}]$.

**Surface Charge Density.** $\sigma$ is the surface density of $\Psi$ and inherits from it the space association, i.e. $\sigma[\tilde{\mathbf{S}}]$. We will discuss the time association subsequently.

### *10.3.3 Birth of Electric Displacement*

**Uniform and Static Electric Field.** Let us consider an electric field which is *uniform* and *static*, such as that between the plates of a parallel plate capacitor (Fig. 10.2). We stated above that the electric flux $\Psi$ collected on a disc is a function of the position **P** of the disc, of area $A$ with a normal **n**, i.e. $\Psi(\mathbf{P}, \mathbf{n}, A)$. In fact, experiments show that

1. The electric flux collected on a face of unit normal **n** is proportional to the area of the disc;
2. There is a preferred direction **N** for which the electric flux is maximum in algebraic value and, as such, positive;
3. The electric flux in any other orientation given by the unit normal **n** equals this maximum flux multiplied by the cosine of the angle $\alpha$ between **n** and **N**.

Property 1 leads us to factorize the electric flux as follows:

$$\Psi(\mathbf{n}, A) = \sigma(\mathbf{n})\, A, \tag{10.5}$$

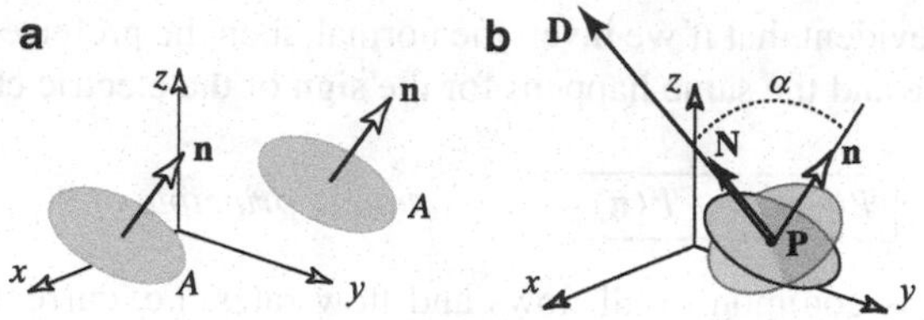


**Fig. 10.2** (**a**) In a uniform and static electric field, the electric flux associated with a flat surface is invariant under space and time translation. (**b**) The direction of **D** is that of maximum $\sigma$

where $\sigma(\mathbf{n})$ is the *surface charge density* which depends on the space orientation of the surface element of area $A$.[15]
Property 2 states that there exists a special direction **N** for which

$$\sigma(\mathbf{N}) = \sigma_{max} \,. \tag{10.6}$$

Property 3 states that

$$\begin{aligned}\sigma(\mathbf{n}) &= \sigma_{max}\cos(\alpha) = \sigma_{max}\,\mathbf{N}\cdot\mathbf{n} \\ &= (\sigma_{max}\,\mathbf{N})\cdot\mathbf{n} = \mathbf{D}\cdot\mathbf{n},\end{aligned} \tag{10.7}$$

where we have introduced the new variable

$$\boxed{\mathbf{D} \overset{\text{def}}{=} \sigma_{max}\,\mathbf{N}} \tag{10.8}$$

called the *electric displacement*. Thus, we can write relation 10.5 in the form $\Psi(\mathbf{n}, A) = \mathbf{D}\cdot\mathbf{n}\,A$. Hence, the electric displacement **D** is orthogonal to the face of the maximum positive flux.[16] The vector **D** inherits from $\Psi$ the space association, i.e. $\mathbf{D}[\widetilde{\mathbf{S}}]$. For the time association we must analyse what happens in a non-static field. Recall that the association with time elements also has meaning in statics because it is involved in the process of measuring the physical variable.
**Electric Field in General.** Let us consider a generic electric field, i.e. one that is neither uniform nor static. This time the electric flux is a function of the time instant $t$ too, i.e. $\Psi(t, \mathbf{P}, \mathbf{n}, A)$. Since the field is no longer uniform, the flux is no longer proportional to the area, and proportionality becomes increasingly true when contracting the plane surface. This leads us to define

$$\sigma(t, \mathbf{P}, \mathbf{n}) \overset{\text{def}}{=} \lim_{A\to 0} \frac{\Psi(t, \mathbf{P}, \mathbf{n}, A)}{A}, \tag{10.9}$$

where $\sigma(t, \mathbf{P}, \mathbf{n}) = \mathbf{D}(t, \mathbf{P})\cdot\mathbf{n}$; see Eq. 10.7. If we add time, it follows that

---

15 Here the term *orientation* refers to the space direction.

16 Rojansky [192, p. 230].

$$\Psi = \int_{\widetilde{\mathbf{S}}} \sigma(t, \mathbf{P}, \mathbf{n})\,\mathrm{d}S = \int_{\widetilde{\mathbf{S}}} \mathbf{D}(t, \mathbf{P}) \cdot \mathbf{n}\,\mathrm{d}S \qquad \longrightarrow \qquad \mathbf{D}[\widetilde{\mathbf{S}}]. \tag{10.10}$$

This shows, even more clearly, that $\mathbf{D}(t, \mathbf{P})$ inherits the space association with $\widetilde{\mathbf{S}}$ from $\Psi$.

**Remark.** Equation 10.10 is often taken as a definition of electric flux, which presupposes that **D** is already defined. In contrast, it is better to start from $\Psi$ because it is a measurable variable and then to deduce **D** by Eq. 10.8.

**Time Association.** The electric flux $\Psi$ is associated with the primal instants $\bar{\mathbf{I}}$, and both $\sigma$ and the vector **D** inherit the same association. In summary we have $\Psi[\bar{\mathbf{I}}, \widetilde{\mathbf{S}}] \longrightarrow \sigma[\bar{\mathbf{I}}, \widetilde{\mathbf{S}}] \longrightarrow \mathbf{D}[\bar{\mathbf{I}}, \widetilde{\mathbf{S}}]$.

### *10.3.4 Critical Remarks*

We must take into account that the electric flux collected on a small metallic disc, considered as a probe, *does not depend* on the medium, whereas the force on a small charge $q$, considered as a test charge, and hence the vector **E**, *does depend* on the medium. Hence, the vectors **E** and **D** *measure two different attributes of an electric field*, like strain and stress in continuum mechanics. Moreover, the electric field strength is a configuration variable, whereas the electric displacement vector is a source variable. This distinction does not depend on the medium; in particular, it is also valid in vacuum. Hence, the relation $\mathbf{D} = \epsilon_0 \mathbf{E}$ is a constitutive, not defining, equation. The situation is similar to the equation $\mathbf{p} = m\mathbf{v}$ of mechanics, which is commonly considered a defining equation for momentum, whereas it is a constitutive equation.[17]

We remark that is not natural to take the flux of **E** just as it is not natural to take the flux of a force from which **E** derives.

### *10.3.5 Electric Charge Flow and Electric Current*

**Space Association.** It is natural to associate the electric charge that flows through a surface with a surface. Since the notion of going *through* a surface presupposes an outer orientation, the surface must be endowed with an outer orientation, i.e. $Q^{\mathrm{f}}[\widetilde{\mathbf{S}}]$.

**Time Association.** To evaluate the amount of electric charge flow, we need to consider a lapse of time, that is a time interval. Since under a reversal of motion

[17] See p. 250.

the sign of an electric charge flow reverses, it follows that the time interval is endowed with an inner orientation, i.e. $Q^{\mathrm{f}}[\overline{\mathbf{T}}]$.

**Electric Current.** The electric current $I$ is the rate of the electric charge flow $Q^{\mathrm{f}}$; hence, it inherits the association with the same time and space elements. In summary we have $Q^{\mathrm{f}}[\overline{\mathbf{T}}, \widetilde{\mathbf{S}}] \longrightarrow I[\overline{\mathbf{T}}, \widetilde{\mathbf{S}}]$.

### *10.3.6 Birth of Electric Current Density Vector*

Let us first consider a *uniform* and *stationary* flow of electric charge. Let us consider an outer oriented surface element $\widetilde{\mathbf{S}}$ and an electric charge $Q^{\mathrm{f}}$ flowing through it. Since the flow is uniform, $Q^{\mathrm{f}}$ is proportional to the area $A$ of the surface element and depends on its space orientation, which is represented by the unit normal $\mathbf{n}$. Moreover, since the flow is also stationary, $Q^{\mathrm{f}}$ is proportional to the elapsed time interval $T$. This double proportionality leads us to write

$$Q^{\mathrm{f}}(T, \mathbf{n}, A) = I(\mathbf{n}, A)T = j(\mathbf{n})\, A\, T, \tag{10.11}$$

where we have introduced the *electric current density* $j(\mathbf{n})$, which depends on the space orientation[18] of the surface element. Changing this orientation we find the orientation for which the electric current density reaches its maximum. Denoting by $\mathbf{N}$ the oniy normal corresponding to this maximum, we write

$$j_{max} = j(\mathbf{N})\ . \tag{10.12}$$

The current density $j(\mathbf{n})$ can be expressed as the product of $j_{max}$ with the cosine of the angle $\alpha$ between $\mathbf{n}$ and $\mathbf{N}$, hence

$$j(\mathbf{n}) = j_{max}\cos(\alpha) = j_{max}(\mathbf{N}\cdot\mathbf{n}) = (j_{max}\mathbf{N})\cdot\mathbf{n}. \tag{10.13}$$

This suggests the need to introduce the vector

$$\boxed{\mathbf{J} \stackrel{\text{def}}{=} j_{max}\mathbf{N}} \tag{10.14}$$

called an *electric current density vector*. We can rewrite Eq. 10.11 as

$$Q^{\mathrm{f}}(T, \mathbf{n}, A) = \mathbf{J}\cdot\mathbf{n}\, A\, T. \tag{10.15}$$

Being a surface density, $\mathbf{J}$ inerits from $Q^{\mathrm{f}}$ the association with a surface endowed with an outer orientation, i.e. $\mathbf{J}[\widetilde{\mathbf{S}}]$. The time association is discussed in what follows.

---

[18] Here the term *orientation* refers to the direction in space.

In a space region in which the electric charge flow is neither uniform in space nor stationary in time, the decomposition which leads to Eq. 10.15 maintains its validity for an infinitesimal time interval $\mathrm{d}t$ and an infinitesimal surface element of area $\mathrm{d}S$, whereas the vector $\mathbf{J}$ depends on the instant $t$ and on the point $\mathbf{P}$ of the region. Let us introduce the infinitesimal electric charge flow $q^{\mathrm{f}}$, defined as

$$q^{\mathrm{f}}(\mathrm{d}t, \mathbf{n}, \mathrm{d}S) = \mathbf{J}(t, \mathbf{P}) \cdot \mathbf{n} \; \mathrm{d}S \; \mathrm{d}t \, . \tag{10.16}$$

The electric charge flow across a surface $\widetilde{\mathbf{S}}$ in a time interval $\overline{\mathbf{T}}$, which is a primal interval, can be expressed in the form

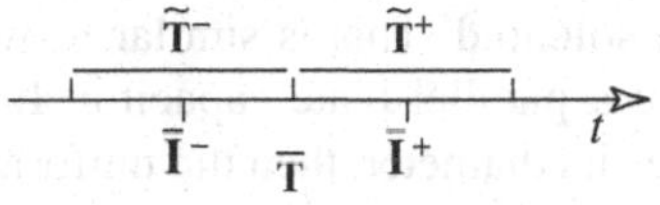


$$\boxed{Q^{\mathrm{f}} = \int_{\overline{\mathbf{T}}} \int_{\widetilde{\mathbf{S}}} \mathbf{J}(t, \mathbf{P}) \cdot \mathbf{n} \, \mathrm{d}S \; \mathrm{d}t \, .} \tag{10.17}$$

Hence, $\mathbf{J}$ inherits from $Q^{\mathrm{f}}$ the space and time association. In summary, we have $Q^{\mathrm{f}}[\overline{\mathbf{T}}, \widetilde{\mathbf{S}}] \longrightarrow I[\overline{\mathbf{T}}, \widetilde{\mathbf{S}}] \longrightarrow \mathbf{J}[\overline{\mathbf{T}}, \widetilde{\mathbf{S}}]$.

### *10.3.7 Electric Vector Potential*

In the theory of antennas[19], it is sometimes useful to introduce the electric pseudovector potential $\check{\mathbf{F}}$ in those regions where $\rho = 0$. Since div $\mathbf{D} = 0$, the vector $\mathbf{D}$ plays a role similar to that of the magnetic vector potential $\mathbf{A}$ for $\check{\mathbf{B}}$ in diagram [ELE2], i.e. $\mathbf{D} = \mathrm{curl}\, \check{\mathbf{F}}$, see diagram [ELE4] (p. 313).[20]

**Space Association.** Since the electric flux $\Psi$ is associated with surfaces endowed with an outer orientation, and since the line integral of the electric pseudovector $\check{\mathbf{F}}$ along the boundary of a surface is equal to the electric flux associated with the surface, as follows from the defining relation $\mathbf{D} = \mathrm{curl}\, \check{\mathbf{F}}$, it follows that $\check{\mathbf{F}}$ is associated with lines endowed with an outer orientation, i.e. $\check{\mathbf{F}}[\widetilde{\mathbf{L}}]$. We remark that the line integral of a vector can be evaluated along a line endowed with an inner orientation (the most common case) but also with lines endowed with an outer orientation, such as in the magnetomotive force.

**Time Association.** Since there is no operational definition of the vector $\check{\mathbf{F}}$, unlike what has been done up to now, we will deduce the association by analysing the equation that defines it. Since the defining relation $\mathbf{D} = \mathrm{curl}\, \check{\mathbf{F}}$ does not in-

---

[19] Milligan [159, p. 49].

[20] These labels refers to the classification diagrams compiled at the end of the chapter. A complete list of the classification diagrams and the corresponding pages can be found on p. xx.

volve a rate or a time derivative, it follows that $\check{\mathbf{F}}$ is associated with the primal time instants, just as the electric flux; hence, we have $\check{\mathbf{F}}[\bar{\mathbf{I}}, \tilde{\mathbf{L}}]$.

### *10.3.8 Magnetic Field Strength*

Let us consider a solenoid carrying a current: we aim to characterize the intensity of the magnetic field inside the solenoid. If we perform measurements using, say, a Hall probe, then we discover that the magnetic field is approximately uniform inside the solenoid and approximately null outside the solenoid. This indicates that the field is almost entirely confined inside the solenoid. This is similar to what happens in the electric field inside and outside a parallel plate capacitor. If we increase the length of the solenoid and decreases its diameter, then the uniformity of the field inside and its cancellation outside become more and more pronounced. This leads us to consider a solenoid as a tube and then to assimilate it to a line.

Since with such a solenoid measurements are impossible, we return to an ordinary solenoid.[21] Let us insert a magnetic needle in the solenoid. In the absence of current, the magnetic needle assumes a south–north direction. Using a torsion balance (Fig. 10.3b) we rotate the thread of the balance to bring it orthogonal to the axis of the solenoid.

At this point, given a current $i$ in the solenoid, the magnetic needle rotates, tending to become parallel to the axis of the solenoid. This is contrasted by the elastic couple generated by the torsion of the thread. Applying a torque we report the magnetic needle in the initial position, a right angle with the axis of the solenoid. The angle of rotation of the thread (to restore the orthogonality of the needle to the coil axis) becomes a measure of the field intensity.

Denoting by $N$ the number of turns of the whole coil, the concentration of the coils is expressed by the ratio $N/L$. Experimentally one sees that taking a first coil of length $L_1$ with $N_1$ turns and current $i_1$ and then a second coil of length $L_2$ with $N_2$ turns and current $i_2$, the same torque (and therefore the same initial rotation of the magnetic needle) is realized when the condition

$$\frac{N_1 i_1}{L_1} = \frac{N_2 i_2}{L_2} \tag{10.18}$$

is satisfied. Clearly, the equality of the rotation, and hence of the torque, indicates the equality of the fields inside the two coils. It follows that the scalar

$$H \stackrel{\text{def}}{=} \frac{N\,i}{L} \tag{10.19}$$

is suitable for characterizing the intensity of the magnetic field.

---

[21] The following description is in Pohl [179, p. 86].

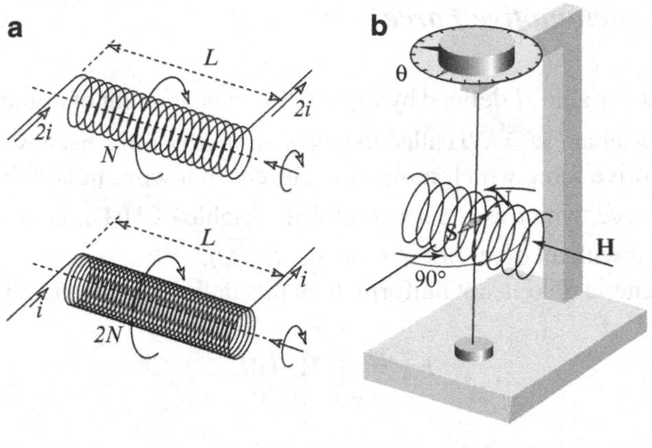


**Fig. 10.3** Torsion balance to measure magnetic field strength. (**a**) what is significant for the intensity of the magnetic field is the product of the current for the number of loops, hence (2i) N = i (2N); (**b**) the operation of measurement of **H**

The scalar variable $H$ is not an exhaustive characterization of the intensity of the magnetic field because the current in the coil may have two opposite directions. Moreover, the axis of the solenoid has a space direction. These two features require the introduction of a vector instead of a scalar. The line of action of the vector is the axis of the solenoid; it remains to find its direction. To this end, we must associate an inner orientation of the axis to the direction of the current flowing in the solenoid, i.e. around the axis.

To determine the side to which the needle rotates, we should choose a direction of the axis of the solenoid (as in the definition of the angular velocity vector in a rotating body). This direction is given by the usual screw rule applied to the direction of the current. Let $\check{\mathbf{t}}$ be the unit pseudovector indicating the oriented direction. This requires the use of the screw rule. In this way, one can introduce a unit vector along the line which, on account of the use of the screw, is an axial vector (≡ pseudovector) $\check{\mathbf{t}}$. This leads us to consider the pseudovector

$$\boxed{\check{\mathbf{H}} \stackrel{\text{def}}{=} \frac{N\,i}{L}\,\check{\mathbf{t}}}, \tag{10.20}$$

which is called the *magnetic field strength*. The magnetic field strength is a physical variable associated with a line endowed with an outer orientation, i.e. $\check{\mathbf{H}}(\widetilde{\mathbf{L}})$.

Since the electric current is associated with time intervals endowed with an inner orientation, i.e. $I[\overline{\mathbf{T}}]$, it follows that the magnetic field strength is also associated with the same time element, hence $\check{\mathbf{H}}[\overline{\mathbf{T}}, \widetilde{\mathbf{L}}]$.

### 10.3.9 Magnetomotive Force

This physical variable $H$ defined by Eq. 10.20 can be thought of as the line density of a global variable $F_m \stackrel{\text{def}}{=} Ni$ called the *magnetomotive force* because it resembles the electromotive force which moves the current in a wire. In addition, the magnetomotive force, which is the space global variable of $\check{\mathbf{H}}(\tilde{\mathbf{L}})$, is associated with lines endowed with an outer orientation, i.e. $F_m(\tilde{\mathbf{L}})$.

If the magnetic field is not uniform, then this definition is generalized:[22]

$$F_m \stackrel{\text{def}}{=} \int_{\tilde{\mathbf{L}}} \check{\mathbf{H}} \cdot \check{\mathbf{t}} \, dL \,. \tag{10.21}$$

Hence, $F_m$ inherits from $\check{\mathbf{H}}$ the space and time associations. In summary, $\check{\mathbf{H}}[\overline{\mathbf{T}}, \tilde{\mathbf{L}}] \longrightarrow F_m[\overline{\mathbf{T}}, \tilde{\mathbf{L}}]$.

**Remark.** A parallel exists between **D** and $\check{\mathbf{H}}$. The analogue of a capacitor with parallel flat faces in electrostatics is a straight solenoid in magnetostatics. Just as in the interior region of a capacitor with sufficiently large flat plates the electric field is essentially uniform, so in the interior of a sufficiently long straight solenoid the magnetic field is essentially uniform. A capacitor with parallel and flat faces was the workhorse of Faraday just as the solenoid was the workhorse of Ampère.

We list here a comparison of the four field vectors of electromagnetism.

| *Electric field* | | *Magnetic field* | |
|---|---|---|---|
| $\mathbf{E} = \dfrac{V_{max}}{L}\mathbf{t}$ | $\mathbf{D} = \dfrac{\Psi_{max}}{A}\mathbf{n}$ | $\check{\mathbf{H}} = \dfrac{F_{max}}{L}\check{\mathbf{t}}$ | $\check{\mathbf{B}} = \dfrac{\Phi_{max}}{A}\check{\mathbf{n}}$ |
| Electromotive force/ length | Electric flux/ area | Magnetomotive force/ length | Magnetic flux/ area |

(10.22)

**Measurement of Magnetomotive Force.** How can we measure the magnetomotive force in a magnetic field? With reference to Fig. 10.4, let us consider the magnetic field generated by a long straight wire carrying a current of intensity $I$. Let us consider a small line segment **L** situated tangentially to the vector lines of the magnetic field. Let us put around **L** a closely wound coil that will be called a *compensating coil*. This causes us to measure the magnetomotive force $F_m$ along the line **L** as the opposite of the total current $n\,i$ which causes the field inside the compensating coil to vanish; hence, $F_m = -n\,i$.

[22] Fleury and Mathieu [71, vol. 6; Sect. 9.8].

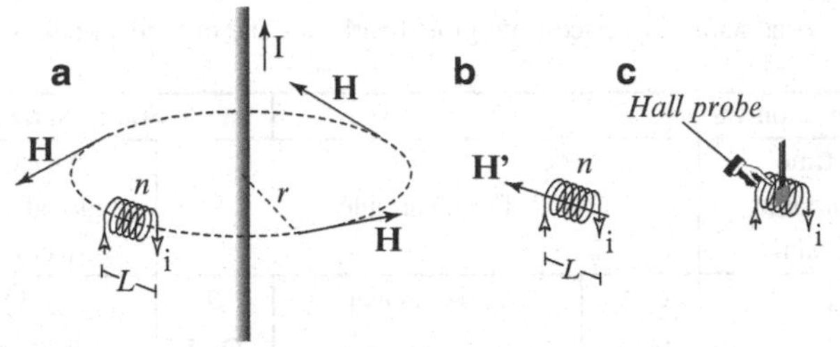


**Fig. 10.4** (**a**) Compensating coil for measuring $F_{\rm m}$ and $\check{\mathbf{H}}$. (**b**) The sign of the current in the compensating coil must be such that the magnetic field inside the coil vanishes. (**c**) This can be tested with a Hall probe

An equivalent way of doing the test is to consider a small tube of superconducting material. The tube is crossed by a uniform current $I' = ni$ which makes the internal field vanish. Experience tells us that the current needed to obtain this result *does not* depend on the medium or on the nature of the wire of the compensating coil.[23] This agrees with the fact that this is a source variable.

**Space Association.** The direction of the current in a solenoid induces an outer orientation on the line segment which we will denote by $\widetilde{\mathbf{L}}$. Hence, we can associate $F_{\rm m}$ with this oriented line segment, i.e. $F_{\rm m}[\widetilde{\mathbf{L}}]$. A further proof of the association with lines endowed with an outer orientation is that the value of the magnetomotive force depends on the direction of the current in the solenoid; hence, it changes sign when we invert the direction of the current. Since in its definition there is no intervention of the screw, it follows that passing from a right-handed to a left-handed screw $F_{\rm m}$ does not change the sign, i.e. it is a true scalar. The Table 10.2 shows the tensorial nature of the variables of electromagnetism.

**Time Association.** As previously stated, the electric current $I$ is the rate of electric charge flow $Q^{\rm f}$, the latter being associated with primal time intervals, $Q^{\rm f}[\overline{\mathbf{T}}]$. The magnetomotive force is essentially a current, hence $F_{\rm m}[\overline{\mathbf{T}}]$.

### *10.3.10 Scalar Magnetic Potential*

Let us consider a region of space where there are no currents and which is simply connected. From Ampère's law the magnetomotive force along any closed line is zero. It follows that we can associate to each point **P** a *magnetic potential* $\check{\phi}_{\rm m}(\mathbf{P})$ defined as the line integral from the point **P** to a fixed point **O** of the region. The line will be endowed with an *inner* orientation so that the unit tangent vector **t** is a

[23] Langevin [127, p. 501], Fouillé [72, p. 224], Pohl [179, p. 66], Schelkunoff [201, p. 41], Hehl and Obukhov [89, p. 137].

**Table 10.2** Tensorial nature of space-time global variables and of field variables of electromagnetism

| Configuration variable | | | | Source variable |
|---|---|---|---|---|
| Space-time global variable (true) scalar | | Field variable | | Space-time global variable (true) scalar |
| $\varphi, \chi$ | $\phi, \chi$ | (True) scalar | $\rho$ | $Q^{\mathrm{c}}$ |
| $\mathscr{E}, a$ | $\mathbf{E}, \mathbf{A}$ | (Polar) vector | $\mathbf{D}, \mathbf{J}$ | $\Psi, Q^{\mathrm{f}}$ |
| $Q^{\mathrm{f}}_{\mathrm{m}}, \Phi$ | $\check{\mathbf{J}}_{\mathrm{m}}, \check{\mathbf{B}}$ | Pseudovector | $\check{\mathbf{F}}, \check{\mathbf{H}}$ | $\mathscr{F}_{\mathrm{m}}$ |
| $Q^{\mathrm{c}}_{\mathrm{m}}$ | $\check{\rho}_{\mathrm{m}}$ | Pseudoscalar | $\check{\phi}_{\mathrm{m}}$ | $\check{\varphi}_{\mathrm{m}}$ |

polar vector. It is convenient to choose the point **O** far enough where the magnetic field vanishes, or, as is commonly said, at 'infinity'. Hence,

$$\check{\phi}_{\mathrm{m}}(\mathbf{P}) \stackrel{\mathrm{def}}{=} \int_{\mathbf{P}}^{\infty} \check{\mathbf{H}} \cdot \mathbf{t}\, \mathrm{d}L \qquad \text{from which} \qquad \check{\mathbf{H}} = -\nabla \check{\phi}_{\mathrm{m}}\,. \tag{10.23}$$

With reference to Fig. 10.5, in the case of a loop crossed by a current, choosing for convenience, for example, a point **P** on the axis of the loop, the computation of the integral Eq. 10.23 yields

$$\check{\phi}_{\mathrm{m}} = -\frac{I}{2}[1 - cos(\theta)] = -I\frac{\check{\Omega}}{4\pi}, \tag{10.24}$$

where $\check{\Omega}$ is a solid angle.

**Remark.** In plane trigonometry, the angles are endowed with a sign depending on their anticlockwise or clockwise orientation. In addition, in spherical trigonometry, we can consider oriented solid angles. It is enough to give an inner or an outer orientation to the spherical surface used to evaluate the solid angle. When endowed with an inner orientation the solid angle $\Omega$ is a pseudoscalar; hence, we put a check on it, i.e. $\check{\Omega}$. The solid angle, when endowed with an inner orientation, is indispensable for giving to the relation of Eq. 10.24 invariance with respect to the inversion of orientation.

**Space Association.** The physical variable $\check{\phi}_{\mathrm{m}}$ is associated with a point and depends on its outer orientation, i.e. $\check{\phi}_{\mathrm{m}}[\widetilde{\mathbf{P}}]$. In fact, with reference to Fig. 10.5, we see that the sign of the magnetic potential due to a loop depends on the direction of the current on the loop. This direction indicates an outer orientation of the point; hence, $\check{\phi}_{\mathrm{m}}$ is a pseudoscalar.[24]

[24] Jouguet [110, vol. II; p. 31].

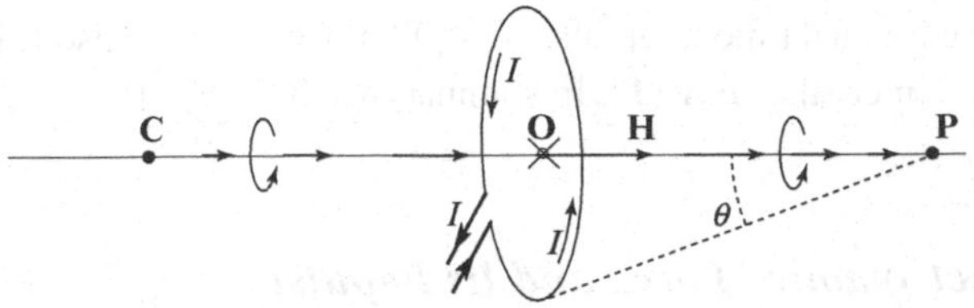


**Fig. 10.5** The scalar magnetic potential at a point depends on the outer orientation of the point

**Time Association.** Since the magnetomotive force $F_{\mathrm{m}}$ is associated with primal time intervals, it follows that the pseudoscalar magnetic potential inherits from it the association with the primal time intervals, i.e. $\breve{\phi}_{\mathrm{m}}[\overline{\mathbf{T}}, \widetilde{\mathbf{P}}]$.

## 10.4 Configuration Variables: Space and Time Classification

These are the *electric field strength*, the *electromotive force*, the *electromotive force impulse*, the *electric potential*, the *electric potential impulse*, the *magnetic flux* and the *magnetic flux density*.

### *10.4.1 Electric Field Strength*

If we insert a small charge $q$ in an electrostatic field, a charge so small that its influence on the charges generating the field can be neglected, it experiences a force that depends on the medium. Let us remark, as stated previously, that the force here is not the source of the field because the source is the electric charge. Experiments show that the force $\mathbf{F}$ is proportional to the charge, i.e. $\mathbf{F} \propto q$. This leads to the introduction of the ratio $\mathbf{E} \stackrel{\text{def}}{=} \mathbf{F}/q$, which is called the *electric field strength*, which no longer depends on the test charge, just as the price of goods no longer depends on the quantity of goods.

**Space Association.** If we perform a line integral of the electric field strength, then we obtain another physical variable called the *electromotive force*. Hence, $\mathbf{E}$ is not a global variable but a line density, and this implies that it must be associated with lines endowed with an inner orientation, hence $\mathbf{E}[\overline{\mathbf{L}}]$, as is shown in more detail in the next section.

**Time Association.** The force does not change sign under a reversal of motion, and since the electric charge $q$ is also invariant under a reversal of motion, it follows that $\mathbf{E}$ is invariant under a reversal of motion, i.e. it is time even. Since the force is associated with dual time intervals, i.e. $\mathbf{F}[\tilde{\mathbf{T}}]$, it follows that also $\mathbf{E}$ is associated with dual time intervals, i.e. $\mathbf{E}[\tilde{\mathbf{T}}]$. In summary we have $\mathbf{F}[\tilde{\mathbf{T}}] \longrightarrow \mathbf{E}[\tilde{\mathbf{T}}]$.

### *10.4.2 Electromotive Force and Its Impulse*

Recall that in a field of force the line integral of the force along a line is the *virtual* work; this has nothing to do with the actual work, which does not presuppose a field of force but a single force and which needs an actual displacement of its point of application. The actual work depends on the motion of the point of application of the force, while the virtual work depends only on the line used to perform the line integral. Performing the line integral of the electric field strength $\mathbf{E}$

$$\boxed{E \stackrel{\text{def}}{=} \int_{\mathbf{L}} \mathbf{E}(t, \mathbf{P}) \cdot \mathbf{t}\, \mathrm{d}L} \tag{10.25}$$

we obtain the virtual work per unit charge.[25] This physical variable is the *electromotive force* and the SI units are joule/coulomb = volt.

**Space Association.** The electromotive force $E$ is associated with a line endowed with an inner orientation, i.e. $E[\overline{\mathbf{L}}]$. Since the electromotive force is a configuration variable, we have here a confirmation that configuration variables are associated with space elements endowed with an inner orientation.

**Time Association.** Since the electric field strength is associated with dual time intervals, the electromotive force is also associated with dual time intervals, hence $E[\tilde{\mathbf{T}}]$. Since it makes sense to compute the definite time integral of a force, obtaining the impulse,[26] it also makes sense to introduce the definite time integral of electromotive force over a dual time interval $\tilde{\mathbf{T}}$, which we call the *electromotive force impulse* and which will be denoted by $\mathscr{E}$.

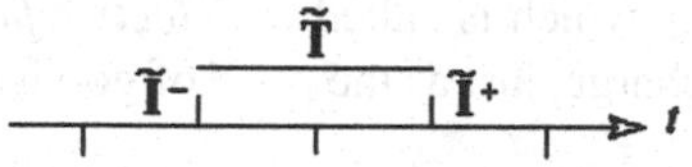


$$\boxed{\mathscr{E} \stackrel{\text{def}}{=} \int_{\tilde{\mathbf{T}}} \int_{\mathbf{L}} \mathbf{E}(t, \mathbf{P}) \cdot \mathbf{t}\, \mathrm{d}L\, \mathrm{d}t\,.} \tag{10.26}$$

The electromotive force impulse is a global variable in both time and space. It can be seen that the electric field strength is a line density and the rate of $\mathscr{E}$; hence, it inherits from it the space and time association that we have just obtained using another method. In summary, we have $\mathscr{E}[\tilde{\mathbf{T}}, \overline{\mathbf{L}}] \longrightarrow E[\tilde{\mathbf{T}}, \overline{\mathbf{L}}] \longrightarrow \mathbf{E}[\tilde{\mathbf{T}}, \overline{\mathbf{L}}]$.

[25] See p. 256 for the role of virtual work in field theories.

[26] See p. 243.

### 10.4.3 Electric Potential

The electrostatic field possesses the property that the electromotive force between any couple of points **A** and **B** does not depend on the line connecting the two points. This suggests that we should consider a reference point **O** and, for every point **P**, the integral

$$\phi(\mathbf{P}) \stackrel{\text{def}}{=} \int_{\mathbf{P}}^{\mathbf{O}} \mathbf{E} \cdot \mathbf{t}\,\mathrm{d}L \tag{10.27}$$

and to call this the *electric potential* of the field at the point **P**. Note the order of points: the initial point is the point **P** and the final point is the reference point **O**. The electromotive force can be expressed as a potential difference,

$$E(t, \mathbf{A}, \mathbf{B}) = \phi(t, \mathbf{A}) - \phi(t, \mathbf{B}), \tag{10.28}$$

in which we must pay attention to the order of the points.[27] From this relation follow $\phi[\widetilde{\mathbf{T}}, \overline{\mathbf{P}}]$ and $\mathbf{E} = -\mathrm{grad}\,\phi$. The electric potential is not global in time: one can introduce the *electric potential impulse* defined as

$$\varphi(t) \stackrel{\text{def}}{=} \int_0^t \phi(t')\,\mathrm{d}t'\,, \tag{10.29}$$

a time and space global variable not commonly used.

### 10.4.4 Magnetic Flux

To present the magnetic flux, we use the electromotive force impulse. Let us consider a region which is the site of a magnetic field generated by an electromagnet. To describe the configuration of the magnetic field at a point and at an instant, let us consider a probe formed by a small coil in the form of a plane loop, as shown in Fig. 10.6. Let us connect the two ends of the wire to a digital voltmeter. When we switch off the field, a short pulse of current is generated in the loop. Registering the electromotive force as a function of time during the shutdown we obtain a curve like the one shown in the bottom part of Fig. 10.6. Experiments show that, while the current in the loop is a function of time and depends on the resistance of the wire, the time integral of the electromotive force during the shutdown, i.e. the electromotive force impulse $\mathscr{E}$, depends neither on the wire resistance nor on the way the electromagnet is switched off. Moreover, the electromotive force impulse *also depends on the medium* in which the flat coil is embedded, and this is a confirmation that it is a configuration variable. Hence, the quantity

[27] Olivieri and Ravelli [170, p. 21].

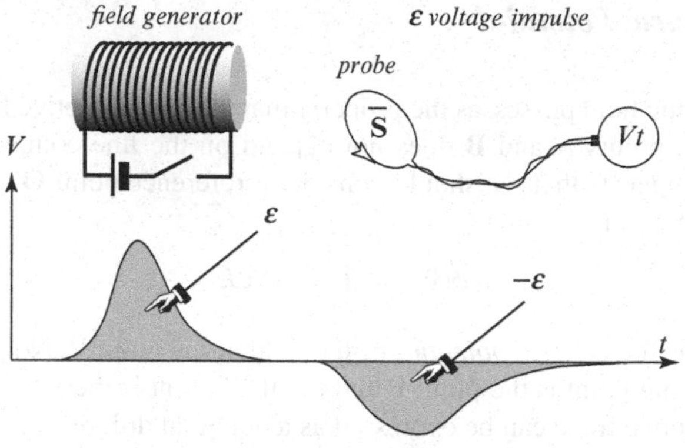


**Fig. 10.6** By removing the magnetic field a short pulse of current is induced in the coil (the probe). This short pulse implies an electromotive force impulse at the terminal of the coil

$$\mathscr{E} \stackrel{\text{def}}{=} \int_0^T E(t)\, \mathrm{d}t = \int_0^T \int_{\partial\overline{\mathbf{S}}} \mathbf{E} \cdot \mathbf{t}\, \mathrm{d}L\, \mathrm{d}t \tag{10.30}$$

evaluated along the loop $\partial\overline{\mathbf{S}}$ characterizes locally the magnetic field. We remark that the interval $[0, T]$ must cover the switching off of the magnetic field. With regard to the sign of the electromotive force impulse, one must choose arbitrarily a positive direction on the loop. If the current generated by the switching goes in the chosen positive direction, then the electromotive force impulse is considered positive; otherwise it is negative.

**Space Association.** What is significant for the classification is that in a region of a *uniform* magnetic field like that inside a solenoid, the electromotive force impulse thus registered is *proportional to the area* enclosed by the loop and does not depend on the shape of the loops enclosing the same area. Moreover, experiments show that the electromotive force impulse depends on the space orientation on the flat coil. The proportionality to the area of the surface indicates that the electromotive force impulse is associated with the surface element **S** enclosed by the flat coil. Hence, the electromotive force impulse is called the *magnetic flux* of the field across the surface **S** and it is denoted by $\Phi$.

As we have seen, in this experiment, there is nothing that goes 'through' the surface of the loop. Instead it is the direction of the current in the coil, hence the boundary of the surface, which determines the sign of the flow. From this fact it follows that the magnetic flux is associated with a surface endowed with an inner orientation, i.e. $\Phi[\overline{\mathbf{S}}]$. The fact that this variable depends on the medium and is associated with an inner orientation is a confirmation that it is a configuration variable.

**Time Association.** The measure of the magnetic flux depends on the existing magnetic field before it is turned off; it does not depend on the duration of the shutdown. If the magnetic field is turned on, then the induced electromotive force impulse is opposite to that registered during the turning off, as shown in the bottom part of Fig. 10.6.

Moreover, the reversal of motion implies that the electric current in the electromagnet is switched on and, hence, the electric current induced in the coil goes in the opposite direction. This means that *the current, the electromotive force and the electromotive force impulse, i.e. the magnetic flux, change sign under a reversal of motion; hence, the electromotive force impulse and* the electric flux are associated with dual time instants $\widetilde{\mathbf{I}}$, i.e. $\mathscr{E}[\widetilde{\mathbf{I}}]$ and $\Phi[\widetilde{\mathbf{I}}]$. In global terms we can write

$$\textit{magnetic flux} \qquad \Phi[\widetilde{\mathbf{I}},\overline{\mathbf{S}}] \stackrel{\text{def}}{=} \mathscr{E}[\widetilde{\mathbf{I}},\partial\overline{\mathbf{S}}] \qquad \textit{turning off the field.} \tag{10.31}$$

## 10.4.5 Magnetic Flux Density

To introduce the magnetic flux density vector $\check{\mathbf{B}}$, we proceed in the same way as we did for the electric displacement $\mathbf{D}$. First, we consider the magnetic flux density $b \stackrel{\text{def}}{=} \Phi/A$ whose dimensions are $\text{Wb/m}^2$, i.e. *tesla*. The magnetic flux density $b$ depends on the point $\mathbf{P}$ of the region and on the space orientation of the surface. This suggests the need to introduce a unit vector $\mathbf{n}$ orthogonal to the surface element, hence $b(\mathbf{P}, \mathbf{n})$. Since the surface element is endowed with an inner orientation, i.e. $\overline{\mathbf{S}}$, we make use of the *screw rule* to assign a direction normal to the surface element. Hence, the vector $\mathbf{n}$ is a pseudovector and will be denoted by $\check{\mathbf{n}}$.[28]

Let us observe that among the space directions sorting from $\mathbf{P}$ there will surely be one, i.e. $\check{\mathbf{N}}$, for which $b$ assumes the maximum value. It is an experimental fact that the value $b(\mathbf{P}, \check{\mathbf{n}})$ for an arbitrary normal from $\mathbf{P}$ is linked to the value $b_{max}$ by the relation

$$b(\mathbf{P},\check{\mathbf{n}}) = b_{max}(\mathbf{P})\cos\alpha = b_{max}(\mathbf{P})\,(\check{\mathbf{N}}\cdot\check{\mathbf{n}}) = [b_{max}(\mathbf{P})\,\check{\mathbf{N}}]\cdot\check{\mathbf{n}}\,. \tag{10.32}$$

This leads us to introduce the vector $\check{\mathbf{B}}$, defined as

$$\boxed{\check{\mathbf{B}}(\mathbf{P}) \stackrel{\text{def}}{=} b_{max}(\mathbf{P})\,\check{\mathbf{N}}} \tag{10.33}$$

called the *magnetic flux density vector*. Let us remark that, since $\check{\mathbf{N}}$ is a pseudovector, then $\check{\mathbf{B}}$ is also pseudo. It is obvious that the magnetic flux $\Phi$ on a surface $\overline{\mathbf{S}}$ can be expressed in the form[29]

---

[28] See p. 145.

[29] Fournet [73, p. 48].

$$\boxed{\Phi = \int_{\overline{\mathbf{S}}} \check{\mathbf{B}}(\mathbf{P}) \cdot \check{\mathbf{n}}\, \mathrm{d}S.} \tag{10.34}$$

This relationship is often taken as a definition of magnetic flux. In contrast, we should take the relationship Eq. 10.33 as a definition of the vector $\check{\mathbf{B}}$ because the flux $\Phi$ is directly measured, whereas $\check{\mathbf{B}}$ is evaluated as the density. The magnetic flux is the overall magnitude associated with a surface, whereas the magnetic flux density vector $\check{\mathbf{B}}(\mathbf{P})$ is a kind of 'price' whose advantage is that it is independent of the area and of the space orientation of the plane surface element $\overline{\mathbf{S}}$.

The magnetic flux $\Phi$ is a quantity associated with a surface and is a domain function, whereas $\check{\mathbf{B}}(\mathbf{P})$ is a point function and inherits from $\Phi$ the association with $\overline{\mathbf{S}}$ and $\widetilde{\mathbf{I}}$. In summary $\Phi[\widetilde{\mathbf{I}}, \overline{\mathbf{S}}] \longrightarrow \mathbf{B}[\widetilde{\mathbf{I}}, \overline{\mathbf{S}}]$.

Figure 10.7 shows a parallel between the measures of $\mathbf{E}$ and of $\check{\mathbf{B}}$.

### *10.4.6 Summary of Physical Variables of Electromagnetism*

Table 10.1 on p. 276 shows the space-time global variables of electromagnetism, the corresponding space global variables and field variables. Table 10.3 summarizes the main global variables of electromagnetism and their link with field variables. In the integrals the inner or outer orientatin of the space element is shown. Thus, the magnetic flux $\Phi$ is associated with a surface endowed with an *inner* orientation, whereas the electric flux $\Psi$ requires an *outer* orientation. The same is true of the impulses of the electromotive force and magnetomotive forces which are associated with lines: the former is referred to lines endowed with an *inner* orientation, whereas the latter is referred to lines endowed with an *outer* orientation.

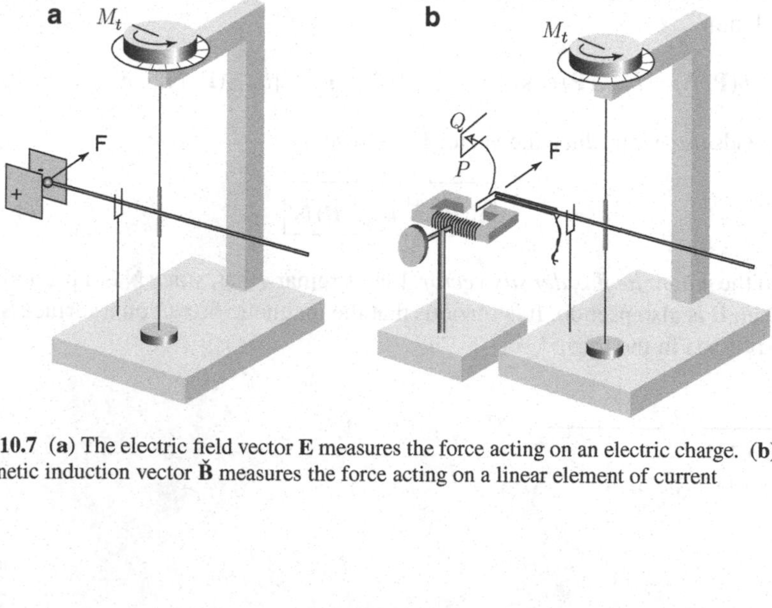


**Fig. 10.7** (**a**) The electric field vector **E** measures the force acting on an electric charge. (**b**) The magnetic induction vector $\check{\mathbf{B}}$ measures the force acting on a linear element of current

We stress that the behaviour of electromagnetic variables under a reversal of motion and space reflection (parity) resulting from their position inside the classification diagram is in full agreement with the behaviour found in the literature.[30]

## 10.5 Field Laws

In the following subsections we will analyse the four equations of the electromagnetic field: Gauss' law, Gauss' law for magnetism, Faraday's electromagnetic induction law, the Ampère–Maxwell law. We want to write these equations in global terms and then deduce them in differential form (see Tables 10.4–10.7).

**Table 10.3** Global physical variables expressed as integrals. The global variable on the left side all have the same physical dimension, that of the magnetic flux; those on the right also have the same physical dimension, that of the charge

| **Configuration variables**<br>Space: *inner orientation*<br>Time: *outer orientation* | **Source variables**<br>Space: *outer orientation*<br>Time: *inner orientation* |
|---|---|
| Gauge function $\chi$ | Electric charge production $Q^p = \int_{\tilde{\mathbf{T}}}\int_{\tilde{\mathbf{V}}} \sigma \, dV \, dt$ |
| Electric potential impulse $\varphi = \int_{\tilde{\mathbf{T}}} \phi \, dt$ | Electric charge content $Q^c = \int_{\tilde{\mathbf{V}}} \rho \, dV$ |
| (Nameless) $a = \int_{\mathbf{L}} \mathbf{A} \cdot \mathbf{t} \, dL$ | Electric charge flow $Q^f = \int_{\mathbf{T}}\int_{\tilde{\mathbf{S}}} \mathbf{J} \cdot \mathbf{n} \, dS \, dt$ |
| Electromotive force impulse $\mathscr{E} = \int_{\tilde{\mathbf{T}}}\int_{\mathbf{L}} \mathbf{E} \cdot \mathbf{t} \, dL \, dt$ | Electric flux $\Psi = \int_{\tilde{\mathbf{S}}} \mathbf{D} \cdot \mathbf{n} \, dS$ |
| Magnetic flux $\Phi = \int_{\mathbf{S}} \check{\mathbf{B}} \cdot \check{\mathbf{n}} \, dS$ | Magnetomotive force impulse $\mathscr{F}_m = \int_{\mathbf{T}}\int_{\tilde{\mathbf{L}}} \check{\mathbf{H}} \cdot \check{\mathbf{t}} \, dL \, dt$ |
| Magnetic charge flow $G^f = \int_{\tilde{\mathbf{T}}}\int_{\mathbf{S}} \check{\mathbf{J}}_m \cdot \check{\mathbf{n}} \, dS \, dt$ | (Nameless) $f = \int_{\tilde{\mathbf{L}}} \check{\mathbf{F}}_m \cdot \check{\mathbf{t}} \, dL$ |
| Magnetic charge content $\check{G}^c = \int_{\mathbf{V}} \check{\rho}_m \, dV$ | Magnetic potential impulse $\check{\varphi}_m = \int_{\mathbf{T}} \check{\phi}_m \, dt$ |
| Magnetic charge production $\check{G}^p = \int_{\tilde{\mathbf{T}}}\int_{\mathbf{V}} \check{\sigma}_m \, dV \, dt$ | Dual gauge function $\check{\eta}$ |
| Global variables, SI unit: *weber* | Global variables, SI unit: *coulomb* |

[30] See Jackson [102, Sect. 6.10].

The global formulation has some advantages: it is more intuitive, it is closer to a verbal formulation; it is also valid when the lines, surfaces and volumes cross or contain different materials, in contrast to the differential formulation, which obligates us to introduce jump conditions.

### *10.5.1 Gauss's Law*

Faraday discovered that if a charge $Q$ is enclosed within a spherical metallic shell, then an equal charge with the same sign appears on the surface of the sphere. He verified that the external field is symmetrical even if the ball is not concentric with the charge (Fig. 10.8). If the external surface charge is removed by momentarily placing the shell in contact with the ground, then an equal surface charge, with opposite sign with respect to the charge inside the case, is collected on the internal surface and can be measured.[31] The charge collected on the external surface of the metal case

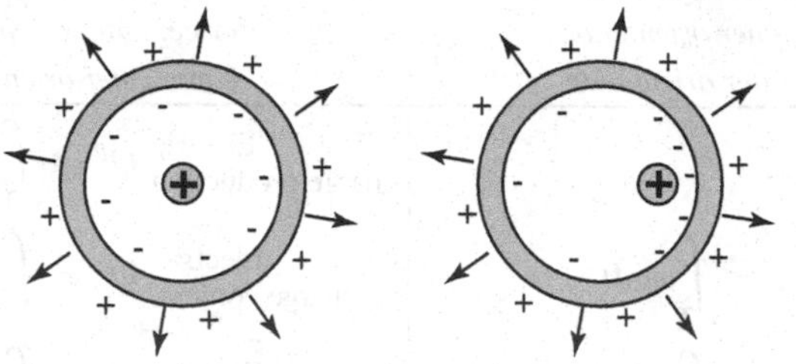

**Fig. 10.8** The charge induced on the external surface of a metal shell is equal to the charge contained inside the surface

- Does not depend on the medium surrounding the charge[32];
- Does not depend on either the shape or the size of the metal.

This law, which is the electrostatic induction law, is the experimental starting point of what is known today as Gauss's law. We want to express in words and in global terms Gauss's law as follows: *at every instant, the electric flux $\Psi$ on the boundary of a volume is equal to the electric charge $Q$ contained in the volume.*

Table 10.4 shows the preceding statement and the law written in a global formulation. The differential form of the first law of electrostatics could be obtained by introducing the densities of previously presented variables. This deduction is also shown in Table 10.4.

[31] Schelkunoff [201, p. 24].

[32] Jordan and Balmain [109, p. 31].

**Table 10.4** From the verbal statement to the differential formulation of Gauss's law

**Gauss's law**

$$\left\{\begin{matrix}\text{at every instant}\\ \text{the electric flux}\\ \text{on the boundary of a volume}\end{matrix}\right\} \overset{\text{law}}{=} \left\{\begin{matrix}\text{electric charge}\\ \text{contaned in the volume}\\ \text{at the instant}\end{matrix}\right\}$$

$$\boxed{\Psi[\bar{\mathbf{I}}, \partial\widetilde{\mathbf{V}}] \overset{\text{law}}{=} Q^c[\bar{\mathbf{I}}, \widetilde{\mathbf{V}}]}$$

$t$ (time axis), $\bar{\mathbf{I}}$

volume $\widetilde{\mathbf{V}}$

boundary $\partial\widetilde{\mathbf{V}}$

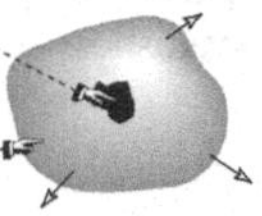

$$\int_{\partial\widetilde{\mathbf{V}}} \mathbf{D}\cdot\mathbf{n}\,\mathrm{d}S \overset{\text{law}}{=} \int_{\widetilde{\mathbf{V}}} \rho\,\mathrm{d}V$$

dividing by $V$ and computing the limit when $\mathbf{V} \to \mathbf{P}$

$$\boxed{\operatorname{div}\mathbf{D} \overset{\text{law}}{=} \rho}$$

## *10.5.2 Gauss's Law for Magnetism*

Gauss's law for magnetism is quickly established experimentally. Let us consider a polyhedron, say a tetrahedron on whose faces are coils, as shown in Fig. 10.9 (right). Since the electromotive force impulse $\mathscr{E}$ along every edge of a coil is opposite to that of the edge of the adjacent coil, the sum of all magnetic fluxes of the faces, automatically vanishes.

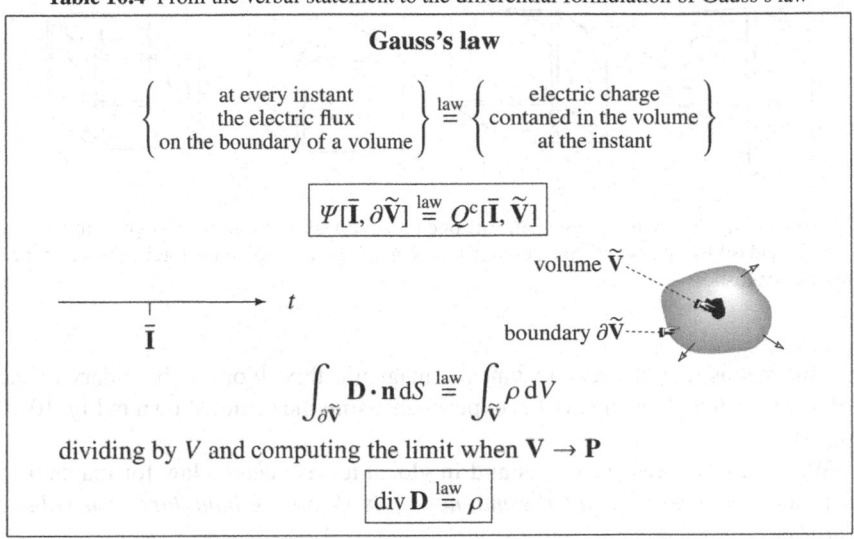

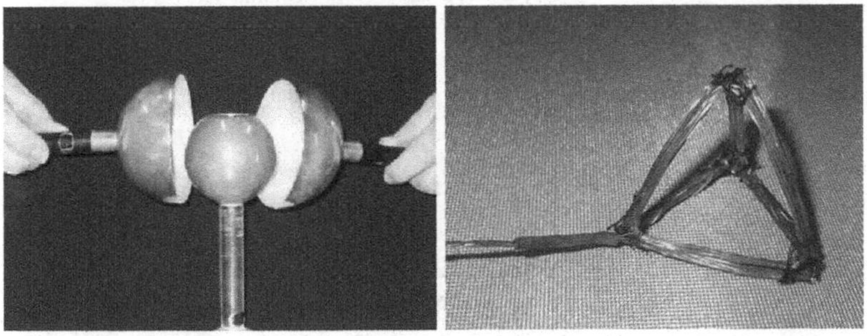

**Fig. 10.9** Devices to prove Gauss' law: (*left*) for electrostatics; (*right*) for the magnetic field

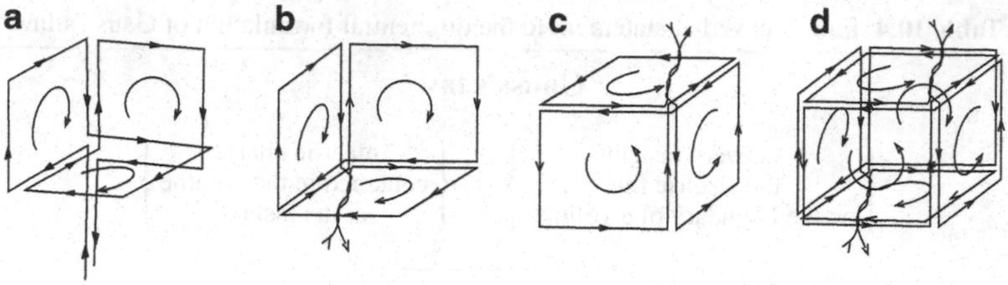


**Fig. 10.10** Gauss's law for magnetism: (**a**) one possible way to form three loops with a single wire; (**b**) and (**c**) two pairs of three loops; (**d**) the sum of two electromotive forces (hence of their impulses) vanishes

This means that at every instant the magnetic flux $\Phi$ on the boundary of the volume vanishes. You can do this experiment using the device shown in Fig. 10.10 (*right*).

We want to express in words and in global terms Gauss's law for magnetism as follows: *at every instant the magnetic flux* $\Phi$ *on the boundary of a volume vanishes*.

This statement and the law written in global terms is shown in Table 10.5. The differential form of Gauss's law of magnetism could be obtained by introducing the densities of previously presented variables. This deduction is also shown in Table 10.5.

**Table 10.5** From the verbal statement to the differential formulation of Gauss's law of magnetism

**Magnetic Gauss's law**

$$\left\{\begin{array}{c}\text{at every instant}\\ \text{the magnetic flux}\\ \text{on the boundary of a volume}\end{array}\right\} \overset{\text{law}}{=} 0$$

$$\boxed{\Phi[\tilde{\mathbf{I}}, \partial\overline{\mathbf{V}}] \overset{\text{law}}{=} 0}$$

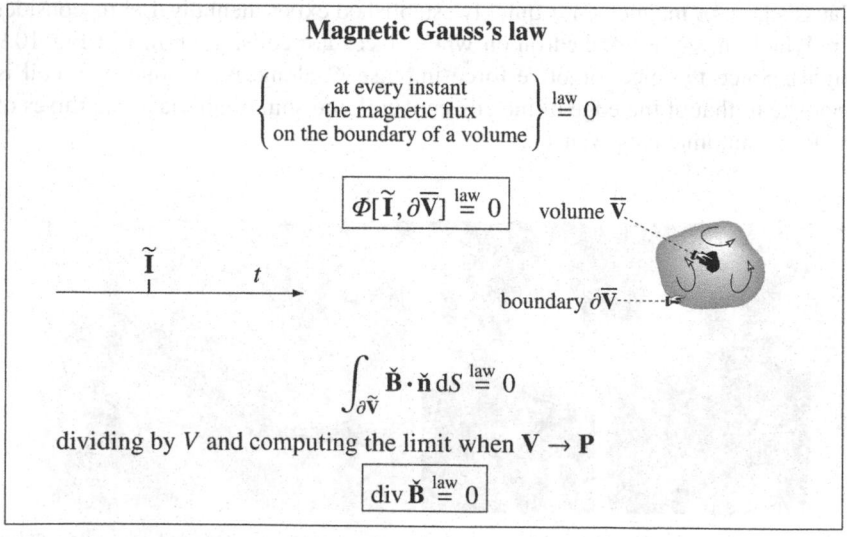

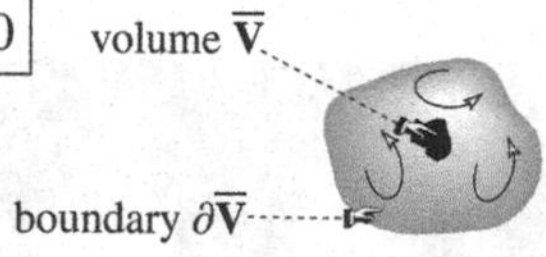


$$\int_{\partial\tilde{\mathbf{V}}} \check{\mathbf{B}} \cdot \check{\mathbf{n}} \, \mathrm{d}S \overset{\text{law}}{=} 0$$

dividing by $V$ and computing the limit when $\mathbf{V} \to \mathbf{P}$

$$\boxed{\operatorname{div} \check{\mathbf{B}} \overset{\text{law}}{=} 0}$$

### *10.5.3 Faraday's Electromagnetic Induction Law*

Experiments tell us that a time variation of a magnetic field implies the birth of an electric field. To describe this phenomenon, we must choose two physical variables, one for the magnetic field and another for the electric field. In line with the viewpoint we have taken in this book, we will choose two global variables, i.e. the magnetic flux $\Phi$ and the electromotive force $E$. We want to express in words and in global terms Faraday's electromagnetic induction law as follows: *the impulse of the electromotive force $E$ along the boundary of a surface in a time interval plus the time variation of the magnetic flux $\Phi$ in the time interval is equal to zero.*

Table 10.6 summarizes this statement and the law written in global terms. The differential form of Faraday's electromagnetic induction law could be obtained by introducing the densities of previously presented variables.

**Table 10.6** From the verbal statement to the differential formulation of Faraday's law

**Faraday's law**

$$\left\{\begin{array}{c}\text{the impulse of the}\\ \text{electromotive force}\\ \text{on the boundary of a surface}\\ \text{during a time interval}\end{array}\right\} + \left\{\begin{array}{c}\text{time variation of}\\ \text{the magnetic flux}\\ \text{in the time interval}\end{array}\right\} \overset{\text{law}}{=} 0$$

$$\boxed{\mathscr{E}[\tilde{\mathbf{T}}, \partial\bar{\mathbf{S}}] + \left\{\Phi[\tilde{\mathbf{I}}^+, \bar{\mathbf{S}}] + \Phi[\tilde{\mathbf{I}}^-, \bar{\mathbf{S}}]\right\} \overset{\text{law}}{=} 0}$$

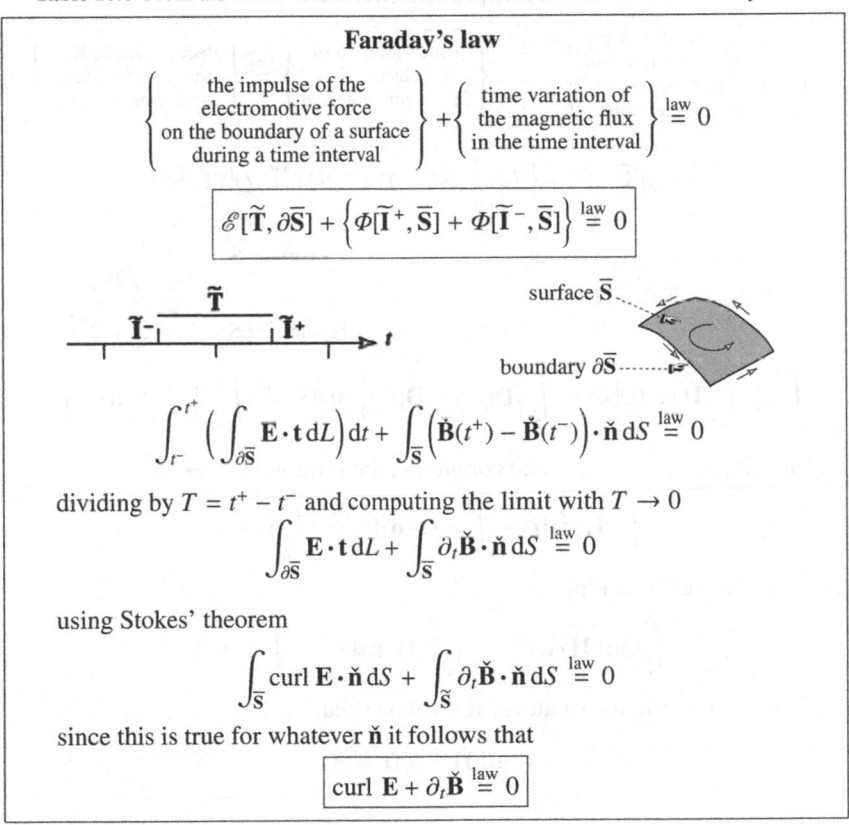

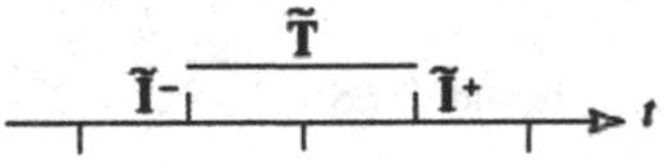


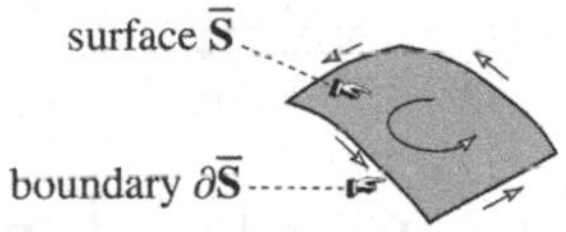


$$\int_{t^-}^{t^+}\left(\int_{\partial\bar{\mathrm{S}}} \mathbf{E}\cdot\mathbf{t}\,\mathrm{d}L\right)\mathrm{d}t + \int_{\bar{\mathrm{S}}}\left(\check{\mathbf{B}}(t^+) - \check{\mathbf{B}}(t^-)\right)\cdot\check{\mathbf{n}}\,\mathrm{d}S \overset{\text{law}}{=} 0$$

dividing by $T = t^+ - t^-$ and computing the limit with $T \to 0$

$$\int_{\partial\bar{\mathrm{S}}} \mathbf{E}\cdot\mathbf{t}\,\mathrm{d}L + \int_{\bar{\mathrm{S}}} \partial_t\check{\mathbf{B}}\cdot\check{\mathbf{n}}\,\mathrm{d}S \overset{\text{law}}{=} 0$$

using Stokes' theorem

$$\int_{\bar{\mathrm{S}}} \operatorname{curl}\mathbf{E}\cdot\check{\mathbf{n}}\,\mathrm{d}S + \int_{\tilde{\mathrm{S}}} \partial_t\check{\mathbf{B}}\cdot\check{\mathbf{n}}\,\mathrm{d}S \overset{\text{law}}{=} 0$$

since this is true for whatever $\check{\mathbf{n}}$ it follows that

$$\boxed{\operatorname{curl}\mathbf{E} + \partial_t\check{\mathbf{B}} \overset{\text{law}}{=} 0}$$

### 10.5.4 Ampère–Maxwell Law

The electromagnetic field arises as an interaction between an electric and a magnetic field. Since the variation of a magnetic field generates an electric field, a variation in an electric field generates a magnetic field. The passage of electric charges across a surface and the time variation of the electric flux on the surface generate an impulse of the magnetomotive force along the boundary of the surface. We want to express in words and in global terms the Ampère–Maxwell law as follows: *the impulse of the magnetomotive force* $\mathscr{F}_{\mathrm{m}}$ *on the boundary of a surface during a time interval minus the time variation of the electric flux* $\Psi$ *in the time interval is equal to electric charge flow across the surface in the time interval.*

**Table 10.7** From the verbal statement to the differential formulation of Ampère's law

**Ampère-Maxwell law**

$$\left\{\begin{array}{c}\text{the impulse of the}\\ \text{magnetomotive force}\\ \text{on the boundary of a surface}\\ \text{during a time interval}\end{array}\right\} - \left\{\begin{array}{c}\text{time variation of}\\ \text{the electric flux}\\ \text{in the time interval}\end{array}\right\} \overset{\text{law}}{=} \left\{\begin{array}{c}\text{electric charge flow}\\ \text{across the surface}\\ \text{in the time interval}\end{array}\right\}$$

$$\boxed{\mathscr{F}_{\mathrm{m}}[\overline{\mathbf{T}}, \partial\widetilde{\mathbf{S}}] - \left\{\Psi[\overline{\mathbf{I}}^{+}, \widetilde{\mathbf{S}}] - \Psi[\overline{\mathbf{I}}^{-}, \widetilde{\mathbf{S}}]\right\} \overset{\text{law}}{=} Q^{\mathrm{f}}[\overline{\mathbf{T}}, \widetilde{\mathbf{S}}]}$$

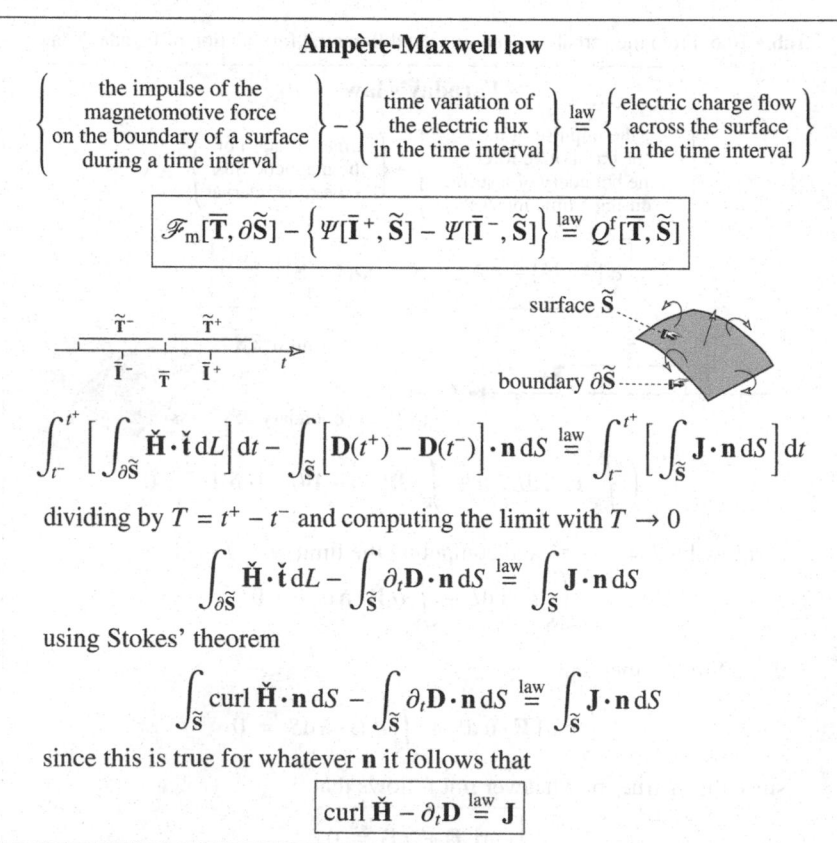

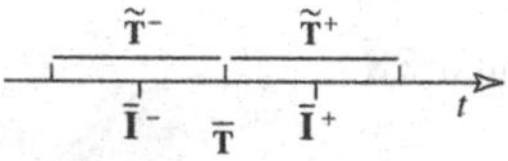


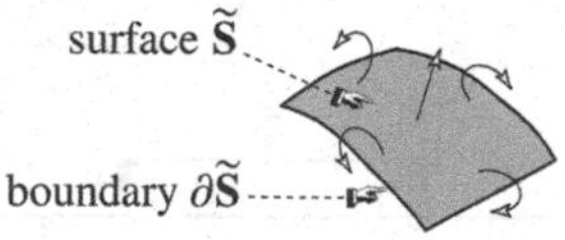


$$\int_{t^-}^{t^+}\left[\int_{\partial\widetilde{\mathbf{S}}} \check{\mathbf{H}}\cdot\check{\mathbf{t}}\,\mathrm{d}L\right]\mathrm{d}t - \int_{\widetilde{\mathbf{S}}}\left[\mathbf{D}(t^+) - \mathbf{D}(t^-)\right]\cdot\mathbf{n}\,\mathrm{d}S \overset{\text{law}}{=} \int_{t^-}^{t^+}\left[\int_{\widetilde{\mathbf{S}}} \mathbf{J}\cdot\mathbf{n}\,\mathrm{d}S\right]\mathrm{d}t$$

dividing by $T = t^+ - t^-$ and computing the limit with $T \to 0$

$$\int_{\partial\widetilde{\mathbf{S}}} \check{\mathbf{H}}\cdot\check{\mathbf{t}}\,\mathrm{d}L - \int_{\widetilde{\mathbf{S}}} \partial_t\mathbf{D}\cdot\mathbf{n}\,\mathrm{d}S \overset{\text{law}}{=} \int_{\widetilde{\mathbf{S}}} \mathbf{J}\cdot\mathbf{n}\,\mathrm{d}S$$

using Stokes' theorem

$$\int_{\widetilde{\mathbf{S}}} \mathrm{curl}\,\check{\mathbf{H}}\cdot\mathbf{n}\,\mathrm{d}S - \int_{\widetilde{\mathbf{S}}} \partial_t\mathbf{D}\cdot\mathbf{n}\,\mathrm{d}S \overset{\text{law}}{=} \int_{\widetilde{\mathbf{S}}} \mathbf{J}\cdot\mathbf{n}\,\mathrm{d}S$$

since this is true for whatever $\mathbf{n}$ it follows that

$$\boxed{\mathrm{curl}\,\check{\mathbf{H}} - \partial_t\mathbf{D} \overset{\text{law}}{=} \mathbf{J}}$$

$$\Phi[\partial\overline{V}] = 0 \qquad E[\partial\widetilde{S}] = -\dot{\Phi}[\overline{S}] \qquad \Psi[\partial\widetilde{V}] = Q[\widetilde{V}] \qquad F_m[\partial\widetilde{S}] = I[\widetilde{S}]$$

inductor        resistor        capacitor

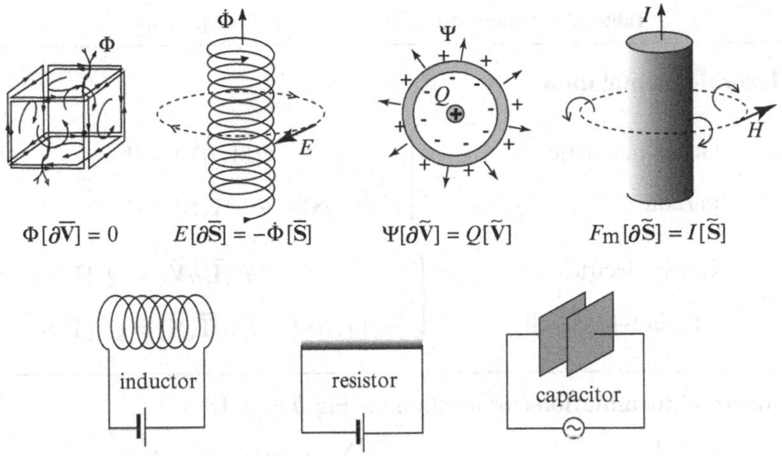

**Fig. 10.11** Simplest devices for illustrating the four Maxwell equations (*top*) and the three constitutive equations (*bottom*)

Table 10.7 summarizes this statement and the law written in global terms. The differential form of the Ampère–Maxwell law could be obtained by introducing the densities of previously presented variables (Fig. 10.11).

Table 10.8 shows different formulations of Maxwell's equations.

## 10.6  Space-Time Representation of Maxwell's Equations

The fact that physical variables are associated with space and time elements implies that the field equations linking them involve space regions, such as squares, cubes and hypercubes. Thus, a circuital equation can be referred to a square and a balance equation can be referred to a cube. Moreover, a circuital equation involving also time, like Ampère's and Faraday's laws, can be referred to a cube obtained by extrusion of a square: the third dimension is the time axis. To put it equivalently in words, these two circuital laws can be associated with a cube in a three-dimensional space-time $x, y, t$. We want to show that the eight scalar field equations can be associated with a cube in space-time, i.e. a hypercube.

**Table 10.8** Various descriptions of Maxwell's equations

**Algebraic formulation**

Gauss magnetic / Faraday

$$\begin{cases} \Phi\,[\tilde{\mathbf{I}}, \partial\overline{\mathbf{V}}] = 0 \\ \mathscr{E}\,[\tilde{\mathbf{T}}, \partial\overline{\mathbf{S}}] + \Phi\,[\partial\tilde{\mathbf{T}}, \overline{\mathbf{S}}] = 0 \end{cases}$$

Gauss electric / Ampère–Maxwell

$$\begin{cases} \Psi\,[\overline{\mathbf{I}}, \partial\tilde{\mathbf{V}}] = Q^{\mathrm{c}}\,[\overline{\mathbf{I}}, \tilde{\mathbf{V}}] \\ \mathscr{F}_{\mathrm{m}}\,[\overline{\mathbf{T}}, \partial\tilde{\mathbf{S}}] - \Psi\,[\partial\overline{\mathbf{T}}, \tilde{\mathbf{S}}] = Q^{\mathrm{f}}\,[\overline{\mathbf{T}}, \tilde{\mathbf{S}}] \end{cases}$$

**Numerical formulation** (for notation see Fig. 7.3, p. 189)

$$\begin{cases} \sum_j \bar{d}_{kj}\ \Phi[\tilde{\mathbf{t}}^n, \bar{\mathbf{s}}^j] = 0 \\ \sum_i \bar{c}_{ji}\ \mathscr{E}[\tilde{\boldsymbol{\tau}}^{n+1}, \bar{\mathbf{l}}_i] + \left\{\Phi[\tilde{\mathbf{t}}^{n+1}, \bar{\mathbf{s}}^j] - \Phi[\tilde{\mathbf{t}}^n, \bar{\mathbf{s}}^j]\right\} = 0 \end{cases}$$

$$\begin{cases} \sum_i \tilde{d}_{hi}\ \Psi[\bar{\mathbf{t}}^n, \tilde{\mathbf{s}}^i] = Q^{\mathrm{c}}[\bar{\mathbf{t}}^n, \tilde{\mathbf{v}}^h] \\ \sum_j \tilde{c}_{ij}\ \mathscr{F}_{\mathrm{m}}[\bar{\boldsymbol{\tau}}^n, \tilde{\mathbf{l}}^j] - \left\{\Psi[\bar{\mathbf{t}}^{n+1}, \tilde{\mathbf{s}}^i] - \Psi[\bar{\mathbf{t}}^n, \tilde{\mathbf{s}}^i]\right\} = Q^{\mathrm{f}}[\bar{\boldsymbol{\tau}}^n, \tilde{\mathbf{s}}^i] \end{cases}$$

**Integral formulation**

$$\begin{cases} \int_{\partial\overline{\mathbf{V}}} \check{\mathbf{B}} \cdot \check{\mathbf{n}}\, \mathrm{d}S = 0 \\ \int_{\tilde{\mathbf{T}}} \int_{\partial\overline{\mathbf{S}}} \mathbf{E} \cdot \mathbf{t}\, \mathrm{d}L\ \mathrm{d}t + \left[\int_{\overline{\mathbf{S}}} \check{\mathbf{B}} \cdot \check{\mathbf{n}}\, \mathrm{d}S\right]_{t^-}^{t^+} = 0 \end{cases}$$

$$\begin{cases} \int_{\partial\tilde{\mathbf{V}}} \mathbf{D} \cdot \mathbf{n}\, \mathrm{d}S = \int_{\tilde{\mathbf{V}}} \rho\, \mathrm{d}V \\ \int_{\overline{\mathbf{T}}} \int_{\partial\tilde{\mathbf{S}}} \check{\mathbf{H}} \cdot \check{\mathbf{t}}\, \mathrm{d}L\ \mathrm{d}t - \left[\int_{\tilde{\mathbf{S}}} \mathbf{D} \cdot \mathbf{n}\, \mathrm{d}S\right]_{t^-}^{t^+} = \int_{\overline{\mathbf{T}}} \int_{\tilde{\mathbf{S}}} \mathbf{J} \cdot \mathbf{n}\, \mathrm{d}S\ \mathrm{d}t \end{cases}$$

**Differential formulation**

$$\begin{cases} \operatorname{div} \check{\mathbf{B}} = 0 \\ \operatorname{curl} \mathbf{E} + \partial_t \check{\mathbf{B}} = 0 \end{cases}$$

$$\begin{cases} \operatorname{div} \mathbf{D} = \rho \\ \operatorname{curl} \check{\mathbf{H}} - \partial_t \mathbf{D} = \mathbf{J} \end{cases}$$

### 10.6.1 Visualizing Space-Time

In physics it is convenient to consider time as a fourth coordinate, added to the three space coordinates, and this becomes space-time. Space-time, which appeared with the advent of relativity, involves *metrical* properties.

We start by posing a question: is it possible to visualize the fourth dimension? The answer is that it is possible using the projection of space-time objects in three-dimensional space. To understand what this means, let us observe that when we draw a cube on a sheet of paper, we represent it by two squares whose homologous vertices are connected by a straight line, as shown in Fig. 10.12. This means that to 'see' a cube, which is an object of three-dimensional space, we project it in two-dimensional space, as shown in Fig. 10.12. Note that the lateral faces of the cube, which are squares, are projected into trapezoids in the perspective projection and in parallelograms in the axonometric projection. Up to now we have shown a *static* view.

Now let us consider a *kinematic* view, so we can visualize the fourth dimension. A cube can be conceived of as being generated by the translation of a square or as a contraction (or dilatation) of a square, as in Fig. 10.13.

By *kinematic* view we mean the registration of different positions of an object at different time instants, as shown in Fig. 10.14. Thus, as a train approaches, its shape expands, but as the train departs, its shape contracts; the rails can be taken as time axes (Fig. 10.15).

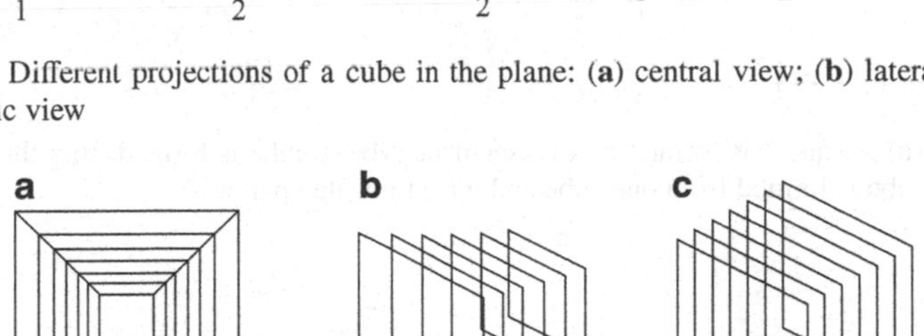

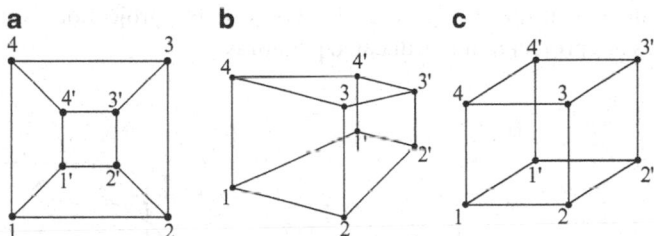

**Fig. 10.12** Different projections of a cube in the plane: (**a**) central view; (**b**) lateral view; (**c**) axonometric view

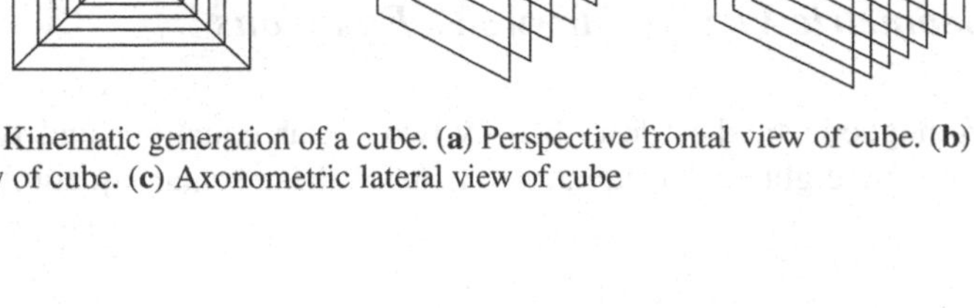


**Fig. 10.13** Kinematic generation of a cube. (**a**) Perspective frontal view of cube. (**b**) Perspective lateral view of cube. (**c**) Axonometric lateral view of cube

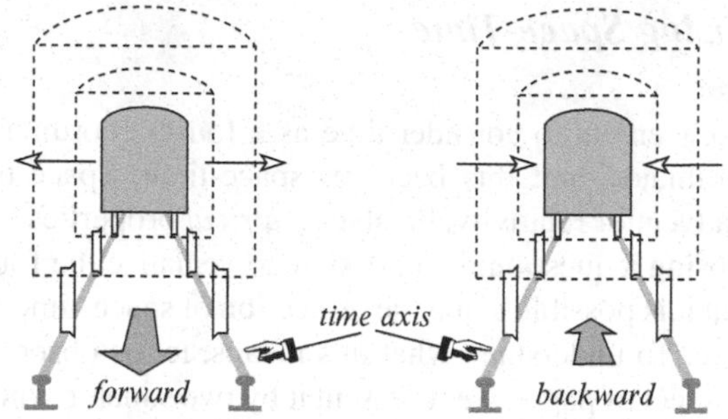


**Fig. 10.14** Kinematic views of back of a moving train. As the train approaches, its shape expands and shrinks as it moves away. A reversal of motion changes an explosion into an implosion

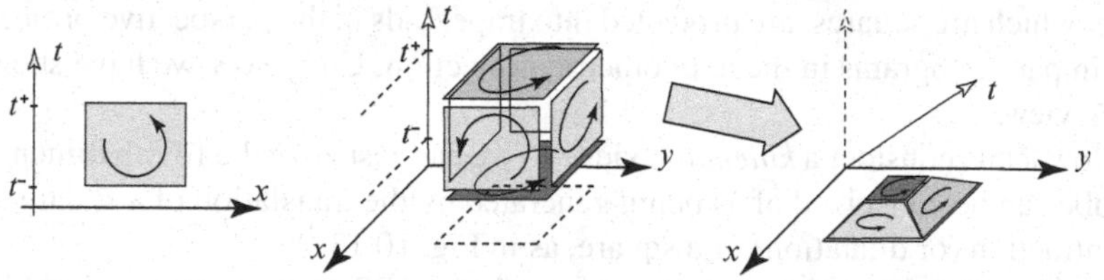


**Fig. 10.15** Geometric support for Faraday's law in a two-dimensional space-time

Figure 10.16 shows how to build a square, a cube and a hypercube starting from line segments, squares and cubes respectively. After these considerations we can visualize a hypercube, i.e. a cube of a four-dimensional space, by its projection in three-dimensional space. Figure 10.17 shows such a projection. Note that the lateral cubes are projected into truncated pyramids.

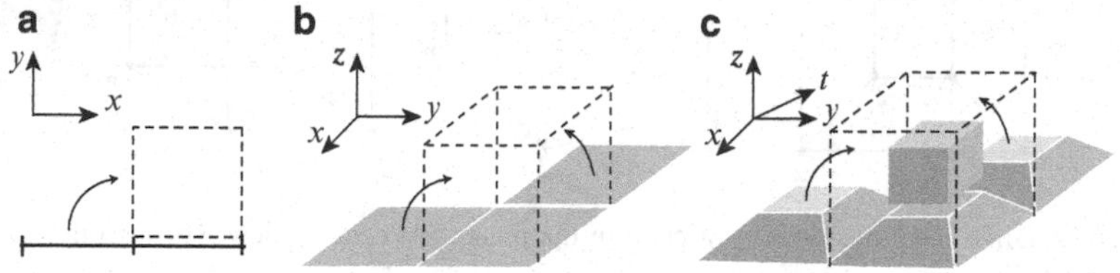


**Fig. 10.16** (**a**) A square is formed by two segments. (**b**) A cube is formed from three squares. (**c**) A hypercube is formed from one cube and three truncated pyramids

## *10.6.2 Geometric View of Maxwell's Equations*

The electromagnetic field is described by two scalar equations and two vector equations, i.e. by eight scalar equations. Four of these scalar equations link the

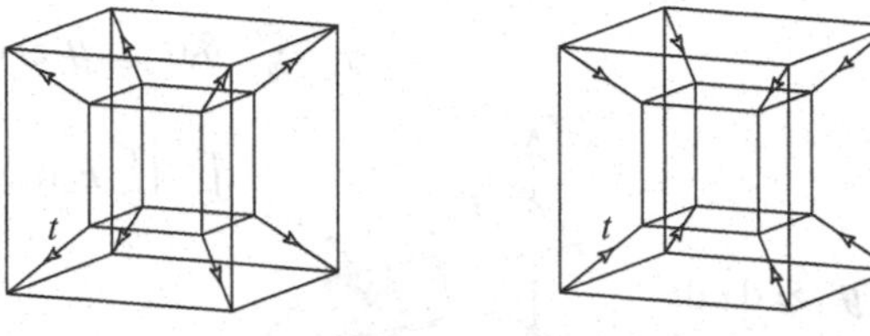


**Fig. 10.17** A hypercube of a four-dimensional space projected into a three-dimensional space and represented in the two-dimensional space of the paper. The *arrows* on the time lines denote the two possible outer orientations of the hypercube

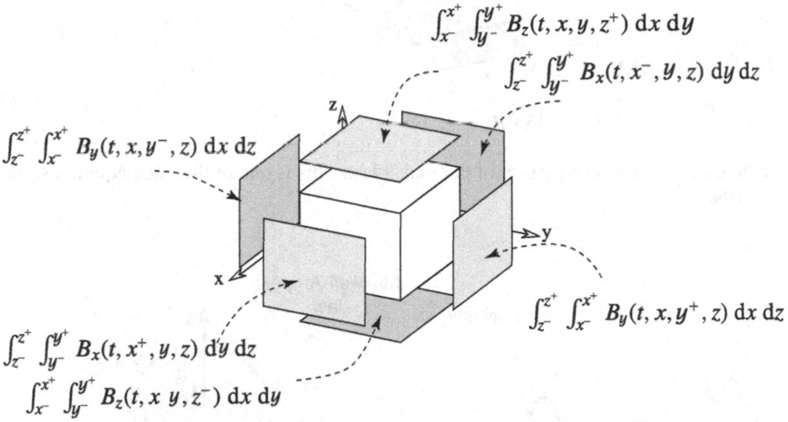


**Fig. 10.18** Geometric interpretation of Gauss's law of magnetism

configuration variables and four link the source variables. We now show that it is possible to produce a beautiful representation of these two sets of equations using a three-dimensional projection of a four-dimensional cube. To this end, let us remark that the two laws of Gauss are static balance equations, while the two circuital laws, those of Faraday and of Maxwell–Ampère, can be conceived as balance equations in three-dimensional space-time cubes, as shown in Figs. 10.18 and 10.19. These figures show only the balances for the first set of Maxwell's equations. Those of the second set are similar and are not included for space reasons only. Equation 10.35 shows the classical differential formulation of Maxwell's equations (left side) and the integral formulation (right). The terms shown in Figs. 10.18 and 10.19 are the pieces which compose the integrals of Eq. 10.35:

$$\begin{cases} \operatorname{div}\check{\mathbf{B}} = 0 \\ \operatorname{curl}\mathbf{E} + \partial_t\check{\mathbf{B}} = 0 \end{cases} \longrightarrow \begin{cases} \displaystyle\int_{\partial\overline{\mathbf{V}}} \check{\mathbf{B}}\cdot\check{\mathbf{n}}\,\mathrm{d}S = 0 \\ \displaystyle\int_{t^-}^{t^+}\int_{\partial\overline{\mathbf{S}}} \mathbf{E}\cdot\mathbf{t}\,\mathrm{d}L\,\mathrm{d}t + \left[\int_{\overline{\mathbf{S}}} \check{\mathbf{B}}\cdot\check{\mathbf{n}}\,\mathrm{d}S\right]_{t^-}^{t^+} = 0 \end{cases} . \tag{10.35}$$

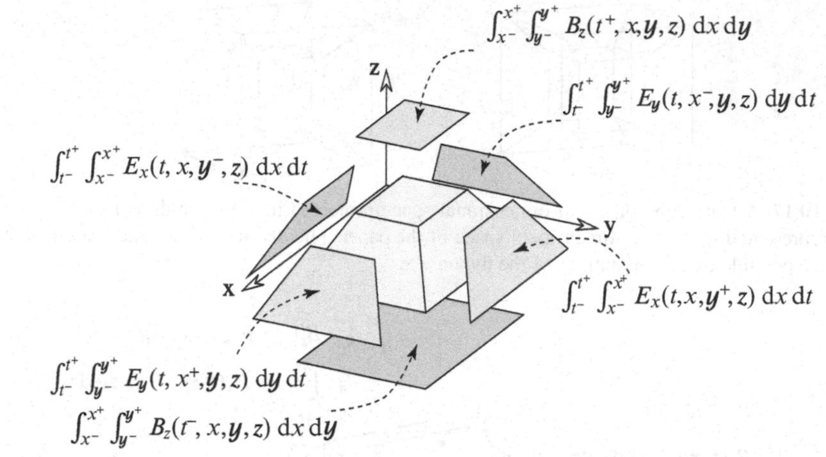


**Fig. 10.19** Geometric interpretation of Faraday's law. This is one of the three equations, the one relative to the $z$-axis

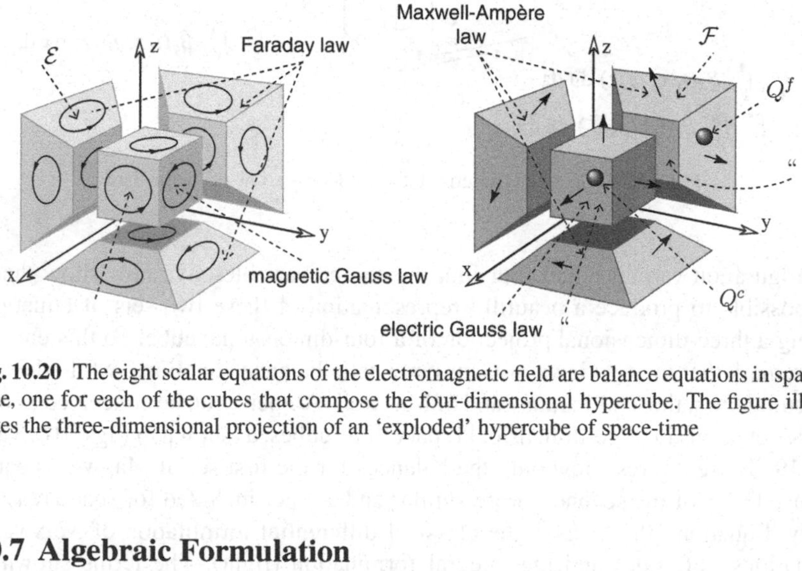


**Fig. 10.20** The eight scalar equations of the electromagnetic field are balance equations in space-time, one for each of the cubes that compose the four-dimensional hypercube. The figure illustrates the three-dimensional projection of an 'exploded' hypercube of space-time

## 10.7 Algebraic Formulation

As far as physics is concerned, it appears natural in a discrete setting to use the discrete form of degree $p$ to describe the global variables associated with $p$-dimensional manifolds (Fig. 10.20). It is enough to cover the space with a cell complex $\overline{\mathbf{K}}$ and its dual $\widetilde{\mathbf{K}}$ and to approximate the $p$-dimensional manifolds with $p$-dimensional chains. In this way, an amount of the global physical quantity is associated with every $p$-chain, and then a discrete form of degree $p$ is obtained.

The association of an amount of a physical variable with every $p$-cell gives rise to a *set function* which is the natural extension of *point functions* used in physics. In Table 10.1 we present six global variables (in space) of electromagnetism and the corresponding space elements. This leads us to the introduction of the six discrete forms defined as follows:

$$\begin{cases} \mathrm{E}^1(l_\alpha) \stackrel{\text{def}}{=} E_\alpha & \text{discrete 1-form of voltage,} \\ \Phi^2(s_\beta) \stackrel{\text{def}}{=} \Phi_\beta & \text{discrete 2-form of magnetic fluxes,} \\ \mathrm{F}^1(\tilde{l}_\alpha) \stackrel{\text{def}}{=} F_\alpha & \text{discrete 1-form of magnetomotive force,} \\ \Psi^2(\tilde{s}_\beta) \stackrel{\text{def}}{=} \Psi_\beta & \text{discrete 2-form of electric fluxes,} \\ \mathrm{I}^2(\tilde{s}_\beta) \stackrel{\text{def}}{=} I_\beta & \text{discrete 2-form of electric currents,} \\ \mathrm{Q}^3(\tilde{v}_h) \stackrel{\text{def}}{=} Q_h & \text{discrete 3-form of electric charge contents.} \end{cases} \tag{10.36}$$

We may write Maxwell's equations in the form

$$\begin{cases} \delta\Phi^2 = 0, \\ \delta\mathrm{E}^1 + \dfrac{\mathrm{d}}{\mathrm{d}t}\Phi^2 = 0, \end{cases} \qquad \begin{cases} \delta\Psi^2 = \mathrm{Q}^3, \\ \delta\mathrm{F}^1 = \dfrac{\mathrm{d}}{\mathrm{d}t}\Psi^2 + \mathrm{I}^2. \end{cases} \tag{10.37}$$

When these equations are applied to the single cells of the two complexes, we obtain a 'local' form of Maxwell's equations in a discrete setting, i.e.

$$\begin{cases} \displaystyle\sum_\alpha d_{h\alpha}\, \Phi_\alpha = 0, \\ \displaystyle\sum_\beta c_{\alpha\beta}\, E_\beta + \frac{\mathrm{d}}{\mathrm{d}t}\Phi_\alpha = 0, \end{cases} \qquad \begin{cases} \displaystyle\sum_\alpha \tilde{d}_{h\alpha}\, \Psi_\alpha = \mathrm{Q}_h, \\ \displaystyle\sum_\beta \tilde{c}_{\alpha\beta}\, F_\beta = \frac{\mathrm{d}}{\mathrm{d}t}\Psi_\alpha + I_\alpha. \end{cases} \tag{10.38}$$

These equations can be used to obtain a numerical solution of electromagnetic problems.[33]

## 10.8 Classification Diagrams of Electromagnetism

We now show how to combine a diagram of electrostatics with a diagram of magnetostatics to obtain a diagram of electromagnetism. Table 10.9 shows that it is appropriate to shift the second and fourth columns backwards with respect to the first and third rows and to shift vertically the first and third rows by half a space. In

[33] See the papers dealing with the cell method for electromagnetism `http://discretephysics.dica.units.it/`.

**Table 10.9** Classification diagram for field variables of electromagnetism: *front*: electrostatics; *rear*: magnetostatics

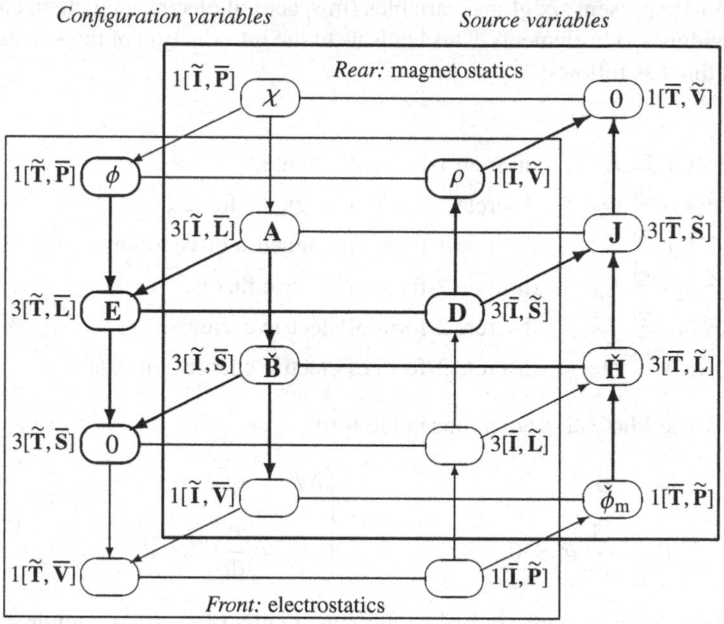


doing this, we assure that the space-time elements which lie in the same row have the same dimension. Thus, $[\tilde{\mathbf{T}}, \overline{\mathbf{L}}]$ has dimension 2 (=1+1) just like $[\tilde{\mathbf{I}}, \overline{\mathbf{S}}]$ (0+2).[34]

The diagrams that follow show the classification of electromagnetic variables and of the equations that link them. In the bottom part of each diagram is drawn a small frame composed of 16 small circles connected by arrows. We have filled the small circles that correspond to a physical variable used in electromagnetism and we have draw in heavy lines the connections corresponding to an existing equation, and it is shown (filled circles and heavy lines) which part of the frame is interested by the diagram.

Diagrams [ELE1] of electrostatics and [ELE2] of magnetostatics can be combined to give a diagram of the electromagnetic field [ELE3]: they are located respectively in the front and back parts. The two fields are coupled by the time derivatives, represented in the horizontal lines connecting the front with the back part of the diagram.

Diagrams [ELE4] and [ELE5] give the tensorial version of [ELE1] and [ELE2] respectively. Diagram [ELE6] gives the space-time version of diagram [ELE3]. Since electromagnetism is a relativistic theory, we see that the frame with 16

[34] The connections among the cells give rise to a sort of 'home' which Bossavit [21, p. 122] was the first to call a 'Maxwell house'. The present book shows that this 'house' is not specific to electromagnetism but is common to all classical physical theories.

boxes is automatically adapted to relativity. Diagram [ELE7] shows how the two-dimensional electric field can be cast in the language of complex variables. Diagram [ELE8] shows how to describe the motion of a charged particle in an electric and magnetic field respectively. Diagram [ELE9] shows how to combine the variables of the electric field to obtain energy variables. Diagram [ELE10] describes electric networks, while [ELE11] describes RLC circuits. Diagram [ELE12] shows the link between the field and the global variables of electrostatics. Diagram [ELE13] describes the motion of a charged particle in an electromagnetic field. This diagram differs from all other diagrams in the book. Diagram [ELE14] shows the link of the electric and magnetic fields with the theory of exterior differential forms. In particular, it shows the difference between the algebraic and metric duals of the second-order covariant tensors $\check{D}_{ij}$ and $B_{hk}$.

Table 10.10 summarizes the link between global variables and field variables of electromagnetism.

**Table 10.10** From field variables to global variables of electromagnetism

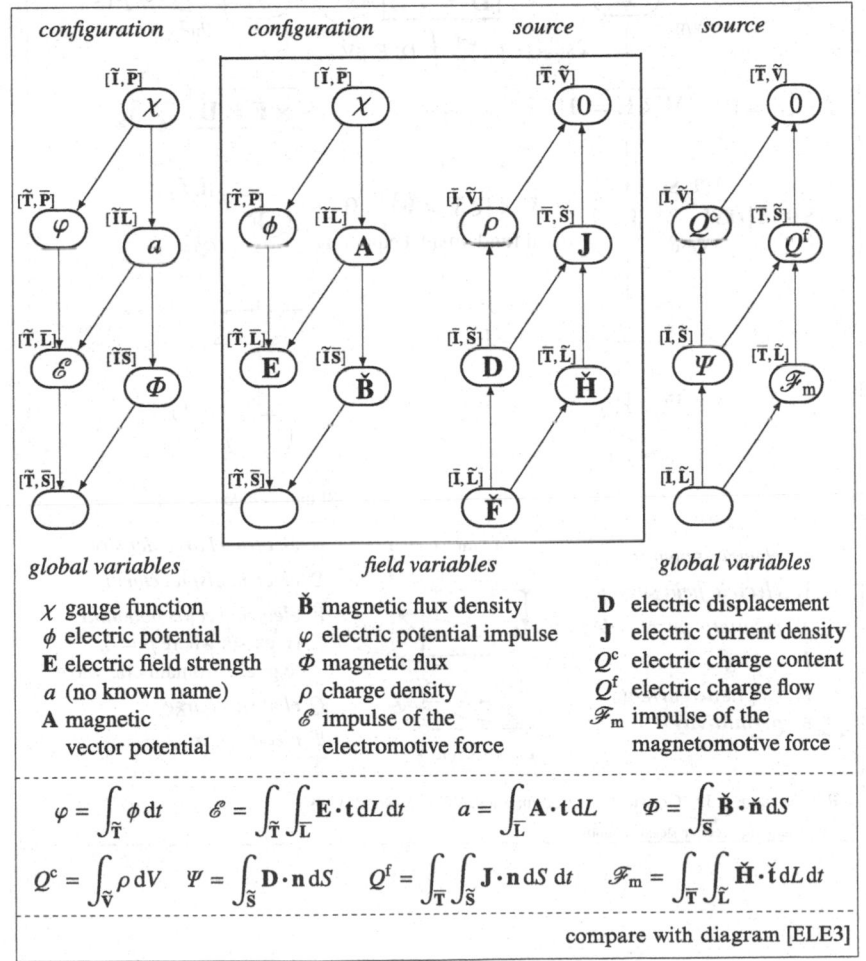

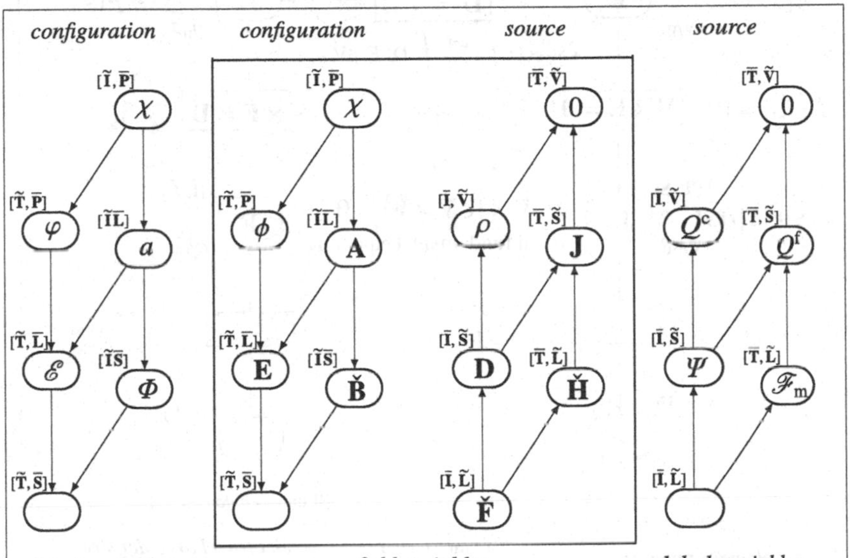


| global variables | field variables | global variables |
|---|---|---|
| $\chi$ gauge function | $\check{\mathbf{B}}$ magnetic flux density | **D** electric displacement |
| $\phi$ electric potential | $\varphi$ electric potential impulse | **J** electric current density |
| **E** electric field strength | $\Phi$ magnetic flux | $Q^c$ electric charge content |
| $a$ (no known name) | $\rho$ charge density | $Q^f$ electric charge flow |
| **A** magnetic vector potential | $\mathscr{E}$ impulse of the electromotive force | $\mathscr{F}_m$ impulse of the magnetomotive force |

$$\varphi = \int_{\tilde{\mathrm{T}}} \phi \, \mathrm{d}t \qquad \mathscr{E} = \int_{\tilde{\mathrm{T}}} \int_{\mathrm{L}} \mathbf{E} \cdot \mathbf{t} \, \mathrm{d}L \, \mathrm{d}t \qquad a = \int_{\mathrm{L}} \mathbf{A} \cdot \mathbf{t} \, \mathrm{d}L \qquad \Phi = \int_{\mathrm{S}} \check{\mathbf{B}} \cdot \check{\mathbf{n}} \, \mathrm{d}S$$

$$Q^c = \int_{\tilde{\mathrm{V}}} \rho \, \mathrm{d}V \qquad \Psi = \int_{\tilde{\mathrm{S}}} \mathbf{D} \cdot \mathbf{n} \, \mathrm{d}S \qquad Q^f = \int_{\bar{\mathrm{T}}} \int_{\tilde{\mathrm{S}}} \mathbf{J} \cdot \mathbf{n} \, \mathrm{d}S \, \mathrm{d}t \qquad \mathscr{F}_m = \int_{\bar{\mathrm{T}}} \int_{\tilde{\mathrm{L}}} \check{\mathbf{H}} \cdot \check{\mathbf{t}} \, \mathrm{d}L \, \mathrm{d}t$$

compare with diagram [ELE3]

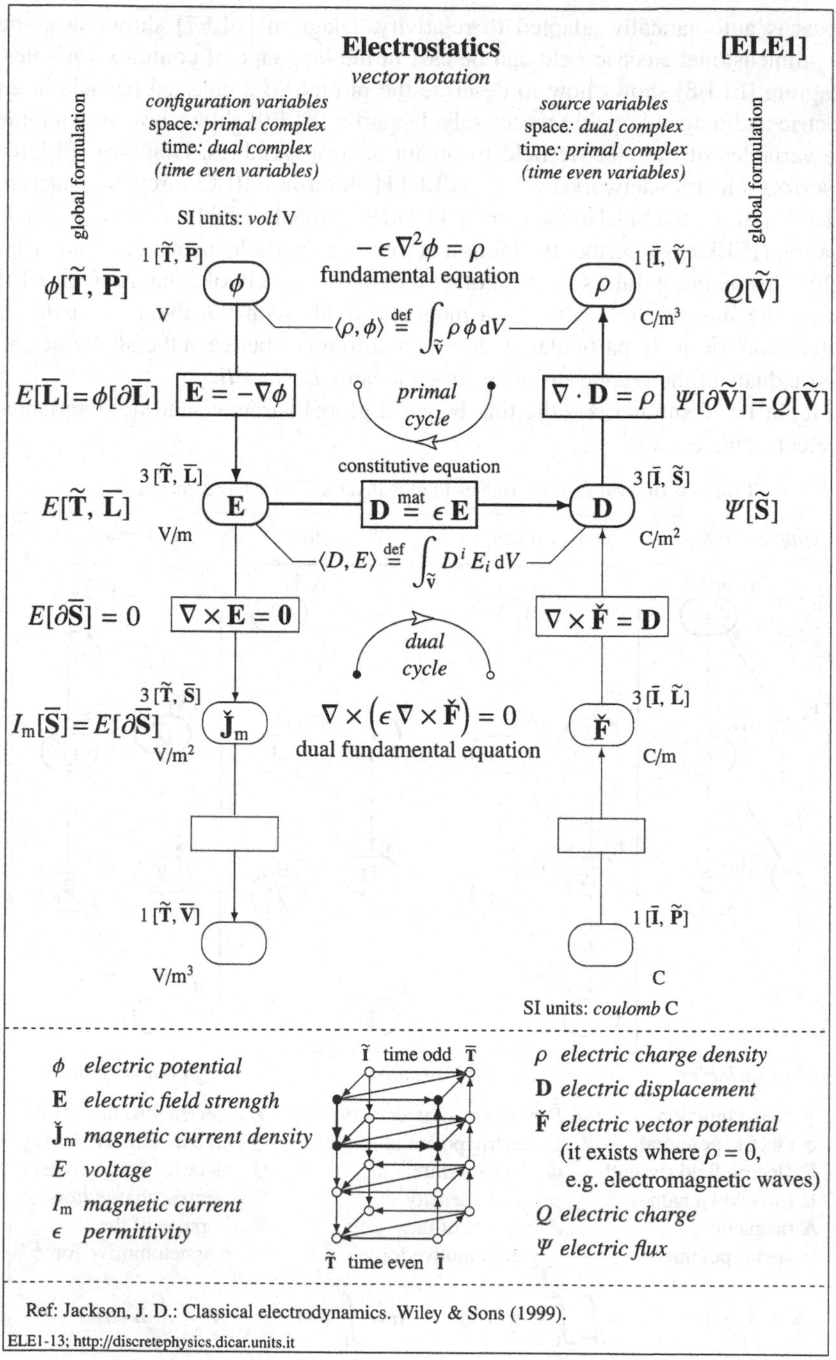
Electrostatics
[ELE1]
vector notation
configuration variables
space: primal complex
time: dual complex
(time even variables)
source variables
space: dual complex
time: primal complex
(time even variables)
global formulation
global formulation
SI units: volt V
$-\epsilon\nabla^2\phi = \rho$
fundamental equation
$1\,[\tilde{\mathbf{T}}, \bar{\mathbf{P}}]$
$\phi[\tilde{\mathbf{T}}, \bar{\mathbf{P}}]$
$\phi$
V
$1\,[\bar{\mathbf{I}}, \tilde{\mathbf{V}}]$
$\rho$
$Q[\tilde{\mathbf{V}}]$
C/m³
$\langle\rho,\phi\rangle \stackrel{\text{def}}{=} \int_{\tilde{\mathbf{V}}} \rho\,\phi\,\mathrm{d}V$
$E[\bar{\mathbf{L}}] = \phi[\partial\bar{\mathbf{L}}]$
$\mathbf{E} = -\nabla\phi$
primal cycle
$\nabla\cdot\mathbf{D} = \rho$
$\Psi[\partial\tilde{\mathbf{V}}] = Q[\tilde{\mathbf{V}}]$
constitutive equation
$3\,[\tilde{\mathbf{T}}, \bar{\mathbf{L}}]$
$E[\tilde{\mathbf{T}}, \bar{\mathbf{L}}]$
$\mathbf{E}$
V/m
$\mathbf{D} \stackrel{\text{mat}}{=} \epsilon\,\mathbf{E}$
$3\,[\bar{\mathbf{I}}, \tilde{\mathbf{S}}]$
$\mathbf{D}$
$\Psi[\tilde{\mathbf{S}}]$
C/m²
$\langle D,E\rangle \stackrel{\text{def}}{=} \int_{\tilde{\mathbf{V}}} D^i E_i\,\mathrm{d}V$
$E[\partial\bar{\mathbf{S}}] = 0$
$\nabla\times\mathbf{E} = \mathbf{0}$
dual cycle
$\nabla\times\check{\mathbf{F}} = \mathbf{D}$
$3\,[\tilde{\mathbf{T}}, \bar{\mathbf{S}}]$
$I_{\mathrm{m}}[\bar{\mathbf{S}}] = E[\partial\bar{\mathbf{S}}]$
$\check{\mathbf{J}}_{\mathrm{m}}$
V/m²
$\nabla\times\left(\epsilon\nabla\times\check{\mathbf{F}}\right) = 0$
dual fundamental equation
$3\,[\bar{\mathbf{I}}, \tilde{\mathbf{L}}]$
$\check{\mathbf{F}}$
C/m
$1\,[\tilde{\mathbf{T}}, \bar{\mathbf{V}}]$
V/m³
$1\,[\bar{\mathbf{I}}, \tilde{\mathbf{P}}]$
C
SI units: coulomb C
$\phi$ electric potential
$\mathbf{E}$ electric field strength
$\check{\mathbf{J}}_{\mathrm{m}}$ magnetic current density
$E$ voltage
$I_{\mathrm{m}}$ magnetic current
$\epsilon$ permittivity
$\tilde{\mathbf{I}}$ time odd $\bar{\mathbf{T}}$
$\tilde{\mathbf{T}}$ time even $\bar{\mathbf{I}}$
$\rho$ electric charge density
$\mathbf{D}$ electric displacement
$\check{\mathbf{F}}$ electric vector potential
(it exists where $\rho = 0$,
e.g. electromagnetic waves)
$Q$ electric charge
$\Psi$ electric flux
Ref: Jackson, J. D.: Classical electrodynamics. Wiley & Sons (1999).
ELE1-13; http://discretephysics.dicar.units.it

## Magnetostatics [ELE2]

*vector notation*

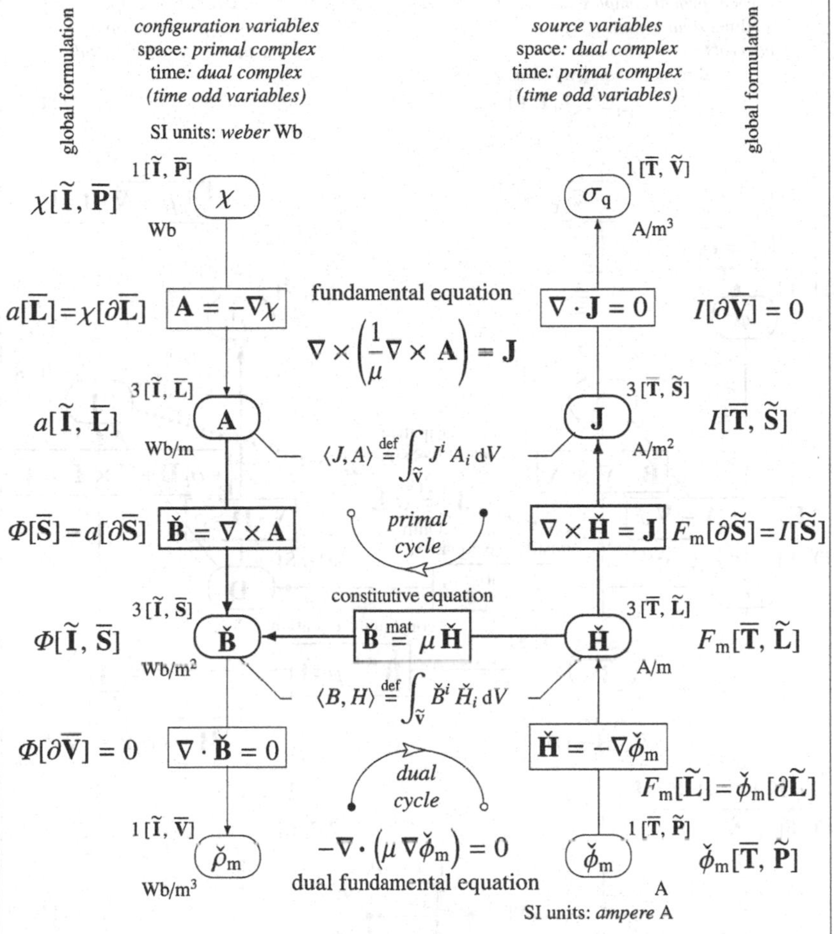


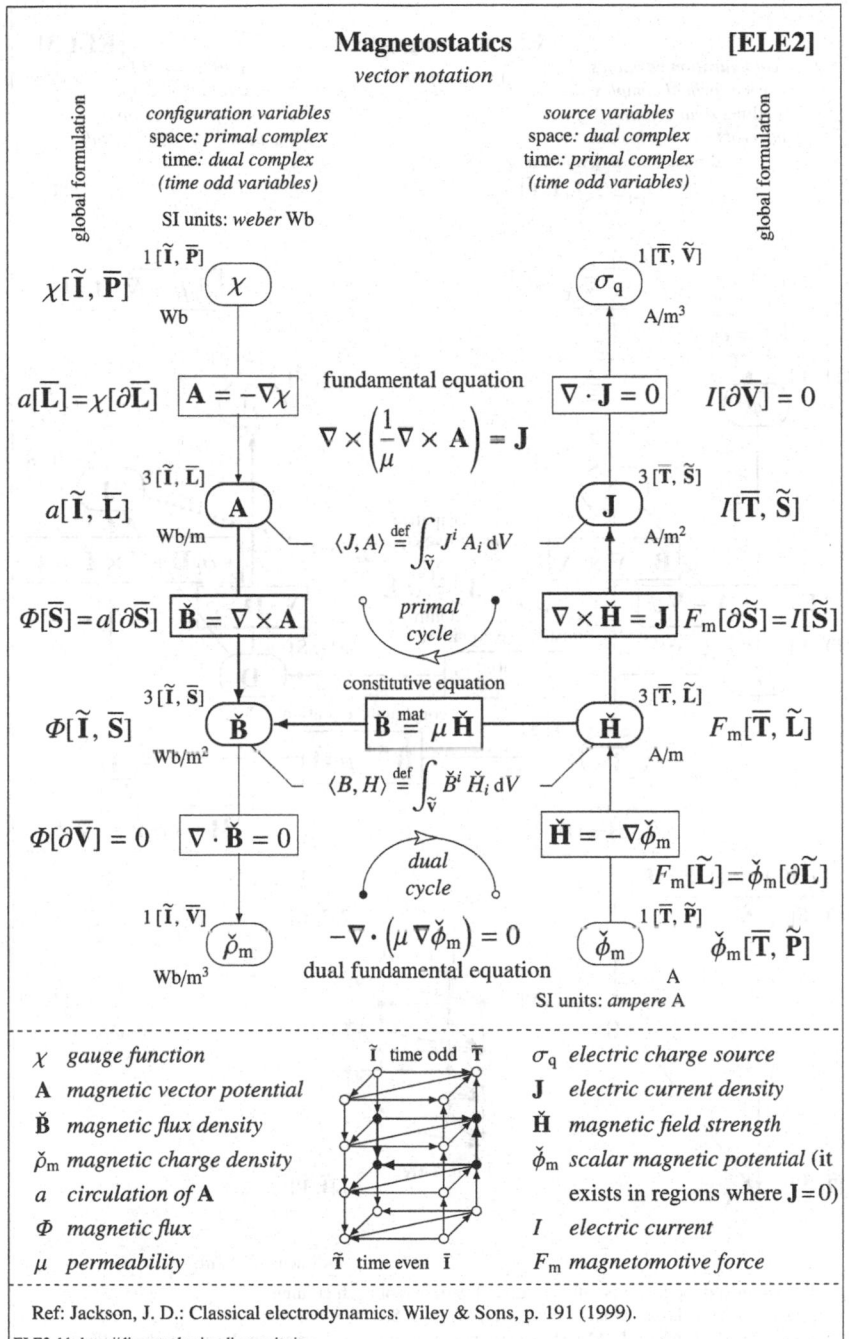

- $\chi$ *gauge function*
- $\mathbf{A}$ *magnetic vector potential*
- $\check{\mathbf{B}}$ *magnetic flux density*
- $\check{\rho}_m$ *magnetic charge density*
- $a$ *circulation of* $\mathbf{A}$
- $\Phi$ *magnetic flux*
- $\mu$ *permeability*

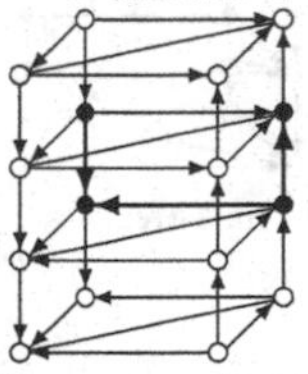


- $\sigma_q$ *electric charge source*
- $\mathbf{J}$ *electric current density*
- $\check{\mathbf{H}}$ *magnetic field strength*
- $\check{\phi}_m$ *scalar magnetic potential* (it exists in regions where $\mathbf{J} = 0$)
- $I$ *electric current*
- $F_m$ *magnetomotive force*

Ref: Jackson, J. D.: Classical electrodynamics. Wiley & Sons, p. 191 (1999).

ELE2-11; http://discretephysics.dicar.units.it

# Electromagnetism [ELE3]

### vector notation

*configuration variables*
space: *primal complex*
time: *dual complex*
intervals          instants
dimension: *action / charge*

*source variables*
space: *dual complex*
time: *primal complex*
instants          intervals

$1\,[\tilde{\mathbf{I}},\,\overline{\mathbf{P}}]$ — $\chi$

$\mathbf{A}=-\nabla\chi$

$\phi=\partial_t\chi$

$1\,[\tilde{\mathbf{T}},\,\overline{\mathbf{P}}]$ — $\phi$

SI units: Wb

SI units: V

$3\,[\tilde{\mathbf{I}},\,\overline{\mathbf{L}}]$ — $\mathbf{A}$

$\check{\mathbf{B}}=\nabla\times\mathbf{A}$

$\mathbf{E}=-\partial_t\mathbf{A}-\nabla\phi$

$3\,[\tilde{\mathbf{T}},\,\overline{\mathbf{L}}]$ — $\mathbf{E}$

dissipative constitutive equation

$$\mathbf{J}\overset{\text{mat}}{=}\sigma\,\mathbf{E}$$

Ohm constitutive equation

$$\mathbf{D}\overset{\text{mat}}{=}\epsilon\,\mathbf{E}$$

$3\,[\tilde{\mathbf{I}},\,\overline{\mathbf{S}}]$ — $\check{\mathbf{B}}$

constitutive equation

$$\check{\mathbf{B}}\overset{\text{mat}}{=}\mu\,\check{\mathbf{H}}$$

$\nabla\cdot\check{\mathbf{B}}=0$

$\partial_t\check{\mathbf{B}}+\nabla\times\mathbf{E}=0$

$3\,[\tilde{\mathbf{T}},\,\overline{\mathbf{S}}]$ — $0$

$1\,[\tilde{\mathbf{I}},\,\overline{\mathbf{V}}]$ — $0$

$1\,[\tilde{\mathbf{T}},\,\overline{\mathbf{V}}]$ —

$1\,[\overline{\mathbf{T}},\,\tilde{\mathbf{V}}]$ — $0$

$\partial_t\rho+\nabla\cdot\mathbf{J}=0$

$1\,[\overline{\mathbf{I}},\,\tilde{\mathbf{V}}]$ — $\rho$

$3\,[\overline{\mathbf{T}},\,\tilde{\mathbf{S}}]$ — $\mathbf{J}$

$-\partial_t\mathbf{D}+\nabla\times\check{\mathbf{H}}=\mathbf{J}$

$\nabla\cdot\mathbf{D}=\rho$

$3\,[\overline{\mathbf{I}},\,\tilde{\mathbf{S}}]$ — $\mathbf{D}$

$3\,[\overline{\mathbf{T}},\,\tilde{\mathbf{L}}]$ — $\check{\mathbf{H}}$

$\check{\mathbf{H}}=+\partial_t\check{\mathbf{F}}-\nabla\check{\phi}_m$

$\mathbf{D}=\nabla\times\check{\mathbf{F}}$

$3\,[\overline{\mathbf{I}},\,\tilde{\mathbf{L}}]$ — $\check{\mathbf{F}}$

SI units: A

SI units: C

$1\,[\overline{\mathbf{T}},\,\tilde{\mathbf{P}}]$ — $\check{\phi}_m$

$1\,[\overline{\mathbf{I}},\,\tilde{\mathbf{P}}]$ —

$\tilde{\mathbf{I}}$ time odd $\overline{\mathbf{T}}$

$\tilde{\mathbf{T}}$ time even $\overline{\mathbf{I}}$

dimension: *charge*

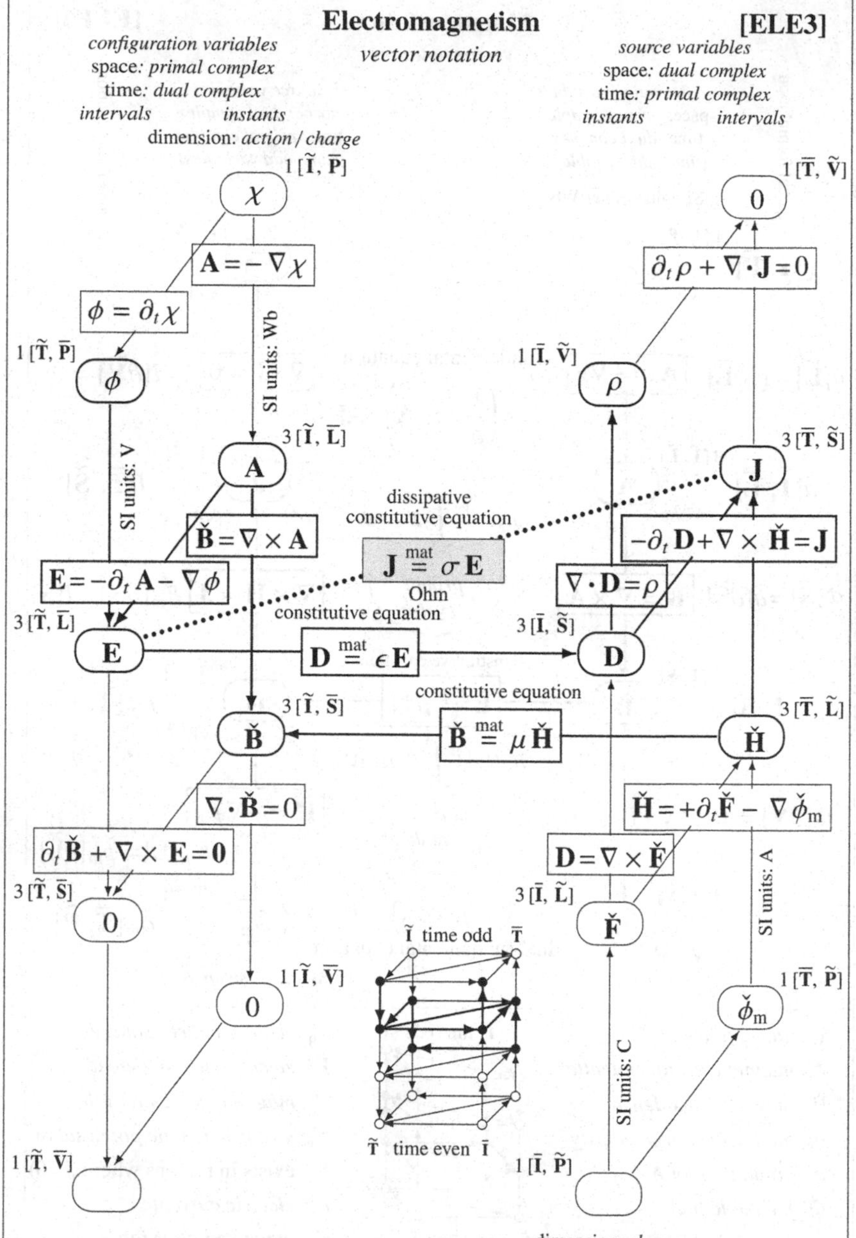

The dimension and the SI-units refers to global variables of each column.
Ref: Jackson, J. D.: Classical electrodynamics. Wiley & Sons (1999).
Ref: Landau, L.D., Lifshitz, E. M.: The classical theory of fields. Pergamon Press, London (1962).

ELE3-9; http://discretephysics.dicar.units.it

**Electrostatics**           **[ELE4]**

*configuration variables*      *tensor notation*        *source variables*
space: *primal complex*                         space: *dual complex*
time: *dual complex*                             time: *primal complex*
*(time even variables)*                          *(time even variables)*

SI units

$1 [\tilde{\mathbf{T}}, \overline{\mathbf{P}}]$    1

V    $\phi$          $\langle \rho, \phi \rangle \overset{\text{def}}{=} \int_{\tilde{\mathbf{v}}} \rho\,\phi\,dV$          $\rho$    C/m³

scalar                scalar

$$E_k = -\partial_k \phi$$    *primal cycle*    $$\frac{1}{\sqrt{g}} \partial_i (\sqrt{g}\, D^i) = \rho$$

constitutive equation

$3 [\tilde{\mathbf{T}}, \overline{\mathbf{L}}]$   1            1   $3 [\overline{\mathbf{I}}, \tilde{\mathbf{S}}]$

V/m    $E_k$    $$D^i \overset{\text{mat}}{=} \epsilon\, g^{ik} E_k$$    $D^i$    C/m²

covariant vector                     contravariant vector

$\langle D, E \rangle \overset{\text{def}}{=} \int_{\tilde{\mathbf{v}}} D^i E_i\,dV$

$$\tilde{\eta}^{ijk} \partial_j E_k = 0$$          $$D^i = \check{\eta}^{ijk} \partial_j \check{F}_k$$

$3 [\tilde{\mathbf{T}}, \overline{\mathbf{S}}]$   sgn($\Delta$)          sgn($\Delta$)   $3 [\overline{\mathbf{I}}, \tilde{\mathbf{L}}]$

V/m²    $\check{J}^i_{\mathrm{m}}$                     $\check{F}_k$    C/m

contravariant pseudovector                     covariant pseudovector

$1 [\tilde{\mathbf{T}}, \overline{\mathbf{V}}]$   sgn($\Delta$)          sgn($\Delta$)   $1 [\overline{\mathbf{I}}, \tilde{\mathbf{P}}]$

V/m³                                     C

pseudoscalar                             pseudoscalar

$\overline{\mathbf{I}}$ time odd   $\overline{\mathbf{T}}$

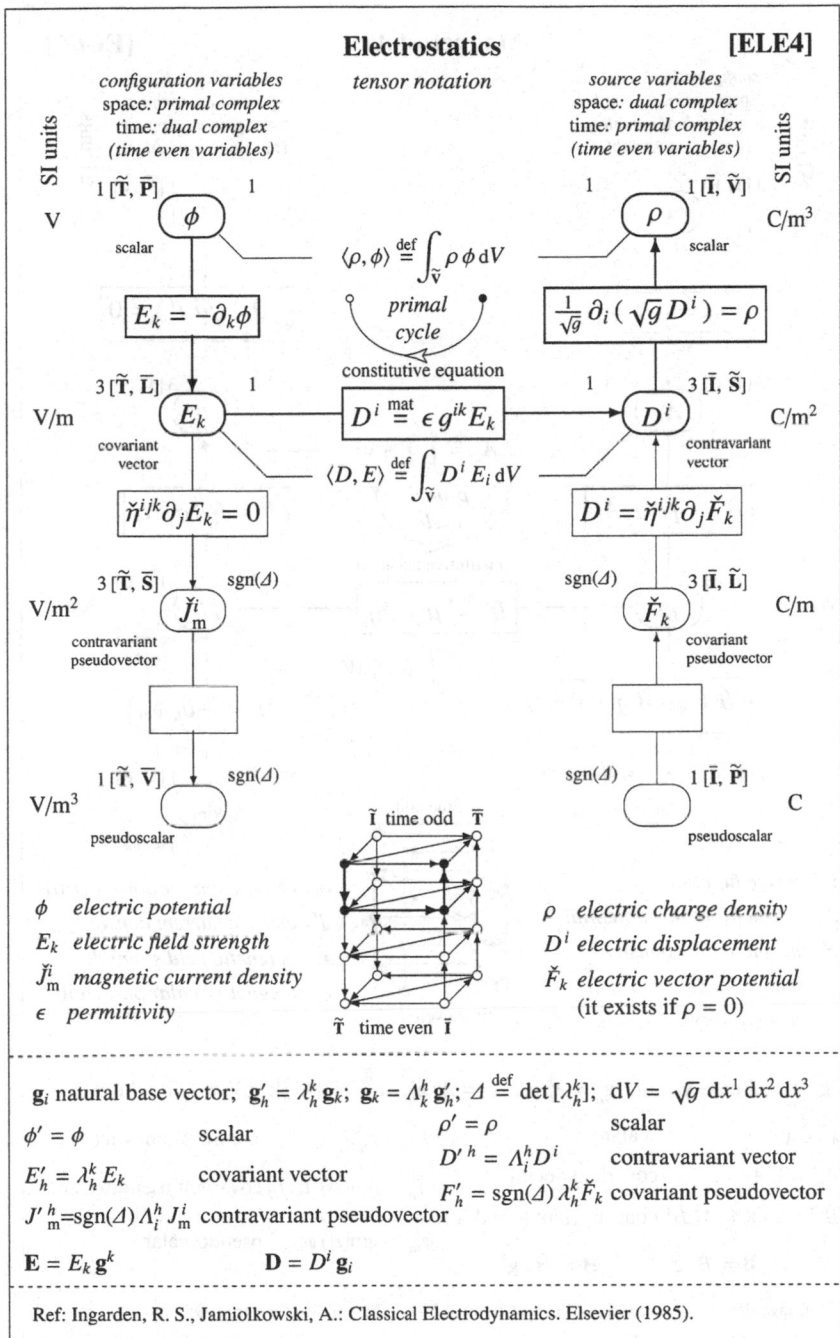

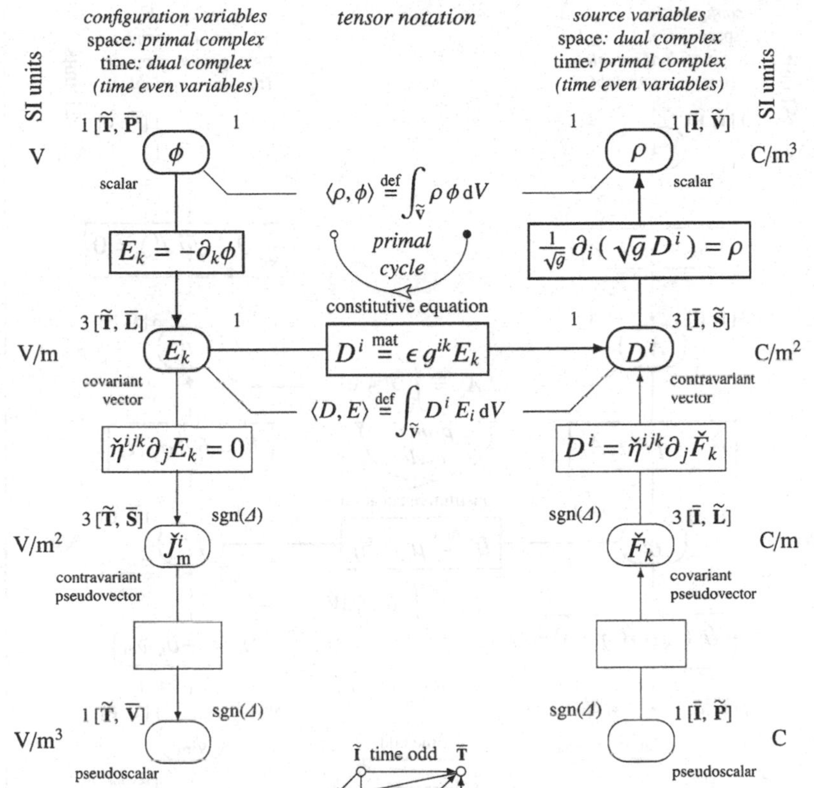

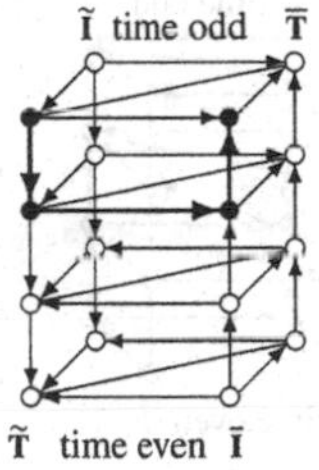

$\phi$   *electric potential*          $\rho$   *electric charge density*
$E_k$   *electric field strength*        $D^i$   *electric displacement*
$\check{J}^i_{\mathrm{m}}$   *magnetic current density*    $\check{F}_k$   *electric vector potential*
$\epsilon$   *permittivity*                           *(it exists if $\rho = 0$)*

$\tilde{\mathbf{T}}$ time even $\overline{\mathbf{I}}$

---

$\mathbf{g}_i$ natural base vector;   $\mathbf{g}'_h = \lambda^k_h \mathbf{g}_k$;   $\mathbf{g}_k = \Lambda^h_k \mathbf{g}'_h$;   $\Delta \overset{\text{def}}{=} \det [\lambda^k_h]$;   $dV = \sqrt{g}\, dx^1\, dx^2\, dx^3$

$\phi' = \phi$          scalar          $\rho' = \rho$          scalar

$E'_h = \lambda^k_h E_k$      covariant vector      $D'^h = \Lambda^h_i D^i$      contravariant vector

$J'^h_{\mathrm{m}} = \mathrm{sgn}(\Delta)\, \Lambda^h_i J^i_{\mathrm{m}}$ contravariant pseudovector    $F'_h = \mathrm{sgn}(\Delta)\, \lambda^k_h \check{F}_k$ covariant pseudovector

$\mathbf{E} = E_k\, \mathbf{g}^k$          $\mathbf{D} = D^i\, \mathbf{g}_i$

---

Ref: Ingarden, R. S., Jamiolkowski, A.: Classical Electrodynamics. Elsevier (1985).

ELE4-16; http://discretephysics.dicar.units.it

**Magnetostatics**      **[ELE5]**

*configuration variables*
space*: primal complex*
time*: dual complex*
*(time even variables)*

*tensor notation*

*source variables*
space*: dual complex*
time*: primal complex*
*(time even variables)*

SI units

Wb $\quad$ 1 $[\tilde{\mathbf{I}}, \overline{\mathbf{P}}]$ $\qquad$ 1

$\chi$

scalar

$$A_k = -\partial_k \chi$$

3 $[\tilde{\mathbf{I}}, \overline{\mathbf{L}}]$ $\qquad$ 1

Wb/m $\quad$ $A_k$

*covariant*
*vector*

$$\breve{B}^i = \breve{\eta}^{ijk} \partial_j A_k$$

3 $[\tilde{\mathbf{I}}, \overline{\mathbf{S}}]$ $\qquad$ sgn($\Delta$)

Wb/m² $\quad$ $\breve{B}^i$

*contravariant*
*pseudovector*

$$\frac{1}{\sqrt{g}} \partial_i (\sqrt{g}\, \breve{B}^i) = 0$$

1 $[\tilde{\mathbf{I}}, \overline{\mathbf{V}}]$ $\qquad$ sgn($\Delta$)

Wb/m³ $\quad$ $\breve{\rho}_m$

*pseudoscalar*

$\langle J, A \rangle \overset{\text{def}}{=} \int_{\tilde{\mathbf{V}}} J^i A_i \, dV$

*primal*
*cycle*

constitutive equation

$$\breve{B}^i \overset{\text{mat}}{=} \mu\, g^{ik} \breve{H}_k$$

$\langle B, H \rangle \overset{\text{def}}{=} \int_{\tilde{\mathbf{V}}} \breve{B}^i \breve{H}_i \, dV$

SI units

1 $[\overline{\mathbf{T}}, \tilde{\mathbf{V}}]$ $\qquad$ A/m³

$\sigma_q$

scalar

$$\frac{1}{\sqrt{g}} \partial_i (\sqrt{g}\, J^i) = 0$$

1 $\qquad$ 3 $[\overline{\mathbf{T}}, \tilde{\mathbf{S}}]$

$J^i$ $\quad$ A/m²

*contravariant*
*vector*

$$J^i = \breve{\eta}^{ijk} \partial_j \breve{H}_k$$

sgn($\Delta$) $\qquad$ 3 $[\overline{\mathbf{T}}, \tilde{\mathbf{L}}]$

$\breve{H}_k$ $\quad$ A/m

*covariant*
*pseudovector*

$$\breve{H}_k = -\partial_k \breve{\phi}_m$$

sgn($\Delta$) $\qquad$ 1 $[\overline{\mathbf{T}}, \tilde{\mathbf{P}}]$

$\breve{\phi}_m$ $\quad$ A

*pseudoscalar*

$\tilde{\mathbf{I}}$ time odd $\overline{\mathbf{T}}$

$\tilde{\mathbf{T}}$ time even $\overline{\mathbf{I}}$

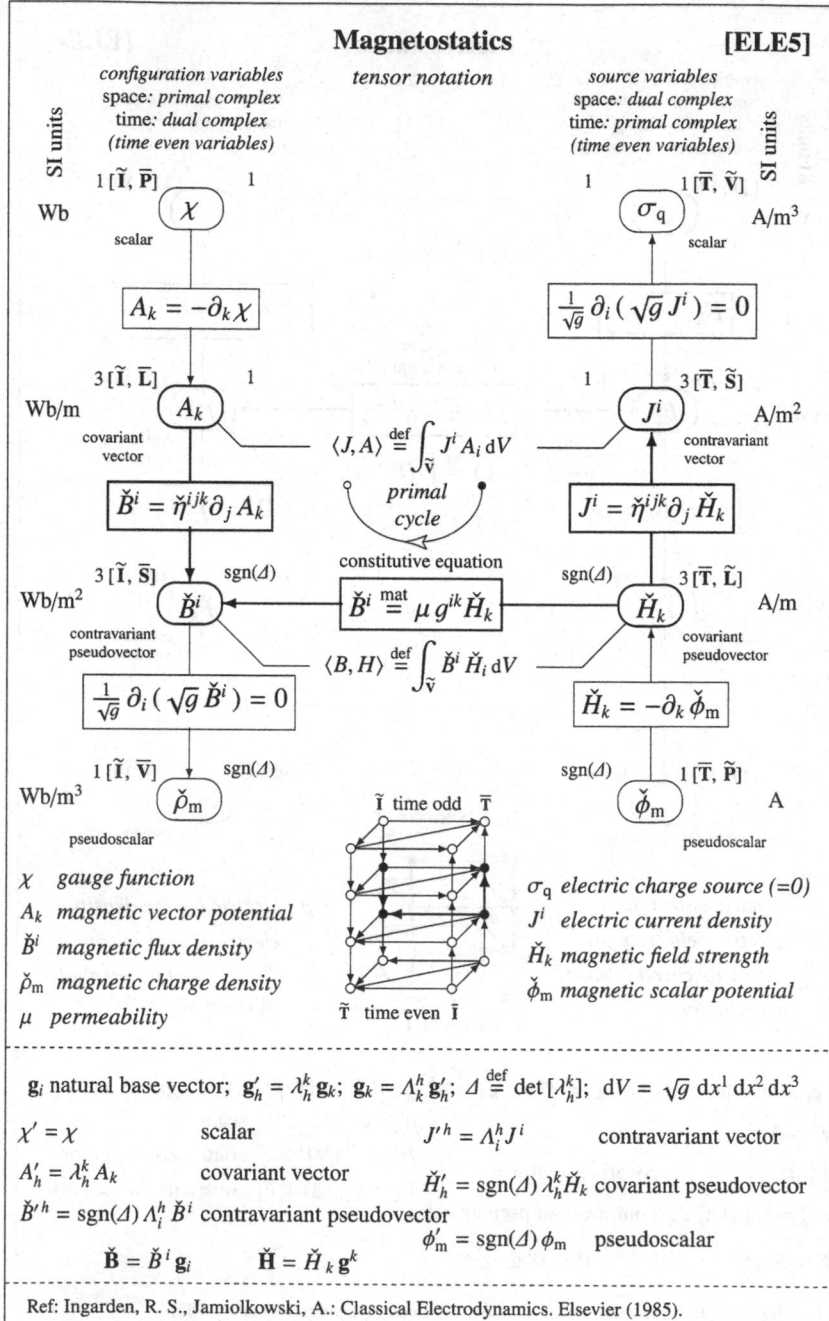

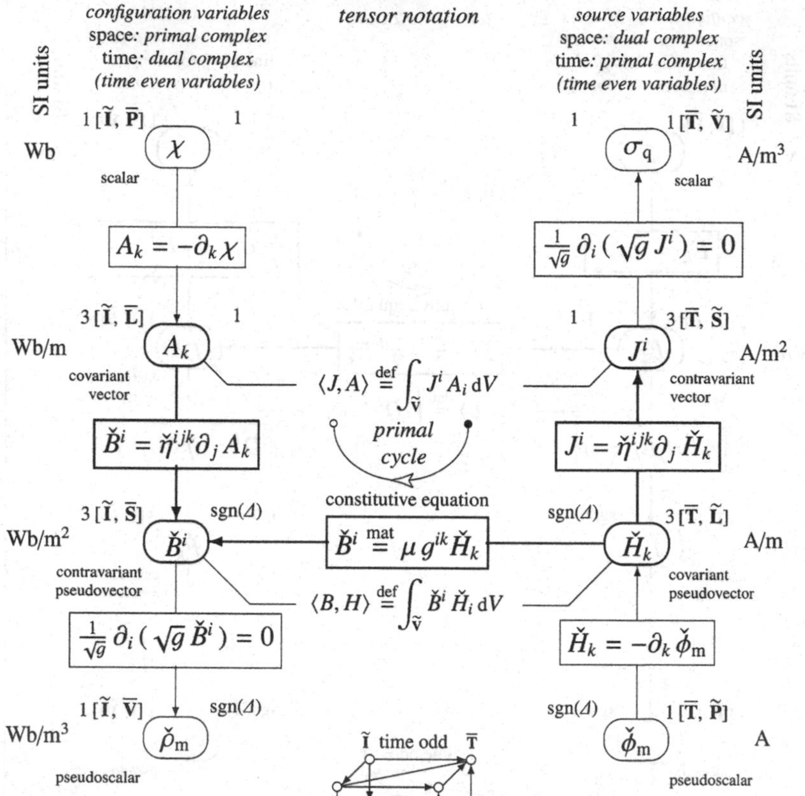

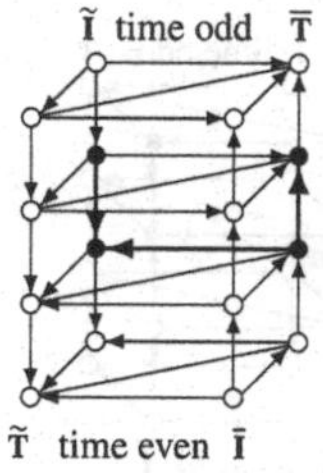

$\chi$   *gauge function*
$A_k$   *magnetic vector potential*
$\breve{B}^i$   *magnetic flux density*
$\breve{\rho}_m$   *magnetic charge density*
$\mu$   *permeability*

$\sigma_q$ *electric charge source (=0)*
$J^i$   *electric current density*
$\breve{H}_k$   *magnetic field strength*
$\breve{\phi}_m$ *magnetic scalar potential*

---

$\mathbf{g}_i$ natural base vector; $\mathbf{g}'_h = \lambda_h^k \mathbf{g}_k$; $\mathbf{g}_k = \Lambda_k^h \mathbf{g}'_h$; $\Delta \overset{\text{def}}{=} \det [\lambda_h^k]$; $dV = \sqrt{g}\, dx^1\, dx^2\, dx^3$

$\chi' = \chi$ $\qquad\qquad$ scalar $\qquad\qquad$ $J'^h = \Lambda_i^h J^i$ $\qquad\qquad$ contravariant vector

$A'_h = \lambda_h^k A_k$ $\qquad$ covariant vector $\qquad$ $\breve{H}'_h = \text{sgn}(\Delta)\, \lambda_h^k \breve{H}_k$ covariant pseudovector

$\breve{B}'^h = \text{sgn}(\Delta)\, \Lambda_i^h \breve{B}^i$ contravariant pseudovector

$\breve{\mathbf{B}} = \breve{B}^i \mathbf{g}_i$ $\qquad$ $\breve{\mathbf{H}} = \breve{H}_k \mathbf{g}^k$ $\qquad$ $\phi'_m = \text{sgn}(\Delta)\, \phi_m$ $\quad$ pseudoscalar

---

Ref: Ingarden, R. S., Jamiolkowski, A.: Classical Electrodynamics. Elsevier (1985).

**Electromagnetism** [ELE6]
*space-time formulation*

*configuration variables*
space: *primal complex*
time: *dual complex*

*source variables*
space: *dual complex*
time: *primal complex*

$1\,[\overline{\mathbf{P}}_4]$

$\chi$

1

$dH \overset{\text{def}}{=} dV\,dt$

$1\,[\widetilde{\mathbf{H}}_4]$

$0$

1

$A_\alpha = \partial_\alpha \chi$

$\langle J, A \rangle \overset{\text{def}}{=} \int_H J^\alpha A_\alpha \, dH$

$\partial_\rho J^\rho = 0$

$4\,[\overline{\mathbf{L}}_4]$

$A_\alpha$

1    four
potential

$\langle G, F \rangle \overset{\text{def}}{=} \int_H \frac{1}{2} G^{\alpha\beta} F_{\alpha\beta} \, dH$

$4\,[\widetilde{\mathbf{V}}_4]$

$J^\rho$

four     1
current

$F_{\alpha\beta} = \partial_\alpha A_\beta - \partial_\beta A_\alpha$

*primal cycle*

$\frac{1}{2}\breve{\eta}^{\rho\gamma\mu\nu} \partial_\gamma \breve{G}_{\mu\nu} = J^\rho$

constitutive equation

$6\,[\overline{\mathbf{S}}_4]$

$F_{\alpha\beta}$

1    field
strength

$\breve{G}_{\mu\nu} \overset{\text{mat}}{=} \frac{1}{2} \breve{\chi}_{\mu\nu}{}^{\alpha\beta} F_{\alpha\beta}$

$\breve{G}_{\mu\nu}$

excitation    $\mathrm{sgn}(\Delta)$
tensor

$6\,[\widetilde{\mathbf{S}}_4]$

$\frac{1}{2}\breve{\eta}^{\gamma\alpha\beta\mu} \partial_\gamma F_{\alpha\beta} = 0$

*dual cycle*

$\breve{G}_{\mu\nu} = \partial_\mu \breve{F}_\nu - \partial_\nu \breve{F}_\mu$

$4\,[\overline{\mathbf{V}}_4]$

$0$

$\mathrm{sgn}(\Delta)$

$\breve{F}_\mu$

$\mathrm{sgn}(\Delta)$

$4\,[\widetilde{\mathbf{L}}_4]$

$1\,[\overline{\mathbf{H}}_4]$

$\mathrm{sgn}(\Delta)$

$1\,[\widetilde{\mathbf{P}}_4]$

$\mathrm{sgn}(\Delta)$

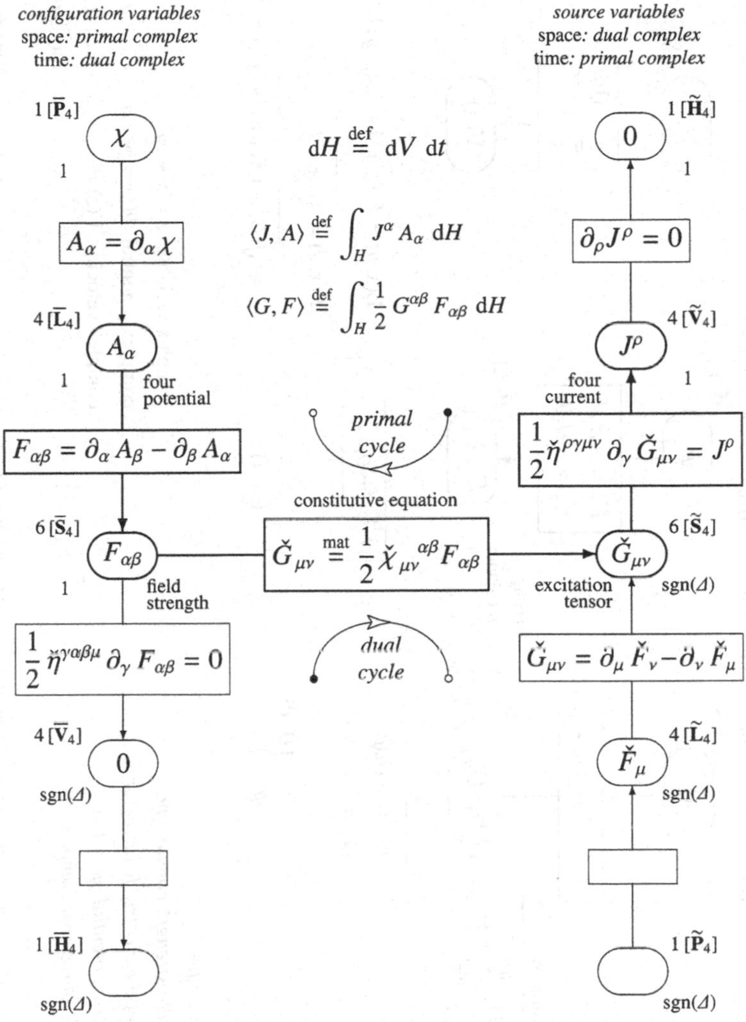

Ref: Post, E.J.: Formal Structure of Electromagnetics; General Covariance and Electromagnetics, Dover (1997)

$\epsilon^{\rho\gamma\mu\nu}$ is the permutation symbol ($\equiv$ Levi–Civita symbol) that is a tensor density
$A_\alpha$ associated with **lines** $\longrightarrow$ spin 1 particle **one** dimensional space-time element.
$\breve{F}_\sigma$ exists in the regions in which $J^\mu = 0$.

ELE6-22; http://discretephysics.dicar.units.it

## Plane electrostatics using complex variables

**[ELE7]**

*use of complex variables for plane electrostatics (only if $\rho = 0$)*

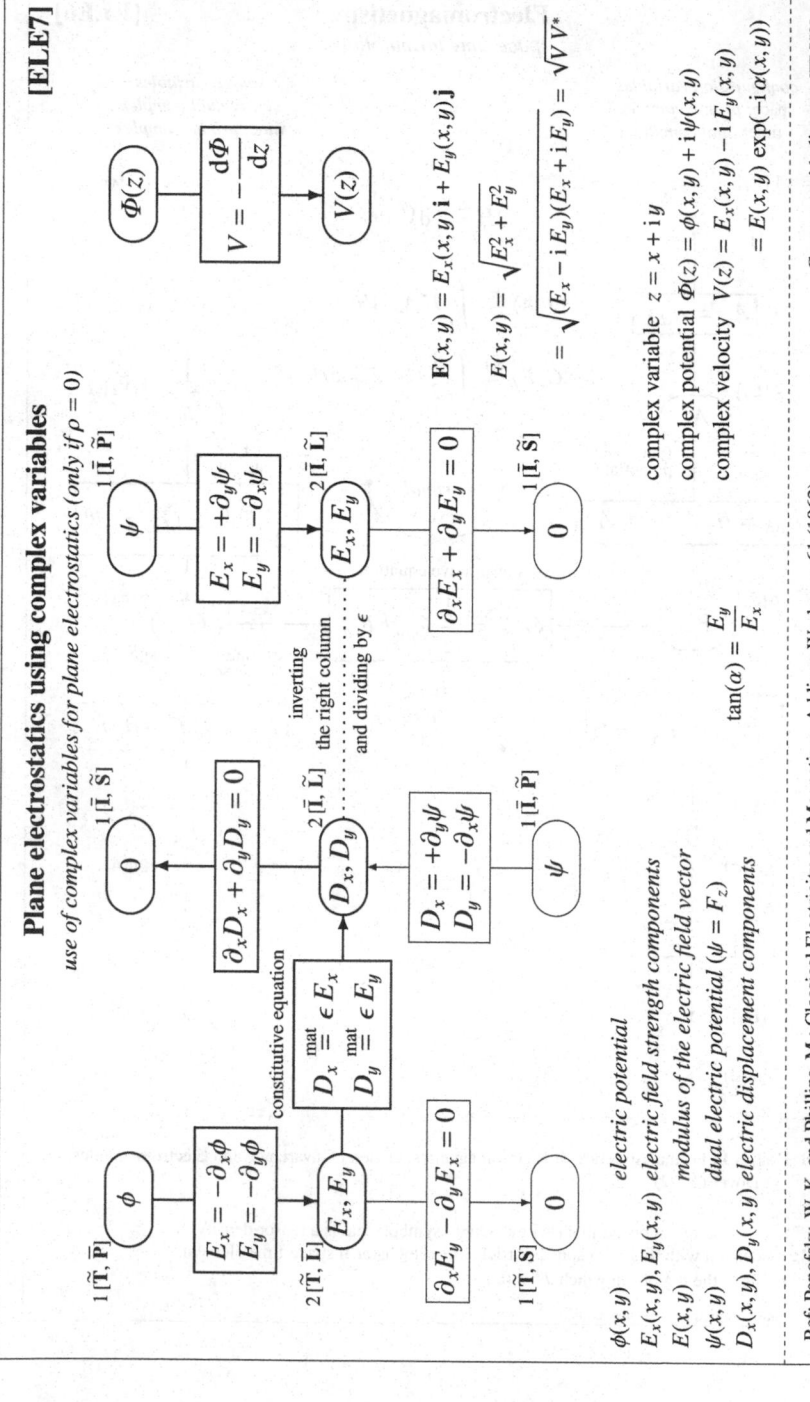

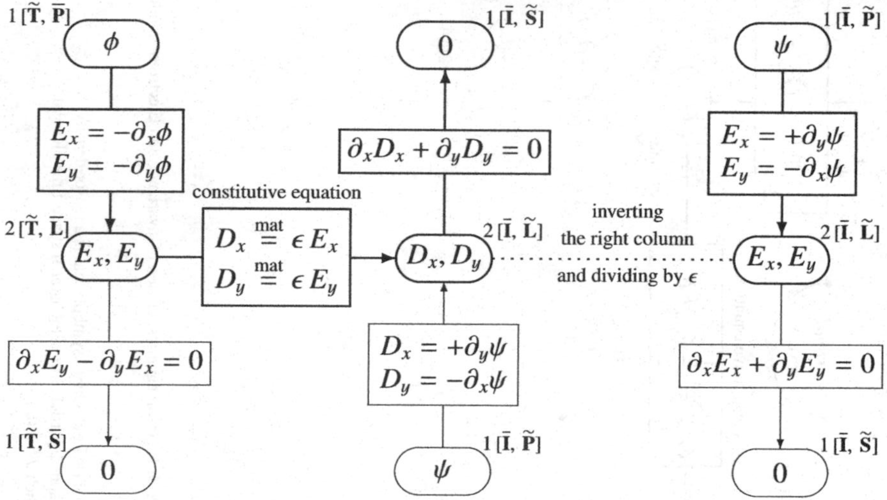


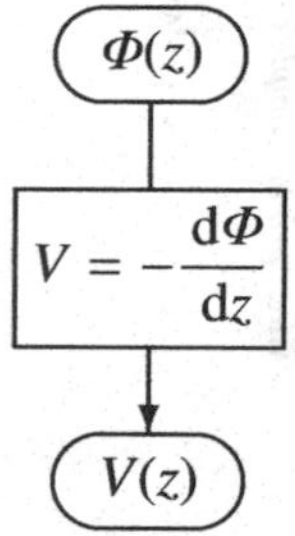


$$\mathbf{E}(x,y) = E_x(x,y)\,\mathbf{i} + E_y(x,y)\,\mathbf{j}$$

$$E(x,y) = \sqrt{E_x^2 + E_y^2} = \sqrt{(E_x - \mathrm{i}\,E_y)(E_x + \mathrm{i}\,E_y)} = \sqrt{V\,V^*}$$

$\phi(x,y)$ *electric potential*

$E_x(x,y), E_y(x,y)$ *electric field strength components*

$E(x,y)$ *modulus of the electric field vector*

$\psi(x,y)$ *dual electric potential* ($\psi = F_z$)

$D_x(x,y), D_y(x,y)$ *electric displacement components*

$$\tan(\alpha) = \frac{E_y}{E_x}$$

complex variable $z = x + \mathrm{i}\,y$

complex potential $\Phi(z) = \phi(x,y) + \mathrm{i}\,\psi(x,y)$

complex velocity $V(z) = E_x(x,y) - \mathrm{i}\,E_y(x,y)$

$\qquad = E(x,y)\,\exp(-\mathrm{i}\,\alpha(x,y))$

Ref: Panofsky, W. K. and Phillips, M.: Classical Electricity and Magnetism. Addison Wesley, p. 61 (1962).

Compare with diagram [FLU2]

ELE7-8; http://discretephysics.dicar.units.it

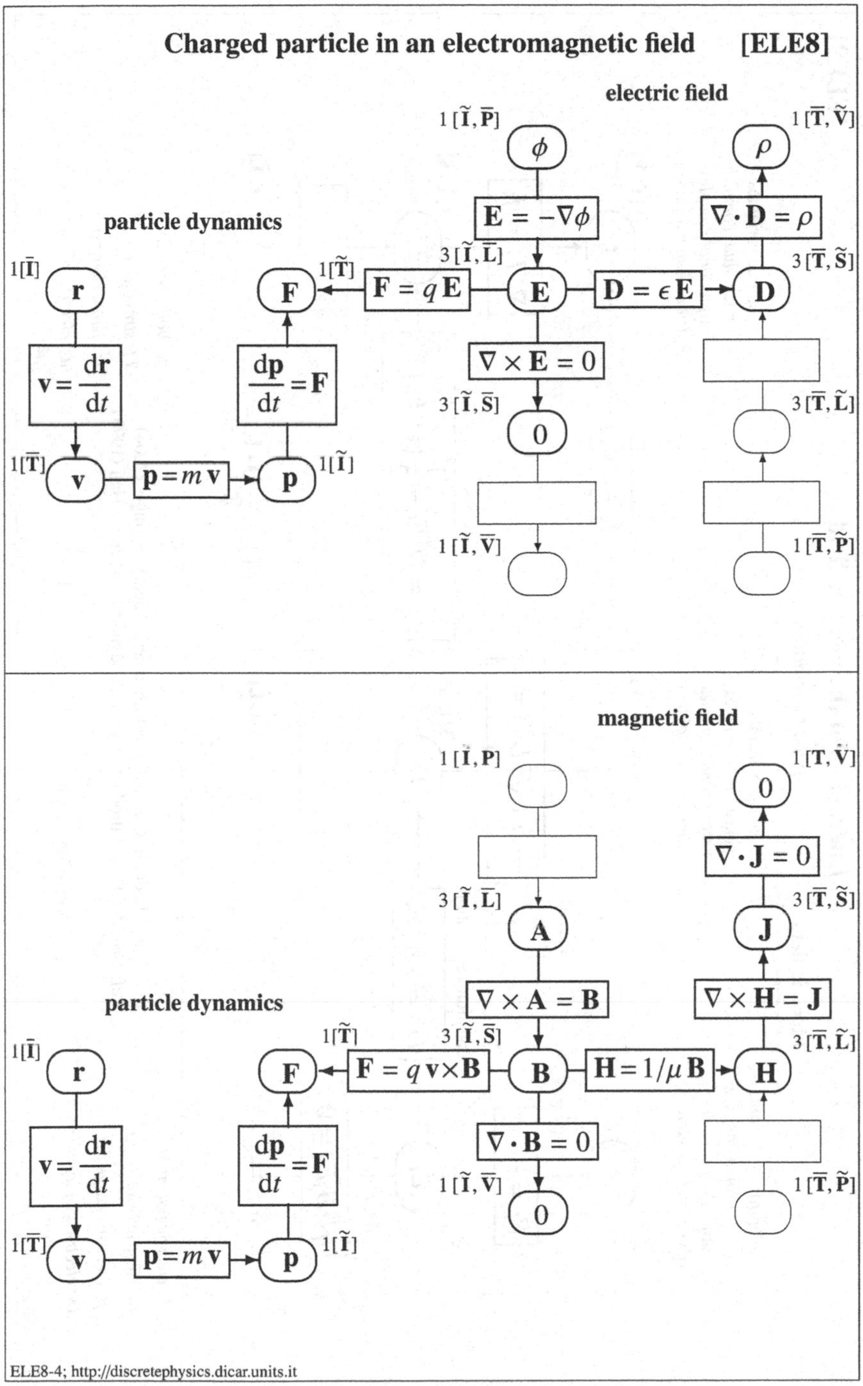

Charged particle in an electromagnetic field [ELE8]
electric field
particle dynamics
r
v = dr/dt
v
p = m v
p
dp/dt = F
F
F = q E
E
E = −∇ϕ
ϕ
∇ × E = 0
0
D = ϵ E
D
∇ · D = ρ
ρ
magnetic field
particle dynamics
r
v = dr/dt
v
p = m v
p
dp/dt = F
F
F = q v×B
B
∇ × A = B
A
∇ · B = 0
0
H = 1/μ B
H
∇ × H = J
J
∇ · J = 0
0
ELE8-4; http://discretephysics.dicar.units.it

## Energetics of the electric field [ELE9]

**Electric field** *space formulation*

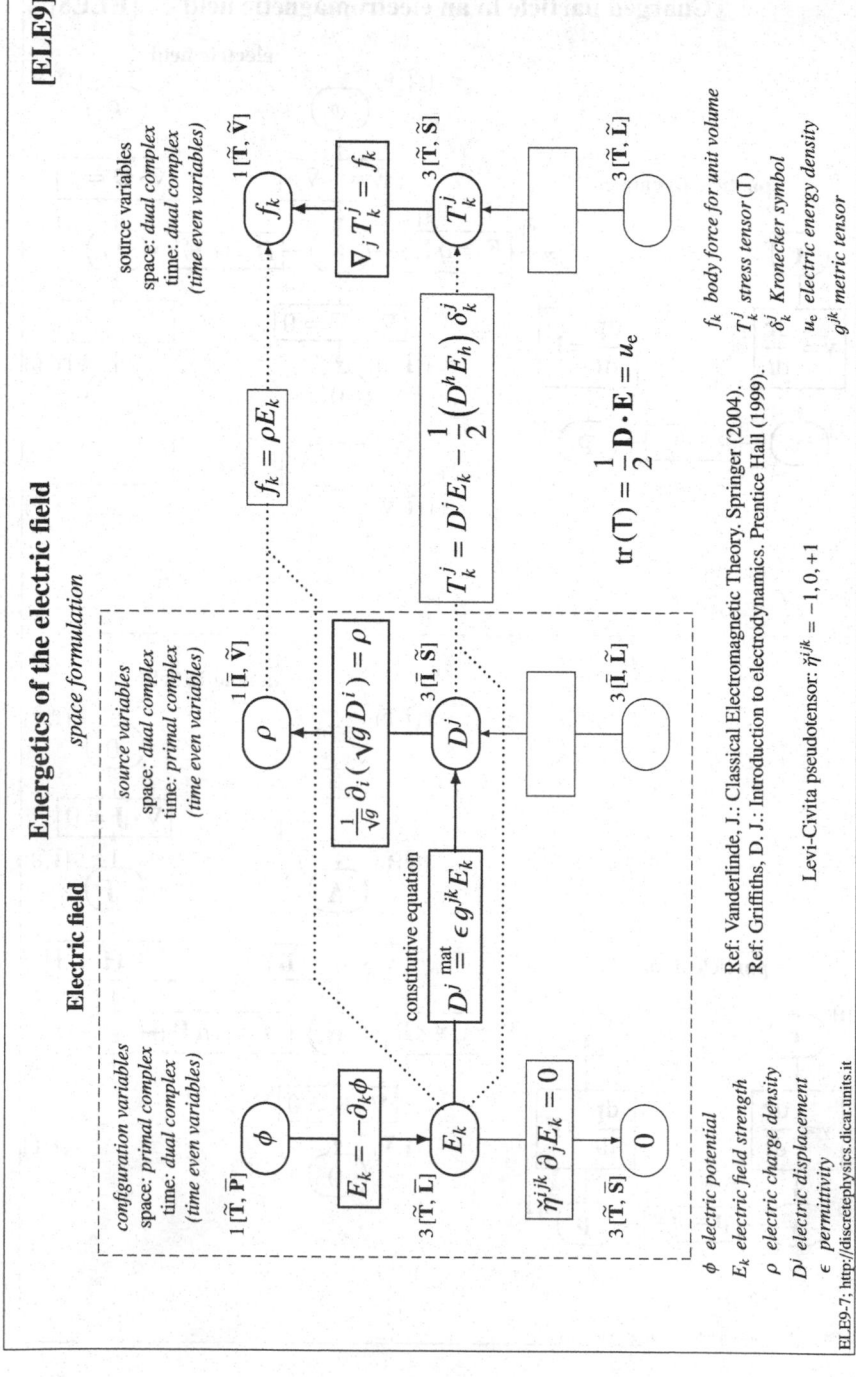

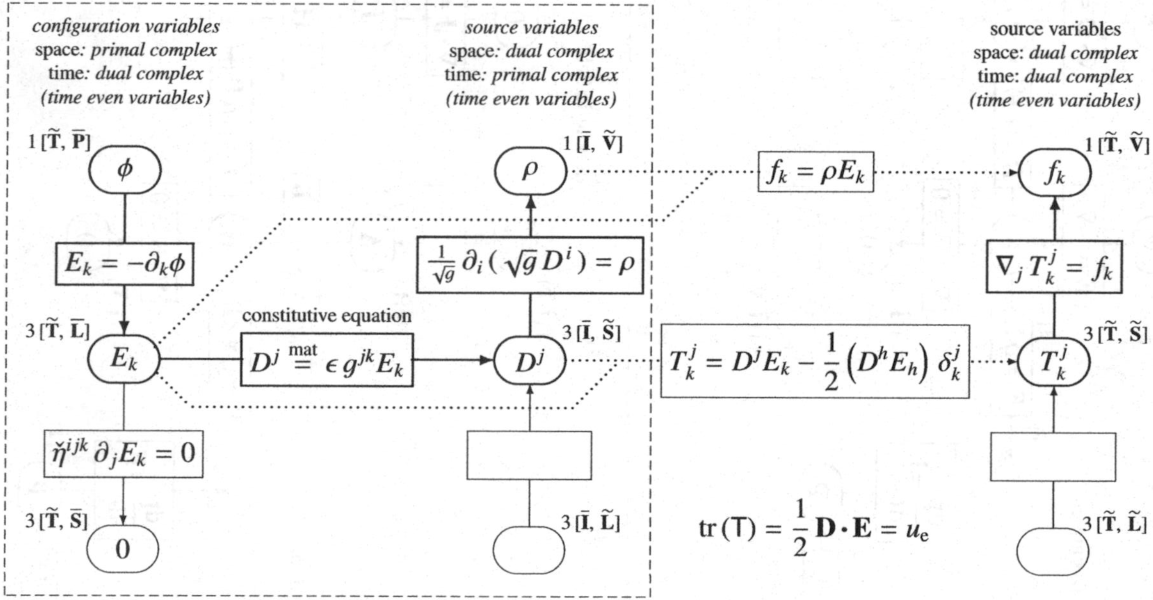


$$\mathrm{tr}\,(\mathsf{T}) = \frac{1}{2}\,\mathbf{D}\cdot\mathbf{E} = u_{\mathrm{e}}$$

$\phi$ *electric potential*
$E_k$ *electric field strength*
$\rho$ *electric charge density*
$D^j$ *electric displacement*
$\epsilon$ *permittivity*

Ref: Vanderlinde, J.: Classical Electromagnetic Theory. Springer (2004).
Ref: Griffiths, D. J.: Introduction to electrodynamics. Prentice Hall (1999).

Levi-Civita pseudotensor: $\check{\eta}^{ijk} = -1, 0, +1$

$f_k$ *body force for unit volume*
$T^j_k$ *stress tensor* (T)
$\delta^j_k$ *Kronecker symbol*
$u_{\mathrm{e}}$ *electric energy density*
$g^{jk}$ *metric tensor*

ELE9-7; http://discretephysics.dicar.units.it

## Network theory                        [ELE10]

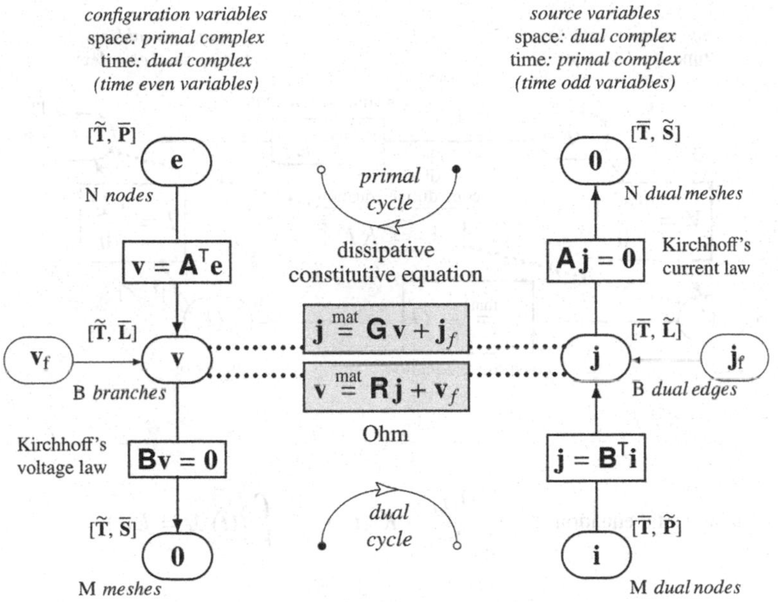

*configuration variables*
space: *primal complex*
time: *dual complex*
*(time even variables)*

*source variables*
space: *dual complex*
time: *primal complex*
*(time odd variables)*

| | | |
|---|---|---|
| N number of nodes | **j** branch current vector | $(j_1, j_2, ..., j_B)$ |
| B number of branches | **i** independent current vector | $(i_1, i_2, ..., i_M)$ |
| M number of meshes | **A** reduced incidence matrix | |

Euler's rule (1752): $N - B + M = 1$ from which $B = (N - 1) + M$

**e** node voltage vector $(e_1, e_2, ..., e_N)$

**v** branch voltages vector $(v_1, v_2, ..., v_B)$

**B** independent mesh matrix

$\mathbf{j}_f$ current source vector

$\mathbf{v}_f$ voltage source vector

$\tilde{\mathbf{I}}$ time odd  $\overline{\mathbf{T}}$

$\tilde{\mathbf{T}}$  time even  $\overline{\mathbf{I}}$

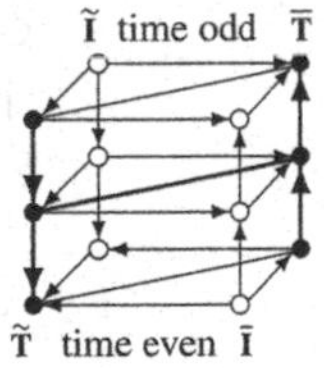

Ref: Reed M.B.,Foundation for Electric Network Theory, Prentice Hall, (1961)

Ref: Balabanian N., Bickart T.A.: Electrical Network Theory, Wiley & Sons, (1969), p. 72

Ref: Desoer, C. A. and Kuh, E. S.: Basic Circuits Theory. McGraw Hill (1969).

ELE10-11; http://discretephysics.dicar.units.it

## RLC circuit **[ELE11]**

configuration variables
time: dual complex
source variables
time: primal complex
$[\tilde{\mathbf{I}}, \mathscr{S}]$
$\Phi$
constitutive equation
$\Phi \overset{\text{mat}}{=} L\,I$
$I$
$[\bar{\mathbf{T}}, \mathscr{S}]$
dissipative
constitutive equation
$V = \dfrac{\mathrm{d}\Phi}{\mathrm{d}t}$
$V \overset{\text{mat}}{=} R\,I$
$I = \dfrac{\mathrm{d}Q}{\mathrm{d}t}$
$[\tilde{\mathbf{T}}, \mathscr{S}]$
$V$
$V \overset{\text{mat}}{=} \dfrac{1}{C}\,Q$
$Q$
$[\bar{\mathbf{I}}, \mathscr{S}]$
$E$
constitutive equation

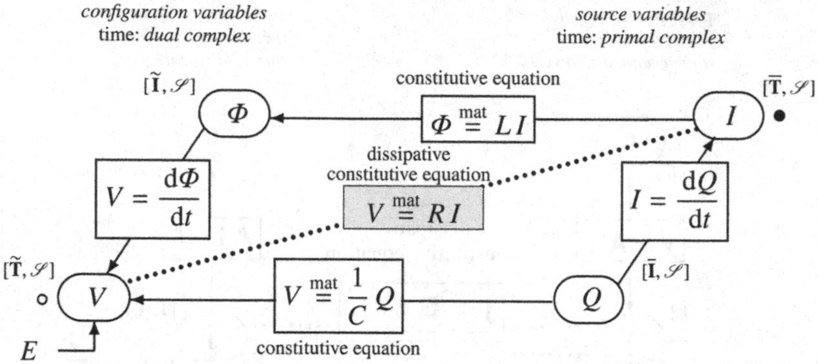

fundamental equation

$$L\frac{\mathrm{d}I(t)}{\mathrm{d}t} + RI(t) + \frac{1}{C}\int I(t)\,\mathrm{d}t = E(t)$$

dual complex
$\tilde{\mathbf{I}}$
$\tilde{\mathbf{T}}$
$\tilde{\mathbf{I}}$
$t$
$\bar{\mathbf{I}}$
$\bar{\mathbf{T}}$
$\bar{\mathbf{I}}$
primal complex

R
C
L
E

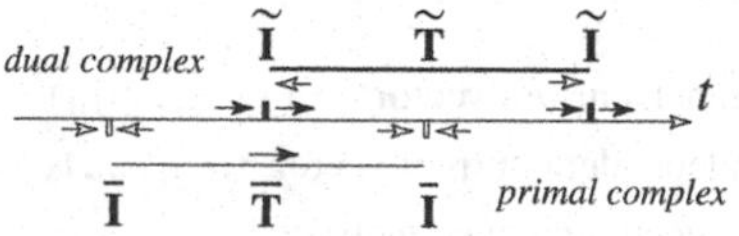

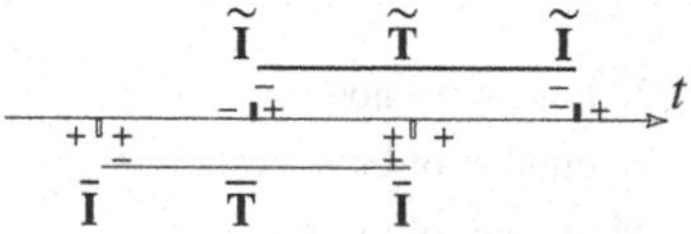

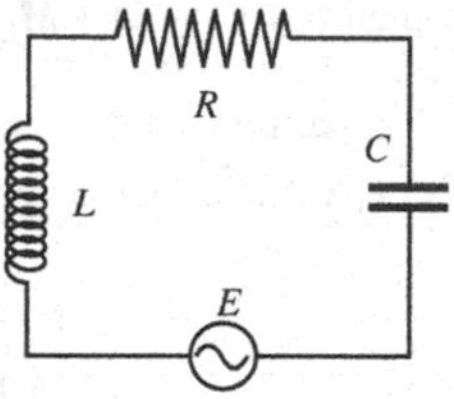

*E impressed voltage*
*R resistance*
*L inductance*
*C capacitance*

*Φ magnetic flux*
*V electromotive force*

$\tilde{\mathbf{I}}$ time odd $\bar{\mathbf{T}}$
$\tilde{\mathbf{T}}$ time even $\bar{\mathbf{I}}$

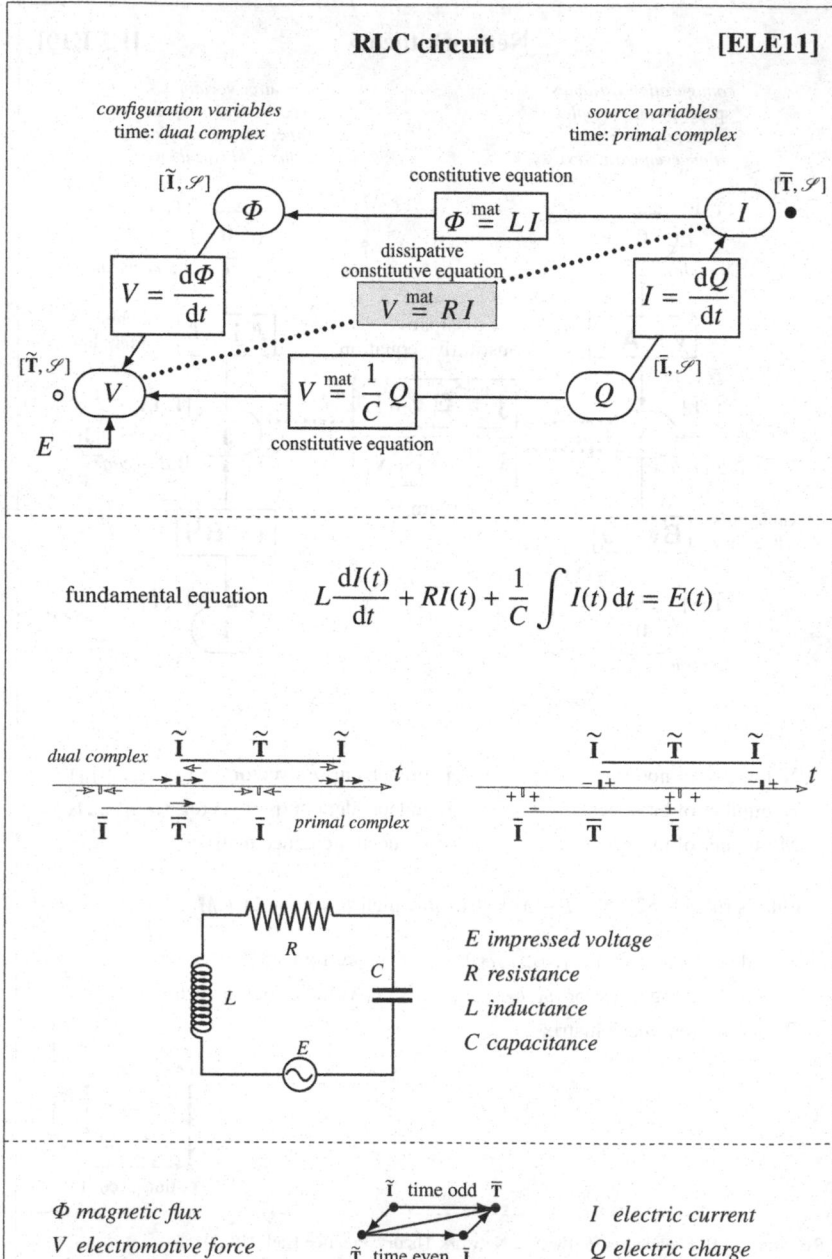

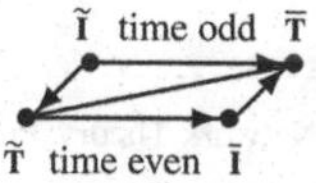

*I electric current*
*Q electric charge*

## Coulomb's law derived from the differential formulation

**[ELE12]**

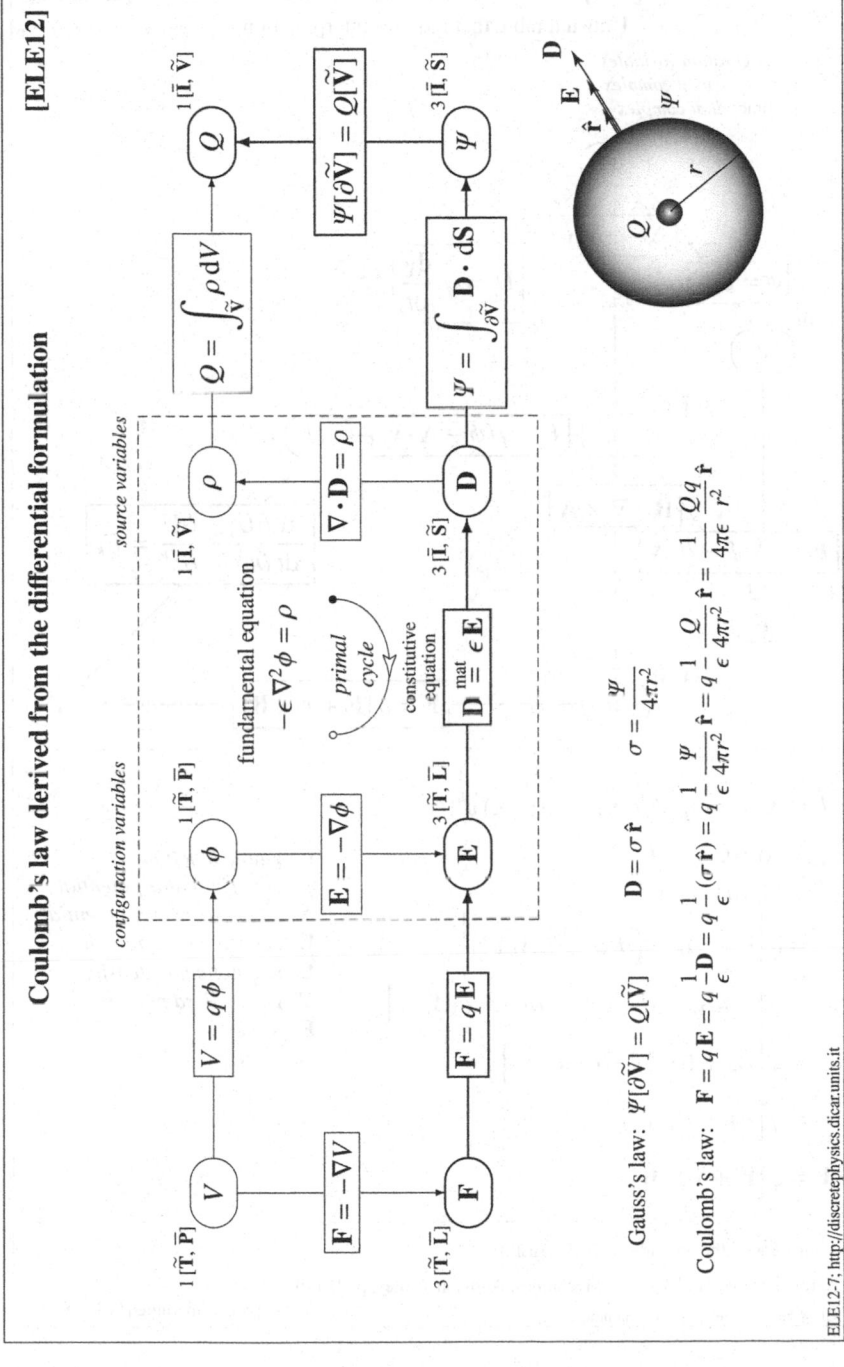

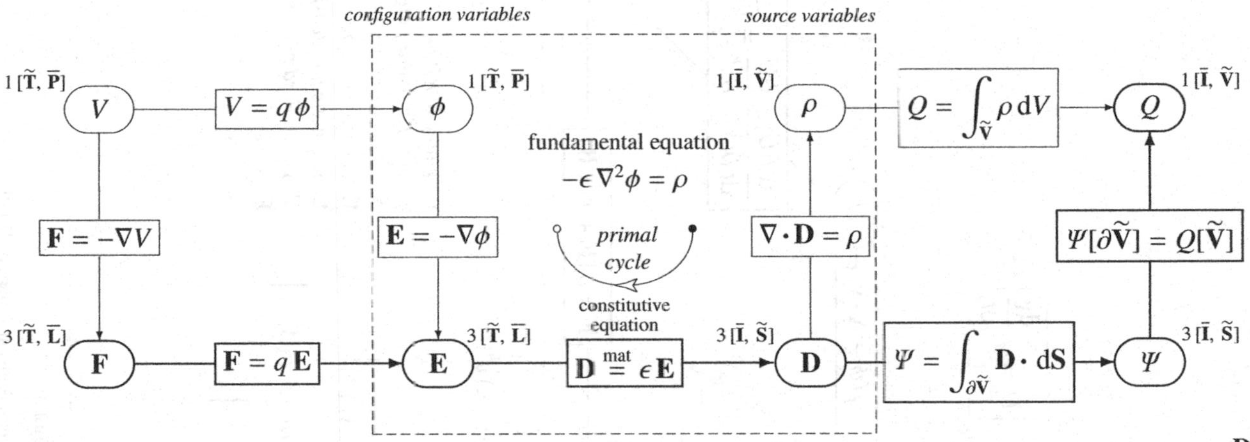


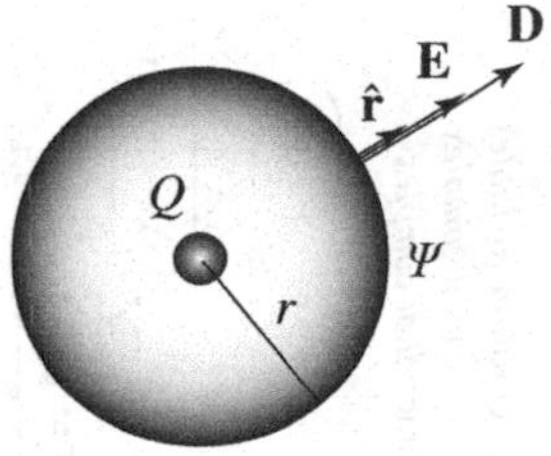


Gauss's law: $\Psi[\partial\tilde{\mathbf{V}}] = Q[\tilde{\mathbf{V}}]$ $\qquad \mathbf{D} = \sigma\,\hat{\mathbf{r}} \qquad \sigma = \dfrac{\Psi}{4\pi r^2}$

Coulomb's law: $\mathbf{F} = q\,\mathbf{E} = q\,\dfrac{1}{\epsilon}\mathbf{D} = q\,\dfrac{1}{\epsilon}(\sigma\,\hat{\mathbf{r}}) = q\,\dfrac{1}{\epsilon}\,\dfrac{\Psi}{4\pi r^2}\,\hat{\mathbf{r}} = q\,\dfrac{1}{\epsilon}\,\dfrac{Q}{4\pi r^2}\,\hat{\mathbf{r}} = \dfrac{1}{4\pi\epsilon}\,\dfrac{Q\,q}{r^2}\,\hat{\mathbf{r}}$

ELE12-7; http://discretephysics.dicar.units.it

## Charged particle in the electromagnetic field [ELE13]

Unusual table that requires interpretation

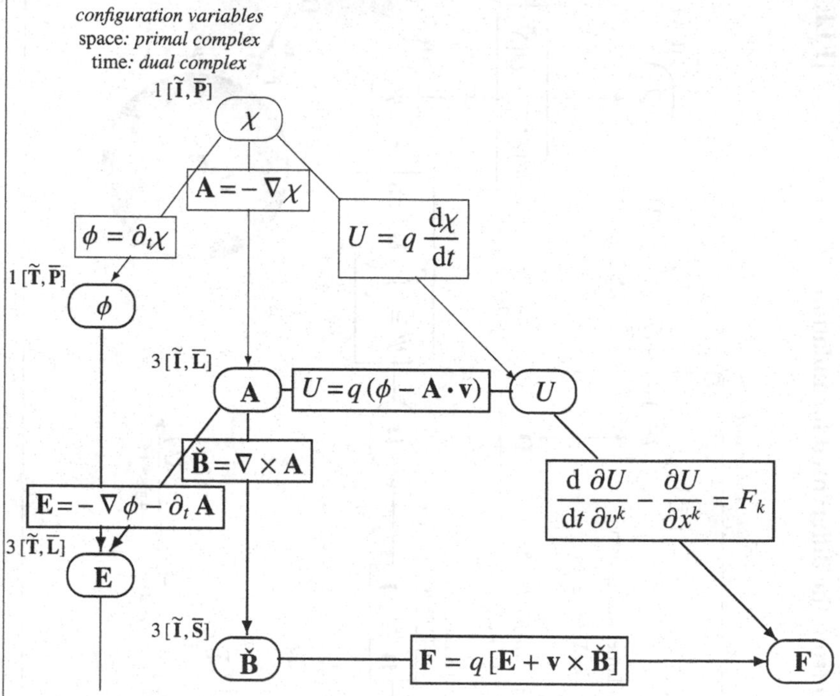


$$U(t,\mathbf{x},\mathbf{v}) = q\left[\phi(t,\mathbf{x}) - A_h(t,\mathbf{x})\, v^h(t)\right]$$

$$\begin{aligned}
F_k &= \frac{\mathrm{d}}{\mathrm{d}t}\frac{\partial U}{\partial v^k} - \frac{\partial U}{\partial x^k} \\
&= q\left[-\frac{\mathrm{d}}{\mathrm{d}t}A_k - \left(\partial_k\phi - \partial_k A_h\, v^h\right)\right] \\
&= q\left[-\partial_t A_k - \partial_h A_k\, v^h - \partial_k\phi + \partial_k A_h\, v^h\right] \\
&= q\left[E_k + (\partial_k A_h - \partial_h A_k)\, v^h\right] \\
&= q\left[E_k + B_{kh}\, v^h\right]
\end{aligned}$$

$$\mathbf{F} = q\left[\mathbf{E} + \mathbf{v}\times\check{\mathbf{B}}\right]$$

$\chi$ *gauge function*
$\phi$ *electric scalar potential*
$\mathbf{A}$ *magnetic vector potential*
$\mathbf{E}$ *electric field strength*
$\check{\mathbf{B}}$ *magnetic flux density*
$U$ *generalized potential*
$\mathbf{F}$ *force*

compare with the left side of Table [ELE3]

Ref: Goldstein, H.: Classical Mechanics. Addison Wesley, p. 21 (1957).

ELE13-6 http://discretephysics.dicar.units.it compare with Table [FLU10]

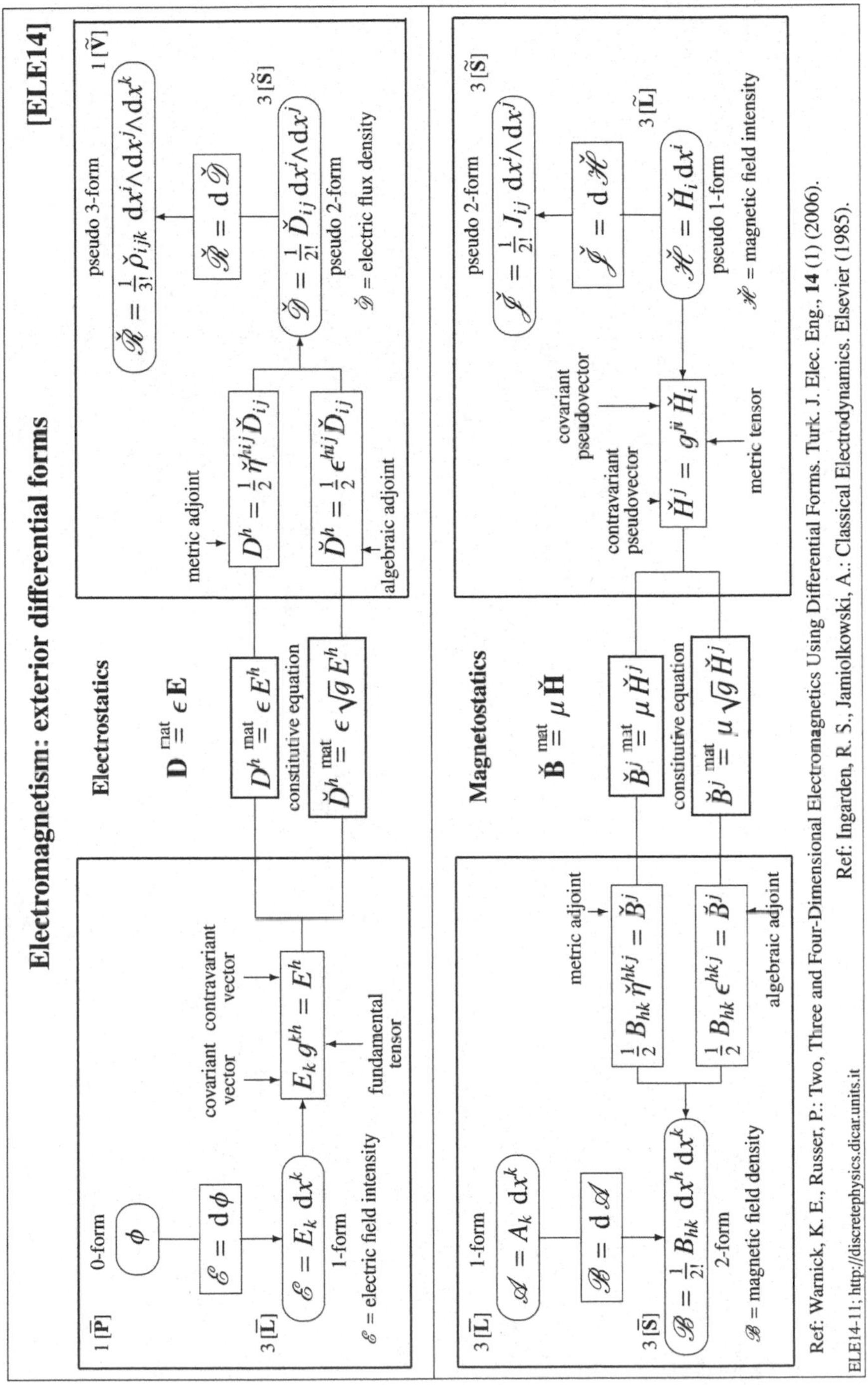
Electromagnetism: exterior differential forms
[ELE14]
Electrostatics
1 [$\bar{P}$]
0-form
$\phi$
$\mathscr{E} = \mathrm{d}\,\phi$
3 [$\bar{L}$]
$\mathscr{E} = E_k\,\mathrm{d}x^k$
1-form
$\mathscr{E}$ = electric field intensity
covariant vector
contravariant vector
$E_k\, g^{kh} = E^h$
fundamental tensor
$\mathbf{D} \overset{\text{mat}}{=} \epsilon\,\mathbf{E}$
$D^h \overset{\text{mat}}{=} \epsilon\, E^h$
constitutive equation
$\check{D}^h \overset{\text{mat}}{=} \epsilon\,\sqrt{g}\,E^h$
metric adjoint
$D^h = \frac{1}{2}\,\check{\eta}^{hij}\,\check{D}_{ij}$
$\check{D}^h = \frac{1}{2}\,\epsilon^{hij}\,\check{D}_{ij}$
algebraic adjoint
pseudo 3-form
$\check{\mathscr{R}} = \frac{1}{3!}\,\check{\rho}_{ijk}\,\mathrm{d}x^i\wedge\mathrm{d}x^j\wedge\mathrm{d}x^k$
1 [$\tilde{V}$]
$\check{\mathscr{R}} = \mathrm{d}\,\check{\mathscr{D}}$
3 [$\tilde{S}$]
$\check{\mathscr{D}} = \frac{1}{2!}\,\check{D}_{ij}\,\mathrm{d}x^i\wedge\mathrm{d}x^j$
pseudo 2-form
$\check{\mathscr{D}}$ = electric flux density
Magnetostatics
$\check{\mathbf{B}} \overset{\text{mat}}{=} \mu\,\check{\mathbf{H}}$
3 [$\bar{L}$]
1-form
$\mathscr{A} = A_k\,\mathrm{d}x^k$
$\mathscr{B} = \mathrm{d}\,\mathscr{A}$
3 [$\bar{S}$]
$\mathscr{B} = \frac{1}{2!}\,B_{hk}\,\mathrm{d}x^h\,\mathrm{d}x^k$
2-form
$\mathscr{B}$ = magnetic field density
metric adjoint
$\frac{1}{2}\,B_{hk}\,\check{\eta}^{hkj} = \check{B}^j$
$\frac{1}{2}\,B_{hk}\,\epsilon^{hkj} = \check{B}^j$
algebraic adjoint
$\check{B}^j \overset{\text{mat}}{=} \mu\,\check{H}^j$
constitutive equation
$\check{B}^j \overset{\text{mat}}{=} \mu\,\sqrt{g}\,\check{H}^j$
pseudo 2-form
$\check{\mathscr{J}} = \frac{1}{2!}\,J_{ij}\,\mathrm{d}x^i\wedge\mathrm{d}x^j$
3 [$\tilde{S}$]
$\check{\mathscr{J}} = \mathrm{d}\,\check{\mathscr{H}}$
3 [$\tilde{L}$]
$\check{\mathscr{H}} = \check{H}_i\,\mathrm{d}x^i$
pseudo 1-form
$\check{\mathscr{H}}$ = magnetic field intensity
covariant pseudovector
contravariant pseudovector
$\check{H}^j = g^{ji}\,\check{H}_i$
metric tensor
Ref: Warnick, K. E., Russer, P.: Two, Three and Four-Dimensional Electromagnetics Using Differential Forms. Turk. J. Elec. Eng., 14 (1) (2006).
Ref: Ingarden, R. S., Jamiolkowski, A.: Classical Electrodynamics. Elsevier (1985).
ELE14-11; http://discretephysics.dicar.units.it

# Chapter 11
# Mechanics of Deformable Solids

## 11.1 Introduction

We assume that the reader is familiar with the main variables of continuum mechanics.[1] We summarize these variables in Tables 11.3 (p. 329) and 11.5 (p. 336), dividing into two categories: source and configuration variables. Our aim is to classify these physical variables on the basis of the oriented spatial and time elements with which each variable is associated. This can be done by starting from the global variables because the densities and the rates inherit the same association of the global variables from which they originated. The first step in the classification is to separate the source variables from the configuration variables. To this end, it is appropriate to explicitly state the fundamental problem of the theory, something which, in traditional expositions, is not usually done initially.

## 11.2 Fundamental Problem

The *fundamental problem* in the mechanics of deformable solids can be stated as follows:

- Given a solid body,
- Given the shape and the nature of the materials which form the body,
- Given the constraints acting on the body,
- Given a time interval,
- Given the initial position and the initial velocity of each point of the body,
- Given the volume forces and the surface forces at the boundary,
- Find the position of each point of the body at every subsequent instant.

---

[1] This chapter presupposes one's having read Chaps. 1–9.

E. Tonti, *The Mathematical Structure of Classical and Relativistic Physics*, Modeling and Simulation in Science, Engineering and Technology,
DOI 10.1007/978-1-4614-7422-7_11, © Springer Science+Business Media New York 2013

The fundamental problem is mathematically expressed by the *fundamental equation* which links source variables with configuration variables via topological and constitutive equations. The founding fathers of the fundamental equations of continuum mechanics are given in Table 11.1, while the corresponding equations are compiled in Table 6.2 (p. 158) (Table 11.2).

**Table 11.1** Fundamental equations of continuum mechanics

| | | | |
|---|---|---|---|
| *Continuum mechanics* | *Solids* | *Perfect* ≡ *rigid* | Newton's equation |
| | | *Deformable* | Navier's equation |
| | *Fluids* | *Perfect* ≡ *inviscid* | Euler's equation |
| | | *Viscous* | Navier–Stokes' equations |

## 11.3 Source Variables

The source variables used in continuum mechanics are listed in Table 11.3 (p. 329). Forces are the common sources of deformations. The *global* source variables are as follows:

1. *Impulse of volume force* $\mathbf{J}^{v}$ equivalent to the *momentum production* $\mathbf{P}^{p}$;
2. Impulse of *surface force* $\mathbf{J}^{s}$;
3. *Momentum content* $\mathbf{P}^{c}$.

The corresponding rates are the volume force $\mathbf{F}^{v}[\widetilde{\mathbf{T}}, \widetilde{\mathbf{V}}]$ and the surface force $\mathbf{F}^{s}[\widetilde{\mathbf{T}}, \widetilde{\mathbf{S}}] \equiv \mathbf{T}[\widetilde{\mathbf{T}}, \widetilde{\mathbf{S}}]$. The corresponding densities are the volume force for unit volume $\mathbf{f}$, the stress vector $\mathbf{t}$ and the momentum density $\mathbf{p}$.[2]

### 11.3.1 Impulse of Volume Forces

Denoting by $\mathbf{F}^{v}$ the volume force, the corresponding impulse $\mathbf{J}^{v}$ during the time interval $(t^{-}, t^{+})$ is

[2] We use capital letters for global variables and lower-case letters for their densities. This is not followed in the literature, where **F** is used to denote the volume force for unit mass; see Prager [188, p. 46], Paterson [172, p. 102], Achenbach [1, p. 51], Batchelor [10, p. 15], Milne [160, p. 78]. Shames [209, p. 154] uses **B** for volume force for unit mass, whereas Chorin and Marsden [41, p. 6] use **B** for volume force and **b** for volume force for unit mass.

**Table 11.2** From global variables to field functions in the mechanics of deformable solids

| | | |
|---|---|---|
| $\mathbf{J}^{\text{v}}[\tilde{\mathbf{T}}, \tilde{\mathbf{V}}]$ | $\mathbf{J}^{\text{s}}[\tilde{\mathbf{T}}, \tilde{\mathbf{S}}]$ | $\mathbf{P}^{\text{c}}[\tilde{\mathbf{I}}, \tilde{\mathbf{V}}]$ |
| Body force impulse | Surface impulse | Momentum |
| ↓ | ↓ | ↓ |
| *Rate* | *Rate* | |
| $\mathbf{F}^{\text{v}}[\tilde{\mathbf{T}}, \tilde{\mathbf{V}}]$ | $\mathbf{T}[\tilde{\mathbf{T}}, \tilde{\mathbf{S}}]$ | |
| Body force | Surface force | |
| ↓ | ↓ | |
| *Density* | *Density* | *Density* |
| $\mathbf{f}[\tilde{\mathbf{T}}, \tilde{\mathbf{V}}] \stackrel{\text{def}}{=} \dfrac{\mathbf{F}^{\text{v}}}{V}$ | $\mathbf{t}[\tilde{\mathbf{T}}, \tilde{\mathbf{S}}] \stackrel{\text{def}}{=} \dfrac{\mathbf{T}}{A}$ | $\mathbf{p}[\tilde{\mathbf{I}}, \tilde{\mathbf{V}}] \stackrel{\text{def}}{=} \dfrac{\mathbf{P}^{\text{c}}}{V}$ |
| Body force density | Surface force density | Momentum density |

$$\mathbf{J}^{\text{v}}(t^-, t^+) \stackrel{\text{def}}{=} \int_{t^-}^{t^+} \mathbf{F}^{\text{v}}(t)\, \mathrm{d}t \quad \longrightarrow \quad \mathbf{J}^{\text{v}}[\tilde{\mathbf{T}}, \tilde{\mathbf{V}}], \tag{11.1}$$

which is associated with a dual time interval and a dual volume.[3]

### 11.3.2 Impulse of Surface Forces

Denoting by $\mathbf{F}^{\text{s}}$ the surface force acting on a surface, the impulse $\mathbf{J}^{\text{s}}$ of this force during the time interval $(t^-, t^+)$ is

$$\mathbf{J}^{\text{s}}(t^-, t^+) \stackrel{\text{def}}{=} \int_{t^-}^{t^+} \mathbf{F}^{\text{s}}(t)\, \mathrm{d}t \quad \longrightarrow \quad \mathbf{J}^{\text{s}}[\tilde{\mathbf{T}}, \tilde{\mathbf{S}}], \tag{11.2}$$

which is associated with a dual interval and with the surface. We remark that in the literature the surface force is often denoted by $\mathbf{T}$.

### 11.3.3 Momentum

Let us consider, for example, a parachutist falling through the atmosphere. His momentum at a given instant $t$ is the time integral of the total force acting on him, i.e. the sum of body and surface forces from the instant of his jump plus the

[3] We showed on p. 242 that the impulse is associated with $\tilde{\mathbf{T}}$. See also p. 131.

momentum at the start $\mathbf{P}_0$ considering the velocity of the aeroplane:

$$\mathbf{P}(t) = \mathbf{P}_0 + \int_0^t [\mathbf{F}^{\mathrm{v}}(t) + \mathbf{F}^{\mathrm{s}}(t)]\, \mathrm{d}t \quad \longrightarrow \quad \mathbf{P}[\tilde{\mathbf{I}}, \mathscr{B}], \tag{11.3}$$

where $\mathscr{B}$ denotes the body. It follows that **P** is associated with dual time instants and with the body.[4]

## *11.3.4 Momentum Content, Flow and Production*

When a body is in motion, all of its particles have a momentum: the sum of the momenta of all particles is the momentum of the body. One theorem states that the momentum of a body is equal to the product of the velocity of the centre of mass for the mass of the whole body. This makes reference to a system description.

In a spatial description, let us consider a fixed control volume and the sum of the momenta of all particles contained in the volume at a given instant; this is the *momentum content* $\mathbf{P}^{\mathrm{c}}$. Considering the momenta of the particles which leave the control volume during a time interval, the sum of these momenta is the *momentum flow* $\mathbf{P}^{\mathrm{f}}$. The action of external forces on the particles contained inside the control volume increase the momentum content, and this is the *momentum production* $\mathbf{P}^{\mathrm{p}}$.

**Space Association.** The momentum content and momentum production are associated with $\tilde{\mathbf{V}}$, the momentum flow with $\tilde{\mathbf{S}}$, as is obvious.

**Time Association.** Since the operation of summing momenta does not change the time association, the momentum content is referred to $\tilde{\mathbf{I}}$, just like the momentum of a particle. The momentum production and the momentum flow are associated with a time interval, necessarily with a dual time interval, i.e. $\tilde{\mathbf{T}}$.

In conclusion, we have $\mathbf{P}^{\mathrm{c}}[\tilde{\mathbf{I}}, \tilde{\mathbf{V}}], \mathbf{P}^{\mathrm{f}}[\tilde{\mathbf{T}}, \tilde{\mathbf{S}}], \mathbf{P}^{\mathrm{p}}[\tilde{\mathbf{T}}, \tilde{\mathbf{V}}]$.

## *11.3.5 Stress Vector, Stress Tensor, Pressure*

Since the stress vector $\mathbf{t}(\mathbf{n})$ referred to a plane areal element of normal $\mathbf{n}$ is the ratio of the surface force $\mathbf{T}(\mathbf{n})$ and the area, it is associated with the same space and time elements of the surface force, hence $\mathbf{t}[\tilde{\mathbf{T}}, \tilde{\mathbf{S}}]$. The traditional condition $\mathbf{T}(-\mathbf{n}) = -\mathbf{T}(\mathbf{n})$ is the expression of the oddness principle and is equivalent to the condition $\mathbf{T}(-\tilde{\mathbf{S}}) = -\mathbf{T}(\tilde{\mathbf{S}})$.

Pressure is used in fluids where the internal surface forces are orthogonal to the plane surface elements on which they act. Denoting by $T$ the modulus of the

[4] We showed on p. 243 that momentum is associated with $\tilde{\mathbf{I}}$.

**Table 11.3** Source variables of continuum mechanics

| | | |
|---|---|---|
| Volume force | $\mathbf{F}^{\mathrm{v}}(\mathbf{V}, t, \mathbf{P})$ | $\mathbf{F}^{\mathrm{v}}\,[\tilde{\mathbf{T}}, \tilde{\mathbf{V}}]$ |
| Force for unit volume | $\mathbf{f}(t, \mathbf{P}) \stackrel{\text{def}}{=} \mathbf{F}^{\mathrm{v}}(\mathbf{V}, t, \mathbf{P})/V$ | $\mathbf{f}\,[\tilde{\mathbf{T}}, \tilde{\mathbf{V}}]$ |
| Mass content | $M^{\mathrm{c}}(\mathbf{V}, t, \mathbf{P})$ | $M^{\mathrm{c}}\,[\bar{\mathbf{I}}, \tilde{\mathbf{V}}]$ |
| Mass density | $\rho(t, \mathbf{P}) \stackrel{\text{def}}{=} M^{\mathrm{c}}(\mathbf{V}, t, \mathbf{P})/V$ | $\rho\,[\bar{\mathbf{I}}, \tilde{\mathbf{V}}]$ |
| Momentum content | $\mathbf{P}^{\mathrm{c}}(\mathbf{V}, t, \mathbf{P})$ | $\mathbf{P}\,[\tilde{\mathbf{I}}, \tilde{\mathbf{V}}]$ |
| Momentum density | $\mathbf{p}(t, \mathbf{P}) \stackrel{\text{def}}{=} \mathbf{P}^{\mathrm{c}}(\mathbf{V}, t, \mathbf{P})/V$ | $\mathbf{p}\,[\tilde{\mathbf{I}}, \tilde{\mathbf{V}}]$ |
| Surface force | $\mathbf{T}(\mathbf{n}A, t, \mathbf{P})$ | $\mathbf{T}\,[\tilde{\mathbf{T}}, \tilde{\mathbf{S}}]$ |
| Stress vector | $\mathbf{t}(\mathbf{n}, t, \mathbf{P}) \stackrel{\text{def}}{=} \mathbf{T}(\mathbf{n}A, t, \mathbf{P})/A$ | $\mathbf{t}\,[\tilde{\mathbf{T}}, \tilde{\mathbf{S}}]$ |
| Pressure | $p(t, \mathbf{P}) \stackrel{\text{def}}{=} F_{\mathrm{n}}/A$ | $p[\tilde{\mathbf{T}}, \tilde{\mathbf{S}}]$ |
| Stress tensor | $\boldsymbol{\tau}(t, \mathbf{P}) \quad \tau_{ik}$<br>$(\mathbf{t} = \mathbf{n} \cdot \boldsymbol{\tau}) \quad (t_k = n_i \tau_{ik})$ | $\boldsymbol{\tau}\,[\tilde{\mathbf{T}}, \tilde{\mathbf{S}}]$ |
| Symmetric stress tensor | $\boldsymbol{\sigma}(t, \mathbf{P}) \stackrel{\text{def}}{=} \frac{1}{2}(\boldsymbol{\tau} + \boldsymbol{\tau}^{\mathrm{T}})$<br>$\sigma_{hk} \stackrel{\text{def}}{=} \frac{1}{2}(\tau_{hk} + \tau_{kh})$ | $\boldsymbol{\sigma}\,[\tilde{\mathbf{T}}, \tilde{\mathbf{S}}]$ |
| Mean normal stress | $\sigma(t, \mathbf{P}) \equiv -p(t, \mathbf{P}) \stackrel{\text{def}}{=} \frac{1}{3}\,\mathrm{tr}(\boldsymbol{\sigma})$ | $p\,[\tilde{\mathbf{T}}, \tilde{\mathbf{S}}]$ |
| Viscous stress (in fluids) | $\boldsymbol{\tau}(t, \mathbf{P}) \stackrel{\text{def}}{=} \boldsymbol{\sigma} - \sigma\,\mathbf{I}$<br>$\tau_{hk} \stackrel{\text{def}}{=} \sigma_{hk} - \sigma\,\delta_{hk}$ | $\boldsymbol{\tau}\,[\tilde{\mathbf{T}}, \tilde{\mathbf{S}}]$ |
| Deviator stress (in solids) | $\boldsymbol{\sigma}'(t, \mathbf{P}) \stackrel{\text{def}}{=} \boldsymbol{\sigma} - \sigma\,\mathbf{I}$<br>$\sigma'_{hk} \stackrel{\text{def}}{=} \sigma_{hk} - \sigma\,\delta_{hk}$ | $\boldsymbol{\sigma}'\,[\tilde{\mathbf{T}}, \tilde{\mathbf{S}}]$ |
| Mass flow<br>Mass current<br>Mass current density | $M^{\mathrm{f}}(\mathbf{n}A, T, \mathbf{P})$<br>$\Phi(\mathbf{n}A, t, \mathbf{P}) \stackrel{\text{def}}{=} M^{\mathrm{f}}(\mathbf{n}A, T, \mathbf{P})/T$<br>$\mathbf{q}(t, \mathbf{P}) \qquad \Phi = \int_A \mathbf{q} \cdot \mathbf{n}\,\mathrm{d}A$ | $M^{\mathrm{f}}[\bar{\mathbf{T}}, \tilde{\mathbf{S}}]$<br>$\Phi\,[\bar{\mathbf{T}}, \tilde{\mathbf{S}}]$<br>$\mathbf{q}\,[\bar{\mathbf{T}}, \tilde{\mathbf{S}}]$ |
| Stream pseudovector<br>Stream function (pseudoscalar) | $\check{\boldsymbol{\psi}}(t, \mathbf{P}) \qquad (\mathbf{q} = \nabla \times \check{\boldsymbol{\psi}})$<br>$\check{\psi}(t, \mathbf{P})$ | $\check{\boldsymbol{\psi}}\,[\bar{\mathbf{T}}, \tilde{\mathbf{L}}]$<br>$\psi\,[\bar{\mathbf{T}}, \tilde{\mathbf{L}}]$ |

normal force the pressure is the ratio $p \stackrel{\text{def}}{=} T/A$, where $A$ is the area of the surface element. Hence, pressure has the same time and space association as the surface force, i.e. $p[\widetilde{\mathbf{T}}, \widetilde{\mathbf{S}}]$.

The stress tensor is a generalization of pressure and, as such, has the same association with space and time elements of the pressure, hence $\tau_{hk}[\widetilde{\mathbf{T}}, \widetilde{\mathbf{S}}]$.

## 11.4 Configuration Variables

The configuration variables used in this chapter are listed in Table 11.5 (p. 336). With reference to Figs. 11.1 and 11.2, the *global* configuration variables are as follows.

**Geometric variables:**

1. The *position vector* $\mathbf{R}(\mathbf{P})$ of a particle $\mathscr{P}$ in the *reference configuration* at the instant $t = 0$ with respect to a fixed coordinate system $\mathbf{O}xyz$;
2. The *position vector* $\mathbf{r}(t, \mathbf{P})$ of the particle $\mathscr{P}$ at an instant $t$;
3. The *relative position* $\mathbf{G}$ of the two points $\mathbf{P}$ and $\mathbf{Q}$ in the reference configuration.

**Kinematic variables:**

1. The *total displacement* $\mathbf{u}(t, \mathbf{P})$ of the particle $\mathscr{P}$ from its position in the reference configuration;
2. The *displacement* $\Delta_t\mathbf{u}$ of a particle $\mathscr{P}$ in an interval (shown in Fig. 11.2).
3. The *relative displacement* $\mathbf{h}(t, \mathbf{G})$ of the two particles $\mathscr{P}$ and $\mathscr{Q}$ at the instant $t$;

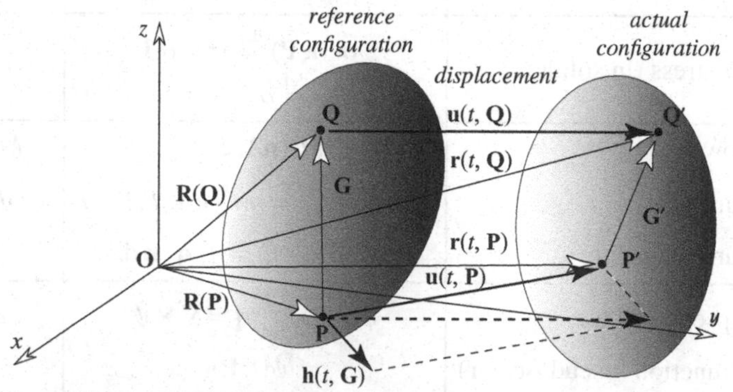


**Fig. 11.1** Two particles $\mathscr{P}$ and $\mathscr{Q}$ in the initial configuration occupy the positions $\mathbf{P}$ and $\mathbf{Q}$. The same particles, at a later instant, occupy the positions $\mathbf{P}'$ and $\mathbf{Q}'$. The vectors in thin lines are position vectors, whereas those in thick lines are the displacements of the two particles $\mathscr{P}$ and $\mathscr{Q}$

In what follows, we give a detailed definition of the configuration variables of continuum mechanics.

### 11.4.1 Initial Position Vector

Let us consider a *reference configuration* of a body. In this configuration, a point **P** can be localized by the initial position vector **R** connecting the origin **O** of a coordinate system with the point, hence $\mathbf{R}[\overline{\mathbf{P}}]$.

### 11.4.2 Relative Position Vector

This is the vector

$$\mathbf{G} \stackrel{\text{def}}{=} \mathbf{R}[\overline{\mathbf{Q}}] - \mathbf{R}[\overline{\mathbf{P}}] \qquad \longrightarrow \qquad \mathbf{G}\,[\overline{\mathbf{L}}] \tag{11.4}$$

and is associated with the line **L** endowed with an inner orientation, i.e. from $\overline{\mathbf{P}}$ to $\overline{\mathbf{Q}}$. Since it is associated with the reference configuration, **G** does not depend on time.

### 11.4.3 Position Vector

This variable is used in both particle dynamics and continuum mechanics. With reference to Fig. 11.2a, let us consider the motion of a particle $\mathscr{P}$. At every instant, the position of the particle is described by the position vector **r**; hence, it is associated with the instant **I**. Is it a primal or a dual instant? Since **r**, by definition, goes from the origin **O** to the point **P** in which the particle lies at the instant $t$, it does not change sign under a reversal of motion; hence, it is associated with the primal instants $\mathbf{r}\,[\overline{\mathbf{I}}, \mathscr{P}]$.

In a spatial description, the position vector, as in Fig. 11.2b, is associated with a point of a body, hence $\mathbf{r}\,[\overline{\mathbf{I}}, \overline{\mathbf{P}}]$. It is appropriate, even if not imperative,[5] to consider **R** as the position vector at the initial instant ($t = 0$), i.e. to set $\mathbf{R}[\overline{\mathbf{P}}] = \mathbf{r}[0, \overline{\mathbf{P}}]$.

### 11.4.4 Displacement

Let us pose the question: *is displacement associated with an instant or an interval?* If you pose this question to different people, you will get different answers. For some it is associated with an instant, for others with an interval. The ambiguity arises because we often use the same term, *displacement*, to denote both the displacement of a particle between two arbitrary instants and the displacement from

[5] Hunter [98, p. 23].

a reference position at a given instant. If the reference configuration is considered the *initial* configuration of a motion, then this displacement can be called *total* displacement. It follows that *total displacement* is associated only with the *final* instant, whereas *relative displacement* between two arbitrary instants is associated with the corresponding *interval* (Table 11.4).

This situation with displacement is similar to that of *impulse*: an impulse is associated with a time interval, but if we consider an impulse given to a particle *starting from rest*, it becomes associated with instants and is properly called *momentum*. Hence we have this analogy:

$$\begin{aligned} &\text{momentum} \quad \mathbf{p}(t) = \int_0^t \mathbf{F}(t')\,\mathrm{d}t' \qquad \text{impulse} \quad \mathbf{J}(t^-,t^+) = \int_{t^-}^{t^+} \mathbf{F}(t)\,\mathrm{d}t, \\ &\text{total displacement} \quad \mathbf{u}(t) = \int_0^t \mathbf{v}(t')\,\mathrm{d}t' \qquad \text{displacement} \quad \boldsymbol{\eta}(t^-,t^+) = \int_{t^-}^{t^+} \mathbf{v}(t)\,\mathrm{d}t. \end{aligned} \tag{11.5}$$

**(a) Particle Mechanics.** Let us consider two instants, $t^-$ and $t^+$: the *displacement* $\boldsymbol{\eta}$ relative to the time interval $(t^-,t^+)$ is the difference between the position at the final and initial instants of a time interval

$$\boldsymbol{\eta}(t^-,t^+) \stackrel{\text{def}}{=} \mathbf{r}(t^+) - \mathbf{r}(t^-) \qquad \longrightarrow \qquad \boldsymbol{\eta}[\overline{\mathbf{T}}]. \tag{11.6}$$

Since the position vector $\mathbf{r}$ is associated with a primal instant $\overline{\mathbf{I}}$, the displacement $\mathbf{u}$ is associated with a primal time interval $\overline{\mathbf{T}}$.

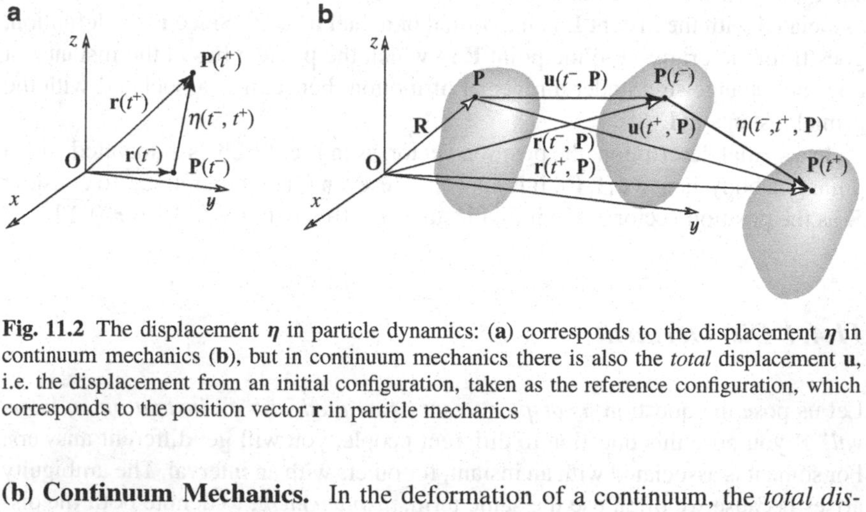


**Fig. 11.2** The displacement $\boldsymbol{\eta}$ in particle dynamics: (**a**) corresponds to the displacement $\boldsymbol{\eta}$ in continuum mechanics (**b**), but in continuum mechanics there is also the *total* displacement $\mathbf{u}$, i.e. the displacement from an initial configuration, taken as the reference configuration, which corresponds to the position vector $\mathbf{r}$ in particle mechanics

**(b) Continuum Mechanics.** In the deformation of a continuum, the *total displacement* $\mathbf{u}$ of a particle is measured from its position in the *reference* configuration to its current position at the instant (Fig. 11.1). We see that $\mathbf{u}$ in continuum mechanics plays the same role as the position vector $\mathbf{r}$ in particle dynamics and is a *function of* the instant. This implies that it does not change sign under a reversal of motion, and for this reason $\mathbf{u}$ is associated with a primal instant, i.e. $\mathbf{u}\,[\bar{\mathbf{I}}]$:

$$\mathbf{u}(t,\mathbf{P}) \stackrel{\text{def}}{=} \mathbf{r}\,(t,\mathbf{P}) - \mathbf{R}\,(\mathbf{P}) \equiv \mathbf{r}\,(t,\mathbf{P}) - \mathbf{r}\,(0,\mathbf{P}) \qquad \longrightarrow \qquad \mathbf{u}\,[\bar{\mathbf{I}},\bar{\mathbf{P}}]\,. \tag{11.7}$$

The *displacement* $\eta$, defined by

$$\eta\,(t^-,t^+,\mathbf{P}) \stackrel{\text{def}}{=} \mathbf{u}\,(t^+,\mathbf{P}) - \mathbf{u}\,(t^-,\mathbf{P}) \qquad \longrightarrow \qquad \eta\,[\bar{\mathbf{T}},\bar{\mathbf{P}}], \tag{11.8}$$

is associated with primal intervals. This implies that it changes sign under a reversal of motion.

**Table 11.4** The term *displacement* and its meanings

| *Particle dynamics* | *Continuum mechanics* | |
|---|---|---|
| Position vector | $\mathbf{R}[\bar{\mathbf{P}}]$ | Initial position vector |
| | $\mathbf{r}[\mathbf{I},\bar{\mathbf{P}}]$ | Position vector |
| Displacement | $\mathbf{u}[\bar{\mathbf{I}},\bar{\mathbf{P}}]$ | Total displacement |
| | $\eta[\bar{\mathbf{T}},\bar{\mathbf{P}}]$ | Displacement |

## *11.4.5 Relative Displacement*

With reference to Fig. 11.1, there is also the *relative displacement* associated with two points, $\mathbf{P}$ and $\mathbf{Q}$, defined as the difference between the total displacements $\mathbf{u}$ associated with the two points at the same instant, i.e.

$$\mathbf{h}(t,\mathbf{P},\mathbf{Q}) \stackrel{\text{def}}{=} \mathbf{u}(t,\mathbf{Q}) - \mathbf{u}(t,\mathbf{P}) \qquad \longrightarrow \qquad \mathbf{h}[\bar{\mathbf{I}},\bar{\mathbf{L}}]. \tag{11.9}$$

Hence, $\mathbf{h}$ is associated with primal instants like $\mathbf{u}$. Since we must choose an order of the points to compute the difference, i.e. $\mathbf{PQ}$ or $\mathbf{QP}$, it follows that $\mathbf{h}(t,\mathbf{P},\mathbf{Q}) = -\mathbf{h}(t,\mathbf{Q},\mathbf{P})$, i.e. the line must be endowed with an inner orientation.

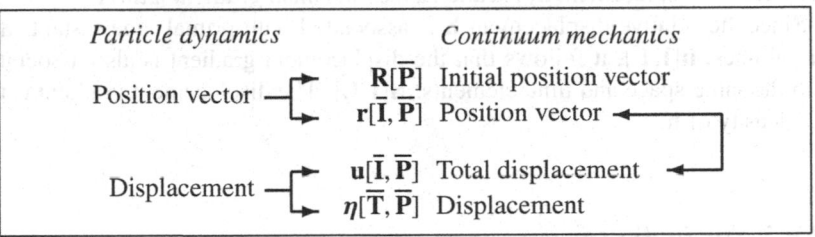

### 11.4.6 Displacement Gradient

In the neighbourhood of a point **P**, the displacement **u** can be approximated by an affine function,[6] i.e.

$$u_i(x_1, x_2, x_3) = a_i + H_{ij}\, x_j \qquad \text{with } a_i, H_{ij} \text{ constants.} \tag{11.10}$$

In this approximation, the relative displacement **h** between a point **P** and a point **Q** is given by

$$h_i = u_i(\mathbf{Q}) - u_i(\mathbf{P}) = H_{ij}\,[x_j(\mathbf{Q}) - x_j(\mathbf{P})] = H_{ij}\, G_j \quad \text{or} \quad \mathbf{h} = \mathsf{H}\,\mathbf{G}\,. \tag{11.11}$$

The matrix $\mathsf{H}$, which links the relative position **G** of two neigbouring points with their relative displacement **h**, is called a *displacement gradient matrix*.

Since the relative displacement **h** is associated with primal time instants and primal lines, $\mathbf{h}[\bar{\mathbf{I}}, \bar{\mathbf{L}}]$, it follows that the displacement gradient is also associated with the same space and time elements: $\mathsf{H}[\bar{\mathbf{I}}, \bar{\mathbf{L}}]$. The displacement gradient is the line density of **h**.

### 11.4.7 Strain Tensor

The strain tensor $\boldsymbol{\varepsilon}$ is the symmetric part of the displacement gradient, whereas the rotation tensor $\boldsymbol{\Omega}$ is the skew-symmetric part of the displacement gradient. The linear invariant of the displacement gradient is the volume dilatation $\Theta$. See Table 11.5.

**Remark.** We emphasize that at first glance one is tempted to associate the volume dilatation to a volume, i.e. to put $\Theta[\mathbf{V}]$, which is wrong. The fact that the volume dilatation comes from the strain tensor, which is associated with lines, implies the inherited association with lines, hence $\Theta[\bar{\mathbf{I}}, \bar{\mathbf{L}}]$. As a second argument, we remark that considering the variation of a volume, we are in a material description, not a spatial one. To see the space element with which $\Theta$ is associated, we must switch to a spatial description. Now it appears natural that the trace of a tensor must be associated with the same space element of the tensor, hence $\Theta[\bar{\mathbf{L}}]$. A third argument is that the constitutive relation $p = -K\Theta$ is reversible; hence, it links two variables which must have the same behaviour under a reversal of motion and the same behaviour for a change in the inner orientation of lines because the outer orientation of a surface coincides with the inner orientation of a line crossing it. Since pressure is associated with $[\tilde{\mathbf{T}}, \tilde{\mathbf{S}}]$, it follows that $\Theta$ must be associated with $[\bar{\mathbf{I}}, \bar{\mathbf{L}}]$.

[6] See Appendix A at p. 457.

### 11.4.8 Velocity

In continuum mechanics, velocity is the *rate* of displacement $\eta$ and, at the same time, the *time derivative* of the *total* displacement $\mathbf{u}$:

$$\mathbf{v}(t,\mathbf{P}) \stackrel{\text{def}}{=} \lim_{\tau \longrightarrow 0} \frac{\eta(\tau,\mathbf{P})}{\tau} = \frac{\mathrm{d}\mathbf{u}(t,\mathbf{P})}{\mathrm{d}t} \qquad \longrightarrow \qquad \mathbf{v}[\overline{\mathbf{T}},\overline{\mathbf{P}}]. \tag{11.12}$$

Since velocity is the rate of displacement, it inherits the association with primal time intervals: $\mathbf{v}[\overline{\mathbf{T}},\overline{\mathbf{P}}]$.

## 11.5 Field Laws

The main equation in the mechanics of deformable solids is the equilibrium equation (in statics), which is a particular case of the momentum balance in dynamics. Since the equation is similar to that used in fluid dynamics, we refer the reader to Table 12.6 (p. 371).

Table 11.6 summarizes the classification of the main physical variables in continuum mechanics, showing their position in the diagram. Moreover, it shows the link between the field variables and the global variables.

## 11.6 Rod Traction

As an example of classification, let us consider the traction of a rod. Let us consider the deformation of a vertical rod under the action of its weight; this is a one-dimensional model. To simplify the description, we consider a hanging rod, like a stalactite, as shown in diagram [SOL0].

We divide the whole rod into equal pieces of thickness $l$ forming a primal cell complex in space and consider their midpoints; they give rise to a dual cell complex because every cell is oriented with an outer orientation.

Let us consider the configuration variables. Because of the action of its weight, the rod extends, and every normal section $s_h$ of the primal subdivision is shifted by a quantity $u_h$. Let $\mathbf{P}_h$ be the intersection of the normal section $s_h$ with the axis of the rod. The longitudinal *displacement* is $u_h = u_h(\overline{\mathbf{P}}_h)$. Under the dilatation the thickness of every piece increases by the amount $\eta_i = u_{h+1} - u_h$; this is the *extension* of the piece. The variable $\eta_i$ is associated with the 1-cell $\overline{\mathbf{L}}_i$.

The two variables $u[\overline{\mathbf{P}}]$ and $\eta[\overline{\mathbf{L}}]$ are global variables which describe the configuration of the deformed rod.

Let us consider the source variables. We must compute the equilibrium on every piece of rod. To this end, we are tempted to apply the equilibrium condition

**Table 11.5** Configuration variables in the mechanics of deformable solids

| | | |
|---|---|---|
| Initial position vector | $\mathbf{R}(\mathbf{P}) \qquad X_h(\mathbf{P})$ | $\mathbf{R}\,[\overline{\mathbf{P}}]$ |
| Position vector | $\mathbf{r}(t,\mathbf{P}) \qquad x_h(t,\mathbf{P})$ | $\mathbf{r}\,[\overline{\mathbf{I}},\overline{\mathbf{P}}]$ |
| Total displacement | $\mathbf{u}(t,\mathbf{P}) \overset{\text{def}}{=} \mathbf{r}(t,\mathbf{P}) - \mathbf{R}(\mathbf{P})$<br>$\eta_h(t,\mathbf{P}) \overset{\text{def}}{=} x_h(t,\mathbf{P}) - X_h(\mathbf{P})$ | $\mathbf{u}\,[\overline{\mathbf{I}},\overline{\mathbf{P}}]$ |
| Displacement | $\boldsymbol{\eta}(T,\mathbf{P}) \overset{\text{def}}{=} \mathbf{u}(t^+,\mathbf{P}) - \mathbf{u}(t^-,\mathbf{P})$<br>$\eta_i(T,\mathbf{P}) \overset{\text{def}}{=} u_i(t^+,\mathbf{P}) - u_i(t^-,\mathbf{P})$ | $\mathbf{u}\,[\overline{\mathbf{T}},\overline{\mathbf{P}}]$ |
| Relative position vector | $\mathbf{G}(\mathbf{L}) \overset{\text{def}}{=} \mathbf{R}(\mathbf{Q}) - \mathbf{R}(\mathbf{P})$<br>$G_k(\mathbf{L}) \overset{\text{def}}{=} X_k(\mathbf{Q}) - X_k(\mathbf{P})$ | $\mathbf{G}\,[\overline{\mathbf{L}}]$ |
| Relative displacement | $\mathbf{h}(t,\mathbf{L}) \overset{\text{def}}{=} \mathbf{u}(t,\mathbf{Q}) - \mathbf{u}(t,\mathbf{P})$<br>$h_k(t,\mathbf{L}) \overset{\text{def}}{=} u_k(t,\mathbf{Q}) - u_k(t,\mathbf{P})$ | $\mathbf{h}\,[\overline{\mathbf{I}},\overline{\mathbf{L}}]$ |
| Displacement gradient | $\mathsf{H}(t,\mathbf{P}) \overset{\text{def}}{=} \operatorname{Grad}\mathbf{u} \qquad \mathrm{d}\mathbf{u} = \mathsf{H}\,\mathrm{d}\mathbf{R}$<br>$H_{hk}(t,\mathbf{P}) \overset{\text{def}}{=} \dfrac{\partial u_h}{\partial X_k} \qquad \mathrm{d}u_h = H_{hk}\,\mathrm{d}X_k$ | $\mathsf{H}\,[\overline{\mathbf{I}},\overline{\mathbf{L}}]$ |
| Strain tensor | $\boldsymbol{\varepsilon}(t,\mathbf{P}) \overset{\text{def}}{=} \frac{1}{2}(\mathsf{H} + \mathsf{H}^\mathsf{T})$<br>$\varepsilon_{hk}(t,\mathbf{P}) \overset{\text{def}}{=} \frac{1}{2}(H_{hk} + H_{kh})$ | $\boldsymbol{\varepsilon}\,[\overline{\mathbf{I}},\overline{\mathbf{L}}]$ |
| Rotation tensor | $\boldsymbol{\Omega}(t,\mathbf{P}) \overset{\text{def}}{=} \frac{1}{2}(\mathsf{H} - \mathsf{H}^\mathsf{T})$<br>$\Omega_{hk}(t,\mathbf{P}) \overset{\text{def}}{=} \frac{1}{2}(H_{hk} - H_{kh})$ | $\boldsymbol{\Omega}\,[\overline{\mathbf{I}},\overline{\mathbf{L}}]$ |
| Volume (or bulk or cubic) dilatation | $\Theta(t,\mathbf{P}) \overset{\text{def}}{=} \operatorname{tr}(\mathsf{H}) = \operatorname{tr}(\boldsymbol{\varepsilon}) = \nabla\cdot\mathbf{u}$ | $\Theta\,[\overline{\mathbf{I}},\overline{\mathbf{L}}]$ |
| Strain deviator | $\boldsymbol{\varepsilon}'(t,\mathbf{P}) \overset{\text{def}}{=} \boldsymbol{\varepsilon} - \frac{1}{3}\Theta\,\mathsf{I}$<br>$\varepsilon'_{hk}(t,\mathbf{P}) \overset{\text{def}}{=} \varepsilon_{hk} - \frac{1}{3}\Theta\,\delta_{hk}$ | $\boldsymbol{\varepsilon}'\,[\overline{\mathbf{I}},\overline{\mathbf{L}}]$ |
| Deformation gradient | $\mathsf{F}(t,\mathbf{P}) \overset{\text{def}}{=} \operatorname{Grad}\mathbf{r} \qquad \mathrm{d}\mathbf{r} = \mathsf{F}\,\mathrm{d}\mathbf{R}$<br>$F_{hk}(t,\mathbf{P}) \overset{\text{def}}{=} \dfrac{\partial x_h}{\partial X_k} \qquad \mathrm{d}x_h = F_{hk}\,\mathrm{d}X_k$ | $\mathsf{F}\,[\overline{\mathbf{I}},\overline{\mathbf{L}}]$ |

**Table 11.6** From field variables to global variables in continuum mechanics

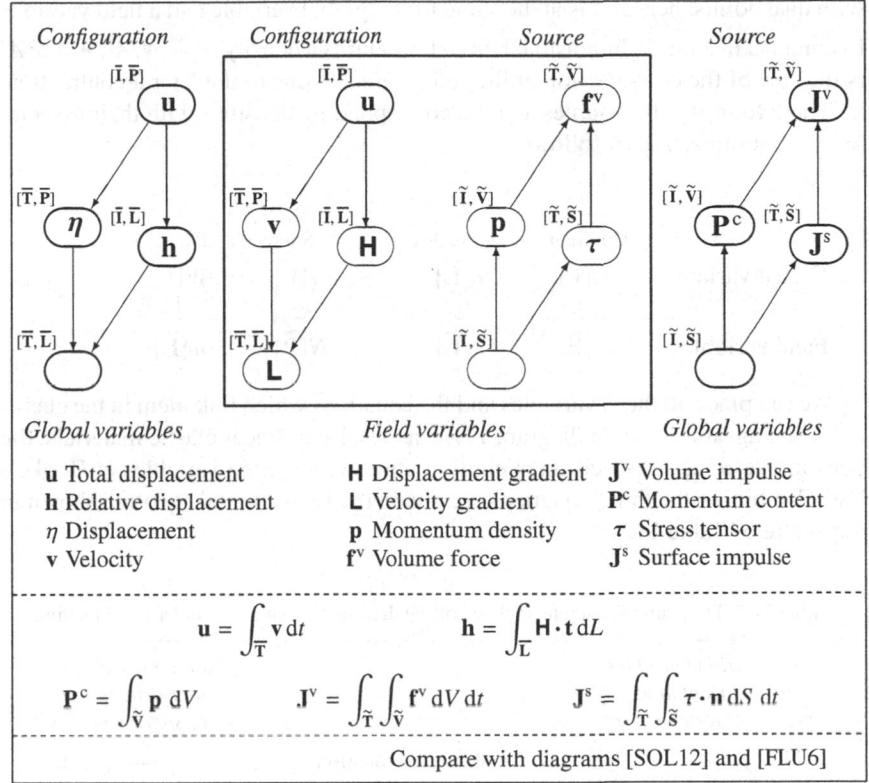

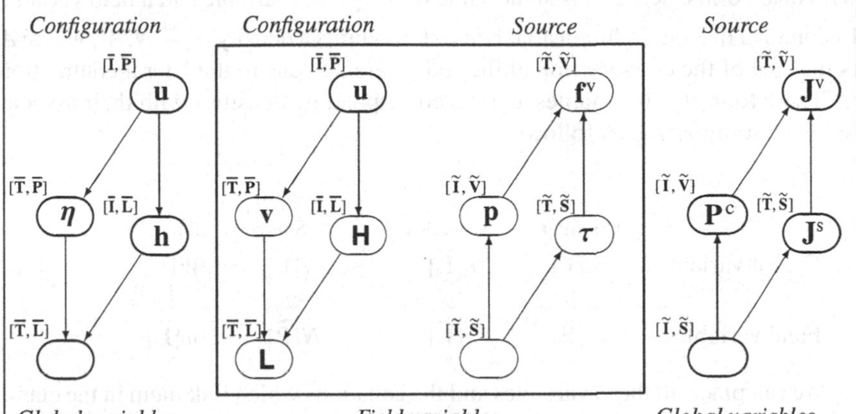


| | | |
|---|---|---|
| **u** Total displacement | **H** Displacement gradient | $\mathbf{J}^{\mathrm{v}}$ Volume impulse |
| **h** Relative displacement | **L** Velocity gradient | $\mathbf{P}^{\mathrm{c}}$ Momentum content |
| $\eta$ Displacement | **p** Momentum density | $\boldsymbol{\tau}$ Stress tensor |
| **v** Velocity | $\mathbf{f}^{\mathrm{v}}$ Volume force | $\mathbf{J}^{\mathrm{s}}$ Surface impulse |

$$\mathbf{u} = \int_{\mathrm{T}} \mathbf{v}\, \mathrm{d}t \qquad \mathbf{h} = \int_{\mathrm{L}} \mathbf{H}\cdot\mathbf{t}\, \mathrm{d}L$$

$$\mathbf{P}^{\mathrm{c}} = \int_{\tilde{\mathrm{V}}} \mathbf{p}\, \mathrm{d}V \qquad \mathbf{J}^{\mathrm{v}} = \int_{\tilde{\mathrm{T}}}\int_{\tilde{\mathrm{V}}} \mathbf{f}^{\mathrm{v}}\, \mathrm{d}V\, \mathrm{d}t \qquad \mathbf{J}^{\mathrm{s}} = \int_{\tilde{\mathrm{T}}}\int_{\tilde{\mathrm{S}}} \boldsymbol{\tau}\cdot\mathbf{n}\, \mathrm{d}S\, \mathrm{d}t$$

Compare with diagrams [SOL12] and [FLU6]

on every piece of the primal complex; this is what is commonly done. However, the equilibrium must be imposed on a dual piece because such a piece is endowed with an outer orientation. In fact, the weight $W_h$ is exerted on the piece from the Earth, i.e. from outside, and this implies that the space element must be endowed with an outer orientation. Moreover, the axial force $N_i$ acting on the section $s_i$ requires an outer orientation of the face, and this is represented in diagram [SOL0] by the white arrows crossing the faces.

The space global variables, which play the role of sources in this theory are, the *traction* $N[\widetilde{\mathbf{P}}]$ acting on a dual section $s_i$ and the *weight* $W[\widetilde{\mathbf{L}}]$ of a dual segment.

The two pieces, the primal and the dual, were made with equal thickness, which we denote by $l$. If we divide the extension of the primal piece by its thickness, then we obtain the *strain* $\varepsilon_i \stackrel{\text{def}}{=} \eta_i/l$; dividing the weight of the dual piece by the thickness we obtain the *weight for unit length*: $w_h \stackrel{\text{def}}{=} W_h/l$.

We remark that, in a one-dimensional model, the axial force $N$ is associated with dual points; hence, it is at the same time a global variable and a field variable. In contrast, in a three-dimensional model, $N$ admits a density $\sigma \stackrel{\text{def}}{=} N/A'$, where $A'$ is the area of the cross section of the rod contracted due to the lateral contraction.

These four global variables and the corresponding densities with their associations are summarized as follows:

$$
\begin{array}{lcccc}
 & \multicolumn{2}{c}{\text{Configuration variables}} & \multicolumn{2}{c}{\text{Source variables}} \\
\text{Global variables} & u[\overline{\mathbf{P}}] & \eta[\overline{\mathbf{L}}] & N[\widetilde{\mathbf{P}}] & W[\widetilde{\mathbf{L}}] \\
 & \downarrow & \downarrow & \downarrow & \downarrow \\
\text{Field variables} & u[\overline{\mathbf{P}}] & \varepsilon[\overline{\mathbf{L}}] & N[\widetilde{\mathbf{P}}] & w[\widetilde{\mathbf{L}}]
\end{array}
\tag{11.13}
$$

We can place all these variables and the equations which link them in the classification diagram shown in diagram 11.7; in the case of linear elastic materials, the constitutive relation which links configuration with source variables is Hooke's law. Combining the three equations we obtain the fundamental equation shown in the centre of Table 11.7.

**Table 11.7** Diagram of discrete analysis of the deformation of a rod under axial loading

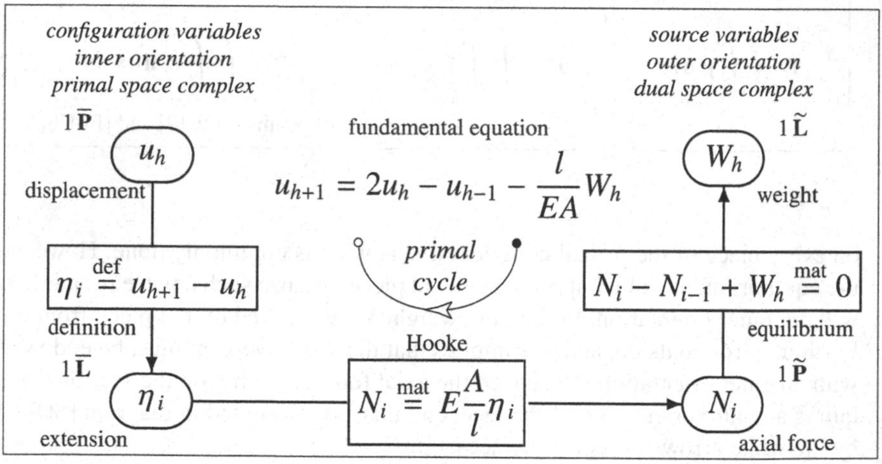


The corresponding diagram for the differential formulation is shown in diagram [SOL0].

## 11.7 Vibrations in One Dimension

**Longitudinal Vibrations of a Rod.** The diagram for the longitudinal vibrations of a rod, [SOL10], can be obtained by a fusion of the diagram of rod traction in [SOL0] with the diagram of particle dynamics in Fig. 8.6 (p. 227). The displacement $u$ of a section of the rod from its reference position depends on the two variables $t$ and $z$, hence $u(t, z)$. As diagram [SOL10] shows, it is convenient to represent the velocity $v(t, z)$ in a shifted position with respect to the column of $u$ and $\varepsilon$ and the momentum $p(t, z)$ in a shifted position with respect to the column of $f$ and $\sigma$. This shift is always made whenever time is used in the theory. In this way, the space-time diagram looks like an axonometric view of a three-dimensional frame or of a building. The advantage of this representation is that it gives separate boxes for the equations containing only space or only time derivatives and, at the same time, gives a clear representation of the equations containing both time and space derivatives.

**Transversal Vibrations of a String.** We showed in diagram [VIB] (p. 228) the composition of the diagram of transversal vibrations of strings starting from the diagram of statics of strings (p. 228) and that of the dynamics of a particle, diagram [PAR1] (p. 268).

## 11.8 Classification Diagrams of Deformable Solids

All diagrams refer to the theory of small displacements from a reference configuration. We start with the simplest cases, the deformation of beams, which is the most treated subject in the theory of the strength of materials. To these simple cases are devoted the diagrams [SOL0]–[SOL5]. Diagram [SOL0] can be compared with Fig. 4.25 (p. 79) in order to understand the need for the dual cell complex in view of a discrete treatment of continuum mechanics. Diagrams [SOL1]–[SOL3] deal with bending, torsion and beam shear. Diagram [SOL4] refers to the classical theory of Bernoulli–Euler beams in which the plane sections which are normal to the axis of the beam remain plane and normal to the deformed axis. Diagram [SOL5] refers to the more sophisticated theory of Timoshenko, in which the plane sections which are normal to the axis of the beam remain plane but no longer normal to the deformed axis.

Diagram [SOL6] deals with plane elasticity, while diagram [SOL7] deals with the theory of plates under plane loads and diagram [SOL8] deals with plates under normal loads.

Diagram [SOL9] deals with three-dimensional elastostatics, while diagram [SOL12] deals with three-dimensional elastodynamics. Diagrams [SOL10] and [SOL11] deal with longitudinal and torsional vibrations of rods. Diagram [SOL13] deals with the statics of strings, whereas diagram [SOL14] deals with the transversal vibrations of strings.

Traction of rods
[SOL0]
configuration variables
space: primal complex
time: primal complex
source variables
space: dual complex
time: dual complex
[Ī, P̄]
u
[T̃, L̃]
w
$\langle w, u\rangle \stackrel{\text{def}}{=} \int_0^L w\,(u\,\mathrm{d}z)$
definition
$\varepsilon = \dfrac{\mathrm{d}u}{\mathrm{d}z}$
primal cycle
$-\dfrac{\mathrm{d}N}{\mathrm{d}z} = w$
equilibrium
constitutive equation
[Ī, L̄]
ε
$N \stackrel{\text{mat}}{=} E\,A\,\varepsilon$
[T̃, P̃]
N
$\langle N, \varepsilon\rangle \stackrel{\text{def}}{=} \int_0^L N\,(\varepsilon\,\mathrm{d}z)$

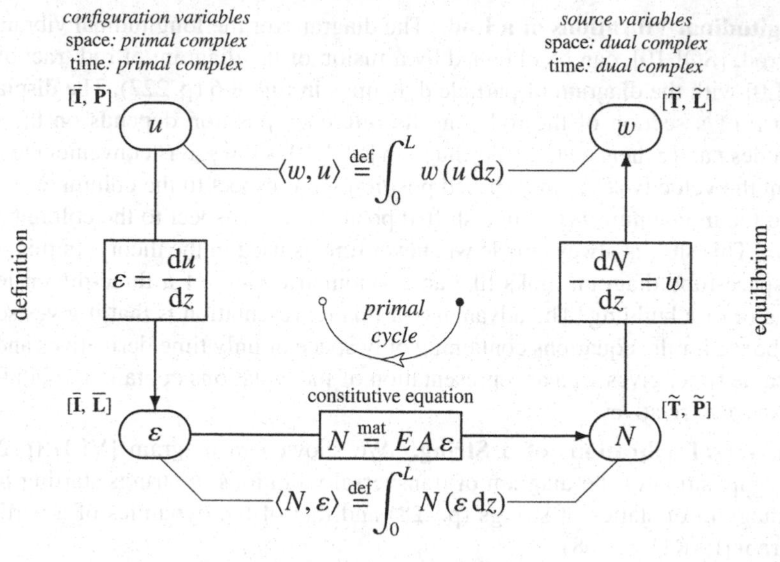

fundamental equation
$-\dfrac{\mathrm{d}}{\mathrm{d}z}\left[E\,A\left(\dfrac{\mathrm{d}u}{\mathrm{d}z}\right)\right] = w$
configuration variables
inner orientation
primal complex
source variables
dual complex
outer orientation
$\eta_i$
$u_h$
$u_{h+1}$
$z_h$
$z_{h+1}$
$\bar{\mathbf{P}}_h$
$\bar{\mathbf{L}}_i$
$\bar{\mathbf{P}}_{h+1}$
$\tilde{\mathbf{P}}_{i-1}$
$\tilde{\mathbf{L}}_h$
$\tilde{\mathbf{P}}_i$
$N_{i-1}$
$z_{i-1}$
$z_i$
$N_i$
$W_h$
z

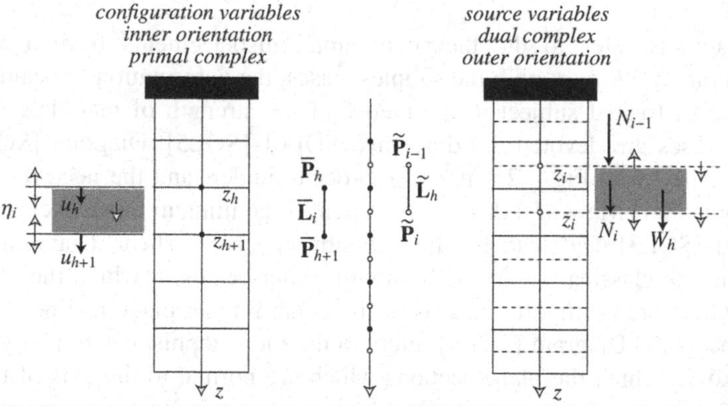

u displacement
ε strain
w axial body force / length
N traction
E elastic modulus
A area of the cross section
Ī time even T̃
T̄ time odd Ĩ

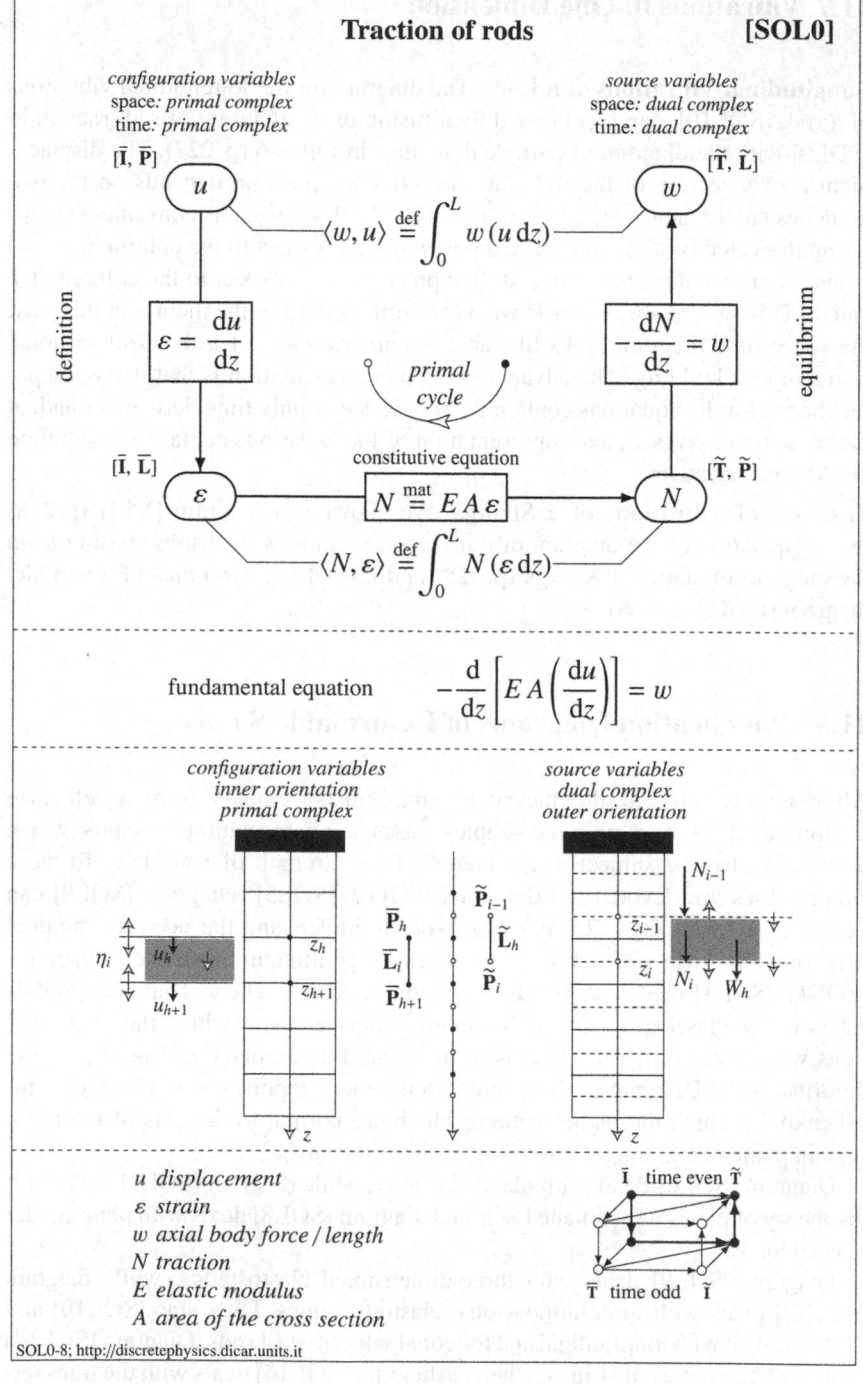

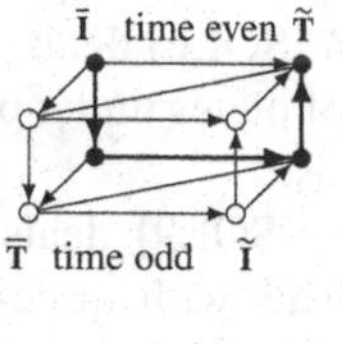

SOL0-8; http://discretephysics.dicar.units.it

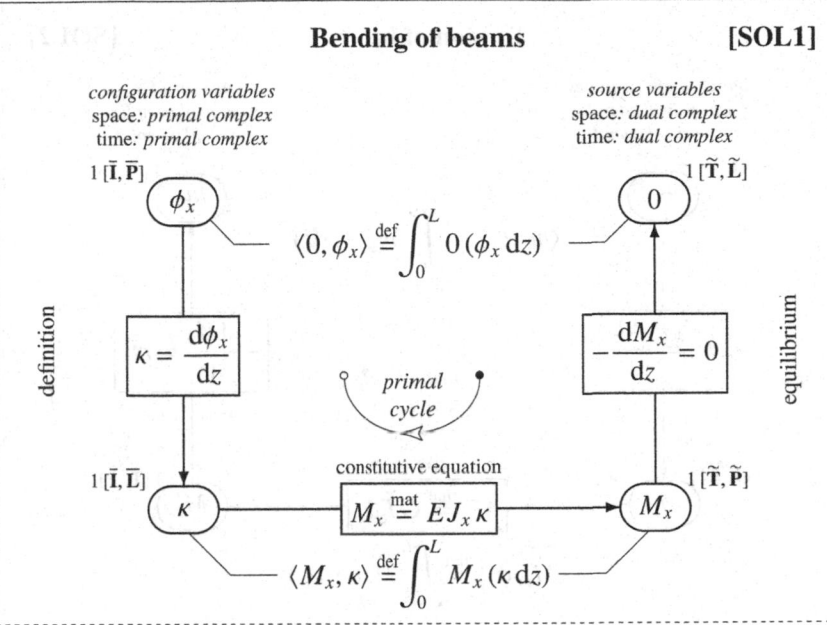

## Bending of beams [SOL1]

*configuration variables*
space: *primal complex*
time: *primal complex*

*source variables*
space: *dual complex*
time: *dual complex*

$1\,[\bar{\mathbf{I}},\bar{\mathbf{P}}]$ $\phi_x$

$1\,[\tilde{\mathbf{T}},\tilde{\mathbf{L}}]$ $0$

$$\langle 0, \phi_x \rangle \stackrel{\text{def}}{=} \int_0^L 0\,(\phi_x\,\mathrm{d}z)$$

definition

$$\kappa = \frac{\mathrm{d}\phi_x}{\mathrm{d}z}$$

equilibrium

$$-\frac{\mathrm{d}M_x}{\mathrm{d}z} = 0$$

*primal cycle*

$1\,[\bar{\mathbf{I}},\bar{\mathbf{L}}]$ $\kappa$

constitutive equation

$$M_x \stackrel{\text{mat}}{=} EJ_x\,\kappa$$

$1\,[\tilde{\mathbf{T}},\tilde{\mathbf{P}}]$ $M_x$

$$\langle M_x, \kappa \rangle \stackrel{\text{def}}{=} \int_0^L M_x\,(\kappa\,\mathrm{d}z)$$

fundamental equation
$$-\frac{\mathrm{d}}{\mathrm{d}z}\left[EJ_x\left(\frac{\mathrm{d}\phi_x}{\mathrm{d}z}\right)\right] = 0$$

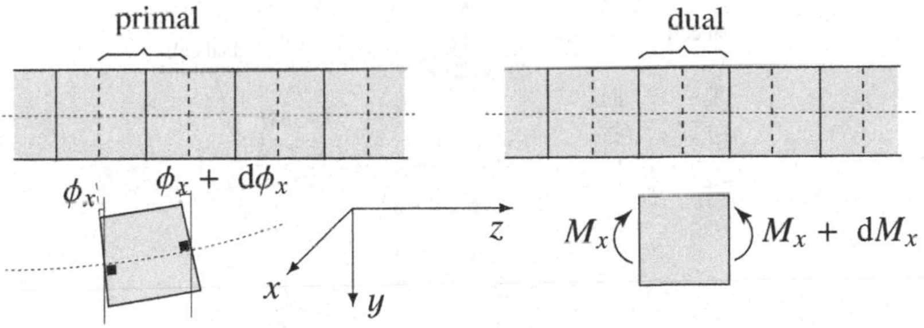


*normal cross sections remain plane and normal to the neutral surface*

$\phi_x$ *angle formed by the tangent and the x-axis*
$\kappa$ *curvature*
$M_x$ *bending moment*
$E$ *elastic modulus*
$J_x$ *second order moment*

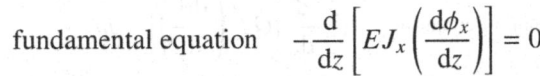

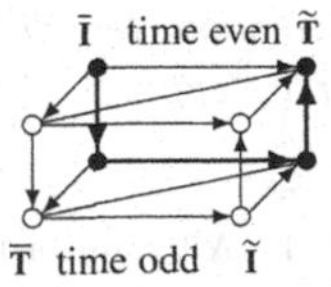


SOL1-7; http://discretephysics.dicar.units.it

## Torsion of beams [SOL2]

*configuration variables*
space: *primal complex*
time: *primal complex*

*source variables*
space: *dual complex*
time: *dual complex*

1 [$\bar{\mathbf{I}}$, $\bar{\mathbf{P}}$] $\theta_z$

1 [$\tilde{\mathbf{T}}$, $\tilde{\mathbf{L}}$] $m_z$

$$\langle m_z, \theta_z \rangle \stackrel{\text{def}}{=} \int_0^L m_z\,(\phi_z\,\mathrm{d}z)$$

definition

$$\phi_z \stackrel{\text{def}}{=} \frac{\mathrm{d}\theta_z}{\mathrm{d}z}$$

*primal cycle*

equilibrium

$$-\frac{\mathrm{d}M_z}{\mathrm{d}z} = m_z$$

1 [$\bar{\mathbf{I}}$, $\bar{\mathbf{L}}$] $\phi_z$

constitutive equation

$$M_z \stackrel{\text{mat}}{=} GJ_z\phi_z$$

1 [$\tilde{\mathbf{T}}$, $\tilde{\mathbf{P}}$] $M_z$

$$\langle M_z, \phi_z \rangle \stackrel{\text{def}}{=} \int_0^L M_z\,(\phi_z\,\mathrm{d}z)$$

fundamental equation

$$-\frac{\mathrm{d}}{\mathrm{d}z}\left[GJ_z\left(\frac{\mathrm{d}\theta_z}{\mathrm{d}z}\right)\right] = m_z$$

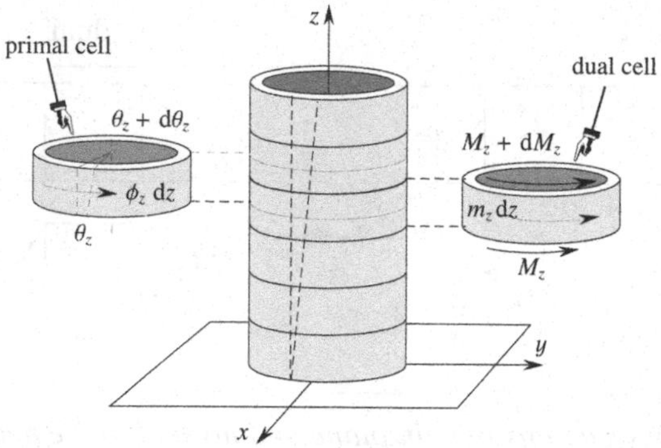


$\theta_z$ *torsional angle*
$\phi_z$ *torsional rotation / length*
$m_z$ *axial couple / length*

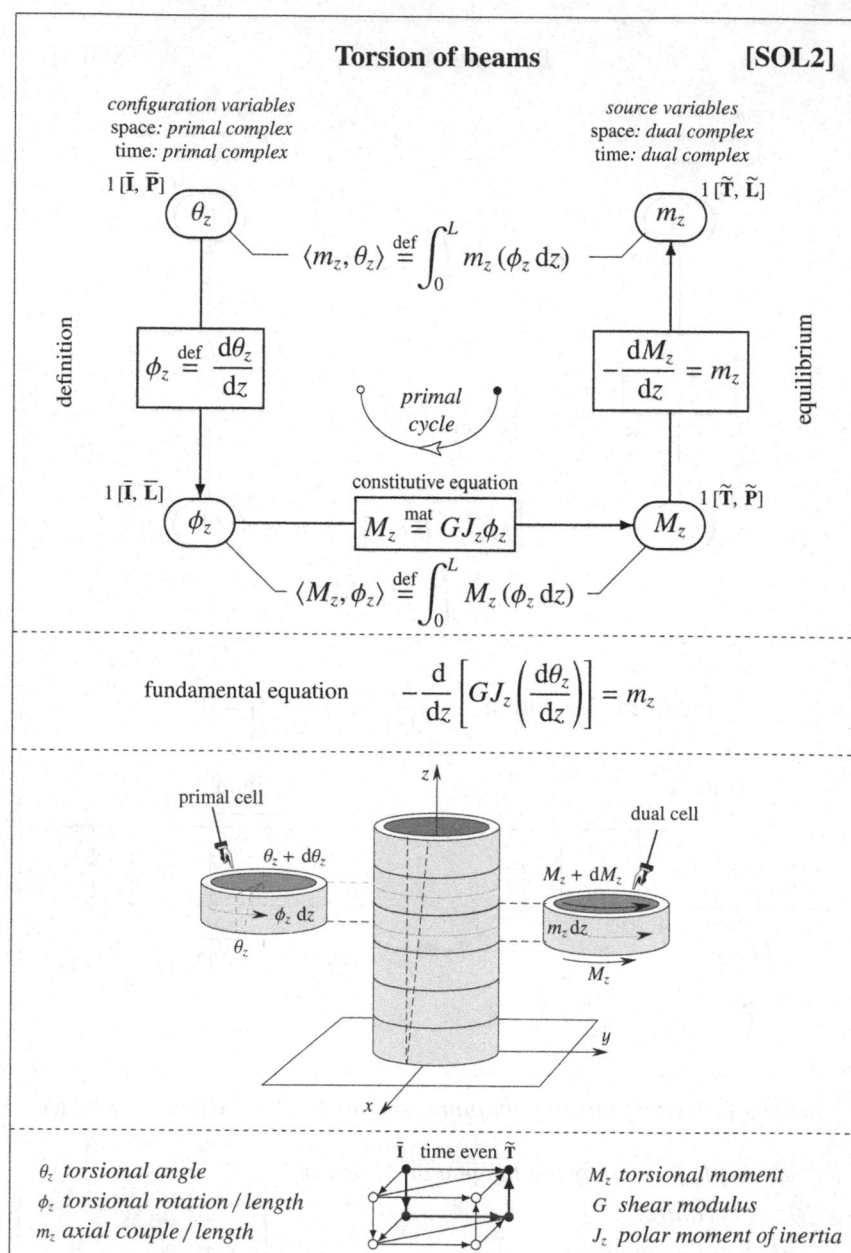

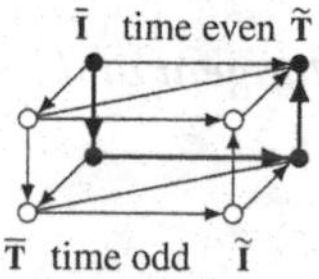


$M_z$ *torsional moment*
$G$ *shear modulus*
$J_z$ *polar moment of inertia*

Ref: Love, A. E. H.: A Treatise on the Mathematical Theory of Elasticity. Dover, Ch. XIV (1944).
Ref: Belluzzi, O.: Scienza delle Costruzioni. Zanichelli, Vol. I, p. 209 (1966).

SOL2-10; http://discretephysics.dicar.units.it

**Shear of beams**      **[SOL3]**

*configuration variables*
space*: primal complex*
time*: primal complex*

*source variables*
space*: dual complex*
time*: dual complex*

$1\,[\bar{\mathbf{I}}, \bar{\mathbf{P}}]$

$1\,[\tilde{\mathbf{T}}, \tilde{\mathbf{L}}]$

$y$

$q$

$\langle q, y \rangle \overset{\text{def}}{=} \int_0^L q\,(y\,\mathrm{d}z)$

definition

$\gamma = \dfrac{\mathrm{d}y}{\mathrm{d}z}$

$-\dfrac{\mathrm{d}V}{\mathrm{d}z} = q$

equilibrium

*primal cycle*

constitutive equation

$1\,[\bar{\mathbf{I}}, \bar{\mathbf{L}}]$

$1\,[\tilde{\mathbf{T}}, \tilde{\mathbf{P}}]$

$\gamma$

$V \overset{\text{mat}}{=} \dfrac{GA}{\chi}\gamma$

$V$

$\langle V, \gamma \rangle \overset{\text{def}}{=} \int_0^L V\,(\gamma\,\mathrm{d}z)$

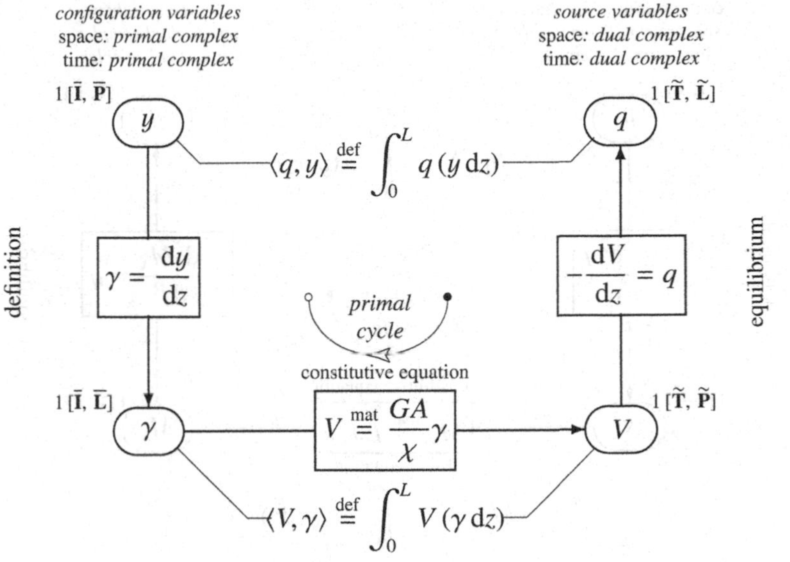

fundamental equation     $-\dfrac{\mathrm{d}}{\mathrm{d}z}\left[\dfrac{GA}{\chi}\left(\dfrac{\mathrm{d}y}{\mathrm{d}z}\right)\right] = q$

$z$

$1\!-\!\chi$

$V$    $V+dV$

$y$

$q\,\mathrm{d}z$

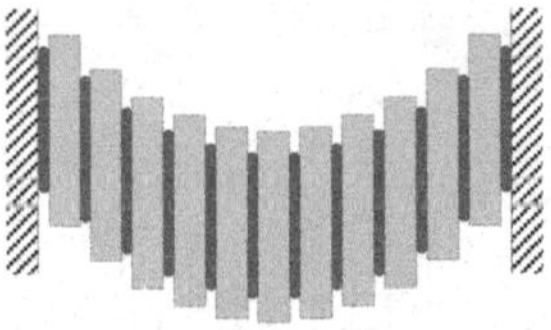

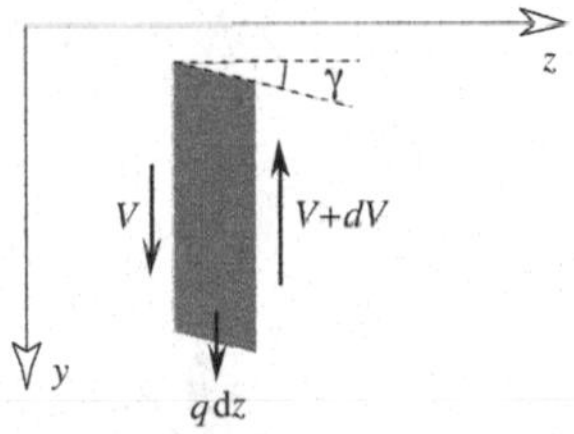

$y$ *transversal displacement*
$\gamma$ *shear angle*
$q$ *transversal force / length*
$V$ *shear force*
$\chi$ *shear factor*
$G$ *shear modulus*
$A$ *area of the cross section*

$\tilde{\mathbf{I}}$ time odd   $\bar{\mathbf{T}}$

$\tilde{\mathbf{T}}$ time even   $\bar{\mathbf{I}}$

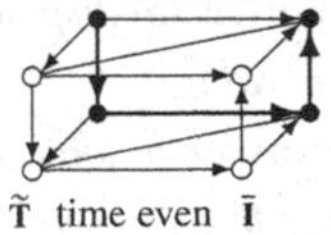

SOL3-6; http://discretephysics.dicar.units.it

## Theory of beams: Bernoulli-Euler [SOL4]

*small displacements*

*configuration variables*
space: *primal complex*
time: *primal complex*

*source variables*
space: *dual complex*
time: *dual complex*

definition

equilibrium

$y$ 1 [$\bar{\mathbf{I}}$, $\bar{\mathbf{P}}$]

$q$ 1 [$\tilde{\mathbf{T}}$, $\tilde{\mathbf{L}}$]

$$\langle q, y\rangle \stackrel{\text{def}}{=} \int_0^L q\,(y\,\mathrm{d}z)$$

$$\kappa = \frac{\mathrm{d}^2 y}{\mathrm{d}z^2}$$

*primal cycle*

$$-\frac{\mathrm{d}^2 M_x}{\mathrm{d}z^2} = q$$

$\kappa$ 1 [$\bar{\mathbf{I}}$, $\bar{\mathbf{L}}$]

$M_x$ 1 [$\tilde{\mathbf{T}}$, $\tilde{\mathbf{P}}$]

constitutive equation

$$M_x \stackrel{\text{mat}}{=} E J_x\,\kappa$$

Euler equation

$$\langle M_x, \kappa\rangle \stackrel{\text{def}}{=} \int_0^L M_x(\kappa\,\mathrm{d}z)$$

fundamental equation $$-\frac{\mathrm{d}^2}{\mathrm{d}z^2}\left[E J_x \frac{\mathrm{d}^2 y}{\mathrm{d}z^2}\right] = q$$

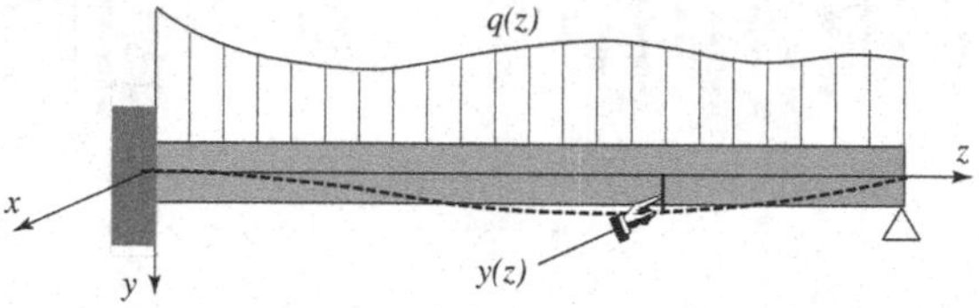


$y$ *transversal displacement*
$\kappa$ *curvature*
$q$ *transversal load / length*
$M_x$ *bending moment*
$E$ *elastic modulus*
$J_x$ *second order moment*

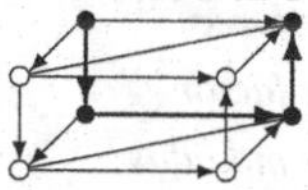


SOL4-7; http://discretephysics.dicar.units.it

## Theory of beams: Timoshenko theory [SOL5]

*configuration variables*
space: *primal complex*
time: *primal complex*

*source variables*
space: *dual complex*
time: *dual complex*

$1\,[\bar{\mathbf{I}}, \bar{\mathbf{P}}]$ $(y, \phi_x)$

$1\,[\tilde{\mathbf{T}}, \tilde{\mathbf{L}}]$ $(q, m_x)$

$$\langle q, y\rangle + \langle m_x, \phi_x\rangle \stackrel{\text{def}}{=} \int_0^L [q(y\,\mathrm{d}z) + m_x(\phi_x\,\mathrm{d}z)]$$

definition

$$\gamma = \frac{\mathrm{d}y}{\mathrm{d}z} - \phi_x \qquad \kappa = \frac{\mathrm{d}\phi_x}{\mathrm{d}z}$$

*primal cycle*

equilibrium

$$\frac{\mathrm{d}M_x}{\mathrm{d}z} - V = m_x \qquad -\frac{\mathrm{d}V}{\mathrm{d}z} = q$$

constitutive equation

$$V \stackrel{\text{mat}}{=} \frac{GA}{\chi}\gamma \qquad M_x \stackrel{\text{mat}}{=} EJ_x\,\kappa$$

$1\,[\bar{\mathbf{I}}, \bar{\mathbf{L}}]$ $(\gamma, \kappa)$

$1\,[\tilde{\mathbf{T}}, \tilde{\mathbf{P}}]$ $(V, M_x)$

$$\langle V, y\rangle + \langle M_x, \phi_x\rangle \stackrel{\text{def}}{=} \int_0^L [V(y\,\mathrm{d}z) + M_x(\phi_x\,\mathrm{d}z)]$$

fundamental equations

$$\frac{\mathrm{d}}{\mathrm{d}z}\left[EJ_x\frac{\mathrm{d}\phi_x}{\mathrm{d}z}\right] - \frac{GA}{\chi}\left[\frac{\mathrm{d}y}{\mathrm{d}z} - \phi_x\right] = m_x$$

$$-\frac{\mathrm{d}}{\mathrm{d}z}\left[\frac{GA}{\chi}\left(\frac{\mathrm{d}y}{\mathrm{d}z} - \phi_x\right)\right] = q$$

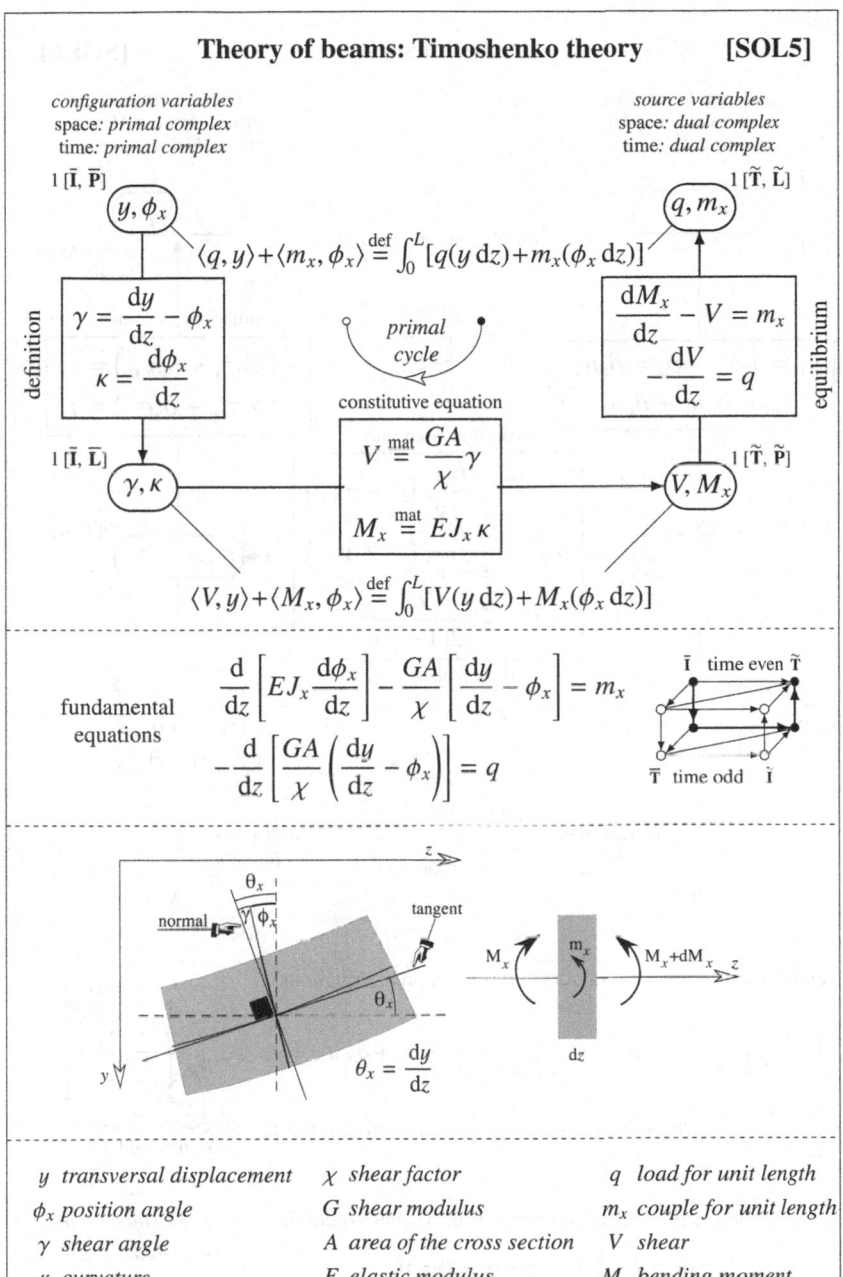

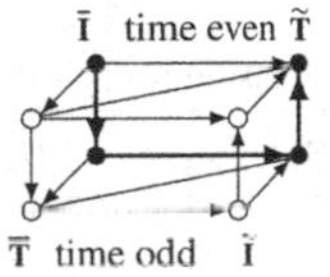


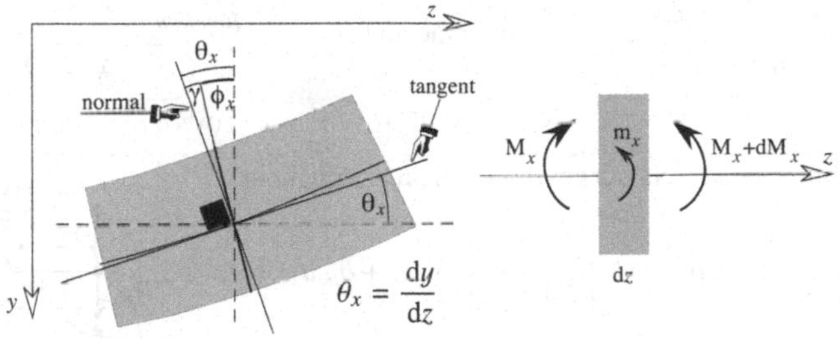


$y$ *transversal displacement*
$\phi_x$ *position angle*
$\gamma$ *shear angle*
$\kappa$ *curvature*
$\chi$ *shear factor*
$G$ *shear modulus*
$A$ *area of the cross section*
$E$ *elastic modulus*
$J_x$ *second order moment*
$q$ *load for unit length*
$m_x$ *couple for unit length*
$V$ *shear*
$M_x$ *bending moment*

SOL5-9; http://discretephysics.dicar.units.it

## Plane Elasticity **[SOL6]**

*configuration variables*
space: *primal complex*
time: *primal complex*

*source variables*
space: *dual complex*
time: *dual complex*

1 [$\bar{\mathbf{I}}$, $\bar{\mathbf{P}}$] $u_x, u_y$ *displacement*

1 [$\tilde{\mathbf{T}}$, $\tilde{\mathbf{S}}$] $f_x, f_y$ *surface force density*

displacement method

*primal cycle*

$$\varepsilon_x = \partial_x u_x \qquad \varepsilon_y = \partial_y u_y$$
$$\gamma_{xy} = \partial_x u_y + \partial_y u_x$$

equilibrium condition

$$-\left(\partial_x \sigma_x + \partial_y \tau_{xy}\right) = f_x$$
$$-\left(\partial_x \tau_{xy} + \partial_y \sigma_y\right) = f_y$$

constitutive equation

$$\sigma_x \overset{\text{mat}}{=} \frac{E}{1-\nu^2}\left(\varepsilon_x + \nu\varepsilon_y\right)$$
$$\sigma_y \overset{\text{mat}}{=} \frac{E}{1-\nu^2}\left(\varepsilon_y + \nu\varepsilon_x\right)$$
$$\tau_{xy} \overset{\text{mat}}{=} \frac{E}{2(1+\nu)}\gamma_{xy}$$

Hooke, plane stress

2 [$\bar{\mathbf{I}}$, $\bar{\mathbf{L}}$] $\varepsilon_x, \varepsilon_y, \gamma_{xy}$ *strain tensor*

2 [$\tilde{\mathbf{T}}$, $\tilde{\mathbf{L}}$] $\sigma_x, \sigma_y, \tau_{xy}$ *stress tensor*

compatibility condition

$$\partial_{yy}\varepsilon_x + \partial_{xx}\varepsilon_y - \partial_{xy}\gamma_{xy} = 0$$

$$\sigma_x = +\partial_{yy}\check{\phi}$$
$$\sigma_y = +\partial_{xx}\check{\phi}$$
$$\tau_{xy} = -\partial_{xy}\check{\phi}$$

*dual cycle*

force method

1 [$\bar{\mathbf{I}}$, $\bar{\mathbf{S}}$] $\check{\eta}$ *incompatibility function*

1 [$\tilde{\mathbf{T}}$, $\tilde{\mathbf{P}}$] $\check{\phi}$ *Airy stress function* (when $f_x = 0$ $f_y = 0$)

---

fundamental equation (Navier) two second order equations

$$\begin{cases} -\left[\dfrac{E}{1-\nu^2}\right](\partial_{xx}u_x + \nu\,\partial_{xy}u_y) - \left[\dfrac{E}{2(1+\nu)}\right](\partial_{xy}u_y + \partial_{yy}u_x) = f_x \\ -\left[\dfrac{E}{2(1+\nu)}\right](\partial_{xx}u_y + \partial_{xy}u_x) - \left[\dfrac{E}{1-\nu^2}\right](\partial_{yy}u_y + \nu\,\partial_{yx}u_x) = f_y \end{cases}$$

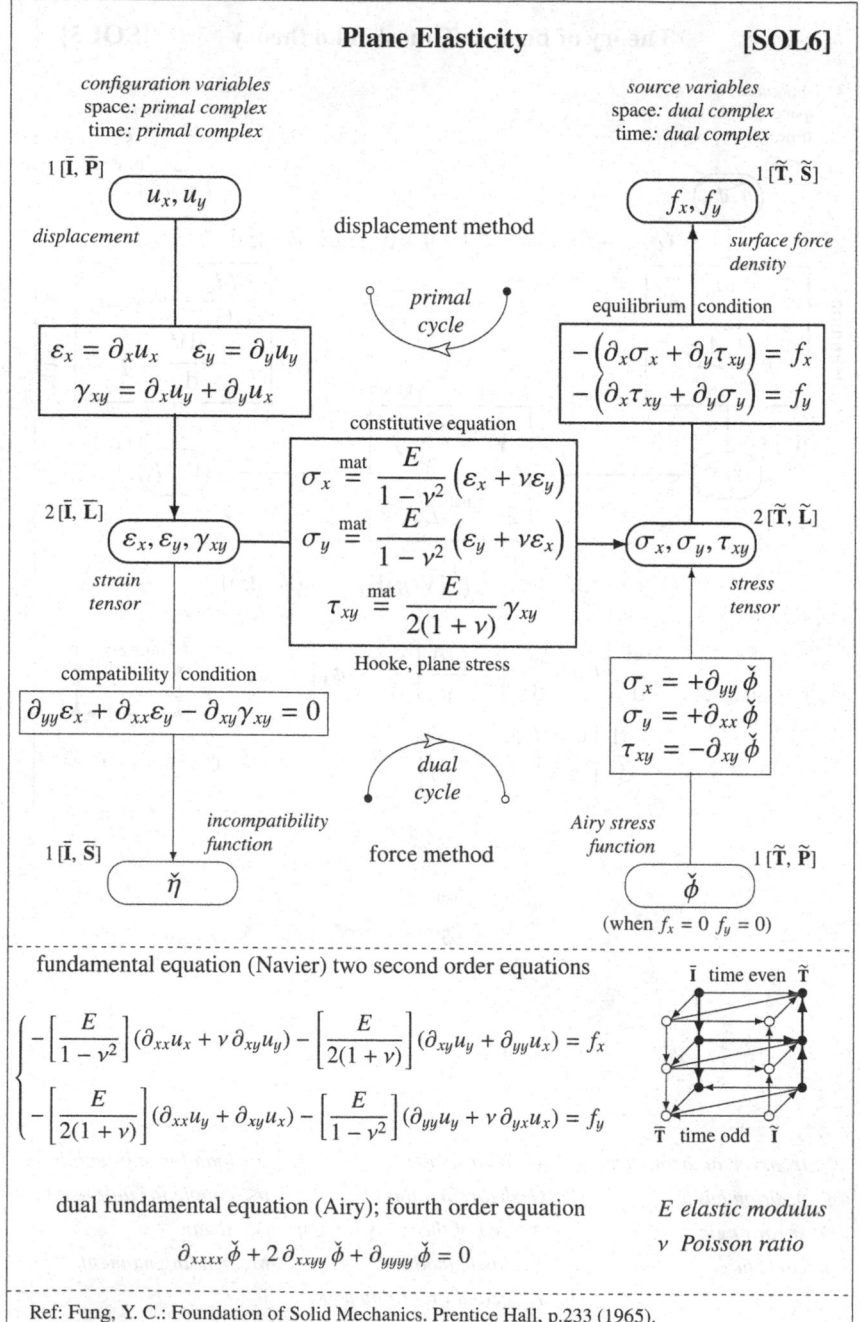

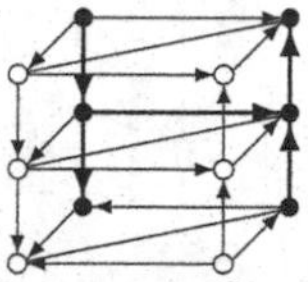


dual fundamental equation (Airy); fourth order equation

$$\partial_{xxxx}\check{\phi} + 2\,\partial_{xxyy}\check{\phi} + \partial_{yyyy}\check{\phi} = 0$$

*E elastic modulus*
*ν Poisson ratio*

---

Ref: Fung, Y. C.: Foundation of Solid Mechanics. Prentice Hall, p.233 (1965).
Ref: Timoshenko, S. and Goodier, J. N.: Theory of Elasticity. Mc Graw Hill, p.11 (1951).
SOL6-8; http://discretephysics.dicar.units.it

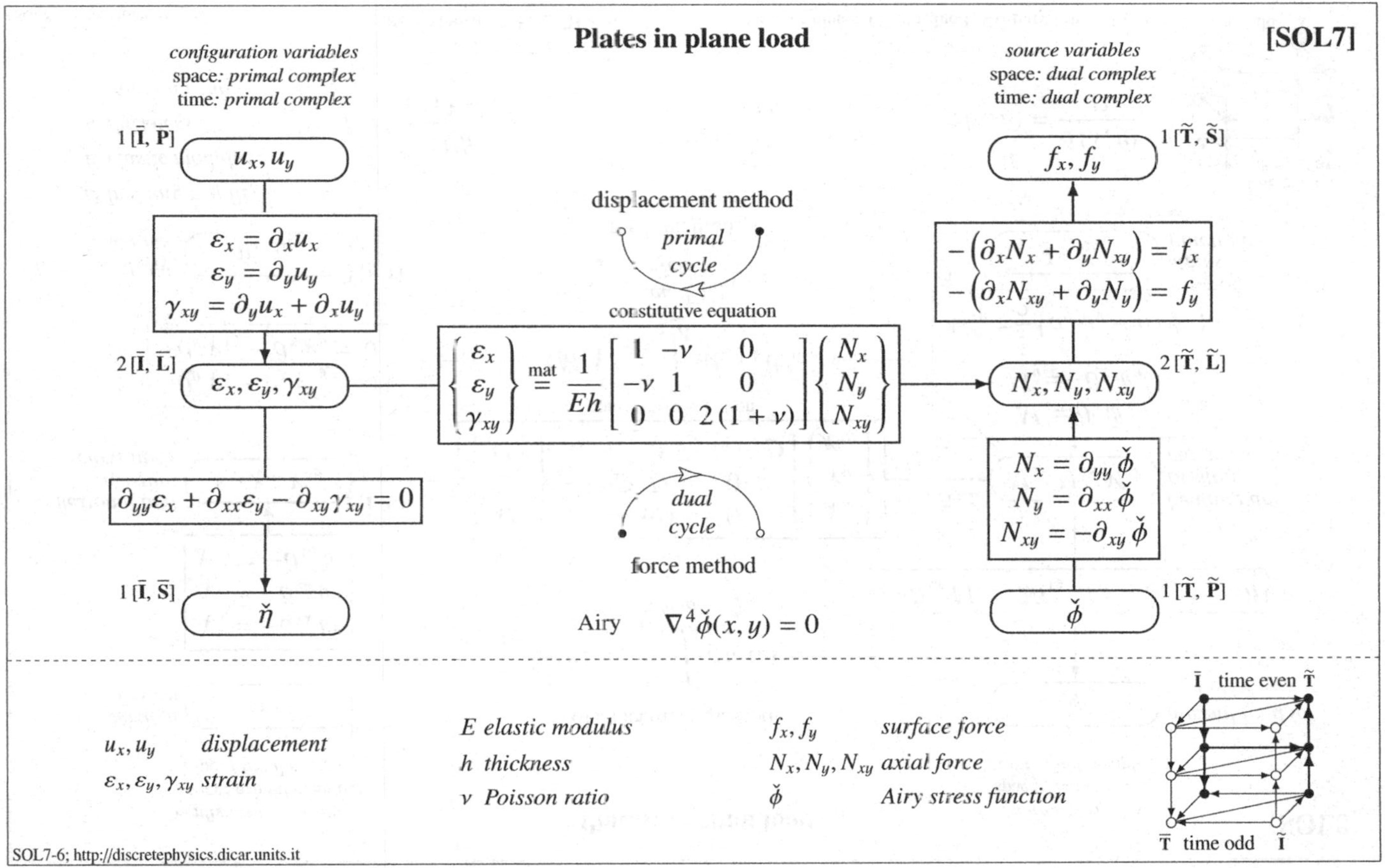
Plates in plane load
[SOL7]
configuration variables
space: primal complex
time: primal complex
source variables
space: dual complex
time: dual complex
1 [Ī, P̄]
$u_x, u_y$
$\varepsilon_x = \partial_x u_x$
$\varepsilon_y = \partial_y u_y$
$\gamma_{xy} = \partial_y u_x + \partial_x u_y$
2 [Ī, L̄]
$\varepsilon_x, \varepsilon_y, \gamma_{xy}$
$\partial_{yy}\varepsilon_x + \partial_{xx}\varepsilon_y - \partial_{xy}\gamma_{xy} = 0$
1 [Ī, S̄]
$\check{\eta}$
displacement method
primal
cycle
constitutive equation
$\begin{Bmatrix} \varepsilon_x \\ \varepsilon_y \\ \gamma_{xy} \end{Bmatrix} \overset{\text{mat}}{=} \frac{1}{Eh} \begin{bmatrix} 1 & -\nu & 0 \\ -\nu & 1 & 0 \\ 0 & 0 & 2(1+\nu) \end{bmatrix} \begin{Bmatrix} N_x \\ N_y \\ N_{xy} \end{Bmatrix}$
dual
cycle
force method
Airy $\nabla^4 \check{\phi}(x, y) = 0$
1 [T̃, S̃]
$f_x, f_y$
$-\left(\partial_x N_x + \partial_y N_{xy}\right) = f_x$
$-\left(\partial_x N_{xy} + \partial_y N_y\right) = f_y$
2 [T̃, L̃]
$N_x, N_y, N_{xy}$
$N_x = \partial_{yy}\check{\phi}$
$N_y = \partial_{xx}\check{\phi}$
$N_{xy} = -\partial_{xy}\check{\phi}$
1 [T̃, P̃]
$\check{\phi}$
$u_x, u_y$ displacement
$\varepsilon_x, \varepsilon_y, \gamma_{xy}$ strain
$E$ elastic modulus
$h$ thickness
$\nu$ Poisson ratio
$f_x, f_y$ surface force
$N_x, N_y, N_{xy}$ axial force
$\check{\phi}$ Airy stress function
Ī time even T̃
T̄ time odd Ĩ
SOL7-6; http://discretephysics.dicar.units.it

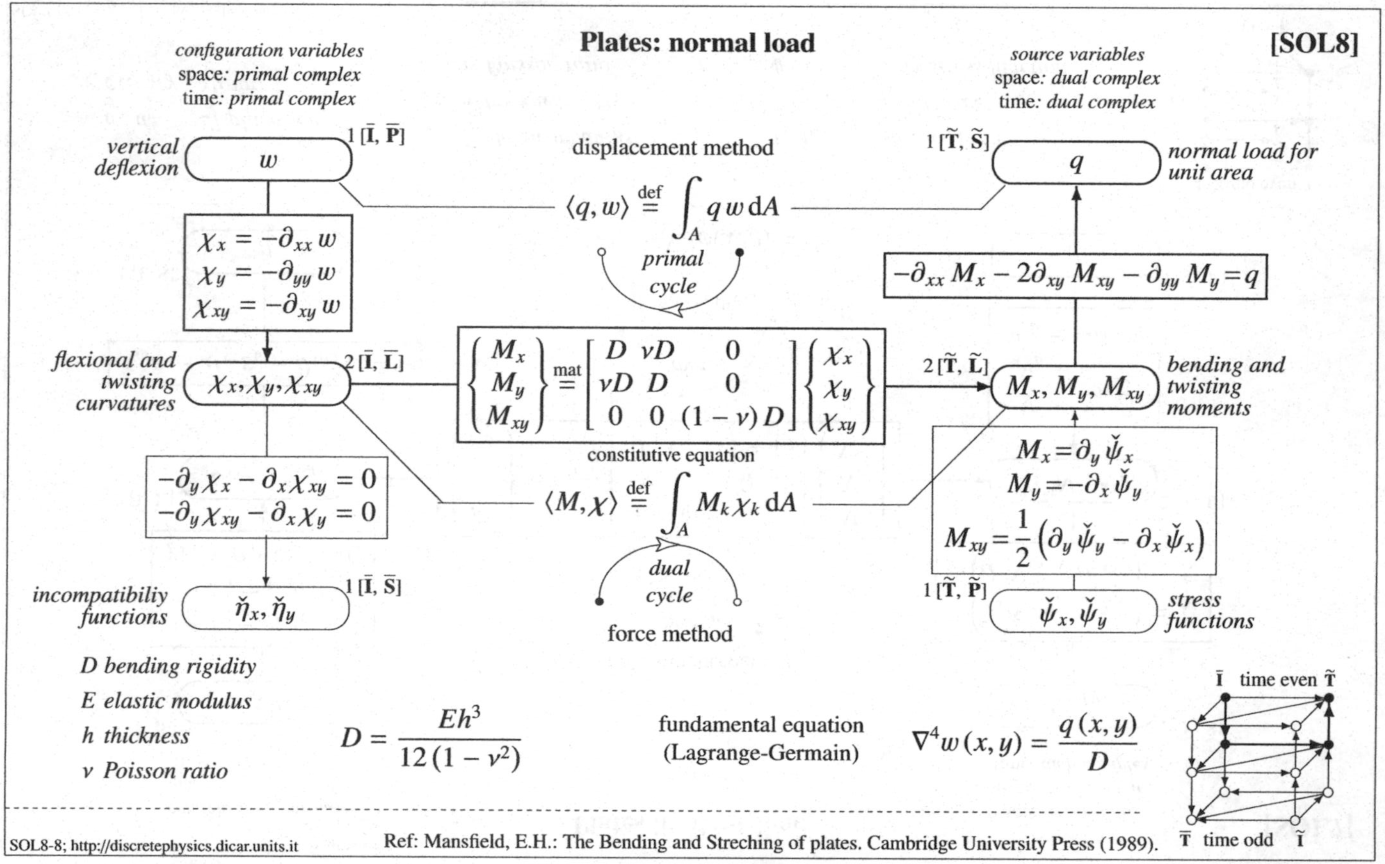
Plates: normal load
[SOL8]
configuration variables
space: primal complex
time: primal complex
source variables
space: dual complex
time: dual complex
vertical deflexion
w
1 [Ī, P̄]
displacement method
$\langle q, w \rangle \stackrel{\text{def}}{=} \int_A q\, w\, \mathrm{d}A$
primal cycle
1 [T̃, S̃]
q
normal load for unit area
$\chi_x = -\partial_{xx} w$
$\chi_y = -\partial_{yy} w$
$\chi_{xy} = -\partial_{xy} w$
$-\partial_{xx} M_x - 2\partial_{xy} M_{xy} - \partial_{yy} M_y = q$
flexional and twisting curvatures
$\chi_x, \chi_y, \chi_{xy}$
2 [Ī, L̄]
$\begin{Bmatrix} M_x \\ M_y \\ M_{xy} \end{Bmatrix} \stackrel{\text{mat}}{=} \begin{bmatrix} D & \nu D & 0 \\ \nu D & D & 0 \\ 0 & 0 & (1-\nu) D \end{bmatrix} \begin{Bmatrix} \chi_x \\ \chi_y \\ \chi_{xy} \end{Bmatrix}$
constitutive equation
2 [T̃, L̃]
$M_x, M_y, M_{xy}$
bending and twisting moments
$-\partial_y \chi_x - \partial_x \chi_{xy} = 0$
$-\partial_y \chi_{xy} - \partial_x \chi_y = 0$
$\langle M, \chi \rangle \stackrel{\text{def}}{=} \int_A M_k \chi_k\, \mathrm{d}A$
$M_x = \partial_y \check{\psi}_x$
$M_y = -\partial_x \check{\psi}_y$
$M_{xy} = \frac{1}{2}\left(\partial_y \check{\psi}_y - \partial_x \check{\psi}_x\right)$
incompatibiliy functions
$\check{\eta}_x, \check{\eta}_y$
1 [Ī, S̄]
dual cycle
force method
1 [T̃, P̃]
$\check{\psi}_x, \check{\psi}_y$
stress functions
D bending rigidity
E elastic modulus
h thickness
ν Poisson ratio
$D = \frac{E h^3}{12(1-\nu^2)}$
fundamental equation
(Lagrange-Germain)
$\nabla^4 w(x,y) = \frac{q(x,y)}{D}$
Ī time even T̃
T̄ time odd Ĩ
SOL8-8; http://discretephysics.dicar.units.it
Ref: Mansfield, E.H.: The Bending and Streching of plates. Cambridge University Press (1989).

## Elastostatics [SOL9]

*Cartesian notation*

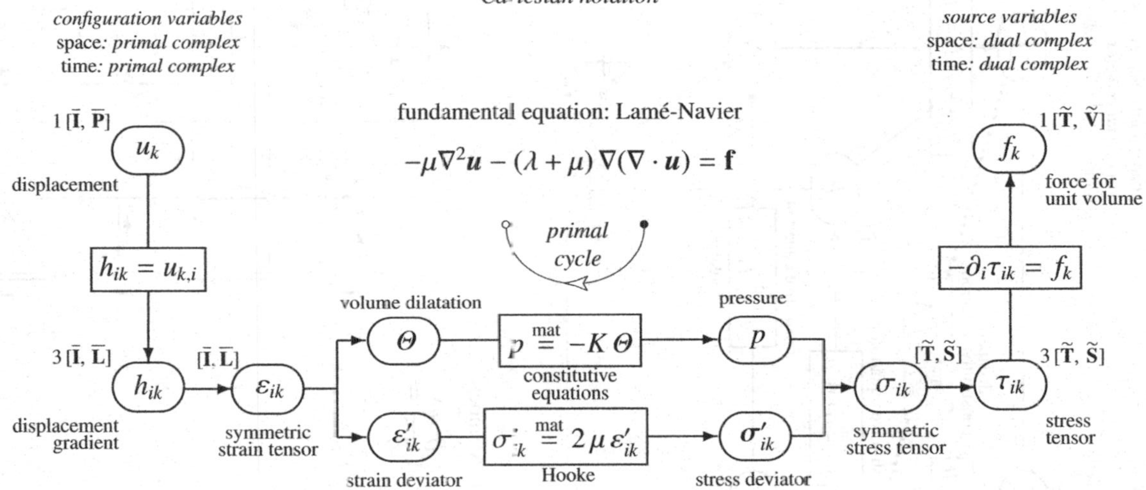


$$\varepsilon_{ik} \overset{\text{def}}{=} \frac{1}{2}(h_{ik} + h_{ik}) \qquad \varepsilon'_{ik} \overset{\text{def}}{=} \varepsilon_{ik} - \frac{1}{3}\Theta\,\delta_{ik}$$

$$\Theta \overset{\text{def}}{=} \sum_k \varepsilon_{kk} = \sum_k \frac{\partial u_k}{\partial x_k} = \nabla\cdot\mathbf{u} = \frac{\Delta V}{V}$$

$K = \frac{2}{3}\mu + \lambda$ *bulk modulus*

$\mu = G$ *shear modulus*

$$p = -\frac{1}{3}\sum_k \sigma_{kk} \qquad \sigma'_{ik} \overset{\text{def}}{=} \tau_{ik} - p\,\delta_{ik}$$

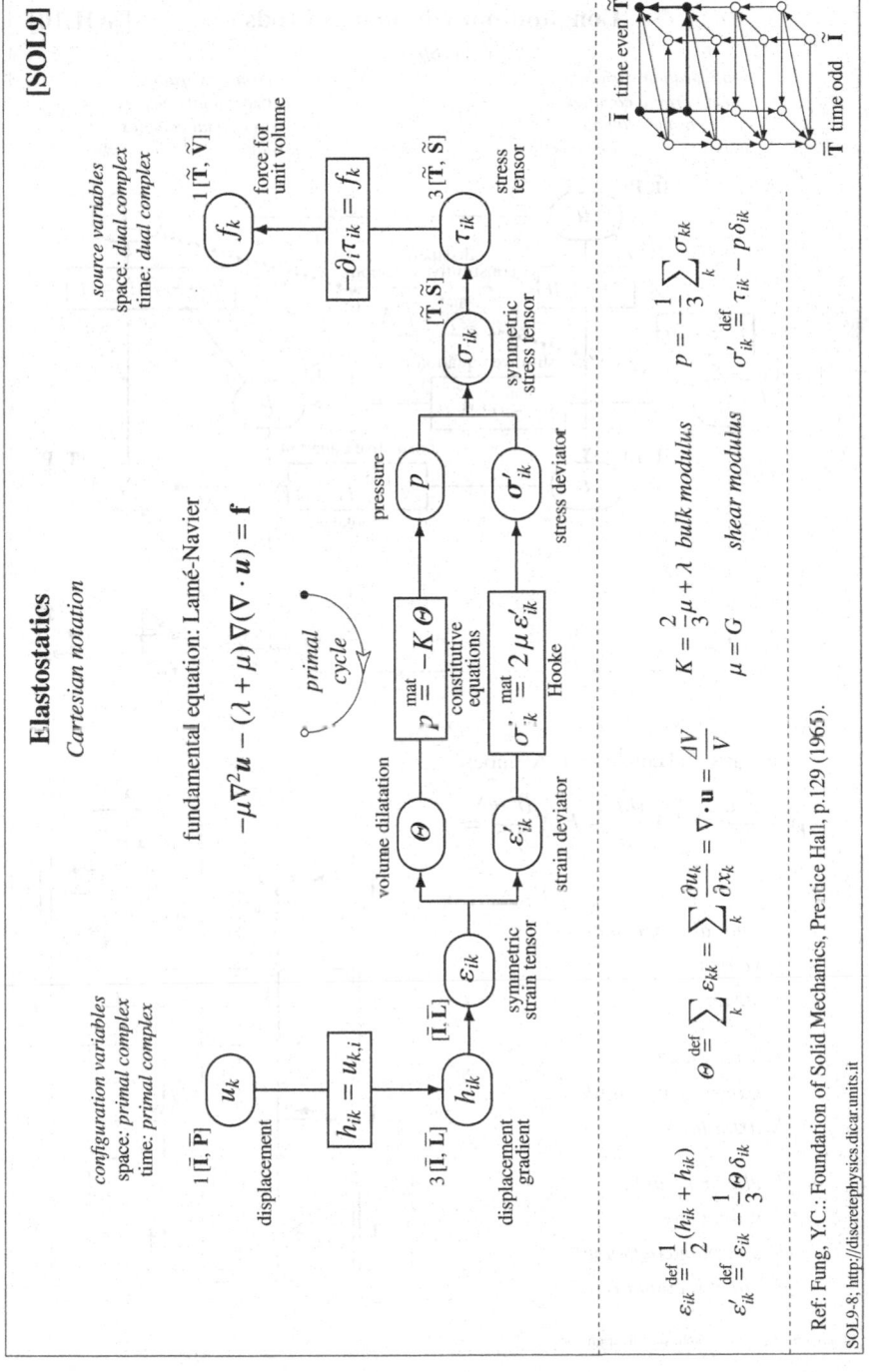

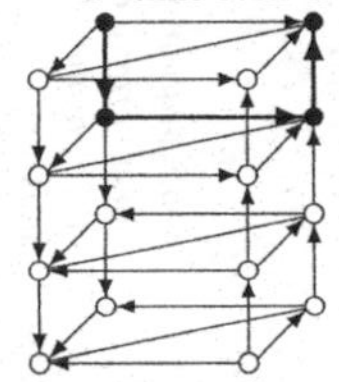


Ref: Fung, Y.C.: Foundation of Solid Mechanics, Prentice Hall, p.129 (1965).

SOL9-8; http://discretephysics.dicar.units.it

## Longitudinal vibrations of rods [SOL10]

*variables: t,z*

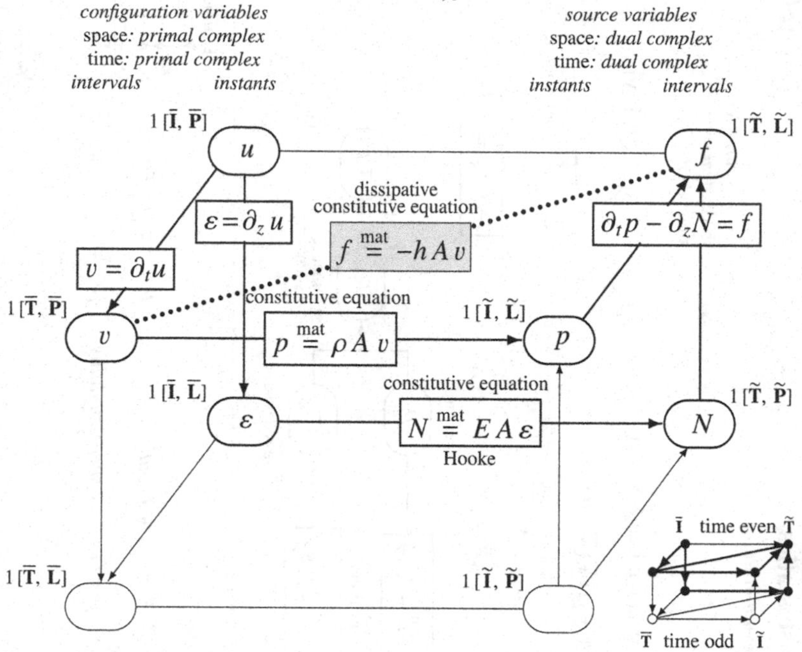


fundamental equation: d'Alembert

$$\rho A \frac{\partial^2 u}{\partial t^2} + hA \frac{\partial u}{\partial t} - E A \frac{\partial^2 u}{\partial z^2} = f$$

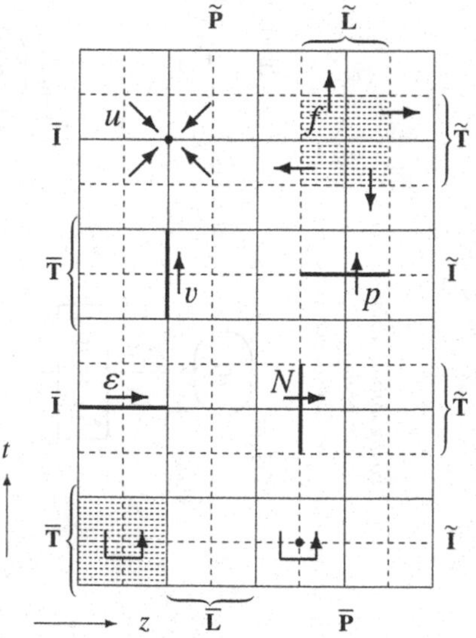


*u* *total displacement*
*v* *velocity*
*ε* *strain*

*f* *force / length*
*p* *momentum / length*
*N* *axial force*

*E* *elastic modulus*
*ρ* *mass density*
*h* *damping coefficient*
*A* *cross-section area*

# Torsional vibrations of rods [SOL11]

### variables: $t,z$

configuration variables
space: primal complex
time: primal complex
intervals        instants

source variables
space: dual complex
time: dual complex
instants        intervals

$1\,[\bar{\mathbf{I}},\bar{\mathbf{P}}]$

$\theta$

$\alpha = \partial_z \theta$

$\omega = \partial_t \theta$

dissipative constitutive equation

$\mu \overset{mat}{=} -h\,\omega$

$1\,[\bar{\mathbf{T}},\bar{\mathbf{P}}]$

$\omega$

constitutive equation

$L \overset{mat}{=} J\omega$

$1\,[\tilde{\mathbf{T}},\tilde{\mathbf{L}}]$

$\mu$

$\partial_t L - \partial_z \tau = \mu$

$1\,[\tilde{\mathbf{I}},\tilde{\mathbf{L}}]$

$L$

$1\,[\bar{\mathbf{I}},\bar{\mathbf{L}}]$

$\alpha$

constitutive equation

$\tau \overset{mat}{=} G\alpha$

$1\,[\tilde{\mathbf{T}},\tilde{\mathbf{P}}]$

$\tau$

$1\,[\bar{\mathbf{T}},\bar{\mathbf{L}}]$

$1\,[\tilde{\mathbf{I}},\tilde{\mathbf{P}}]$

$\bar{\mathbf{I}}$ time even $\tilde{\mathbf{T}}$

$\bar{\mathbf{T}}$ time odd $\tilde{\mathbf{I}}$

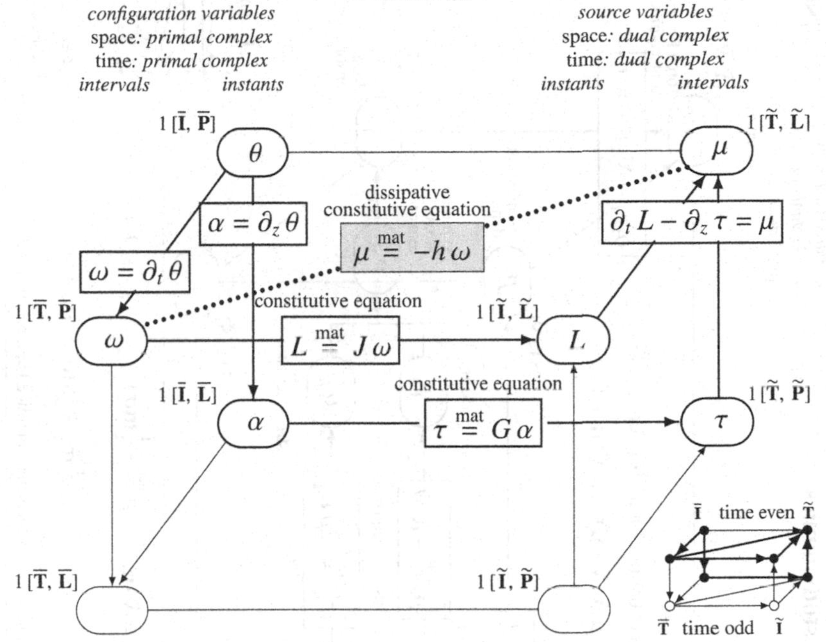

fundamental equation: d'Alembert

$$J\frac{\partial^2 \theta}{\partial t^2} - G\frac{\partial^2 \theta}{\partial z^2} = -h\frac{\partial \theta}{\partial t}$$

$\theta$ angle of torsion
$\omega$ angular velocity
$\alpha$ rotation gradient

$\mu$ body couple for unit length
$L$ angular momentum for unit length
$\tau$ torque for unit length

$J$ moment of inertia for unit length
$G$ modulus of rigidity
$h$ torsional damping coefficient

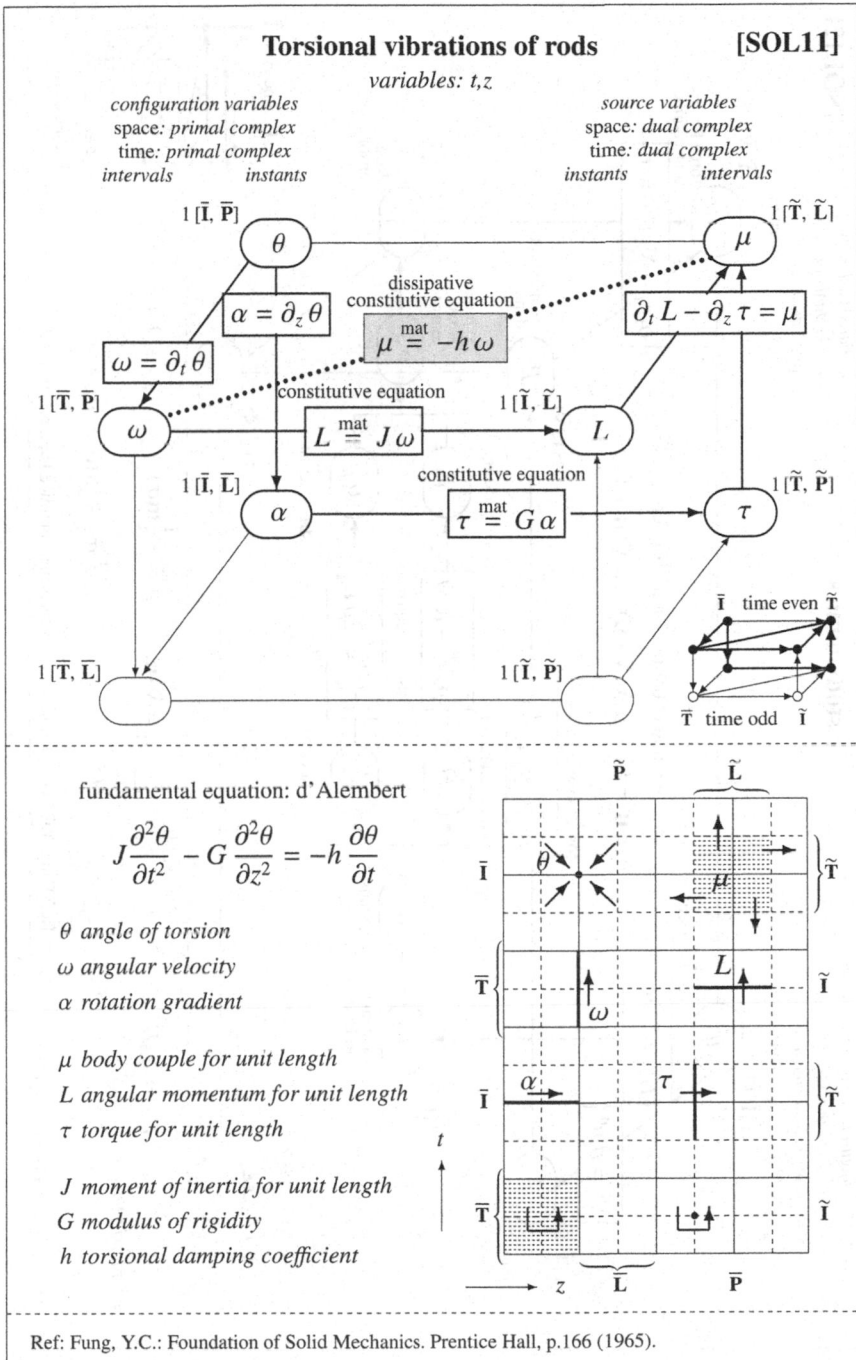

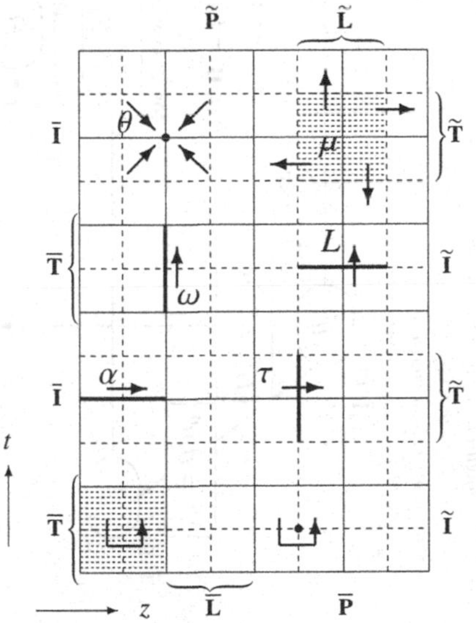

Ref: Fung, Y.C.: Foundation of Solid Mechanics. Prentice Hall, p.166 (1965).

Elastodynamics
[SOL12]
configuration variables
space: primal complex
time: primal complex
intervals instants
source variables
space: dual complex
time: dual complex
instants intervals
1 [$\bar{\mathbf{I}}$, $\bar{\mathbf{P}}$]
$u^h$
total displacement
$H^h_k = \nabla_k u^h$
$v^h = \partial_t u^h$
1 [$\bar{\mathbf{T}}$, $\bar{\mathbf{P}}$]
$v^h$
velocity
fundamental equation: Lamé-Navier
$\rho\,\partial_{tt}\mathbf{u} = (\lambda + \mu)\nabla(\nabla \cdot \mathbf{u}) + \mu\,\nabla^2\mathbf{u} + \mathbf{f}$
constitutive equation
$p_i \overset{\text{mat}}{=} g_{ih}\,\rho\,v^h$
1 [$\tilde{\mathbf{I}}$, $\tilde{\mathbf{V}}$]
$p_i$
momentum density
1 [$\tilde{\mathbf{T}}$, $\tilde{\mathbf{V}}$]
$f_i$
force for unit volume
$\partial_t p_i - \nabla_h \tau_i{}^h = f_i$
3 [$\bar{\mathbf{I}}$, $\bar{\mathbf{L}}$]
$H^h_k$
displacement gradient
$\varepsilon_{ik}$
strain tensor
volume dilatation
$\Theta$
$p \overset{\text{mat}}{=} -K\,\Theta$
pressure
$p$
constitutive equations
$\varepsilon'_{ik}$
strain deviator
$\sigma'_{ik} \overset{\text{mat}}{=} 2\mu\,\varepsilon'_{ik}$
Hooke
$\sigma'_{ik}$
stress deviator
symmetric stress tensor
$\sigma_{ik}$
3 [$\tilde{\mathbf{T}}$, $\tilde{\mathbf{S}}$]
$\tau_i{}^h$
stress tensor
3 [$\bar{\mathbf{T}}$, $\bar{\mathbf{L}}$]
3 [$\tilde{\mathbf{I}}$, $\tilde{\mathbf{S}}$]

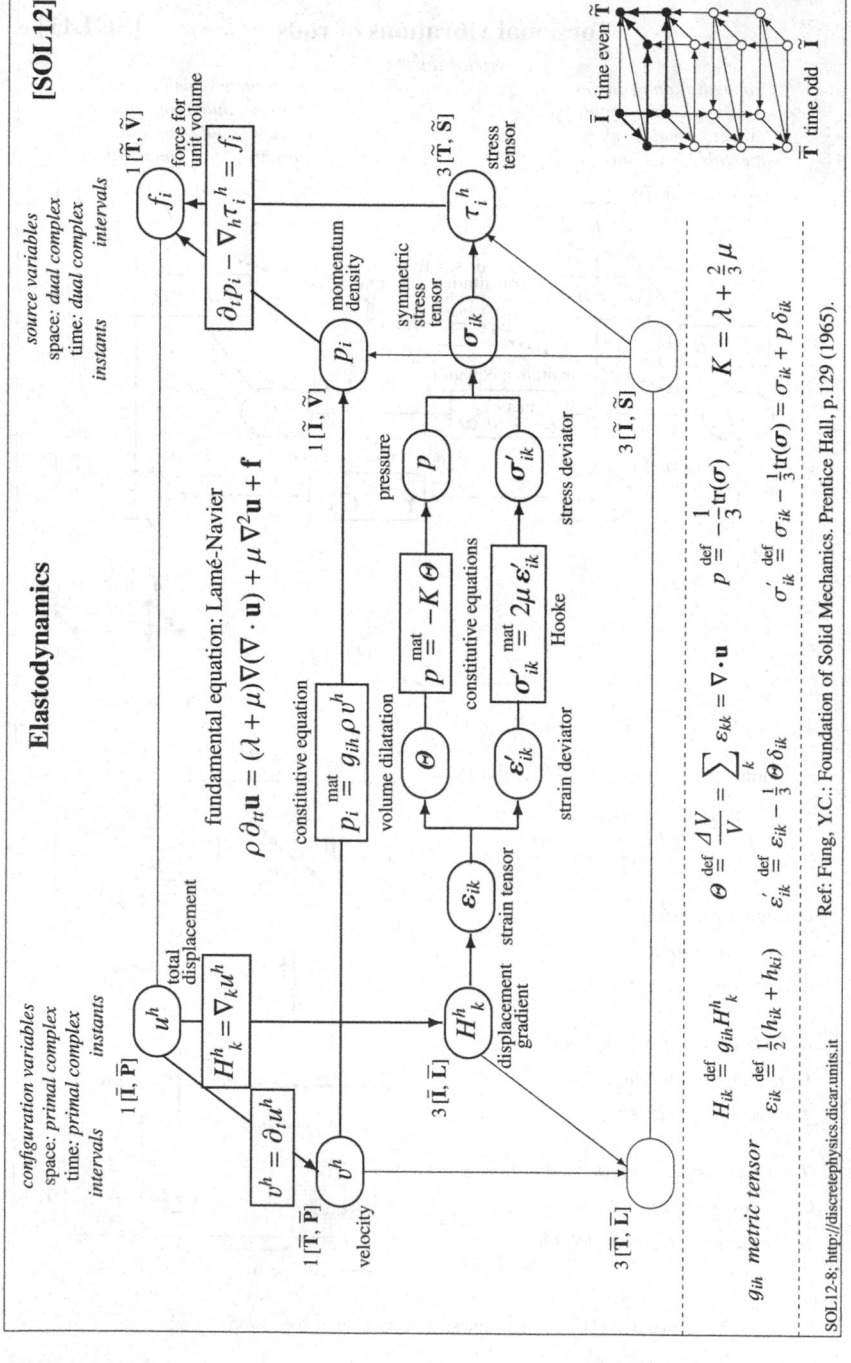

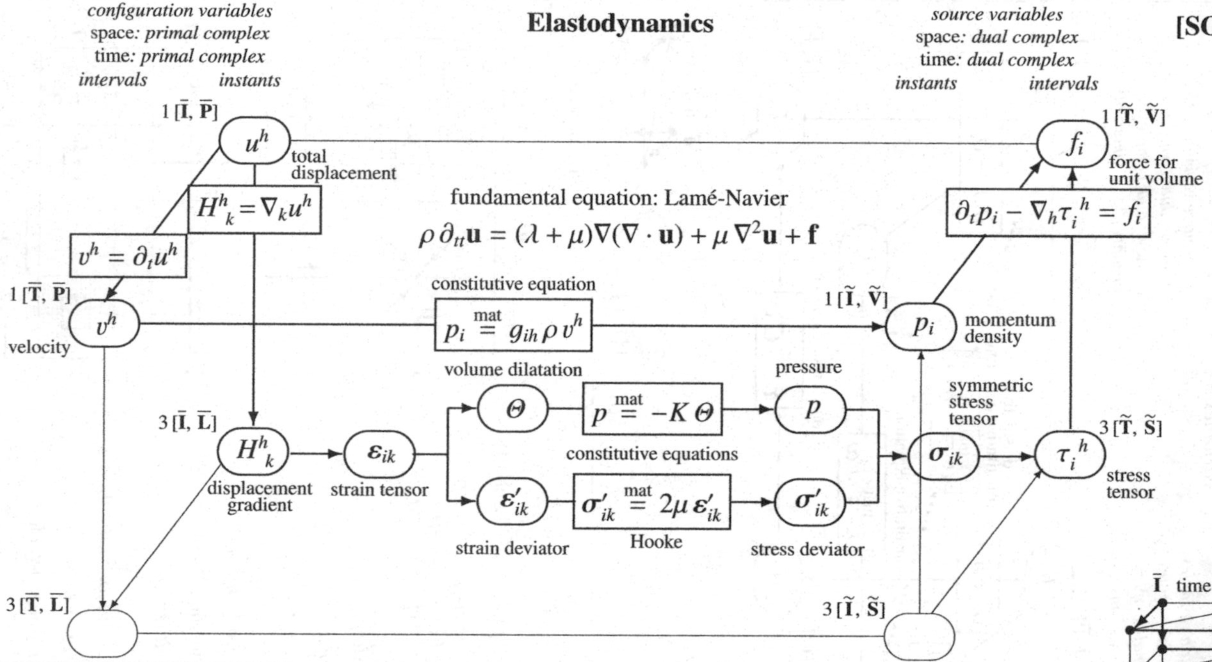

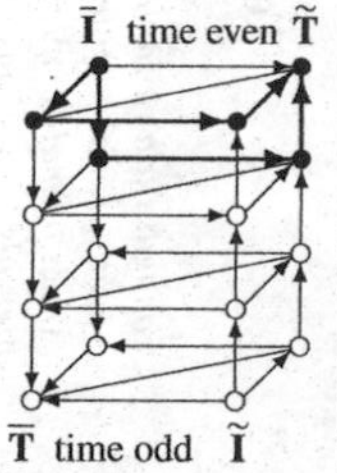

$g_{ih}$ metric tensor
$H_{ik} \overset{\text{def}}{=} g_{ih} H^h_k$
$\varepsilon_{ik} \overset{\text{def}}{=} \frac{1}{2}(h_{ik} + h_{ki})$
$\Theta \overset{\text{def}}{=} \frac{\Delta V}{V} = \sum_k \varepsilon_{kk} = \nabla \cdot \mathbf{u}$
$\varepsilon'_{ik} \overset{\text{def}}{=} \varepsilon_{ik} - \frac{1}{3}\,\Theta\,\delta_{ik}$
$p \overset{\text{def}}{=} -\frac{1}{3}\mathrm{tr}(\boldsymbol{\sigma})$
$\sigma'_{ik} \overset{\text{def}}{=} \sigma_{ik} - \frac{1}{3}\mathrm{tr}(\boldsymbol{\sigma}) = \sigma_{ik} + p\,\delta_{ik}$
$K = \lambda + \frac{2}{3}\mu$
$\bar{\mathbf{I}}$ time even $\tilde{\mathbf{T}}$
$\bar{\mathbf{T}}$ time odd $\tilde{\mathbf{I}}$
SOL12-8; http://discretephysics.dicar.units.it
Ref: Fung, Y.C.: Foundation of Solid Mechanics. Prentice Hall, p.129 (1965).

## Statics of strings [SOL13]

*inextensible, small deflections*

*configuration variables*
space*: primal complex*
time*: primal complex*

fundamental equation

$$-\frac{\mathrm{d}}{\mathrm{d}x}\left(T_x \frac{\mathrm{d}}{\mathrm{d}x}\, y\right) = q_y(x)$$

*source variables*
space*: dual complex*
time*: dual complex*

1 $[\bar{\mathbf{I}}, \bar{\mathbf{P}}]$ $y$ — 1 $[\tilde{\mathbf{T}}, \tilde{\mathbf{L}}]$ $q_y$

$$\langle q_y, y\rangle \stackrel{\mathrm{def}}{=} \int_0^L q_y\, y\, \mathrm{d}x$$

$$\alpha = \frac{\mathrm{d}y}{\mathrm{d}x} \qquad -\frac{\mathrm{d}T_y}{\mathrm{d}x} = q_y$$

1 $[\bar{\mathbf{I}}, \bar{\mathbf{L}}]$ $\alpha$ — 1 $[\tilde{\mathbf{T}}, \tilde{\mathbf{P}}]$ $T_y$

constitutive equation

$$T_y \stackrel{\mathrm{mat}}{=} T_x\, \alpha$$

$$\langle T_y, \alpha\rangle \stackrel{\mathrm{def}}{=} \int_0^L T_y\, \alpha\, \mathrm{d}x$$

$y(x)$ *height*
$\alpha(x)$ *slope*
$T_x$ *traction (constant)*
$q_y(x)$ *vertical load for unit length*
$T_y(x)$ *vertical component of traction*

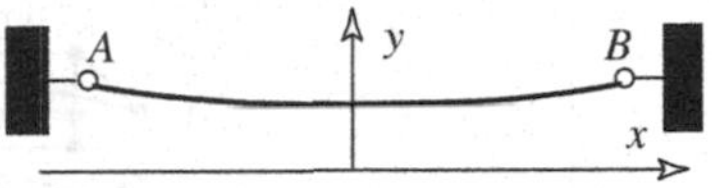


---

*configuration variables*
space*: primal complex*
time*: primal complex*

*large deflections*
*inextensible*

*source variables*
space*: dual complex*
time*: dual complex*

1 $[\bar{\mathbf{I}}, \bar{\mathbf{P}}]$ $\mathbf{r}$ — 1 $[\tilde{\mathbf{T}}, \tilde{\mathbf{L}}]$ $\mathbf{q}$

fundamental equation

$$\frac{\mathrm{d}}{\mathrm{d}s}\left(\mathbf{T}\, \frac{\mathrm{d}\mathbf{r}}{\mathrm{d}s}\right) = \mathbf{q}$$

$$\mathbf{t} = \frac{\mathrm{d}\mathbf{r}}{\mathrm{d}s} \qquad -\frac{\mathrm{d}\mathbf{T}}{\mathrm{d}s} = \mathbf{q}$$

1 $[\bar{\mathbf{I}}, \bar{\mathbf{L}}]$ $\mathbf{t}$ — 1 $[\tilde{\mathbf{T}}, \tilde{\mathbf{P}}]$ $\mathbf{T}$

constitutive equation

$$\mathbf{T} \stackrel{\mathrm{mat}}{=} T\, \mathbf{t}$$

$\mathbf{r}(s)$ *radius vector*
$\mathbf{t}(s)$ *unit tangent vector*
$T(s)$ *modulus of tension*
$\mathbf{q}(s)$ *vertical load for unit length*
$\mathbf{T}(s)$ *tension*

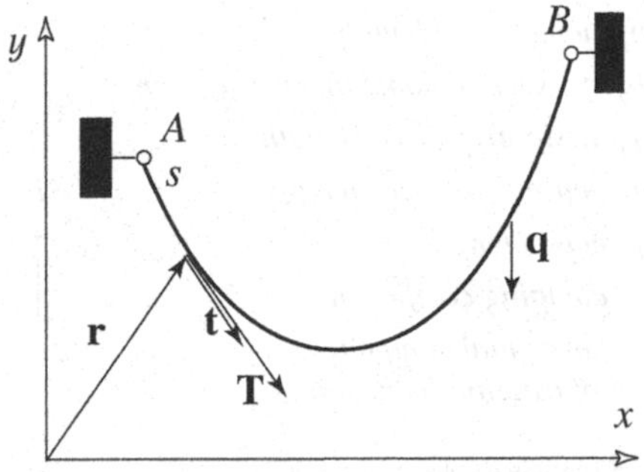


SOL13-7; http://discretephysics.dicar.units.it

## Transverse vibrations of strings [SOL14]

*variables: t, x*

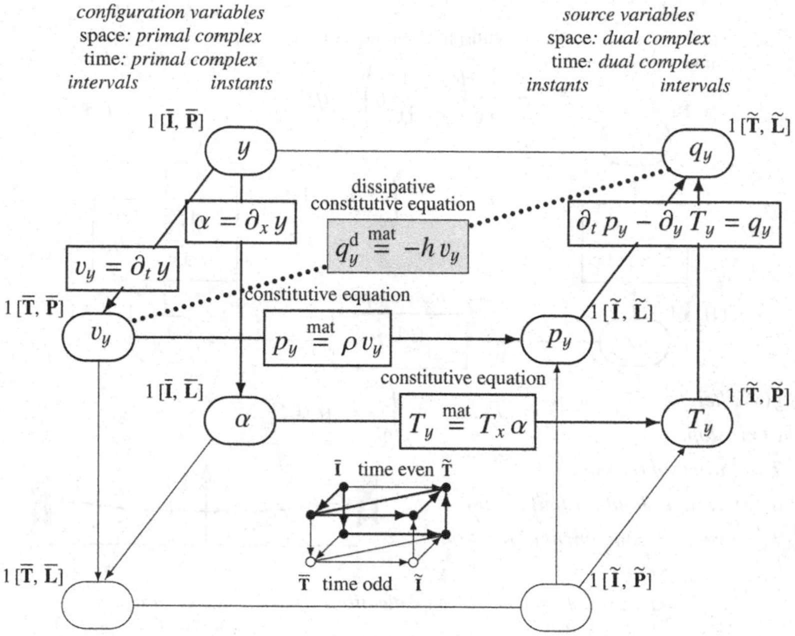


fundamental equation: d'Alembert

$$\rho \frac{\partial^2 y}{\partial t^2} - T_x \frac{\partial^2 y}{\partial x^2} = q_y - h \frac{\partial y}{\partial t}$$

$y$ *transversal displacement*
$v_y$ *transversal velocity*
$\alpha$ *slope*
$q_y$ *body force / length*
$p_y$ *momentum / length*
$T_y$ *transversal component of tension*
$q^{d}_y$ *dissipative force / length*
$q_y$ *impressed force / length*
$\rho$ *mass / length*
$h$ *damping coefficient*
$T_x$ *horizontal component of traction (constant)*

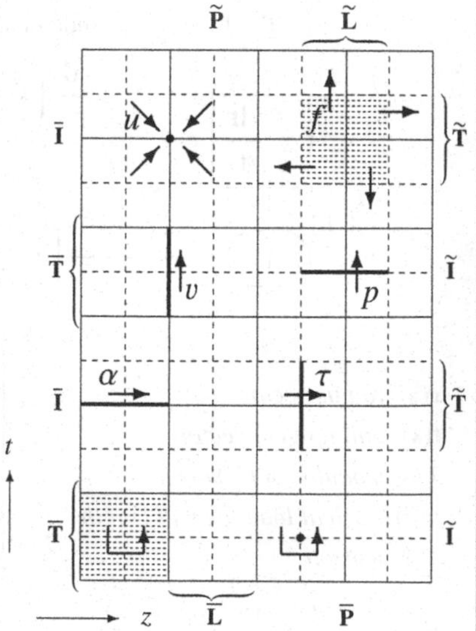


SOL14-6; http://discretephysics.dicar.units.it

# Chapter 12
# Mechanics of Fluids

## 12.1 Particles and Points

In mechanics, we often use the terms *particle* and *point* interchangeably.[1] The tendency to abuse the term *point* instead of *particle* is evidenced by the use of the term *material point*. This custom may generate ambiguities in the description of fluid dynamics as well as in the notion of work. For the present classification we must carefully distinguish between these two notions; more precisely, we must consider the following concepts:[2]

1. A reference system $\mathbf{O}xyz$ with its fixed space points which we denote in boldface, i.e. $\mathbf{A}, \mathbf{B}, \mathbf{P}, \mathbf{Q}, \ldots$;
2. A body $\mathscr{B}$ whose elements are called *particles*;
3. A particle $\mathscr{P}$ of the body that moves with respect to the reference system.

A particle $\mathscr{P}$ is in motion when it passes through different points of a reference system. The mass and the charge of a particle can be denoted by $m(\mathscr{P})$ and $q(\mathscr{P})$. A force acting on a particle will be denoted by $\mathbf{F}(t, \mathscr{P})$. Velocity and acceleration are attributes of particles, i.e. $\mathbf{v}(t, \mathscr{P})$ and $\mathbf{a}(t, \mathscr{P})$.

**Spatial Description.** When we switch from a *material* description to a *spatial* description, the velocity of a particle passing through a (fixed) point $\mathbf{P}$ at the instant $t$ can be referred directly to the point $\mathbf{P}$ at that instant. Hence, writing $\mathbf{v}(t, \mathbf{P})$ does *not* mean that the point $\mathbf{P}$ has a velocity $\mathbf{v}$ at the instant $t$ but that the *particle* passing through the point $\mathbf{P}$ at the instant $t$ has a velocity $\mathbf{v}$.

[1] This chapter presupposes a reading of Chaps. 1–9.

[2] Chadwick [39, p. 50].

E. Tonti, *The Mathematical Structure of Classical and Relativistic Physics*, Modeling and Simulation in Science, Engineering and Technology, DOI 10.1007/978-1-4614-7422-7_12, 

## 12.2 Some Peculiarities of the Fluid Field

A first peculiarity lies in the fact that in some cases velocity is associated with dual instants and primal lines, i.e. $\mathbf{v}[\tilde{\mathbf{I}}, \overline{\mathbf{L}}]$, as shown in diagrams [FLU3]–[FLU5] whereas in other cases it is associated with primal time intervals and primal points, i.e. $\mathbf{v}[\overline{\mathbf{T}}, \overline{\mathbf{P}}]$, as shown in diagrams [FLU6] and [FLU9]. This double association of fluid velocity makes velocity unique among the approximately 180 physical variables of many physical theories classified in this book. How to explain this ambiguity?

Well, this double association seems to be the consequence of a double description used in fluid dynamics. When we consider a fluid particle in a material (= Lagrangian) description, velocity $\mathbf{v}(t, \mathscr{P})$ is associated with a time interval, hence $\mathbf{v}[\overline{\mathbf{T}}, \mathscr{P}]$. In the transition to a spatial (= Eulerian) description, we consider the point $\mathbf{P}$ through which the particle $\mathscr{P}$ goes to the instant $t$ and attribute its velocity to that point. This shift *particle* → *point* implies a forgetting of the history of the particle's motion and to consider the velocity as a field vector applied to the point $\mathbf{P}$ at the instant $t$ instead of to the particle $\mathscr{P}$. The association with time is changed, and the velocity is associated with the instant considered.

However, since velocity, due to its physical meaning, must change sign under a reversal of motion, it follows that the instant must be the dual. Hence, the passage from a material to a spatial description implies not only the passage $\mathscr{P} \to \mathbf{P}$ but also $\mathbf{v}[\overline{\mathbf{T}}] \to \mathbf{v}[\tilde{\mathbf{I}}]$.

Once the velocity has resulted in a vector field it is natural to compute the integral along lines. The same thing happens with the electric field vector $\mathbf{E}$, the acceleration of gravity $\mathbf{g}$ and a force $\mathbf{F}$ in a force field. It follows that velocity refers to lines, i.e. $\mathbf{v}[\tilde{\mathbf{I}}, \overline{\mathbf{L}}]$.

There is another feature of fluid motion which is linked to the transition from one description to another. This feature can be grasped through the following question: *What is the meaning of the product* $\rho\mathbf{v}$*?* We can see that there are two possible answers to this question:

- It could be the *momentum density*, denoted by $\mathbf{p} = \rho\mathbf{v}$. In this case, it is associated with volumes and time instants: $\mathbf{p}[\tilde{\mathbf{I}}, \tilde{\mathbf{V}}]$. The association with $\tilde{\mathbf{I}}$ is required by the fact that momentum, like velocity, changes sign under a reversal of motion; the association with $\tilde{\mathbf{V}}$ is due to the fact that we must consider the momentum of the fluid contained in a volume.
- It could be the *mass current density*, denoted by $\mathbf{q} = \rho\mathbf{v}$. In this case it is associated with surfaces and time intervals: $\mathbf{q}[\overline{\mathbf{T}}, \tilde{\mathbf{S}}]$. The association with $\overline{\mathbf{T}}$ is required by the fact that mass current density, like velocity, changes sign under a reversal of motion; the association with surfaces endowed with an outer orientation is due to the fact that we must consider the mass current through a surface.

**Table 12.1** The two possible associations of velocity in fluid dynamics

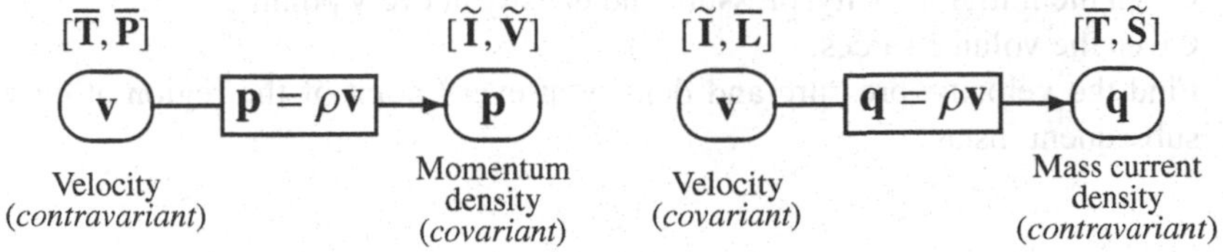


As we see, the variable $\rho\mathbf{v}$ has two different interpretations; this ambiguity is linked to the ambiguity of the velocity just discussed, as shown in diagram 12.1. In the first case, the dual of $[\tilde{\mathbf{I}}, \tilde{\mathbf{V}}]$ is $[\bar{\mathbf{T}}, \bar{\mathbf{P}}]$, hence $\mathbf{v}[\bar{\mathbf{T}}, \bar{\mathbf{P}}]$. In the second case, the dual of $[\bar{\mathbf{T}}, \tilde{\mathbf{S}}]$ is $[\tilde{\mathbf{I}}, \bar{\mathbf{L}}]$, hence $\mathbf{v}[\tilde{\mathbf{I}}, \bar{\mathbf{L}}]$. It is interesting to remark that while momentum is a covariant vector, velocity must be a contravariant vector. In the second case, since the mass current density is a contravariant vector, velocity must be a covariant vector.

Since velocity is the rate of a displacement, that is a contravariant vector, it follows that velocity is also a contravariant vector. To reduce velocity to being a covariant vector, we must make use of the metric tensor $v_h = g_{hk}\, v^k$.

In the first case, where $\mathbf{v}[\bar{\mathbf{T}}, \bar{\mathbf{P}}]$, velocity is the time rate of a global variable in time and is a global variable in space; in the second case, where $\mathbf{v}[\tilde{\mathbf{I}}, \bar{\mathbf{L}}]$, it is a global variable in time and a line density in space.

When velocity is a covariant vector, as in diagram [FLU3], one can perform its line integral along a line, like the vector $\mathbf{E}$ of electromagnetism; this shows that velocity is associated with lines. The velocity line integral is homologous to voltage in electrostatics. Moreover, when a velocity line integral along all reducible closed lines vanishes, one can introduce the velocity potential $\phi$, and the homologous variable in electrostatics is the electric potential $\phi$. We have in both theories $\phi[\tilde{\mathbf{I}}, \bar{\mathbf{P}}]$: compare diagram [FLU3] with diagram [ELE1].

## 12.3 Fundamental Problem

Let us consider a fluid at rest or in motion. The forces exerted on it from outside generate a pressure at every point of the fluid; this may change its density and, if the fluid is free to move, also generate a velocity at every point.

The *fundamental problem* of fluid mechanics can be stated as follows:

- Given a fluid,
- Given the shape and the dimension of the region in which it moves,
- Given a time interval,
- Given the type of fluid,

- Given the boundary conditions,
- Given the initial velocity, pressure and density at every point,
- Given the volume forces,
- Find the velocity, pressure and density at every point of the region at every subsequent instant.

## 12.4 Fluids and Flows

Let us start with a definition of *fluid*:[3]

DEFINITION. *A* fluid *is a continuum medium which has the property that when it is in equilibrium, the stress at every point is one of compression.*

It is essential to distinguish between the properties of a *fluid* and the properties of a *flow*.[4]

### 12.4.1 Kinds of Fluids

Fluids can be classified as follows:

- *A* perfect *or* inviscid *or* non-viscous *or* ideal *fluid is one for which the stress at every point is one of compression also during motion; in Cartesian notation,* $t_{hk} = -p(t, \mathbf{x})\delta_{hk}$.[5] A perfect fluid can be compressible or incompressible: recall that a fluid can be a gas or a liquid, and a perfect gas is surely compressible. Its complementary class is that of *viscous* or *real* fluids.
- *A* Newtonian *fluid is a* viscous *fluid for which the viscous stress tensor is a linear function of the strain rate tensor.*[6] The complementary class is that of *non-Newtonian* fluids.
- *An* incompressible *fluid is one with constant density.* The complementary class is that of *compressible* fluids.

---

[3] Lai et al. [117, p. 176], McLeod [155, p. 3].

[4] Aris [6, p. 124].

[5] Chorin and Marsden [41, p. 5], Aris [6, p. 105], McLeod [155, p. 7]. We remark that Meyer [158, p. 64] distinguishes between *perfect* and *ideal* fluids: an ideal fluid is a perfect fluid in isochoric motion.

[6] Batchelor [10, p. 146], Yih [255, p. 31].

**Table 12.2** The many kinds of fluids and flows

| *Fluid* | | *Flow* | |
|---|---|---|---|
| Liquid | Gas | Steady | Unsteady |
| Perfect | Viscous | Irrotational | Rotational |
| Incompressible | Compressible | Laminar | Turbulent |
| Newtonian | Non-Newtonian | Isochoric | Non-isochoric |
| Piezotropic | Non-piezotropic | Barotropic | Non-barotropic |

- *A* piezotropic *fluid is one for which the pressure depends only on the density.* The complementary class is that of *non-piezotropic* fluids for which the pressure depends not only on the density but also on the temperature.

## *12.4.2 Kind of Flows*

Fluid flows can be classified as follows (Table 12.2):

- A *stationary* or *steady* or *permanent* flow is one for which the velocity at every point does not change with time. The complementary class is that of *unsteady* flows.
- An *irrotational* flow is one for which the line integral of the velocity along any reducible closed line vanishes. The complementary class is that of *rotational* flows.
- A *turbulent* flow *is an irregular condition of flow in which the various quantities show a random variation with time and space coordinates, so that statistically distinct average values can be discerned.*[7] The complementary class is that of *laminar* or *regular* flows.
- An *isochoric* flow is one for which the volume of a fluid body does not change during the flow: $\operatorname{div} \mathbf{v} = 0$.[8] An incompressible fluid undergoes only an isochoric flow, but a compressible fluid, like air, can also move without changing its volume.
- A *barotropic* flow is one for which, during motion, the pressure depends only on the density. A piezotropic fluid undergoes only barotropic flows but a non-piezotropic fluid can also undergo a barotropic flow.
- A *creeping* flow is one for which the velocity is so small that one can neglect its square by comparing it with the velocity itself.

[7] Cebeci and Smith [38, p. 2].

[8] Billington and Tate [15, p. 96].

## 12.5 Variables Used in the Steady Motion of a Perfect Fluid

In this brief summary, we analyse the main variables of perfect fluid dynamics in steady motion for the purpose of finding the space elements with which they are associated.

The space global variables of the theory of perfect fluids are (diagram 12.2) the *velocity potential* $\phi[\overline{\mathbf{P}}]$, *velocity line integral* $\Gamma[\overline{\mathbf{L}}]$, *vortex flux* $W[\overline{\mathbf{S}}]$, *mass production* $M^{\mathrm{p}}[\widetilde{\mathbf{V}}]$, *mass flow* $M^{\mathrm{f}}[\widetilde{\mathbf{S}}]$ and *line integral* $\Psi[\widetilde{\mathbf{L}}]$ *of the stream vector* $\check{\psi}[\widetilde{\mathbf{L}}]$. These six global variables and the corresponding densities are as follows:

$$\begin{array}{r|ccc|ccc} & \multicolumn{3}{c|}{\textit{Configuration variables}} & \multicolumn{3}{c}{\textit{Source variables}} \\ \textit{Global variables} & \phi[\overline{\mathbf{P}}] & \Gamma[\overline{\mathbf{L}}] & W[\overline{\mathbf{S}}] & M^{\mathrm{p}}[\widetilde{\mathbf{V}}] & M^{\mathrm{f}}[\widetilde{\mathbf{S}}] & \Psi[\widetilde{\mathbf{L}}] \\ & \downarrow & \downarrow & \downarrow & \downarrow & \downarrow & \downarrow \\ \textit{Field variables} & \phi[\overline{\mathbf{P}}] & \mathbf{v}[\overline{\mathbf{L}}] & \check{\mathbf{w}}[\overline{\mathbf{S}}] & \sigma_{\mathrm{m}}[\widetilde{\mathbf{V}}] & \mathbf{q}[\widetilde{\mathbf{S}}] & \check{\psi}[\widetilde{\mathbf{L}}] \end{array} \tag{12.1}$$

## 12.6 Source Variables

The source variables in fluid mechanics are essentially the following (diagram 12.3): *volume forces*, *surface forces*, *mass content* and *mass density*, *mass flow*, *mass current*, *mass current density*, *stream function*, *stream vector*, *momentum content*, *momentum flow*, *momentum production*.

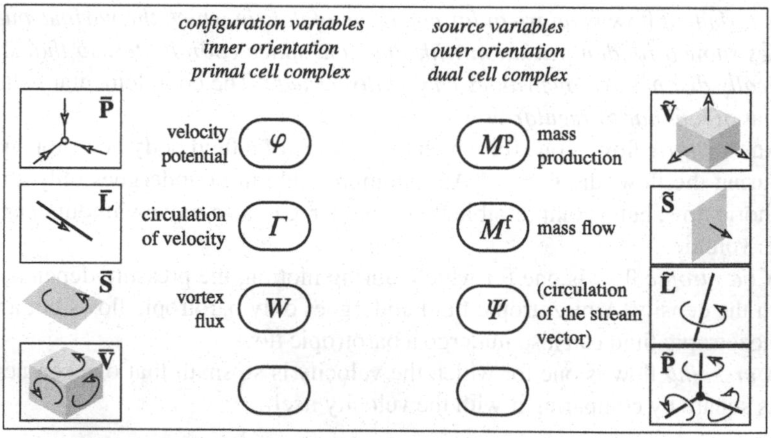


**Fig. 12.1** Perfect fluid motion: classification of space global variables

**Table 12.3** Perfect fluid motion: classification of field variables

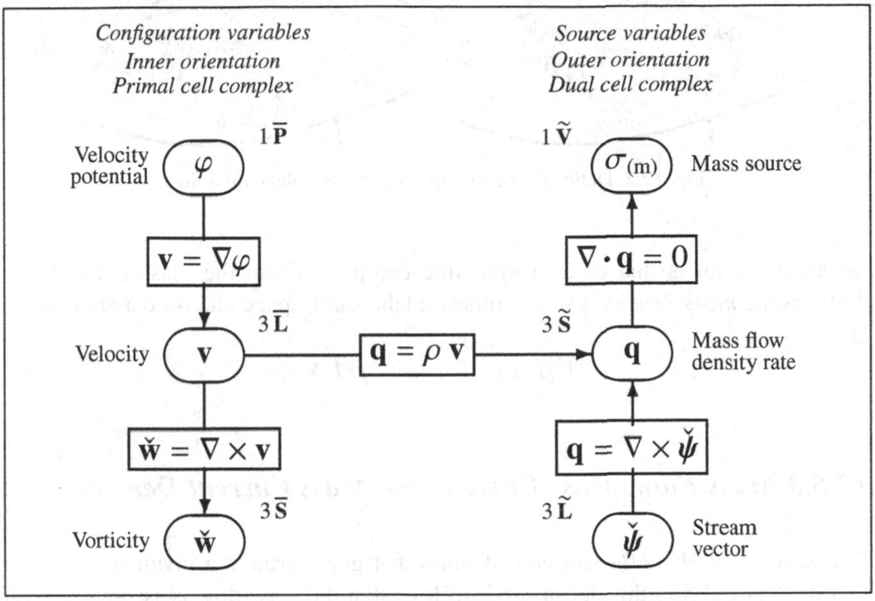


## 12.6.1 Volume and Surface Forces and Their Impulses

The impulse of a volume force $\mathbf{J}^{\mathrm{v}}$ is defined as the definite time integral of the volume force $\mathbf{F}^{\mathrm{v}}$ over a time interval. Hence, it is associated with time intervals and with volumes. Since the force does not change sign under a reversal of motion (Chap. 9), it follows that $\mathbf{F}^{\mathrm{v}}[\widetilde{\mathbf{T}}, \widetilde{\mathbf{V}}]$. In conclusion,

$$\mathbf{J}^{\mathrm{v}}\,[\widetilde{\mathbf{T}}, \widetilde{\mathbf{V}}] \quad \longrightarrow \quad \mathbf{F}^{\mathrm{v}}\,[\widetilde{\mathbf{T}}, \widetilde{\mathbf{V}}]\,.$$

The impulse of a surface force $\mathbf{J}^{\mathrm{s}}$ is defined as the definite time integral of the surface force $\mathbf{F}^{\mathrm{s}}$ over a time interval. Hence,

$$\mathbf{J}^{\mathrm{s}}\,[\widetilde{\mathbf{T}}, \widetilde{\mathbf{S}}] \quad \longrightarrow \quad \mathbf{F}^{\mathrm{s}}\,[\widetilde{\mathbf{T}}, \widetilde{\mathbf{S}}]\,.$$

## 12.6.2 Mass Content and Mass Density

The *mass content* is the amount of mass contained in a fixed volume at a given time instant. It is associated with volumes endowed with an outer orientation. Since the mass content has no reason to change sign under a reversal of motion,

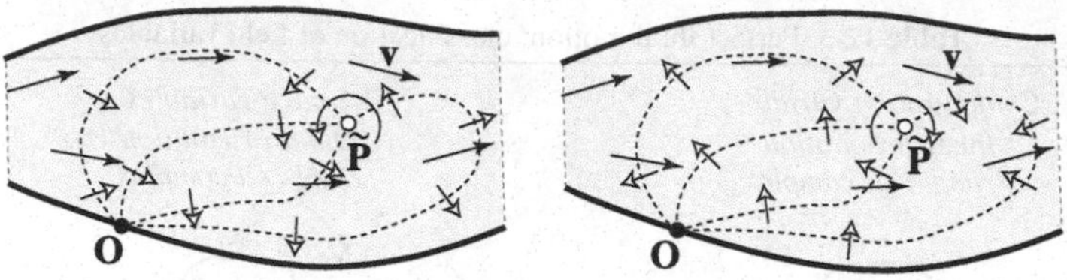


**Fig. 12.2** Definition of stream function in a plane fluid flow

the time instant is that of a primal time complex. From the mass content one deduces the *mass density* $\rho$ which inherited the same space and time associations. Hence,

$$M^{c}[\bar{\mathbf{I}}, \tilde{\mathbf{V}}] \quad \longrightarrow \quad \rho[\bar{\mathbf{I}}, \tilde{\mathbf{V}}]\,.$$

### 12.6.3 Mass Flow, Mass Current and Mass Current Density

The *mass flow* $M^{f}$ is the amount of mass that goes through a given surface in a given interval. From this definition it follows that the mass flow is associated with a surface endowed with an outer orientation, because the surface is crossed by the mass, and with a time interval, the time during which the mass flows. Under a reversal of motion, the flow changes sign; hence, the interval is one which is endowed with an inner orientation. The *mass current* $\Phi$ is the mass flow rate; it inherits the same space and time associations. The *mass current density* $\mathbf{q}$ is the mass current for a unit area. In summary,

$$M^{f}[\bar{\mathbf{T}}, \tilde{\mathbf{S}}] \quad \longrightarrow \quad \Phi[\bar{\mathbf{T}}, \tilde{\mathbf{S}}] \quad \longrightarrow \quad \mathbf{q}[\bar{\mathbf{T}}, \tilde{\mathbf{S}}]\,.$$

### 12.6.4 Stream Function

The mass conservation law, when applied to an incompressible fluid or to a compressible fluid in a stationary flow, states that the total flow going out from any closed surface which does not enclose sources or sinks vanishes. In this case, the fluid flow is said to be *solenoidal*. If we divide a closed surface $S$ into two parts, $S'$ and $S''$, by a closed line which belongs to the surface, the fluid flow entering into $S'$ is equal to the flow exiting $S''$.

**Two-dimensional flow.** The most useful application of this property concerns two-dimensional fluid flows. In this case, the traces of the surface are lines, which we represent by dotted lines in Fig. 12.2. Let us consider a fixed point **O**, taken in any position inside a fluid. For every point **P** of the fluid we can consider the lines going from **O** to **P**. We give an outer orientation to the lines (indicated by white arrows in the figure), so that all outer orientations are compatible. Due to the solenoidal character of the flow, the mass current (i.e. mass flow rate) through each line is the same. Since the mass current density is $\mathbf{q} = \rho\mathbf{v}$, the mass current that goes through each line is

$$\Phi = \int_{\mathbf{O}}^{\mathbf{P}} \mathbf{q} \cdot \mathbf{n} \, dL \quad \text{(SI unit: kg/s).} \tag{12.2}$$

Since this mass current does not depend on a line, we can assign to every point **P** a scalar variable,

$$\psi(\mathbf{P}) \stackrel{\text{def}}{=} \int_{\mathbf{O}}^{\mathbf{P}} \mathbf{q} \cdot \mathbf{n} \, dL \quad \text{placing} \quad \psi(\mathbf{O}) = 0, \tag{12.3}$$

called a *stream function*.[9]

If we invert the outer orientation of the lines, as shown on the right side of Fig. 12.2, the mass current changes sign; hence, the stream function changes sign. If we compare the left with the right side of the figure, we see that the outer orientation of the lines induces an outer orientation to the point **P**. It follows that the point is endowed with a natural outer orientation which, in the plane, is represented by a curved arrow around it. Such a point is denoted by $\widetilde{\mathbf{P}}$. Hence, the stream function changes sign under a reversal of the outer orientation of the point, i.e. it is a pseudoscalar; we will write $\check{\psi}[\widetilde{\mathbf{P}}]$. Hence, in a two-dimensional fluid flow, the stream function is associated with points, but in a three-dimensional motion, we must consider the stream vector.

### 12.6.5 Stream Vector

In three dimensions, a solenoidal flow leads us to introduce a *stream pseudovector* $\check{\boldsymbol{\psi}}$ whose common definition is $\mathbf{q} = \nabla \times \check{\boldsymbol{\psi}}$.[10] This definition is equivalent to the relation

$$\Phi = \int_{\widetilde{\mathbf{S}}} \mathbf{q} \cdot \mathbf{n} \, dS = \int_{\partial\widetilde{\mathbf{S}}} \check{\boldsymbol{\psi}} \cdot \check{\mathbf{t}} \, dL \, . \tag{12.4}$$

Since the mass current density vector **q** is associated with primal time intervals, it follows that the stream vector is also associated with the same time element, hence $\check{\boldsymbol{\psi}}[\overline{\mathbf{T}}, \widetilde{\mathbf{L}}]$. The stream function $\psi$ in the plane motion is the normal component of the stream vector $\check{\boldsymbol{\psi}}$.

[9] For this physical interpretation of the stream function, which is usually presented by the purely mathematical relations $q_x = \partial_y \psi$, $q_y = -\partial_x \psi$, see Milne and Thomson [160, p. 476].

[10] Kundu et al. [116, p. 99].

### 12.6.6 Mass Production, Mass Source

In a nuclear reactor and in chemical reactions, we must consider the *mass production* $M^{\text{p}}$ and the *mass source* $\sigma_{\text{m}}$. The mass production is associated with time intervals, and, in particular, since it changes sign under a reversal of motion, it is associated with primal time intervals. We have the following association:

$$M^{\text{p}}[\overline{\mathbf{T}}, \widetilde{\mathbf{V}}] \quad \longrightarrow \quad \sigma_{\text{m}}[\overline{\mathbf{T}}, \widetilde{\mathbf{V}}]\,.$$

### 12.6.7 Momentum Content and Momentum Density

As explained in Chap. 9, the momentum $\mathbf{p}$ of a particle is, by definition, the indefinite time integral of the force $\mathbf{F}$ acting on the particle, starting from rest. Since momentum changes sign under a reversal of motion, it is associated with a time instant endowed with an outer orientation. The momentum of a body is the sum of the momenta of all its particles. The momentum content $\mathbf{P}^{\text{c}}$ is the momentum of the fluid contained in a fixed volume at a given instant. Moreover, since momentum is involved in a balance, it must be associated with volumes endowed with an outer orientation. The corresponding momentum density $\mathbf{p}$ inherits the same space and time associations. In conclusion,

$$\mathbf{P}^{\text{c}}[\widetilde{\mathbf{I}}, \widetilde{\mathbf{V}}] \quad \longrightarrow \quad \mathbf{p}[\widetilde{\mathbf{I}}, \widetilde{\mathbf{V}}]\,.$$

### 12.6.8 Momentum Flow and Momentum Current

The *momentum flow* is the momentum that crosses a given surface in a given time interval; it will be denoted by $\mathbf{P}^{\text{f}}$. It does not change sign under a reversal of motion like an impulse. Moreover, since momentum flow is involved in a balance, it must be associated with surfaces endowed with an outer orientation. The *momentum current* is, by definition, the rate of the momentum flow; hence it inherits the same space and time associations, hence $\mathbf{P}^{\text{f}}[\widetilde{\mathbf{T}}, \widetilde{\mathbf{S}}]$.

### 12.6.9 Momentum Production and Momentum Source

The *momentum production* $\mathbf{P}^{\text{p}}$ is associated with a time interval and with a volume. The momentum production is an alternate name for the *impulse of volume force*: $\mathbf{P}^{\text{p}} \equiv \mathbf{J}^{\text{v}}$. Since the impulse of volume forces does not change sign under a reversal of motion, it follows that the momentum production does the same. Moreover, since momentum production is involved in a balance, it must be associated with volumes endowed with an outer orientation. The *momentum production rate* is simply another name for *volume force*. *Momentum source* is another name for *momentum production density rate*, which is another name for *volume force density* **f**. In conclusion,

$$\mathbf{P}^{\mathrm{p}}[\widetilde{\mathbf{T}},\widetilde{\mathbf{V}}] \equiv \mathbf{J}^{\mathrm{v}}[\widetilde{\mathbf{T}},\widetilde{\mathbf{V}}] \quad \longrightarrow \quad \mathbf{F}^{\mathrm{v}}[\widetilde{\mathbf{T}},\widetilde{\mathbf{V}}] \quad \longrightarrow \quad \mathbf{f}[\widetilde{\mathbf{T}},\widetilde{\mathbf{V}}]\,.$$

## 12.7 Configuration Variables

The configuration variables in fluid dynamics are those linked to velocity (Table 12.4): the *velocity* **v**, the line integral of *velocity* $\Gamma$, *velocity potential* $\phi$, *vortex flux* $W$, *vorticity* **w**, *velocity gradient* **L**, *strain rate tensor* **D**, and *volume dilatation rate* $\theta$.[11]

### *12.7.1 Line Integral of Velocity and Velocity Potential*

As was stated previously, velocity has a double association with space and time elements. When it is associated with lines, it behaves as a covariant vector. In this case, we are led to consider the integral of the velocity along a line endowed with an inner orientation, i.e.

$$\Gamma(t) \stackrel{\text{def}}{=} \int_{\mathbf{L}} \mathbf{v}(t,\mathbf{P})\cdot\mathbf{t}\,\mathrm{d}L \qquad \longrightarrow \qquad \Gamma[\widetilde{\mathbf{I}},\overline{\mathbf{L}}]\,. \tag{12.5}$$

In this case, velocity is not a global variable in space but a line density, and it is associated with dual time instants, i.e. $\mathbf{v}[\widetilde{\mathbf{I}},\overline{\mathbf{L}}]$. The reason for this time association is not clear to the author: it can be inferred from classification diagrams such as [FLU3], [FLU4] or [FLU5].

Taking into account the one-to-one correspondence which exists between the association of a physical variable with a space-time element and its position in a classification diagram, it is natural to infer the association of the variable from its location in the diagram.

When the velocity line integral along all *reducible* closed lines vanishes,[12] the field is called *irrotational*, and we write this condition in differential terms, i.e.

[11] Recall the peculiar role of velocity in fluid dynamics, discussed in Sect. 12.2.

[12] Recall that a closed line is said to be *reducible* when it can be contracted to a point by a continuous deformation, without passing outside the fluid region. See Batchelor [10, p. 92].

**Table 12.4** Configuration variables in fluid dynamics

| | | |
|---|---|---|
| Velocity *(contravariant)* | $\mathbf{v}(t,\mathbf{P}) \stackrel{\text{def}}{=} \dfrac{d\mathbf{r}(t,\mathbf{P})}{dt} = \dfrac{d\eta(t,\mathbf{P})}{dt}$ $\quad v^h(t,\mathbf{P}) = \dfrac{\partial \mathbf{r}}{\partial x^h}$ | $\mathbf{v}\,[\overline{\mathbf{T}},\overline{\mathbf{P}}]$ |
| Relative velocity | $\mathbf{V}(t,\mathbf{PQ}) \stackrel{\text{def}}{=} \mathbf{v}(t,\mathbf{Q}) - \mathbf{v}(t,\mathbf{P})$ | $\mathbf{V}\,[\overline{\mathbf{T}},\overline{\mathbf{L}}]$ |
| Velocity gradient tensor | $\mathsf{L}(t,\mathbf{P}) \stackrel{\text{def}}{=} \text{grad}\,\mathbf{v}(t,\mathbf{P})$ $\quad L^h_k \stackrel{\text{def}}{=} \dfrac{\partial v^h}{\partial x^k}$ <br> $\mathrm{d}\mathbf{v} = \mathsf{L}\,\mathrm{d}\mathbf{r}$ $\quad \mathrm{d}v^h = L^h_k\,\mathrm{d}x^k$ | $\mathsf{L}\,[\overline{\mathbf{T}},\overline{\mathbf{L}}]$ |
| Strain rate tensor (≡ rate of deformation) | $\mathsf{D}(t,\mathbf{P}) \stackrel{\text{def}}{=} \frac{1}{2}(\mathsf{L}+\mathsf{L}^{\mathsf{T}})$ $\quad D_{hk} \stackrel{\text{def}}{=} \frac{1}{2}(L_{hk}+L_{hk})$ | $\mathsf{D}\,[\overline{\mathbf{T}},\overline{\mathbf{L}}]$ |
| Spin tensor | $\mathsf{W}(t,\mathbf{P}) \stackrel{\text{def}}{=} \frac{1}{2}(\mathsf{L}-\mathsf{L}^{\mathsf{T}})$ $\quad W_{hk} \stackrel{\text{def}}{=} \frac{1}{2}(L_{hk}-L_{hk})$ | $\mathsf{W}\,[\overline{\mathbf{T}},\overline{\mathbf{L}}]$ |
| Cubic (or bulk or volume) dilatation rate | $\theta(t,\mathbf{P}) \stackrel{\text{def}}{=} \text{tr}\,(\mathsf{L}) = \text{tr}\,(\mathsf{D}) = \nabla\cdot\mathbf{v}$ | $\theta\,[\overline{\mathbf{T}},\overline{\mathbf{L}}]$ |
| Deviator strain rate tensor | $\mathsf{D}'(t,\mathbf{P}) \stackrel{\text{def}}{=} \mathsf{D}(t,\mathbf{P}) - \frac{1}{3}\theta(t,\mathbf{P})\,\mathsf{I}$ <br> $D'_{hk} \stackrel{\text{def}}{=} D_{hk} - \frac{1}{3}\theta\delta_{hk}$ | $\mathsf{D}'\,[\overline{\mathbf{T}},\overline{\mathbf{L}}]$ |
| Velocity *(covariant)* | $\mathbf{v}(t,\mathbf{P})$ $\quad v_h(t,\mathbf{P})$ | $\mathbf{v}\,[\widetilde{\mathbf{I}},\overline{\mathbf{L}}]$ |
| Velocity line integral | $\Gamma \stackrel{\text{def}}{=} \int_{\overline{\mathbf{L}}} \mathbf{v}(t,\mathbf{P})\cdot\mathbf{t}\,\mathrm{d}L$ | $\Gamma\,[\widetilde{\mathbf{I}},\overline{\mathbf{L}}]$ |
| Velocity potential (≡ kinetic potential) | $\phi(t,\mathbf{P}) \stackrel{\text{def}}{=} \int_{\mathbf{O}}^{\mathbf{P}} \mathbf{v}(t,\mathbf{P})\cdot\mathbf{t}\,\mathrm{d}L$ | $\phi\,[\widetilde{\mathbf{I}},\overline{\mathbf{P}}]$ |
| Vorticity | $\check{\mathbf{w}}(t,\mathbf{P}) \stackrel{\text{def}}{=} \text{curl}\,\mathbf{v}(t,\mathbf{P})$ | $\check{\mathbf{w}}\,[\widetilde{\mathbf{I}},\overline{\mathbf{S}}]$ |
| Vortex flux | $W \stackrel{\text{def}}{=} \int_{\partial\overline{\mathbf{S}}} \mathbf{v}\cdot\mathbf{t}\,\mathrm{d}L = \int_{\overline{\mathbf{S}}} \check{\mathbf{w}}\cdot\check{\mathbf{n}}\,\mathrm{d}S$ | $W\,[\widetilde{\mathbf{I}},\overline{\mathbf{S}}]$ |

curl $\mathbf{v} = 0$. In this case, we can choose an origin and evaluate the line integral of velocity along a line starting at **O** and ending at any point **P**. To every point **P** we can associate the line integral from **O** to **P** which does not depend on the line connecting the two points. This is a scalar $\phi$ defined as

$$\phi(t) \stackrel{\text{def}}{=} \int_{\mathbf{O}}^{\mathbf{P}} \mathbf{v}(t,\mathbf{P})\cdot\mathbf{t}\,\mathrm{d}L \qquad \longrightarrow \qquad \phi[\widetilde{\mathbf{I}},\overline{\mathbf{P}}]\,. \tag{12.6}$$

To explain why the point must be endowed with an inner orientation, let us remark that one can decide to compute the line integral from the point **P** to the origin **O**, as was done in early books.[13] In this case, we would obtain

$$\phi'(t) \stackrel{\text{def}}{=} \int_{\mathbf{P}}^{\mathbf{O}} \mathbf{v}(t, \mathbf{P}) \cdot \mathbf{t}\, \mathrm{d}L = -\phi(t), \tag{12.7}$$

Thus, the sign of the velocity potential changes when the direction of the lines passing through **P** is reversed. This means that the point, instead of being oriented as a sink, would be oriented as a source.

### 12.7.2 Vortex Flux and Vorticity

The vortex flux $W$ across a surface can be defined as the *line integral of the velocity along the boundary of the surface*, i.e.

$$W(t) \stackrel{\text{def}}{=} \int_{\partial\overline{\mathbf{S}}} \mathbf{v}(t, \mathbf{P}) \cdot \mathbf{t}\, \mathrm{d}L \quad \longrightarrow \quad W[\widetilde{\mathbf{I}}, -\overline{\mathbf{S}}] = -W[\widetilde{\mathbf{I}}, \overline{\mathbf{S}}]. \tag{12.8}$$

Since the sign of the line integral depends on the direction chosen on the boundary, the vortex flux is associated with surfaces endowed with an inner orientation, hence $W[\overline{\mathbf{S}}]$. Thus, the vortex flux changes sign when the inner orientation of the surface is changed (oddness principle). Since $\Gamma$ is associated with dual time instants, it follows that $W$ is also associated with dual time instants, hence $W[\widetilde{\mathbf{I}}, \overline{\mathbf{S}}]$.

The introduction of the vorticity pseudovector $\check{\mathbf{w}}$ is analogous to the introduction of the magnetic flux density $\check{\mathbf{B}}$ in the magnetic field.[14] One obtains the relation

$$W(t) = \int_{\overline{\mathbf{S}}} \check{\mathbf{w}}(t, \mathbf{P}) \cdot \check{\mathbf{n}}\, \mathrm{d}S \quad \longrightarrow \quad \check{\mathbf{w}}[\widetilde{\mathbf{I}}, \overline{\mathbf{S}}]\,. \tag{12.9}$$

### 12.7.3 Relative Velocity and Velocity Gradient

When velocity is considered as being associated with points, it is the partial derivative of the position vector; hence, it behaves as a contravariant vector: $\mathbf{v}[\overline{\mathbf{T}}, \overline{\mathbf{P}}]$. In this case, we introduce the *relative velocity*:

$$\mathbf{V}(t, \mathbf{PQ}) \stackrel{\text{def}}{=} \mathbf{v}(t, \mathbf{Q}) - \mathbf{v}(t, \mathbf{P}) \quad \longrightarrow \quad \mathbf{V}[\overline{\mathbf{T}}, \overline{\mathbf{L}}] \tag{12.10}$$

---

[13] See for example Milne and Thomson [160, p. 53], Lamb [119, p. 17]. In the first case, one can write $\mathbf{v} = \nabla\phi$, whereas with the old convention $\mathbf{v} = -\nabla\phi'$.

[14] See p. 293.

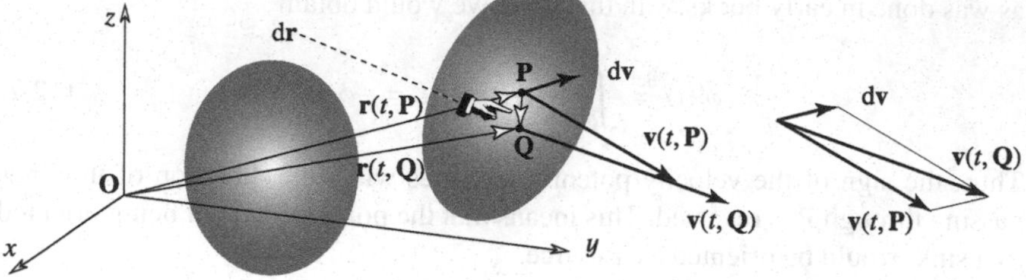


**Fig. 12.3** Velocity of two particles $\mathscr{P}$ and $\mathscr{Q}$, which at the instant $t$ are in the points **P** and **Q**. *Thick lines*: velocities; *thin lines*: position vectors

Consider the motion of a fluid (Fig. 12.3). In the neighbourhood of each point of the region in which the flow is regular (continuous and with continuous variation), the velocity can be considered an affine function (Appendix A) of the position vector, namely

$$\begin{cases} v_x = a_x + L_{xx}x + L_{xy}y + L_{xz}z \\ v_y = a_y + L_{yx}x + L_{yy}y + L_{yz}z \\ v_z = a_z + L_{zx}x + L_{zy}y + L_{zz}z \end{cases} \longrightarrow \begin{array}{c} \mathbf{v} = \mathbf{a} + \mathsf{L}\,\mathbf{r} \\ \\ v^h = a^h + L^h_k x^k\ . \end{array} \tag{12.11}$$

The matrix $\mathsf{L}$ is called the *velocity gradient* at the point $x, y, z$ considered.

Since velocity is, in general, dependent on time and space, the matrix is also dependent on time and space. Hence, we must write $\mathsf{L}(t, \mathbf{P}) \equiv \mathsf{L}(t, \mathbf{r}) \equiv \mathsf{L}(t, \mathbf{x})$. Its components $L^h_k$ depend on space coordinates, and they change according to the law of a mixed second-order tensor:

$$\mathrm{d}v^h(t, \mathbf{x}) = L^h_k(t, \mathbf{x})\,\mathrm{d}x^k \quad \longrightarrow \quad L^h_k = \frac{\partial v^h}{\partial x^k}\ . \tag{12.12}$$

While $\mathsf{L}$ is locally a matrix, its dependence on the coordinate system makes it a second-order tensor and is called the *velocity gradient tensor*.[15]

## 12.7.4 Strain Rate Tensor and Volume Dilatation Rate

From the velocity gradient tensor one introduces the *strain rate tensor* ≡ *rate of deformation*[16]

[15] Chadwick [39, p. 65], Billington and Tate [15, p. 44], Jaunzemis [106, p. 207], Lai et al. [117, p. 76].

[16] Billington and Tate [15, p. 47], Prager [188, p. 64], Aris [6, p. 84], Hunter [98, p. 111], Chadwick [39, p. 74], Jaunzemis [106, p. 208].

$$\mathbf{D} \stackrel{\text{def}}{=} \frac{1}{2}(\mathbf{L} + \mathbf{L}^{\mathsf{T}}) \qquad D_{hk} \stackrel{\text{def}}{=} \frac{1}{2}(L_{hk} + L_{hk}), \tag{12.13}$$

the *spin tensor*

$$\mathbf{W} \stackrel{\text{def}}{=} \frac{1}{2}(\mathbf{L} - \mathbf{L}^{\mathsf{T}}) \qquad W_{hk} \stackrel{\text{def}}{=} \frac{1}{2}(L_{hk} - L_{hk}) \tag{12.14}$$

and the *cubic (or bulk or volume) dilatation rate*

$$\theta(t, \mathbf{P}) \stackrel{\text{def}}{=} \mathrm{tr}\,(\mathbf{L}) = \mathrm{tr}\,(\mathbf{D}) = \nabla \cdot \mathbf{v} \tag{12.15}$$

**Space and Time Associations.** Since the strain rate tensor and the spin tensor are the symmetric and skew-symmetric parts of the velocity gradient tensor, they maintain the same association: $\mathbf{D}\,[\overline{\mathbf{T}}, \overline{\mathbf{L}}]$, $\mathbf{W}\,[\overline{\mathbf{T}}, \overline{\mathbf{L}}]$. The same thing happens for the cubic dilatation rate: $\theta[\overline{\mathbf{T}}, \overline{\mathbf{L}}]$.

### 12.7.5 Global Form of Mass Balance

Temple wrote: 'The fundamental equations of hydrodynamics should therefore be first formulated in global form, for a finite, extended mass of fluid.' Moreover, 'To emphasize this point we can employ the expression a fluid body, in contrast to the familiar rigid body, to designate a mass of fluid which always consists of the same particles, like the yolk of an egg oozing through a crack in the shell without breaking its sac'.[17]

Mass content is associated with primal time instants because it does not change sign under a reversal of motion. It follows that the mass balance must be computed on primal time intervals.

**Material Description.** Let us consider a fluid body $\mathscr{B}$ as a 'system'. Considering two arbitrary instants the mass balance requires

$$M[\overline{\mathbf{I}}^{+}, \mathscr{B}] - M[\overline{\mathbf{I}}^{-}, \mathscr{B}] \stackrel{\text{law}}{=} 0, \tag{12.16}$$

that reduces to a mass *conservation*.

**Spatial Description.** Considering a fixed control volume $\widetilde{\mathbf{V}}$ endowed with an outer orientation, the mass conservation becomes: *the increase in mass content in a volume in a time interval plus the mass which flowed out across the boundary of the volume is zero.* Table 12.5 summarizes this law.

---

[17] Temple [223, p. 3]. The expression *fluid body* is also used by Meyer [158, p. 3].

**Table 12.5** From verbal statement to differential formulation via global formulation

**Mass conservation law**

*(spatial description)*

- *verbal formulation:*

$$\begin{Bmatrix} \text{mass} \\ \text{storage} \end{Bmatrix} + \begin{Bmatrix} \text{mass} \\ \text{outflow} \end{Bmatrix} \overset{\text{law}}{=} 0$$

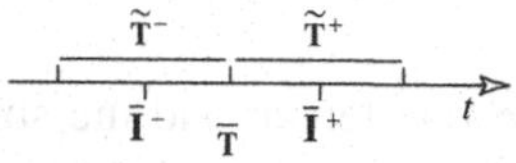


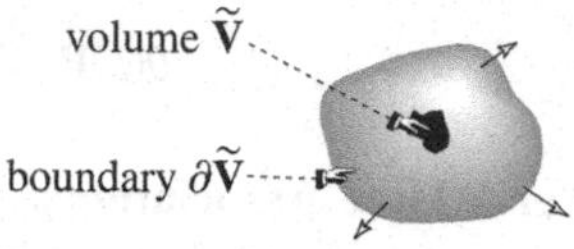


- *global formulation:*

$$\underbrace{M^{\mathrm{c}}[\bar{\mathbf{I}}^+, \widetilde{\mathbf{V}}] - M^{\mathrm{c}}[\bar{\mathbf{I}}^-, \widetilde{\mathbf{V}}]}_{\text{storage}} + \underbrace{M^{\mathrm{f}}[\bar{\mathbf{T}}, \partial\widetilde{\mathbf{V}}]}_{\text{outflow}} \overset{\text{law}}{=} 0$$

- *integral formulation:*

$$\int_{\widetilde{\mathbf{V}}} \left[\rho(t^+) - \rho(t^-)\right] \mathrm{d}V + \int_{t^-}^{t^+} \left[ \int_{\partial\widetilde{\mathbf{V}}} \mathbf{q} \cdot \mathbf{n} \, \mathrm{d}S \right] \mathrm{d}t \overset{\text{law}}{=} 0$$

dividing by $T = t^+ - t^-$ and performing the limit when $T \to 0$

$$\int_{\widetilde{\mathbf{V}}} \partial_t \rho \, \mathrm{d}V + \int_{\partial\widetilde{\mathbf{V}}} \mathbf{q} \cdot \mathbf{n} \, \mathrm{d}S \overset{\text{law}}{=} 0$$

dividing by $V$ and performing the limit when $\mathbf{V} \to \mathbf{P}$ we obtain the

- *differential formulation:*

$$\partial_t \rho + \operatorname{div} \mathbf{q} \overset{\text{law}}{=} 0 \qquad \text{equivalent to} \qquad \partial_t \rho + \nabla_k q^k \overset{\text{law}}{=} 0$$

## *12.7.6 Global Form of Momentum Balance*

Momentum is associated with dual time instants because it changes sign under a reversal of motion. Moreover, momentum production (≡ impulse) is associated with dual time intervals because it does not change sign under a reversal of motion. It follows that the momentum balance must be computed on dual time intervals.

**Material Description.** Let us consider a fluid body $\mathscr{B}$. The impulse communicated to the body is conveniently divided into a volume impulse $\mathbf{J}^{\mathrm{v}}$, like gravity, and a surface impulse $\mathbf{J}^{\mathrm{s}}$, like pressure and viscous impulses. Since the impulse is associated with dual time intervals, the momentum balance becomes

**Table 12.6** From verbal statement to differential formulation via global formulation

**Momentum balance law**

*(spatial description)*

- *verbal formulation:*

$$\left\{\begin{matrix}\text{momemtum}\\ \text{storage}\end{matrix}\right\} + \left\{\begin{matrix}\text{momentum}\\ \text{outflow}\end{matrix}\right\} \overset{\text{law}}{=} \left\{\begin{matrix}\text{impulse of}\\ \text{volume forces}\end{matrix}\right\} + \left\{\begin{matrix}\text{impulse of}\\ \text{surface forces}\end{matrix}\right\}$$

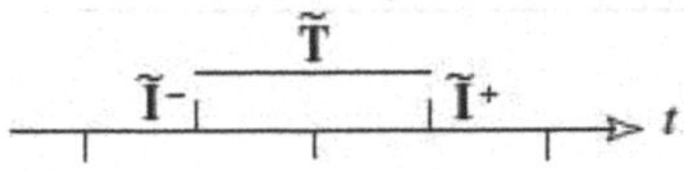


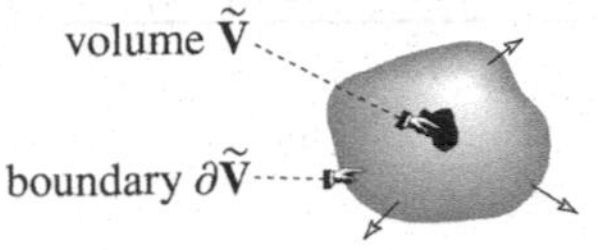


- *global formulation:*

$$\underbrace{\mathbf{P}^{\mathrm{c}}[\widetilde{\mathbf{I}}^+,\widetilde{\mathbf{V}}] - \mathbf{P}^{\mathrm{c}}[\widetilde{\mathbf{I}}^-,\widetilde{\mathbf{V}}]}_{\text{momentum storage}} + \underbrace{\mathbf{P}^{\mathrm{f}}[\widetilde{\mathbf{T}},\partial\widetilde{\mathbf{V}}]}_{\text{momentum outflow}} \overset{\text{law}}{=} \underbrace{\mathbf{J}^{\mathrm{v}}[\widetilde{\mathbf{T}},\widetilde{\mathbf{V}}]}_{\text{volume forces}} + \underbrace{\mathbf{J}^{\mathrm{s}}[\widetilde{\mathbf{T}},\partial\widetilde{\mathbf{V}}]}_{\text{surface forces}}$$

- *integral formulation:*

$$\int_{\widetilde{\mathbf{V}}} \left[\mathbf{p}(t^+) - \mathbf{p}(t^-)\right] \mathrm{d}V + \int_{t^-}^{t^+} \left[\int_{\partial\widetilde{\mathbf{V}}} \mathbf{p}\,(\mathbf{v}\cdot\mathbf{n})\,\mathrm{d}S\right] \mathrm{d}t \overset{\text{law}}{=} \int_{t^-}^{t^+} (\mathbf{F}^{\mathrm{V}} + \mathbf{F}^{\mathrm{S}})\,\mathrm{d}t$$

dividing by $T = t^+ - t^-$ and performing the limit when $T \to 0$

$$\int_{\mathbf{V}} \partial_t p_h \,\mathrm{d}V + \int_{\partial\mathbf{V}} p_h\,(v^k n_k)\,\mathrm{d}S \overset{\text{law}}{=} \int_{\mathbf{V}} f_h \,\mathrm{d}V + \int_{\partial\mathbf{V}} \sigma_h^{\ k} n_k \,\mathrm{d}S$$

dividing by $V$ and performing the limit when $\mathbf{V} \to \mathbf{P}$ we obtain the

- *differential formulation:*

$$\partial_t p_h + \nabla_k (p_h v^k) \overset{\text{law}}{=} f_h + \nabla_k \sigma_h^{\ k}$$

$$\mathbf{P}[\widetilde{\mathbf{I}}^+,\mathscr{B}] - \mathbf{P}[\widetilde{\mathbf{I}}^-,\mathscr{B}] \overset{\text{law}}{=} \mathbf{J}^{\mathrm{v}}[\widetilde{\mathbf{T}},\mathscr{B}] + \mathbf{J}^{\mathrm{s}}[\widetilde{\mathbf{T}},\partial\mathscr{B}]. \tag{12.17}$$

**Spatial Description.** Let us consider a fixed control volume $\widetilde{\mathbf{V}}$ endowed with an outer orientation. The momentum balance becomes: *the increase in momentum content in a volume in a time interval plus the momentum which flowed out across the boundary of the volume is equal to the sum of the impulses of volume forces and surface forces.*

Table 12.6 summarizes this law. Note that while mass balance is computed on primal time intervals, momentum balance must be computed on dual time intervals. This distinction, which is not at all evident in the differential formulation, may have a role to play in computational fluid dynamics.

## 12.8 Constitutive Laws

The most widely used constitutive equations in continuum mechanics are as follows:[18]

| | Ideal | Real |
|---|---|---|
| Incompressible | $\sigma_{ij} \overset{\text{mat}}{=} -p\,\delta_{ij}$ | $\sigma_{ij} \overset{\text{mat}}{=} -p\,\delta_{ij} + 2\mu\,D_{ij}$ |
| Compressible | $\sigma_{ij} \overset{\text{mat}}{=} -p\,\delta_{ij} + \lambda\,\theta\,\delta_{ij}$ | $\sigma_{ij} \overset{\text{mat}}{=} -p\,\delta_{ij} + 2\mu\,D_{ij} + \lambda\,\theta\,\delta_{ij}$ |

(12.18)

## 12.9 Classification Diagrams of Fluid Dynamics

There are two classes of diagrams: those which have as the source variables

- The mass, such as [FLU1]–[FLU5];
- The force, such as [FLU6], [FLU8], [FLU9], [FLU11].

The diagrams of the first class are typical of a field theory, whereas those of the second class are typical of a mechanical theory. The main difference between the two classes is that in a field theory velocity is associated with lines and time instants, whereas in a mechanical theory velocity is associated with points and time intervals.

Diagram [FLU10] differs from the others because it shows the link with the variables entering diagrams of the first kind (those with masses as source variables) and forces. Even if correct, the diagram needs an interpretation. It is similar to the diagram [ELE13] of a charged particle in an electromagnetic field.

[18] O'Neil [171, p. 55], Chorin and Marsden [41, p. 32].

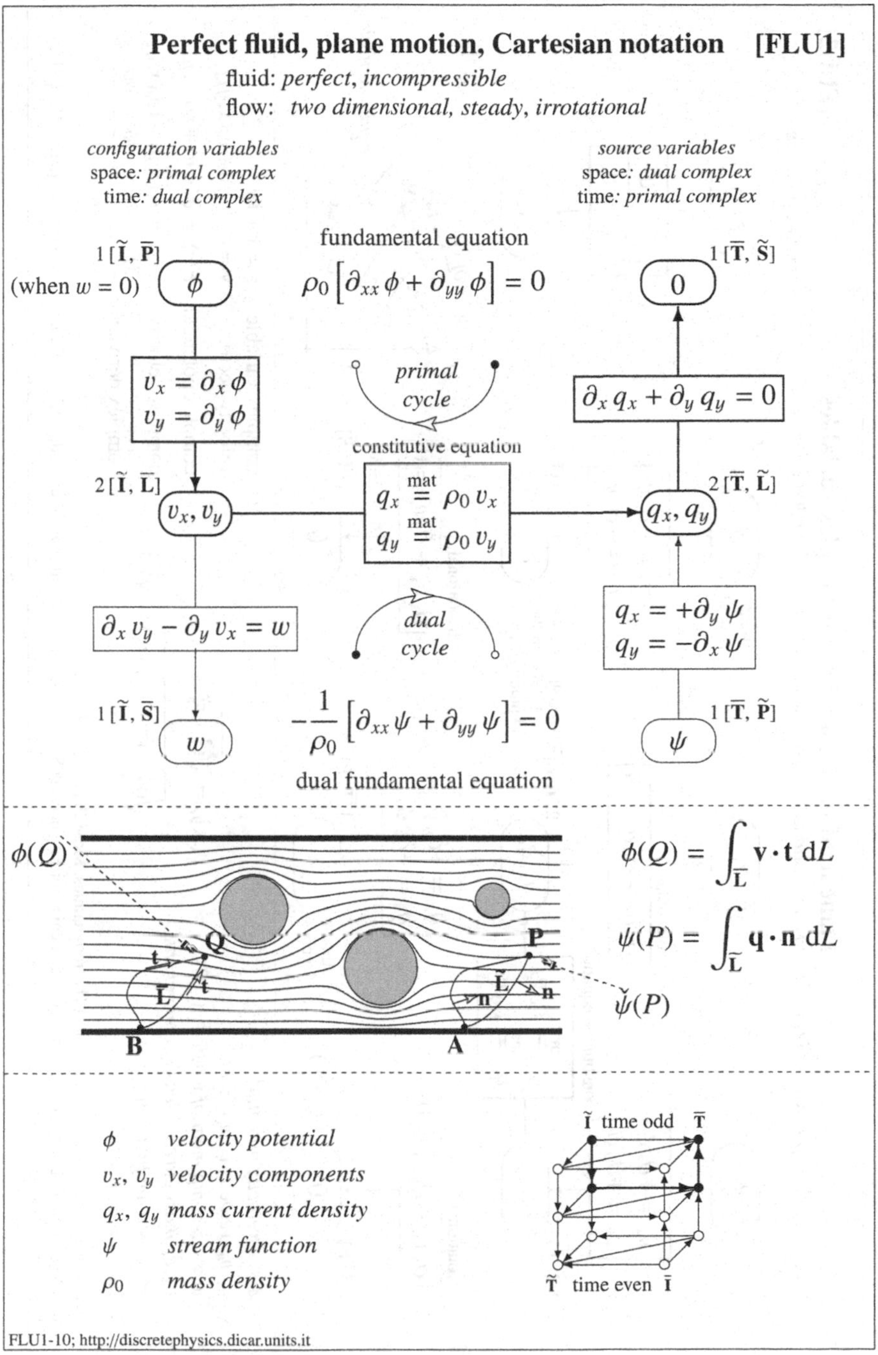
**Perfect fluid, plane motion, Cartesian notation [FLU1]**
fluid: *perfect, incompressible*
flow: *two dimensional, steady, irrotational*
*configuration variables*
space: *primal complex*
time: *dual complex*
*source variables*
space: *dual complex*
time: *primal complex*
fundamental equation
$\rho_0 \left[\partial_{xx}\phi + \partial_{yy}\phi\right] = 0$
1 [$\tilde{\mathbf{I}}$, $\bar{\mathbf{P}}$]
(when $w = 0$) $\phi$
1 [$\bar{\mathbf{T}}$, $\tilde{\mathbf{S}}$]
0
$v_x = \partial_x \phi$
$v_y = \partial_y \phi$
*primal cycle*
$\partial_x q_x + \partial_y q_y = 0$
constitutive equation
2 [$\tilde{\mathbf{I}}$, $\bar{\mathbf{L}}$]
$v_x, v_y$
$q_x \overset{\text{mat}}{=} \rho_0 v_x$
$q_y \overset{\text{mat}}{=} \rho_0 v_y$
2 [$\bar{\mathbf{T}}$, $\tilde{\mathbf{L}}$]
$q_x, q_y$
$\partial_x v_y - \partial_y v_x = w$
*dual cycle*
$q_x = +\partial_y \psi$
$q_y = -\partial_x \psi$
1 [$\tilde{\mathbf{I}}$, $\bar{\mathbf{S}}$]
$w$
$-\frac{1}{\rho_0}\left[\partial_{xx}\psi + \partial_{yy}\psi\right] = 0$
1 [$\bar{\mathbf{T}}$, $\tilde{\mathbf{P}}$]
$\psi$
dual fundamental equation
$\phi(Q)$
Q
P
t
$\bar{\mathbf{L}}$
$\tilde{\mathbf{L}}$
n
B
A
$\check{\psi}(P)$
$\phi(Q) = \int_{\bar{L}} \mathbf{v}\cdot\mathbf{t}\, dL$
$\psi(P) = \int_{\tilde{L}} \mathbf{q}\cdot\mathbf{n}\, dL$
$\phi$ *velocity potential*
$v_x$, $v_y$ *velocity components*
$q_x$, $q_y$ *mass current density*
$\psi$ *stream function*
$\rho_0$ *mass density*
$\tilde{\mathbf{I}}$ time odd $\bar{\mathbf{T}}$
$\tilde{\mathbf{T}}$ time even $\bar{\mathbf{I}}$
FLU1-10; http://discretephysics.dicar.units.it

## Plane motion of a fluid using complex variables [FLU2]

*The use of complex variables for steady, irrotational, perfect, incompressible, plane fluid flow*

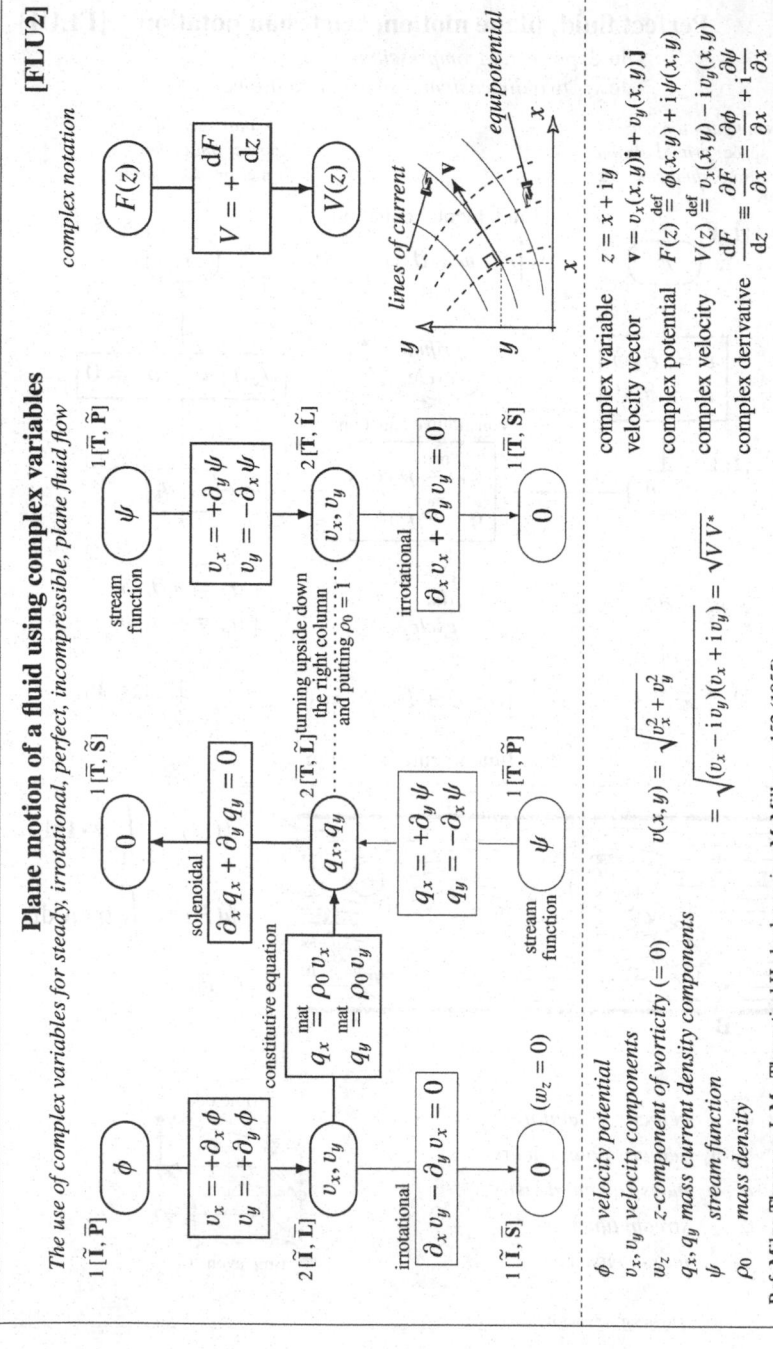

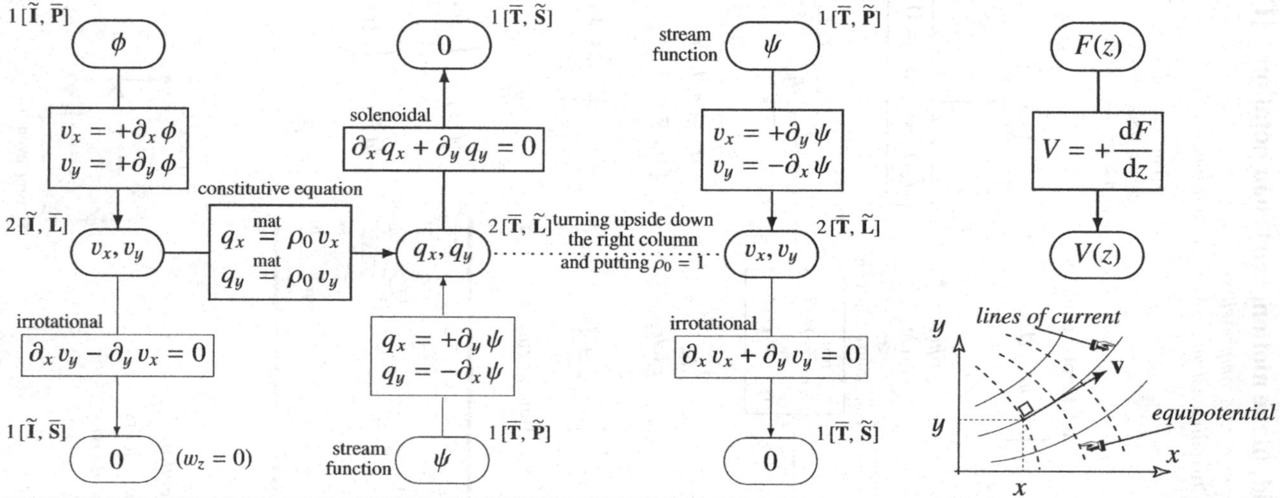


| | |
|---|---|
| $\phi$ | *velocity potential* |
| $v_x, v_y$ | *velocity components* |
| $w_z$ | *z-component of vorticity* (= 0) |
| $q_x, q_y$ | *mass current density components* |
| $\psi$ | *stream function* |
| $\rho_0$ | *mass density* |

$$v(x,y) = \sqrt{v_x^2 + v_y^2} = \sqrt{(v_x - \mathrm{i}\, v_y)(v_x + \mathrm{i}\, v_y)} = \sqrt{V\, V^*}$$

| | |
|---|---|
| complex variable | $z = x + \mathrm{i}\, y$ |
| velocity vector | $\mathbf{v} = v_x(x,y)\mathbf{i} + v_y(x,y)\mathbf{j}$ |
| complex potential | $F(z) \overset{\text{def}}{=} \phi(x,y) + \mathrm{i}\, \psi(x,y)$ |
| complex velocity | $V(z) \overset{\text{def}}{=} v_x(x,y) - \mathrm{i}\, v_y(x,y)$ |
| complex derivative | $\frac{dF}{dz} \equiv \frac{\partial F}{\partial x} = \frac{\partial \phi}{\partial x} + \mathrm{i}\frac{\partial \psi}{\partial x}$ |

Ref: Milne-Thomson, L.M.: Theoretical Hydrodynamics. McMillan, p.152 (1955).
Ref: Houghton, E. L. and Brock, A.E.: Aerodynamics for Engineering Students. Edward Arnold (Publisher) Ltd, Ch.11 (1972).

Compare with table [ELE7]

FLU2-9; http://discretephysics.dicar.units.it

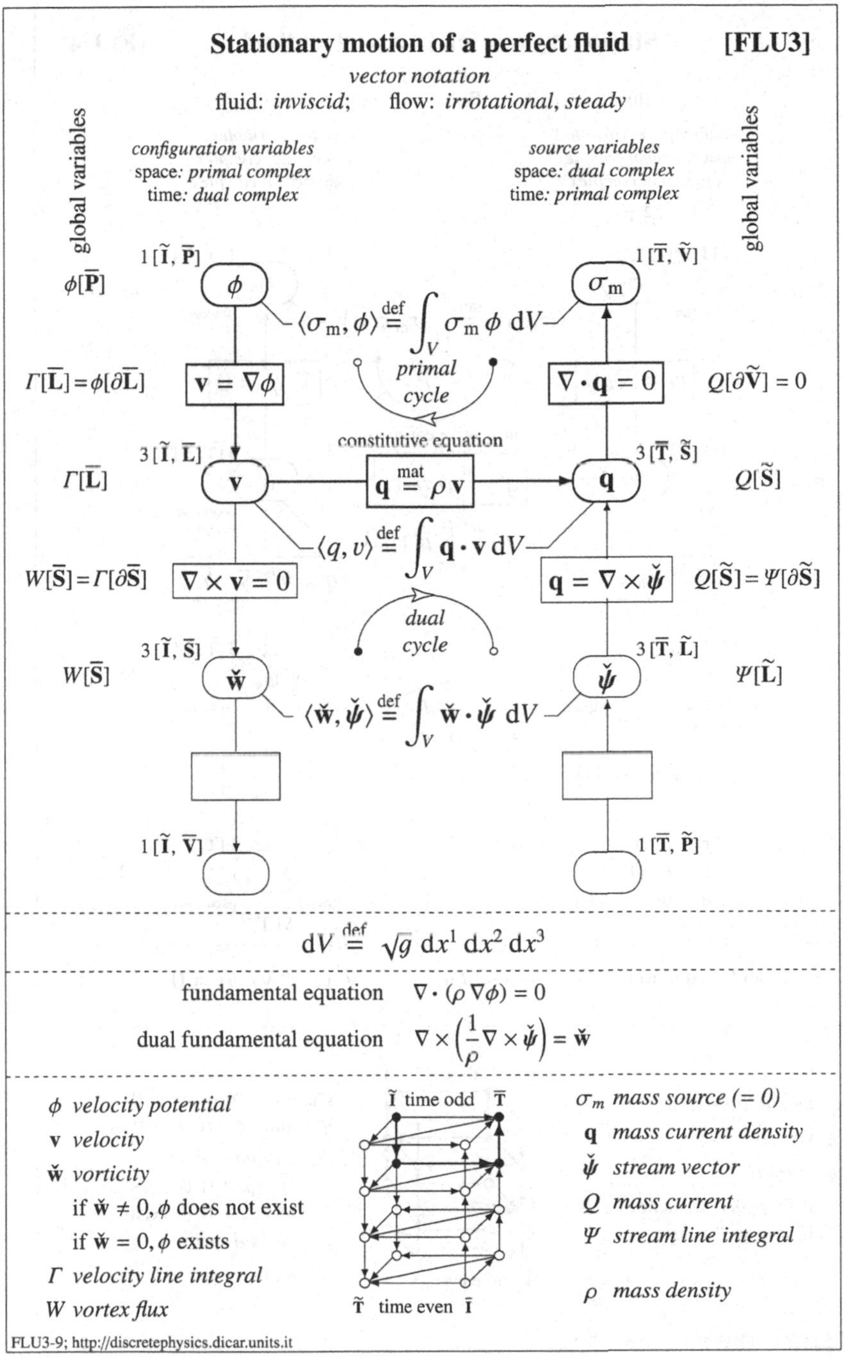
Stationary motion of a perfect fluid [FLU3]
vector notation
fluid: inviscid; flow: irrotational, steady
global variables
configuration variables
space: primal complex
time: dual complex
source variables
space: dual complex
time: primal complex
global variables
$\phi[\overline{\mathbf{P}}]$
$1\,[\tilde{\mathbf{I}}, \overline{\mathbf{P}}]$
$\phi$
$\langle \sigma_{\rm m}, \phi \rangle \stackrel{\rm def}{=} \int_V \sigma_{\rm m}\, \phi \, {\rm d}V$
$1\,[\overline{\mathbf{T}}, \tilde{\mathbf{V}}]$
$\sigma_{\rm m}$
$\Gamma[\overline{\mathbf{L}}] = \phi[\partial \overline{\mathbf{L}}]$
$\mathbf{v} = \nabla \phi$
primal cycle
$\nabla \cdot \mathbf{q} = 0$
$Q[\partial \tilde{\mathbf{V}}] = 0$
constitutive equation
$3\,[\tilde{\mathbf{I}}, \overline{\mathbf{L}}]$
$3\,[\overline{\mathbf{T}}, \tilde{\mathbf{S}}]$
$\Gamma[\overline{\mathbf{L}}]$
$\mathbf{v}$
$\mathbf{q} \stackrel{\rm mat}{=} \rho\, \mathbf{v}$
$\mathbf{q}$
$Q[\tilde{\mathbf{S}}]$
$\langle q, v \rangle \stackrel{\rm def}{=} \int_V \mathbf{q} \cdot \mathbf{v}\, {\rm d}V$
$W[\overline{\mathbf{S}}] = \Gamma[\partial \overline{\mathbf{S}}]$
$\nabla \times \mathbf{v} = 0$
$\mathbf{q} = \nabla \times \check{\boldsymbol{\psi}}$
$Q[\tilde{\mathbf{S}}] = \Psi[\partial \tilde{\mathbf{S}}]$
dual cycle
$3\,[\tilde{\mathbf{I}}, \overline{\mathbf{S}}]$
$3\,[\overline{\mathbf{T}}, \tilde{\mathbf{L}}]$
$W[\overline{\mathbf{S}}]$
$\check{\mathbf{w}}$
$\check{\boldsymbol{\psi}}$
$\Psi[\tilde{\mathbf{L}}]$
$\langle \check{\mathbf{w}}, \check{\boldsymbol{\psi}} \rangle \stackrel{\rm def}{=} \int_V \check{\mathbf{w}} \cdot \check{\boldsymbol{\psi}}\, {\rm d}V$
$1\,[\tilde{\mathbf{I}}, \overline{\mathbf{V}}]$
$1\,[\overline{\mathbf{T}}, \tilde{\mathbf{P}}]$
${\rm d}V \stackrel{\rm def}{=} \sqrt{g}\, {\rm d}x^1\, {\rm d}x^2\, {\rm d}x^3$
fundamental equation $\nabla \cdot (\rho\, \nabla \phi) = 0$
dual fundamental equation $\nabla \times \left( \frac{1}{\rho} \nabla \times \check{\boldsymbol{\psi}} \right) = \check{\mathbf{w}}$
$\phi$ velocity potential
$\mathbf{v}$ velocity
$\check{\mathbf{w}}$ vorticity
if $\check{\mathbf{w}} \neq 0$, $\phi$ does not exist
if $\check{\mathbf{w}} = 0$, $\phi$ exists
$\Gamma$ velocity line integral
$W$ vortex flux
$\tilde{\mathbf{I}}$ time odd $\overline{\mathbf{T}}$
$\tilde{\mathbf{T}}$ time even $\overline{\mathbf{I}}$
$\sigma_m$ mass source (= 0)
$\mathbf{q}$ mass current density
$\check{\boldsymbol{\psi}}$ stream vector
$Q$ mass current
$\Psi$ stream line integral
$\rho$ mass density
FLU3-9; http://discretephysics.dicar.units.it

# Stationary motion of a perfect fluid  [FLU4]

*tensor notation*
fluid: *perfect*;   flow: *irrotational, steady*

*configuration variables*
space: *primal complex*
time: *dual complex*

*source variables*
space: *dual complex*
time: *primal complex*

SI units

SI units

$L^2 T^{-1}$

$m^2/s$

$1\,[\tilde{\mathbf{I}}, \overline{\mathbf{P}}]$   1

$\phi$

scalar

$\langle \sigma_m, \phi \rangle \overset{\text{def}}{=} \int_A \sigma_m \, \phi \, dV$

*primal cycle*

constitutive equation

$v_k = \nabla_k \phi$

$3\,[\tilde{\mathbf{I}}, \overline{\mathbf{L}}]$   1

$v_k$

covariant vector

$q^i \overset{\text{mat}}{=} \rho \, g^{ik} v_k$

$\langle q, v \rangle \overset{\text{def}}{=} \int_A q^i v_i \, dV$

$\breve{\eta}^{ihk} \nabla_h v_k = \breve{w}^i$

$3\,[\tilde{\mathbf{I}}, \overline{\mathbf{S}}]$

$\breve{w}^i$

contravariant pseudovector        $\text{sgn}(\Delta)$

*dual cycle*

$\nabla_i \breve{w}^i = 0$

$1\,[\tilde{\mathbf{I}}, \overline{\mathbf{V}}]$

$0$

pseudoscalar        $\text{sgn}(\Delta)$

1   $1\,[\overline{\mathbf{T}}, \tilde{\mathbf{V}}]$

$\sigma_m$

scalar

$kg/s/m^3$

$\nabla_i q^i = 0$

1   $3\,[\overline{\mathbf{T}}, \tilde{\mathbf{S}}]$

$q^i$

contravariant vector

$kg/s/m^2$

$q^i = \breve{\eta}^{ihk} \nabla_h \breve{\psi}_k$

$3\,[\overline{\mathbf{T}}, \tilde{\mathbf{L}}]$

$\breve{\psi}_k$

$\text{sgn}(\Delta)$   covariant pseudovector

$kg/s/m$

$1\,[\overline{\mathbf{T}}, \tilde{\mathbf{P}}]$

$kg/s$

$\text{sgn}(\Delta)$   pseudoscalar

$M\,T^{-1}$

m/s

1/s

---

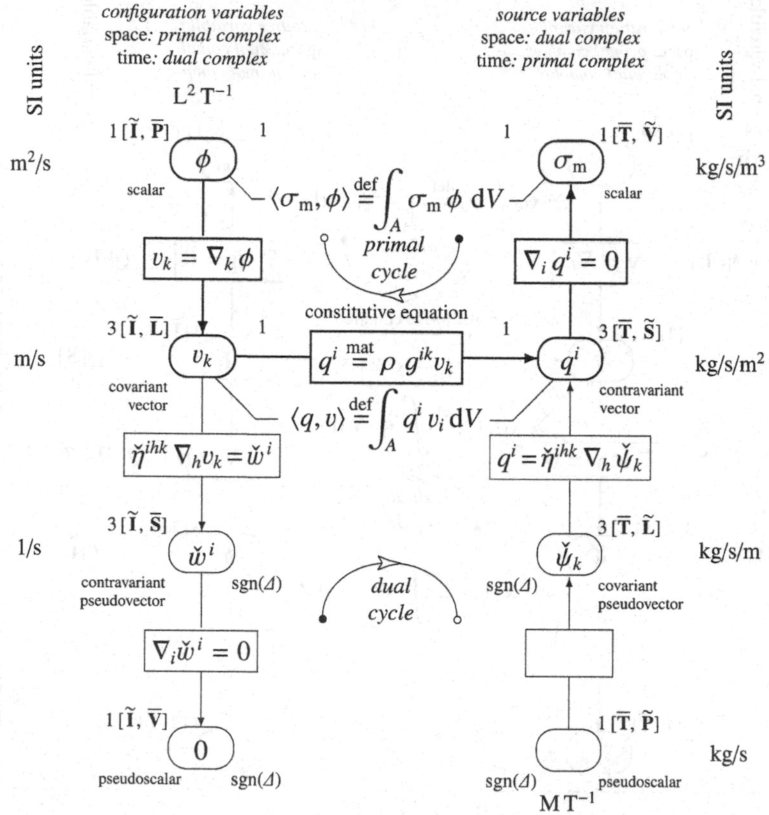

fundamental equation: mass conservation        $\nabla_i (\rho \, g^{ik} \nabla_k \phi) = 0$

---

$\phi$  *kinetic potential*
$v_k$  *velocity*
$\breve{w}^i$  *vorticity*
   if $\breve{w}^i \neq 0, \phi$ does not exist
   if $\breve{w}^i = 0, \phi$ exists

$\tilde{\mathbf{I}}$ time odd  $\overline{\mathbf{T}}$

$\overline{\mathbf{T}}$ time even  $\tilde{\mathbf{I}}$

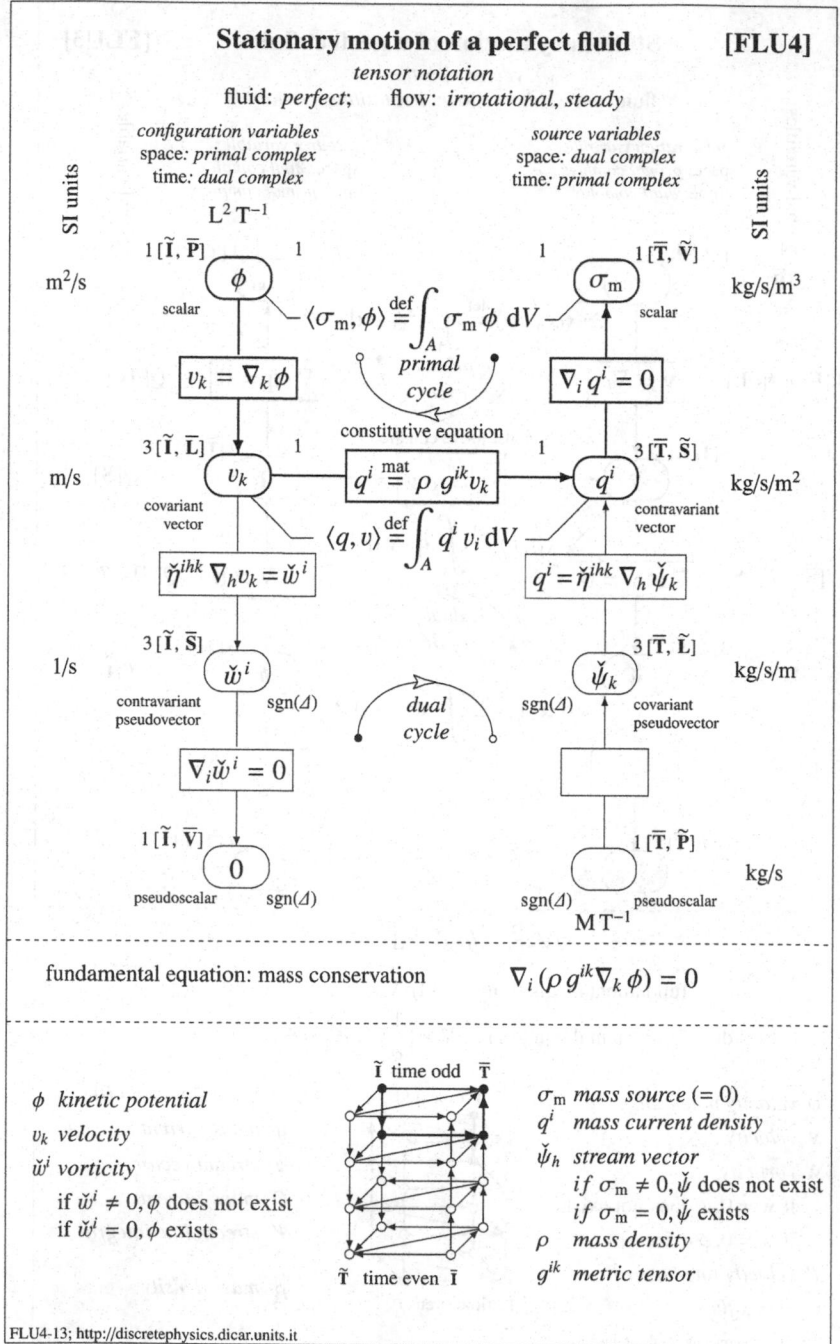

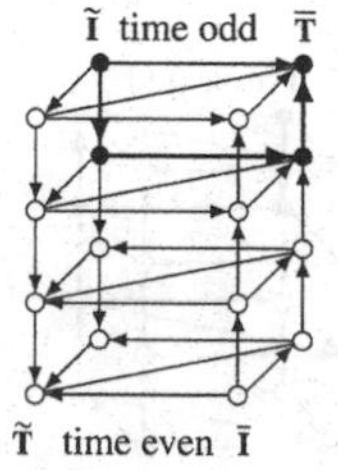

$\sigma_m$  *mass source* $(= 0)$
$q^i$  *mass current density*
$\breve{\psi}_h$  *stream vector*
   *if* $\sigma_m \neq 0, \breve{\psi}$ does not exist
   *if* $\sigma_m = 0, \breve{\psi}$ exists
$\rho$  *mass density*
$g^{ik}$  *metric tensor*

FLU4-13; http://discretephysics.dicar.units.it

## Acoustics in fluids [FLU5]

fluid: *perfect, barotropic*
flow: *irrotational, small velocities*

configuration variables
space: *primal complex*
time: *dual complex*
intervals   instants
dimension: *action / mass*

source variables
space: *dual complex*
time: *primal complex*
instants   intervals

$1\,[\tilde{\mathbf{I}}, \overline{\mathbf{P}}]$

$\phi$

$\mathbf{v} = \nabla\phi$

$\chi = -\partial_t\phi$

$1\,[\tilde{\mathbf{T}}, \overline{\mathbf{P}}]$

$\chi$

$3\,[\tilde{\mathbf{I}}, \overline{\mathbf{L}}]$

$\mathbf{v}$

$3\,[\tilde{\mathbf{T}}, \overline{\mathbf{L}}]$

constitutive equation

$$\rho \overset{\text{mat}}{=} \rho_o\left[1 + \frac{\chi}{c^2}\right]$$

Bernoulli

$$\mathbf{q} \overset{\text{mat}}{=} \rho_o\mathbf{v}$$

constitutive equation

$1\,[\overline{\mathbf{T}}, \tilde{\mathbf{V}}]$

$\sigma_{\mathrm{m}}$

$$\partial_t\rho + \nabla\cdot\mathbf{q} = \sigma_{\mathrm{m}}$$

$1\,[\overline{\mathbf{I}}, \tilde{\mathbf{V}}]$

$\rho$

$3\,[\overline{\mathbf{T}}, \tilde{\mathbf{S}}]$

$\mathbf{q}$

$3\,[\overline{\mathbf{I}}, \tilde{\mathbf{S}}]$

dimension: *mass*

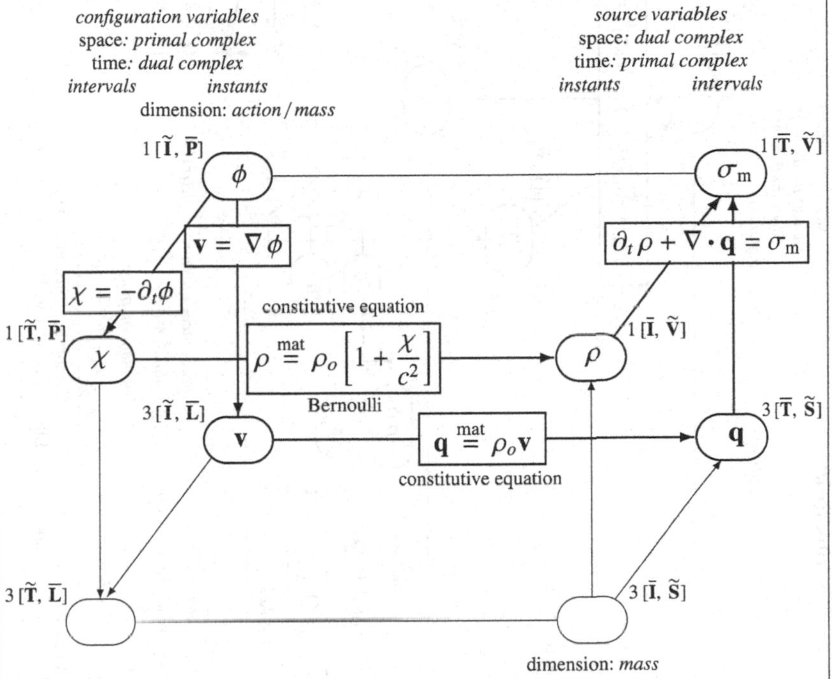

fundamental equation
D'Alembert

$$-\frac{1}{c^2}\frac{\partial^2\phi}{\partial t^2} + \nabla^2\phi = \frac{1}{\rho_0}\sigma_{\mathrm{m}}$$

$$c \overset{\text{def}}{=} \sqrt{\gamma\,\frac{p_0}{\rho_o}}$$

$\phi$  *kinetic potential*
$\mathbf{v}$  *velocity*
$\chi$  *no known name*
$\rho$  *mass density*
$\mathbf{q}$  *mass current density*
$\sigma_{\mathrm{m}}$  *mass production density*
$\rho_o$  *rest mass density*
$p_o$  *static pressure*
$c$  *sound speed*
$\gamma$  *polytropic index*

$\tilde{\mathbf{I}}$ time odd $\overline{\mathbf{T}}$

$\tilde{\mathbf{T}}$ time even $\overline{\mathbf{I}}$

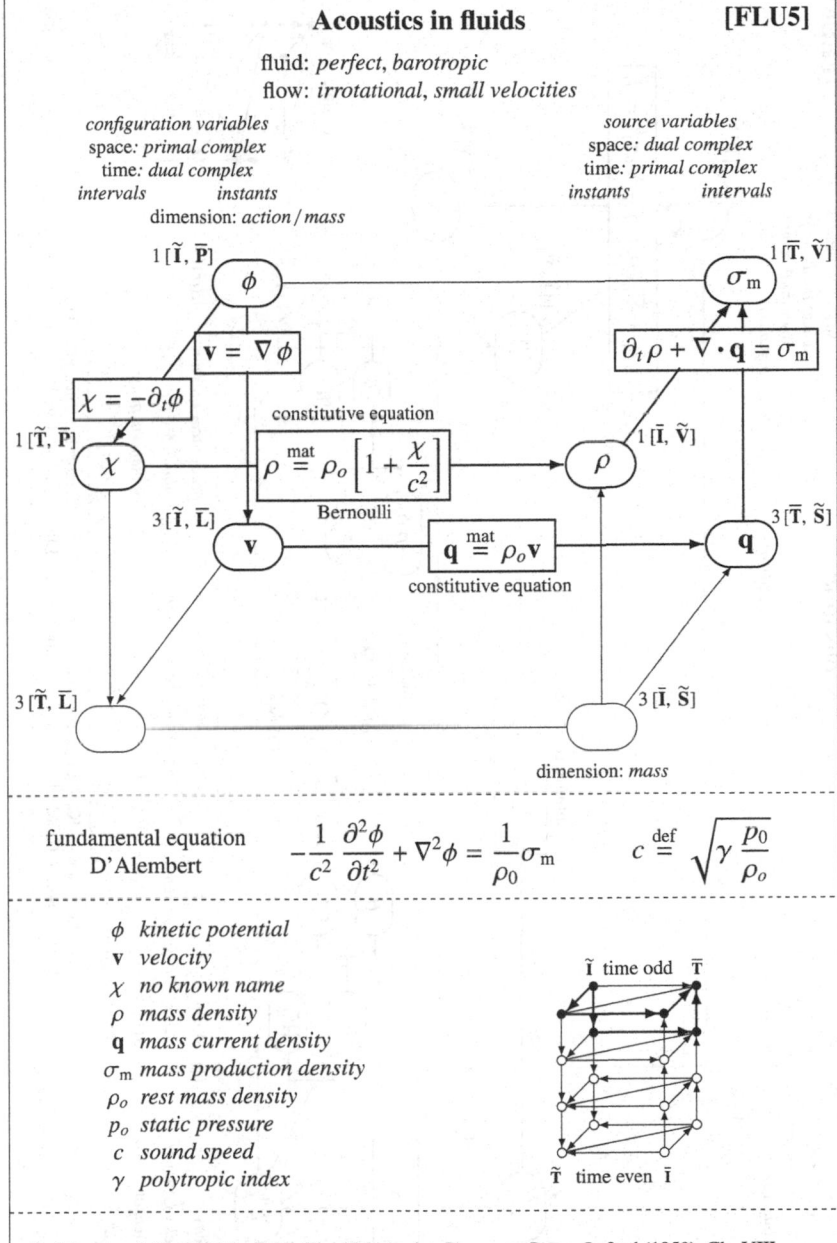

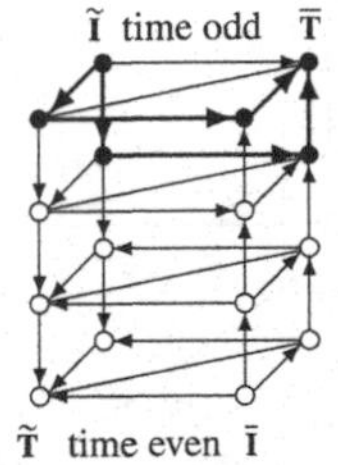

Ref: Landau, L.D., Lifshitz, E.M.: Fluid Mechanics. Pergamon Press, Oxford (1958), Ch. VIII
Ref: Paterson, A.R.: A First Course in Fluid Dynamics. Cambridge University Press, (1997) p. 250.

FLU5-8; http://discretephysics.dicar.units.it

related diagrams: FLU1, FLU4

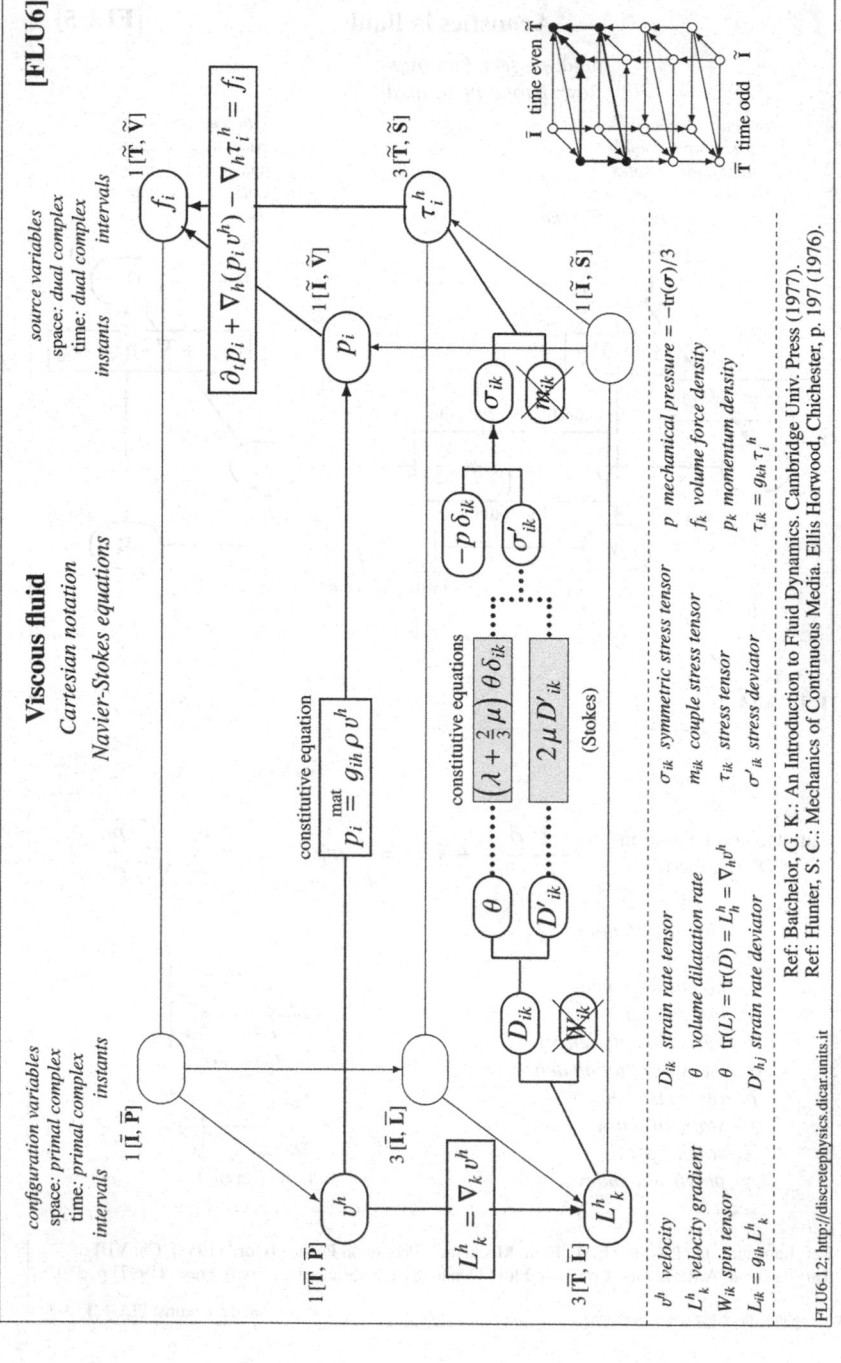

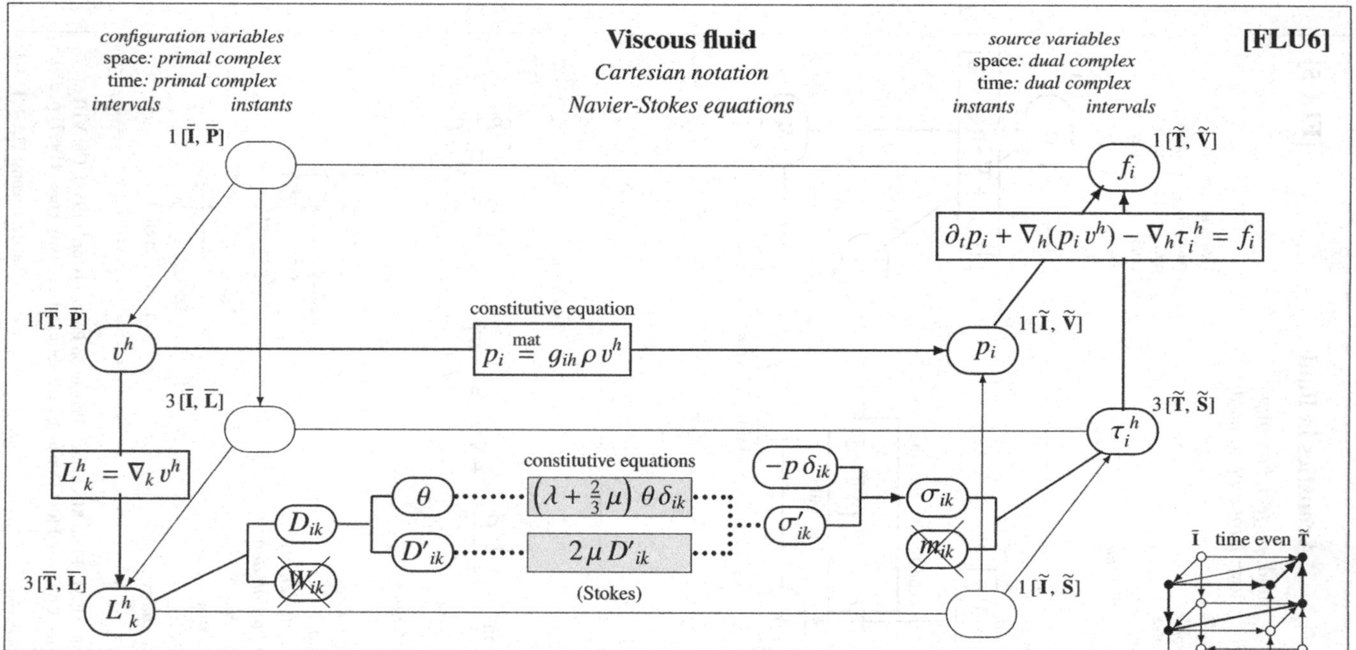

[FLU6]

**Viscous fluid**
*Cartesian notation*
*Navier-Stokes equations*

*source variables*
*space: dual complex*
*time: dual complex*
*instants          intervals*

$$\partial_t p_i + \nabla_h(p_i v^h) - \nabla_h \tau_i^h = f_i$$

$1[\tilde{\mathbf{T}}, \tilde{\mathbf{V}}]$

$3[\tilde{\mathbf{T}}, \tilde{\mathbf{S}}]$

$1[\tilde{\mathbf{I}}, \tilde{\mathbf{V}}]$

$1[\tilde{\mathbf{I}}, \tilde{\mathbf{S}}]$

$\tilde{\mathbf{I}}$ time even $\tilde{\mathbf{T}}$
$\bar{\mathbf{T}}$ time odd $\tilde{\mathbf{I}}$

*constitutive equation*

$$p_i \overset{\text{mat}}{=} g_{ih} \rho v^h$$

*constitutive equations*

$\left(\lambda + \frac{2}{3}\mu\right)\theta\,\delta_{ik}$

$2\mu D'_{ik}$

(Stokes)

$-p\,\delta_{ik}$          $\sigma'_{ik}$

$\sigma_{ik}$          $m_{ik}$

$\theta$          $D'_{ik}$

$D_{ik}$          $W_{ik}$

*configuration variables*
*space: primal complex*
*time: primal complex*
*intervals          instants*

$1[\bar{\mathbf{I}}, \bar{\mathbf{P}}]$

$3[\bar{\mathbf{I}}, \bar{\mathbf{L}}]$

$1[\bar{\mathbf{T}}, \bar{\mathbf{P}}]$

$3[\bar{\mathbf{T}}, \bar{\mathbf{L}}]$

$v^h$

$$L^h_k = \nabla_k v^h$$

$L^h$

$v^h$  *velocity*                      $D_{ik}$  *strain rate tensor*          $\sigma_{ik}$  *symmetric stress tensor*          $p$  *mechanical pressure* $= -\mathrm{tr}(\sigma)/3$
$L^h_k$  *velocity gradient*          $\theta$  *volume dilatation rate*          $m_{ik}$  *couple stress tensor*          $f_k$  *volume force density*
$W_{ik}$  *spin tensor*                $\theta = \mathrm{tr}(L) = \mathrm{tr}(D) = L^h_h = \nabla_h v^h$          $\tau_{ik}$  *stress tensor*          $p_k$  *momentum density*
$L_{ik} = g_{ih} L^h_k$              $D'_{hj}$  *strain rate deviator*          $\sigma'_{ik}$  *stress deviator*          $\tau_i^h = g_{ih}\tau_i^h$

Ref: Batchelor, G. K.: An Introduction to Fluid Dynamics. Cambridge Univ. Press (1977).
Ref: Hunter, S. C.: Mechanics of Continuous Media. Ellis Horwood, Chichester, p. 197 (1976).

FLU6-12: http://discretephysics.dicar.units.it

**Hydraulics: percolation in porous media     [FLU7]**

*configuration variables*
space: *primal complex*
time: *dual complex*
intervals          instants
$1\,[\tilde{\mathbf{I}}, \bar{\mathbf{P}}]$

*source variables*
space: *dual complex*
time: *primal complex*
instants          intervals
$1\,[\bar{\mathbf{T}}, \tilde{\mathbf{V}}]$

$$\nabla \cdot \mathbf{q} = 0$$

$1\,[\tilde{\mathbf{T}}, \bar{\mathbf{P}}]$   $\phi$

$1\,[\bar{\mathbf{I}}, \tilde{\mathbf{V}}]$

$3\,[\tilde{\mathbf{I}}, \bar{\mathbf{L}}]$

$3\,[\bar{\mathbf{T}}, \tilde{\mathbf{S}}]$   $\mathbf{q}$

$$\mathbf{u} = \nabla\phi$$

constitutive equation

$$\mathbf{q} = -\frac{k}{\mu}\,\mathbf{u}$$

Darcy

$3\,[\tilde{\mathbf{T}}, \bar{\mathbf{L}}]$   $\mathbf{u}$

$3\,[\bar{\mathbf{I}}, \tilde{\mathbf{S}}]$

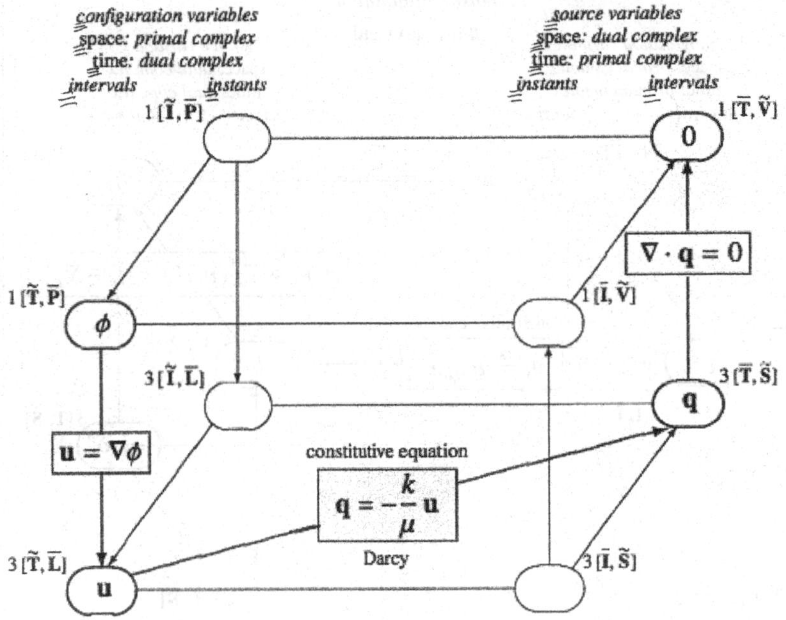

$$\phi \stackrel{\text{def}}{=} p + \rho g z \qquad H \stackrel{\text{def}}{=} \frac{\phi}{\rho g} = \frac{p}{\rho g} + z \qquad \text{total head}$$

$p$ *pressure* (Pa)

$z$ *elevation* (m)

$g$ *acceleration of gravity* $(\mathrm{m\,s^{-2}})$

$\phi$ *fluid potential* (Pa)

$\mathbf{u}$ *pressure gradient* $(\mathrm{Pa\,m^{-1}})$

$\mathbf{q}$ *volumetric current density* $(\mathrm{m\,s^{-1}})$

$\rho$ *density* $(\mathrm{kg\,m^{-3}})$

$\mu$ *absolute viscosity* $(\mathrm{N\,s\,m^{-2}})$

$k$ *permeability* $(\mathrm{m^2})$

$\tilde{\mathbf{I}}$ time odd $\bar{\mathbf{T}}$

$\bar{\mathbf{T}}$ time even $\tilde{\mathbf{I}}$

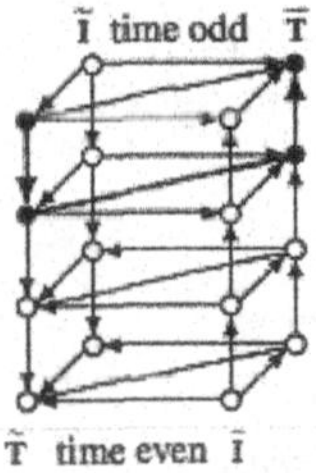

Ref: Hubbert M.K., The Theory of Ground-Water Motion, The Journal of Geology,
    vol. XLVIII, No. 8, Part 1. pp. 785-944, (1940). This paper is a masterpiece!

Ref: Henry P.G. Darcy and other pioneer: in hydraulics : contributions in celebration of the
    200th birthday of Henry Philibert Gaspard Darcy. June 23-26, 2003. Philadelphia, ASCE

**Perfect fluid: Euler equation**      **[FLU8]**

*tensorial notation*

*fluid*: inviscid

*configuration variables*
space*: primal complex*
time*: primal complex*
*intervals*     *instants*

*source variables*
space*: dual complex*
time*: dual complex*
*instants*     *intervals*

$1\,[\overline{\mathbf{I}},\,\overline{\mathbf{P}}]$      $1\,[\widetilde{\mathbf{T}},\,\widetilde{\mathbf{V}}]$

$f_i$

$$\partial_t\, p_i + \nabla_h\,(p_i\, v^h) = f_i - \nabla_i\, p$$

$1\,[\overline{\mathbf{T}},\,\overline{\mathbf{P}}]$    *constitutive equation*    $1\,[\widetilde{\mathbf{I}},\,\widetilde{\mathbf{V}}]$

$v^h$      $p_i \overset{\text{mat}}{=} \rho\, g_{ih}\, v^h$      $p_i$

$3\,[\overline{\mathbf{I}},\,\overline{\mathbf{L}}]$      $3\,[\widetilde{\mathbf{T}},\,\widetilde{\mathbf{S}}]$

$-p\,\delta_i^h$

$3\,[\overline{\mathbf{T}},\,\overline{\mathbf{L}}]$      $3\,[\widetilde{\mathbf{I}},\,\widetilde{\mathbf{S}}]$

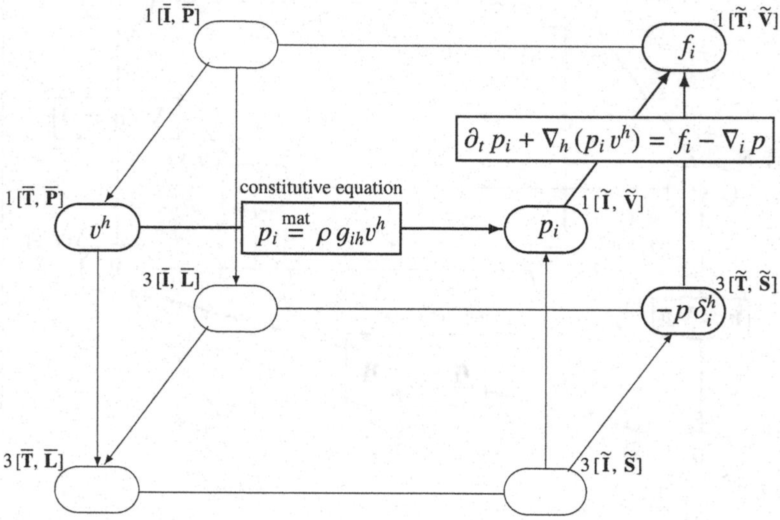

fundamental equation
Euler equation      $\partial_t\,(\rho\, g_{ih} v^h) + \nabla_h\,(\rho\, v_i\, v^h) = f_i - \nabla_i\, p$

$v^h$ *velocity*
$\rho$ *density*
$g_{ih}$ *metric tensor*
$f_i$ *force for unit volume*
$p_i$ *momentum density*
$\delta_i^h$ *Kronecker tensor*

$\overline{\mathbf{I}}$ time even $\widetilde{\mathbf{T}}$

$\overline{\mathbf{T}}$ time odd $\widetilde{\mathbf{I}}$

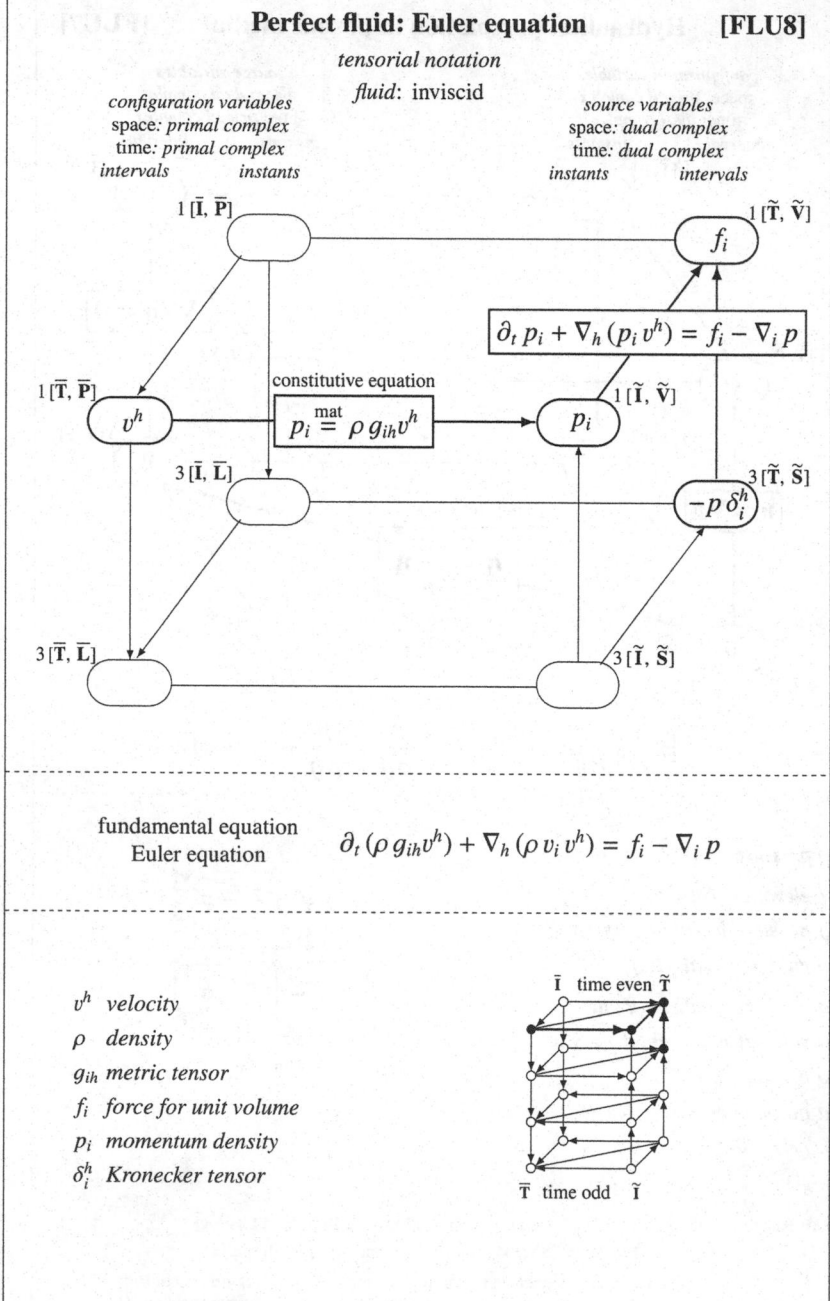

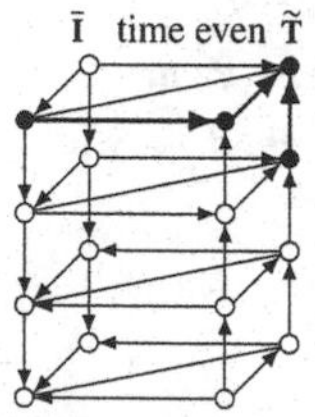

FLU8-5; http://discretephysics.dicar.units.it

## Viscous fluid motion in a pipe: Poiseuille law [FLU9]

fluid: *viscous*
flow: *steady, incompressible, laminar in a horizontal pipe*

*configuration variables*
space*: primal complex*
time*: primal complex*

*source variables*
space*: dual complex*
time*: dual complex*

$1\,[\bar{\mathbf{T}}, \bar{\mathbf{P}}]$ $v$

$\gamma = \dfrac{\mathrm{d}v}{\mathrm{d}r}$

$1\,[\bar{\mathbf{T}}, \bar{\mathbf{L}}]$ $\gamma$

fundamental equation

$$-\mu \frac{1}{r}\frac{\mathrm{d}}{\mathrm{d}r}\left[r\,\frac{\mathrm{d}v}{\mathrm{d}r}\right] = G$$

*primal cycle*

constitutive equation

$$\tau \overset{\text{mat}}{=} -\mu\gamma$$

Newton

$1\,[\tilde{\mathbf{T}}, \tilde{\mathbf{L}}]$ $G$

$\dfrac{1}{r}\dfrac{\mathrm{d}}{\mathrm{d}r}[r\,\tau] = G$

$1\,[\tilde{\mathbf{T}}, \tilde{\mathbf{P}}]$ $\tau$

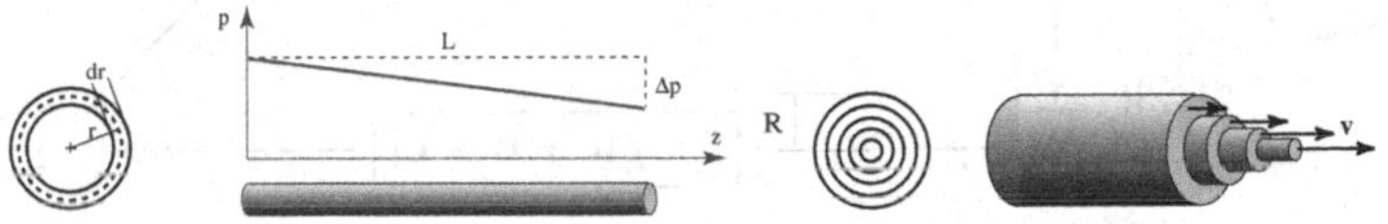


$$G \overset{\text{def}}{=} \frac{p_{\text{out}} - p_{\text{in}}}{L} \qquad v(r) = \frac{1}{4\mu}\,G\,(R^2 - r^2) \qquad Q = \int_0^R v(r)\,(2\pi r\,\mathrm{d}r)$$

$$\mathscr{R} \overset{\text{def}}{=} \frac{8\,\mu\,L}{\pi\,R^4} \qquad Q = \frac{\pi\,R^4}{8\,\mu}G \qquad p_{\text{out}} - p_{\text{in}} = -\mathscr{R}\,Q$$

$p_{\text{in}}$ *inlet pressure*
$p_{\text{out}}$ *outlet pressure*
$L$ *length of the pipe*
$G$ *pressure gradient*
$v(r)$ *velocity*
$\mu$ *viscosity coefficient*
$R$ *radius of the pipe*
$\mathscr{R}$ *hydraulic resistance*
$\gamma(r)$ *shear strain rate*
$\tau(r)$ *shear stress*
$Q$ *volumetric flow rate*

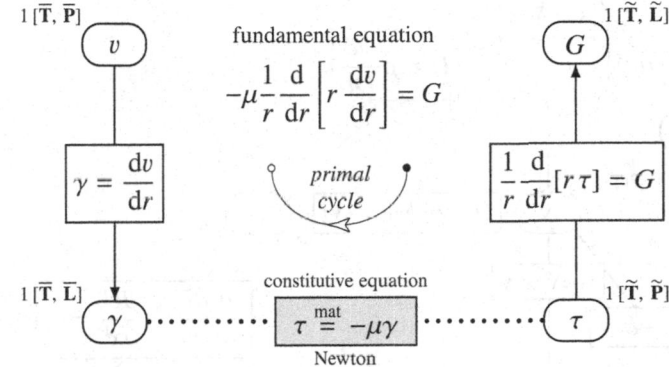

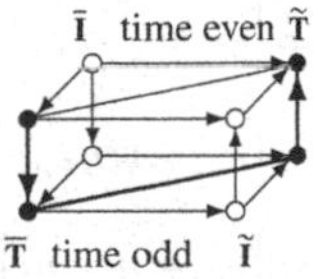


Ref: Bird, R. B., Stewart, W. E. and Lightfoot, E.N.: Transport Phenomena. Whiley, (1960).
Ref: Paterson, A. R.: A First Course in Fluid Dynamics. Cambridge University Press, p. 143 (1997).
Ref: Perucca, E.: Fisica generale e sperimentale. UTET, Vol. I, p. 444 (1940).

FLU9-12; http://discretephysics.dicar.units.it

## Fluid motion: convective term [FLU10]

Unusual table that requires interpretation

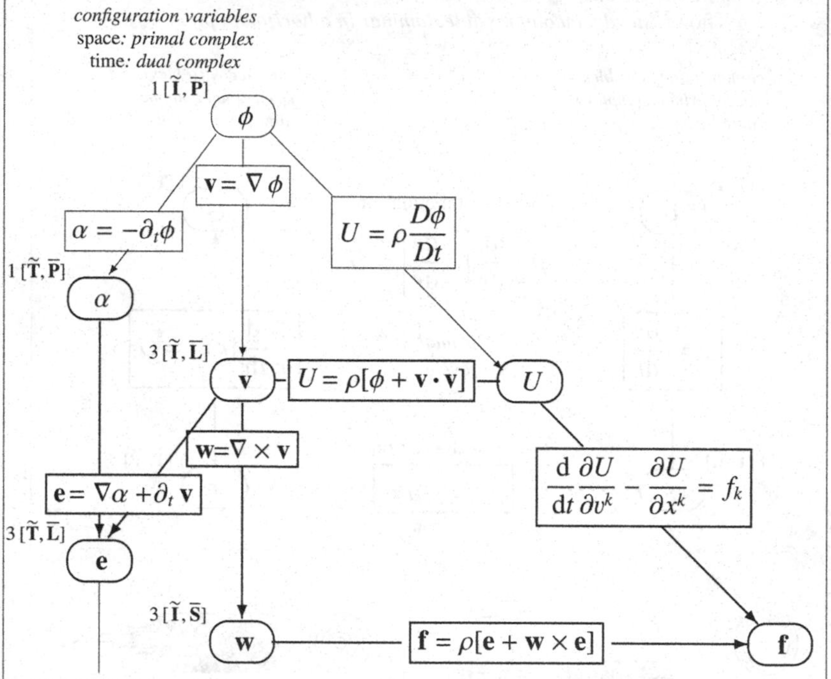


$$U(t, \mathbf{x}, \mathbf{v}) = \rho \left[\alpha(t, \mathbf{x}) + v_h(t, \mathbf{x})\, v^h(t)\right]$$

$$\begin{aligned}
f_k &= \frac{\mathrm{d}}{\mathrm{d}t}\frac{\partial U}{\partial v^k} - \frac{\partial U}{\partial x^k} \\
&= \left[-\frac{\mathrm{d}}{\mathrm{d}t} v_k - \left(\partial_k \phi - \partial_k v_h\, v^h\right)\right] \\
&= \left[-\partial_t v_k - \partial_h v_k\, v^h - \partial_k \phi + \partial_k v_h\, v^h\right] \\
&= \left[e_k + (\partial_k v_h - \partial_h v_k)\, v^h\right] \\
&= \left[e_k + w_{kh}\, v^h\right]
\end{aligned}$$

$$\mathbf{f} = \rho\,[\mathbf{e} + \mathbf{v} \times \mathbf{e}]$$

$\phi$ *velocity potential*
$\alpha$ $= v^2/2$
$\mathbf{v}$ *velocity*
$\mathbf{w}$ *vorticity*
$e_k$ $= -\partial_t v_k - \partial_k \phi$
$\rho$ *mass density*
$U$ *generalized potential*
$\mathbf{f}$ *force for unit volume*

compare with the left side of diagram [FLU3]
compare with diagram [ELE13]

FLU10-7; http://discretephysics.dicar.units.it

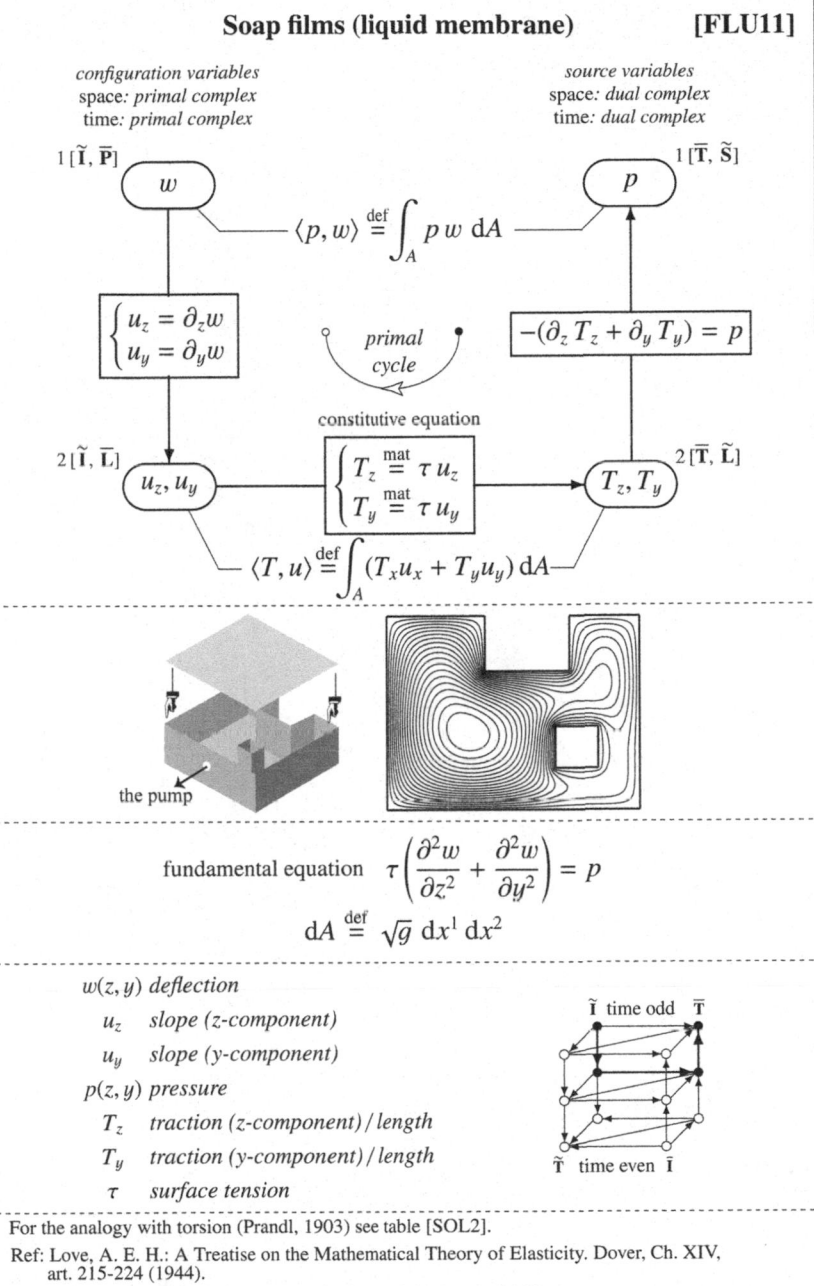

## Soap films (liquid membrane) [FLU11]

*configuration variables*
space: *primal complex*
time: *primal complex*

*source variables*
space: *dual complex*
time: *dual complex*

$1\,[\tilde{\mathbf{I}}, \bar{\mathbf{P}}]$ $w$

$1\,[\bar{\mathbf{T}}, \tilde{\mathbf{S}}]$ $p$

$$\langle p, w\rangle \stackrel{\text{def}}{=} \int_A p\, w\, \mathrm{d}A$$

$$\begin{cases} u_z = \partial_z w \\ u_y = \partial_y w \end{cases}$$

*primal cycle*

$$-(\partial_z T_z + \partial_y T_y) = p$$

constitutive equation

$$\begin{cases} T_z \stackrel{\text{mat}}{=} \tau\, u_z \\ T_y \stackrel{\text{mat}}{=} \tau\, u_y \end{cases}$$

$2\,[\tilde{\mathbf{I}}, \bar{\mathbf{L}}]$ $u_z, u_y$

$2\,[\bar{\mathbf{T}}, \tilde{\mathbf{L}}]$ $T_z, T_y$

$$\langle T, u\rangle \stackrel{\text{def}}{=} \int_A (T_x u_x + T_y u_y)\, \mathrm{d}A$$

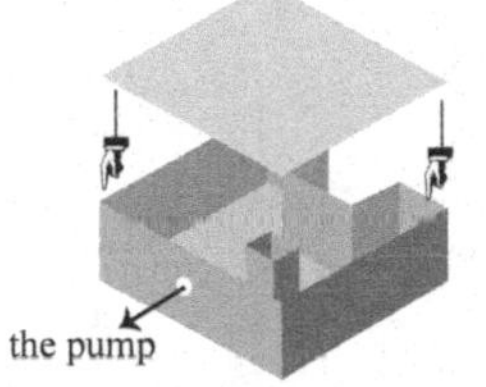


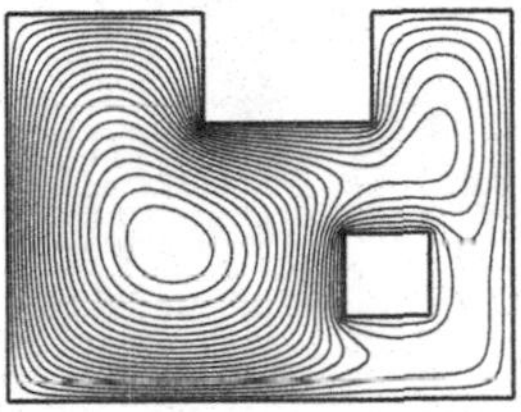

fundamental equation $$\tau\left(\frac{\partial^2 w}{\partial z^2} + \frac{\partial^2 w}{\partial y^2}\right) = p$$

$$\mathrm{d}A \stackrel{\text{def}}{=} \sqrt{g}\, \mathrm{d}x^1\, \mathrm{d}x^2$$

$w(z, y)$ *deflection*
$u_z$ *slope (z-component)*
$u_y$ *slope (y-component)*
$p(z, y)$ *pressure*
$T_z$ *traction (z-component) / length*
$T_y$ *traction (y-component) / length*
$\tau$ *surface tension*

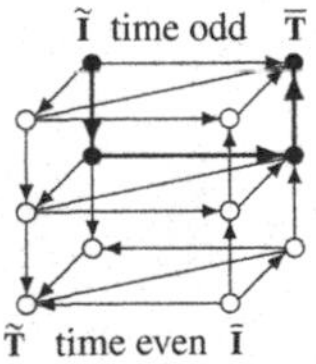


For the analogy with torsion (Prandl, 1903) see table [SOL2].

Ref: Love, A. E. H.: A Treatise on the Mathematical Theory of Elasticity. Dover, Ch. XIV, art. 215-224 (1944).

Ref: Den Hartog, J. P.: Avanced Strength of materials. Dover (1952).

FLU11-10; http://discretephysics.dicar.units.it

# Chapter 13
# Other Physical Theories

## 13.1 Equilibrium Thermodynamics

The thermodynamics of equilibrium, also called simply *thermodynamics*, is entirely based on a *material* description.[1] In fact, the subject of thermodynamics is a thermodynamic *system*, and one of its main physical variables is the volume $\mathcal{V}$.[2] The fact that in a material description the volume is a variable, like for example pressure, temperature and entropy, implies that there is no reason to introduce space elements. In truth, the consideration of space elements arises in a *spatial* description, which deals with control volumes and control surfaces where these elements do not change with time.

The thermodynamic variables, such as volume, mass, number of moles, entropy and the various forms of energy (internal, enthalpy, Helmholtz and Gibbs free energies), relate to the whole system and are additive on the subsystems; for this reason they are called *extensive variables*. In contrast, variables such as temperature, chemical potential, pressure and concentration are *uniform* in the system and are called *intensive variables*.

In thermodynamics a so-called fundamental problem does not exist because the distinction between source and configuration variables need not be introduced. For these reasons the diagrams of thermodynamics differ from the usual diagrams of this book.

**Remark.** It is important not to confuse *fundamental equation*, which we have used in this book as expressing the fundamental problem, with the *fundamental equation* used in thermodynamics, which gives the internal energy $U$ as a function of the extensive parameters such as entropy, volume and mole numbers, i.e. $S, V, N_1, N_2, \cdots N_r$. See Callen [33, p. 31].

[1] This chapter presupposes a previous reading of Chaps. 1–9.

[2] We use the calligraphic letter $\mathcal{V}$ to avoid confusion with the control volume $V$ of the spatial description. See pp. 21 and 92.

E. Tonti, *The Mathematical Structure of Classical and Relativistic Physics*, Modeling and Simulation in Science, Engineering and Technology,
DOI 10.1007/978-1-4614-7422-7_13, 

Diagrams [THE1], [THE2] and [THE3] show an ordered disposition of the variables of thermodynamics, demonstrating Maxwell's reciprocity relations, which are the analogues of the 'curl' in field theories. Moreover, these diagrams show the localization of the Legendre duality transform in the diagram which can be compared with diagrams [PAR3] and [PAR4] of particle and analytical mechanics.

## 13.2 Non-equilibrium Thermodynamics

Non-equilibrium thermodynamics, also called *irreversible thermodynamics*, is entirely based on a *spatial* description. Here the temperature becomes a point function, like pressure, chemical potential and density, and hence they are not uniform as in equilibrium thermodynamics. This association with points implies the consideration of the space differences of these variables, and in this way one introduces new variables associated with lines. The line densities of these variables are the so-called *thermodynamic forces*, i.e. the gradients of temperature, of chemical potential, and so forth. In non-equilibrium thermodynamics, it is possible to distinguish the configuration variables from the source variables, and the fundamental problem makes sense. Hence, for non-equilibrium thermodynamics we build a classification diagram. The various *productions*, like entropy production and molar production, with their corresponding *fluxes*, are source variables. On the other hand, temperature, chemical potentials and their gradients are configuration variables.

In non-equilibrium thermodynamics, additivity remains for those variables associated with volumes, and these can be called extensive, but additivity has a meaning also for variables associated with surfaces and lines, even if calling them *extensive* might generate confusion.

### 13.2.1 Internal Energy

Internal energy, as with every form of energy, is *associated* with $\tilde{\mathbf{T}}$ but is a *function* of primal instants $\bar{\mathbf{I}}$. This is expressed by the two notations $U[\tilde{\mathbf{T}}]$ and $U(\bar{\mathbf{I}})$. This is similar to velocity, which is *associated* with $\bar{\mathbf{T}}$ but is a *function* of $\tilde{\mathbf{I}}$. This is expressed by the two notations $\mathbf{v}[\bar{\mathbf{T}}]$ and $\mathbf{v}(\tilde{\mathbf{I}})$. The same holds for temperature, $T[\tilde{\mathbf{T}}]$ and $T(\bar{\mathbf{I}})$, and for force, $\mathbf{F}[\tilde{\mathbf{T}}]$ and $\mathbf{F}(\bar{\mathbf{I}})$.

**Table 13.1** First principle of thermodynamics

*The increase of internal energy in a volume in a time interval, is equal to the heat and work entering the volume in the time interval.*

$$\underbrace{U[\tilde{\mathbf{T}}^+,\tilde{\mathbf{V}}]-U[\tilde{\mathbf{T}}^-,\tilde{\mathbf{V}}]}_{\text{increase of internal energy}}=\underbrace{Q[\bar{\mathbf{T}},\partial\tilde{\mathbf{V}}]}_{\text{heat}}+\underbrace{W[\bar{\mathbf{T}},\partial\tilde{\mathbf{V}}]}_{\text{work}}\,.$$

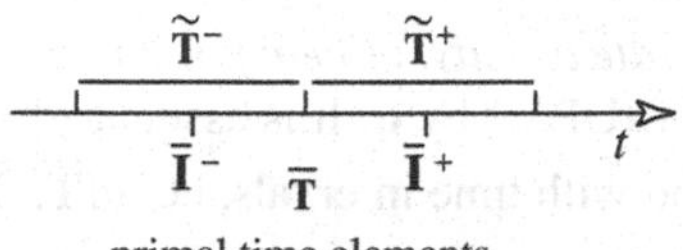

primal time elements

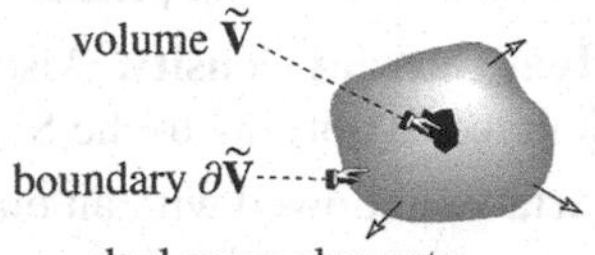

dual space elements

# 13.3 Thermal Conduction

This is a physical theory which belongs to non-equilibrium thermodynamics. The protagonists of thermal conduction are the two physical variables *temperature* and *heat*. Heat is the source of the thermal field, while temperature describes the thermal configuration of a system.

## *13.3.1 Fundamental Problem*

The *fundamental problem* of thermal conduction can be stated as follows:

- Given a solid body,
- Given a time interval,
- Given the shapes and the nature of the materials which fill the region,
- Given the boundary conditions,
- Given the initial temperature at every point of the body,
- Given the distribution and the intensities of heat generators in space and in time,
- Find the temperature at every point of the body at every subsequent instant.

## *13.3.2 Source Variables: Space and Time Classification*

**Heat.** Heat is a particular form of energy transfer through a surface (usually the boundary of a region) during a time interval. Since a reversal of motion changes the direction of heat (heat coming into a region is transformed into heat coming out of the region), it follows that time is endowed with an inner orientation, hence $Q[\overline{\mathbf{T}}, \widetilde{\mathbf{S}}]$.[3]

**Heat Current.** Also called *heat flow rate*, or *thermal flux*, this is the rate of heat crossing a surface in a time interval. The SUN commission of IUPAP say that it must be denoted by $\Phi$. It is associated with surfaces endowed with an outer orientation and with primal time intervals, i.e. $\Phi[\overline{\mathbf{T}}, \widetilde{\mathbf{S}}]$.

**Heat Current Density.** Also called *heat flow rate density*, or *heat flux*,[4] it must be denoted, according to the SUN commission of IUPAP, by $\mathbf{q}$. It is associated with surfaces endowed with an outer orientation and with time intervals, i.e. $\mathbf{q}[\overline{\mathbf{T}}, \widetilde{\mathbf{S}}]$.

### *13.3.3 Configuration Variables: Space and Time Classifications*

**Temperature.** Temperature is invariant under a reversal of motion; one of the reasons for this is that the indication of a mercury thermometer does not change with reversals of motion. Hence temperature can be associated with $\overline{\mathbf{I}}$ or with $\widetilde{\mathbf{T}}$. To decide which one is correct, we remark that the measurement of temperature implies a time interval to reach thermal equilibrium between the body and the thermometer bulb, hence $T[\widetilde{\mathbf{T}}]$.

Another proof stems from the relation between heat and temperature. Denoting by $C$ the thermal capacity of a body, the constitutive relation

$$Q = C\,\Delta_t T \tag{13.1}$$

is valid for both heating and cooling of the body. Hence it must be invariant under a reversal of motion. We now show that, as a consequence, temperature does not change sign under a reversal of motion. In fact,

$$\text{since} \quad \mathscr{R}\,Q = -Q \quad \text{then} \quad \mathscr{R}\,(C\,\Delta_t T) = -(C\Delta_t T) \tag{13.2}$$

since $\mathscr{R}\,(\Delta_t T) = -\Delta_t(\mathscr{R}T)$ then $\mathscr{R}\,T = T$. Hence, $T$ is associated with dual time intervals, $T[\widetilde{\mathbf{T}}]$.[5]

A further proof lies in the fact that temperature is a measure of the internal energy of a body; in particular, for a perfect gas the reversible constitutive equation $U = U_0 + C\,T$ is valid. It follows that, since $U[\widetilde{\mathbf{T}}]$, we have $T[\widetilde{\mathbf{T}}]$.

---

[3] We know that heat and work are two forms of energy flow. Thus, the use of the term *heat flow* is equivalent to *flow of energy flow*. This is incorrect because in this expression the term *flow* is repeated: heat is already a flow!

[4] See p. 30 for a critique of the term *flux* in this context.

[5] See p. 128.

**Thermacy.** Since energy can be integrated in time to give action,[6] temperature can also be integrated in time: its indefinite time integral is called *thermacy* and will be denoted by $\mathscr{T}[\tilde{\mathbf{I}}, \overline{\mathbf{P}}]$.[7] The definite time integral, which we will call the *impulse of thermodynamic temperature*, will be denoted by $\mathcal{T}$. Moreover, since temperature is not the density of another variable, it is associated with points, hence $T[\tilde{\mathbf{T}}, \overline{\mathbf{P}}]$.

**Temperature Gradient.** The gradient of temperature, $\mathbf{g} = \text{grad}\, T$, is associated with lines endowed with an inner orientation and with dual time intervals, like temperature, hence $\mathbf{g}[\tilde{\mathbf{T}}, \overline{\mathbf{L}}]$.

Let us consider the conduction of heat in solids. The space global physical variables are the *temperature* $T[\overline{\mathbf{P}}]$, *temperature difference* between two points $G[\overline{\mathbf{L}}]$, *heat current* $\Phi[\tilde{\mathbf{S}}]$ and *heat generation rate* $P[\tilde{\mathbf{V}}]$. The corresponding field variables are

| | Configuration variables | | Source variables | |
|---|---|---|---|---|
| Global variables | $T[\overline{\mathbf{P}}]$ | $G[\overline{\mathbf{L}}]$ | $\Phi[\tilde{\mathbf{S}}]$ | $P[\tilde{\mathbf{V}}]$ |
| | $\downarrow$ | $\downarrow$ | $\downarrow$ | $\downarrow$ |
| Field variables | $T[\overline{\mathbf{P}}]$ | $\mathbf{g}[\overline{\mathbf{L}}]$ | $\mathbf{q}[\tilde{\mathbf{S}}]$ | $\sigma_{\mathrm{q}}[\tilde{\mathbf{V}}]$ |

(13.3)

Table 13.2 shows how space global variables are organized in the diagram, and the diagram 13.3 shows the corresponding densities.

Lastly, diagram 13.3 adds the relations between the variables: on the left is the definition of temperature gradient $\mathbf{g}$ in terms of the temperature $T$; on the right is the balance equation which links the heat current density $\mathbf{q}$ with the heat source $\sigma_q$. The horizontal link represents the Fourier constitutive relation, which links the configuration variable $\mathbf{g}$ with the source variable $\mathbf{q}$. Diagrams [TCO2] and [TCO3] show unsteady heat conduction using an entropy or energy representation.[8] Diagram [TCO4] shows an algebraic formulation of steady thermal conduction, which is useful for a numerical solution.[9]

## 13.4 Gravitational Field

The *fundamental problem* of gravitation can be stated as follows:

- Given the space and time distribution of masses,
- Given the initial position and initial velocity of the masses,

---

[6] See p. 127.

[7] Maugin and Berezowsky [151, p. 433], De Broglie [47, p. 95], Max von Laue [128].

[8] Callen [33, p. 26].

[9] Cell method, p. 111.

**Table 13.2** Steady thermal conduction: classification of space global variables

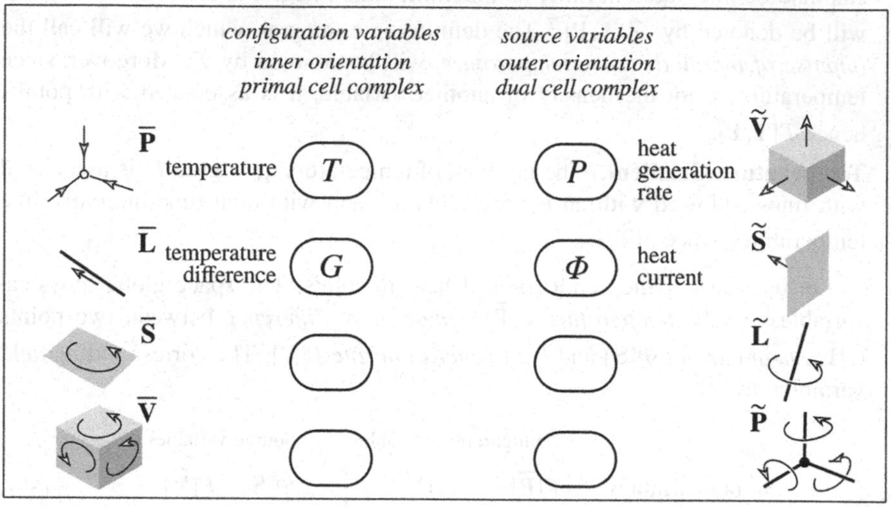


**Table 13.3** Steady thermal conduction: classification of field variables

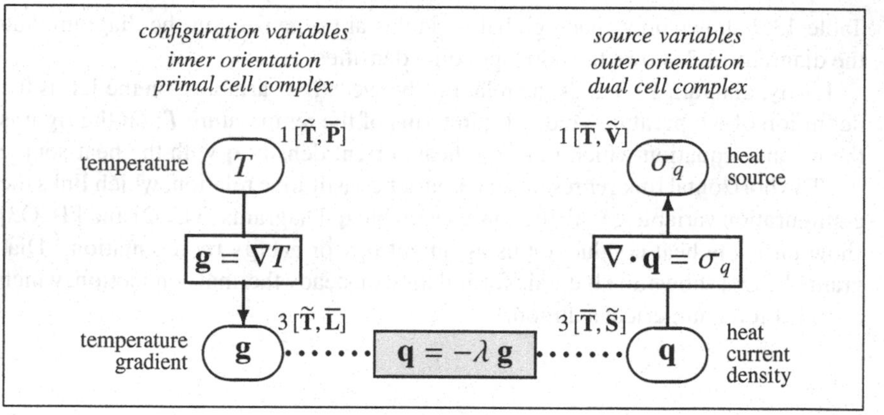


- Find the gravitational potential at every point at all subsequent instants.

In the differential formulation, the fundamental equation links the mass density $\rho$ with the gravitational potential $U_g$. The equation is

$$-\frac{1}{4\pi G}\nabla^2 U_g = \rho. \tag{13.4}$$

In general, as we have seen, in physical theories, one can distinguish quantities which depend on the medium from those quantities which do not depend on it.

**Table 13.4** From field variables to global variables in thermal conduction

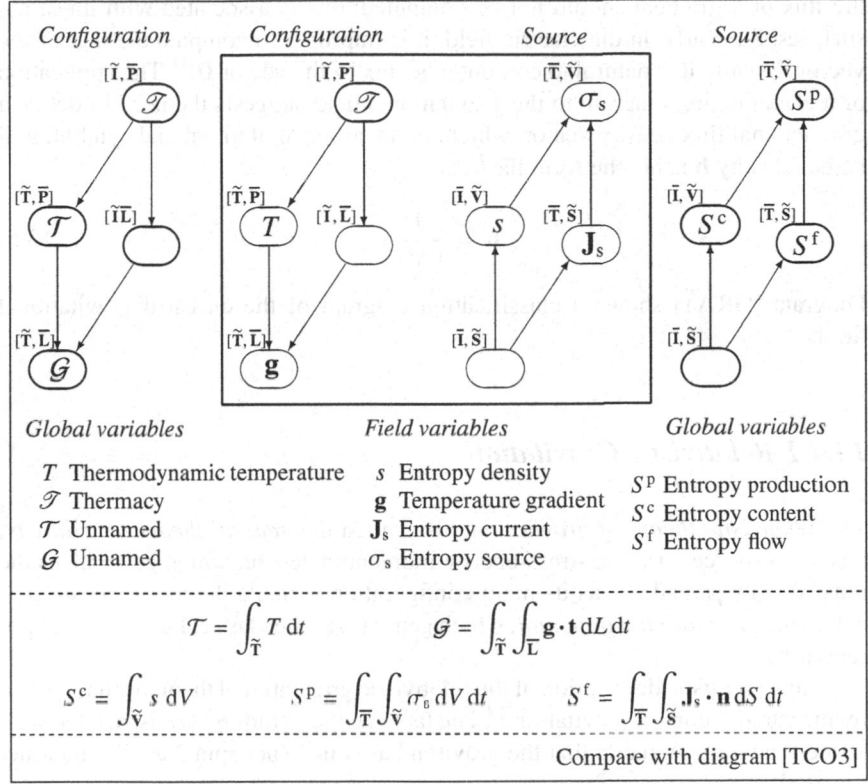

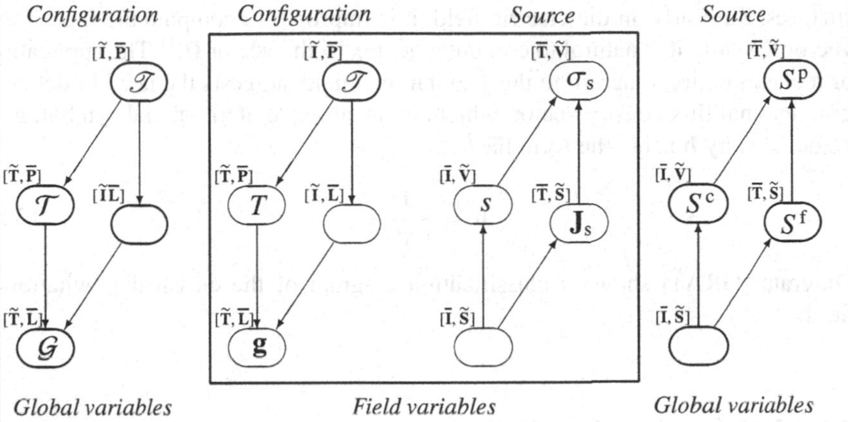


| Global variables | Field variables | Global variables |
|---|---|---|
| $T$ Thermodynamic temperature | $s$ Entropy density | $S^{p}$ Entropy production |
| $\mathcal{T}$ Thermacy | $\mathbf{g}$ Temperature gradient | $S^{c}$ Entropy content |
| $\mathcal{T}$ Unnamed | $\mathbf{J}_s$ Entropy current | $S^{f}$ Entropy flow |
| $\mathcal{G}$ Unnamed | $\sigma_s$ Entropy source | |

$$\mathcal{T} = \int_{\tilde{T}} T \, dt \qquad \mathcal{G} = \int_{\tilde{T}} \int_{\tilde{L}} \mathbf{g} \cdot \mathbf{t} \, dL \, dt$$

$$S^{c} = \int_{\tilde{V}} s \, dV \qquad S^{p} = \int_{\bar{T}} \int_{\tilde{V}} \sigma_s \, dV \, dt \qquad S^{f} = \int_{\bar{T}} \int_{\tilde{S}} \mathbf{J}_s \cdot \mathbf{n} \, dS \, dt$$

Compare with diagram [TCO3]

This is the case with the vectors **E** and **B** of the electromagnetic field, which depend on the medium, and with the vectors **D** and **H**, which do not. The same happens in the mechanics of deformable solids with the deformation tensor $\boldsymbol{\varepsilon}$, which depends on the material, and with the stress $\boldsymbol{\sigma}$, which does not depend on it.

Gravity is a peculiar exception because the gravitational attraction between two masses is not affected by the medium in which the masses are placed; this is summarized in the gravitational constant $G$, which is a *universal* constant. Since the acceleration of gravity $\mathbf{g}$ is the gradient of the gravitational potential, it is associated with a line, as are all gradients since they stem from the difference between the values of a function at two points.

Since there is no distinction between properties which depend on the medium and properties which do not, the gravitational field does not suggest the need to introduce a vector which describes the *gravitational flux*. Despite this, some authors[10] call *gravitational flux* the flux of the vector **g**; this is improper because the flux of a gradient should not be computed; this is associated with lines, not surfaces. Similarly, in the electric field, it is improper to compute the flux of the vector **E**, while it is natural to compute the flux of the vector **D**.[11] The application of a classification diagram to the gravitational field suggests the need to define a gravitational flux density vector, which, in the absence of an official symbol, will be denoted by **h** using the formula[12]

$$\mathbf{h} \stackrel{\text{def}}{=} \frac{1}{4\pi G}\,\mathbf{g}\,. \tag{13.5}$$

Diagram [GRA1] shows a classification diagram of the classical gravitational field.

### 13.4.1 Relativistic Gravitation

The *relativistic theory of gravitation*, also called the *general theory of relativity*, has as its source term the stress-energy-momentum tensor, which generalizes the mass density $\rho$, and has as its main configuration variable the metric tensor $g_{\alpha\beta}$, called the *gravitational potential*, which generalizes the classical gravitational potential $U_{\mathrm{g}}$.

A more realistic description of the relativistic gravitational theory is that offered by the tetrad theory of gravitation.[13] The fact that the tetrads $h_\alpha^A$ are associated with space-time *lines* suggests that the graviton has spin 1 (not spin 2 as is commonly asserted).[14]

## 13.5 Quantum Mechanics

The author has no particular knowledge on quantum mechanics; hence it would be inappropriate to insert diagrams here related to this field of physics. Nonetheless, the relativistic diagrams of particles of integer spin show some impressive analogies with classical physical theories which, in the hands of theoretical physicists, can be interpreted, justified or modified.

---

[10] Kaufman [111, p. 179].

[11] Which some authors do! See Jackson [103, Sect. 1.3], Lorrain et al. [143, p. 50], Akhiezer [5, p. 17].

[12] Melehy [157, p. 49], Mansfield and O'Sullivan [150, p. 90].

[13] Treder [236].

[14] This statement is in accordance with Treder [236, p. 82].

The reader is encouraged to compare diagram [QME4] that deals with the Proca equation for particles of spin 1 and non-zero rest mass, such as mesons, with diagram [ELE6] on electromagnetism, whose associated particle is the photon, also of spin 1 but zero rest mass. Both particles are described by real vector valued functions and are associated with space-time *lines*, i.e. with space-time elements of dimension 1 (as the spin).

Another meaningful comparison is that between the Klein–Gordon equation, see diagram [QME3], relative to particles of spin 0, and diagram [QME4] relative to the Proca equation on particles of spin 1.

The diagram [QME3], which concerns the Klein–Gordon equation, shows that the wave function of particles of spin 0 are associated with space-time points, i.e. with space-time elements of dimension 0 (as the spin). This is in accordance with Schönberg:[15]

> Fields with spin 0 particles ought to be associated with the simplest three dimensional affine objects, the point of space. The spin 1 fields should be associated with the straight lines.

The same equation, decomposed into its space and time components, shown in diagram [QME2], has a striking similarity with the acoustic equation in [FLU5].

Diagram [QME1] shows the analogy between classical particle mechanics and Heisenberg's matrix mechanics.

Diagram [QME5] shows the real physical variables that can be obtained from the complex wave function $\psi$; they are five *bilinear covariants*, each being associated with a space time element.

The author has found that the paper by Schönberg, just cited, contains a description of quantum mechanics which agrees with the analysis given in this book. Among the notions of interest there is the *spherical orientation* of points, which coincides with our notion of source and sink.[16] Schönberg states that '*charge conjugation is thus related to the reversal of the spherical orientation*'.

An interesting remark is that in four-dimensional diagrams, such as that of electromagnetism, diagrams [ELE6], [GEN9], [GEN10], [GEN11] and [GEN12], the boxes in the bottom part contain pseudoscalar and pseudovector functions; these functions can describe pseudoscalar and pseudovector particles.

In conclusion, we stress that these few diagrams on quantum mechanics are diagrams of trials and are offered here as a stimulus for theoretical physicists to investigate their link with classical field theories.

~ * ~

[15] Schönberg [202, p. 326].

[16] Schönberg [202, p. 323].

**Thermodynamics**                                            **[THE1]**
*Legendre's transform*

$(S, V, N)$ = thermodynamic configuration space

**Fundamental equation**

$U(S, V, N)$                                    U = **Internal energy**

$$\begin{cases} T = \dfrac{\partial U}{\partial S} \\[2mm] -p = \dfrac{\partial U}{\partial V} \\[2mm] \mu = \dfrac{\partial U}{\partial N} \end{cases}$$

$$\begin{cases} T \overset{\text{law}}{=} T(S, V, N) \\[1mm] p \overset{\text{law}}{=} p(S, V, N) \\[1mm] \mu \overset{\text{law}}{=} \mu(S, V, N) \end{cases}$$

$S, V, N$                    **Equations of state**                    $T, -p, \mu$

$$\begin{cases} S \overset{\text{law}}{=} S(T, p, \mu) \\[1mm] V \overset{\text{law}}{=} V(T, p, \mu) \\[1mm] N \overset{\text{law}}{=} N(T, p, \mu) \end{cases}$$

Legendre's transform

$$\begin{cases} S = -\dfrac{\partial G}{\partial T} \\[2mm] V = \dfrac{\partial G}{\partial p} \\[2mm] N = \dfrac{\partial G}{\partial \mu} \end{cases}$$

$G(T, p, \mu)$                                    G = **Gibbs function**

**Dual fundamental equation**

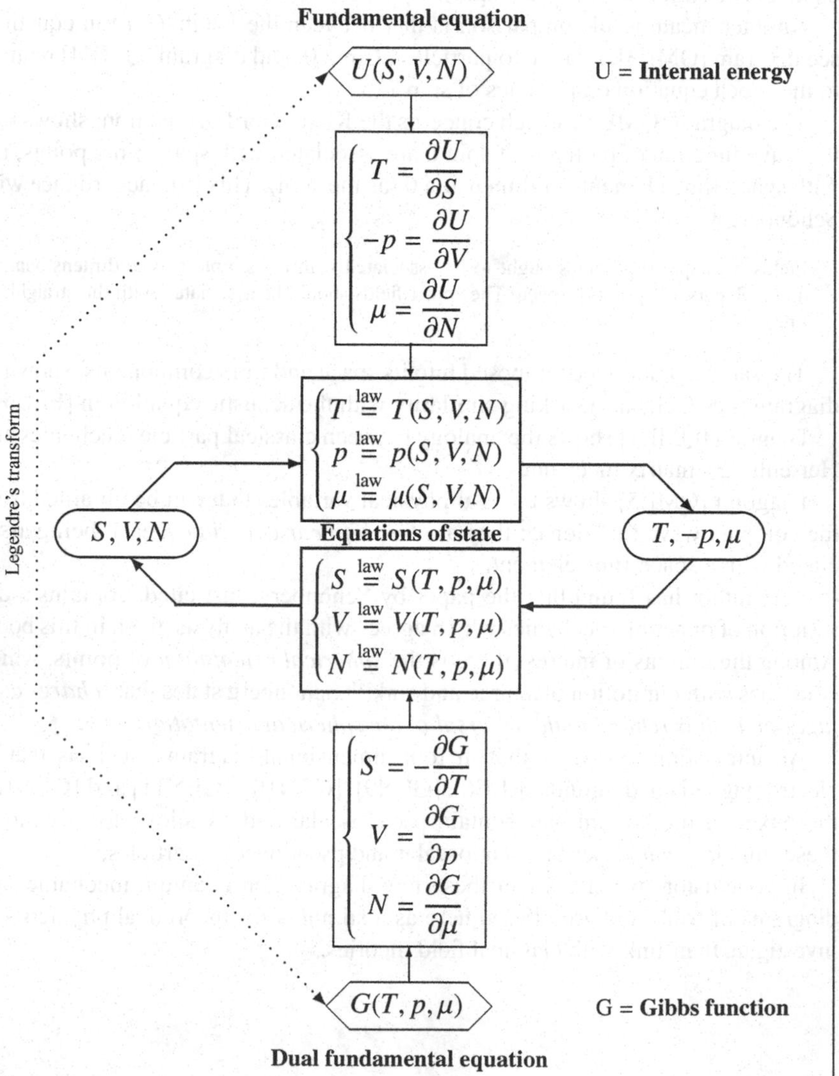

Ref: Callen, H.B.: Thermodynamics. Wiley, New York, pp. 33, 100, 118 (1960)

THE1-8; http://discretephysics.dicar.units.it

**Thermodynamics** [THE2]
*Entropy representation*

S = **Entropy**

**Fundamental equation**

$$S(U, V, N)$$

$$\begin{cases} \dfrac{1}{T} = \dfrac{\partial S}{\partial U} \\[2mm] \dfrac{p}{T} = \dfrac{\partial S}{\partial V} \\[2mm] -\dfrac{\mu}{T} = \dfrac{\partial S}{\partial N} \end{cases}$$

Gradient

**Equations of state**

Extensive
entropy variables

$$U, V, N$$

$$\begin{cases} \dfrac{1}{T} \overset{\text{law}}{=} f(U, V, N) \\[2mm] \dfrac{p}{T} \overset{\text{law}}{=} g(U, V, N) \\[2mm] -\dfrac{\mu}{T} \overset{\text{law}}{=} h(U, V, N) \end{cases}$$

Intensive
entropy variables

$$\dfrac{1}{T}, \dfrac{p}{T}, -\dfrac{\mu}{T}$$

**Maxwell reciprocity relations**

$$\frac{\partial}{\partial V}\left(\frac{1}{T}\right) = -\frac{\partial}{\partial U}\left(\frac{p}{T}\right) = 0$$

$$\frac{\partial}{\partial N}\left(\frac{p}{T}\right) = -\frac{\partial}{\partial V}\left(-\frac{\mu}{T}\right) = 0$$

$$\frac{\partial}{\partial U}\left(-\frac{\mu}{T}\right) = -\frac{\partial}{\partial N}\left(\frac{1}{T}\right) = 0$$

Curl

$$0$$

$$d\mathbf{r} = [\,dU, \, dV, \, dN\,]$$

$$\mathbf{v} = \left[\frac{1}{T}, \frac{p}{T}, -\frac{\mu}{T}\right]$$

$$dS = \mathbf{v} \cdot d\mathbf{r}$$

$$dS = \frac{1}{T}\,dU + \frac{p}{T}\,dV - \frac{\mu}{T}\,dN$$

$$\nabla \times \mathbf{v} = 0$$

Ref: Callen, H. B.: Thermodynamics. Wiley, New York, pp. 33, 118 (1960)
THE2-9; http://discretephysics.dicar.units.it

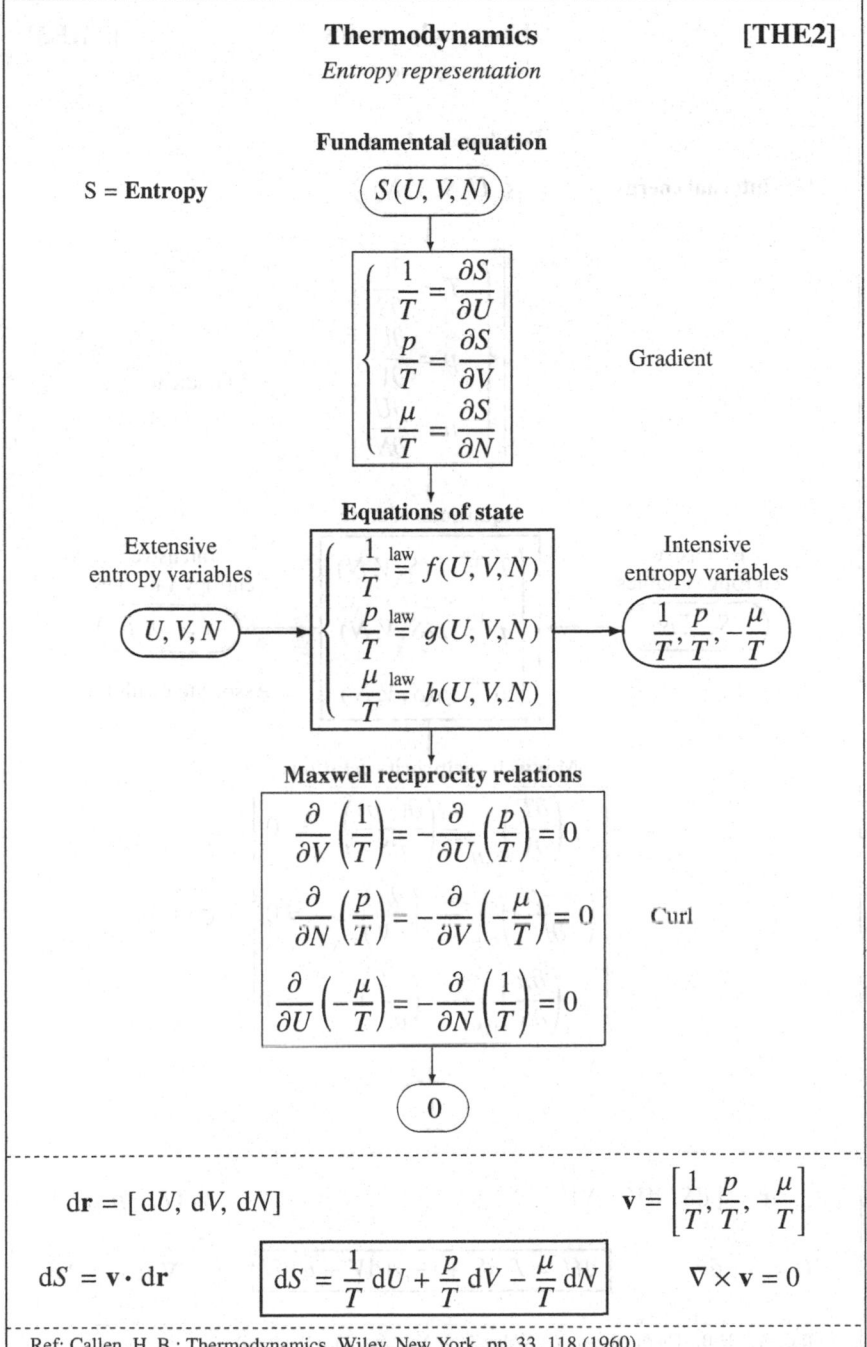

## Thermodynamics [THE3]

*Energy representation*

**Fundamental equation**

U = **Internal energy**

$$U(S, V, N)$$

$$\begin{cases} T = \dfrac{\partial U}{\partial S} \\ -p = \dfrac{\partial U}{\partial V} \\ \mu = \dfrac{\partial U}{\partial N} \end{cases}$$

Gradient

**Equations of state**

Extensive entropy variables: $S, V, N$

$$\begin{cases} T \overset{\text{law}}{=} T(S, V, N) \\ p \overset{\text{law}}{=} p(S, V, N) \\ \mu \overset{\text{law}}{=} \mu(S, V, N) \end{cases}$$

Intensive entropy variables: $T, -p, \mu$

Associated with lines

**Maxwell reciprocity relations**

$$\left(\frac{\partial T}{\partial V}\right)_{S,N} - \left(\frac{\partial(-p)}{\partial S}\right)_{V,N} = 0$$

$$\left(\frac{\partial(-p)}{\partial N}\right)_{S,V} - \left(\frac{\partial \mu}{\partial V}\right)_{S,N} = 0$$

$$\left(\frac{\partial \mu}{\partial S}\right)_{V,N} - \left(\frac{\partial T}{\partial N}\right)_{S,V} = 0$$

Curl

$$0$$

---

$$\mathrm{d}\mathbf{r} = [\,\mathrm{d}S,\ \mathrm{d}V,\ \mathrm{d}N] \qquad \mathbf{v} = [T, -p, \mu]$$

$$\mathrm{d}U = \mathbf{v}\cdot \mathrm{d}\mathbf{r} \qquad \boxed{\mathrm{d}U = T\,\mathrm{d}S + (-p)\,\mathrm{d}V + \mu\,\mathrm{d}N} \qquad \nabla \times \mathbf{v} = 0$$

---

Ref: Callen, H.B.: Thermodynamics. Wiley, New York, pp. 33, 118 (1960)

THE3-8; http://discretephysics.dicar.units.it

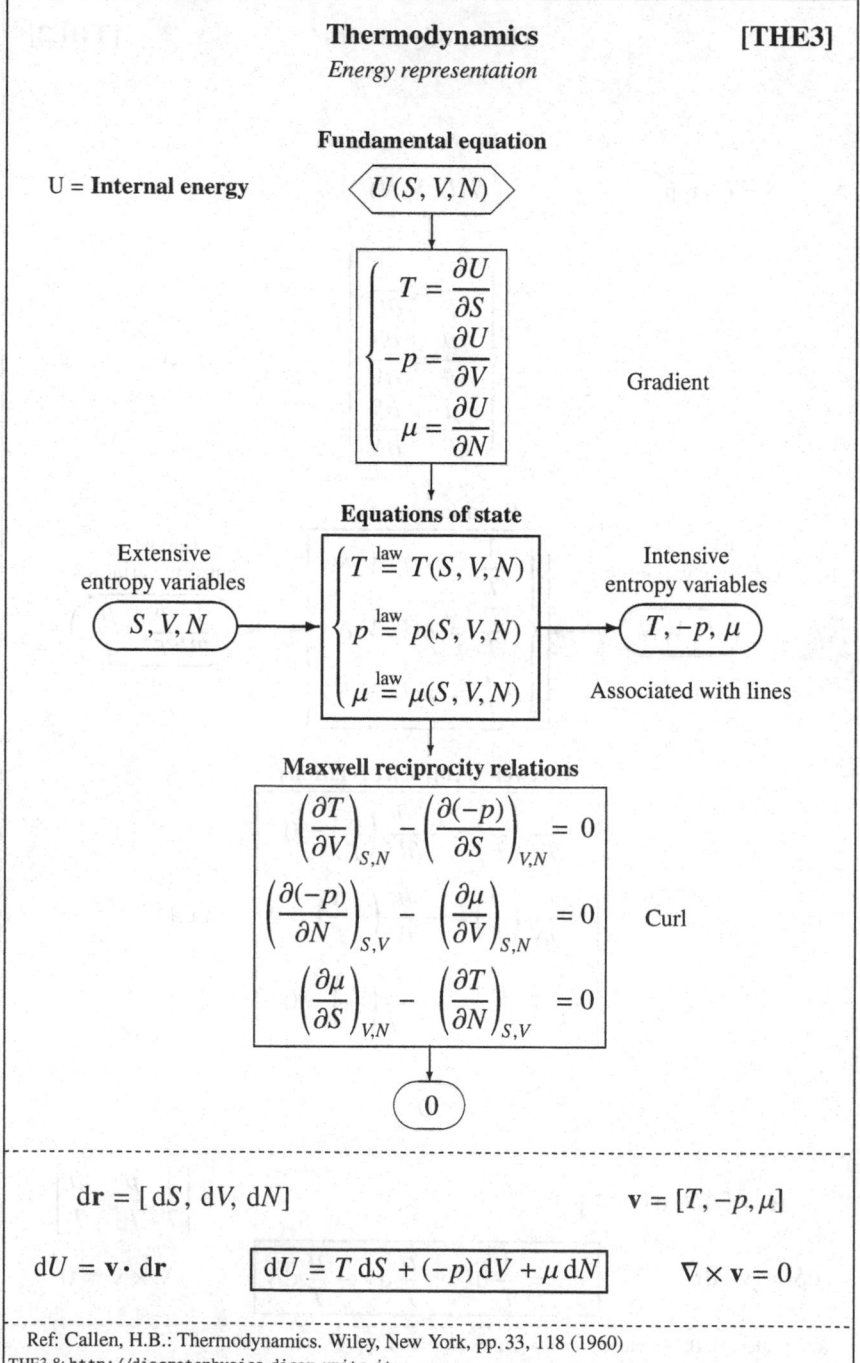

## Steady thermal conduction in solids [TCO1]

*configuration variables*
space: *primal complex*
time: *dual complex*
*(time even variables)*

*source variables*
space: *dual complex*
time: *primal complex*
*(time odd variables)*

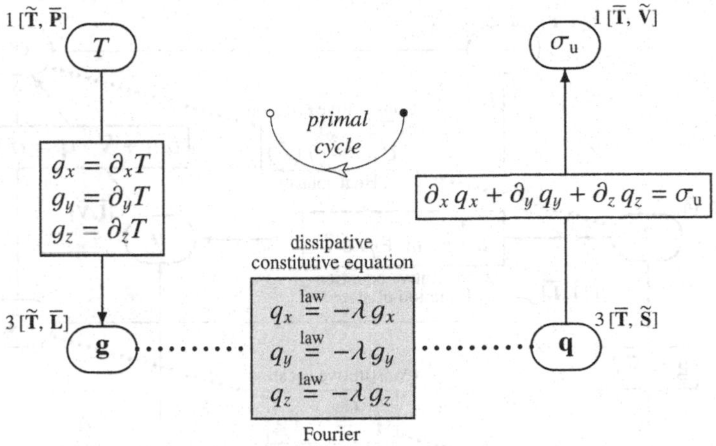


Fundamental equation
Poisson

$$-\lambda \left( \frac{\partial^2 T}{\partial x^2} + \frac{\partial^2 T}{\partial y^2} + \frac{\partial^2 T}{\partial z^2} \right) = \sigma_{\mathrm{u}}$$

$T$ *temperature*
$\mathbf{g}$ *temperature gradient*
$\sigma_{\mathrm{u}}$ *heat source*
$\mathbf{q}$ *heat current density*
$\lambda$ *thermal conductivity*

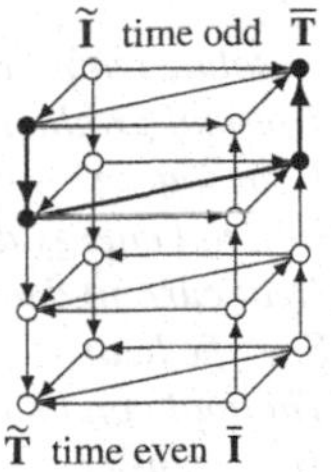


TCO1-6; http://discretephysics.dicar.units.it

# Thermal conduction in solids [TCO2]

*Entropy representation*

*configuration variables*
space: *primal complex*
time: *dual complex*
*intervals         instants*

*source variables*
space: *dual complex*
time: *primal complex*
*instants         intervals*

$1\,[\widetilde{\mathbf{I}},\overline{\mathbf{P}}]$

$1\,[\overline{\mathbf{T}},\widetilde{\mathbf{V}}]$

$\sigma_{\mathrm{u}}$

Constitutive equation

$$\boxed{\sigma_{\mathrm{u}} \overset{law}{=} f(T)}$$

Heat source

$$\boxed{\partial_t u + \nabla \cdot \mathbf{q} = \sigma_{\mathrm{u}}}$$

$1\,[\widetilde{\mathbf{T}},\overline{\mathbf{P}}]$

$T$

$$u \overset{law}{=} u_0 + \rho\,c\,T$$

$1\,[\overline{\mathbf{I}},\widetilde{\mathbf{V}}]$

$u$

$1\,[\widetilde{\mathbf{I}},\overline{\mathbf{L}}]$

Constitutive equation
Equation of state

$1\,[\overline{\mathbf{T}},\widetilde{\mathbf{S}}]$

$\mathbf{q}$

$$\boxed{\mathbf{g} = \nabla T}$$

Constitutive equation

$$\boxed{\mathbf{q} \overset{law}{=} -\lambda\,\mathbf{g}}$$

Fourier

$1\,[\widetilde{\mathbf{T}},\overline{\mathbf{L}}]$

$\mathbf{g}$

$1\,[\overline{\mathbf{I}},\widetilde{\mathbf{S}}]$

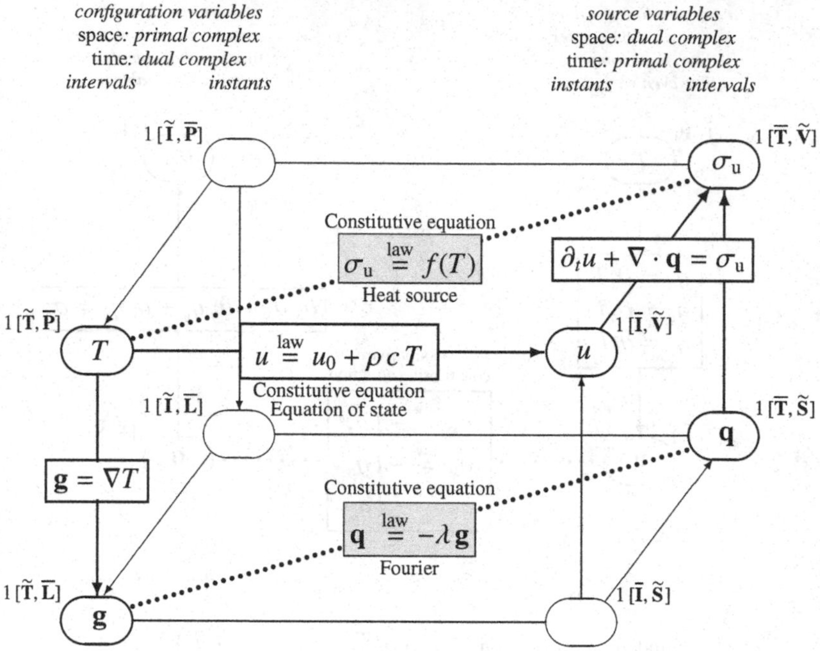

Fundamental equation:
Fourier

$$\rho\,c\frac{\partial T}{\partial t} - \lambda\left(\frac{\partial^2 T}{\partial x^2} + \frac{\partial^2 T}{\partial y^2} + \frac{\partial^2 T}{\partial z^2}\right) = \sigma_{\mathrm{u}}(T)$$

$T$  *Absolute temperature*
$\mathbf{g}$  *Thermal gradient*
$\sigma_{\mathrm{u}}$ *Heat source*
$u$  *Internal energy density*
$\mathbf{q}$  *Heat current density*
$c$  *Specific heat*
$\lambda$  *Thermal conductivity*
$\rho$  *Mass density*

$\widetilde{\mathbf{I}}$ Time odd $\overline{\mathbf{T}}$

$\widetilde{\mathbf{T}}$ Time even $\overline{\mathbf{I}}$

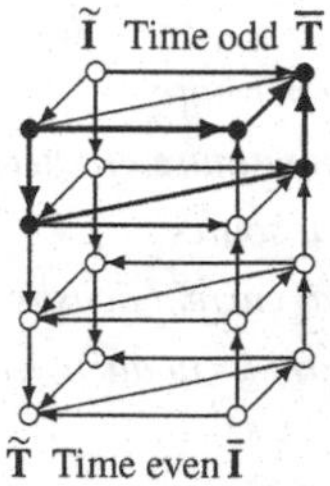

In this diagram, the internal energy density, which is associated with $\widetilde{\mathbf{T}}$, is inserted in the box of $\overline{\mathbf{I}}$.

The best diagram is that in energy representation: see diagram [TCO3].

TCO2-8; http://discretephysics.dicar.units.it

## Thermal conduction in solids [TCO3]

*configuration variables* — space: *primal complex* — time: *dual complex* — *intervals* / *instants*

*Energy representation*

*source variables* — space: *dual complex* — time: *primal complex* — *instants* / *intervals*

Column dimension: action/entropy

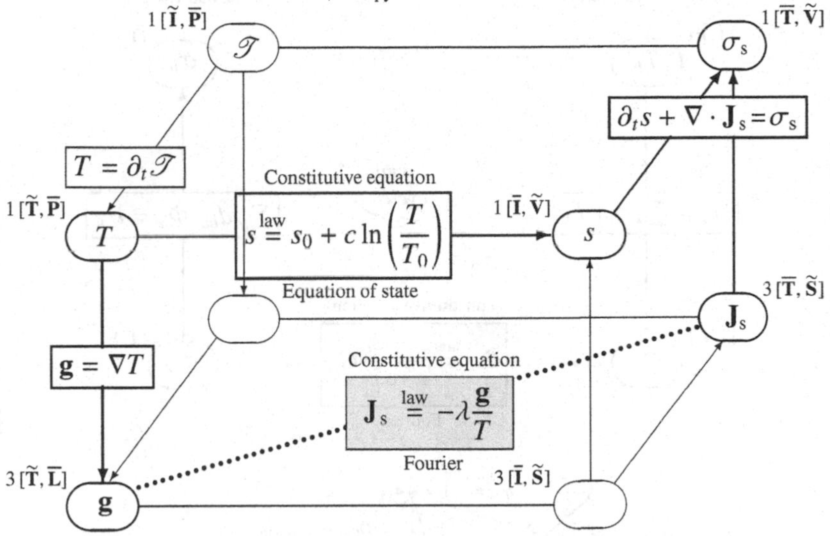


Column dimension: entropy

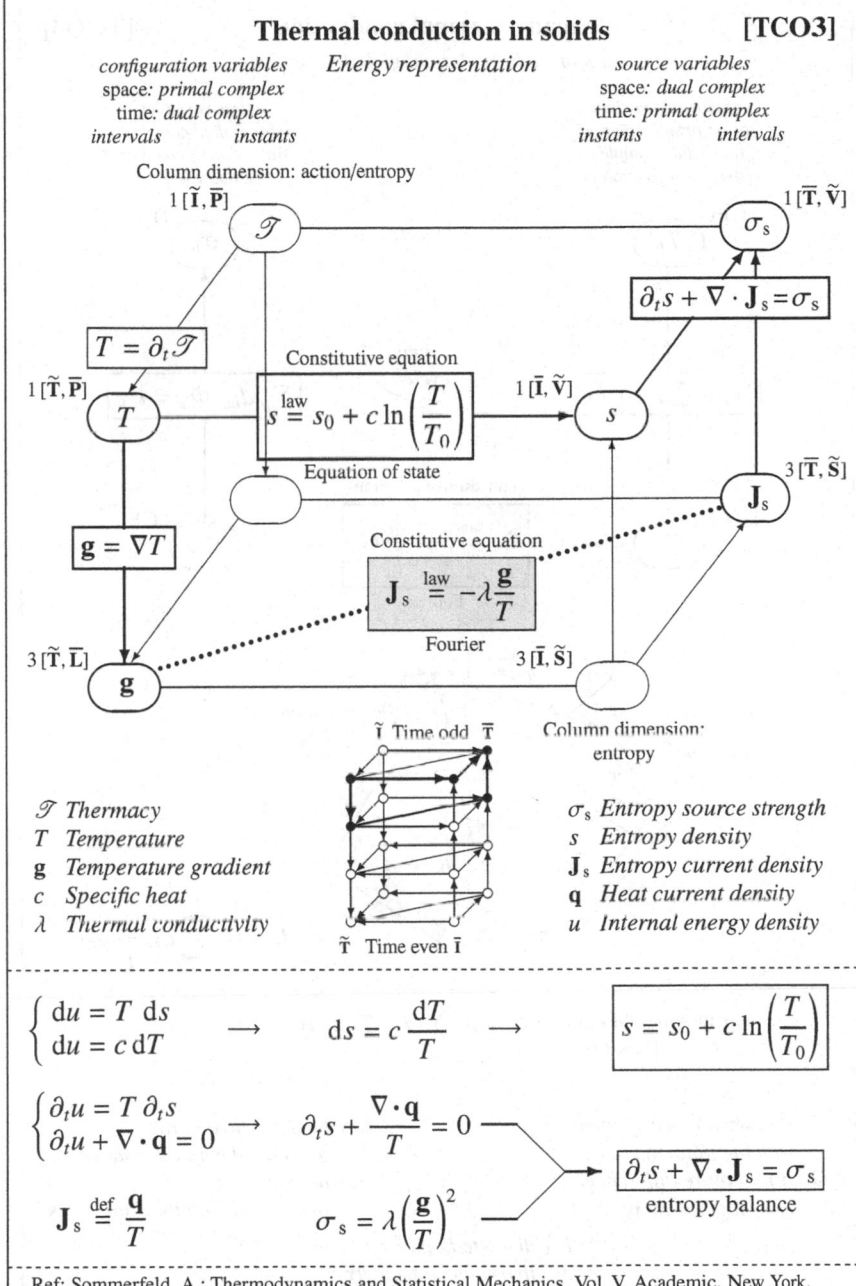

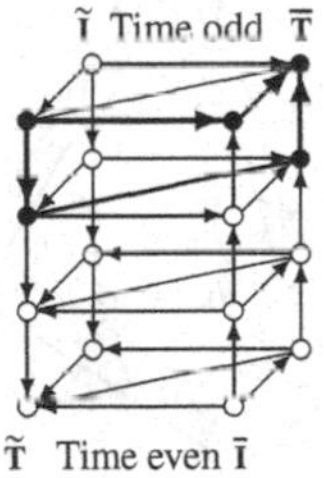


$\mathscr{T}$ *Thermacy*
$T$ *Temperature*
$\mathbf{g}$ *Temperature gradient*
$c$ *Specific heat*
$\lambda$ *Thermal conductivity*

$\sigma_s$ *Entropy source strength*
$s$ *Entropy density*
$\mathbf{J}_s$ *Entropy current density*
$\mathbf{q}$ *Heat current density*
$u$ *Internal energy density*

---

$$\begin{cases} du = T\, ds \\ du = c\, dT \end{cases} \longrightarrow \quad ds = c\frac{dT}{T} \quad \longrightarrow \quad \boxed{s = s_0 + c \ln\left(\frac{T}{T_0}\right)}$$

$$\begin{cases} \partial_t u = T\, \partial_t s \\ \partial_t u + \nabla \cdot \mathbf{q} = 0 \end{cases} \longrightarrow \quad \partial_t s + \frac{\nabla \cdot \mathbf{q}}{T} = 0$$

$$\mathbf{J}_s \overset{\text{def}}{=} \frac{\mathbf{q}}{T} \qquad \sigma_s = \lambda\left(\frac{\mathbf{g}}{T}\right)^2$$

$$\longrightarrow \boxed{\partial_t s + \nabla \cdot \mathbf{J}_s = \sigma_s}$$

entropy balance

---

Ref: Sommerfeld, A.: Thermodynamics and Statistical Mechanics, Vol. V. Academic, New York,
Ref: p. 152 (1952) for thermacy see: Maugin, G.A.: Towards an Analytical Mechanics of Dissipative Materials, Rend. Sem. Mat. Univ. Pol. Torino, Vol. 58 (2), (2000) Callen,
Ref: H.B.: Thermodynamics. Wiley, New York, p. 36 (1960)

TCO3-9; http://discretephysics.dicar.units.it

## Steady thermal conduction [TCO4]

*algebraic formulation, global variables*

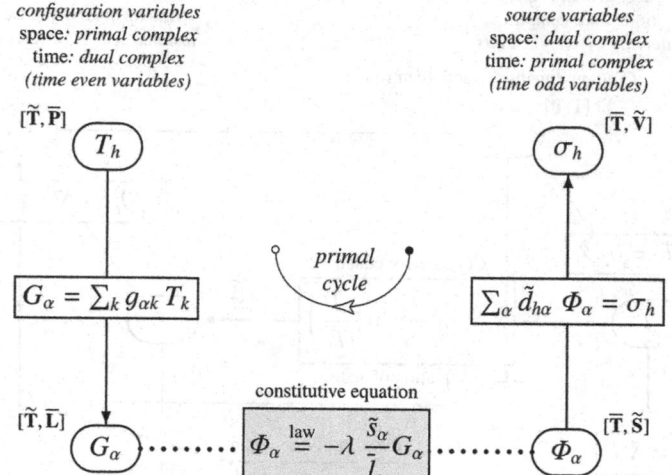

*configuration variables*
space: *primal complex*
time: *dual complex*
*(time even variables)*

*source variables*
space: *dual complex*
time: *primal complex*
*(time odd variables)*

$[\tilde{\mathbf{T}}, \bar{\mathbf{P}}]$ $T_h$

$$G_\alpha = \sum_k g_{\alpha k}\, T_k$$

$[\tilde{\mathbf{T}}, \bar{\mathbf{L}}]$ $G_\alpha$

*primal cycle*

$[\bar{\mathbf{T}}, \tilde{\mathbf{V}}]$ $\sigma_h$

$$\sum_\alpha \tilde{d}_{h\alpha}\, \Phi_\alpha = \sigma_h$$

$[\bar{\mathbf{T}}, \tilde{\mathbf{S}}]$ $\Phi_\alpha$

constitutive equation

$$\Phi_\alpha \overset{\text{law}}{=} -\lambda \frac{\tilde{s}_\alpha}{\bar{l}_\alpha} G_\alpha$$

Fourier

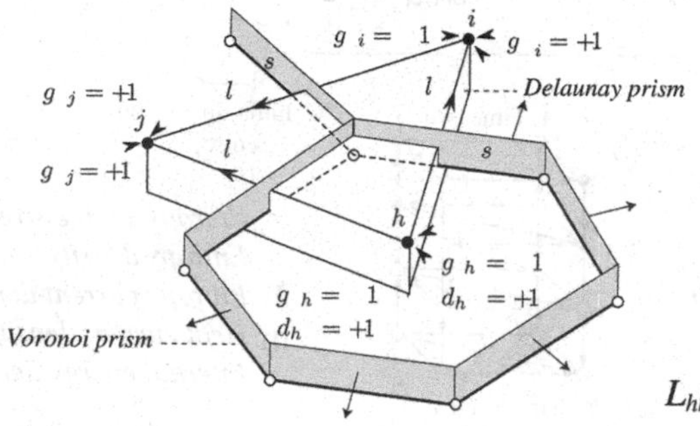


$$L_{hk} \overset{\text{def}}{=} \sum_\alpha \tilde{d}_{h\alpha} \frac{\tilde{s}_\alpha}{\bar{l}_\alpha} g_{\alpha k}$$

fundamental equation: Poisson

$$-\lambda \sum_k L_{hk}\, T_k = \sigma_h$$

$T_h$ *temperature associated with the point* $\bar{\mathbf{p}}_h$

$G_\alpha$ *temperature difference associated with* $\bar{\mathbf{l}}_\alpha$

$\sigma_h$ *heat production rate associated with the dual cell* $\tilde{\mathbf{v}}_h$

$\Phi_\alpha$ *heat current associated with the dual cell* $\tilde{\mathbf{s}}_\alpha$

$L_{hk}$ *discrete Laplacian*

$\lambda$ *thermal conductivity*

TCO4-4; http://discretephysics.dicar.units.it

## Classical gravitational field [GRA1]

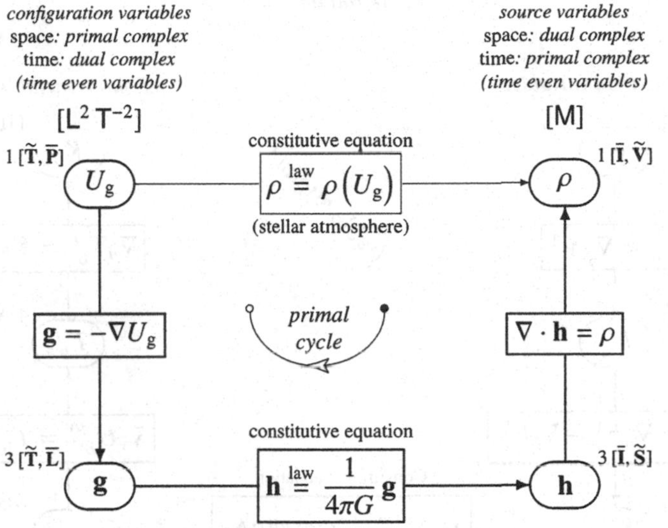
configuration variables
space: primal complex
time: dual complex
(time even variables)
[L² T⁻²]
source variables
space: dual complex
time: primal complex
(time even variables)
[M]
1 [T̃, P̄]
$U_g$
constitutive equation
$\rho \overset{\text{law}}{=} \rho(U_g)$
(stellar atmosphere)
$\rho$
1 [Ī, Ṽ]
$\mathbf{g} = -\nabla U_g$
primal cycle
$\nabla \cdot \mathbf{h} = \rho$
3 [T̃, L̄]
$\mathbf{g}$
constitutive equation
$\mathbf{h} \overset{\text{law}}{=} \frac{1}{4\pi G}\mathbf{g}$
$\mathbf{h}$
3 [Ī, S̃]

Newton: $\mathbf{g}(r) = -G\frac{M}{r^2}\mathbf{e}_r$ $\qquad U_g(r) = -G\frac{M}{r}$

Poisson: $-\frac{1}{4\pi G}\nabla^2 U_g = \rho$ $\qquad \mathbf{g} = -\nabla U_g$

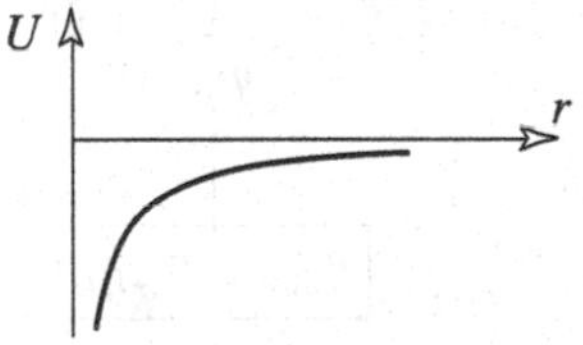
U
r

$$U_g(r) \overset{\text{def}}{=} -\int_{\infty}^{r} \mathbf{g}(r) \cdot \mathbf{e}_r \, dr$$

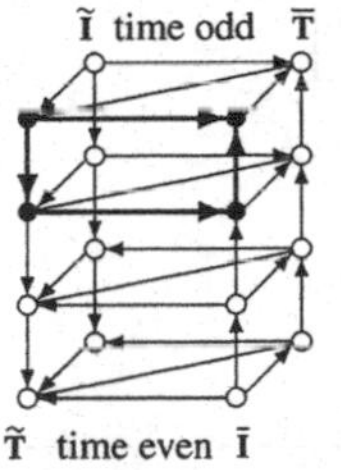
Ĩ time odd T̄
T̃ time even Ī

$U_g$ *gravitational potential*
$\mathbf{g}$ *acceleration of gravity*
$\rho$ *gravitational mass density*
$\mathbf{h}$ *gravitational flux density*
$G$ *gravitational constant*

Ref: Melehy, M. A. in: Foundation of the Thermodynamic Theory of Generalized Fields. Mono Book Corp, p.49 (1973), introduces the vector $\mathbf{D}_g = \mathbf{g}/(4\pi G)$
Ref: Mansfield, M., O'Sullivan, C: Understanding Physics, Wiley, p.90 (2010), $\boldsymbol{\Gamma}_g = \mathbf{g}/(4\pi G)$.
Some authors uses $\mathbf{g} = +\nabla U_g$. See:
Ref: Kellog, O. D.: Foundation of Potential Theory. Ungar Publishing Co. (1929).
Ref: Blakely, R. J.: Potential Theory in Gravity & Magnetic Applications. Cambridge University Press, p.45 (1996).

GRA1-7; http://discretephysics.dicar.units.it

## Relativistic gravitation [GRA2]

*Tetrad theory*

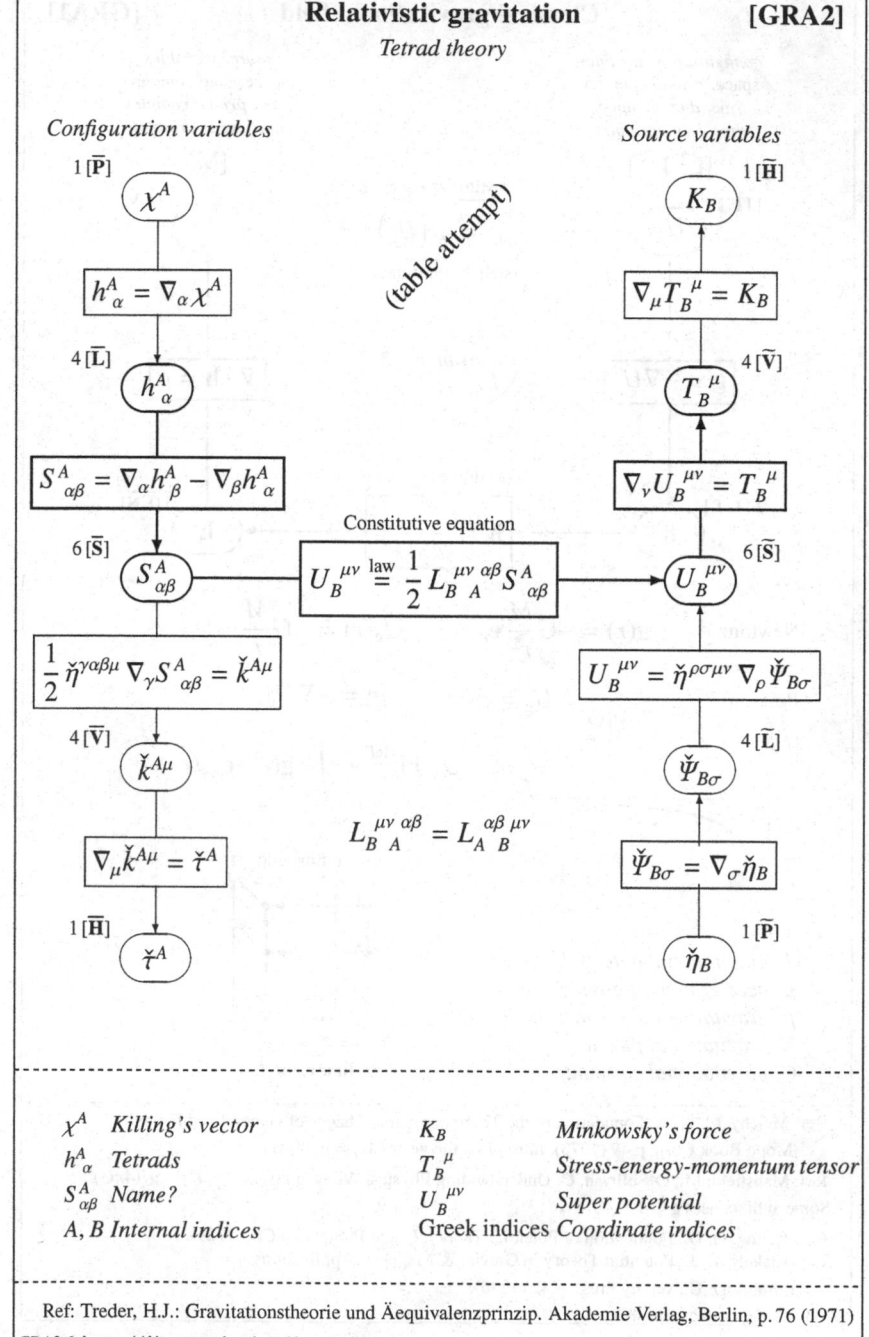

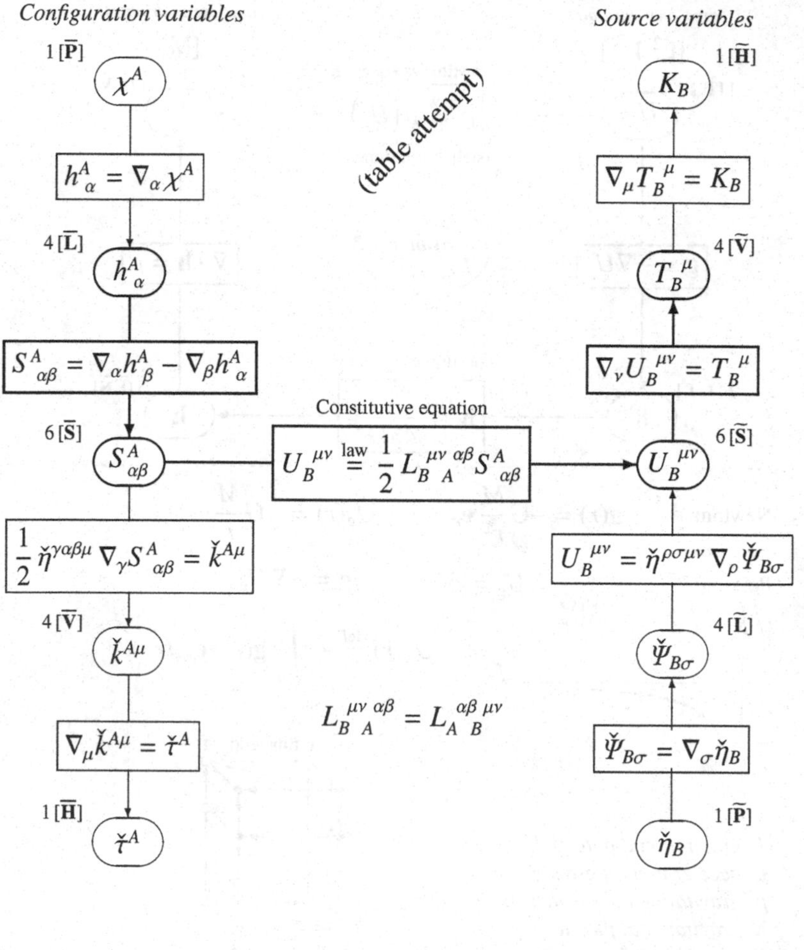


| | | | |
|---|---|---|---|
| $\chi^A$ | *Killing's vector* | $K_B$ | *Minkowsky's force* |
| $h^A_{\ \alpha}$ | *Tetrads* | $T_B^{\ \mu}$ | *Stress-energy-momentum tensor* |
| $S^A_{\ \alpha\beta}$ | *Name?* | $U_B^{\ \mu\nu}$ | *Super potential* |
| $A, B$ | *Internal indices* | Greek indices | *Coordinate indices* |

Ref: Treder, H.J.: Gravitationstheorie und Äequivalenzprinzip. Akademie Verlag, Berlin, p. 76 (1971)

GRA2-6; http://discretephysics.dicar.units.it

## Particle dynamics in conservative fields: Jacobi [ANA1]

$$p(\mathbf{r}) = \sqrt{2m\,[E - V(\mathbf{r})]}$$

trajectory equation : given $V(\mathbf{r})$, $\mathbf{r}(0)$, $\mathbf{p}(0)$, to find $\mathbf{r}(s)$

*kinematic variables* *dynamic variables*

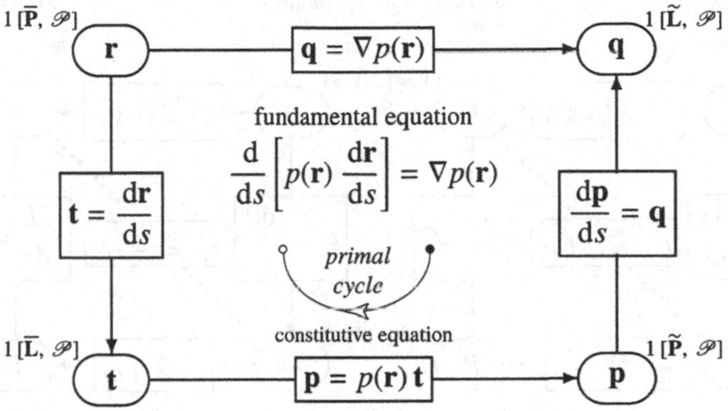


$\mathbf{r}(s)$ *radius vector*
$\mathbf{t}(s)$ *unit tangent vector*
$m$ *mass*

$V$ potential energy
$T$ *kinetic energy*
$E$ *total energy*

$\mathbf{q}(s)$ *unnamed*
$\mathbf{p}(s)$ *momentum*

## Geometrical optics: Fermat

ray equation : given $n(\mathbf{r})$, $\mathbf{r}(0)$, $\mathbf{p}(0)$, to find $\mathbf{r}(s)$

*kinematic variables* *dynamic variables*

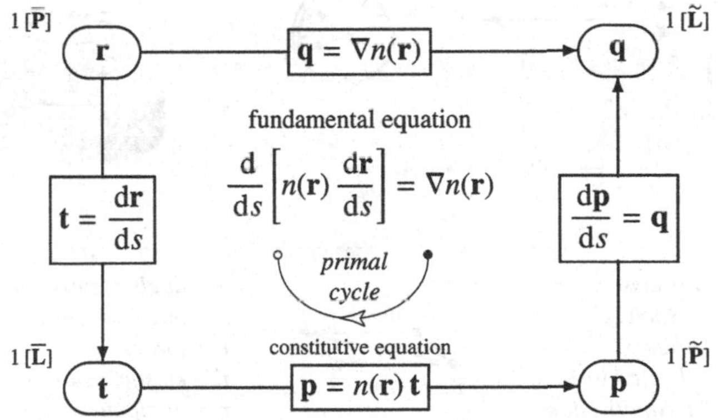


$s$ *arc parameter*
$\mathbf{r}$(s) *radius vector*
$\mathbf{t}$(s) *unit tangent vector*

$\mathbf{q}(s)$ *unnamed*
$\mathbf{p}(s)$ *ray vector*
$n(\mathbf{r})$ *refractive index*

Ref: Novozilov, J. V., Jappa, J. A.: Elektrodynamika. § 21, Nakua, Moscow (1978).
Ref: Born, M. and Wolf, E.: Principles of optics. Pergamon Press. Oxford, p. 122 (1975).
Ref: Joas, C. and Lehner C.:The classical roots of wave mechanics. Preprint 2009, (from internet).
ANA1-7; http://discretephysics.dicar.units.it

## Analogy between translatory and rotatory motions  [ANA2]

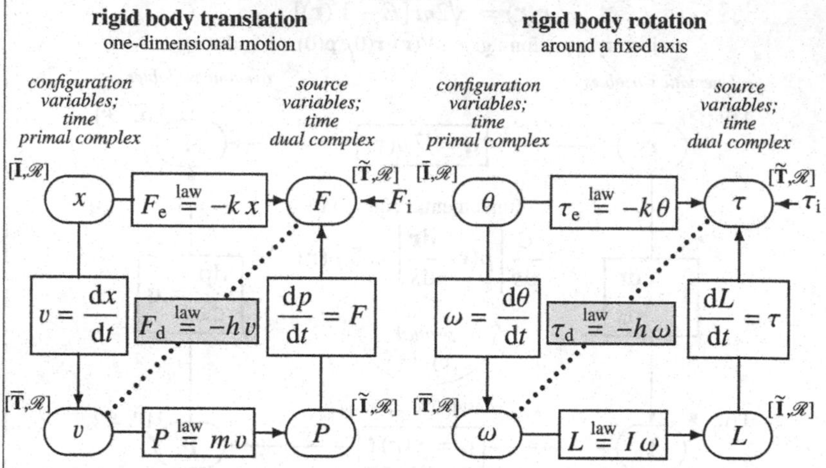

**rigid body translation**
one-dimensional motion

**rigid body rotation**
around a fixed axis

*configuration variables; time primal complex*

*source variables; time dual complex*

*configuration variables; time primal complex*

*source variables; time dual complex*

$[\bar{\mathbf{I}},\mathscr{R}]$

$x$  →  $F_e \overset{law}{=} -k\,x$  →  $F$  ←  $F_i$   $[\tilde{\mathbf{T}},\mathscr{R}]$

$v = \dfrac{dx}{dt}$   $F_d \overset{law}{=} -h\,v$   $\dfrac{dp}{dt} = F$

$[\bar{\mathbf{T}},\mathscr{R}]$   $v$  →  $P \overset{law}{=} m\,v$  →  $P$   $[\tilde{\mathbf{I}},\mathscr{R}]$

$[\bar{\mathbf{I}},\mathscr{R}]$

$\theta$  →  $\tau_e \overset{law}{=} -k\,\theta$  →  $\tau$  ←  $\tau_i$   $[\tilde{\mathbf{T}},\mathscr{R}]$

$\omega = \dfrac{d\theta}{dt}$   $\tau_d \overset{law}{=} -h\,\omega$   $\dfrac{dL}{dt} = \tau$

$[\bar{\mathbf{T}},\mathscr{R}]$   $\omega$  →  $L \overset{law}{=} I\,\omega$  →  $L$   $[\tilde{\mathbf{I}},\mathscr{R}]$

fundamental equation

$$m\,\frac{d^2 r}{dt^2} = -k\,r - h\,\frac{dx}{dt} + F_i(t)$$

fundamental equation

$$I\,\frac{d^2 \theta}{dt^2} = -k\,\theta - h\,\frac{d\theta}{dt} + \tau_i(t)$$

$hv$
$kx$
$F$
$h$
$m$
$k$
$x$

$\mathbf{I}^-$ instant
$\mathbf{T}$ interval
$\mathbf{I}^+$ instant

$\tau$
$L$
$\omega$
$\theta$

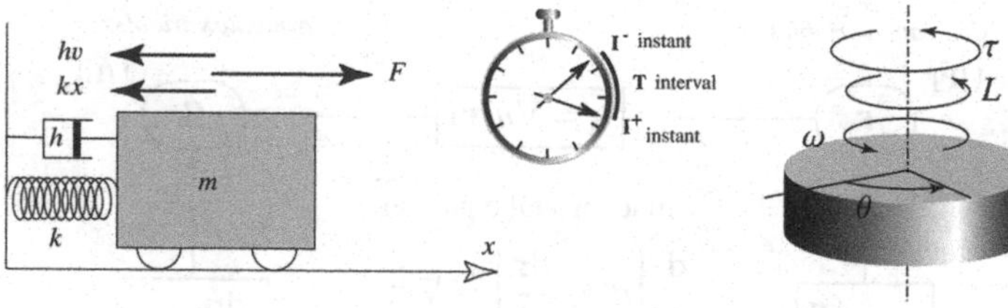

| $x$ | abscissa | | $\theta$ | angle of position |
|---|---|---|---|---|
| $v$ | velocity | $\bar{\mathbf{I}}$ time even $\tilde{\mathbf{T}}$ | $\omega$ | angular velocity |
| $F$ | force | | $\tau$ | torque |
| $F_e$ | elastic force | $\bar{\mathbf{T}}$ time odd $\tilde{\mathbf{I}}$ | $\tau_e$ | elastic restoring torque |
| $F_d$ | dissipative force | | $\tau_d$ | dissipative torque |
| $F_i$ | impressed force | | $\tau_i$ | impressed torque |
| $P$ | momentum | | $L$ | angular momentum |
| $k$ | stiffness | | $k$ | angular stiffness |
| $h$ | damping coefficient | | $h$ | damping coefficient |
| $m$ | mass | | $I$ | moment of inertia |

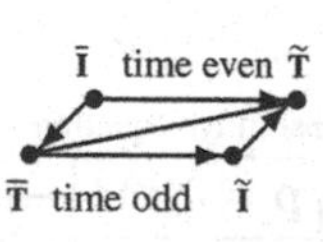

ANA2-7; http://discretephysics.dicar.units.it

# Analogies among different physical theories [ANA3]

**Waves** $1[\tilde{\mathbf{I}}, \bar{\mathbf{P}}]$

$\phi^-$    $S = \hbar \phi$

$\mathbf{k} = -\nabla \phi$    $E = \hbar \omega$    Einstein

$\omega = \partial_t \phi$

$\omega^+$ $1[\tilde{\mathbf{T}}, \bar{\mathbf{P}}]$

$\mathbf{k}^-$ $3[\tilde{\mathbf{I}}, \bar{\mathbf{L}}]$

$3[\tilde{\mathbf{T}}, \bar{\mathbf{L}}]$

**Analytical mechanics** $1[\tilde{\mathbf{I}}, \bar{\mathbf{P}}]$

$S^-$

$\mathbf{p} = -\nabla S$    $E = \partial_t S$    $E^+$ $1[\tilde{\mathbf{T}}, \bar{\mathbf{P}}]$

$E = e\phi$

$\mathbf{p} = \hbar \mathbf{k}$    De Broglie

$\mathbf{p}^-$ $3[\tilde{\mathbf{I}}, \bar{\mathbf{L}}]$

$3[\tilde{\mathbf{T}}, \bar{\mathbf{L}}]$

**Electromagnetism**

$\chi^-$ $1[\tilde{\mathbf{I}}, \bar{\mathbf{P}}]$

$\mathbf{A} = -\nabla \chi$    $\phi = \partial_t \chi$    $\phi^+$ $1[\tilde{\mathbf{T}}, \bar{\mathbf{P}}]$

$\mathbf{p} = e\mathbf{A}$

$\mathbf{A}^-$ $3[\tilde{\mathbf{I}}, \bar{\mathbf{L}}]$

$\mathbf{B} = \nabla \times \mathbf{A}$    $\mathbf{E} = -\partial_t \mathbf{A} - \nabla \phi$

$\mathbf{E}^+$ $3[\tilde{\mathbf{T}}, \bar{\mathbf{L}}]$

$\mathbf{B}^-$ $3[\tilde{\mathbf{I}}, \bar{\mathbf{S}}]$

The + and − signs as upper indices refer to behaviour under reversal of motion.

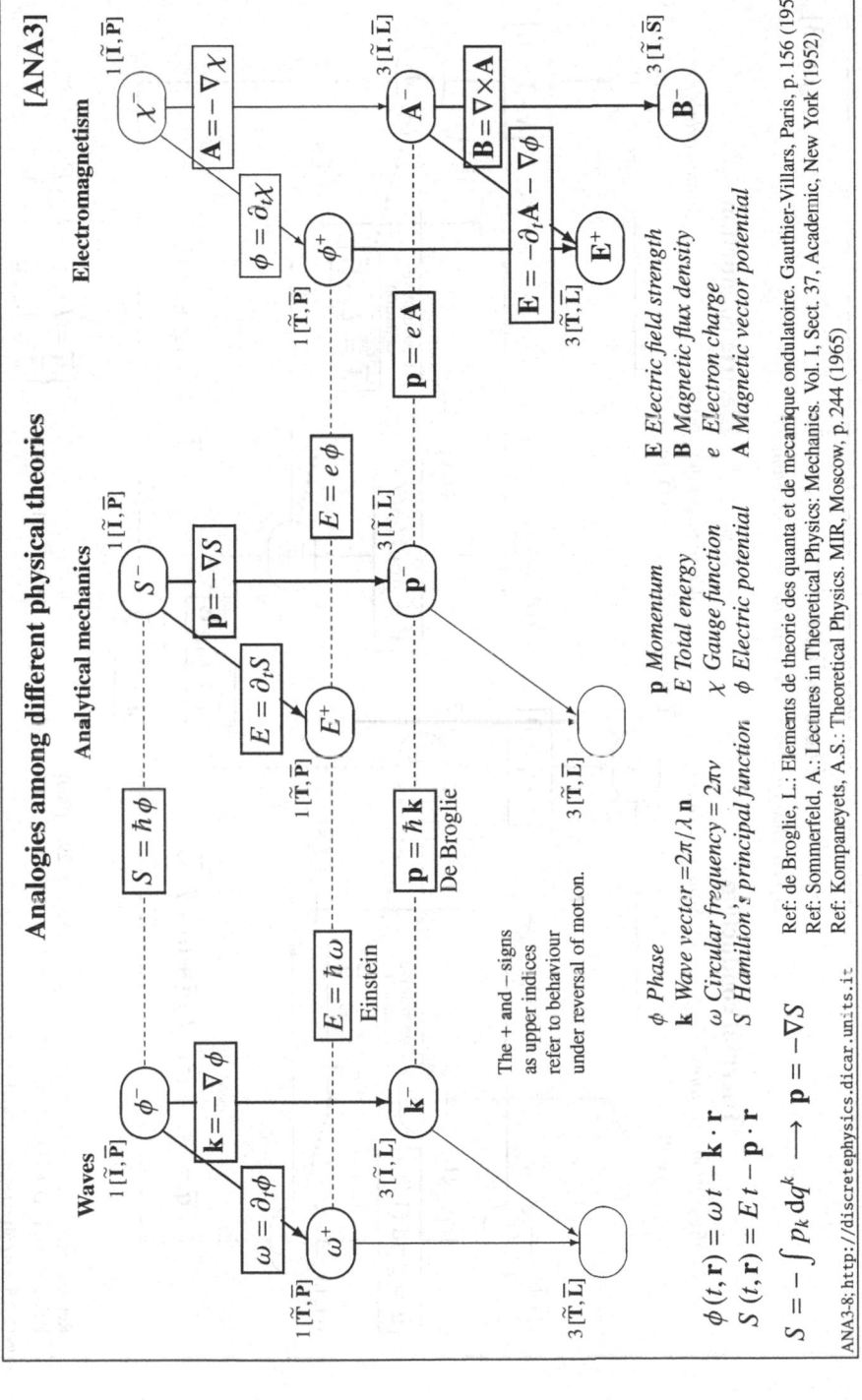

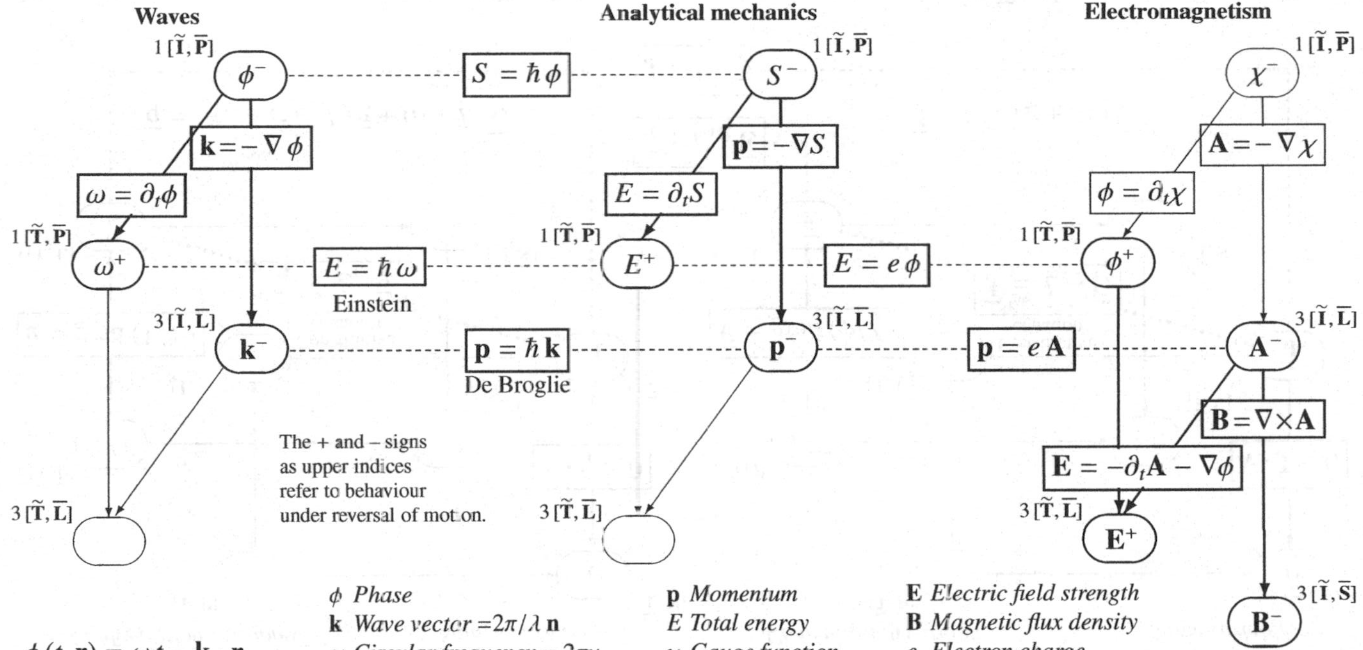

$\phi$ *Phase*
$\mathbf{k}$ *Wave vector* $= 2\pi/\lambda \, \mathbf{n}$
$\omega$ *Circular frequency* $= 2\pi \nu$
$S$ *Hamilton's principal function*

$\mathbf{p}$ *Momentum*
$E$ *Total energy*
$\chi$ *Gauge function*
$\phi$ *Electric potential*

$\mathbf{E}$ *Electric field strength*
$\mathbf{B}$ *Magnetic flux density*
$e$ *Electron charge*
$\mathbf{A}$ *Magnetic vector potential*

$$\phi(t, \mathbf{r}) = \omega t - \mathbf{k} \cdot \mathbf{r}$$
$$S(t, \mathbf{r}) = E t - \mathbf{p} \cdot \mathbf{r}$$
$$S = -\int p_k \, dq^k \longrightarrow \mathbf{p} = -\nabla S$$

Ref: de Broglie, L.: Elements de theorie des quanta et de mecanique ondulatoire. Gauthier-Villars, Paris, p. 156 (1959)
Ref: Sommerfeld, A.: Lectures in Theoretical Physics: Mechanics. Vol. I, Sect. 37, Academic, New York (1952)
Ref: Kompaneyets, A.S.: Theoretical Physics. MIR, Moscow, p. 244 (1965)

ANA3-8: http://discretephysics.dicar.units.it

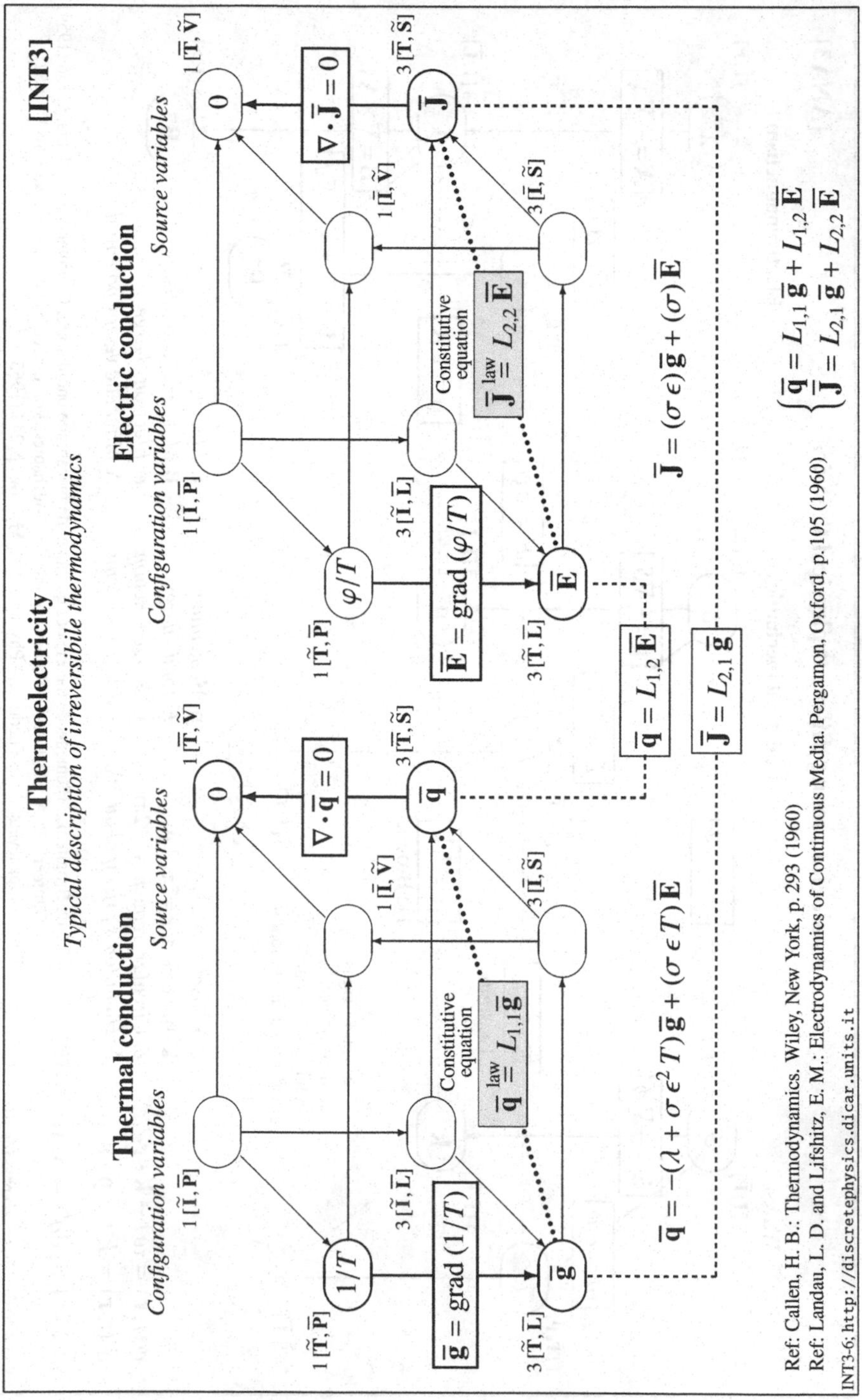
[INT3]
Thermoelectricity
Typical description of irreversibile thermodynamics
Thermal conduction
Electric conduction
Configuration variables
Source variables
Configuration variables
Source variables
$1[\tilde{\mathbf{T}},\bar{\mathbf{P}}]$
$1[\tilde{\mathbf{I}},\bar{\mathbf{P}}]$
$3[\tilde{\mathbf{I}},\bar{\mathbf{L}}]$
$3[\tilde{\mathbf{T}},\bar{\mathbf{L}}]$
$1[\bar{\mathbf{I}},\tilde{\mathbf{V}}]$
$3[\bar{\mathbf{I}},\tilde{\mathbf{S}}]$
$1[\bar{\mathbf{T}},\tilde{\mathbf{V}}]$
$3[\bar{\mathbf{T}},\tilde{\mathbf{S}}]$
$1/T$
$\bar{\mathbf{g}}$
$\bar{\mathbf{g}} = \text{grad}\,(1/T)$
$\bar{\mathbf{q}}$
0
$\nabla\cdot\bar{\mathbf{q}} = 0$
Constitutive equation
$\bar{\mathbf{q}} \overset{\text{law}}{=} L_{1,1}\bar{\mathbf{g}}$
$\varphi/T$
$\bar{\mathbf{E}}$
$\bar{\mathbf{E}} = \text{grad}\,(\varphi/T)$
$\bar{\mathbf{J}}$
$\nabla\cdot\bar{\mathbf{J}} = 0$
Constitutive equation
$\bar{\mathbf{J}} \overset{\text{law}}{=} L_{2,2}\,\bar{\mathbf{E}}$
$\bar{\mathbf{q}} = L_{1,2}\,\bar{\mathbf{E}}$
$\bar{\mathbf{J}} = L_{2,1}\,\bar{\mathbf{g}}$
$\bar{\mathbf{q}} = -(\lambda + \sigma\epsilon^2 T)\,\bar{\mathbf{g}} + (\sigma\epsilon T)\,\bar{\mathbf{E}}$
$\bar{\mathbf{J}} = (\sigma\epsilon)\,\bar{\mathbf{g}} + (\sigma)\,\bar{\mathbf{E}}$
$\begin{cases}\bar{\mathbf{q}} = L_{1,1}\,\bar{\mathbf{g}} + L_{1,2}\,\bar{\mathbf{E}} \\ \bar{\mathbf{J}} = L_{2,1}\,\bar{\mathbf{g}} + L_{2,2}\,\bar{\mathbf{E}}\end{cases}$
Ref: Callen, H. B.: Thermodynamics. Wiley, New York, p. 293 (1960)
Ref: Landau, L. D. and Lifshitz, E. M.: Electrodynamics of Continuous Media. Pergamon, Oxford, p.105 (1960)
INT3-6; http://discretephysics.dicar.units.it

# The quantization process     [QME1]

## Classical mechanics

*configuration variables*          *source variables*

$1\,[\bar{\mathbf{I}}]$        constitutive equation      $1\,[\tilde{\mathbf{T}}]$

$q^k$    $\boxed{Q_h = [p_h, H]}$    $Q_h$

$[...] = $ *Poisson brackets*

$$v^k = \frac{dq^k}{dt}$$     *primal cycle*     $$Q_h = \frac{dp_h}{dt}$$

$1\,[\tilde{\mathbf{I}}]$        constitutive equation      $1\,[\tilde{\mathbf{I}}]$

$v^k$    $\boxed{v^k = [q^k, H]}$    $p_h$

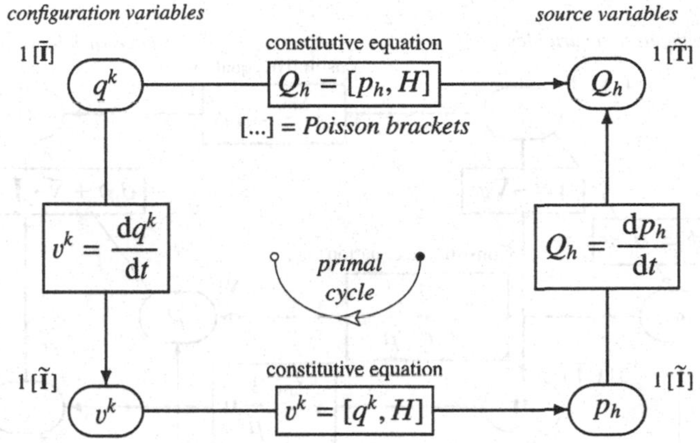

$q^k(t)$ *generalized coordinate*    *H Hamiltonian*    $Q_h(t)$ *generalized forces*
$v^k(t)$ *generalized velocities*                 $p_h(t)$ *generalized momenta*

Ref: Ter Haar, Goldstein

## Quantum mechanics (Heisenberg formulation)

*configuration variables*          *source variables*

$1\,[\bar{\mathbf{I}}]$        constitutive equation      $1\,[\tilde{\mathbf{T}}]$

$\mathbf{q}$    $\boxed{\mathbf{Q} = \frac{i}{\hbar}[\mathbf{H}, \mathbf{p}]}$    $\mathbf{Q}$

$[...] = $ *commutator*

$$\mathbf{v} = \frac{d\mathbf{q}}{dt}$$     *primal cycle*     $$\mathbf{Q} = \frac{d\mathbf{p}}{dt}$$

$1\,[\tilde{\mathbf{I}}]$        constitutive equation      $1\,[\tilde{\mathbf{I}}]$

$\mathbf{v}$    $\boxed{\mathbf{v} = \frac{i}{\hbar}[\mathbf{H}, \mathbf{q}]}$    $\mathbf{p}$

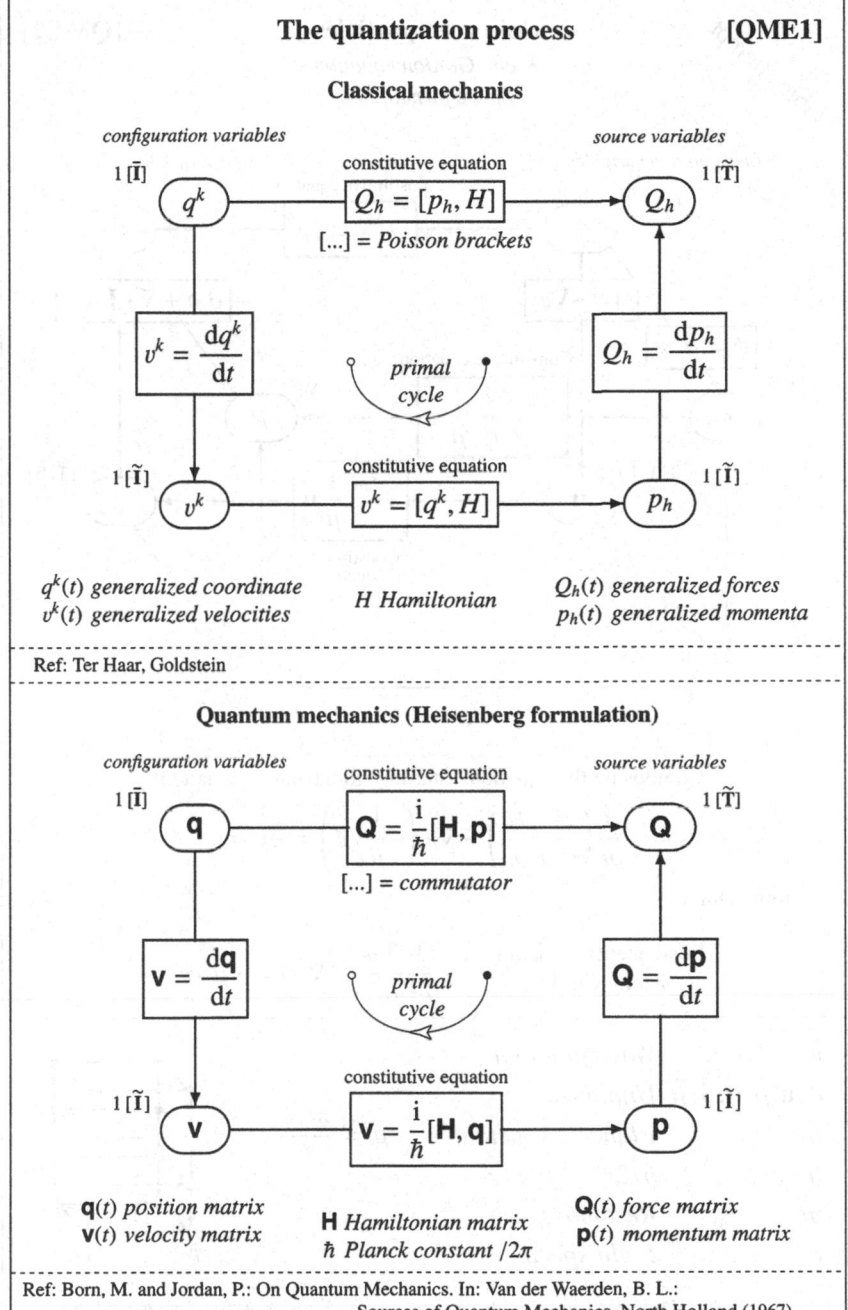

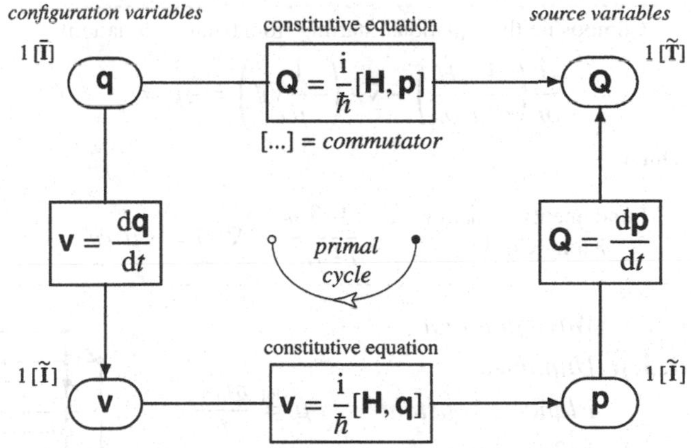

$\mathbf{q}(t)$ *position matrix*    **H** *Hamiltonian matrix*    $\mathbf{Q}(t)$ *force matrix*
$\mathbf{v}(t)$ *velocity matrix*    $\hbar$ *Planck constant* $/2\pi$    $\mathbf{p}(t)$ *momentum matrix*

Ref: Born, M. and Jordan, P.: On Quantum Mechanics. In: Van der Waerden, B. L.:
Sources of Quantum Mechanics. North Holland (1967).

QME1-3; http://discretephysics.dicar.units.it

table attempt

## Spin zero particles [QME2]

*Klein–Gordon equation*
*Space formulation*

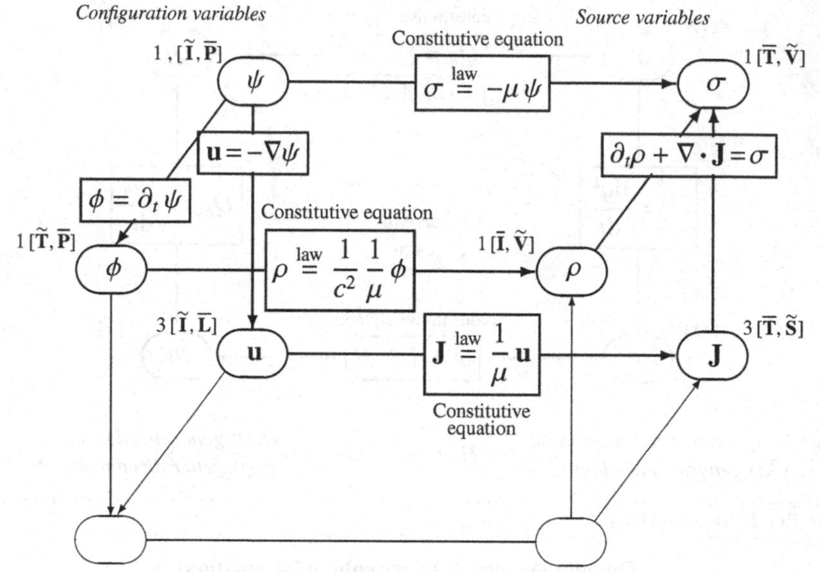


Composing the equations, starting from balance equation:

$$\frac{\partial}{\partial t}\left(\frac{1}{c^2\mu}\frac{\partial\psi}{\partial t}\right)+\nabla\cdot\left(-\frac{1}{\mu}\nabla\psi\right)=-\mu\,\psi$$

from which

Fundamental equation
Klein–Gordon
$$\frac{1}{c^2}\frac{\partial^2\psi}{\partial t^2}-\nabla^2\psi=-\mu^2\psi$$

| | |
|---|---|
| $\psi$ | *Wave function* |
| $\phi,\mathbf{u},\rho,\sigma,\mathbf{J},\mu$ | *Unnamed* |
| $h$ | *Planck's constant* |
| $\hbar$ | $h/2\pi$ |
| $m_o$ | *Rest mass* |
| $c$ | *Light speed* |

$$\mu \overset{\text{def}}{=} \frac{m_o\,c}{\hbar}$$

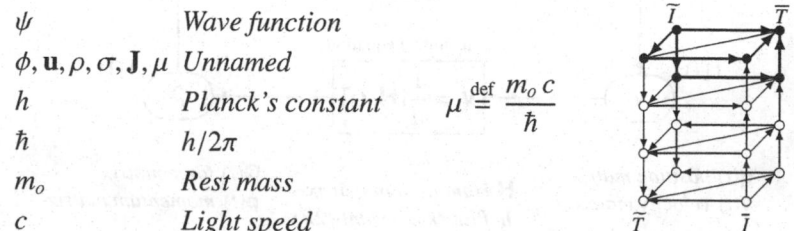

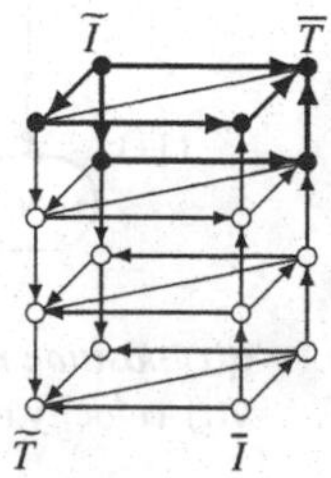


Ref: De Broglie, L.: The Vigier Theory of Elementary Particles. Elsevier, Amsterdam, p. 33 (1963)
Ref: Wichmann, E.H.: Quantum Physics, Berkeley Physics Course, Vol. 4. McGraw-Hill, New York, p. 205

QME2-11; http://discretephysics.dicar.units.it

**Spin zero particles**                                          **[QME3]**
*Klein-Gordon equation*
*space-time formulation*

*configuration variables ?*                                    *source variables ?*
*primal space-time complex*                                    *dual space-time complex*
*inner orientation*                                            *outer orientation*

$1\,[\overline{\mathbf{P}}_4]$                                 $1\,[\widetilde{\mathbf{H}}_4]$

$\psi$        constitutive equation        $\sigma$
$\sigma \overset{\text{law}}{=} -\mu\,\psi$

$U_\alpha = \partial_\alpha \psi$     *primal cycle*     $\nabla_\rho J^\rho = \sigma$

constitutive equation

$4\,[\overline{\mathbf{L}}_4]$                                 $4\,[\widetilde{\mathbf{V}}_4]$

$U_\alpha$     $J^\rho \overset{\text{law}}{=} \dfrac{1}{\mu} g^{\rho\alpha} U_\alpha$     $J^\rho$

$\partial_\beta U_\alpha - \partial_\alpha U_\beta = 0$     $\dfrac{1}{2!}\check{\eta}^{\rho\gamma\mu\nu}\,\partial_\gamma \check{G}_{\mu\nu} = J^\rho$

$6\,[\overline{\mathbf{S}}_4]$                                 $6\,[\widetilde{\mathbf{S}}_4]$

$0$     $\mu \overset{\text{def}}{=} \dfrac{m_o c}{\hbar} \qquad \hbar \overset{\text{def}}{=} \dfrac{h}{2\pi}$     $\check{G}_{\mu\nu}$

$\psi$  *wave function*
$g_{\alpha\beta}$  *metric tensor*
$c$  *light speed*

$4\,[\overline{\mathbf{V}}_4]$                                 $4\,[\widetilde{\mathbf{L}}_4]$

$m_o$  *rest mass*
$\sigma$  *unnamed*

$g^{\mu\nu}\partial_\mu\partial_\nu\psi + \mu^2\psi = 0$

$1\,[\overline{\mathbf{H}}_4]$                                 $1\,[\widetilde{\mathbf{P}}_4]$

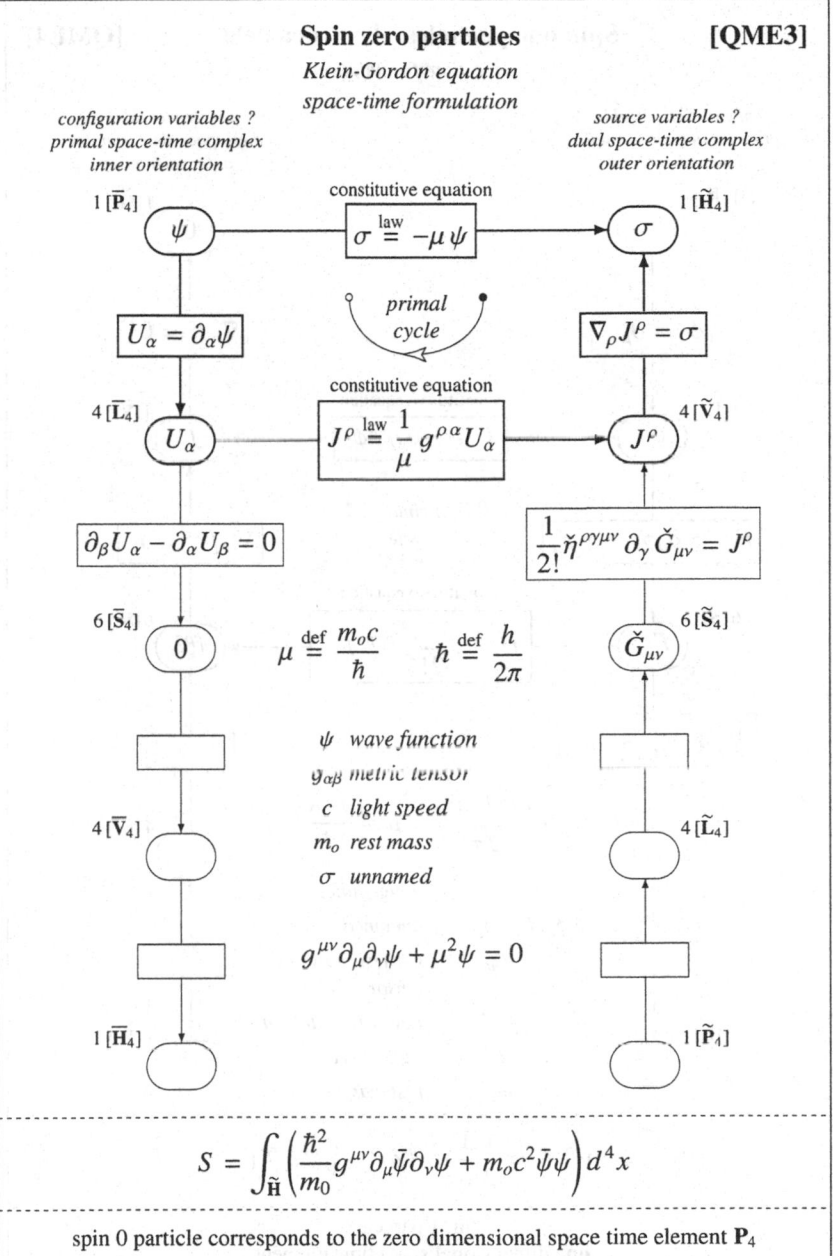

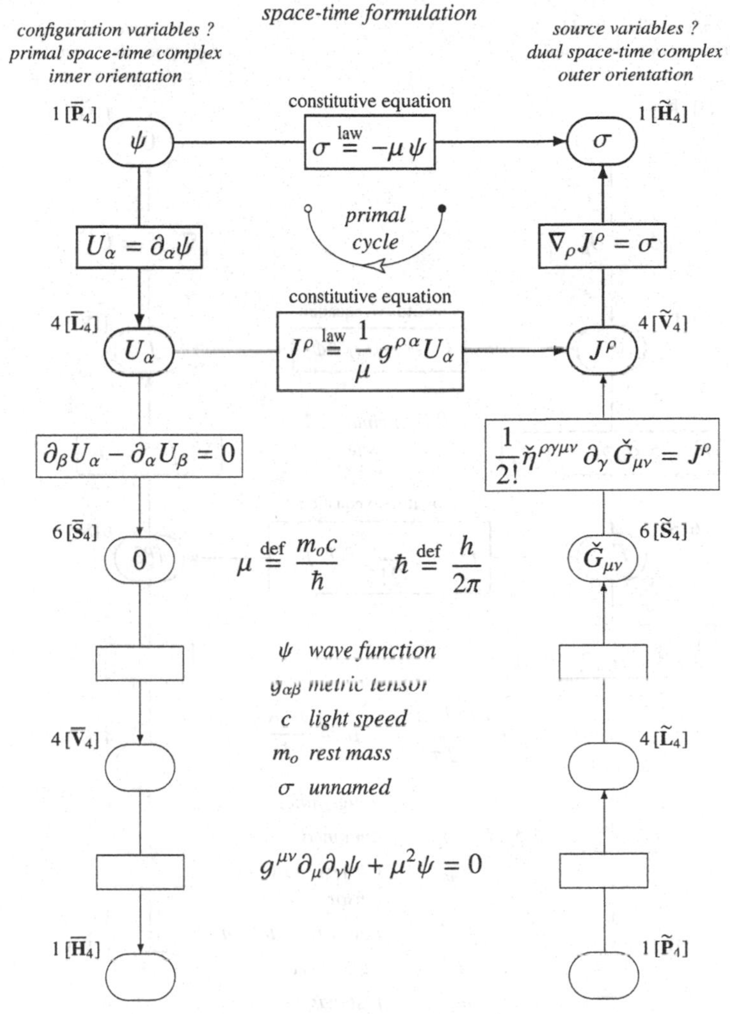

$$S = \int_{\widetilde{\mathbf{H}}} \left( \frac{\hbar^2}{m_0} g^{\mu\nu}\partial_\mu\bar{\psi}\partial_\nu\psi + m_o c^2 \bar{\psi}\psi \right) d^4 x$$

spin 0 particle corresponds to the zero dimensional space time element $\mathbf{P}_4$

Ref: Bjorken, J. D. and Drell, S. D.: Relativistic quantum mechanics. McGraw-Hill, p. 5 (1964).
Ref: McMahon, D.: Quantum Field Theory Demystified. McGraw-Hill, p. 110 (2008).

QME3-10; http://discretephysics.dicar.units.it

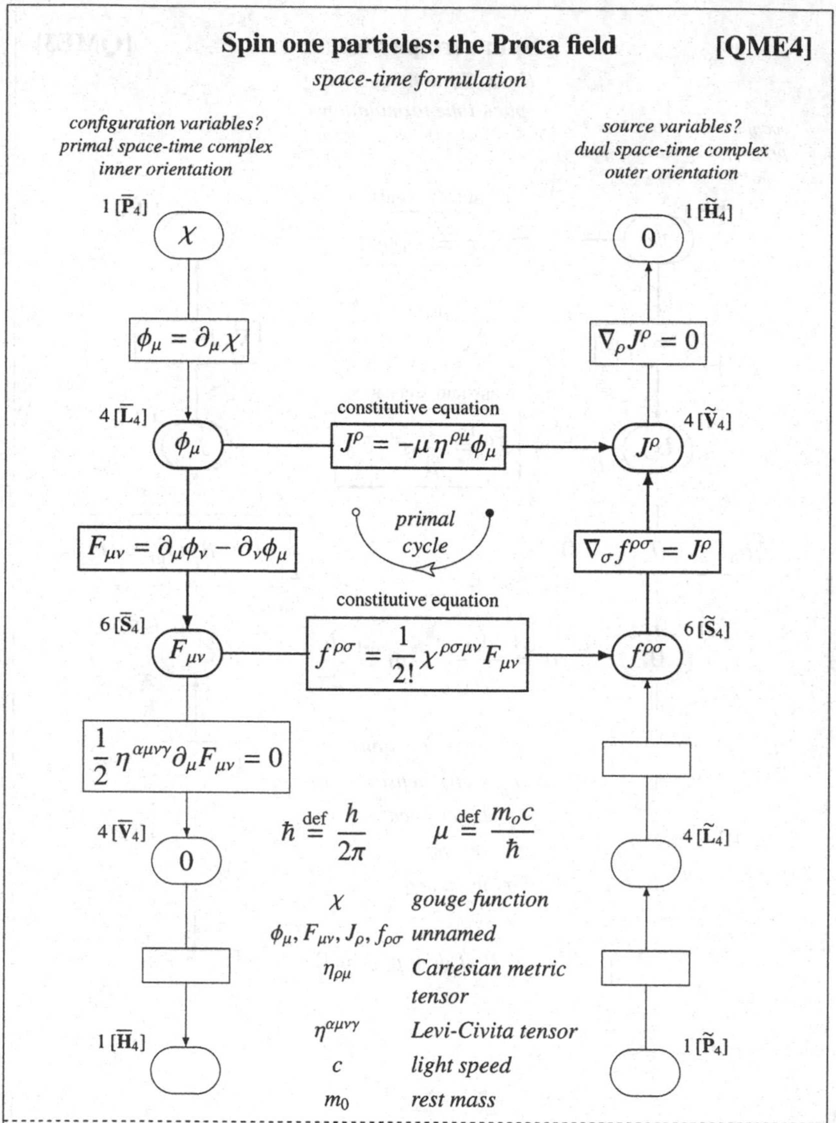
Spin one particles: the Proca field
[QME4]
space-time formulation
configuration variables?
primal space-time complex
inner orientation
source variables?
dual space-time complex
outer orientation
1 [$\bar{\mathbf{P}}_4$]
$\chi$
$\phi_\mu = \partial_\mu \chi$
4 [$\bar{\mathbf{L}}_4$]
$\phi_\mu$
constitutive equation
$J^\rho = -\mu\, \eta^{\rho\mu} \phi_\mu$
$F_{\mu\nu} = \partial_\mu \phi_\nu - \partial_\nu \phi_\mu$
primal
cycle
constitutive equation
6 [$\bar{\mathbf{S}}_4$]
$F_{\mu\nu}$
$f^{\rho\sigma} = \frac{1}{2!} \chi^{\rho\sigma\mu\nu} F_{\mu\nu}$
$\frac{1}{2} \eta^{\alpha\mu\nu\gamma} \partial_\mu F_{\mu\nu} = 0$
4 [$\bar{\mathbf{V}}_4$]
0
1 [$\bar{\mathbf{H}}_4$]
1 [$\tilde{\mathbf{H}}_4$]
0
$\nabla_\rho J^\rho = 0$
4 [$\tilde{\mathbf{V}}_4$]
$J^\rho$
$\nabla_\sigma f^{\rho\sigma} = J^\rho$
6 [$\tilde{\mathbf{S}}_4$]
$f^{\rho\sigma}$
4 [$\tilde{\mathbf{L}}_4$]
1 [$\tilde{\mathbf{P}}_4$]
$\hbar \overset{\text{def}}{=} \frac{h}{2\pi}$
$\mu \overset{\text{def}}{=} \frac{m_o c}{\hbar}$
$\chi$ gouge function
$\phi_\mu, F_{\mu\nu}, J_\rho, f_{\rho\sigma}$ unnamed
$\eta_{\rho\mu}$ Cartesian metric tensor
$\eta^{\alpha\mu\nu\gamma}$ Levi-Civita tensor
$c$ light speed
$m_0$ rest mass

$$\chi^{\rho\sigma\mu\nu} = \frac{1}{2}\frac{1}{\mu}\left(g^{\rho\mu} g^{\sigma\nu} - g^{\rho\nu} g^{\sigma\mu}\right)$$

spin 1 particle

**one** dimensional space time element

QME4-5; http://discretephysics.dicar.units.it

(table attempt)

## Bilinear covariants [QME5]

For Dirac theory

**1 P**
Scalar $S = \overline{\psi}\,\psi$

**4 L**
Vector $V_\alpha = \overline{\psi}\,\gamma_\alpha \psi$

**6 S**
Skew-symmetric tensor $T^{\alpha\beta} = \mathrm{i}\overline{\psi}\,[\gamma^\alpha \gamma^\beta - \gamma^\beta \gamma^\alpha]\psi$

**4 V**
Pseudovector $A^\mu = \overline{\psi}\,\mathrm{i}\,\gamma^\mu \gamma_5 \psi$

**1 H**
Pseudoscalar $P = \mathrm{i}\,\overline{\psi}\,\gamma_5 \psi$

Pauli spin matrices

$$\sigma_0 \overset{\text{def}}{=} \begin{bmatrix} 1 & 0 \\ 0 & 1 \end{bmatrix} \quad \sigma_1 \overset{\text{def}}{=} \begin{bmatrix} 0 & 1 \\ 1 & 0 \end{bmatrix}$$

$$\sigma_2 \overset{\text{def}}{=} \begin{bmatrix} 0 & -\mathrm{i} \\ \mathrm{i} & 0 \end{bmatrix} \quad \sigma_3 \overset{\text{def}}{=} \begin{bmatrix} 1 & 0 \\ 0 & -1 \end{bmatrix}$$

$$\gamma_0 \overset{\text{def}}{=} \begin{bmatrix} 0 & \sigma_0 \\ \sigma_0 & 0 \end{bmatrix} \quad \gamma_k \overset{\text{def}}{=} \begin{bmatrix} 0 & \sigma_k \\ \sigma_k & 0 \end{bmatrix}$$

$$\gamma_5 \overset{\text{def}}{=} \gamma_0\, \gamma_1\, \gamma_2\, \gamma_3$$

$$\gamma^0 \overset{\text{def}}{=} \ldots . \quad \gamma^k \overset{\text{def}}{=} \ldots .$$

| | |
|---|---|
| $\psi^*$ | Complex conjugate |
| $\widetilde{\psi}$ | Transposed (?) |
| $\psi^\dagger = \widetilde{\psi}^*$ | Adjoint |
| $\overline{\psi} = \psi^\dagger \gamma^0$ | Pauli adjoint |

Ref: de Broglie, L.: La theorie des particules de spin 1/2. Gauthier-Villars, Paris, p. 76 (1952) (signature +—)
Ref: Berestetskii, V.B, Lifshitz, E.M. and Pitaevskii, L.P.: Relativistic Quantum Theory. Pergamon, Oxford, p. 85 (1971) (signature +—)
Ref: Feynman, R.P.: Theory of Fundamental Processes. Benjamin, p. 115 (1962) (signature +—).
Ref: Rzewusky, J.: Field Theory, Part I, Classical Theory. Polish Scientific Publisher p. 63 (1964) (signature —+) covariance and contravariance not respected.
Ref: Yilmaz, H.: Introduction to the Theory of Relativity and the Principles of Modern Physics. Blaisdell, p. 121 (1965) (signature —+) covariance and contravariance not respected.
Ref: Sakurai, J.J.: Invariance Principles and Elementary Particles. Princeton University Press, Princeton p. 27 (1964)

QME5-7; http://discretephysics.dicar.units.it

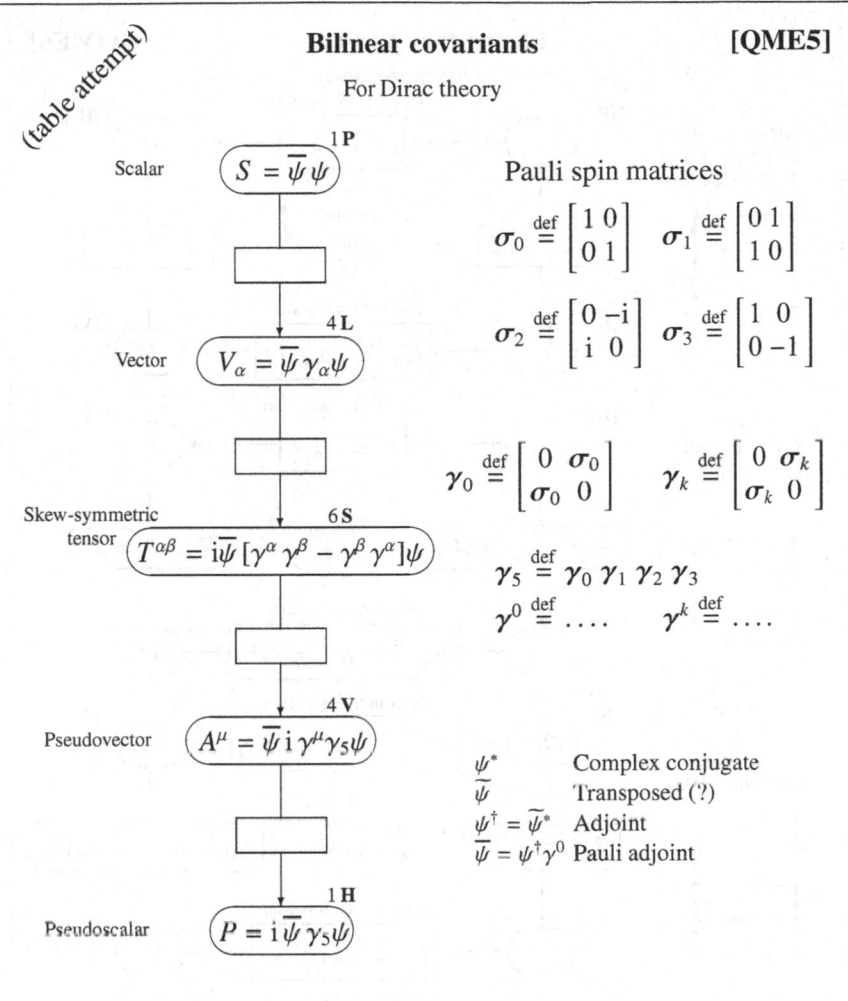

**Elementary particles**                                              **[QME6]**

*relativistic formulation*

Klein-Gordon scalar particle [Corson, p.98]

Proca vector particle [Corson, p.100]

Proca pseudovector particle [Corson, p.103]

Klein-Gordon pseudoscalar particle $\pi$-meson [Corson, p.98] [Ylmaz, p.121]

$1\,[\overline{\mathbf{P}}_4]$ $\phi$ — constitutive equation $\sigma \overset{law}{=} -k\,\phi$ — $\sigma$ $1\,[\widetilde{\mathbf{H}}_4]$

$U_\alpha = \partial_\alpha \phi$ $\qquad$ *primal cycle* $\qquad$ $\nabla_\mu J^\mu = \sigma$

$4\,[\overline{\mathbf{L}}_4]$ $U_\alpha$ — constitutive equation $U_\alpha \overset{law}{=} k\,g_{\alpha\mu}\,J^\mu$ — $J^\mu$ $4\,[\widetilde{\mathbf{V}}_4]$

$4\,[\overline{\mathbf{L}}_4]$ $U_\alpha$ — constitutive equation $\sigma \overset{law}{=} -k\,\psi$ — $J^\mu$ $4\,[\widetilde{\mathbf{V}}_4]$

$F_{\alpha\beta} = \partial_\beta U_\alpha - \partial_\alpha U_\beta$ $\qquad$ *primal cycle* $\qquad$ $\nabla_\nu G^{\mu\nu} = J^\mu$

$6\,[\overline{\mathbf{S}}_4]$ $F_{\alpha\beta}$ — constitutive equation $F_{\alpha\beta} \overset{law}{=} k\,g_{\alpha\mu}\,g_{\beta\nu}\,G^{\mu\nu}$ — $G^{\mu\nu}$ $6\,[\widetilde{\mathbf{S}}_4]$

$1\,[\overline{\mathbf{S}}_4]$ $\breve{G}^{\mu\nu}$ — constitutive equation $\breve{G}^{\mu\nu} \overset{law}{=} k\,g^{\mu\alpha}\,g^{\nu\beta}\,\breve{\phi}_{\alpha\beta}$ — $\breve{\phi}_{\alpha\beta}$ $1\,[\widetilde{\mathbf{S}}_4]$

$\nabla_\nu \breve{G}^{\mu\nu} = \breve{\chi}^\mu$ $\qquad$ *dual cycle* $\qquad$ $\phi_{\alpha\beta} = \partial_\alpha \breve{V}_\beta - \partial_\beta \breve{V}_\alpha$

$4\,[\overline{\mathbf{V}}_4]$ $\breve{\chi}^\mu$ — constitutive equation $\breve{\chi}^\mu \overset{law}{=} -k\,g^{\mu\beta}\,\breve{V}_\beta$ — $\breve{V}_\beta$ $4\,[\widetilde{\mathbf{L}}_4]$

$1\,[\overline{\mathbf{V}}_4]$ $\breve{\chi}^\mu$ — constitutive equation $\breve{\chi}^\mu \overset{law}{=} k\,g^{\mu\beta}\,\breve{V}_\beta$ — $\breve{V}_\beta$ $1\,[\widetilde{\mathbf{L}}_4]$

$\nabla_\mu \breve{\chi}^\mu = \breve{\tau}$ $\qquad$ *dual cycle* $\qquad$ $\breve{V}_\beta = \partial_\beta \breve{\psi}$

$4\,[\overline{\mathbf{H}}_4]$ $\breve{\tau}$ — constitutive equation $\breve{\tau} \overset{law}{=} -k\,\breve{\psi}$ — $\breve{\psi}$ $4\,[\widetilde{\mathbf{P}}_4]$

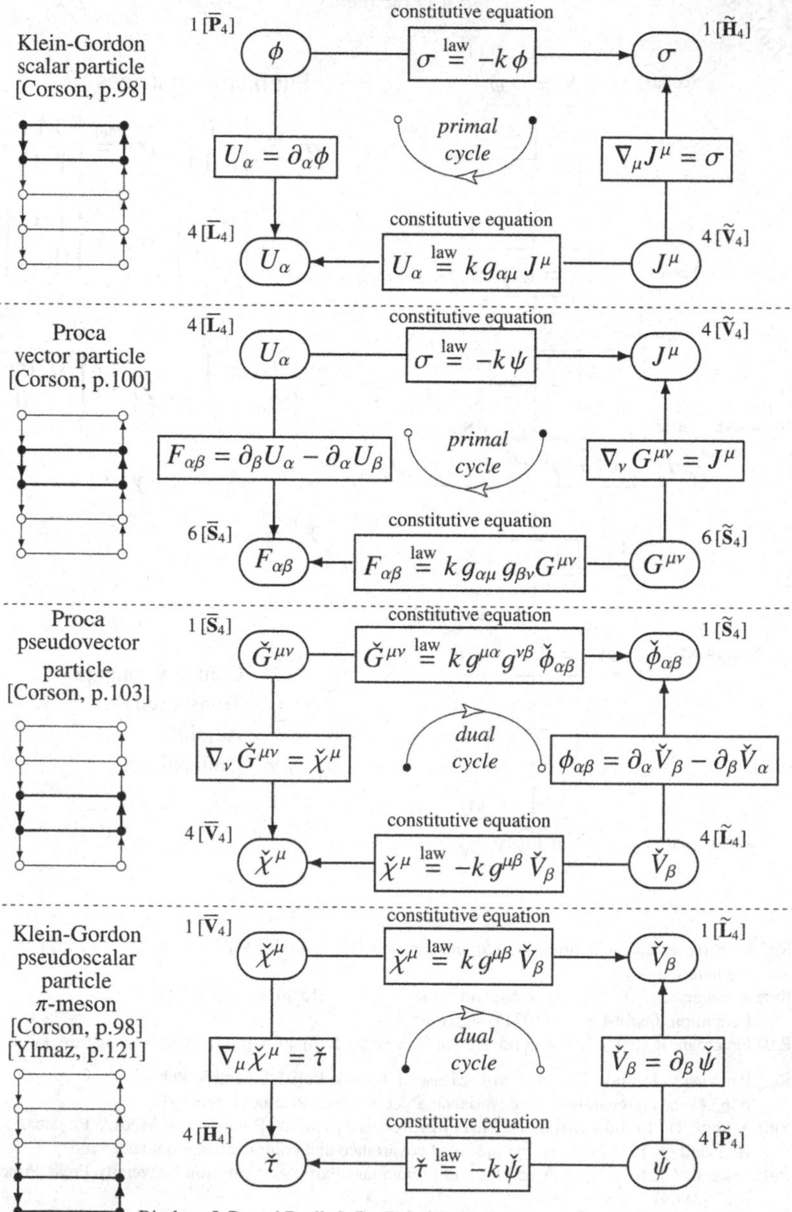

Bjorken, J. D. and Drell, S. D.: Relativistic quantum mechanics. McGraw-Hill, p. 5 (1964).

# Part III
# Advanced Analysis

# Chapter 14
# General Structure of the Diagrams

## 14.1 Introduction

In this chapter we analyse the general structure of a classification diagram.[1]

We start with diagram [GEN1], which shows the relation of exterior differential forms with the classification diagram. The physical variables associated with space elements endowed with an *inner* orientation are described by multicovectors and, hence, by *even* differential forms. In contrast, the physical variables associated with space elements endowed with an *outer* orientation are described by pseudo multicovectors and, hence, by *odd* differential forms.

Diagram [GEN2] shows the passage from the even differential forms to the tensor notation up to the formation of the 'standard column', which is the one employed in all classification diagrams. Diagram [GEN3] shows the analogous passage from the odd differential forms to the pseudomultivector. Diagram [GEN4] shows the relation between discrete forms on primal and dual complexes and even and odd differential forms and from these to the standard columns. Diagram [GEN5] shows the various notations for the standard columns of the two sides of the diagram. Diagram [GEN6] shows a comparison between the standard columns and the de Rham complexes. Diagram [GEN7] makes the same comparison using the language of discrete forms. Diagrams [GEN8]–[GEN11] deal with space-time and must be compared with diagrams [GEN1]–[GEN4]. Diagram [GEN12] shows the standard diagrams in space of one, two, three and four dimensions.

Diagram 14.1 shows the strict link which exists between the de Rham theory and the classification diagram presented in this book. We have added the affine scalar and vector fields and the uniform scalar and vector fields which enjoy many simple and beautiful mathematical properties (Appendix A).

---

[1] This chapter presupposes previous reading of Chaps. 1–9. The tensorial notation which we will use is summarized in Appendix B.

E. Tonti, *The Mathematical Structure of Classical and Relativistic Physics*, Modeling and Simulation in Science, Engineering and Technology,
DOI 10.1007/978-1-4614-7422-7_14, 

**Table 14.1** Link between present classification diagram and de Rham theory

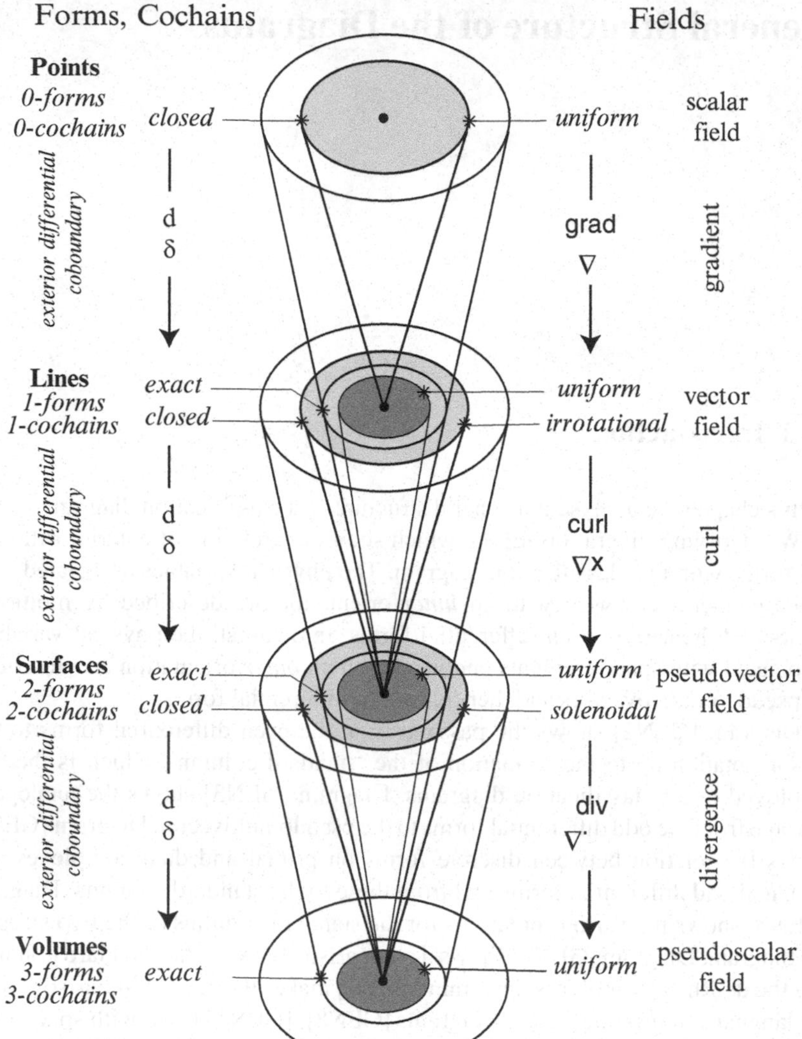

## Even and odd differential forms in three dimensions [GEN1]

*No metric, no connection*

| *Configuration variables* | *Source variables* |
|---|---|
| **Even** *differential forms* | **Odd** *differential forms* |
| *Inner orientation* | *Outer orientation* |

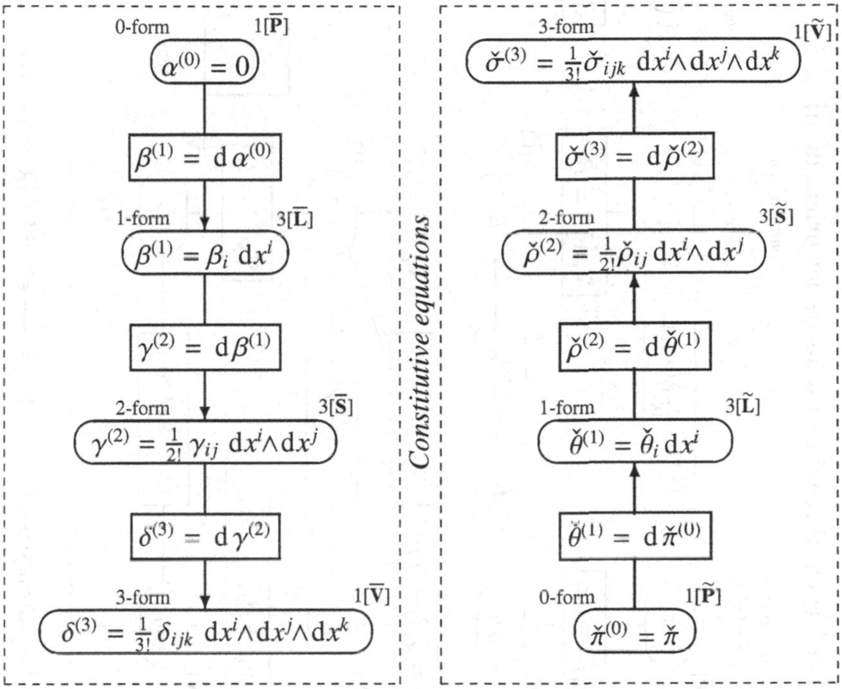


REMARK: we put the degree of the form inside rounded brackets to distinguish them from discrete forms.

$\mathbf{g}_k$ natural base vectors $\quad \mathbf{g}'_h = \lambda^k_h \mathbf{g}_k \quad \mathbf{g}_k = \Lambda^h_k \mathbf{g}'_h \quad \Delta \stackrel{\text{def}}{=} \det[\lambda^k_h]$

| Multicovectors | Pseudomulticovectors |
|---|---|
| $\alpha' = 1\,\alpha$ | $\breve{\sigma}'_{pqr} = \text{sgn}(\Delta)\ \lambda^i_p \lambda^j_q \lambda^k_r \breve{\sigma}_{ijk}$ |
| $\beta'_p = 1\,\lambda^i_p \beta_i$ | $\breve{\rho}'_{pq} = \text{sgn}(\Delta)\ \lambda^i_p \lambda^j_q \breve{\rho}_{ij}$ |
| $\gamma'_{pq} = 1\,\lambda^i_p \lambda^j_q \gamma_{ij}$ | $\breve{\theta}'_p = \text{sgn}(\Delta)\ \lambda^i_p \breve{\theta}_i$ |
| $\delta'_{pqr} = 1\,\lambda^i_p \lambda^j_q \lambda^k_r \delta_{ijk}$ | $\breve{\pi}' = \text{sgn}(\Delta)\ \breve{\pi}$ |

GEN1-12; http://discretephysics.dicar.units.it

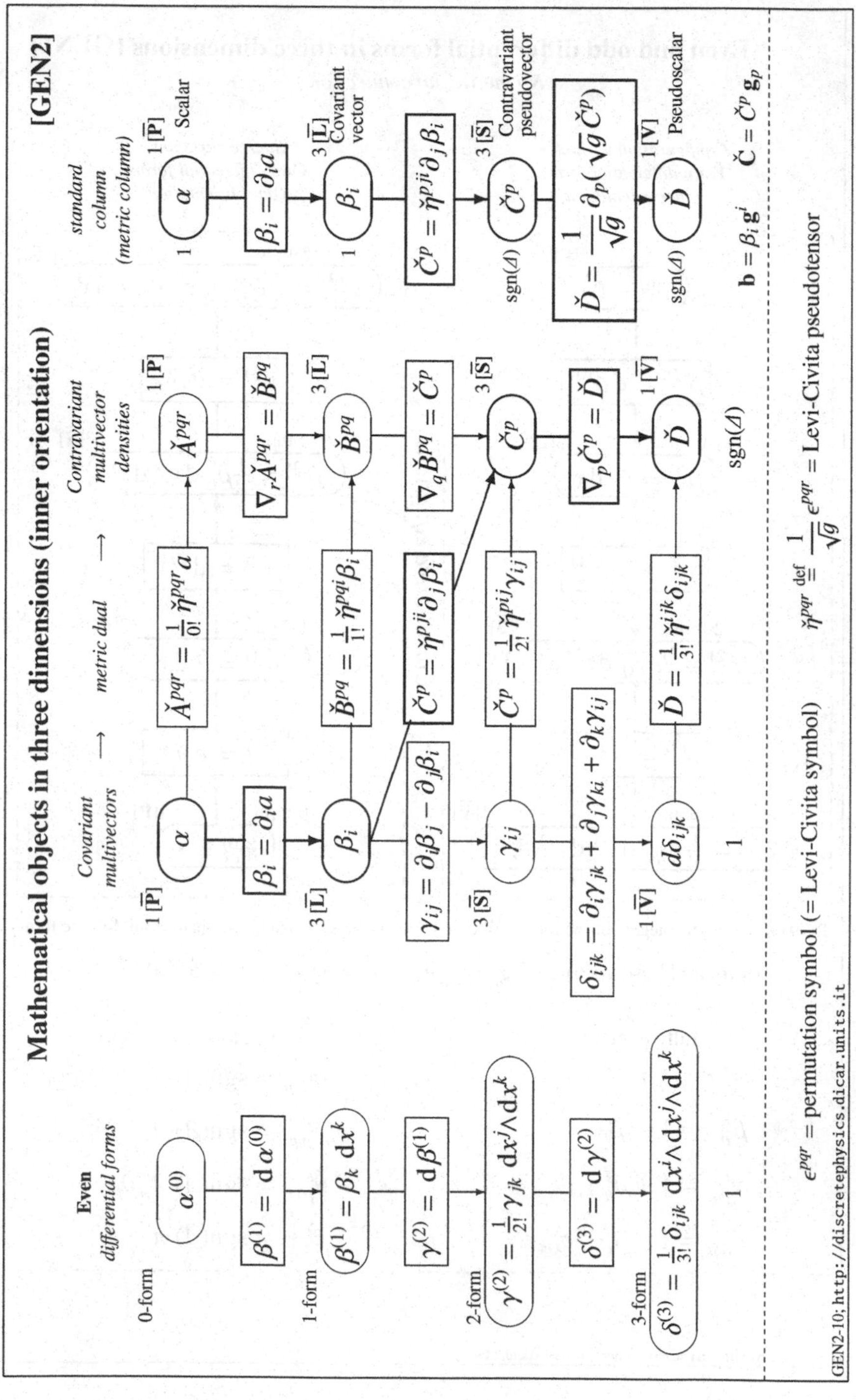

[GEN2]
Mathematical objects in three dimensions (inner orientation)
Even differential forms
Covariant multivectors → metric dual → Contravariant multivector densities
standard column (metric column)
0-form
$\alpha^{(0)}$
$\beta^{(1)} = \mathrm{d}\,\alpha^{(0)}$
1-form
$\beta^{(1)} = \beta_k\,\mathrm{d}x^k$
$\gamma^{(2)} = \mathrm{d}\,\beta^{(1)}$
2-form
$\gamma^{(2)} = \frac{1}{2!}\gamma_{jk}\,\mathrm{d}x^j\wedge\mathrm{d}x^k$
$\delta^{(3)} = \mathrm{d}\,\gamma^{(2)}$
3-form
$\delta^{(3)} = \frac{1}{3!}\delta_{ijk}\,\mathrm{d}x^i\wedge\mathrm{d}x^j\wedge\mathrm{d}x^k$
1
1 [$\overline{\mathbf{P}}$]
$\alpha$
$\beta_i = \partial_i a$
3 [$\overline{\mathbf{L}}$]
$\beta_i$
$\gamma_{ij} = \partial_i\beta_j - \partial_j\beta_i$
3 [$\overline{\mathbf{S}}$]
$\gamma_{ij}$
$\delta_{ijk} = \partial_i\gamma_{jk} + \partial_j\gamma_{ki} + \partial_k\gamma_{ij}$
1 [$\overline{\mathbf{V}}$]
$d\delta_{ijk}$
1
$\check{A}^{pqr} = \frac{1}{0!}\check{\eta}^{pqr}a$
$\check{B}^{pq} = \frac{1}{1!}\check{\eta}^{pqi}\beta_i$
$\check{C}^p = \check{\eta}^{pji}\partial_j\beta_i$
$\check{C}^p = \frac{1}{2!}\check{\eta}^{pij}\gamma_{ij}$
$\check{D} = \frac{1}{3!}\check{\eta}^{ijk}\delta_{ijk}$
1 [$\overline{\mathbf{P}}$]
$\check{A}^{pqr}$
$\nabla_r\check{A}^{pqr} = \check{B}^{pq}$
3 [$\overline{\mathbf{L}}$]
$\check{B}^{pq}$
$\nabla_q\check{B}^{pq} = \check{C}^p$
3 [$\overline{\mathbf{S}}$]
$\check{C}^p$
$\nabla_p\check{C}^p = \check{D}$
1 [$\overline{\mathbf{V}}$]
$\check{D}$
sgn($\Delta$)
1
$\alpha$
1 [$\overline{\mathbf{P}}$] Scalar
$\beta_i = \partial_i a$
1
$\beta_i$
3 [$\overline{\mathbf{L}}$] Covariant vector
$\check{C}^p = \check{\eta}^{pji}\partial_j\beta_i$
sgn($\Delta$)
$\check{C}^p$
3 [$\overline{\mathbf{S}}$] Contravariant pseudovector
$\check{D} = \frac{1}{\sqrt{g}}\,\partial_p(\sqrt{g}\,\check{C}^p)$
sgn($\Delta$)
$\check{D}$
1 [$\overline{\mathbf{V}}$] Pseudoscalar
$\mathbf{b} = \beta_i\,\mathbf{g}^i$ $\check{\mathbf{C}} = \check{C}^p\,\mathbf{g}_p$
$\epsilon^{pqr}$ = permutation symbol (= Levi-Civita symbol) $\check{\eta}^{pqr} \overset{\text{def}}{=} \frac{1}{\sqrt{g}}\epsilon^{pqr}$ = Levi-Civita pseudotensor
GEN2-10; http://discretephysics.dicar.units.it

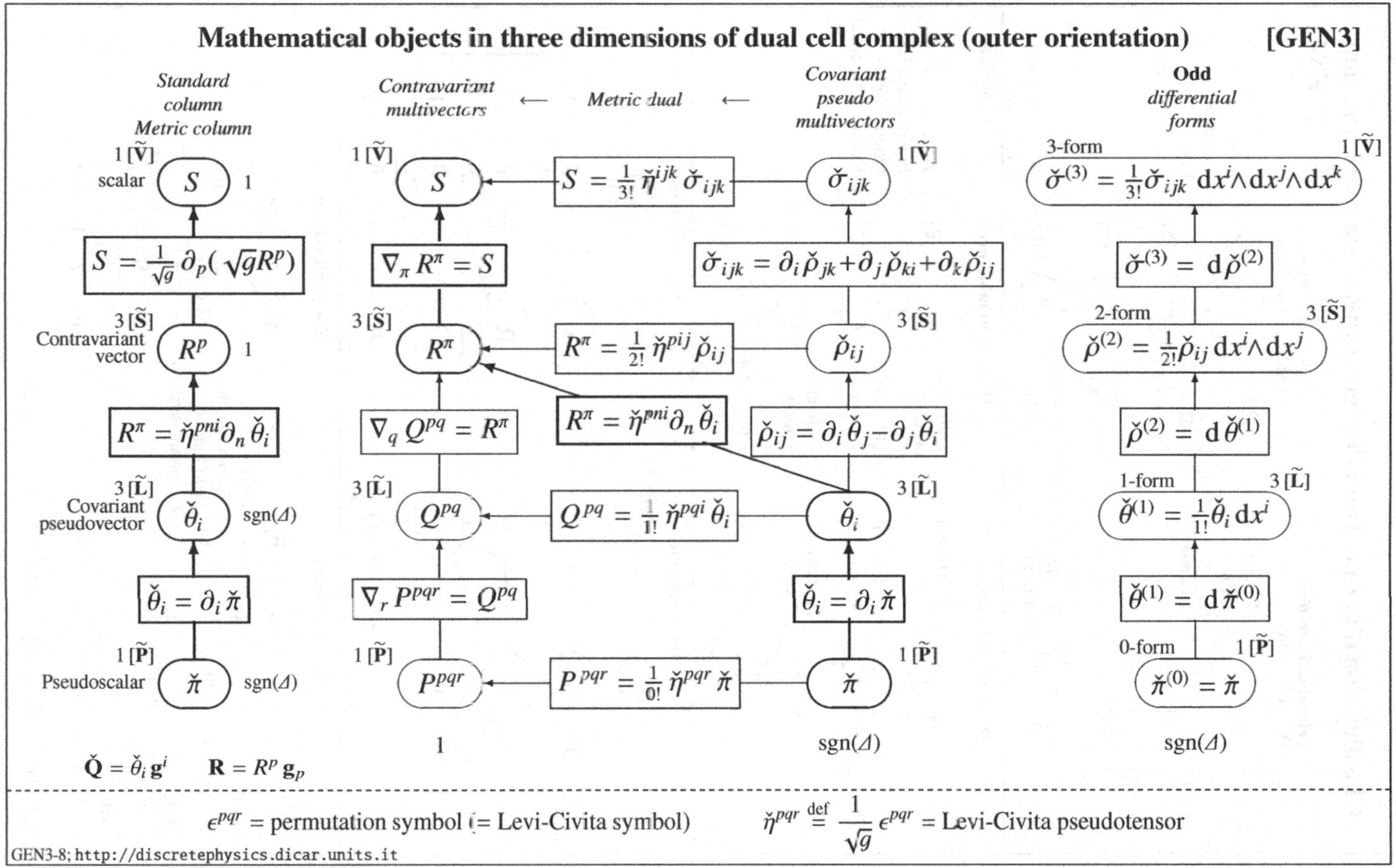
Mathematical objects in three dimensions of dual cell complex (outer orientation) [GEN3]
Standard column Metric column
Contravariant multivectors
Metric dual
Covariant pseudo multivectors
Odd differential forms
1 [$\tilde{\mathbf{V}}$] scalar $S$ 1
$S = \frac{1}{\sqrt{g}}\,\partial_p(\sqrt{g}R^p)$
3 [$\tilde{\mathbf{S}}$] Contravariant vector $R^p$ 1
$R^\pi = \check{\eta}^{pni}\partial_n\check{\theta}_i$
3 [$\tilde{\mathbf{L}}$] Covariant pseudovector $\check{\theta}_i$ sgn($\Delta$)
$\check{\theta}_i = \partial_i\check{\pi}$
1 [$\tilde{\mathbf{P}}$] Pseudoscalar $\check{\pi}$ sgn($\Delta$)
$\check{\mathbf{Q}} = \check{\theta}_i\,\mathbf{g}^i$
$\mathbf{R} = R^p\,\mathbf{g}_p$
1 [$\tilde{\mathbf{V}}$] $S$
$\nabla_\pi R^\pi = S$
3 [$\tilde{\mathbf{S}}$] $R^\pi$
$\nabla_q Q^{pq} = R^\pi$
3 [$\tilde{\mathbf{L}}$] $Q^{pq}$
$\nabla_r P^{pqr} = Q^{pq}$
1 [$\tilde{\mathbf{P}}$] $P^{pqr}$
1
$S = \frac{1}{3!}\check{\eta}^{ijk}\check{\sigma}_{ijk}$
$R^\pi = \frac{1}{2!}\check{\eta}^{pij}\check{\rho}_{ij}$
$R^\pi = \check{\eta}^{pni}\partial_n\check{\theta}_i$
$Q^{pq} = \frac{1}{1!}\check{\eta}^{pqi}\check{\theta}_i$
$P^{pqr} = \frac{1}{0!}\check{\eta}^{pqr}\check{\pi}$
$\check{\sigma}_{ijk}$ 1 [$\tilde{\mathbf{V}}$]
$\check{\sigma}_{ijk} = \partial_i\check{\rho}_{jk}+\partial_j\check{\rho}_{ki}+\partial_k\check{\rho}_{ij}$
$\check{\rho}_{ij}$ 3 [$\tilde{\mathbf{S}}$]
$\check{\rho}_{ij} = \partial_i\check{\theta}_j-\partial_j\check{\theta}_i$
$\check{\theta}_i$ 3 [$\tilde{\mathbf{L}}$]
$\check{\theta}_i = \partial_i\check{\pi}$
$\check{\pi}$ 1 [$\tilde{\mathbf{P}}$]
sgn($\Delta$)
3-form 1 [$\tilde{\mathbf{V}}$]
$\check{\sigma}^{(3)} = \frac{1}{3!}\check{\sigma}_{ijk}\,\mathrm{d}x^i\wedge\mathrm{d}x^j\wedge\mathrm{d}x^k$
$\check{\sigma}^{(3)} = \mathrm{d}\,\check{\rho}^{(2)}$
2-form 3 [$\tilde{\mathbf{S}}$]
$\check{\rho}^{(2)} = \frac{1}{2!}\check{\rho}_{ij}\,\mathrm{d}x^i\wedge\mathrm{d}x^j$
$\check{\rho}^{(2)} = \mathrm{d}\,\check{\theta}^{(1)}$
1-form 3 [$\tilde{\mathbf{L}}$]
$\check{\theta}^{(1)} = \frac{1}{1!}\check{\theta}_i\,\mathrm{d}x^i$
$\check{\theta}^{(1)} = \mathrm{d}\,\check{\pi}^{(0)}$
0-form 1 [$\tilde{\mathbf{P}}$]
$\check{\pi}^{(0)} = \check{\pi}$
sgn($\Delta$)
$\epsilon^{pqr}$ = permutation symbol (= Levi-Civita symbol)
$\check{\eta}^{pqr} \overset{\text{def}}{=} \frac{1}{\sqrt{g}}\,\epsilon^{pqr}$ = Levi-Civita pseudotensor
GEN3-8; http://discretephysics.dicar.units.it

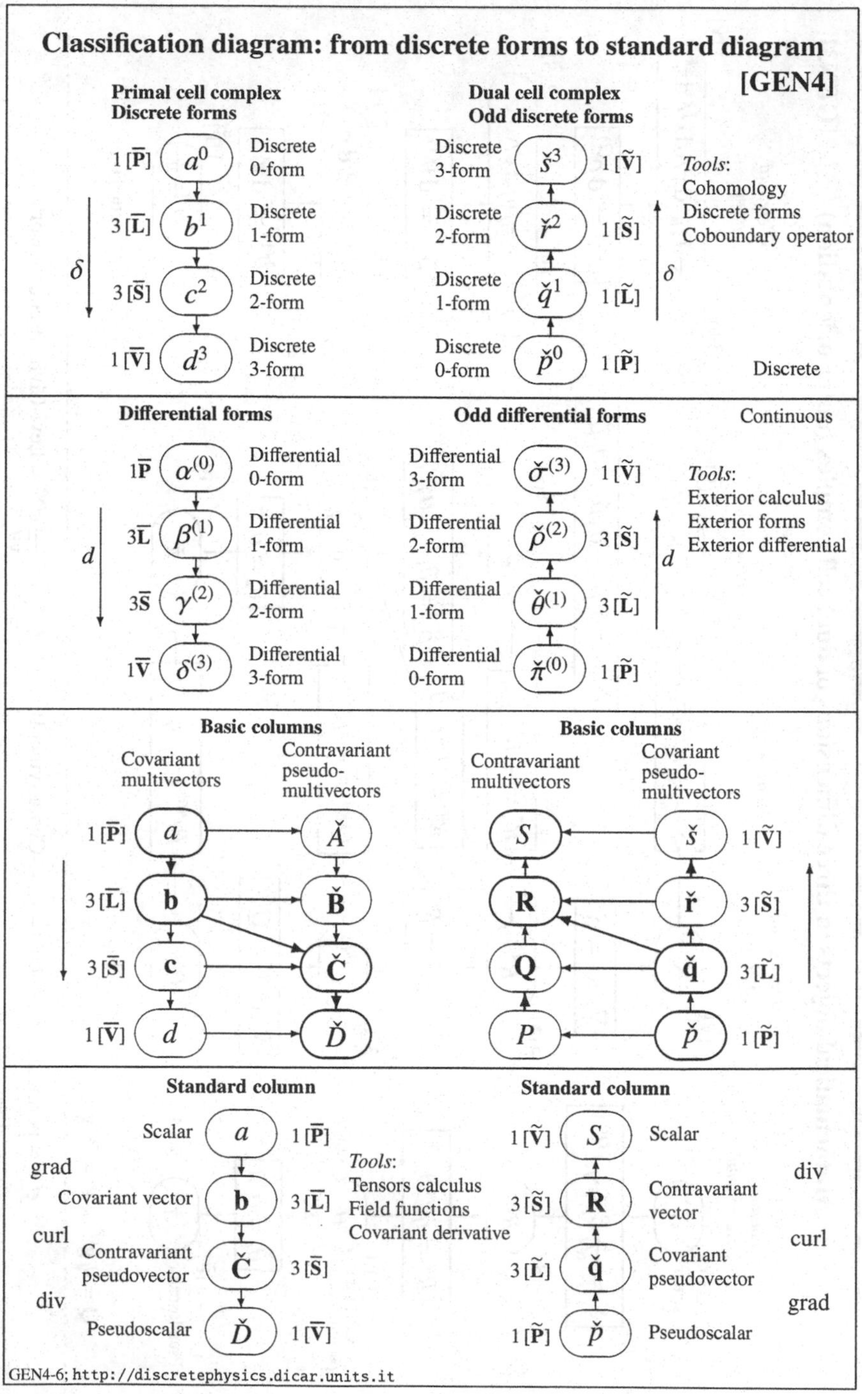
Classification diagram: from discrete forms to standard diagram
[GEN4]
Primal cell complex
Discrete forms
Dual cell complex
Odd discrete forms
1 [$\bar{\mathbf{P}}$] $a^0$ Discrete 0-form
3 [$\bar{\mathbf{L}}$] $b^1$ Discrete 1-form
3 [$\bar{\mathbf{S}}$] $c^2$ Discrete 2-form
1 [$\bar{\mathbf{V}}$] $d^3$ Discrete 3-form
$\delta$
Discrete 3-form $\check{s}^3$ 1 [$\tilde{\mathbf{V}}$]
Discrete 2-form $\check{r}^2$ 1 [$\tilde{\mathbf{S}}$]
Discrete 1-form $\check{q}^1$ 1 [$\tilde{\mathbf{L}}$]
Discrete 0-form $\check{p}^0$ 1 [$\tilde{\mathbf{P}}$]
$\delta$
Tools:
Cohomology
Discrete forms
Coboundary operator
Discrete
Differential forms
Odd differential forms
Continuous
1$\bar{\mathbf{P}}$ $\alpha^{(0)}$ Differential 0-form
3$\bar{\mathbf{L}}$ $\beta^{(1)}$ Differential 1-form
3$\bar{\mathbf{S}}$ $\gamma^{(2)}$ Differential 2-form
1$\bar{\mathbf{V}}$ $\delta^{(3)}$ Differential 3-form
$d$
Differential 3-form $\check{\sigma}^{(3)}$ 1 [$\tilde{\mathbf{V}}$]
Differential 2-form $\check{\rho}^{(2)}$ 3 [$\tilde{\mathbf{S}}$]
Differential 1-form $\check{\theta}^{(1)}$ 3 [$\tilde{\mathbf{L}}$]
Differential 0-form $\check{\pi}^{(0)}$ 1 [$\tilde{\mathbf{P}}$]
$d$
Tools:
Exterior calculus
Exterior forms
Exterior differential
Basic columns
Covariant multivectors
Contravariant pseudo-multivectors
1 [$\bar{\mathbf{P}}$] $a$ $\check{A}$
3 [$\bar{\mathbf{L}}$] $\mathbf{b}$ $\check{\mathbf{B}}$
3 [$\bar{\mathbf{S}}$] $\mathbf{c}$ $\check{\mathbf{C}}$
1 [$\bar{\mathbf{V}}$] $d$ $\check{D}$
Basic columns
Contravariant multivectors
Covariant pseudo-multivectors
$S$ $\check{s}$ 1 [$\tilde{\mathbf{V}}$]
$\mathbf{R}$ $\check{\mathbf{r}}$ 3 [$\tilde{\mathbf{S}}$]
$\mathbf{Q}$ $\check{\mathbf{q}}$ 3 [$\tilde{\mathbf{L}}$]
$P$ $\check{p}$ 1 [$\tilde{\mathbf{P}}$]
Standard column
Scalar $a$ 1 [$\bar{\mathbf{P}}$]
grad
Covariant vector $\mathbf{b}$ 3 [$\bar{\mathbf{L}}$]
curl
Contravariant pseudovector $\check{\mathbf{C}}$ 3 [$\bar{\mathbf{S}}$]
div
Pseudoscalar $\check{D}$ 1 [$\bar{\mathbf{V}}$]
Tools:
Tensors calculus
Field functions
Covariant derivative
Standard column
1 [$\tilde{\mathbf{V}}$] $S$ Scalar
div
3 [$\tilde{\mathbf{S}}$] $\mathbf{R}$ Contravariant vector
curl
3 [$\tilde{\mathbf{L}}$] $\check{\mathbf{q}}$ Covariant pseudovector
grad
1 [$\tilde{\mathbf{P}}$] $\check{p}$ Pseudoscalar
GEN4-6; http://discretephysics.dicar.units.it

**Mathematical objects in three dimensions: various notations** **[GEN5]**

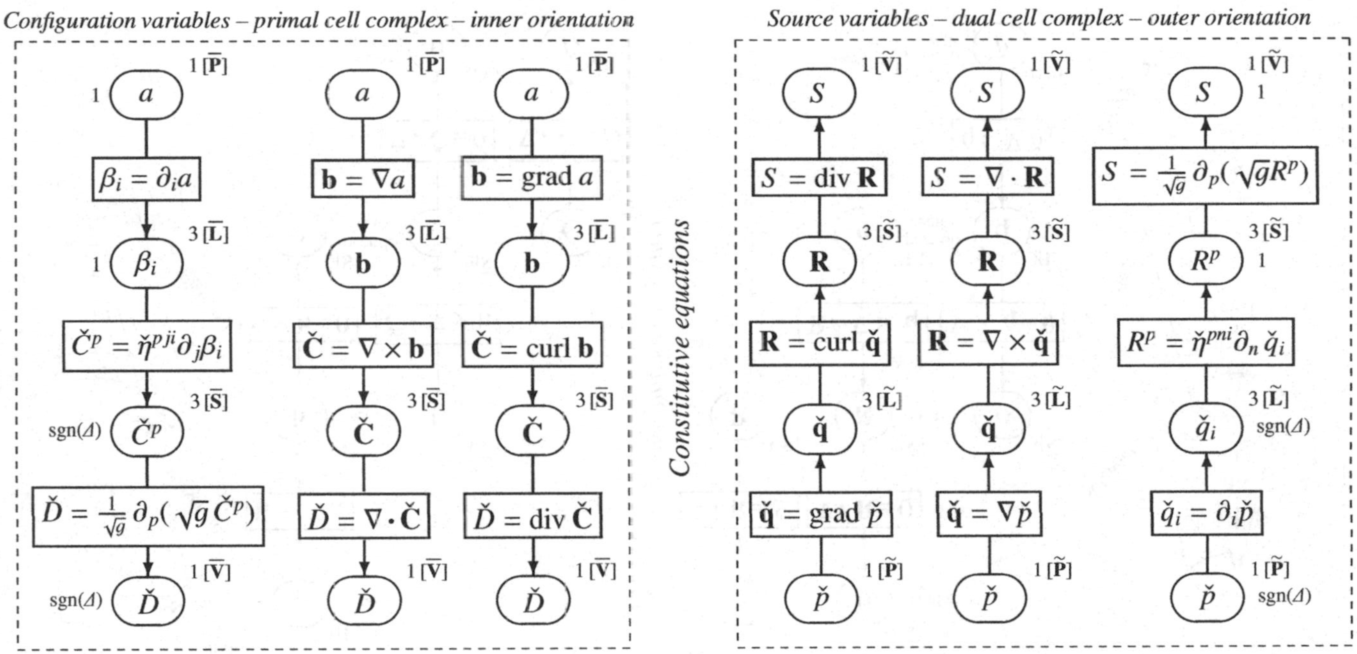


$\epsilon^{pqr}$ = permutation symbol (= Levi-Civita symbol) $\quad \check{\eta}^{pqr} \stackrel{\text{def}}{=} \frac{1}{\sqrt{g}}\epsilon^{pqr}$ = Levi-Civita pseudotensor

GEN5-6; http://discretephysics.dicar.units.it

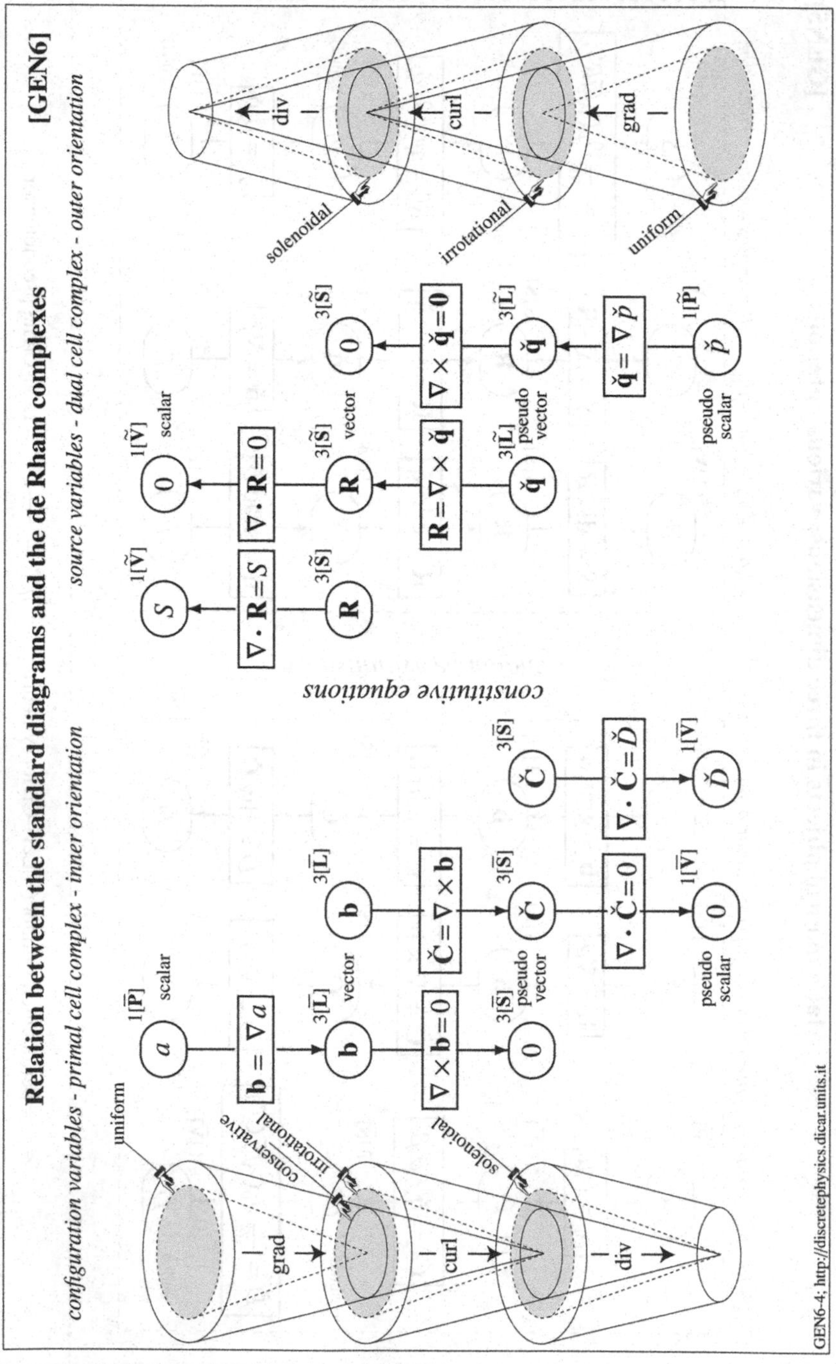
Relation between the standard diagrams and the de Rham complexes
[GEN6]
configuration variables - primal cell complex - inner orientation
source variables - dual cell complex - outer orientation
constitutive equations
uniform
conservative
irrotational
solenoidal
grad
curl
div
scalar
vector
pseudo vector
pseudo scalar
$\mathbf{b} = \nabla a$
$\nabla \times \mathbf{b} = 0$
$\check{\mathbf{C}} = \nabla \times \mathbf{b}$
$\nabla \cdot \check{\mathbf{C}} = 0$
$\nabla \cdot \check{\mathbf{C}} = \check{D}$
$\nabla \cdot \mathbf{R} = S$
$\nabla \cdot \mathbf{R} = 0$
$\mathbf{R} = \nabla \times \check{\mathbf{q}}$
$\nabla \times \check{\mathbf{q}} = 0$
$\check{\mathbf{q}} = \nabla \check{p}$
GEN6-4; http://discretephysics.dicar.units.it

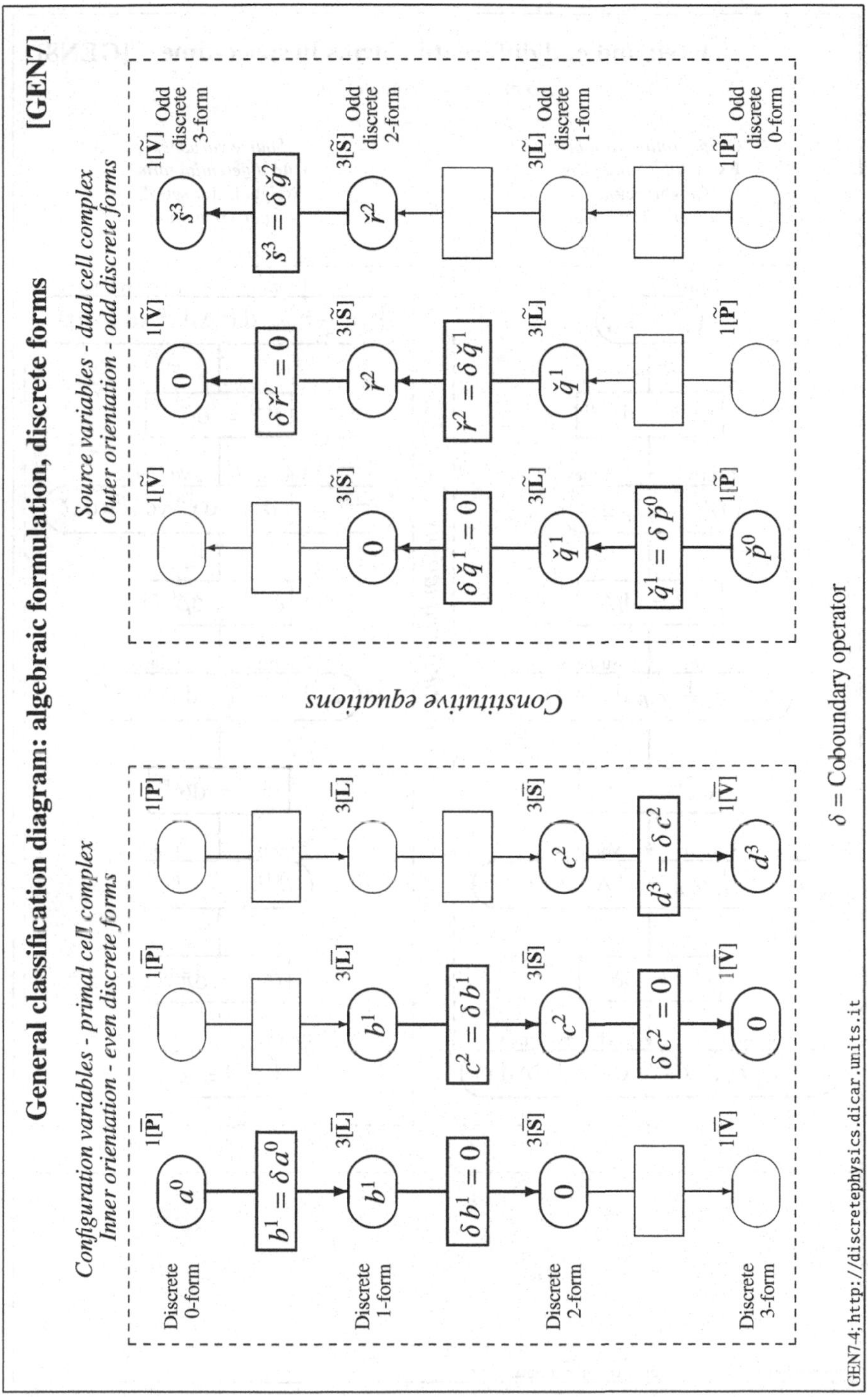
General classification diagram: algebraic formulation, discrete forms
[GEN7]
Configuration variables - primal cell complex
Inner orientation - even discrete forms
Source variables - dual cell complex
Outer orientation - odd discrete forms
Constitutive equations
Discrete 0-form
Discrete 1-form
Discrete 2-form
Discrete 3-form
Odd discrete 3-form
Odd discrete 2-form
Odd discrete 1-form
Odd discrete 0-form
$b^1 = \delta a^0$
$\delta b^1 = 0$
$c^2 = \delta b^1$
$\delta c^2 = 0$
$d^3 = \delta c^2$
$\check{q}^1 = \delta \check{p}^0$
$\delta \check{q}^1 = 0$
$\check{r}^2 = \delta \check{q}^1$
$\delta \check{r}^2 = 0$
$\check{s}^3 = \delta \check{g}^2$
$\delta$ = Coboundary operator
GEN7-4; http://discretephysics.dicar.units.it

## Even and odd differential forms in space-time [GEN8]

*No metric, no connection*

*Configuration variables*
**Even** *differential forms*
Weight factor: 1
*Inner orientation*

0-form — Points

$$a^{(0)} = a$$

$$\beta^{(1)} = \mathrm{d}a^{(0)}$$

1-form — Lines

$$\beta^{(1)} = \beta_\alpha \ \mathrm{d}x^\alpha$$

$$\gamma^{(2)} = \mathrm{d}\beta^{(1)}$$

2-form — Surfaces

$$\gamma^{(2)} = \frac{1}{2!}\,\gamma_{\alpha\beta}\ \mathrm{d}x^\alpha\wedge\mathrm{d}x^\beta$$

$$\delta^{(3)} = \mathrm{d}\gamma^{(2)}$$

3-form — Volumes

$$\delta^{(3)} = \frac{1}{3!}\,\delta_{\alpha\beta\gamma}\ \mathrm{d}x^\alpha\wedge\mathrm{d}x^\beta\wedge\mathrm{d}x^\gamma$$

$$\epsilon^{(4)} = \mathrm{d}\delta^{(3)}$$

4-form — Hypervolumes

$$\epsilon^{(4)} = \frac{1}{4!}\,\epsilon_{\alpha\beta\gamma\delta}\,\mathrm{d}x^\alpha\wedge\mathrm{d}x^\beta\wedge\mathrm{d}x^\gamma\wedge\mathrm{d}x^\delta$$

*Constitutive equations*

*Source variables*
**Odd** *differential forms*
Weight factor sgn($\Delta$)
*Outer orientation*

4-form — Hypervolumes

$$\check{\tau}^{(4)} = \frac{1}{4!}\,\check{\tau}_{\alpha\beta\gamma\delta}\,\mathrm{d}x^\alpha\wedge\mathrm{d}x^\beta\wedge\mathrm{d}x^\gamma\wedge\mathrm{d}x^\delta$$

$$\check{\tau}^{(4)} = \mathrm{d}\check{\sigma}^{(3)}$$

3-form — Volumes

$$\check{\sigma}^{(3)} = \frac{1}{3!}\,\check{\sigma}_{\alpha\beta\gamma}\ \mathrm{d}x^\alpha\wedge\mathrm{d}x^\beta\wedge\mathrm{d}x^\gamma$$

$$\check{\sigma}^{(3)} = \mathrm{d}\check{\rho}^{(2)}$$

2-form — Surfaces

$$\check{\rho}^{(2)} = \frac{1}{2!}\,\check{\rho}_{\alpha\beta}\ \mathrm{d}x^\alpha\wedge\mathrm{d}x^\beta$$

$$\check{\rho}^{(2)} = \mathrm{d}\check{\theta}^{(1)}$$

1-form — Lines

$$\check{\theta}^{(1)} = \check{\theta}_\alpha\ \mathrm{d}x^\alpha$$

$$\check{\theta}^{(1)} = \mathrm{d}\check{\pi}^{(0)}$$

0-form — Points

$$\check{\pi}^{(0)} = \check{\pi}$$

GEN8-4; http://discretephysics.dicar.units.it

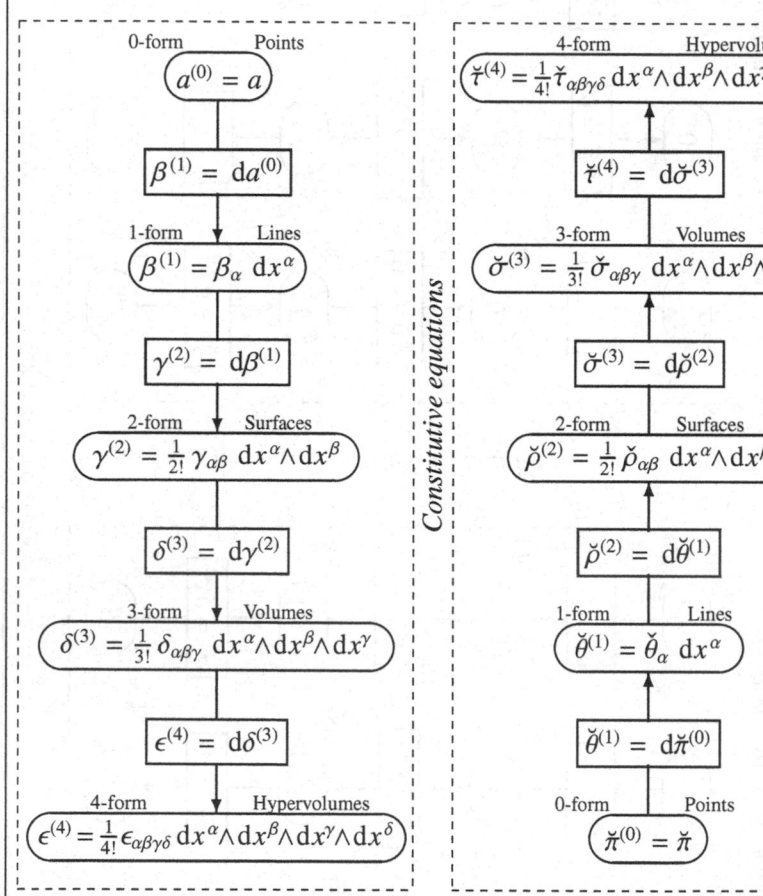

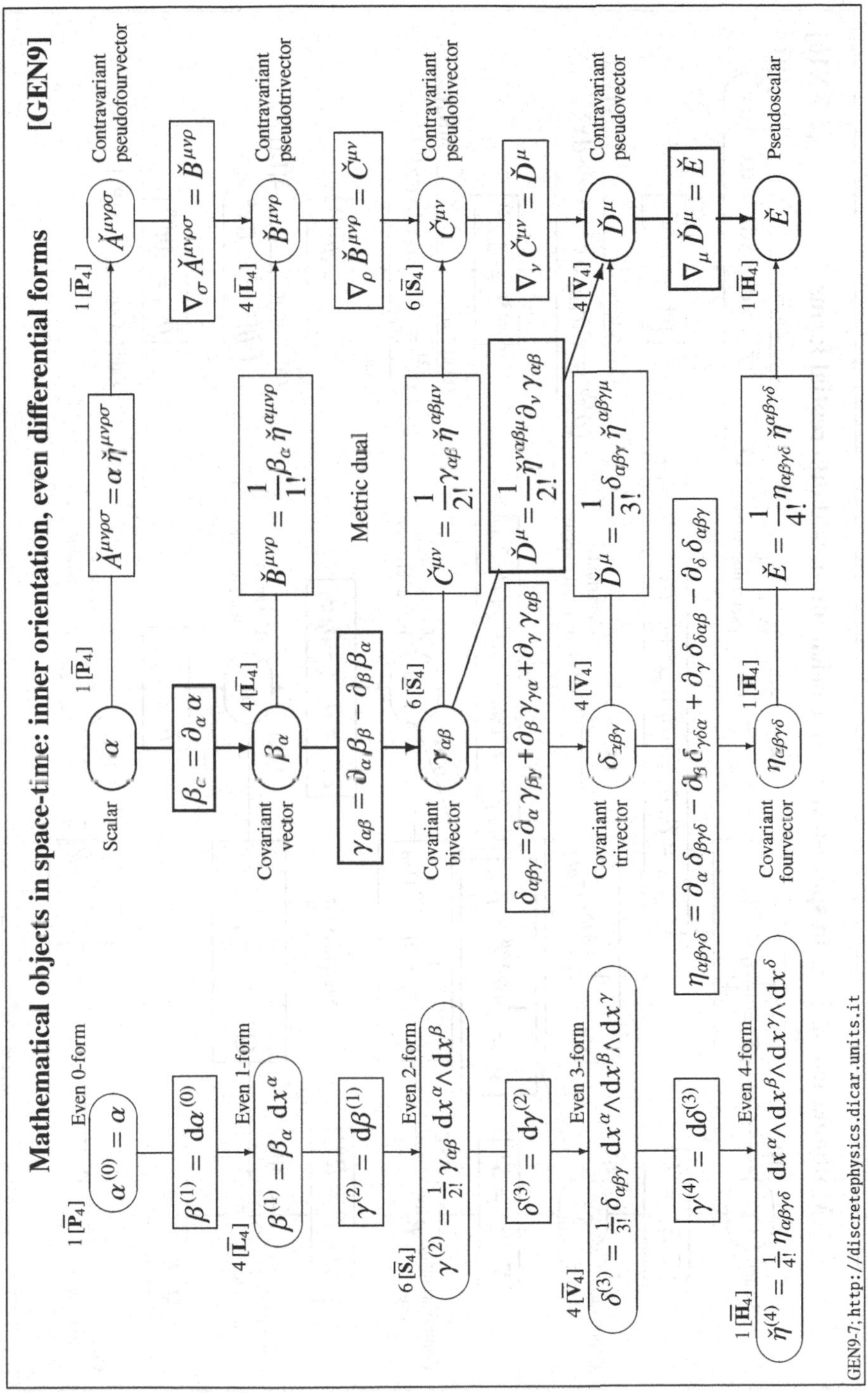
Mathematical objects in space-time: inner orientation, even differential forms
[GEN9]
Even 0-form
1 $[\overline{\mathbf{P}}_4]$
$\alpha^{(0)} = \alpha$
$\beta^{(1)} = \mathrm{d}\alpha^{(0)}$
Even 1-form
4 $[\overline{\mathbf{L}}_4]$
$\beta^{(1)} = \beta_\alpha\, \mathrm{d}x^\alpha$
$\gamma^{(2)} = \mathrm{d}\beta^{(1)}$
Even 2-form
6 $[\overline{\mathbf{S}}_4]$
$\gamma^{(2)} = \frac{1}{2!}\gamma_{\alpha\beta}\, \mathrm{d}x^\alpha \wedge \mathrm{d}x^\beta$
$\delta^{(3)} = \mathrm{d}\gamma^{(2)}$
Even 3-form
4 $[\overline{\mathbf{V}}_4]$
$\delta^{(3)} = \frac{1}{3!}\delta_{\alpha\beta\gamma}\, \mathrm{d}x^\alpha \wedge \mathrm{d}x^\beta \wedge \mathrm{d}x^\gamma$
$\gamma^{(4)} = \mathrm{d}\delta^{(3)}$
Even 4-form
1 $[\overline{\mathbf{H}}_4]$
$\breve{\eta}^{(4)} = \frac{1}{4!}\eta_{\alpha\beta\gamma\delta}\, \mathrm{d}x^\alpha \wedge \mathrm{d}x^\beta \wedge \mathrm{d}x^\gamma \wedge \mathrm{d}x^\delta$
Scalar
1 $[\overline{\mathbf{P}}_4]$
$\alpha$
$\beta_c = \partial_\alpha \alpha$
Covariant vector
4 $[\overline{\mathbf{L}}_4]$
$\beta_\alpha$
$\gamma_{\alpha\beta} = \partial_\alpha \beta_\beta - \partial_\beta \beta_\alpha$
Covariant bivector
6 $[\overline{\mathbf{S}}_4]$
$\gamma_{\alpha\beta}$
$\delta_{\alpha\beta\gamma} = \partial_\alpha \gamma_{\beta\gamma} + \partial_\beta \gamma_{\gamma\alpha} + \partial_\gamma \gamma_{\alpha\beta}$
Covariant trivector
4 $[\overline{\mathbf{V}}_4]$
$\delta_{\alpha\beta\gamma}$
$\eta_{\alpha\beta\gamma\delta} = \partial_\alpha \delta_{\beta\gamma\delta} - \partial_\beta \delta_{\gamma\delta\alpha} + \partial_\gamma \delta_{\delta\alpha\beta} - \partial_\delta \delta_{\alpha\beta\gamma}$
Covariant fourvector
1 $[\overline{\mathbf{H}}_4]$
$\eta_{\alpha\beta\gamma\delta}$
$\breve{A}^{\mu\nu\rho\sigma} = \alpha\, \breve{\eta}^{\mu\nu\rho\sigma}$
$\breve{B}^{\mu\nu\rho} = \frac{1}{1!}\beta_\alpha\, \breve{\eta}^{\alpha\mu\nu\rho}$
Metric dual
$\breve{C}^{\mu\nu} = \frac{1}{2!}\gamma_{\alpha\beta}\, \breve{\eta}^{\alpha\beta\mu\nu}$
$\breve{D}^{\mu} = \frac{1}{2!}\breve{\eta}^{\nu\alpha\beta\mu}\partial_\nu \gamma_{\alpha\beta}$
$\breve{D}^{\mu} = \frac{1}{3!}\delta_{\alpha\beta\gamma}\, \breve{\eta}^{\alpha\beta\gamma\mu}$
$\breve{E} = \frac{1}{4!}\eta_{\alpha\beta\gamma\delta}\, \breve{\eta}^{\alpha\beta\gamma\delta}$
1 $[\overline{\mathbf{P}}_4]$
$\breve{A}^{\mu\nu\rho\sigma}$
Contravariant pseudofourvector
$\nabla_\sigma \breve{A}^{\mu\nu\rho\sigma} = \breve{B}^{\mu\nu\rho}$
4 $[\overline{\mathbf{L}}_4]$
$\breve{B}^{\mu\nu\rho}$
Contravariant pseudotrivector
$\nabla_\rho \breve{B}^{\mu\nu\rho} = \breve{C}^{\mu\nu}$
6 $[\overline{\mathbf{S}}_4]$
$\breve{C}^{\mu\nu}$
Contravariant pseudobivector
$\nabla_\nu \breve{C}^{\mu\nu} = \breve{D}^{\mu}$
4 $[\overline{\mathbf{V}}_4]$
$\breve{D}^{\mu}$
Contravariant pseudovector
$\nabla_\mu \breve{D}^{\mu} = \breve{E}$
1 $[\overline{\mathbf{H}}_4]$
$\breve{E}$
Pseudoscalar
GEN9-7; http://discretephysics.dicar.units.it

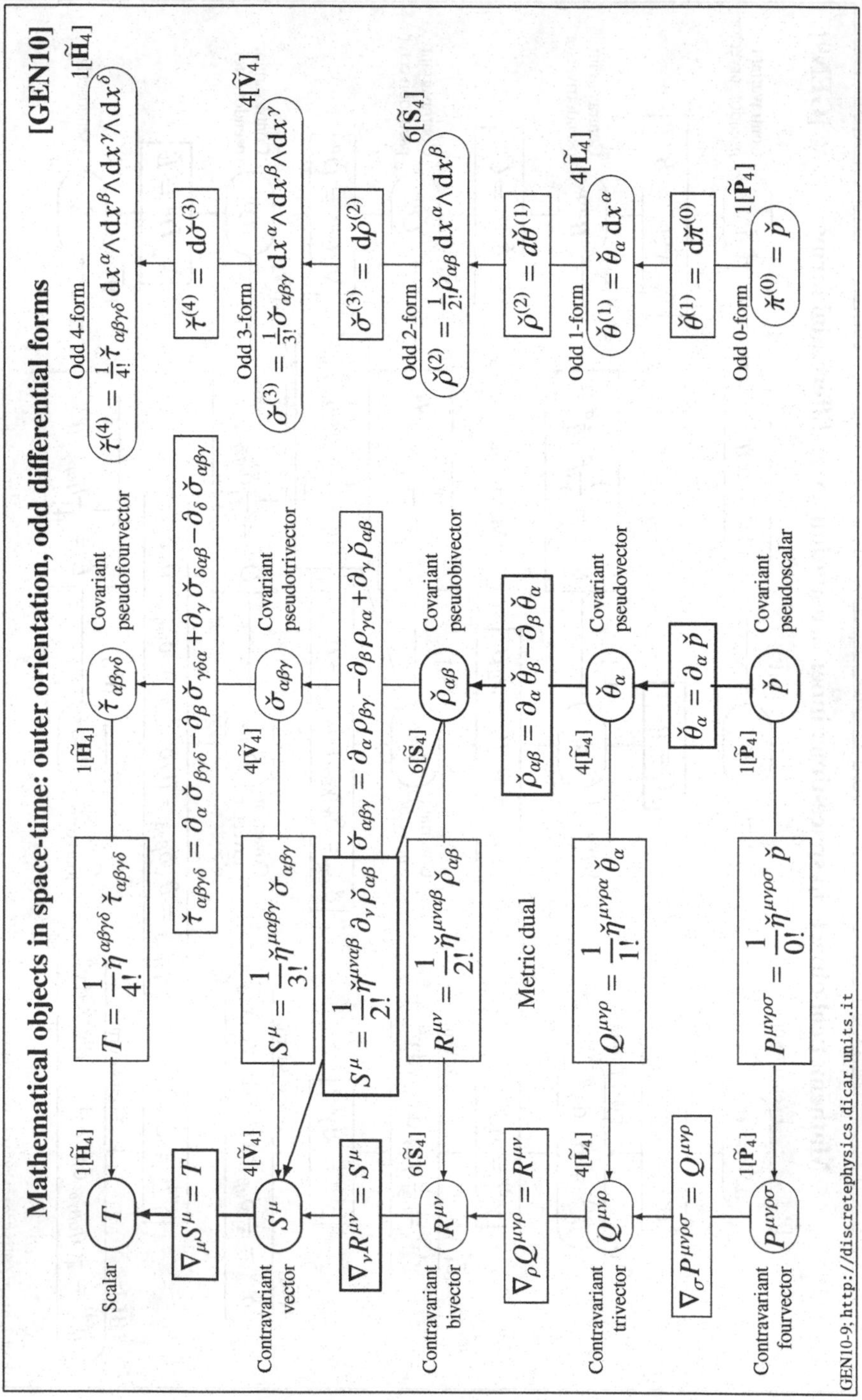

Mathematical objects in space-time: outer orientation, odd differential forms
[GEN10]
Scalar
$T$
1[$\tilde{\mathbf{H}}_4$]
$\nabla_\mu S^\mu = T$
Contravariant vector
$S^\mu$
4[$\tilde{\mathbf{V}}_4$]
$\nabla_\nu R^{\mu\nu} = S^\mu$
Contravariant bivector
$R^{\mu\nu}$
6[$\tilde{\mathbf{S}}_4$]
$\nabla_\rho Q^{\mu\nu\rho} = R^{\mu\nu}$
Contravariant trivector
$Q^{\mu\nu\rho}$
4[$\tilde{\mathbf{L}}_4$]
$\nabla_\sigma P^{\mu\nu\rho\sigma} = Q^{\mu\nu\rho}$
Contravariant fourvector
$P^{\mu\nu\rho\sigma}$
1[$\tilde{\mathbf{P}}_4$]
$T = \frac{1}{4!}\breve{\eta}^{\alpha\beta\gamma\delta}\,\breve{\tau}_{\alpha\beta\gamma\delta}$
$S^\mu = \frac{1}{3!}\breve{\eta}^{\mu\alpha\beta\gamma}\,\breve{\sigma}_{\alpha\beta\gamma}$
$S^\mu = \frac{1}{2!}\breve{\eta}^{\mu\nu\alpha\beta}\,\partial_\nu\breve{\rho}_{\alpha\beta}$
$R^{\mu\nu} = \frac{1}{2!}\breve{\eta}^{\mu\nu\alpha\beta}\,\breve{\rho}_{\alpha\beta}$
Metric dual
$Q^{\mu\nu\rho} = \frac{1}{1!}\breve{\eta}^{\mu\nu\rho\alpha}\,\breve{\theta}_\alpha$
$P^{\mu\nu\rho\sigma} = \frac{1}{0!}\breve{\eta}^{\mu\nu\rho\sigma}\,\breve{p}$
1[$\tilde{\mathbf{H}}_4$]
$\breve{\tau}_{\alpha\beta\gamma\delta}$
Covariant pseudofourvector
$\breve{\tau}_{\alpha\beta\gamma\delta} = \partial_\alpha\breve{\sigma}_{\beta\gamma\delta} - \partial_\beta\breve{\sigma}_{\gamma\delta\alpha} + \partial_\gamma\breve{\sigma}_{\delta\alpha\beta} - \partial_\delta\breve{\sigma}_{\alpha\beta\gamma}$
4[$\tilde{\mathbf{V}}_4$]
$\breve{\sigma}_{\alpha\beta\gamma}$
Covariant pseudotrivector
$\breve{\sigma}_{\alpha\beta\gamma} = \partial_\alpha\rho_{\beta\gamma} - \partial_\beta\rho_{\gamma\alpha} + \partial_\gamma\breve{\rho}_{\alpha\beta}$
6[$\tilde{\mathbf{S}}_4$]
$\breve{\rho}_{\alpha\beta}$
Covariant pseudobivector
$\breve{\rho}_{\alpha\beta} = \partial_\alpha\breve{\theta}_\beta - \partial_\beta\breve{\theta}_\alpha$
4[$\tilde{\mathbf{L}}_4$]
$\breve{\theta}_\alpha$
Covariant pseudovector
$\breve{\theta}_\alpha = \partial_\alpha\breve{p}$
1[$\tilde{\mathbf{P}}_4$]
$\breve{p}$
Covariant pseudoscalar
Odd 4-form
$\breve{\tau}^{(4)} = \frac{1}{4!}\breve{\tau}_{\alpha\beta\gamma\delta}\,dx^\alpha\wedge dx^\beta\wedge dx^\gamma\wedge dx^\delta$
1[$\tilde{\mathbf{H}}_4$]
$\breve{\tau}^{(4)} = d\breve{\sigma}^{(3)}$
Odd 3-form
$\breve{\sigma}^{(3)} = \frac{1}{3!}\breve{\sigma}_{\alpha\beta\gamma}\,dx^\alpha\wedge dx^\beta\wedge dx^\gamma$
4[$\tilde{\mathbf{V}}_4$]
$\breve{\sigma}^{(3)} = d\breve{\rho}^{(2)}$
Odd 2-form
$\breve{\rho}^{(2)} = \frac{1}{2!}\breve{\rho}_{\alpha\beta}\,dx^\alpha\wedge dx^\beta$
6[$\tilde{\mathbf{S}}_4$]
$\breve{\rho}^{(2)} = d\breve{\theta}^{(1)}$
Odd 1-form
$\breve{\theta}^{(1)} = \breve{\theta}_\alpha\,dx^\alpha$
4[$\tilde{\mathbf{L}}_4$]
$\breve{\theta}^{(1)} = d\breve{\pi}^{(0)}$
Odd 0-form
$\breve{\pi}^{(0)} = \breve{p}$
1[$\tilde{\mathbf{P}}_4$]
GEN10-9; http://discretephysics.dicar.units.it

**Space-time diagram**                                    **[GEN11]**

*Tensorial notation*

| | |
|---|---|
| *Configuration variables*<br>*Primal space-time complex*<br>*Inner orientation* | *Source variables*<br>*Dual space-time complex*<br>*Outer orientation* |

$1\,[\overline{\mathbf{P}}_4]$   $\alpha$   Scalar

1

$$\beta_\alpha = \partial_\alpha\,\alpha$$

$4\,[\overline{\mathbf{L}}_4]$   $\beta_\alpha$   Covariant vector

1

$$\gamma_{\alpha\beta} = \partial_\alpha\beta_\beta - \partial_\beta\beta_\alpha$$

$6\,[\overline{\mathbf{S}}_4]$   $\gamma_{\alpha\beta}$   Covariant bivector

1

$$\frac{1}{2!}\,\breve{\eta}^{\gamma\alpha\beta\mu}\,\partial_\gamma\,\gamma_{\alpha\beta} = \breve{D}^\mu$$

$4\,[\overline{\mathbf{V}}_4]$   $\breve{D}^\mu$   Contravariant pseudovector

sgn($\Delta$)

$$\frac{1}{\sqrt{g}}\,\partial_\mu(\sqrt{g}\,\breve{D}^\mu) = \breve{E}$$

$1\,[\overline{\mathbf{H}}_4]$   $\breve{E}$   Pseudoscalar

sgn($\Delta$)

*Constitutive equations*

Scalar   $T$   $1\,[\widetilde{\mathbf{H}}_4]$

1

$$\frac{1}{\sqrt{g}}\,\partial_\nu(\sqrt{g}\,S^\mu) = T$$

Contravariant vector   $S^\mu$   $4\,[\widetilde{\mathbf{V}}_4]$

1

$$\frac{1}{2!}\,\breve{\eta}^{\mu\nu\alpha\beta}\,\partial_\nu\,\breve{\rho}_{\alpha\beta} = S^\mu$$

Convariant pseudobivector   $\breve{\rho}_{\alpha\beta}$   $6\,[\widetilde{\mathbf{S}}_4]$

sgn($\Delta$)

$$\breve{\rho}_{\alpha\beta} = \partial_\alpha\,\breve{\theta}_\beta - \partial_\beta\,\breve{\theta}_\alpha$$

Covariant pseudovector   $\breve{\theta}_\sigma$   $4\,[\widetilde{\mathbf{L}}_4]$

sgn($\Delta$)

$$\breve{\theta}_\sigma = \partial_\sigma\breve{\pi}$$

Pseudoscalar   $\breve{\pi}$   $1\,[\widetilde{\mathbf{P}}_4]$

sgn($\Delta$)

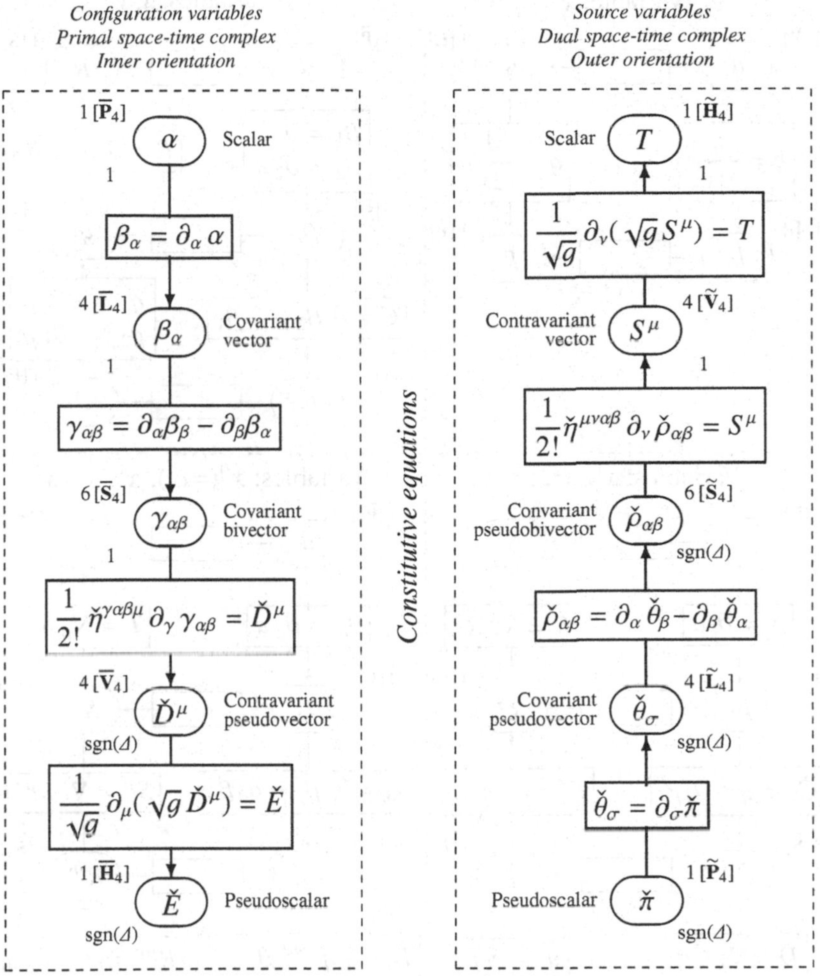

The prototype of this diagram is that of electromagnetism; see diagram [ELE6].
It is also the diagram of the Klein–Gordon field for particles of spin 0; see diagram [QME3]
and of Proca equation for particles of spin 1; see diagram [QME4].

$x^0 = c\,t$      Greek index runs from 0 to 3      $\mathrm{d}s^2 = g_{\alpha\beta}\,\mathrm{d}x^\alpha\,\mathrm{d}x^\beta$

**Standard diagrams for** $n = 1, 2, 3, 4$      **[GEN12]**
*Tensorial notation*

Variable: $x$

1 $[\overline{\mathbf{P}}]$    $a$      $q$    1 $[\widetilde{\mathbf{L}}]$

$b = \dfrac{\mathrm{d}}{\mathrm{d}x}a$      $q = \dfrac{\mathrm{d}}{\mathrm{d}x}p$

1 $[\overline{\mathbf{L}}]$    $b$      $p$    1 $[\widetilde{\mathbf{P}}]$

Variables: $x, y$

1 $[\overline{\mathbf{P}}]$    $a$      $R$    1 $[\widetilde{\mathbf{S}}]$

$\begin{aligned}\beta_1 &= \partial_1 a\\ \beta_2 &= \partial_2 a\end{aligned}$      $R = \partial_1 q_1 + \partial_2 q_2$

2 $[\mathbf{L}]$    $\beta_i$      $q_i$    2 $[\widetilde{\mathbf{L}}]$

$C = \partial_1 \beta_2 - \partial_2 \beta_1$      $\begin{aligned}q_1 &= +\partial_2 p\\ q_2 &= -\partial_1 p\end{aligned}$

1 $[\overline{\mathbf{S}}]$    $C$      $p$    1 $[\widetilde{\mathbf{P}}]$

Variables: $x^1, x^2, x^3$

1 $[\overline{\mathbf{P}}]$    $a$      $S$    1 $[\widetilde{\mathbf{V}}]$

$\beta_i = \partial_i a$      $S = \nabla_h R^h$

3 $[\overline{\mathbf{L}}]$    $\beta_i$      $R^h$    3 $[\widetilde{\mathbf{S}}]$

$\check{C}^p = \check{\eta}^{pji}\,\partial_j \beta_i$      $R^h = \check{\eta}^{hji}\,\partial_j \check{q}_i$

3 $[\overline{\mathbf{S}}]$    $\check{C}^p$      $\check{q}_i$    3 $[\widetilde{\mathbf{L}}]$

$\check{D} = \nabla_p \check{C}^p$      $\check{q}_i = \partial_i \check{p}$

1 $[\overline{\mathbf{V}}]$    $\check{D}$      $\check{p}$    1 $[\widetilde{\mathbf{P}}]$

Variables: $x^0 (=ct),\ x^1,\ x^2,\ x^3$

1 $[\overline{\mathbf{P}}_4]$    $a$      $T$    1 $[\widetilde{\mathbf{H}}_4]$

$\beta_\alpha = \partial_\alpha a$      $T = \nabla_\mu S^\mu$

4 $[\overline{\mathbf{L}}_4]$    $\beta_\alpha$      $S^\mu$    4 $[\widetilde{\mathbf{V}}_4]$

$\gamma_{\alpha\beta} = \partial_\alpha \beta_\beta - \partial_\beta \beta_\alpha$      $S^\mu = \nabla_\nu R^{\mu\nu}$

6 $[\overline{\mathbf{S}}_4]$    $\gamma_{\alpha\beta}$      $R^{\mu\nu}$    6 $[\widetilde{\mathbf{S}}_4]$

$\check{D}^\mu = \tfrac{1}{2!}\,\check{\eta}^{\mu\gamma\alpha\beta}\,\partial_\gamma \gamma_{\alpha\beta}$      $R^{\mu\nu} = \check{\eta}^{\mu\nu\beta\alpha}\,\partial_\beta \check{q}_\alpha$

4 $[\overline{\mathbf{V}}_4]$    $\check{D}^\mu$      $\check{q}_\alpha$    4 $[\widetilde{\mathbf{L}}_4]$

$\check{E} = \nabla_\mu \check{D}^\mu$      $\check{q}_\alpha = \partial_\alpha \check{p}$

1 $[\overline{\mathbf{H}}_4]$    $\check{E}$      $\check{p}$    1 $[\widetilde{\mathbf{P}}_4]$

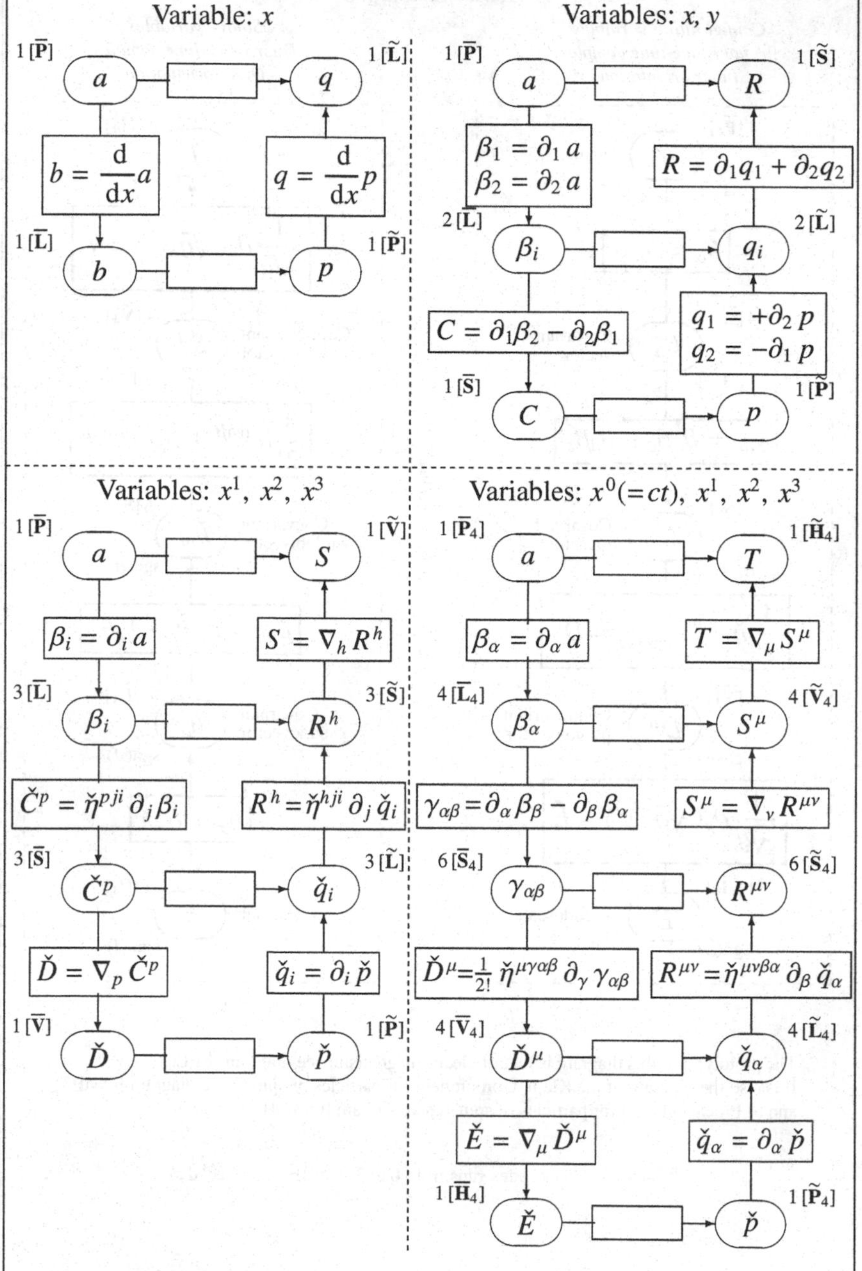

GEN12-7; http://discretephysics.dicar.units.it

# Chapter 15
# The Mathematical Structure

## 15.1 Introduction

In this chapter, we will show that for a mathematical description of physics it is much better to start with algebra than with differential calculus. An algebraic description has some advantages over the differential formulation:

1. The first advantage is that the algebraic description is enriched by the geometric interpretation given by algebraic topology. The analogous description in the differential formulation is that of exterior differential forms, which add a geometric description to the traditional differential formulation.
2. A second advantage of an algebraic description is that it involves using global variables, which are often the ones we measure directly.
3. A third advantage is that the global variables are not subjected to the regularity requirements which the differential formulation imposes, such as the derivability of field functions. For example, temperature is continuous across the separation surface of two material media, but its derivatives are not continuous. Hence, the differential formulation cannot be applied to such a separation surface where we must introduce the 'jump conditions' which are expressed by algebraic equations.
4. A fourth advantage is that an algebraic formulation leads directly to a numerical formulation of physical problems. In contrast, to apply a numerical analysis to differential equations, we must discretize these equations, i.e. come back to algebraic equations.

We want to show that it is more natural to describe physics starting from geometric aspects, in particular from topological aspects, and then translate them into algebraic terms and, finally, into differential terms, instead of proceeding along the traditional road (Fig. 15.1).

Up to now the classification diagram has been made up of rounded boxes containing physical variables and of rectangular boxes containing the equations

E. Tonti, *The Mathematical Structure of Classical and Relativistic Physics*, Modeling and Simulation in Science, Engineering and Technology,
DOI 10.1007/978-1-4614-7422-7_15, 

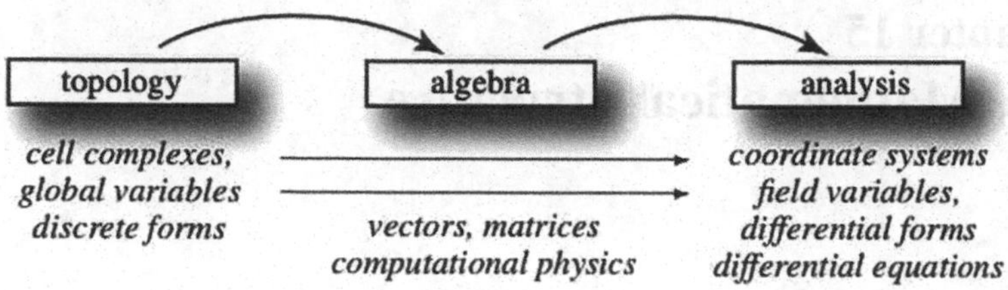


**Fig. 15.1** Three stages in description of physical theories

linking them. Now we can go a step further and consider each rounded box as a vector space; it can be a finite-dimensional vector space or a function space, depending on whether we use global variables or field functions. It follows that when global variables are used, each rectangular box contains an algebraic operator, usually a matrix, while when field functions are used, each box contains a differential operator. The use of vector spaces and operators makes it possible to build up a mathematical model that can be applied to various physical theories.[1]

This mathematical model is made possible by the fact that despite the different physical meaning of the variables which reside in the corresponding boxes of two theories, despite their different tensorial natures, and hence the different number of components, the equations that connect the vertical boxes have the same mathematical structure. In fact, we will show that the operators in the left and right columns that lie on the same level are one adjoint of the other and that the constitutive equations, when they are linear, are formed by symmetric operators. The adjointness of the operators on the two sides and the symmetry of the constitutive operator represent the necessary and sufficient condition for a variational formulation of the fundamental problem. Herein resides the root of the Lagrangians, which dominates all physical theories dealing with conservative phenomena.

To start the construction of the model, we must introduce some mathematical tools, such as the notion of pairs of vector spaces placed in duality by nondegenerate bilinear forms and the notion of adjoint operators. To this we add the link between a purely algebraic description and an algebraic topological description using the notions of discrete forms and coboundary operator. As we will see, the rudiments of the cell complexes presented in Chap. 4 and the notions of algebraic topology presented in Chap. 7 are enough to explain the reason for the unitary structure underlying the theories of physics.

---

[1] To make clear when the rounded boxes are considered vector spaces, we draw them as circles, as shown in Table 15.5 (p. 441).

### 15.1.1 Discovery of Adjointness

Let us consider two adjacent rooms (Fig. 15.2) within which there is heat generation through radiators. Suppose the ceiling, the floor and the side walls are insulated so that heat can only pass horizontally: the two extreme walls are in contact with two thermostats whose temperatures are 20° on the left and 5° on the right. Let us denote by $P_1$ and $P_2$ the heat generation rates (in watt), which we assume are assigned, and by $T_1$ and $T_2$ the unknown average temperatures (in degrees Celsius) measured at the centres of the two rooms.

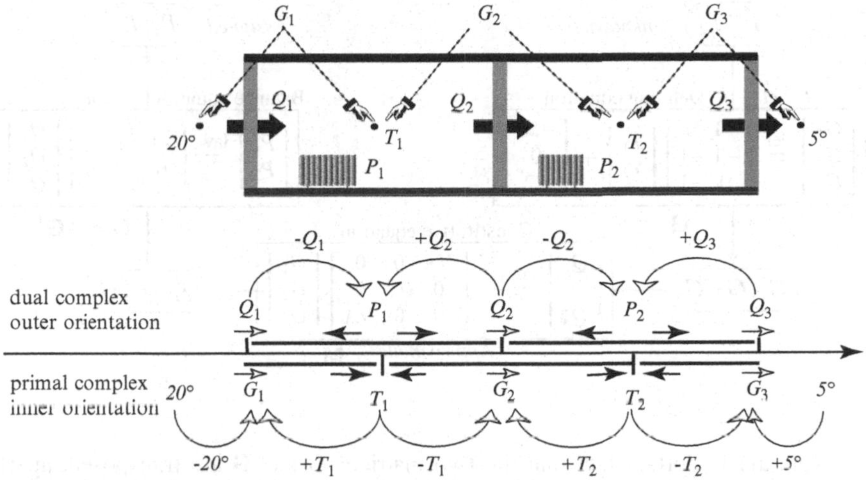


**Fig. 15.2** Heat flows horizontally only. The defining equations and the equation of balance perform the coboundary process on a one-dimensional cell complex and its dual

We number 1, 2 and 3 the three transverse walls crossed by the heat currents $Q_1, Q_2, Q_3$ (in watt) and $G_1, G_2, G_3$ the temperature differences across the walls. Let us denote by $U$ the thermal transmittance of the walls and by $A$ the area of the walls. We can write the three relations as follows:

$$\overbrace{\begin{cases} G_1 = T_1 - 20 \\ G_2 = T_2 - T_1 \\ G_3 = 5 - T_2 \end{cases}}^{definition}, \quad \overbrace{\begin{cases} Q_1 = -U A G_1 \\ Q_2 = -U A G_2 \\ Q_3 = -U A G_3 \end{cases}}^{constitutive\ equation}, \quad \overbrace{\begin{cases} P_1 = Q_2 - Q_1 \\ P_2 = Q_3 - Q_2 \end{cases}}^{balance}. \tag{15.1}$$

Recall that the balances must be made on dual cells endowed with an outer orientation, in this case the rooms. The (mean) temperatures are those measured at the centres of the cells, that is at the vertices of the primal complex, which have an inner orientation (oriented as sinks). These equations can be organized in matrix notation, as shown in Table 15.1. For the considerations that follow we should use the following matrices: **G** for the defining equation, **M** for the constitutive equation and **D** for the balance equation.

**Remark.** The letter **M**, or the constitutive matrix, is suggested by the fact that it is the initial of the words *material* and *medium*. The use of the letter **C** for the constitutive matrix would conflict with the use of the same letter to denote the matrix corresponding to the *curl* of the differential formulation. Recall the symbols **G**, **C**, **D** in Chap. 7.

**Table 15.1** The balance matrix is the transpose of the definition matrix with a minus sign

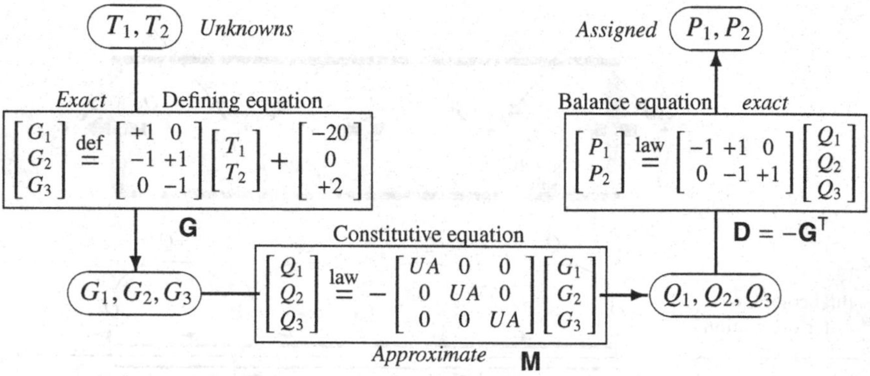


Note the important fact that the two matrices **D** and **G** are transposed, apart from the change in sign. The minus sign in the relation $\mathbf{D} = -\mathbf{G}^{\mathsf{T}}$ depends on the fact that an outer orientation of dual cells (outer normals, i.e. arrows coming out of the volume) is the opposite of the orientation induced by an inner orientation of the vertices of the primal complex (oriented as a sink, so the arrows are coming into the vertex).

What we have described above is an application of the *coboundary process*. Recall that the coboundary process allows us to switch from a discrete $p$-form to a discrete $(p+1)$-form, as mentioned in Chap. 7.

Let us consider the defining equation. Let us consider the discrete 0-form of the nodal temperatures $T^0 = [T_1, T_2]$ and apply to it the coboundary process, which consists of two steps. The first step is to transfer the value associated with each node (0-cell) to the two incident edges, applying the + or − sign depending on the mutual incidence edges/nodes. Given the orientations shown in Fig. 15.2, the value transferred to the left has a + sign while the value transferred to the right has a − sign. The second step is to add, for each edge, the two values which derive from the extreme nodes of the segment. In this way, we obtain the discrete 1-form $G^1 = [G_1, G_2, G_3]$ whose elements are given by the second equation in Eq. 15.1. The result thus obtained can be expressed with the matrix **G** of the previous section, which describes the coboundary process performed on the discrete 0-form $T^0$ of the temperatures and which gives rise to the discrete 1-form $G^1$ of the temperature differences. In the language of algebraic topology, we can write $G^1 = \delta T^0$, where $\delta$ denotes the coboundary operator.

Let us consider the balance equation. The balance is also formed by two steps. The first step consists in transferring the values of the heat currents $Q_k$ which pass through each wall (0-cells of the dual complex) to the 1-cells (the rooms) which form the coboundary of the walls, each multiplied by the incidence number between the faces and the cells. The second step consists in adding, for each 1-cell, the heat currents coming from its faces and making this sum equal to the rate of heat generation within each 1-cell.

The reason of the transposition lies in the fact that *the same* incidence number (apart from the sign) appears in the balance equation, where we pass from the faces of a dual cell to the dual cell itself, and in the defining equation, where we pass from the primal vertices to the primal edges. We are lead to introduce the discrete 1-form $P^1 \stackrel{\text{def}}{=} [P_1, P_2]$ and the discrete 0-form $Q^0 \stackrel{\text{def}}{=} [Q_1, Q_2.Q_3]$.

Hence in the balance equation we pass from a discrete 0-form to a discrete 1-form, i.e. from a discrete form of lower dimension to one of higher dimension. In the language of algebraic topology, we can write $P^1 = \delta Q^0$, where $\delta$ denotes the coboundary operator. However, since the dimensions of the dual cells increase in inverse order of the cells of the primal complex, we have $\tilde{d}_{k\alpha} = -g_{\alpha k}$, i.e. $\widetilde{\mathbf{D}} = -\mathbf{G}^{\mathsf{T}}$, which is the adjointness property, as was already shown.[2]

The fact that this property of adjointness is systematic in the diagram and is valid for all physical theories is so amazing that we pose the question: *why is this so? We will show in what follows that the reason must be found in the nature of the coboundary process applied to the discrete forms of the primal and the dual cell complexes.*

To explain why, we will consider a simple one-dimensional problem which can be treated in terms of algebraic equations.

### *15.1.2 Topological Equivalent of Algebraic Formulation*

We now show how to give a geometric interpretation of the algebraic formulation described previously. With reference to Fig. 15.3, the values $(\varphi_1, \varphi_2, \varphi_3, \varphi_4)$ of the set which is associated with the points of the primal complex give rise to a discrete 0-form $\varphi^0$, whereas the values $(u_1, u_2, u_3)$ of the set which is associated with the 1-cells of the primal complex give rise to a discrete 1-form $u^1$. The first relation of Eq. 15.18 expresses the coboundary process performed on the discrete 0-form $\varphi^0$,

[2] See Eq. 7.1 on p. 189.

**Table 15.2** One dimensional, stationary, fluid flow

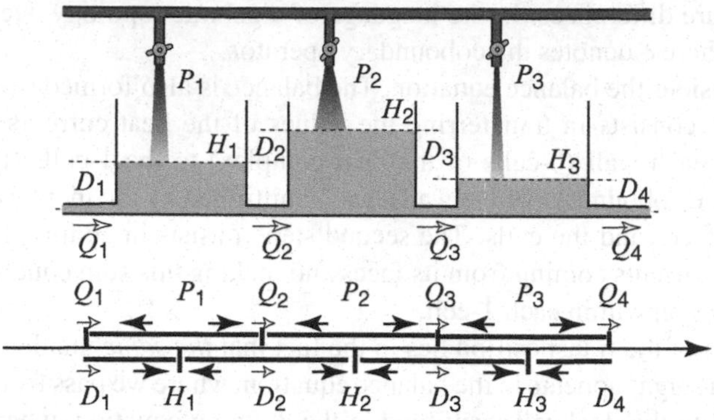


Piezometric differences between tanks

$$\begin{cases} D_1 = H_1 - H_0 \\ D_2 = H_2 - H_1 \\ D_3 = H_3 - H_2 \\ D_4 = H_4 - H_3 \end{cases}$$

Constitutive equations

$$\begin{cases} Q_1 = -\lambda \frac{A}{L} D_1 \\ Q_2 = -\lambda \frac{A}{L} D_2 \\ Q_3 = -\lambda \frac{A}{L} D_3 \\ Q_4 = -\lambda \frac{A}{L} D_4 \end{cases}$$

Fluid flow between tanks

$$\begin{cases} Q_2 - Q_1 = P_1 \\ Q_3 - Q_2 = P_2 \\ Q_4 - Q_3 = P_3 \end{cases}$$

which, in matrix notation, can be written

$$\begin{bmatrix} D_1 \\ D_2 \\ D_3 \\ D_4 \end{bmatrix} = \begin{bmatrix} +1 & 0 & 0 \\ -1 & +1 & 0 \\ 0 & -1 & +1 \\ 0 & 0 & -1 \end{bmatrix} \begin{bmatrix} H_1 \\ H_2 \\ H_3 \end{bmatrix} + \begin{bmatrix} -H_0 \\ 0 \\ 0 \\ +H_4 \end{bmatrix} \qquad \begin{bmatrix} -1 & +1 & 0 & 0 \\ 0 & -1 & +1 & 0 \\ 0 & 0 & -1 & +1 \end{bmatrix} \begin{bmatrix} Q_1 \\ Q_2 \\ Q_3 \\ Q_4 \end{bmatrix} = \begin{bmatrix} P_1 \\ P_2 \\ P_3 \end{bmatrix}$$

Note that the two matrices are adjoint to one another: from here

$$\mathbf{D} = \mathbf{G}\,\mathbf{H} + \overline{\mathbf{H}} \qquad \mathbf{Q} = \mathbf{M}\,\mathbf{D} \qquad -\mathbf{G}^{\mathsf{T}}\,\mathbf{Q} = \mathbf{P}$$

which give rise to the fundamental problem

$$-\mathbf{G}^{\mathsf{T}}\,\mathbf{M}\,\mathbf{G}\,\mathbf{H} = \mathbf{P} + \mathbf{G}^{\mathsf{T}}\,\mathbf{M}\,\overline{\mathbf{H}}$$

$$u^1 = \delta\varphi^0 \,, \tag{15.2}$$

and can be written in the equivalent matrix notation:

$$\begin{bmatrix} u_1 \\ u_2 \\ u_3 \end{bmatrix} = \begin{bmatrix} -1 & +1 & 0 & 0 \\ 0 & -1 & +1 & 0 \\ 0 & 0 & -1 & +1 \end{bmatrix} \begin{bmatrix} \varphi_1 \\ \varphi_2 \\ \varphi_3 \\ \varphi_4 \end{bmatrix} \quad or \quad \mathbf{u} = \mathbf{G}\boldsymbol{\varphi}\,. \tag{15.3}$$

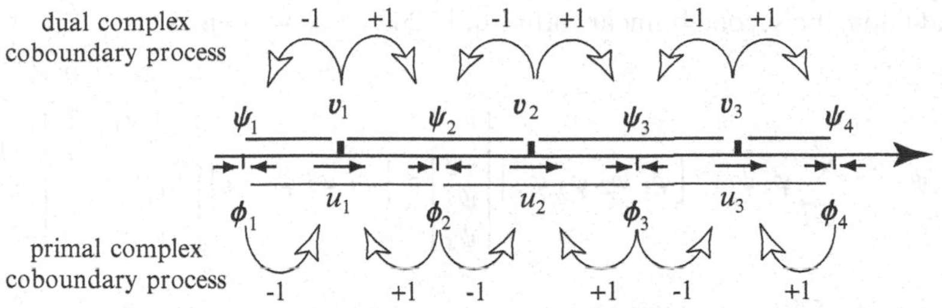


**Fig. 15.3** Coboundary process on primal and dual cell complexes in one dimension

Hence, the coboundary process between discrete 0-forms and discrete 1-forms on the primal cell complex of Fig. 15.3, which is endowed with an inner orientation, is described by the matrix **G** (Chap. 7).

Still, with reference to Fig. 15.3, the set of values $(v_1, v_2, v_3)$ associated with the 0-cells of the dual complex give rise to a discrete 0-form $v^0$, whereas the set of values $(\psi_1, \psi_2, \psi_3, \psi_4)$ associated with the 1-cells of the dual complex give rise to a discrete 1-form $\psi^1$. The second relation of Eq. 15.18 between the two discrete forms can be considered the coboundary process

$$\psi^1 = \delta\, v^0 \,. \tag{15.4}$$

This relation can also be expressed in matrix form as follows:

$$\begin{bmatrix} \psi_1 \\ \psi_2 \\ \psi_3 \\ \psi_4 \end{bmatrix} = \begin{bmatrix} -1 & 0 & 0 \\ +1 & -1 & 0 \\ 0 & +1 & -1 \\ 0 & 0 & +1 \end{bmatrix} \begin{bmatrix} v_1 \\ v_2 \\ v_3 \end{bmatrix} \qquad or \qquad \boldsymbol{\psi} = \mathbf{G}^{\mathsf{T}} \mathbf{v} \,. \tag{15.5}$$

Hence, the matrix $\mathbf{G}^{\mathsf{T}}$ describes the coboundary process between the discrete 0-form and the discrete 1-form on the dual cell complex of Fig. 15.3. Thus, we see that the same coboundary process performed on a primal or on a dual cell complex is described by a matrix and by its transpose respectively.

The first bilinear form Eq. 15.20 can be written as

$$\langle v,\, u\rangle_2 \stackrel{\text{def}}{=} \sum_{h=1}^{3} v_h\, u_h = \begin{bmatrix} v_1 & v_2 & v_3 \end{bmatrix} \begin{bmatrix} u_1 \\ u_2 \\ u_3 \end{bmatrix} = \begin{bmatrix} v_1 & v_2 & v_3 \end{bmatrix} \begin{bmatrix} -1 & +1 & 0 & 0 \\ 0 & -1 & +1 & 0 \\ 0 & 0 & -1 & +1 \end{bmatrix} \begin{bmatrix} \varphi_1 \\ \varphi_2 \\ \varphi_3 \\ \varphi_4 \end{bmatrix} . \tag{15.6}$$

In addition, the second bilinear form Eq. 15.20 can be written as

$$\langle \varphi , \psi \rangle_1 \stackrel{\text{def}}{=} \sum_{k=1}^{4} \varphi_k \psi_k = \begin{bmatrix} \varphi_1 & \varphi_2 & \varphi_3 & \varphi_4 \end{bmatrix} \begin{bmatrix} \psi_1 \\ \psi_2 \\ \psi_3 \\ \psi_4 \end{bmatrix} = \begin{bmatrix} \varphi_1 & \varphi_2 & \varphi_3 & \varphi_4 \end{bmatrix} \begin{bmatrix} -1 & 0 & 0 \\ +1 & -1 & 0 \\ 0 & +1 & -1 \\ 0 & 0 & +1 \end{bmatrix} \begin{bmatrix} v_1 \\ v_2 \\ v_3 \end{bmatrix} . \tag{15.7}$$

Hence, Eq. 15.21 can be written as

$$\overbrace{\mathbf{v}^{\mathsf{T}}\,\mathbf{G}\varphi \equiv \varphi^{\mathsf{T}}\,\mathbf{G}^{\mathsf{T}}\,\mathbf{v}}^{\text{matrix notation}} \qquad \text{or} \qquad \overbrace{\langle v,\, G\varphi\rangle_2 \equiv \langle \varphi , G^{\mathsf{T}}\, v\rangle_1 .}^{\text{operatorial notation}} \tag{15.8}$$

This shows that bilinear forms are connected with the coboundary process.

However, to help the reader who is more familiar with the differential approach, we will do the reverse process: we will start from the differential formulation to arrive at the algebraic formulation.

The algebraic description possesses the advantage of not having the conditionings of the differential formulation, primarily those conditionings imposed by the conditions of derivability. These conditions are specifically required by the differential formulation to use the theorems which the differential formulation provides. Algebraic topology makes use of global variables and therefore does not use those field functions which are obtained by the density of global variables.

Since we presume that our reader is more familiar with the traditional differential formulation of physical laws than with the algebraic-topological formulation, we will start by showing how to pass from the differential formulation to the algebraic one and from this to the formulation offered by algebraic topology.

## 15.2 From Differential Operators to Algebraic Operators

The ‘discovery’ that global physical variables have a natural association with space and time elements, endowed with an inner or outer orientation, leads us to make use of cell complexes instead of coordinate systems to describe physical theories. The proper tools for this algebraic description are algebraic topology and the notions of chains and discrete forms.[3] This leads us to use algebra instead of infinitesimal calculus. Since the physical literature has been modelled on the differential formulation since Newton’s time, it is imperative to analyse the link between algebraic operators and differential operators. This is the main purpose of the following sections.

---

[3] See Chap. 7.

### 15.2.1 Differential Operators: Some Specifications

The traditional description of physical theories is based on the differential formulation. For this reason the classification diagrams contained in this book were written in the differential formulation. Hence the mathematical properties are contained in the differential operators which appear in these equations. In common language, expressions like

$$\frac{\mathrm{d}}{\mathrm{d}x}, \quad \frac{\partial}{\partial t}, \quad \frac{\partial^2}{\partial t^2}, \quad \text{div}, \quad \text{grad} \quad \text{curl}\ , \quad \nabla^2, \quad \mathrm{i}\hbar\,\partial \tag{15.9}$$

are called *differential operators*. This terminology conflicts with the terminology of functional analysis, where one is interested in notions like, for example, adjointness, eigenvalues, inverse of an operator and variational formulation. Properly speaking, the expressions (15.9) are *formal differential operators*[4] which become differential operators only when the set of functions on which they must operate is assigned. This set is called the *domain* of the operator $L$ and is denoted by $\mathsf{D}(L)$. To define the domain, one must give *additional conditions* composed of the following elements:

- The *functional class*, i.e. the regularity requirement of the functions and of its derivatives, say $C^1[a,b], C^2[a,b], C^\infty[\Omega], C_0^\infty[\Omega]$, etc.
- The *boundary* conditions for boundary value problems and the *initial* conditions for evolution problems.

The functional class is largely dictated by mathematical convenience and is not imposed by physics, apart from a minimum of derivability conditions; for this reason we will omit specifying it.[5] The same *formal* differential operator gives rise to many differential operators, one for every domain selected. In summary:

$$\textit{differential operator} = \textit{formal differential operator} + \textit{domain}. \tag{15.10}$$

The following relations are examples of differential operators:

$$\begin{cases} L_1 = \left\{\dfrac{\mathrm{d}}{\mathrm{d}x};\ x \in [a,b];\ u(a) = 0\right\}, \\[2ex] L_2 = \left\{\dfrac{\mathrm{d}^2}{\mathrm{d}x^2};\ x \in [a,b];\ u(a) = 0;\ u(b) = 0\right\}, \\[2ex] L_3 = \left\{\nabla^2;\ \mathbf{x} \in [\Omega];\ \left.\dfrac{\partial u}{\partial n}\right|_{\partial\Omega} = 0\right\}. \end{cases} \tag{15.11}$$

4 Stone [220, p. 116, 123], Locker [141, p. iv, 26], Kato [112, p. 146], Dass and Sharma [45, p. 346], Dunford and Schwartz [61], Goldberg [79]. Naimark [167, p. 3] uses the term *linear differential expression* instead of *linear formal differential operator.*

5 ... as many authors do.

Why should we add the domain? One reason is that while for a square $n \times n$ matrix $\mathbf{Q}$ the eigenvalues $\lambda_1, \lambda_2, \ldots, \lambda_n$ are the numbers $\lambda_k$ for which the system

$$\mathbf{Q}\,\mathbf{u} = \lambda\,\mathbf{u} \tag{15.12}$$

admits non-zero solutions, such a requirement for the 'operator' $\nabla^2$ is meaningless because the eigenvalues depend on the shape of the domain and on the kind of boundary conditions. Thus the normal modes of a membrane blocked on its boundary depends on the shape of the boundary. The following two operators, having the same formal differential operator but different additional conditions

$$\begin{cases} L_4 = \left\{ \dfrac{\mathrm{d}^2}{\mathrm{d}x^2};\ x \in [0, \pi];\ u(0) = 0;\ u(\pi) = 0 \right\}, \\ L_5 = \left\{ \dfrac{\mathrm{d}^2}{\mathrm{d}x^2};\ x \in [0, \pi];\ u(0) = 0;\ u'(\pi) = 0 \right\}, \end{cases} \tag{15.13}$$

have different eigenvalues $\lambda_n = -n^2$ and $\lambda_n = -(n - 1/2)^2$ respectively. Hence, the eigenvalues are different only because they have different boundary conditions at $\pi$.

It would be meaningless to ask for the eigenvalues of the 'operator' $\mathrm{d}/\mathrm{d}x$. Moreover, in contrast to a common statement in books on quantum mechanics, it is not true that the 'operator' $i\hbar\,\mathrm{d}/\mathrm{d}x$ is Hermitian, simply because this property depends on the kind of boundary conditions. Another reason is that it would be meaningless to ask for the inverse of the 'operator' $\nabla^2$ or $\mathrm{d}^2/\mathrm{d}x^2$. These are meaningless statements simply because these 'operators' are only *formal* differential operators; these statements become meaningful only when the additional conditions are taken into account, i.e. for a *true* differential operator.

Hence, if we want to extend the differential calculus concepts, which are rooted in matrix calculus, we must consider the additional conditions as being an essential part of a differential operator and, consequently, make a clear distinction between a *formal differential operator* and a *differential operator*.

It is also important to make a clear distinction between the terms *equation* and *problem*. While $-\epsilon\nabla^2\phi = \rho$ is an equation, the sets

$$\begin{cases} -\epsilon\nabla^2\phi = \rho & in\ \Omega \\ \phi\big|_{\partial\Omega} = assigned \end{cases} \qquad \begin{cases} -\epsilon\nabla^2\phi = \rho & in\ \Omega \\ \dfrac{\partial\phi}{\partial n}\bigg|_{\partial\Omega} = assigned \end{cases} \tag{15.14}$$

are the Dirichlet and the Neumann *problems* respectively. An equation contains a *formal* differential operator, say $\mathscr{L}$, whereas a problem has a differential operator, say $L$. We will distinguish the two notions by writing

$$\left\{\begin{matrix}\textbf{problem}\\ L\,u=v\end{matrix}\right\}=\left\{\begin{matrix}\textbf{equation}\\ \mathscr{L}\,u=v\end{matrix}\right\}+\left\{\begin{matrix}\textbf{additional conditions}\\ \text{initial/boundary/regularity}\end{matrix}\right\}. \tag{15.15}$$

We will denote a linear operator by the letter $L$ and a non-linear operator by $N$. The kernel of a linear operator will be denoted by $\mathsf{K}(L)$ and the range of a linear and a non-linear operator by $\mathsf{R}(L)$ and $\mathsf{R}(N)$ respectively.

### 15.2.2 Algebraic Equivalent of Differential Formulation

Our aim is to show the link between the linear *differential* operators and the corresponding linear *algebraic* operators. To this end, the following integral identity plays a pivotal role:[6]

$$\boxed{\int_a^b v(x)\,\mathrm{d}\varphi(x) \equiv [v(b)\,\varphi(b) - v(a)\,\varphi(a)] - \int_a^b \varphi(x)\,\mathrm{d}v(x)\,.} \tag{15.16}$$

Let us subdivide the interval $[a,b]$ into subintervals, for example, in three subintervals of equal length, as shown in Fig. 15.4. Let us number the subdivision points from 1 to 4 and the subintervals from 1 to 3. Equation 15.16 becomes

$$\begin{aligned} &v_1\,(\varphi_2-\varphi_1)+v_2\,(\varphi_3-\varphi_2)+v_3\,(\varphi_4-\varphi_3)\equiv\\ &\qquad [\varphi_4 v_3-\varphi_1 v_1]-\varphi_2\,(v_2-v_1)-\varphi_3\,(v_3-v_2)\,. \end{aligned} \tag{15.17}$$

Notice that the number of components of $\varphi$ is four, whereas that of $v$ is three: this obligates us to approximate the expression $v(a)\,\varphi(a)$ with $\varphi_1\,v_1$ and $v(b)\,\varphi(b)$ with $\varphi_4\,v_3$. The last formula suggests the need to introduce two new sets of numbers:

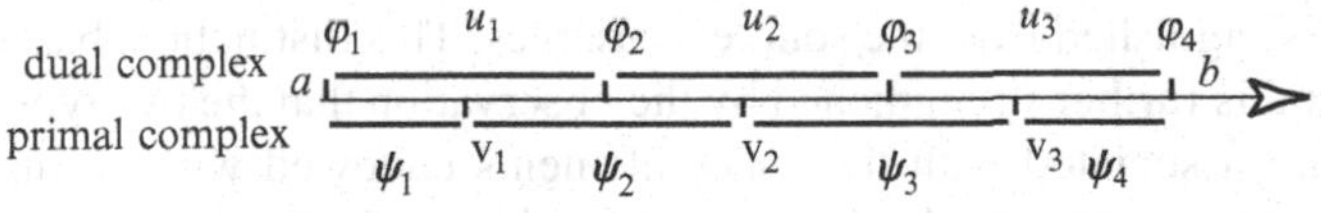


**Fig. 15.4** Primal and dual subdivision of an interval [a,b]

[6] Lanczos [121, p. 152].

$$\begin{cases} u_1 \stackrel{\text{def}}{=} \varphi_2 - \varphi_1 \\ u_2 \stackrel{\text{def}}{=} \varphi_3 - \varphi_2 \\ u_3 \stackrel{\text{def}}{=} \varphi_4 - \varphi_3 \end{cases} \qquad \begin{cases} \psi_1 \stackrel{\text{def}}{=} 0 - v_1 \\ \psi_2 \stackrel{\text{def}}{=} v_1 - v_2 \\ \psi_3 \stackrel{\text{def}}{=} v_2 - v_3 \\ \psi_4 \stackrel{\text{def}}{=} v_3 - 0 \end{cases} \tag{15.18}$$

so that Eq. 15.17 can be written as

$$\boxed{\sum_{h=1}^{3} v_h\, u_h \equiv \sum_{k=1}^{4} \varphi_k\, \psi_k\,.} \tag{15.19}$$

This relation is the algebraic equivalent of the differential identity of Eq. 15.16. This notation suggests the need to introduce an appropriate tool, the *bilinear forms*, which will be explicitly presented shortly. Hence, when we introduce the bilinear forms

$$\langle v,\, u\rangle_2 \stackrel{\text{def}}{=} \sum_{h=1}^{3} v_h\, u_h \qquad \langle \varphi\,, \psi\rangle_1 \stackrel{\text{def}}{=} \sum_{k=1}^{4} \varphi_k\, \psi_k, \tag{15.20}$$

Eq. 15.19 becomes

$$\langle v,\, u\rangle_2 \equiv \langle \varphi\,, \psi\rangle_1\,. \tag{15.21}$$

This shows that the formula of integration by parts of Eq. 15.16 becomes the equality of two bilinear forms in an algebraic setting. As we will see, in the sequel, this equality plays a pivotal role in the mathematical structure of physical theories.

## 15.3 Physics Needs Couples of Vector Spaces

The analysis we have done so far, which has led to the construction of a classification scheme for the variables and equations of physical theories of the macrocosm, has demonstrated a truth: in all the theories examined, the configuration variables must be distinguished from the sources variables. This distinction, based on physical content, is further strengthened by the observation that the two types of global variable are associated with the space elements endowed with an inner and an outer orientation respectively. The reason for this marked correspondence

**Configuration variables ⟷ Inner orientation**
**Source variables ⟷ Outer orientation**

remains mysterious to the present author.

This distinction, between configuration and source variables, is absent in traditional descriptions of physics, which, rather than two separate function spaces, use a single function space.

For example, functional analysis, when dealing with Poisson's equation of electrostatics $-\epsilon\nabla^2\phi = \rho$, omits the physical constant $\epsilon$ and (implicitly!) introduces the function $f = \rho/\epsilon$. In this way, Poisson's equation is reduced to $-\nabla^2\phi = f$, which links two functions, $\phi$ and $f$, of the *same* function space. This reduction of equations describing physical laws to purely mathematical equations hides information on the physical phenomenon described.

Furthermore, this single space is endowed with a scalar product ($\equiv$ inner product), creating a Hilbert space. We will now show that it is possible and convenient to maintain the two distinct spaces, equipping them with a tool similar to a scalar product in a Hilbert space. This requires the introduction of a bilinear form between the two spaces; in this way, the two spaces are said to be 'placed in duality'. This bilinear forms plays the role of scalar products between the elements of the two vector spaces, as shown in the left part of Fig. 15.5.

$$\langle v, L\phi\rangle_2 = \langle L^*v, \phi\rangle_1$$
*adjoint*

$$\langle Cu, u'\rangle_2 = \langle Cu', u\rangle_2$$
*selfadjoint*

$$\langle L^*CL\phi, \phi'\rangle_1 = \langle L^*CL\phi', \phi\rangle_1$$
*selfadjoint*

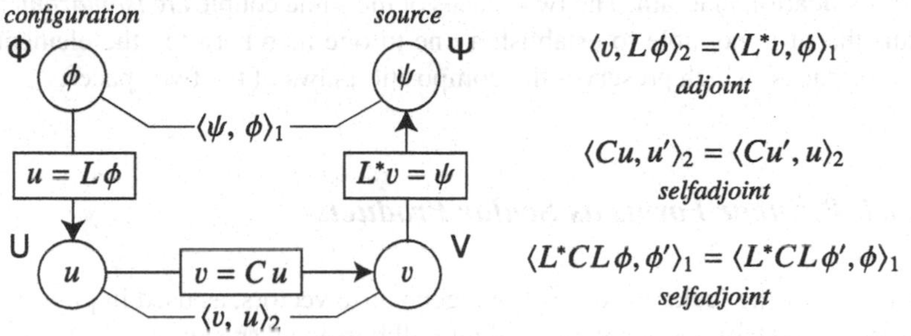

**Fig. 15.5** Two couples of spaces placed in duality by non-degenerate bilinear forms. The formulae on the right part will be explained later

**REMARK.** It may seem strange to talk about a pair of vector spaces and of a 'scalar product' between entities of two different spaces; but just consider the fact that the first scalar product of history, which is also the most used in physics, is that between a force and a displacement which gives rise to work. Forces and displacements are members of two different communities, they do not have the same physical dimensions and therefore cannot be added to one another: how can they reasonably coexist in the same function space? This cannot be: *it is better to consider two vector spaces placed in duality by a bilinear form* instead of a single vector space endowed with a scalar product. The same remark is valid for all scalar products used in physics, such as, for example, electric displacement **D** and electric field strength **E** in the electric field, magnetic induction $\check{\mathbf{B}}$ and magnetic field strength $\check{\mathbf{H}}$ in the magnetic field; stress $\sigma$ and strain $\varepsilon$ in the mechanics of deformable solids, and momentum **p** and velocity **v** in particle mechanics. Kinetic energy, which uses the scalar product of velocity by itself, seems to be an exception. But we must remember that it comes from the scalar product $\mathbf{v}\cdot d\mathbf{p}$ and is reduced to the product $\mathbf{v}\cdot d\mathbf{v}$ only after having linked the momentum and the velocity with the constitutive relation $\mathbf{p} = m\mathbf{v}$.

Another mathematical reason for maintaining the distinction between the two sets of variables is that the source of a physical field can be a highly discontinuous function, up to a distribution, like Dirac's delta function, whereas the configuration variable, typically a potential, is always continuous and derivable whenever there are no material discontinuities. How can they reasonably coexist in the same space?

A further mathematical reason for this distinction is that the variables of the two spaces are conjugated, i.e. their product is an energy.[7] They have an opposed tensorial variance: if one variable is covariant, then the corresponding conjugate variable is contravariant. Since the product of two conjugate variables gives an energy, this assures the invariance of the energy with respect to a change in coordinates.

We can say that these two function spaces, such as the spaces $\Phi$ and $\Psi$ and the spaces $\mathsf{U}$ and $\mathsf{V}$ of Fig. 15.5, correspond to the rounded boxes of each level of the classification diagram. The two spaces of the same couple are *isomorphic*: this means that it is possible to establish a one-to-one map between the elements of the two spaces, which preserves the composition laws of the two spaces.

### 15.3.1 Bilinear Forms as Scalar Products

The notion of the scalar product of two geometric vectors, as used in physics, can be generalized to algebraic vectors of an $n$-dimensional space:

$$\overbrace{\mathbf{u}\cdot\mathbf{v} = u_x\,v_x + u_y\,v_y + u_z\,v_z}^{vectorial} \;\Rightarrow\; \overbrace{\langle u, v\rangle}^{operatorial} \overset{\text{def}}{=} \overbrace{\sum_{k=1}^{n} u_k\,v_k = [\,u_1\;u_2\;\ldots\;u_n\,]\begin{bmatrix} v_1 \\ v_2 \\ \ldots \\ v_n \end{bmatrix} = \mathbf{u}^{\mathsf{T}}\mathbf{v}}^{matrix}\;. \tag{15.22}$$

The scalar product is an application which associates a number to two vectors, and this association is linear in both the first and second terms; hence, it is a *bilinear* form. Let us recall some definitions involving the term *bilinear*:

- Bilinear *map* denotes three vector spaces[8] with $\mathsf{U}, \mathsf{V}$ and $\mathsf{W}$, a bilinear map $w = B(u, v)$ is an application between the Cartesian product of the first two spaces and the third space. Hence $B : \mathsf{U} \times \mathsf{V} \mapsto \mathsf{W}$.
- Bilinear **form** or bilinear *functional*[9] is a particular bilinear map, in which the third space is $\mathbb{R}$. In this case $B : \mathsf{U} \times \mathsf{V} \mapsto \mathbb{R}$.

---

[7] See p. 149.

[8] Recall that the terms *vector space* and *linear space* are synonymous.

[9] Halmos [88, p. 36].

The extension of the notion of scalar product to a function space can be done as follows. Let $u(x)$ and $v(x)$ belong to two isomorphic function spaces U and V. We can use the notation

$$\langle u, v\rangle \stackrel{\text{def}}{=} \int_a^b u(x)\, v(x)\, dx \qquad with \qquad u \in \mathsf{U},\ v \in \mathsf{V} \tag{15.23}$$

and consider this relation as the definition of the *scalar product* of the two functions. If $\mathbf{u}(\mathbf{x})$ and $\mathbf{v}(\mathbf{x})$ are two vector-valued functions which belong to two isomorphic function spaces U and V, then a useful scalar product is the following:

$$\langle u, v\rangle \stackrel{\text{def}}{=} \int_{\mathsf{V}} \mathbf{u}(\mathbf{x}) \cdot \mathbf{v}(\mathbf{x})\, dV \qquad with \qquad \mathbf{u} \in \mathsf{U},\ \mathbf{v} \in \mathsf{V}\,. \tag{15.24}$$

Let us give the following definition:[10]

DEFINITION. *Given two vector spaces* U *and* V *over the reals, a* bilinear form *$B$ is an application which to every couple of elements $u \in \mathsf{U}$ and $v \in \mathsf{V}$ associates a real number, i.e. $\mathsf{U} \times \mathsf{V} \mapsto \mathbb{R}$, and which is linear in both arguments:*

$$\begin{aligned} B(u + u', v) &= B(u, v) + B(u', v),\\ B(u, v + v') &= B(u, v) + B(u, v'),\\ B(\lambda u, v) &= B(u, \lambda v) = \lambda B(u, v)\,. \end{aligned} \tag{15.25}$$

*A bilinear form is called* non-degenerate *if*

$$\begin{aligned} &B(u, v) = 0 \text{ for all } u \in \mathsf{U} \text{ implies } v = 0 \text{ and}\\ &B(u, v) = 0 \text{ for all } v \in \mathsf{V} \text{ implies } u = 0 \end{aligned} \tag{15.26}$$

*and the two spaces* U *and* V *are said to be* placed in duality. *The space* V *is called* dual *of the space* U *and is denoted by* $\mathsf{U}^*$.

The notation we will use is $\langle v, u\rangle \equiv B(u, v)$, i.e. we take as the first elements in the notation $\langle v, u\rangle$ the element of the second space.[11]

**REMARK.** The reason for the order $\langle v, u\rangle$, instead of the more common one $\langle u, v\rangle$, is suggested by the differential of a function $v = f(\mathbf{u})$, which is a bilinear form $dv = (\partial f / \partial u^k)\, du^k \equiv \langle \nabla f, d\mathbf{u}\rangle$. In physics, the source variable may depend on the configuration variable; for example, in solid mechanics the force density $\mathbf{f}$ (source) depends on the displacement $\eta$ (configuration), and this is written as $\mathbf{f}(\eta)$. This implies that the elementary work of a force is written as $w = \mathbf{f} \cdot d\eta \equiv \langle f, d\eta\rangle$; hence, the source variable $\mathbf{f}$ precedes the configuration variable $\eta$.

[10] Chevalley [40, p. 106], Schaefer [200, p. 123], Bourbaki [23, p. II.43].

[11] As used by Kato [112, p. 12].

Since the scalar product of two vectors is a real number, we see that the bilinear form is an extension of the notion of scalar product. The only request we must add is that the bilinear form must be non-degenerate. With this requirement it is possible and appropriate to extend the name *scalar product* to a non-degenerate bilinear form. Two elements $u \in \mathsf{U}$ and $v \in \mathsf{V}$ for which $\langle v, u \rangle = 0$ are called *orthogonal*. With the notion of orthogonality the condition of being non-degenerate can be restated as follows: only the null vector of one space can be orthogonal to all vectors of the other space.

Table 15.3 gives a list of bilinear forms for various physical theories. What is remarkable is that all these bilinear forms have the dimension of an *energy*, while those dealing with space-time, i.e. with relativity, have the dimension of an *action*. One can easily recognize that when dealing with finite-dimensional vector spaces, two spaces can be placed in duality only if they have the same dimension. The simplest counterexample is the case $\mathsf{U}^2$ and $\mathsf{V}^1$ on which the most general bilinear form is $\langle v, u \rangle = au_1v_1 + bu_2v_1 \equiv (au_1 + bu_2)\,v_1$: all vectors such that $au_1 + bu_2 = 0$ are orthogonal to any vector $v$. For function spaces the condition is that the two spaces $\mathsf{U}$ and $\mathsf{V}$ must be isomorphic.

The scalar product is the natural tool for introducing the notions of the transpose of a matrix and the adjoint of a linear operator. We will introduce these notions by distinguishing between two cases:

- When the operator works between two isomorphic spaces;
- When the two spaces are not isomorphic.

### *15.3.2 From Transposed Matrix to Adjoint Operator*

We want to show that between the formal differential operators which lie on the same level of the classification diagrams, such as '-grad' and 'div', one is the adjoint of the other. To this end, we summarize the notion of adjoint of an operator starting from the familiar notion of transpose of a matrix.

Let us start with a *square* matrix $\mathbf{L}$ operating on the vectors of a single space $\mathsf{U}$ endowed with a non-degenerate inner product $(u, u')$ or with two spaces of the same dimension $\mathsf{U}$ and $\mathsf{V}$ *placed in duality* by a non-degenerate bilinear form $\langle v, u \rangle$. These two cases are represented in the left and right parts of Fig. 15.6. The transpose of a matrix $\mathbf{L}$ is commonly defined as the matrix $\mathbf{H}$ obtained by interchanging the rows and columns of the matrix. Taking this interchange as the *definition* of the transpose is not a good idea: it is better to define the transposed matrix $\mathbf{H}$ as the matrix which satisfies the relation

$$\underbrace{(L\,u, u') \stackrel{\text{def}}{=} (H\,u', u)}_{\text{when } L:\,\mathsf{U}\mapsto\mathsf{U}}, \qquad \underbrace{\langle L\,u, u'\rangle \stackrel{\text{def}}{=} \langle H\,u', u\rangle}_{\text{when } L:\,\mathsf{U}\mapsto\mathsf{V}}. \tag{15.27}$$

**Table 15.3** Main bilinear forms

| Field theories | | | |
|---|---|---|---|
| Electrostatics (energy) | $\langle\rho,\phi\rangle \stackrel{\text{def}}{=} \int_{\mathrm{V}} \rho\,\phi\,\mathrm{dV}$ | $\langle D,E\rangle \stackrel{\text{def}}{=} \int_{\mathrm{V}} D^k\,E_k\,\mathrm{dV}$ | [ELE1] |
| Magnetostatics (energy) | $\langle J,A\rangle \stackrel{\text{def}}{=} \int_{\mathrm{V}} J^k\,A_k\,\mathrm{dV}$ | $\langle H,B\rangle \stackrel{\text{def}}{=} \int_{\mathrm{V}} \frac{1}{2} H^{hk}\,B_{hk}\,\mathrm{dV}$ | [ELE2] |
| Electromagnetism (action) | $\langle J,A\rangle \stackrel{\text{def}}{=} \int_{H} J^\alpha A_\alpha\,\mathrm{d}H$ | $\langle G,F\rangle \stackrel{\text{def}}{=} \int_{H} \frac{1}{2} G^{\alpha\beta} F_{\alpha\beta}\,\mathrm{d}H$ | [ELE6] |
| Gravitation (energy) | $\langle\rho,U\rangle \stackrel{\text{def}}{=} \int_{\mathrm{V}} \rho\,U\,\mathrm{dV}$ | $\langle g,h\rangle \stackrel{\text{def}}{=} \int_{\mathrm{V}} h^k\,g_k\,\mathrm{dV}$ | [GRA1] |
| Mechanical theories | | | |
| Particle mechanics (energy) | $\langle F,r\rangle \stackrel{\text{def}}{=} \int_{T} F_k\,x^k\,\mathrm{d}t$ | $\langle p,v\rangle \stackrel{\text{def}}{=} \int_{T} p_k\,v^k\,\mathrm{d}t$ | [PAR1] |
| Acoustics in fluids (energy) | $\langle\sigma_{\mathrm{m}},\phi\rangle \stackrel{\text{def}}{=} \int_{\mathrm{V}} \sigma\,\phi\,\mathrm{dV}$ | $\langle q,v\rangle \stackrel{\text{def}}{=} \int_{\mathrm{V}} q_k\,v^k\,\mathrm{dV}$ | [FLU5] |
| Analytical dynamics (energy) | $\langle Q,q\rangle \stackrel{\text{def}}{=} \int_{T} Q_i\,q^i\,\mathrm{d}t$ | $\langle p,v\rangle \stackrel{\text{def}}{=} \int_{T} p_i\,v^i\,\mathrm{d}t$ | [PAR3] |
| Relativistic dynamics (action) | $\langle K,x\rangle \stackrel{\text{def}}{=} \int_{T} K_\alpha\,x^\alpha\,\mathrm{d}\tau$ | $\langle p,u\rangle \stackrel{\text{def}}{=} \int_{T} p_\alpha\,u^\alpha\,\mathrm{d}\tau$ | [PAR3] |
| Statics of continua (energy) | $\langle f,\eta\rangle \stackrel{\text{def}}{=} \int_{\mathrm{V}} f_k\,\eta^k\,\mathrm{dV}$ | $\langle\sigma,\varepsilon\rangle \stackrel{\text{def}}{=} \int_{\mathrm{V}} \sigma_{hk}\,\varepsilon^{hk}\,\mathrm{dV}$ | [SOL9] |

Note: the last column indicates a reference diagram

This relation can also be written as

$$\sum_{h,k=1}^{n} u'_k\,L_{kh}\,u_h = \sum_{h,k=1}^{n} u_h\,H_{hk}\,u'_k \tag{15.28}$$

from which it follows that $H_{hk} = L_{kh}$, i.e. from the definition of Eq. 15.28 we *deduce* that the transposed matrix has the rows exchanged with the columns of the original matrix. The matrix $\mathbf{H}$ is denoted by $\mathbf{L}^{\mathsf{T}}$. Hence the relation Eq. 15.28 can also be written in the form

$$\mathbf{u}'^{\mathsf{T}}\mathbf{L}\,\mathbf{u} = \mathbf{u}^{\mathsf{T}}\mathbf{L}^{\mathsf{T}}\mathbf{u}'. \tag{15.29}$$

### *15.3.3 Inhomogeneous Boundary Condition: Convex Set*

A region $\Omega$ of $\mathbb{R}^2$ or of $\mathbb{R}^3$ is said to be *convex* if, for every pair of points within the region, all points on the straight line segment which joins them is also within the region. Denoting by $P_1$ and $P_2$ two points of $\Omega$ and by $\mathbf{r}_1$ and $\mathbf{r}_2$ their radius

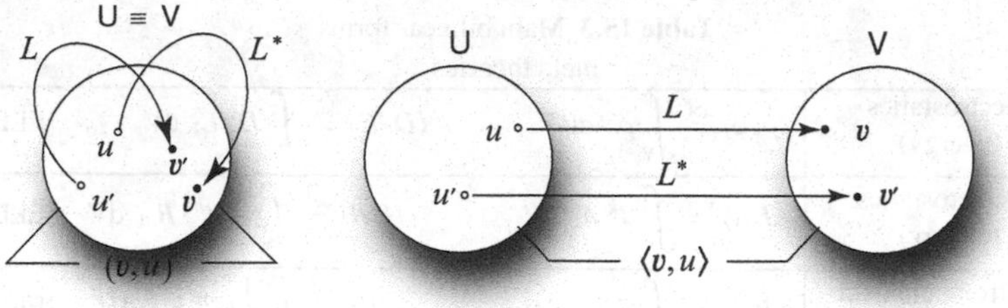


**Fig. 15.6** (*Left*): linear operator $L : \mathsf{U} \mapsto \mathsf{U}$; (*right*) linear operator $L$ between two isomorphic spaces $L : \mathsf{U} \mapsto \mathsf{V}$

vectors from an origin $O$, and denoting by $\lambda$ a real parameter with the condition $0 \leq \lambda \leq 1$, the straight line segment has the equation $\mathbf{r}(\lambda) = (1-\lambda)\mathbf{r}_1 + \lambda\mathbf{r}_2$. The convexity of the region requires that $\mathbf{r}(\lambda)$ is the radius vector of a point $P \in \Omega$.

What is beautiful is that the notion of convex set can be extended to a function space with the same definition. Let us consider the domain of a differential operator formed by the functions which satisfy inhomogeneous boundary or initial conditions.

Let $u_1(x)$ and $u_2(x)$ be two functions of this set with the conditions $u_1(a) = u_2(a) = A$ and $u_1(b) = u_2(b) = B$. All the functions of the family $u(x;\lambda) = (1-\lambda)u_1(x) + \lambda u_2(x)$ satisfy the same inhomogeneous conditions. Since the family is a linear function of $\lambda$ with $0 \leq \lambda \leq 1$, all these functions define a straight 'segment' in function space connecting the 'points' $u_1$ and $u_2$. Since all the 'points' of this segment belong to the domain, we can say that the set of functions is **convex** (Fig. 15.7).

## *15.3.4 Adjoint Operator*

The notion of the transpose of a matrix can be generalized to a linear operator, in particular to a linear *differential* operator. We remark that the linearity of a

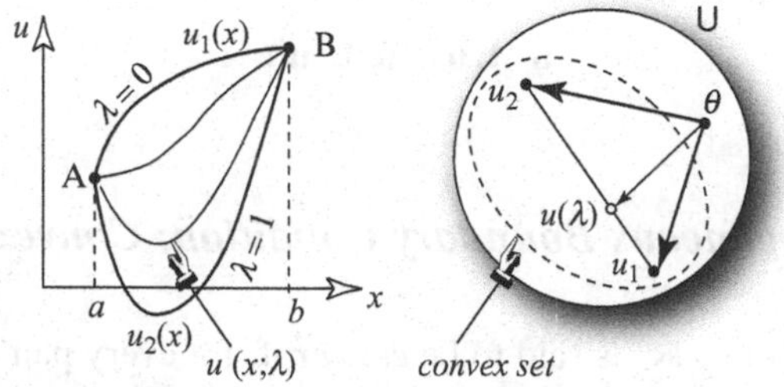


**Fig. 15.7** Straight line segment connecting two elements in a function space

differential operator implies the linearity of its formal part and the linearity of its domain; this means that the initial or boundary conditions must be homogeneous.

The generalization runs as follows: let us consider, at first, a linear operator $L$ operating between two *isomorphic* vector spaces $\mathsf{U}$ and $\mathsf{V}$ placed in duality by a non-degenerate bilinear form $\langle v, u\rangle$ called a *scalar product*, as shown in Fig. 15.6. Hence, $L : \mathsf{U} \mapsto \mathsf{V}$. Let us consider the linear operator $H : \mathsf{U} \mapsto \mathsf{V}$, which satisfy the equality

$$\langle L\,u, u'\rangle \stackrel{\text{def}}{=} \langle H\,u', u\rangle\,. \tag{15.30}$$

The operator $H$ is called the *adjoint* of the operator $L$ and is denoted by $L^*$. The equality Eq. 15.30 can be written as

$$\boxed{\langle Lu, u'\rangle \stackrel{\text{def}}{=} \langle L^* u', u\rangle}\,. \tag{15.31}$$

This is the definition of adjoint of an operator when it operates between two isomorphic spaces. Equation 15.31 presupposes, obviously, that $u' \in \mathsf{D}(L^*)$ and is valid not for all elements of $\mathsf{U}$ but only for some subsets of $\mathsf{U}$: the larger of these subsets is the domain of the adjoint operator, i.e. $u' \in \mathsf{D}(L^*) \subseteq \mathsf{U}$.[12]

Given a differential operator $L$, let us denote by $\mathscr{L}$ its formal part. Let us denote by $L^*$ the adjoint operator and by $\mathscr{L}^*$ its formal part. One says that the two formal differential operators $\mathscr{L}$ and $\mathscr{L}^*$ are *formally adjoint*.

Example 1. The adjoint of the linear operator $L_1$ of Eq. 15.11 with respect to the bilinear form Eq. 15.23 can be obtained with an integration by parts as follows:

$$\int_a^b \frac{\mathrm{d}u(x)}{\mathrm{d}x} v(x)\,\mathrm{d}x \equiv \Big[u(x)\,v(x)\Big]_a^b + \int_a^b -\frac{\mathrm{d}v(x)}{\mathrm{d}x} u(x)\,\mathrm{d}x. \tag{15.32}$$

Since $u(a) = 0$, to eliminate the boundary term, the *minimum* requirement is to impose the condition $v(b) = 0$. Hence, the adjoint is

$$L_1^* \stackrel{\text{def}}{=} \left\{-\frac{\mathrm{d}}{\mathrm{d}x};\; x \in [a, b];\; v(b) = 0\right\}. \tag{15.33}$$

The formal differential operators $\mathscr{L}_1 = \mathrm{d}/\mathrm{d}x$ and $\mathscr{L}_1^* = -\,\mathrm{d}/\mathrm{d}x$ are formally adjoint.

### 15.3.5 Further Extension of Notion of Adjoint Operator

In the preceding section, we considered linear operators working between two isomorphic spaces placed in duality by a non-degenerate bilinear form. In this

---

[12] One criticism to the definition of the transpose of a matrix (i.e. a matrix with rows exchanged with columns) is that such a definition cannot be extended to other kinds of linear operators. Be careful: the transpose of a matrix must be not called 'adjoint' because this term is already used with a different meaning for matrices.

section, we present the notion of adjoint for linear operators working between two spaces which are *not* isomorphic; hence, they *cannot be* placed in duality. An example is the operator 'grad', which is applied to a scalar-valued function and gives rise to a vector-valued function. Refer to the diagram of Fig. 15.5.

For a better understanding of this impossibility, let us return to matrix calculus and let us consider a rectangular matrix, say $m \times n$ matrix. This works on vectors of $n$ components and produces vectors of $m$ components, i.e. it works from a space $\mathsf{U}^n$ and a space $\mathsf{V}^m$. The different dimensions of the spaces implies that they are not isomorphic; hence, they cannot be placed in duality.

To introduce the notion of adjoint of a linear operator $L : \Phi \mapsto \mathsf{U}$ when the two spaces are not isomorphic, we introduce two other spaces $\Psi$ and $\mathsf{V}$ so that $\Psi$ can be placed in duality with $\Phi$ and $\mathsf{V}$ with $\mathsf{U}$, as shown in Fig. 15.5. This allows us to introduce two non-degenerate bilinear forms $\langle \psi, \phi \rangle_1$ and $\langle v, u \rangle_2$. The defining relation of the adjoint is

$$\boxed{\langle v, L\phi \rangle_1 \overset{\text{def}}{=} \langle L^* v, \phi \rangle_2}. \tag{15.34}$$

What is interesting is that the mathematical description of physical theories spontaneously shows couples of isomorphic spaces! These are the function spaces which lie on the same level in the classification diagram. In the following sections, we will show that the relation expressed by Eq. 15.34 plays a pivotal role in the mathematical description of physical theories.[13]

### *15.3.6 Role of Boundary Conditions*

In the example shown in Sect. 15.1.1, we have assigned two boundary conditions, the temperatures at the extremes. With reference to Table 15.4, let us examine now the following three cases in which are assigned

1. The temperatures $T_A$ and $T_B$,
2. The temperature $T_A$ and the heat $Q_D$ going out on the right side, and
3. The heat $Q_C$ entering the left side and the heat $Q_D$ going out on the right side.

These three cases are the most simple algebraic, one-dimensional version of the classical Dirichlet problem, mixed problem and Neumann's problem for Poisson's equation used in field theories. In the third problem, the data must satisfy the compatibility condition $P_1 + P_2 = Q_D - Q_C$ and the temperature is defined up to an additive constant. In the three cases, we consider the constitutive equation $Q_i = -U A D_i$. Table 15.4 shows that in all these cases the matrix of the balance problem is (minus) the transpose of the matrix of the definition problem.

[13] For the notion of adjoint see Kato [112, p. 167].

**Table 15.4** Algebraic version of three kinds of boundary value problem

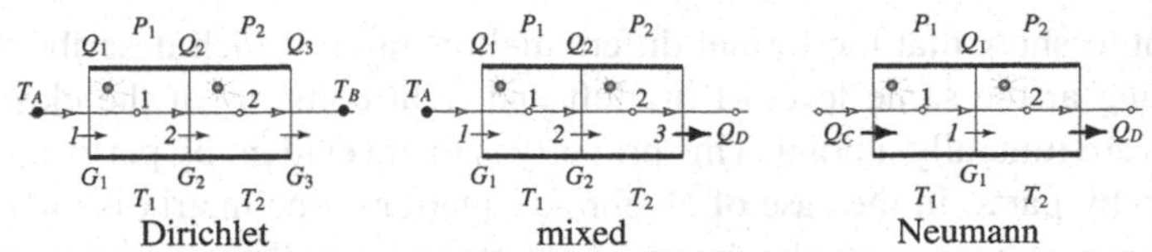


algebraic version of the Dirichlet's problem

$$\begin{bmatrix} G_1 \\ G_2 \\ G_3 \end{bmatrix} = \begin{bmatrix} +1 & 0 \\ -1 & +1 \\ 0 & -1 \end{bmatrix} \begin{bmatrix} T_1 \\ T_2 \end{bmatrix} + \begin{bmatrix} -T_A \\ 0 \\ +T_B \end{bmatrix} \qquad \begin{bmatrix} -1 & +1 & 0 \\ 0 & -1 & +1 \end{bmatrix} \begin{bmatrix} Q_1 \\ Q_2 \\ Q_3 \end{bmatrix} = \begin{bmatrix} P_1 \\ P_2 \end{bmatrix}$$

algebraic version of the mixed problem

$$\begin{bmatrix} G_1 \\ G_2 \end{bmatrix} = \begin{bmatrix} +1 & 0 \\ -1 & +1 \end{bmatrix} \begin{bmatrix} T_1 \\ T_2 \end{bmatrix} + \begin{bmatrix} -T_A \\ 0 \end{bmatrix} \qquad \begin{bmatrix} -1 & +1 \\ 0 & -1 \end{bmatrix} \begin{bmatrix} Q_1 \\ Q_2 \end{bmatrix} + \begin{bmatrix} 0 \\ +Q_D \end{bmatrix} = \begin{bmatrix} P_1 \\ P_2 \end{bmatrix}$$

algebraic version of Neumann's problem

$$\begin{bmatrix} G_1 \end{bmatrix} = \begin{bmatrix} -1 & +1 \end{bmatrix} \begin{bmatrix} T_1 \\ T_2 \end{bmatrix} \qquad \begin{bmatrix} +1 \\ -1 \end{bmatrix} \begin{bmatrix} Q_1 \end{bmatrix} + \begin{bmatrix} -Q_C \\ +Q_D \end{bmatrix} = \begin{bmatrix} P_1 \\ P_2 \end{bmatrix}$$

definition | balance

Example 2. We encourage the reader to perform the calculations of the temperatures $T_1$ and $T_2$ once $P_1$ and $P_2$ are assigned. As a check, we suggest the following values: $P_1 = 60\,W; P_2 = 30\,W; T_A = 20°; T_B = 5°; UA = 2\,W/°C$. The calculated values are $T_1 = 39°; T_2 = 28°; Q_1 = -38\,W; Q_2 = -22\,W; Q_3 = +52\,W$.

### *15.3.7 Symmetric and Self-Adjoint Operators*

When the operator $L$ operates between two spaces $\mathsf{U}$ and $\mathsf{V}$ placed in duality, as seen in Fig. 15.6, its adjoint $L^*$ has its domain in the same space $\mathsf{U}$ and its range in the same space $\mathsf{V}$. In this case it happens that the two operators coincide, i.e. $L = L^*$. This means that the formal operators and the domains coincide: $\mathscr{L} = \mathscr{L}^*$ and $\mathsf{D}(L) = \mathsf{D}(L^*)$. When this happens, the operator $L$ is said to be *self-adjoint*.

It happens more frequently that the domain of the adjoint contains the domain of the operator, i.e. $\mathsf{D}(L) \subseteq \mathsf{D}(L^*)$, always with the condition $\mathscr{L} = \mathscr{L}^*$. In this case, the operator $L$ is said to be *symmetric*.[14]

[14] Naimark [168, p. 13].

### 15.3.8 Operators at the Same Level Are Mutually Adjoint

We want to show that the formal differential operators which describe the equations lying at the same level of the left and right columns of the classification diagram are mutually adjoint. This property is made evident by performing an integration by parts. In the case of algebraic equations, one matrix is the transpose of the other, as happens in the theory of electrical networks.

We refer to two diagrams, one of perfect fluid dynamics, [FLU1], and one of electrostatics, [ELE1]. Diagram 15.5 shows the first two boxes of the left columns of the two diagrams. In the same diagram, the last column on the right shows the notation we will use for the operators. In this notation, the boxes are regarded as spaces of functions. In diagram 15.5 the operator $L$ operates between the spaces $\Phi$ and $\mathsf{U}$. In diagram 15.6, the operator $B$ operates between the spaces $V$ and $\Sigma$. We have chosen the letter $B$, the first letter of the word 'balance', because, in general, the equation $B\,v = \sigma$ is a balance equation. Diagram 15.6 shows the last two boxes of the right columns of the two diagrams.

**Table 15.5** Configuration side of diagrams [FLU1] and [ELE1]

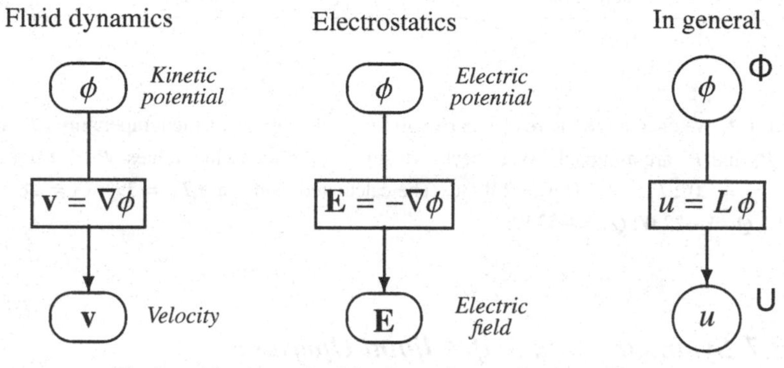


The last column on the right shows the notation we will use.

Let us consider now the far right columns of the two diagrams 15.5 and 15.6. We aim to show that the formal part of the differential operator $B$ is the adjoint of the formal part of the operator $L$, i.e. $\mathcal{B} = \mathcal{L}^*$.

Denoting by $\mathbf{S}_1$ and $\mathbf{S}_2$ the two parts of the boundary of the region $\mathbf{V}$, such that $\mathbf{S}_1 \cup \mathbf{S}_2 = \partial\mathbf{V}$, in electrostatics the main problems are

$$\begin{cases} E_k = -\nabla_k\,\phi \\ \phi = 0 \quad \text{on } \mathbf{S}_1 \end{cases}, \qquad \begin{cases} \nabla_k\,D^k = \rho \\ n_k\,D^k = 0 \quad \text{on } \mathbf{S}_2 \end{cases}, \tag{15.35}$$

**Table 15.6** Source side of diagrams [FLU1] and [ELE1]

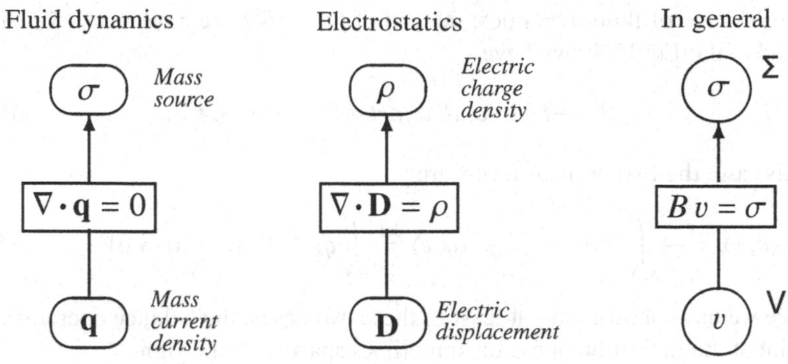


The last column on the right shows the notation we will use.

which can also be written in vector form as

$$\begin{cases} \mathbf{E} = -\nabla\phi \\ \phi = 0 \quad \textit{on } \mathbf{S}_1 \end{cases}, \qquad \begin{cases} \nabla\cdot\mathbf{D} = \rho \\ \mathbf{n}\cdot\mathbf{D} = 0 \quad \textit{on } \mathbf{S}_2 \end{cases}. \tag{15.36}$$

These two problems contain the linear operators

$$L \stackrel{\text{def}}{=} \left\{ -\nabla\,;\phi|_{\mathbf{S}_1} = 0 \right\}, \qquad B \stackrel{\text{def}}{=} \left\{ \nabla\cdot\,;(\mathbf{n}\cdot\mathbf{D})|_{\mathbf{S}_2} = 0 \right\}. \tag{15.37}$$

A relation between these two operators can be found by performing an integration by parts as follows:

$$\int_{\mathbf{V}} \rho\,\phi\,\mathrm{d}V = \int_{\mathbf{V}} (\nabla_k D^k)\,\phi\,\mathrm{d}V \equiv \cancel{\int_{\partial\mathbf{V}} (n_k D^k)\,\phi\,\mathrm{d}S} - \int_{\mathbf{V}} D^k\,(\nabla_k\phi)\,\mathrm{d}V = \int_{\mathbf{V}} D^k\,E_k\,\mathrm{d}V \tag{15.38}$$

in which the boundary term vanishes on account of the given boundary conditions. It follows that the operators $B$ and $L$ adjoint to one another with respect to the bilinear forms

$$\langle\rho,\phi\rangle \stackrel{\text{def}}{=} \int_{\mathbf{V}} \rho\,\phi\,\mathrm{d}V, \qquad \langle D,E\rangle \stackrel{\text{def}}{=} \int_{\mathbf{V}} \mathbf{D}\cdot\mathbf{E}\,\mathrm{d}V\,. \tag{15.39}$$

Hence, we have the property

$$B = L^* \quad \textit{and, a fortiori,} \quad \mathscr{B} = \mathscr{L}^* \tag{15.40}$$

i.e. the operator of the balance equation is the adjoint of the operator of the defining equation. We have stressed the fact that the formal differential operators are also adjoint because this property is valid even when the boundary conditions are not considered. We remark that the so-called operators -grad , curl , div are, more properly, *formal* differential operators.

In the case of fluid dynamics, instead of $\mathbf{E} = -\nabla\phi$ we have $\mathbf{v} = +\nabla\phi$, hence instead of the Eq. 15.40 we have

$$B = -L^* \qquad \textit{and, a fortiori} \qquad \mathscr{B} = -\mathscr{L}^* \,. \tag{15.41}$$

In this case, the two bilinear forms are

$$\langle\rho,\phi\rangle \stackrel{\text{def}}{=} -\int_{\mathbf{V}} \sigma\,\phi\,\mathrm{d}V \qquad \langle q,v\rangle \stackrel{\text{def}}{=} \int_{\mathbf{V}} q_k\,v^k\,\mathrm{d}V \equiv \int_{\mathbf{V}} \mathbf{q}\cdot\mathbf{v}\,\mathrm{d}V\,. \tag{15.42}$$

Hence we have shown that, at least in these two cases, the balance operator is the adjoint of the definition operator, sometimes apart from the sign.

We can conclude that the *formal* operators div and -grad are adjoint to one another. An analogous property is enjoyed by the two problems of statics of continua, which in Cartesian notation are

$$\begin{cases} \varepsilon_{hk} = \dfrac{1}{2}(\partial_h\,\eta_k + \partial_k\,\eta_h) \\ \eta_h = 0 \qquad \text{on } \mathbf{S}_1 \end{cases}, \qquad \begin{cases} -\partial_k\sigma_{hk} = f_h \\ n_k\sigma_{hk} = 0 \qquad \text{on } \mathbf{S}_2 \end{cases}. \tag{15.43}$$

The two operators are adjoint to one another with respect to the two bilinear forms $\langle f,\eta\rangle$ and $\langle\sigma,\varepsilon\rangle$ of Table 15.3 on p. 445. In fact,

$$\begin{aligned} \langle f\,,\eta\rangle &\stackrel{\text{def}}{=} \int_{\mathbf{V}} f_h\,\eta_h\,\mathrm{d}V = \\ &= \int_{\mathbf{V}} \Big(-\partial_k\sigma_{hk}\Big)\eta_h\,\mathrm{d}V \equiv -\int_{\partial V} \sigma_{hk}\,\eta_h\,n_k\,\mathrm{d}S + \int_{\mathbf{V}} \sigma_{hk}\Big(\partial_k\eta_h\Big)\mathrm{d}V \equiv \\ &\equiv \int_{\mathbf{V}} \sigma_{hk}\,\frac{1}{2}(\partial_h\,\eta_k + \partial_k\,\eta_h)\,\mathrm{d}V = \int_{\mathbf{V}} \sigma_{hk}\,\varepsilon_{hk}\,\mathrm{d}V \stackrel{\text{def}}{=} \langle\sigma\,,\varepsilon\rangle, \end{aligned} \tag{15.44}$$

where the boundary term vanishes on account of the given boundary conditions. Hence, also for the statics of continua, we have the relation Eq. 15.40.

### 15.3.9 Formal Operator 'curl' is Self-Adjoint

We refer to two diagrams, one of perfect fluid dynamics, [FLU2], and one of magnetostatics, [ELE3]. The diagram 15.7 shows the first two boxes of the left columns of the two diagrams, while the last column on the right shows the notation we will use. In this notation, the boxes are regarded as spaces of functions and the operator $R$ as operating between the spaces U and T.

**Table 15.7** Configuration side of diagrams [FLU2] and [ELE3]

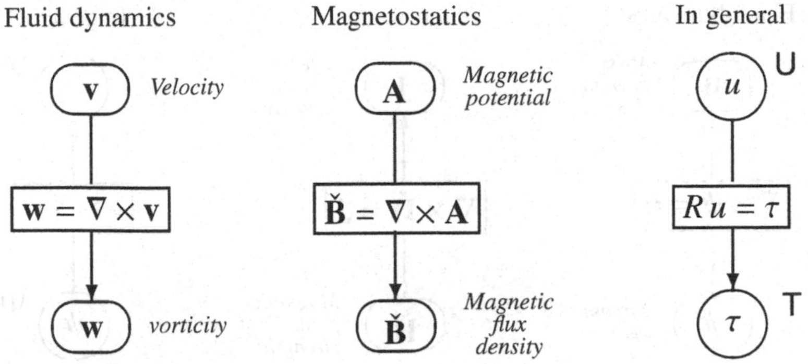


The last column on the right shows the notation we will use.

The diagram 15.8 shows the last two boxes of the right columns of the two diagrams. We aim to show that the differential operator $R$ is the adjoint of the operator $S$ and, at the same time, that the two formal differential operators coincide, i.e. $\mathcal{R} = \mathcal{S}$. Denoting by $\mathbf{S}_1$ and $\mathbf{S}_2$ the two parts of the boundary of the region $\mathbf{V}$, such that $\mathbf{S}_1 \cup \mathbf{S}_2 = \partial\mathbf{V}$, the main problems of magnetostatics are[15]

$$\begin{cases} B^k = \eta^{hki}\nabla_k A_i \\ A_i = 0 \quad \text{on } \mathbf{S}_1, \end{cases} \qquad \begin{cases} \eta^{pqr}\nabla_q H_r = J^p \\ \eta^{pqr} n_q H_r = 0 \quad \text{on } \mathbf{S}_2, \end{cases} \tag{15.45}$$

which can also be written in vector form as follows:

$$\begin{cases} \check{\mathbf{B}} = \nabla \times \mathbf{A} \\ \mathbf{A} = 0 \quad \text{on } \mathbf{S}_1, \end{cases} \qquad \begin{cases} \nabla \times \check{\mathbf{H}} = \mathbf{J} \\ \mathbf{n} \times \check{\mathbf{H}} = 0 \quad \text{on } \mathbf{S}_2, \end{cases} \tag{15.46}$$

and they are formed with the two linear operators

$$R \stackrel{\text{def}}{=} \left\{ \nabla\times \,;\, \mathbf{A}\big|_{\mathbf{S}_1} = 0 \right\}, \qquad S \stackrel{\text{def}}{=} \left\{ \nabla\times \,;\, (\mathbf{n}\cdot\mathbf{J})\big|_{\mathbf{S}_2} = 0 \right\}. \tag{15.47}$$

Taking account of the assigned boundary conditions, the boundary term in the following integration by parts vanishes, hence

---

[15] The symbol $\eta^{hki}$ denotes the Levi-Civita pseudotensor $\epsilon^{hki}/\sqrt{g}$ where $\epsilon^{hki}$ is the permutation symbol and $\sqrt{g}$ is the square root of the determinant of the metric tensor. See Appendix B.

**Table 15.8** Source side of diagrams [FLU2] and [ELE3]

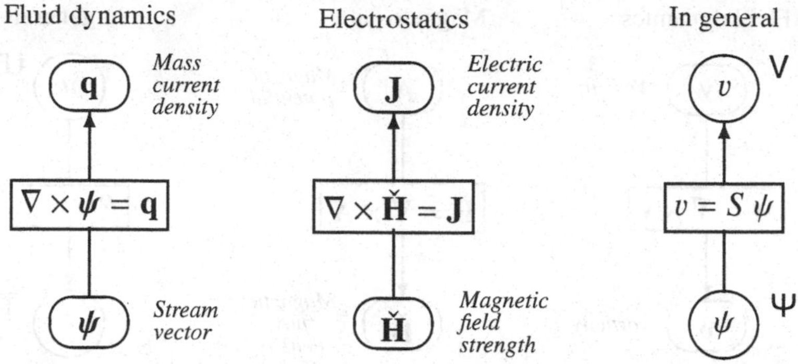


The last column on the right shows the notation we will use.

$$
\begin{aligned}
&\int_{\mathbf{V}} \mathbf{J}\cdot\mathbf{A}\,\mathrm{d}V \equiv \int_{\mathbf{V}} J^p A_p\,\mathrm{d}V = \int_{\mathbf{V}} (\eta^{pqr}\nabla_q H_r) A_p\,\mathrm{d}V \equiv \int_{\mathbf{V}} (\nabla_q \eta^{pqr} H_r) A_p\,\mathrm{d}V \equiv \\
&\equiv \int_{\mathbf{V}} \nabla_q(\eta^{pqr} H_r A_p)\,\mathrm{d}V - \int_{\mathbf{V}} (\nabla_q \eta^{pqr} A_p) H_r\,\mathrm{d}V \equiv \int_{\partial\mathbf{V}} n_q(\eta^{qrp} H_r A_p)\,\mathrm{d}S + \\
&+ \int_{\mathbf{V}} (\eta^{rqp}\nabla_q A_p) H_r\,\mathrm{d}V \equiv \int_{\partial\mathbf{V}} \mathbf{n}\cdot(\check{\mathbf{H}}\times\mathbf{A})\,\mathrm{d}S + \int_{\mathbf{V}} (\nabla\times\mathbf{A})\cdot\check{\mathbf{H}}\,\mathrm{d}V = \\
&= \int_{\partial\mathbf{V}} \mathbf{A}\cdot(\mathbf{n}\times\check{\mathbf{H}})\,\mathrm{d}S + \int_{\mathbf{V}} \check{\mathbf{B}}\cdot\check{\mathbf{H}}\,\mathrm{d}V = \int_{\mathbf{V}} \check{\mathbf{B}}\cdot\check{\mathbf{H}}\,\mathrm{d}V,
\end{aligned}
\tag{15.48}
$$

and it follows that the operators $R$ and $S$ are adjoint to one another with respect to the bilinear forms

$$
\langle J, A\rangle \overset{\text{def}}{=} \int_{\mathbf{V}} \mathbf{J}\cdot\mathbf{A}\,\mathrm{d}V, \qquad \langle B, H\rangle \overset{\text{def}}{=} \int_{\mathbf{V}} \check{\mathbf{B}}\cdot\check{\mathbf{H}}\,\mathrm{d}V\,. \tag{15.49}
$$

Hence we have the property

$$
S = R^* \quad \textit{and, a fortiori,} \quad \mathscr{S} = \mathscr{R}^*\,, \tag{15.50}
$$

i.e. the operator of the dual balance equation is the adjoint of the operator of the dual defining equation. We have stressed the fact that the formal differential operators are also adjoint. We can conclude that the *formal* operator 'curl ' is adjoint to itself, i.e. it is self-adjoint.

A great advantage of the introduction of the notion of formal differential operator arises when we consider non-homogeneous boundary or initial conditions. In fact, if the formal differential operator is linear, when we add homogeneous boundary or initial conditions, the differential operator is also linear, while by adding non-homogeneous boundary conditions we obtain a nonlinear operator.

Table [PDE] shows how the three kinds of partial differential equations, elliptic, hyperbolic and parabolic are located in the classification diagram.

## 15.4 The Three Kinds of Partial Differential Equations

Partial differential equations are divided into three classes: elliptic, hyperbolic and parabolic. Diagram [PDE] shows how these three classes arise in a classification diagram.

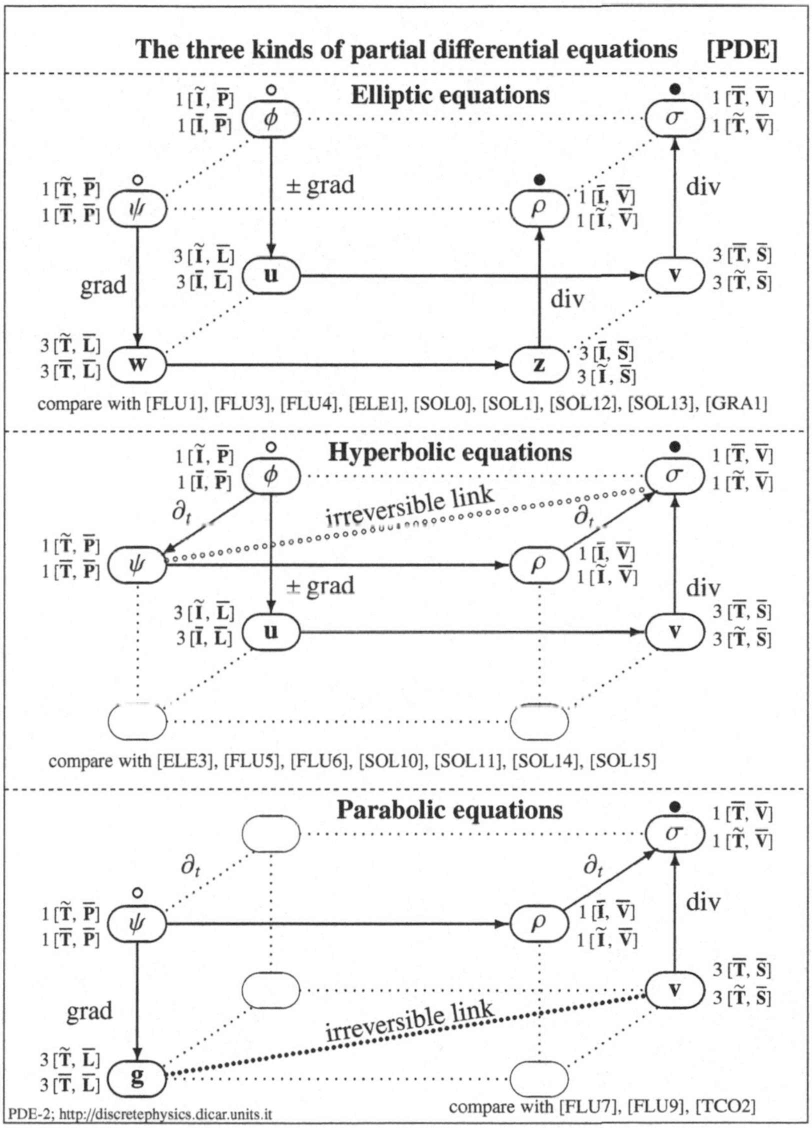

# Appendix A
# Affine Vector Fields

## A.1 Affine Fields

Given a scalar function of one variable $u = f(x)$, which we assume to be regular, i.e. continuous and with continuous derivative, the tangent to its representative curve at a point $x_0$ is the straight line of the equation

$$u = f(x_0) + \left.\frac{\mathrm{d}f}{\mathrm{d}x}\right|_{x_0} (x - x_0)\,. \tag{A.1}$$

In an analogous way, given a regular function of two variables $u = f(x, y)$, the tangent plane to its representative surface in the point of coordinates $(x_0, y_0)$ has the equation

$$u = f(x_0, y_0) + \left.\frac{\partial f}{\partial x}\right|_{(x_0,y_0)} (x - x_0) + \left.\frac{\partial f}{\partial y}\right|_{(x_0,y_0)} (y - y_0)\,. \tag{A.2}$$

These functions are of the kind

$$\begin{aligned} u &= a + b\,(x - x_0), \\ u &= a + b\,(x - x_0) + c\,(y - y_0), \\ u &= a + b\,(x - x_0) + c\,(y - y_0) + d\,(z - z_0) \end{aligned} \tag{A.3}$$

and have a linear behaviour: they are called *affine* (scalar) functions.

The notion of affine function can be extended to a vector-valued function as follows: denoted by $\mathbf{v}$, a vector field, with $\mathbf{O}$ a point of the three-dimensional region and with $\mathbf{v}_O$ the vector in such a point, we will call an *affine vector field* one that satisfies the following relation:

$$\mathbf{v} = \mathbf{v}_O + \mathsf{H}\,(\mathbf{r} - \mathbf{r}_O), \tag{A.4}$$

E. Tonti, *The Mathematical Structure of Classical and Relativistic Physics*, Modeling and Simulation in Science, Engineering and Technology, DOI 10.1007/978-1-4614-7422-7, where we have denoted by $\mathbf{r}$ the position vector and by $\mathbf{H}$ a $3 \times 3$ matrix with constant coefficients.

### A.1.1 Affine Scalar Field

An affine scalar field is described in Cartesian coordinates by the equation

$$\phi = a + h_x x + h_y y + h_z z \qquad \longrightarrow \qquad \phi = a + \mathbf{h} \cdot \mathbf{r}, \tag{A.5}$$

where $a, h_x, h_y, h_z$ are constants. The presence of the constant $a$ permits us to say that Eq. A.5 is an affine equation with a *linear* behaviour and *not* a linear equation. In the sequel, it is appropriate to call this function a *potential*. In fact, in physical theories such as electrostatics, the function $\phi$ indicates the electric potential, in fluid dynamics the velocity potential, in gravitation the gravitational potential, and so forth.

Every *regular* scalar field in a region can be approximated in the neighbourhood of a point with an affine scalar field.

The surfaces along which the function $\phi$ has a constant value, i.e. the equipotential surfaces, are parallel planes (Fig. A.1). In particular, denoting by $\phi_P$ and $\phi_Q$ the values of the function at two points arbitrarily chosen $\mathbf{P}$ and $\mathbf{Q}$ and considering the equipotential planes passing through $\mathbf{P}$ and $\mathbf{Q}$, all the planes which match the values $\phi = \phi_P + k[\phi_Q - \phi_P]$, where $k$ is a positive or a negative integer, are equidistant.

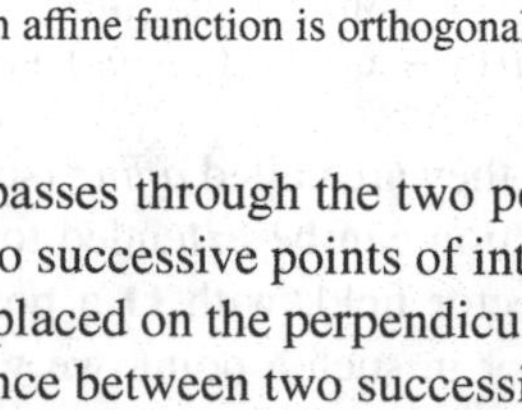

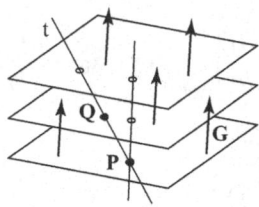

**Fig. A.1** The gradient of an affine function is orthogonal to the equipotential planes

In fact, the line $t$ which passes through the two points intersects these planes, and the distance between two successive points of intersection is constant. In particular, when the point $\mathbf{Q}$ is placed on the perpendicular to the equipotential plane passing through $\mathbf{P}$, the distance between two successive intersections is minimal.

It is obvious that the difference $u_Q - u_P$ is proportional to the length of the segment $(\mathbf{Q} - \mathbf{P})$. Therefore, the ratio $[u_Q - u_P]/||\mathbf{Q} - \mathbf{P}||$ becomes significant because it is the same for all the straight lines parallel to $t$. This ratio is called the *gradient in the direction t* and is denoted by $G_t$. In particular, $G_t$ is at its maximum when the straight line $t$ is perpendicular to the equipotential planes, and it is the same regardless of the point from which it is calculated. This suggests the need to establish a representative vector of this maximum ratio: this vector, denoted by **G**, goes towards the values where the potential increases and its modulus is equal to the maximum value of $G_t$. The vector thus defined is called *gradient* of the affine function $\phi$. Since for a scalar-valued affine function the gradient is independent of the point at which it is calculated, the vector field generated by the gradient vector field is a *uniform* vector field.

EXAMPLE 1. An example of affine scalar field is the gravitational potential near the ground (assumed plane); another is the field of pressure of the atmosphere at low altitudes (the pressure decreases linearly with altitude). A third is the field of temperatures at low altitudes.

An interesting property of affine scalar fields is the following:

**Theorem A1:** *the value of an affine function in the middle point of a segment is the average of the values of the function at the two extremes.* This property is easily proved.

### A.1.2 Affine Vector Field

An affine vector field in Cartesian coordinates has the form

$$\begin{cases} v_x = a_x + h_{xx}x + h_{xy}y + h_{xz}z \\ v_y = a_y + h_{yx}x + h_{yy}y + h_{yz}z \\ v_z = a_z + h_{zx}x + h_{zy}y + h_{zz}z \end{cases} \longrightarrow \qquad \mathbf{v} = \mathbf{a} + \mathbf{H}\,\mathbf{r}\,. \tag{A.6}$$

The matrix $\mathbf{H}$, whose elements are $h_{hk} = \partial v_h/\partial x_k$, is called *gradient* of the vector $\mathbf{v}$. Every *regular* vector field in a region can be approximated in the neighbourhood of a point with an affine vector field.

EXAMPLE 2. An affine vector field is a vector field of the motion of a *rigid* body. At each *instant* the velocity of a point $P$ is given by the equation

$$\begin{bmatrix} v_x \\ v_y \\ v_z \end{bmatrix} = \begin{bmatrix} c_x \\ c_y \\ c_z \end{bmatrix} + \begin{bmatrix} 0 & -\omega_z & \omega_y \\ \omega_z & 0 & -\omega_x \\ -\omega_y & \omega_x & 0 \end{bmatrix} \begin{bmatrix} x \\ y \\ z \end{bmatrix}, \tag{A.7}$$

where $c_x, c_y, c_z$ are the velocity components of one of its points $O$; $\omega_x, \omega_y, \omega_z$ are the components of the angular velocity with respect to a Cartesian system with origin in the point $O$; and $x, y, z$ are the coordinates of $P$. From the rotation of a rigid body originates the term 'rot', which is the European equivalent of the term 'curl'.

Among the useful properties of affine vector fields we mention the following, which are easily proved:[1]

[1] For the proofs see, for example, Tonti and Nuzzo [235, p. 22].

**Theorem A2:** *the value of an affine vector-valued function in the midpoint of a segment is the mean of the values of the affine vector-valued function at the two extremes.* This property is evident from Fig. A.2.

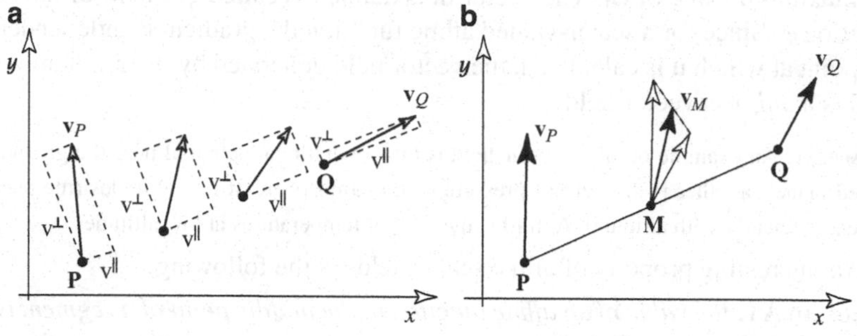


**Fig. A.2** The variation of a vector field along a straigth line. (**a**) The tangential and the normal components vary linearly; (**b**) The vector in the midpoint of a line segment is the mean of the vectors at its ends

**Theorem A3:** *the line integral of an affine vector field along a straight line segment is equal to the scalar product of the vector, evaluated in the midpoint of the segment, for that vector which describes the oriented segment.* In short,

$$\int_{\mathbf{P}}^{\mathbf{Q}} \mathbf{v}\cdot\mathbf{t}\,\mathrm{d}L = \mathbf{v}(\mathbf{M})\cdot(\mathbf{Q}-\mathbf{P})\,. \tag{A.8}$$

Proof. Placing the $x$-axis along the line segment $\mathbf{L}$ and putting the origin in the point $\mathbf{P}$, since the points of the segment have $y = 0$, we have $\mathbf{v}\cdot\mathbf{t} = v_x = a + bx$. The integral from 0 to $L$ is $(a + bL/2)L$. This is simply the product of the length $L$ for the component of the vector, evaluated in the midpoint of the segment, in the direction of the line. □

**Theorem A4:** *the circulation of a vector along the boundary of a plane surface is proportional to the area of the surface* [2]

$$\Gamma \stackrel{\text{def}}{=} \int_{\partial S} \mathbf{v}\cdot\mathbf{t}\,\mathrm{d}L \,\propto\, A\,. \tag{A.9}$$

Proof. For simplicity we consider a plane affine field and a triangle. With reference to the left part of Fig. A.3, let us consider three points $\mathbf{P}, \mathbf{Q}, \mathbf{R}$ and three vectors $\mathbf{v}_P, \mathbf{v}_Q, \mathbf{v}_R$. We aim to find the vector $\mathbf{v}$ at each point of the region using the formulas of the affine field, i.e.

$$v_x = a + bx + cy \qquad v_y = d + ex + fy\,. \tag{A.10}$$

[2] See, for example, Tonti and Nuzzo [235, p. 154].

Let us evaluate the circulation of the vector along the boundary of the triangle **PQR**. Setting $\mathbf{L}_1 = \mathbf{Q}-\mathbf{P}, \mathbf{L}_2 = \mathbf{R}-\mathbf{Q}, \mathbf{L}_3 = \mathbf{P}-\mathbf{R}$ and remarking that $\mathbf{L}_1+\mathbf{L}_2+\mathbf{L}_3 = 0$ we have

$$\begin{aligned} \Gamma &= +\frac{1}{2}[\mathbf{v}_P \cdot (\mathbf{L}_1 + \mathbf{L}_3) + \mathbf{v}_Q \cdot (\mathbf{L}_1 + \mathbf{L}_2) + \mathbf{v}_R \cdot (\mathbf{L}_2 + \mathbf{L}_3)] \\ &= -\frac{1}{2}[\mathbf{v}_P \cdot \mathbf{L}_2 + \mathbf{v}_Q \cdot \mathbf{L}_3 + \mathbf{v}_R \cdot \mathbf{L}_1] \, . \end{aligned} \tag{A.11}$$

Since $v_{Px} = a + bx_P + cy_P$, $v_{Py} = d + ex_P + fy_P$ and similarly for the other two points, developing the calculations we obtain

$$-2\Gamma = (c - e)(x_R y_P - x_Q y_P + x_P y_Q - x_R y_Q + x_Q y_R - x_P y_R) \, . \tag{A.12}$$

Denoting by $A$ the area of the triangle we have

$$\begin{aligned} 2A &= \begin{bmatrix} 1 & x_P & y_P \\ 1 & x_Q & y_Q \\ 1 & x_R & y_R \end{bmatrix} = \begin{bmatrix} 1 & x_P & y_P \\ 0 & (x_Q - x_P) & (y_Q - y_P) \\ 0 & (x_R - x_Q) & (y_R - y_Q) \end{bmatrix} \\ &= (x_Q y_R - x_P y_R + x_P y_Q - x_R y_Q + x_R y_P - x_Q y_P), \end{aligned} \tag{A.13}$$

from which

$$-2\Gamma = (c - e)2A \quad \longrightarrow \quad \frac{\Gamma}{A} = (e - c), \tag{A.14}$$

which shows that for a triangle the ratio between the circulation and the area is a constant, characteristic of the affine field. With reference to the right part of Fig. A.3, let us now consider a polygon of arbitrary shape and divide it into triangles. Considering two adjacent triangles the line integral along the edge of one triangle is opposite to the same line integral along the edge of the adjacent triangle. Since $\Gamma_k = (e - c)A_k$, it follows that $\Gamma = \int_k \Gamma_k = (e - c) \int_k A_k = (e - c)A$, as the theorem states. □

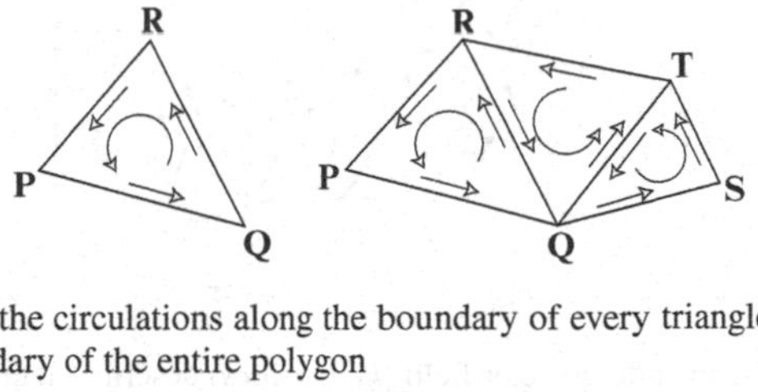


**Fig. A.3** The sum of the circulations along the boundary of every triangle is equal to the circulation along the boundary of the entire polygon

As a particular case of this theorem follows that the circulation along the boundary of different regions but with equal areas are equal, as shown in Fig. A.4.

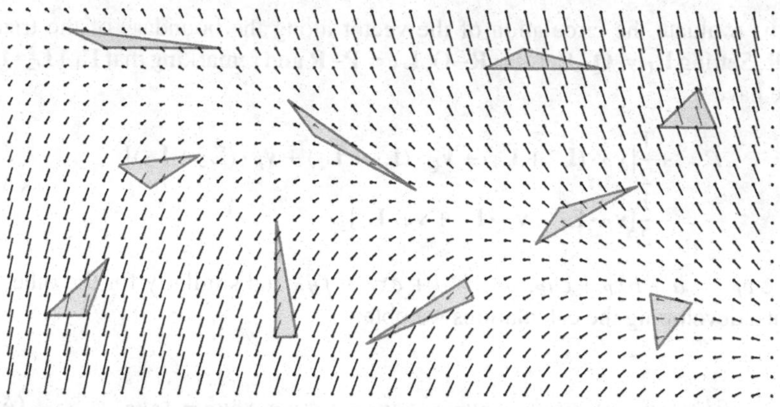

**Fig. A.4** In an affine vector field the circulations of the vector along the boundary of all *triangles* of equal area are equal. In another words, the circulation of the vector along a closed line is proportional to the area enclosed. The *small filled circles* denote the application points of vectors, while the line segments, without arrows, denote the vectors

This proportionality suggests the need to calculate the ratio between the circulation and the area obtaining a variable which depends on the space inclination of the plane and does not depend on the point. This ratio can be called *the curl in the considered direction*. With reference to a three-dimensional affine vector field, such as that shown in Fig. A.5, by changing the space inclination, the ratio changes. It is natural to consider as privileged the space inclination along which this ratio is at its maximum.

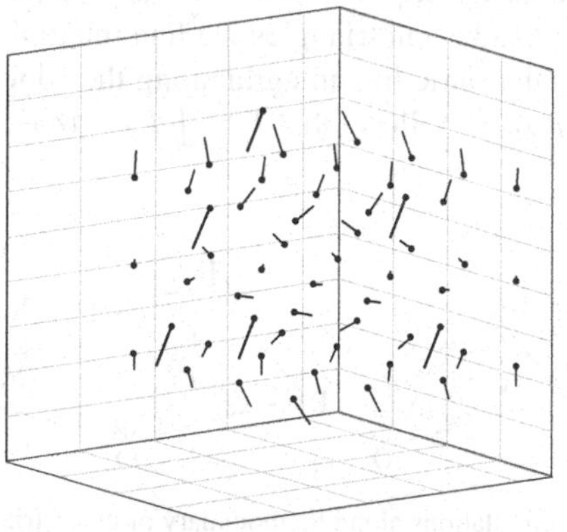

**Fig. A.5** The curl of an affine vector field (*heavy lines*) describes a uniform vector field

We can create a vector whose direction is perpendicular to the plane for which we have the maximum value of this ratio, whose modulus is the value of this maximum ratio and whose direction is the one obtained by applying the screw rule to the circuit along which the circulation is evaluated. This vector is called the curl of the affine vector field. The field of the 'curl' of an affine vector field is therefore a *uniform* vector field. In particular, the curl of an affine plane vector field is orthogonal to the plane.

Recall that the centroid of a plane surface is the point of coordinates

$$x_C \overset{\text{def}}{=} \frac{\sum_i x_i A_i}{\sum_i A_i}, \qquad y_C \overset{\text{def}}{=} \frac{\sum_i y_i A_i}{\sum_i A_i}, \qquad x_C \overset{\text{def}}{=} \frac{\int_S x\,\mathrm{d}S}{\int_S \mathrm{d}S}, \qquad y_C \overset{\text{def}}{=} \frac{\int_S y\,\mathrm{d}S}{\int_S \mathrm{d}S}. \tag{A.15}$$

If the plane surface is a strip made of homogeneous material, then the centroid coincides with the centre of mass.

**Theorem A5:** *the flux of an affine vector field across a plane surface of area A and unit normal* **n** *is equal to the scalar product of the vector evaluated in the centroid of the plane surface for the vector* **n** *A which describes the surface.*

Proof. With reference to Fig. A.6a, let us consider a plane surface element of area $A$. Let us consider a Cartesian system of axes with the $z$-axis orthogonal to the plane. Taking into account Eq. A.6, the flux $\Phi$ of the affine vector field on the surface is

$$\Phi = \int_{\mathrm{S}} \mathbf{v}_P \cdot \mathbf{n}\,\mathrm{d}S = \int_{\mathrm{S}} v_z\,\mathrm{d}S = \int_{\mathrm{S}} a_z\,\mathrm{d}S + h_{zx} \int_{\mathrm{S}} x_P\,\mathrm{d}S + h_{zy} \int_{\mathrm{S}} y_P\,\mathrm{d}S\,. \tag{A.16}$$

Using the centroid given by Eq. A.15 we can write

$$\Phi = a_z A + h_{zx}\, x_C\, A + h_{zy}\, y_C\, A = A\, v_{Cz} = \mathbf{v}_C \cdot \mathbf{n} A\,. \qquad \square \tag{A.17}$$

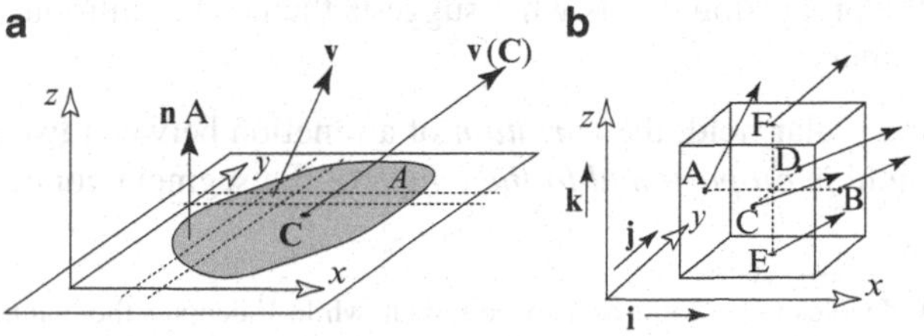


**Fig. A.6** (**a**) Flux across a plane surface. (**b**) Flux across boundary of a cube

**Theorem A6:** *the flux of the vector on a closed surface is proportional to the volume enclosed by the surface*[3]

$$\Phi \stackrel{\text{def}}{=} \int_{\partial \mathbf{V}} \mathbf{v} \cdot \mathbf{n}\, dS \propto V\,. \tag{A.18}$$

PROOF. With reference to Fig. A.6b, let us consider a cube of side $a$ and the fluxes across its six faces. Since every face is plane, we can use the result of Theorem A5 and compute the scalar products of the vectors in the centroids of the faces for the vectors $\mathbf{n}A$ relative to every face. We have

$$\begin{aligned}\Phi &= a^2\,[\mathbf{v}_A \cdot (-\mathbf{i}) + \mathbf{v}_C \cdot \mathbf{i} + \mathbf{v}_C \cdot (-\mathbf{j}) + \mathbf{v}_D \cdot \mathbf{j} + \mathbf{v}_E \cdot (-\mathbf{k}) + \mathbf{v}_F \cdot \mathbf{k}] \\ &= a^2\,[(\mathbf{v}_C - \mathbf{v}_A) \cdot \mathbf{i} + (\mathbf{v}_D - \mathbf{v}_C) \cdot \mathbf{j} + (\mathbf{v}_E - \mathbf{v}_F) \cdot \mathbf{k}] \\ &= a^3\,[h_{xx} + h_{yy} + h_{zz}],\end{aligned} \tag{A.19}$$

where we have used Eq. A.6, so that

$$(\mathbf{v}_C - \mathbf{v}_A) \cdot \mathbf{i} = v_{Bx} - v_{Ax} = h_{xx}(x_C - x_A) = a\,h_{xx} \tag{A.20}$$

and so forth. □

This proportionality suggests the need to compute the ratio between the flux on a generic closed surface and the volume enclosed in the region. This ratio has the same value at any point in the affine vector field: it is called a *divergence* of the affine vector field. The divergence of an affine vector field is a *uniform* scalar field.

In short, we have

| | | | |
|---|---|---|---|
| gradient (vector) | $\mathbf{G} = \dfrac{\phi_Q - \phi_P}{L}\mathbf{N}$ | $\mathbf{N}$ is the unit vector in the direction of maximum ratio (increment/distance) | |
| curl (vector) | $\mathbf{C} = \dfrac{\Gamma[\partial \mathbf{S}]}{A}\mathbf{N}$ | $\mathbf{N}$ is the unit vector in the direction orthogonal to the plane for which we have the maximum ratio circulation/area | (A.21) |
| divergence (scalar ) | $D = \dfrac{\Phi[\partial \mathbf{V}]}{V}$ | | |

As we see, the proportionality is what suggests the need to introduce meaningful ratios. In summary:

- In an affine scalar field, the *variation* of a function between two points of the physical field is *proportional to the length* of the segment connecting the two points.

[3] Recall that $\mathbf{V}$ denotes the volume as a space region, while $V$ denotes the volume as a measure of the extension of the space region: it is an unfortunate coincidence of terminology that we try to alleviate by using the symbol in bold. This ambiguity does not happen for lines and surfaces.

- In an affine vector field,
  - The *circulation* of a vector along a plane closed line is *proportional to the area* enclosed by the line;
  - The *flux* of a vector through a closed surface is *proportional to the volume* of the region enclosed by the surface.

1. A first property of affine vector fields is that the line integral of the vector along an oriented line segment joining two points of $P$ and $Q$ is equal to the scalar product between the vector in the midpoint of the segment and the vector representing the oriented line segment.
2. A second property is that the flux of the vector through an element of a plane surface is equal to the scalar product of the vector in the centroid of the surface element and the vector which describes the surface element.

# Appendix B
# Tensorial Notation

## B.1 Summary of Tensorial Notation Used in This Book

Let us start by recalling the role of coordinate systems in geometry. Before the introduction of Cartesian coordinates, geometry and algebra were two distinct disciplines. After the idea of Descartes to associate with every point a set of numbers (two in the plane and three in space) and vice versa, every line and every surface in space could be described analytically by functions giving the coordinates of their points as functions of one parameter (for one-dimensional lines) or two parameters (for two-dimensional surfaces). This was the birth of analytic geometry. The role of a coordinate system is precisely to build a bridge between geometry and analysis.

When in physics new entities called vectors, characterized by an oriented direction in space, a magnitude and an application point, were introduced, it was natural to seek an analytical representation of such entities. This is the case of displacement, velocity, force, momentum and many others physical variables. Vector calculus, making use of geometric objects, which are segments provided with an arrow and a point of application, allows a purely geometrical description. We can apply the parallelogram rule to compute the sum of two vectors which have the same application point, and we can multiply a vector by a real number to obtain another vector with the same application point. In short, vector calculus is an extension of classical geometry.

Faced with the great utility of analytic geometry for physics, it is natural to search for a corresponding analytical treatment of vector calculus. This requires associating some kind of 'coordinates' with every vector. This is realized by introducing a basis at every space point and considering the vector at every point as a linear combination of the base vectors at the point. The coefficients of this linear combination are called the *components* of the vector at the point.

The basis vectors at each point in space can, in principle, be chosen at will, but it is spontaneous to choose vectors which are tangent to the coordinate lines

E. Tonti, *The Mathematical Structure of Classical and Relativistic Physics*,
Modeling and Simulation in Science, Engineering and Technology,
DOI 10.1007/978-1-4614-7422-7, © Springer Science+Business Media New York 2013

passing through the point. We can choose base vectors of unit length; a basis so obtained is called a *physical basis*. Another choice is that of the *natural basis*, which we present subsequently.

Due to its geometrical nature, vector calculus does not require the use of coordinate systems to express physical laws, i.e. it gives rise to an intrinsic formulation. This is the case of the fundamental equation of motion for a particle, i.e. $m\,\mathbf{a} = \mathbf{F}$.

Nevertheless, performing calculations necessitates working with components instead of with vectors themselves. This prompts us to inquire as to how components are transformed in passing from one basis to another. The transformation formulae are the subject of *tensor analysis*.

The aim of tensor analysis is to introduce notations and rules to write variables and equations in a way which is independent of the coordinate system chosen and of the basis chosen.

In dealing with tensors, some authors start with tensor *calculus*, others with tensor *algebra*. Authors of the first group start with coordinate systems, while those of the second group start with the basis in a vector space.[1]

In algebra, the starting point is vector spaces and base vectors, while in physics the starting point is vector fields, hence coordinate systems, natural bases and metric tensors. In algebra, one considers the transition from an old base $\mathbf{e}_1, \mathbf{e}_3, \mathbf{e}_3$ to a new base $\mathbf{e}'_1, \mathbf{e}'_3, \mathbf{e}'_3$, while in physics one considers the transition from an old coordinate system $x^1, x^2, x^3$ and its natural basis to a new coordinate system $x'^1, x'^2, x'^3$ and its natural basis.

In algebra, given two bases of a vector space, one can always link the vectors of the two bases by a linear combination (recall that *vector* space is synonymous with *linear* space):

$$\mathbf{e}'_h = \lambda^k_h\, \mathbf{e}_k, \tag{B.1}$$

in which the coefficients $\lambda^k_h$ are constants.

In physics, one is led to consider the transition from two coordinate systems

$$x'^h = f^h(x^k) \qquad \text{from which} \qquad \mathrm{d}x'^h = \frac{\partial x'^h}{\partial x^k}\,\mathrm{d}x^k\,. \tag{B.2}$$

Denoting by $\mathbf{P}(x)$ a point, we will denote by $\mathbf{g}_k$ the *natural* base vectors $\partial\mathbf{P}/\partial x^k$. The transformation law of the base vectors is

$$\mathbf{g}_k(x) \stackrel{\text{def}}{=} \frac{\partial\mathbf{P}(x)}{\partial x^k}, \qquad \mathbf{g}'_h(x) = \frac{\partial\mathbf{P}(x)}{\partial x'^h} = \frac{\partial\mathbf{P}(x)}{\partial x^k}\frac{\partial x^k(x')}{\partial x'^h} = \mathbf{g}_k(x)\frac{\partial x^k(x')}{\partial x'^h}\,. \tag{B.3}$$

[1] From space coordinates: Levi-Civita [136, p. 2], Synge and Schild [221, p. 3], Laugwitz [129, p. 94], Eisenhart [62, p. 1]. From basis in algebra: Willmore [252, p. 166], Lichnerowicz [138, p. 8].

Comparing Eq. B.3 with Eq. B.1 we see that the coefficients of the change in the natural basis are

$$\lambda^k_h(x') \stackrel{\text{def}}{=} \frac{\partial x^k(x')}{\partial x'^h}, \tag{B.4}$$

and here the coefficients $\lambda^k_h$ are no longer constants but functions of the new coordinates. In particular, the infinitesimal vector $\mathbf{dP}$ can be written as $\mathbf{dP}(x) = \mathrm{d}x^k\,\mathbf{g}_k(x)$.

Vectors which are transformed like base vectors are called *covariant* because they vary in the same way. Under a change in basis, the components $\mathrm{d}x'^h$ of the infinitesimal vector $\mathbf{dP}$ change according to the relation

$$\mathrm{d}x'^h = \frac{\partial x'^h(x)}{\partial x^k}\,\mathrm{d}x^k = \Lambda^h_k(x)\,\mathrm{d}x^k, \qquad \text{where we have set} \qquad \Lambda^h_k(x) \stackrel{\text{def}}{=} \frac{\partial x'^h(x)}{\partial x^k}, \tag{B.5}$$

that is, the differentials of the coordinates change in accordance with the inverse transformation and for this reason are said to be *contravariant.*

A matrix for a change in basis whose elements are $\lambda^k_h$ is called a *transition matrix.*[2] The two indices of the elements of the transition matrix can be written on the same vertical[3] simply because there is no reason to put them in two distinct columns, like $\lambda_h{}^k$ or $\lambda^k{}_h$.

The determinant of the transition matrix, commonly denoted by $\Delta$, plays a special role in tensor calculus:

$$\overbrace{\Delta \stackrel{\text{def}}{=} \det(\lambda^k_h),}^{\text{tensor algebra}} \qquad \overbrace{\Delta(x') \stackrel{\text{def}}{=} \det(\lambda^k_h(x'))}^{\text{tensor calculus}}\ . \tag{B.6}$$

## *B.1.1 Generalized Kronecker Delta*

As is well known, by definition $\delta^p_q = 1$ for $p = q$ and $\delta^p_q = 0$ for $p \neq q$ in *every* basis. The transformation law for $\delta^p_q$ is the law of a *mixed tensor*, as can be seen from the identity

$$\delta'^p_q = \frac{\partial x'^p}{\partial x'^q} \equiv \frac{\partial x'^p}{\partial x^i}\frac{\partial x^i}{\partial x'^q} \equiv \delta^j_i \frac{\partial x'^p}{\partial x^j}\frac{\partial x^i}{\partial x'^q} \equiv \delta^j_i \Lambda^p_j\,\lambda^i_q\ . \tag{B.7}$$

[2] Postnikov [180, vol. 1, p. 97], Godement [78, p. 206], Brillouin [30, p. 46].

[3] This rule is followed by most authors: see Schouten [203, p. 1], Postnikov [180, p. 97], Lichnerowicz [138, p. 11], Brillouin [30, p. 25], Levi-Civita [136, p. 69], Weyl [247, p. 35]. Nevertheless, some authors put the indices in two distinct columns: see Laugwitz [129, p. 69], Schutz [205, p. 61], Corson [43, p. 9], Hehl and Obukhov [89, p. 20].

The Kronecker *delta* is a particular case of the *generalized Kronecker delta* $\delta^p_h$, $\delta^{pq}_{hi}$, $\delta^{pqr}_{hij}$ whose definition is[4]

$$\delta^{pqr}_{ijk} \overset{\text{def}}{=} \begin{cases} +1 \text{ if } i,j,k \text{ form an even permutation of } p,q,r; \\ -1 \text{ if } i,j,k \text{ form an odd permutation of } p,q,r; \\ \;\;0 \text{ if two upper or lower indices are equal,} \end{cases} \tag{B.8}$$

and similarly for $\delta^{pq}_{hi}$ and $\delta^p_h$. Its transformation law is also one of mixed tensors. We have

$$\delta^{pq}_{hi} \overset{\text{def}}{=} \begin{vmatrix} \delta^p_h & \delta^q_h \\ \delta^p_i & \delta^q_i \end{vmatrix} = \delta^p_h\,\delta^q_i - \delta^p_i\,\delta^q_h, \qquad \delta^{pqr}_{hij} \overset{\text{def}}{=} \begin{vmatrix} \delta^p_h & \delta^q_h & \delta^r_h \\ \delta^p_i & \delta^q_i & \delta^r_i \\ \delta^p_j & \delta^q_j & \delta^r_j \end{vmatrix} = \delta^p_h\,\delta^{qr}_{ij} - \delta^p_i\,\delta^{qr}_{hj} + \delta^p_j\,\delta^{qr}_{hi}\,. \tag{B.9}$$

## B.1.2 Permutation Symbol

This is defined as

$$\epsilon_{hij} \overset{\text{def}}{=} \delta^{123}_{hij}, \qquad \epsilon^{hij} \overset{\text{def}}{=} \delta^{hij}_{123}, \qquad \epsilon_{hijk} \overset{\text{def}}{=} \delta^{1234}_{hijk}, \qquad \epsilon^{hijk} \overset{\text{def}}{=} \delta^{hijk}_{1234}, \tag{B.10}$$

$$\epsilon_{ijk} = \begin{cases} +1 \text{ if } h,i,j \text{ are a cyclic permutation of } 1,2,3; \\ -1 \text{ if } h,i,j \text{ are a cyclic permutation of } 2,1,3; \\ \;\;0 \text{ otherwise;} \end{cases} \tag{B.11}$$

and similarly for $\epsilon^{ijk}$. To maintain the same values, i.e. $-1, 0, +1$, in *all* coordinate systems, the transformation law of these symbols is that of a tensor *capacity* and tensor *density* respectively,[5] i.e.

$$\underbrace{\epsilon'_{pqr} = \frac{1}{\Delta}\lambda^h_p\,\lambda^i_q\lambda^j_r\epsilon_{hij},}_{\text{tensor capacity}} \qquad \underbrace{\epsilon'^{pqr} = \Delta\,\Lambda^p_h\,\Lambda^q_i\Lambda^r_j\epsilon^{hij}}_{\text{tensor density}}\,. \tag{B.12}$$

Despite the use of the same letter $\epsilon$, once with subscripts and once with superscripts, the two symbols are *not* the covariant and contravariant components of a single object; hence the indices cannot be lowered or raised with the metric tensor. Expressions like $\epsilon^{pq}_r$, $\epsilon^p_{qr}$, etc. have no meaning in tensor calculus.

**Remark.** Schouten [203, p. 29] uses two different letters for the covariant and contravariant components: the letters $e$ and $E$ to stress that the indices cannot be raised and lowered with the

[4] Synge [221, p. 242], Eriksen [63, p. 38].

[5] Brillouin [30, Chap. III, Sect. 4].

fundamental tensor. Synge [221, p. 250] uses the same letter $\epsilon$ as we do but advises that, in the presence of a metric, the covariant and controvariant forms are *not* obtained from one another by raising or lowering suffixes with the metric tensor. He advises that this is a violation of the rules of tensor calculus, a violation commonly committed in the literature.

The term *tensor capacity* was introduced by Brillouin.[6] It is well chosen because it refers to the volume considered as a recipe which can be filled with matter which has a certain density. This can be seen from the fundamental relation[7]

$$\begin{aligned} M &= \int_{\mathbf{V}} \rho\, \mathrm{d}V = \int_{\mathbf{V}} \rho\,(\sqrt{g}\,\mathrm{d}x^1 \wedge \mathrm{d}x^2 \wedge \mathrm{d}x^3) = \int_{\mathbf{V}} (\rho\sqrt{g})\,(\mathrm{d}x^1 \wedge \mathrm{d}x^2 \wedge \mathrm{d}x^3) \\ &= \int_{\mathbf{V}} \rho_{123}\, \mathrm{d}x^1 \wedge \mathrm{d}x^2 \wedge \mathrm{d}x^3 = \int_{\mathbf{V}} \rho_{123}\, \mathrm{d}\tau. \end{aligned} \tag{B.13}$$

A tensor capacity is also called a relative tensor of weight $-1$, while a tensor density is a relative tensor of weight $+1$.[8]

**Levi-Civita Pseudotensor.** The element of volume in general coordinates has the form $\mathrm{d}V = \sqrt{g}\,\mathrm{d}x^1 \wedge \mathrm{d}x^2 \wedge \mathrm{d}x^3$, where $g \stackrel{\text{def}}{=} \det(g_{hk})$ is the determinant of the metric tensor which obeys the transformation law

$$\sqrt{g'} = |\varDelta|\,\sqrt{g}\,. \tag{B.14}$$

The permutation symbols and $\sqrt{g}$ combine, giving rise to two other quantities:

$$\boxed{\check{\eta}^{hij} \stackrel{\text{def}}{=} \frac{1}{\sqrt{g}}\,\epsilon^{hij}, \qquad \check{\eta}_{hij} \stackrel{\text{def}}{=} \sqrt{g}\,\epsilon_{hij}} \tag{B.15}$$

which are called the *Levi-Civita pseudotensor*.[9] The law of transformation of the Levi-Civita pseudotensor is as follows: denoted by $\mathrm{sgn}(\varDelta) \equiv \varDelta/|\varDelta| \equiv |\varDelta|/\varDelta$, the sign of the determinant of the transition matrix, it can be shown that

$$\check{\eta}'^{pqr} = \mathrm{sgn}(\varDelta)\,\Lambda^p_h\,\Lambda^q_i\Lambda^r_j\,\check{\eta}^{hij}, \qquad \check{\eta}'_{pqr} = \mathrm{sgn}(\varDelta)\,\lambda^h_p\,\lambda^i_q\lambda^j_r\,\check{\eta}_{hij}\,. \tag{B.16}$$

The first is a contravariant pseudotensor, while the second is a covariant pseudotensor. This transformation law is similar to that of contravariant and covariant

6 Brillouin [30, Chap. III, Sect. 4].

7 Brillouin [30, p. 109]; Ingarden and Jamiolkowski [100, p. 44].

8 Schouten [204, p. 29].

9 The letter $\eta$ is used by Synge and Schild [221, p. 249], Zund and Brown [258], Burke [32, p. 182], Ingarden and Jamiolkowski [100], Von Westenholz [243, vol. I, p. 184].

tensors respectively but differs by the factor sgn($\Delta$), which is $\pm 1$ and $-1$ when we pass from a right-handed to a left-handed basis. For this reason they are called pseudotensors.[10] We will put a hacek over the letter $\eta$.[11]

### B.1.3  Main Use of Levi-Civita Pseudotensor

The main use of the Levi-Civita pseudotensor is in the evaluation of the components of the vector product of two vectors, i.e.

$$\check{\mathbf{w}} = \mathbf{u} \times \mathbf{v}, \qquad \check{w}^h = \check{\eta}^{hij} u_i v_j, \qquad \check{w}_h = \check{\eta}_{hij} u^i v^j . \qquad (B.17)$$

The hacek over the letter $\mathbf{w}$ recalls the 'pseudo' nature of the vector, hence the transformation law

$$\check{w}'^p = \text{sgn}(\Delta)\, \Lambda_h^p\, \check{w}^h, \qquad \check{w}'_p = \text{sgn}(\Delta)\, \lambda_p^h\, \check{w}_h . \qquad (B.18)$$

### B.1.4  Vector Components

In physics, one deals frequently with different coordinate systems and with different basis vectors. There are three main sets of basis vectors:

- *Natural basis vectors* $\mathbf{g}_k \overset{\text{def}}{=} \dfrac{\partial \mathbf{P}}{\partial x^k}$, from which $g_{hk} \overset{\text{def}}{=} \mathbf{g}_h \cdot \mathbf{g}_k$;
- *Reciprocal basis vectors* $\mathbf{g}^h$, defined as $\mathbf{g}^h \overset{\text{def}}{=} \check{\eta}^{hij} \mathbf{g}_i \times \mathbf{g}_j$, from which $\mathbf{g}^h \cdot \mathbf{g}_k = \delta_k^h$;
- *Physical basis vectors* $\mathbf{e}_{(k)} \overset{\text{def}}{=} \dfrac{\mathbf{g}_k}{\sqrt{g_{kk}}}$:

$$
\begin{array}{ccccccc}
\text{vector} & \text{basis vectors} & & \begin{array}{c}\text{reciprocal}\\\text{basis vectors}\end{array} & & \begin{array}{c}\text{physical}\\\text{basis vectors}\end{array} & \\
\downarrow & \downarrow & & \downarrow & & \downarrow & \\
\mathbf{v} & = \quad v^h\, \mathbf{g}_h & = & v_h\, \mathbf{g}^h & = & v_{(h)}\, \mathbf{e}_{(h)} & ; \\
& \uparrow & & \uparrow & & \uparrow & \\
& \begin{array}{c}\text{contravariant}\\\text{components}\end{array} & & \begin{array}{c}\text{covariant}\\\text{components}\end{array} & & \begin{array}{c}\text{physical}\\\text{components}\end{array} &
\end{array} \qquad (B.19)
$$

---

[10] Nowadays also called a *twisted* tensor. For the present author, the term *twisted* is inappropriate when compared with its use in everyday life: just think of *a twisted sentence*. What changes here is simply a sign in passing from a dextrorotatory coordinate system to a levorotatory one; nothing is 'twisted'!

[11] As is done by Corson [43, pp. 8, 98].

the $v_{(k)}$ are called *physical components*.[12] The covariant, contravariant and physical components are given by

$$v_h = \mathbf{v} \cdot \mathbf{g}_h, \qquad v^h = \mathbf{v} \cdot \mathbf{g}^h, \qquad v_{(h)} = v^h \sqrt{g_{hh}} \quad \text{(not summed)} . \qquad \text{(B.20)}$$

**Table B.1** The six kinds of scalars

| **Densities** | | **Capacities** |
|---|---|---|
| Ordinary density<br>$S' = \lvert\Delta\rvert\, S$<br>$\sqrt{g},\ \rho_{123}$ | Scalar<br>$S' = S$<br>Mass, electric charge, energy, work, heat electric potential, gauge function, action electromotive force, phase, entropy, pressure magnetic flux, electric flux, temperature particle number, charge flow, mass density | Ordinary capacity<br>$S' = \dfrac{1}{\lvert\Delta\rvert} S$<br>Volume with outer orientation<br>$\dfrac{1}{\sqrt{g}}$, $\mathrm{d}\tau = \mathrm{d}x^1\,\mathrm{d}x^2\,\mathrm{d}x^3$ |
| Density<br>$S' = \Delta\, S$ | Pseudoscalar<br>$S' = \dfrac{\lvert\Delta\rvert}{\Delta} S$<br>Magnetic scalar potential (magnetic charge) | Capacity<br>$S' = \dfrac{1}{\Delta} S$<br>Volume with inner orientation |

## B.2 Algebraic and Metric Duals

Given a second-order tensor, one may consider its dual (Table B.1). There are two kinds of dual, the *algebraic dual* made with the permutation symbol and the *metric dual* made with the Levi-Civita pseudotensor. Following Corson[13] we will place a breve (like a hacek but with a rounded bottom) over the letter to denote the *algebraic* dual and a hacek to denote the *metric* dual. Thus, for the magnetic induction tensor $B_{ij}$ we have

$$\overbrace{\breve{B}^k \stackrel{\text{def}}{=} \epsilon^{kij} B_{ij}}^{\text{algebraic dual}}, \qquad \overbrace{\check{B}^k \stackrel{\text{def}}{=} \check{\eta}^{kij} B_{ij}}^{\text{metric dual}}, \qquad \text{hence} \qquad \check{B}^k = \frac{1}{\sqrt{g}} \breve{B}^k . \qquad \text{(B.21)}$$

---

[12] Truesdell [238, p. 347]; Spiegel [217, p. 172].

[13] Corson [43, p. 6, 8].

In tensor calculus, the most important dual is the metric dual, which is useful for the vector product of two vectors and for the curl of a vector:

$$\check{\mathbf{w}} = \text{curl}\,\mathbf{v} \quad \longrightarrow \quad \check{w}^h = \check{\eta}^{kij}\partial_i v_j \equiv \check{\eta}^{kij}\nabla_i v_j. \tag{B.22}$$

The last equality is due to the symmetry of the Christoffel symbols entering the covariant derivative.

**Table B.2** Main formulae involving gradient, curl and divergence

| | Gradient | Curl | Divergence |
|---|---|---|---|
| Cartesian | $G_x = -\partial_x\phi;\ldots$ | $C_x = \partial_y u_z - \partial_z u_y;\ldots$ | $D = \partial_x v_x + \partial_y v_y + \partial_z v_z$ |
| General | $G_k = -\partial_k\phi$ | $\check{C}^m = \check{\eta}^{mhk}\partial_k u_k$ | $D = \frac{1}{\sqrt{g}}\partial_k(\sqrt{g}\,v^k)$ |
| Symbolic | $\mathbf{G} = -\nabla\phi \equiv -\text{grad}\,\phi$ | $\mathbf{C} = \nabla\times\mathbf{u} \equiv \text{curl}\,\mathbf{u}$ | $D = \nabla\cdot\mathbf{v} \equiv \text{div}\,\mathbf{v}$ |
| Integral | $\mathbf{G} = \lim_{\mathbf{V}\to\mathbf{P}}\frac{1}{V}\int_{\partial\mathbf{V}}\phi\,\mathbf{n}\,\mathrm{d}S$ | $\mathbf{C} = \lim_{\mathbf{V}\to\mathbf{P}}\frac{1}{V}\int_{\partial\mathbf{V}}\mathbf{n}\times\mathbf{u}\,\mathrm{d}S$ | $D = \lim_{\mathbf{V}\to\mathbf{P}}\frac{1}{V}\int_{\partial\mathbf{V}}\mathbf{v}\cdot\mathbf{n}\,\mathrm{d}S$ |

## B.3 Bivectors

The notion of *exterior product* between two vectors is usually described in a purely algebraic way which deprives the reader of its intuitive geometric meaning and makes difficult its introduction in courses on experimental physics. Elie Cartan, the creator of the theory of differential forms, wrote: '*We call* bivector *the figure formed by two vectors* **x**, **y** *kept in a certain order.*' Of course, '*This definition does not make sense before we define the equality of two bivectors*'. [14]

According to its general meaning, the term *product of two objects* denotes the result of an operation performed on two objects. For extension of the term it also denotes the operation itself. Thus, the *scalar product* of two vectors is a number; the *vector product* of two vectors is another vector; the *cartesian product* of two sets $X$ and $Y$ is another set formed by all pairs of elements $(x, y)$, with $x \in X$ and $y \in Y$; the *matrix product* of two matrices $P$ and $Q$ is another matrix $R$ resulting from the successive application of the two matrices; and so forth.

Let us consider two vectors **u** and **v** in Euclidean space. Arranging the vectors in sequence, as shown in Fig. B.1*left*, we may consider the oriented parallelograms formed by them. There are two possible configuration according to whether **u** precedes **v** or **v** precedes **u**. Even if the parallelogram is the same, the two parallelograms have different inner orientation. This is a first stage. Going one step further,

[14] Cartan [34, p. 5].

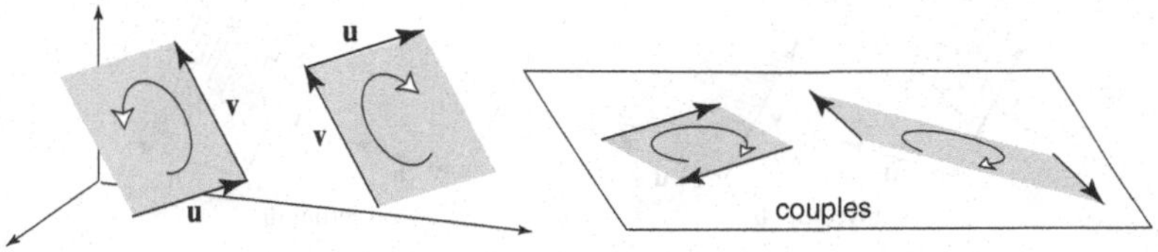


**Fig. B.1** *Left*: A bivector is a parallelogram with an inner orientation; *right*: a couple (mechanics) can be described by a bivector

we will consider two parallelograms to be equivalent when they satisfy three properties: (1) they lie on the same plane or on parallel planes, (2) have equal areas, and (3) have the same orientation. This suggests that all oriented parallelograms which are equivalent to a given one may be regarded as the same object. This equivalence class is called a *bivector*, and every parallelogram of this class is a *representation* of the bivector.

The reader will certainly recall the notion of couple in mechanics: two forces which are parallel, of equal modulus but with opposite orientations form a couple. They give rise to an oriented parallelogram, as shown in Fig. B.1 *right*. The oriented area of this parallelogram is equal, up to a constant factor (due to the different scales of lengths and forces), to the moment of the couple. All pairs of forces which give rise to equivalent parallelograms constitute the same couple. The couple is the equivalence class of all pairs of opposite forces with the same moment. Commonly speaking, we say that two opposite forces *form* a couple even if we are conscious that they only *represent* a couple. A couple is a more general concept which gives us the freedom to choose a representative pair of opposite forces among an infinity of them. In the same sense, one may say that an oriented parallelogram *is* a bivector even if it is better to say that it *represents* a bivector.

## *B.3.1 Exterior Product of Two Vectors*

Let us introduce the notion of the *exterior product* of two vectors. We may consider an oriented parallelogram as being generated by the translation of the first vector along the second one or by the translation of the second vector along the first one, as shown in Fig. B.1. One can consider this operation, the generation of an oriented parallelogram by the translation of one vector along another, as a new kind of 'product' between vectors. The oriented parallelogram thus generated, i.e. the bivector, is called the *exterior product* of two vectors. If $\mathbf{u}$ and $\mathbf{v}$ are the two vectors and if we denote by $\mathbf{b}$ the bivector, we may introduce the symbol $\wedge$, called a wedge, to denote this kind of product and write (Fig. B.2)

$$\mathbf{b} = \mathbf{u} \wedge \mathbf{v} \qquad -\mathbf{b} = \mathbf{v} \wedge \mathbf{u}; \tag{B.23}$$

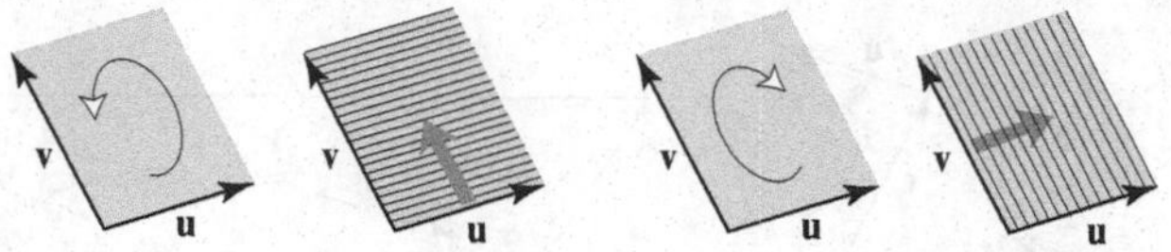


**Fig. B.2** The exterior product of two vectors generates a bivector. Inverting the order of the two vectors changes the sign of the bivector

hence, the exterior product is skew-symmetric. We may say that the bivector $\mathbf{b} = \mathbf{u} \wedge \mathbf{v}$ is the entity which permits the algebraic description of an oriented parallelogram and then of any oriented polygon equivalent to it.

Regarding the tendency to make a purely algebraic formulation of the geometry, we quote the words of Hilbert [91, Preface] (emphasis in the original):

> *In mathematics, as in any scientific research, we find two tendencies present. On the one hand, the tendency toward* abstraction *seeks to crystallize the* logical *relations inherent in the maze of material that is being studied, and to correlate the material in a systematic and orderly manner. On the other hand, the tendency toward* intuitive understanding *fosters a more immediate grasp of the objects one studies, a live* rapport *with them, so to speak, which stresses the concrete meaning of their relations.*
>
> *As to geometry, in particular, the abstract tendency has here led to the magnificent systematic theories of Algebraic Geometry, of Riemannian Geometry, and of Topology; these theories make extensive use of abstract reasoning and symbolic calculation in the sense of algebra. Notwithstanding this, it is still as true today as it ever was that* intuitive understanding *plays a major role in geometry. And such concrete intuition is of great value not only for the research worker, but also for anyone who wishes to study and appreciate the results of research in geometry.*

# Appendix C
# On Observable Quantities

Many physicists argue that physics should not use physical quantities which are not directly measurable. Nothing is more absurd than this taboo.

The same authors then use the magnetic vector potential **A**, the entropy $S$, the scalar magnetic potential $\varphi_{\mathrm{m}}$, the stress potentials $\chi_{hk}$ in continuum mechanics, the Airy function $\phi$ and the wave function $\psi$ of quantum mechanics, which are not measurable quantities. It is a misconception, mistakenly attributed to Heisenberg.

It is obvious that physics should start from measurable quantities, but in the course of its treatment one is free to introduce quantities which are not directly measurable but which can serve as bridges to other measurable quantities. Thus, the magnetic vector potential **A**, defined up to the gradient of the (gauge) function $\chi$, is not measurable, but its 'curl' is the magnetic induction vector **B**, which *is* measurable.

Similarly, the wave function $\psi$, defined up to a factor of phase $\exp(i\phi)$, is not measurable, but the product $\psi\psi^*$ integrated on a region of space gives the probability of finding a particle within that region, and it is a measurable quantity.

To remove this taboo, we quote the opinions of some authoritative physicists.

**Louis de Broglie:** 'The quantities which physicists use in their reasoning are not all observable and measurable. Certain of these serve only as intermediaries; they come into the calculations but are eliminated when comparison is made with experiment. We have attempted, by taking a purely phenomenological point of view, to eject all the non-measurable quantities from physical theories. The doctrine of energy and, more recently, the doctrine of quantal mechanics by Heisenberg are remarkable examples of this kind of effort. But these attempts have never completely succeeded; there always exist in the theories quantities which are non-measurable, and for example, in wave mechanics the well-known wave function $\psi$ belongs to this category. Nevertheless the measurable quantities retain a basic importance for it is through them that the indispensable experimental control of the consequences of the theory is carried out.' (de Broglie [48, Chap. 5])

E. Tonti, *The Mathematical Structure of Classical and Relativistic Physics*, Modeling and Simulation in Science, Engineering and Technology,
DOI 10.1007/978-1-4614-7422-7, 

**Richard Feynman:** 'It is not true that we can pursue science completely by using only those concepts which are directly subject to experiment.' (Feynman et al. [69, vol. III, pp. 2–9])

**Arnold Sommerferld:** 'Now we do not accept the "positivistic" standpoint, according to which only observables may be employed in theoretical physics; but instead are of the opinion that the introduction of not directly observable quantities is justified whenever the resulting conclusions agree with experiment (as in the kinetic theory of gases). Nevertheless we demand that the concepts introduced in a hypothesis may be based at least on an imaginary experiment, i.e. an observational method, even if it cannot be carried out in practice.' (Sommerferld [215, p. 72])

**Max Born:** 'It is often said that it was a metaphysical idea which led Heisenberg to the principle of matrix mechanics, and this statement is used by the believers in the power of pure reason as an example in their favour. Well, if you were to ask Heisenberg, he would strongly oppose this view. As we worked together I think I know what was going on in his mind. At that time we were all convinced that the new mechanics must be based on new concepts having only a loose connection with classical concepts, as expressed in Bohr's postulate of correspondence. Heisenberg felt that quantities which had no direct relation to experiment ought to be eliminated. He wished to found the new mechanics as directly as possible on experience. If this is a "metaphysical" principle, well, I cannot contradict; I only wish to say that it is exactly the fundamental principle of modern science as a whole, that which distinguishes it from scholasticism and dogmatic systems of philosophy. But if it is taken (as many have taken it) to mean the elimination of all non-observables from theory, it leads to nonsense. For instance, Schrödinger's wave function $\psi$ is such a non-observable quantity, but it was of course later accepted by Heisenberg as a useful concept. He stated not a dogmatic, but a heuristic principle. He found by an act of scientific intuition the spurious conceptions that have to be eliminated.' (Born [28, p. 18])

**Max Planck:** 'It is absolutely false, although it is often asserted, that the world picture of physics contains, or may contain, directly observable magnitudes only. On the contrary, directly observable magnitudes are not found at all in the world picture. It contains symbols only.' (Planck [177, p. 129])

**Richard Tolman:** '... the probability amplitudes $\psi$ and $\phi$ will in general actually turn out to be complex numbers consisting of a real and imaginary part, and are not themselves measurable but are to be regarded as summarizing the directly observable properties of the system. For example, the squares of their absolute magnitudes are real quantities equal to the probability densities which can be empirically observed. This is in agreement with the idea that the equations of mathematical physics are to provide a formalism for computation which leads to results capable of empirical determination even though the formalism itself contains symbols which have no direct reference to observable physical quantities. . .This is contrary to the apparently pleasing but somewhat unfortunate statement that the equations of mathematical physics should contain only quantities which are susceptible of direct measurement.' (Tolman [226, p. 192])

# Appendix D
# History of the Diagram

## D.1 Historical Remarks

The classification diagram of physical variables presented in this book evolved from a similar diagram originating in electrical networks. A network can be described using a *graph*, and the theory of graphs is a subject of algebraic topology. Since other physical theories make use of the notions of graph theory, such as chemistry, biology, economics, operations research and sociology, it follows that the use of graph theory makes it easy to grasp the similarities between the various physical theories.

**Electrical Networks.** The story begins in the years 1944–1953, with Gabriel Kron in the field of electrical circuit theory. It was followed in 1955–1959 by the works of Paul Roth, who introduced the first notions of algebraic topology.[1] Great progress was made in the years 1966–1977 by Franklin Branin, Jr.[2]

Gabriel Kron, an engineer at General Electric, published many papers showing the analogy between electric circuit theory and electromagnetic field, elasticity, fluid dynamics, vibrations of polyatomic molecules and the Schrödinger equation. Unhappily, his target was always to reduce every one of these fields to a cumbersome combination of RLC circuits.

**Graph Theory.** Paul Roth, a mathematician, realized the role of algebraic topology in network analysis.[3] Figure D.1 shows his diagram for an electrical circuit.[4] Quoting Roth[5]

> 'The electrical network problem comes up in several branches of physics and engineering. In this presentation the mathematical aspects of the pure problem itself will be treated.

---

[1] Roth [195, pp. 518–521].

[2] Branin [29, pp. 453–487].

[3] Branin [29, p. 454].

[4] Grady and Polimeni [83, p. 64].

[5] Roth [196, p. 1].

E. Tonti, *The Mathematical Structure of Classical and Relativistic Physics*, Modeling and Simulation in Science, Engineering and Technology,
DOI 10.1007/978-1-4614-7422-7, 

The first comprehensive treatment of the electrical network problem is due to *Kirchhoff in 1847*. He argued for the existence of a solution to the problem in probably the first paper to introduce the equivalent of incidence matrices. *Maxwell* systematized this approach in *1892* and introduced the concept of potential, introducing what are now known as Maxwell's mesh and node methods for solving electrical networks. A complete proof for the existence of a solution was given by *Hermann Weyl, in 1923*, for the case of a purely resistive network when sources of electromotive force are placed in series with the branches. In *1953 Kron* introduced the method of tearing as another means of solving the electrical network problem. In *1955* the author [*Roth*] gave a proof for what is probably the most general case wherein the network is "ohmic" as well as a proof of the validity of the method of tearing. In *1966 Branin* proposed a considerably more general algebraic topological model for Maxwell's electromagnetic equations.'

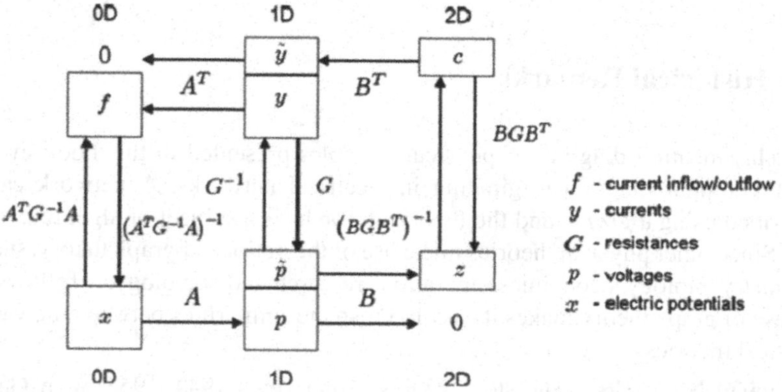


**Fig. D.1** The Roth diagram of electrical circuits

Franklin Branin, an IBM engineer, made a crucial contribution when he discovered the relationship between physical variables and space elements, as this quote shows:[6]

> 'There are several significant features of the evidently intimate relation between algebraic topology, network theory, and the vector calculus which are worth noting. First, algebraic topology deals with very simple but fundamental properties of the space in which physical phenomena happen. Second, network theory is founded directly on the most elemental principles of algebraic topology. Third, the very same principles inevitably come into play in the derivation of the vector calculus, which starts by considering certain numerical quantities associated with what amounts to a network of discrete points, lines, surface and volume elements interconnected with each other.'

Branin, moreover, was the first to describe the electromagnetic field, introducing a cell complex in space and its dual, extending the duality which exists in graph theory between a graph and its dual. This led him to make use of two cochain

6 Branin [29, p. 454].

sequences[7] instead of one chain sequence and one cochain sequence, as can be seen by a comparison of Fig. D.2 with Fig. D.3.

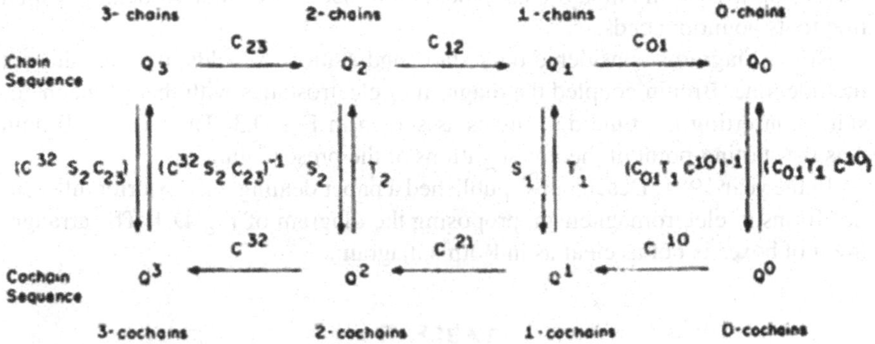


**Fig. D.2** 1966: Branis uses Roth's diagram formed by chains and cochains

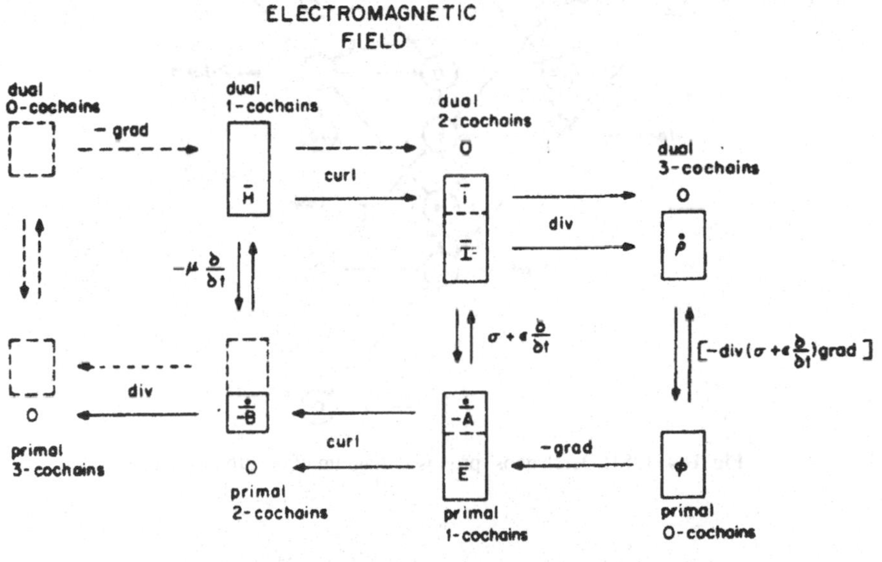


**Fig. D.3** 1966: Branin's diagram for electromagnetic field

The ubiquitous gradient operator arises in descriptions of physics because in all physical theories we find it appropriate to compute the difference between the values of a scalar function between two points in space and to assign this difference

[7] Recall that the term *cochain* can be replaced by *discrete form*; see the remark on p. 196.

to the line connecting them. This process is well described by the coboundary operator, which transforms a 0-cochain into a 1-cochain. In contrast, the gradient operator cannot be described by the boundary operator on chains because it transforms a 1-chain into a 0-chain, that is it transfers the value associated with a line to its bounding ends.

Roth's diagrams considered only static and stationary fields, i.e. they did not include time. Branin coupled the diagram of electrostatics with that of magnetostatics, inserting the time derivatives as shown in Fig. D.3. The work of Branin was the starting point of the investigations of the present author.

In the year 1981, Deschamps[8] published a paper dealing with exterior differential forms in electromagnetism, proposing the diagram of Fig. D.4. The arrangement of boxes is not as clear as in Roth's diagram.

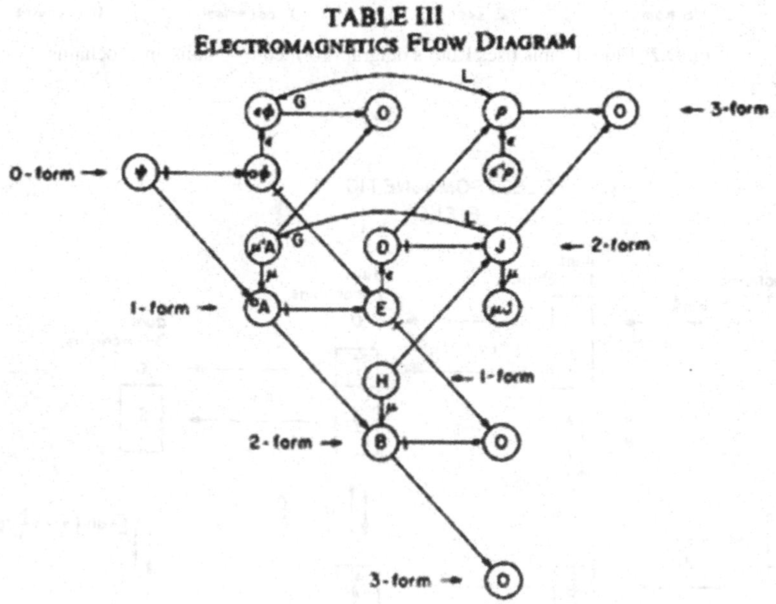


Fig. D.4 1981: Deschamps' proposed diagram of electromagnetism

[8] Deschamps [53].

# Appendix E
# List of Physical Variables

In this list we use 'C' to denote a configuration variable, 'S' a source variable, 'E' an energy variable, $\mathscr{S}$ a system, $\mathscr{P}$ a particle, and $\mathscr{R}$ a rigid body.

| *Name* | *Time-space* | | *dim* | *SI unit* |
|---|---|---|---|---|
| **A** | | | | |
| Acceleration (of a particle) | $\mathbf{a}\,[\tilde{\mathbf{T}}, \mathscr{P}]$ | C | $\mathsf{L}\,\mathsf{T}^{-2}$ | $\mathrm{m\,s^{-2}}$ |
| Acceleration of gravity | $\mathbf{g}\,[\tilde{\mathbf{T}}, \bar{\mathbf{L}}]$ | C | $\mathsf{L}\,\mathsf{T}^{-2}$ | $\mathrm{m\,s^{-2}}$ |
| Action of a system (Hamiltonian) | $A_H\,[\tilde{\mathbf{T}}, \mathscr{S}]$ | E | $\mathsf{L}^2\,\mathsf{M}\,\mathsf{T}^{-1}$ | $\mathrm{J\,s}$ |
| Action of a system (Lagrangian) | $A_L\,[\tilde{\mathbf{T}}, \mathscr{S}]$ | E | $\mathsf{L}^2\,\mathsf{M}\,\mathsf{T}^{-1}$ | $\mathrm{J\,s}$ |
| Airy stress function | $\phi\,[\bar{\mathbf{T}}, \tilde{\mathbf{P}}]$ | S | $\mathsf{L}\,\mathsf{M}\,\mathsf{T}^{-2}$ | $\mathrm{N\,s^{-2}}$ |
| Angle of rotation | $\alpha\,[\bar{\mathbf{I}}, \mathscr{R}]$ | C | 1 | rad |
| Angular acceleration (rigid body) | $\boldsymbol{\alpha}\,[\tilde{\mathbf{T}}, \mathscr{R}]$ | C | $\mathsf{T}^{-2}$ | $\mathrm{rad\,s^{-2}}$ |
| Angular frequency | $\omega\,[\bar{\mathbf{T}}]$ | C | $\mathsf{T}^{-1}$ | $\mathrm{rad\,s^{-1}}$ |
| Angular velocity (rigid body) | $\boldsymbol{\omega}\,[\bar{\mathbf{T}}, \mathscr{R}]$ | C | $\mathsf{T}^{-1}$ | $\mathrm{rad\,s^{-1}}$ |
| Angular momentum (of a system) | $\mathbf{L}\,[\tilde{\mathbf{I}}, \mathscr{S}]$ | S | $\mathsf{L}^2\,\mathsf{M}\,\mathsf{T}^{-1}$ | $\mathrm{kg\,m^2\,s^{-1}}$ |
| Angular momentum content | $\mathbf{L}^c\,[\tilde{\mathbf{I}}, \tilde{\mathbf{V}}]$ | S | $\mathsf{L}^2\,\mathsf{M}\,\mathsf{T}^{-1}$ | $\mathrm{kg\,m^2\,s^{-1}}$ |
| Angular momentum density | $\mathbf{l}\,[\tilde{\mathbf{I}}, \tilde{\mathbf{V}}]$ | S | $\mathsf{L}^{-1}\,\mathsf{M}\,\mathsf{T}^{-1}$ | $\mathrm{kg\,m^{-1}\,s^{-1}}$ |
| Angular momentum flow | $\mathbf{L}^f\,[\tilde{\mathbf{T}}, \tilde{\mathbf{S}}]$ | S | $\mathsf{L}^2\,\mathsf{M}\,\mathsf{T}^{-1}$ | $\mathrm{kg\,m^2\,s^{-1}}$ |
| Angular momentum current (= torque) | $\mathbf{T}\,[\tilde{\mathbf{T}}, \tilde{\mathbf{S}}]$ | S | $\mathsf{L}^2\,\mathsf{M}\,\mathsf{T}^{-2}$ | $\mathrm{kg\,m^2\,s^{-2}}$ |

E. Tonti, *The Mathematical Structure of Classical and Relativistic Physics*, Modeling and Simulation in Science, Engineering and Technology,
DOI 10.1007/978-1-4614-7422-7, 

| *Name* | *Time-space* | | *dim* | *SI unit* |
|---|---|---|---|---|
| **B** | | | | |
| Bending moment (beams) | $M\,[\widetilde{\mathbf{T}},\widetilde{\mathbf{S}}]$ | S | $\mathsf{L^2\,M\,T^{-2}}$ | N m |
| Burgers vector (dislocations) | $\mathbf{b}\,[\overline{\mathbf{I}},\overline{\mathbf{S}}]$ | C | $\mathsf{L}$ | m |
| **C** | | | | |
| Chemical potential | $\mu\,[\widetilde{\mathbf{T}},\overline{\mathbf{P}}]$ | C | $\mathsf{L^2\,M\,T^{-2}}$ mol$^{-1}$ | J mol$^{-1}$ |
| Concentration (molarity) | $c\,[\overline{\mathbf{I}},\widetilde{\mathbf{V}}]$ | C | mol $\mathsf{L^{-3}}$ | mol m$^{-3}$ |
| Couple stress tensor | $\mu\,[\widetilde{\mathbf{T}},\mathbf{S}]$ | S | $\mathsf{M\,T^{-2}}$ | N m$^{-1}$ |
| Cubic dilatation (bulk or volume) | $\Theta\,[\overline{\mathbf{T}},\widetilde{\mathbf{V}}]$ | C | 1 | – |
| Cubic dilatation rate | $\theta\,[\overline{\mathbf{T}},\widetilde{\mathbf{V}}]$ | C | $\mathsf{T^{-1}}$ | s$^{-1}$ |
| **D** | | | | |
| Deformation gradient | $\mathsf{F}\,[\overline{\mathbf{I}},\mathbf{L}]$ | C | 1 | – |
| Displacement (incremental) | $\boldsymbol{\eta}\,[\overline{\mathbf{T}},\overline{\mathbf{P}}]$ | C | $\mathsf{L}$ | m |
| Displacement (initial) | $\mathbf{u}\,[\overline{\mathbf{I}},\overline{\mathbf{P}}]$ | C | $\mathsf{L}$ | m |
| Displacement (particle dynamics) | $\mathbf{u}\,[\overline{\mathbf{T}},\overline{\mathbf{P}}]$ | C | $\mathsf{L}$ | m |
| Displacement (relative) | $\mathbf{h}\,[\overline{\mathbf{I}},\overline{\mathbf{L}}]$ | C | $\mathsf{L}$ | m |
| Displacement gradient tensor | $\mathsf{H}\,[\overline{\mathbf{I}},\mathbf{L}]$ | C | 1 | – |
| **E** | | | | |
| Eikonal | $S\,[\overline{\mathbf{I}},\overline{\mathbf{P}}]$ | C | 1 | – |
| Electric charge (of a system) | $Q\,[\overline{\mathbf{I}},\mathscr{S}]$ | S | $\mathsf{T\,I}$ | C |
| Electric charge content | $Q^{\mathrm{c}}\,[\overline{\mathbf{I}},\widetilde{\mathbf{V}}]$ | S | $\mathsf{T\,I}$ | C |
| Electric charge density | $\rho\,[\overline{\mathbf{I}},\widetilde{\mathbf{V}}]$ | S | $\mathsf{L^{-3}\,T\,I}$ | C m$^{-3}$ |
| Electric charge flow | $Q^{\mathrm{f}}\,[\overline{\mathbf{T}},\widetilde{\mathbf{S}}]$ | S | $\mathsf{I\,T}$ | C |
| Electric current | $I\,[\overline{\mathbf{T}},\widetilde{\mathbf{S}}]$ | S | $\mathsf{I}$ | A |
| Electric current density | $\mathbf{J}\,[\overline{\mathbf{T}},\widetilde{\mathbf{S}}]$ | S | $\mathsf{L^{-2}\,I}$ | A m$^{-2}$ |
| Electric displacement | $\mathbf{D}\,[\overline{\mathbf{I}},\widetilde{\mathbf{S}}]$ | S | $\mathsf{L^{-2}\,T\,I}$ | C m$^{-2}$ |
| Electric energy | $U_{\mathrm{e}}\,[\widetilde{\mathbf{T}},\widetilde{\mathbf{V}}]$ | E | $\mathsf{L^2\,M\,T^{-2}}$ | J |
| Electric energy density | $u_{\mathrm{e}}\,[\widetilde{\mathbf{T}},\widetilde{\mathbf{V}}]$ | E | $\mathsf{L^{-1}\,M\,T^{-2}}$ | J m$^{-3}$ |
| Electric field strength | $\mathbf{E}\,[\widetilde{\mathbf{T}},\overline{\mathbf{L}}]$ | C | $\mathsf{L\,M\,T^{-3}\,I^{-1}}$ | V m$^{-1}$ |
| Electric flux | $\Psi\,[\overline{\mathbf{I}},\widetilde{\mathbf{S}}]$ | S | $\mathsf{T\,I}$ | C |
| Electric flux density | $\sigma\,[\overline{\mathbf{I}},\widetilde{\mathbf{S}}]$ | S | $\mathsf{L^{-2}\,T\,I}$ | C m$^{-2}$ |
| Electric potential | $\phi\,[\widetilde{\mathbf{T}},\overline{\mathbf{P}}]$ | C | $\mathsf{L^2\,M\,T^{-3}\,I^{-1}}$ | V |
| Electric potential impulse | $\varphi\,[\widetilde{\mathbf{T}},\overline{\mathbf{P}}]$ | C | $\mathsf{L^2\,M\,T^{-2}\,I^{-1}}$ | Wb |
| Electric vector potential | $\check{\mathbf{F}}\,[\overline{\mathbf{T}},\widetilde{\mathbf{L}}]$ | S | $\mathsf{L^{-1}\,T\,I}$ | C m$^{-1}$ |
| Electric voltage | $V\,[\widetilde{\mathbf{T}},\overline{\mathbf{L}}]$ | C | $\mathsf{L^2\,M\,T^{-3}\,I^{-1}}$ | V |

| *Name* | *Time-space* | | *dim* | *SI unit* |
|---|---|---|---|---|
| Electromotive force | $E\,[\tilde{\mathbf{T}},\bar{\mathbf{L}}]$ | C | $\mathsf{L^2MT^{-3}I^{-1}}$ | V |
| Electromotive force impulse | $\mathcal{E}\,[\tilde{\mathbf{T}},\bar{\mathbf{L}}]$ | C | $\mathsf{L^2MT^{-2}I^{-1}}$ | Wb |
| Energy (of a system) | $E\,[\tilde{\mathbf{T}},\mathscr{S}]$ | E | $\mathsf{L^2MT^{-2}}$ | J |
| Energy content | $E^c\,[\tilde{\mathbf{T}},\tilde{\mathbf{V}}]$ | E | $\mathsf{L^2MT^{-2}}$ | J |
| Energy content density | $e\,[\tilde{\mathbf{T}},\tilde{\mathbf{V}}]$ | E | $\mathsf{L^{-1}MT^{-2}}$ | $\mathrm{J\,m^{-3}}$ |
| Energy current (= power) | $I_e\,[\bar{\mathbf{T}},\tilde{\mathbf{S}}]$ | E | $\mathsf{L^2MT^{-3}}$ | W |
| Energy current density | $\mathbf{J}_e\,[\bar{\mathbf{T}},\tilde{\mathbf{S}}]$ | E | $\mathsf{MT^{-3}}$ | $\mathrm{W\,m^{-2}}$ |
| Energy flow (= work, heat) | $E^f\,[\tilde{\mathbf{T}},\tilde{\mathbf{S}}]$ | E | $\mathsf{L^2MT^{-2}}$ | J |
| Enthalpy (of a system) | $H\,[\tilde{\mathbf{T}},\mathscr{S}]$ | E | $\mathsf{L^2MT^{-2}}$ | J |
| Entropy (of a system) | $S\,[\bar{\mathbf{I}},\mathscr{S}]$ | S | $\mathsf{L^2MT^{-2}\Theta^{-1}}$ | $\mathrm{J\,K^{-1}}$ |
| Entropy content | $S^c\,[\bar{\mathbf{I}},\tilde{\mathbf{V}}]$ | S | $\mathsf{L^2MT^{-2}\Theta^{-1}}$ | $\mathrm{J\,K^{-1}}$ |
| Entropy current | $I_s\,[\bar{\mathbf{T}},\tilde{\mathbf{S}}]$ | S | $\mathsf{L^2MT^{-3}\Theta^{-1}}$ | $\mathrm{W\,K^{-1}}$ |
| Entropy current density | $\mathbf{J}_s\,[\bar{\mathbf{T}},\tilde{\mathbf{S}}]$ | S | $\mathsf{MT^{-3}\Theta^{-1}}$ | $\mathrm{WK^{-1}m^{-2}}$ |
| Entropy density | $s\,[\bar{\mathbf{I}},\tilde{\mathbf{V}}]$ | S | $\mathsf{L^{-1}MT^{-2}\Theta^{-1}}$ | $\mathrm{J\,K^{-1}m^{-3}}$ |
| Entropy flow | $S^f\,[\bar{\mathbf{T}},\tilde{\mathbf{S}}]$ | S | $\mathsf{L^2MT^{-2}\Theta^{-1}}$ | $\mathrm{J\,K^{-1}}$ |
| Entropy production (in a system) | $S^p\,[\bar{\mathbf{T}},\mathscr{S}]$ | S | $\mathsf{L^2MT^{-2}\Theta^{-1}}$ | $\mathrm{J\,K^{-1}}$ |
| Entropy source | $\sigma_S\,[\bar{\mathbf{T}},\tilde{\mathbf{V}}]$ | S | $\mathsf{L^{-1}MT^{-3}\Theta^{-1}}$ | $\mathrm{WK^{-1}m^{-3}}$ |
| Extension | $e\,[\bar{\mathbf{I}},\bar{\mathbf{L}}]$ | C | $\mathsf{L}$ | m |
| **F** | | | | |
| Force (on a system) | $\mathbf{F}\,[\tilde{\mathbf{T}},\mathscr{S}]$ | S | $\mathsf{LMT^{-2}}$ | N |
| Force (line) | $\mathbf{F}^l\,[\tilde{\mathbf{T}},\bar{\mathbf{L}}]$ | S | $\mathsf{LMT^{-2}}$ | N |
| Force (surface) | $\mathbf{F}^s\,[\tilde{\mathbf{T}},\tilde{\mathbf{S}}]$ | S | $\mathsf{LMT^{-2}}$ | N |
| Force (volume) | $\mathbf{F}^v\,[\tilde{\mathbf{T}},\tilde{\mathbf{V}}]$ | S | $\mathsf{LMT^{-2}}$ | N |
| Force impulse (surface) | $\mathbf{J}^s\,[\tilde{\mathbf{T}},\tilde{\mathbf{S}}]$ | S | $\mathsf{LMT^{-1}}$ | N s |
| Force impulse (volume) | $\mathbf{J}^v\,[\tilde{\mathbf{T}},\tilde{\mathbf{V}}]$ | S | $\mathsf{LMT^{-1}}$ | N s |
| Frequency | $f\,[\tilde{\mathbf{T}}]$ | C | $\mathsf{T^{-1}}$ | $\mathrm{s^{-1}}$ |
| **G** | | | | |
| Gauge function (electromagnetism) | $\chi\,[\bar{\mathbf{I}},\bar{\mathbf{P}}]$ | C | $\mathsf{L^2MT^{-2}I^{-1}}$ | Wb |
| Gibbs function (of a system) | $G\,[\tilde{\mathbf{T}},\mathscr{S}]$ | E | $\mathsf{L^2MT^{-2}}$ | J |
| Gravitational potential | $U_g\,[\tilde{\mathbf{T}},\bar{\mathbf{P}}]$ | C | $\mathsf{L^2T^{-2}}$ | J/kg |
| **H** | | | | |
| Hamilton principal function (system) | $W\,[\tilde{\mathbf{I}},\mathscr{S}]$ | E | $\mathsf{L^2MT^{-1}}$ | J s |
| Hamiltonian (of a system) | $H\,[\tilde{\mathbf{T}},\mathscr{S}]$ | E | $\mathsf{L^2MT^{-2}}$ | J |

| *Name* | *Time-space* | | *dim* | *SI unit* |
|---|---|---|---|---|
| Hamiltonian density | $\mathcal{H}\,[\widetilde{\mathbf{T}},\widetilde{\mathbf{V}}]$ | E | $\mathsf{L^{-1}\,M\,T^{-2}}$ | $\mathrm{J\,m^{-3}}$ |
| Heat (crossing a surface) | $Q\,[\overline{\mathbf{T}},\widetilde{\mathbf{S}}]$ | E | $\mathsf{L^{2}\,M\,T^{-2}}$ | J |
| Heat current | $\Phi\,[\overline{\mathbf{T}},\widetilde{\mathbf{S}}]$ | E | $\mathsf{L^{2}\,M\,T^{-3}}$ | W |
| Heat current density | $\mathbf{J}_{\mathrm{q}}\,[\overline{\mathbf{T}},\widetilde{\mathbf{S}}]$ | E | $\mathsf{M\,T^{-3}}$ | $\mathrm{W\,m^{-2}}$ |
| Heat production | $Q^{\mathrm{p}}[\overline{\mathbf{T}},\widetilde{\mathbf{V}}]$ | E | $\mathsf{L^{2}\,M\,T^{-2}}$ | J |
| Heat production rate | $P\,[\overline{\mathbf{T}},\widetilde{\mathbf{V}}]$ | E | $\mathsf{L^{2}\,M\,T^{-3}}$ | W |
| Heat source | $\sigma_{\mathrm{q}}[\overline{\mathbf{T}},\widetilde{\mathbf{V}}]$ | E | $\mathsf{L^{-1}\,M\,T^{-3}}$ | $\mathrm{W\,m^{-3}}$ |
| Helmholtz free energy (system) | $F\,[\widetilde{\mathbf{T}},\mathscr{S}]$ | E | $\mathsf{L^{2}\,M\,T^{-2}}$ | J |
| **I** | | | | |
| Impulse of a force (of a system) | $\mathbf{J}\,[\widetilde{\mathbf{T}},\mathscr{S}]$ | S | $\mathsf{L\,M\,T^{-1}}$ | N s |
| Impulse of a volume force | $\mathbf{J}^{\mathrm{v}}[\widetilde{\mathbf{T}},\widetilde{\mathbf{V}}]$ | S | $\mathsf{L^{2}\,M\,T^{-1}}$ | N s |
| Impulse of a surface force | $\mathbf{J}^{\mathrm{s}}[\widetilde{\mathbf{T}},\widetilde{\mathbf{S}}]$ | S | $\mathsf{L^{2}\,M\,T^{-1}}$ | N s |
| Internal energy (of a system) | $U\,[\widetilde{\mathbf{T}},\mathscr{S}]$ | E | $\mathsf{L^{2}\,M\,T^{-2}}$ | J |
| Internal energy content | $U^{\mathrm{c}}[\widetilde{\mathbf{T}},\widetilde{\mathbf{V}}]$ | E | $\mathsf{L^{2}\,M\,T^{-2}}$ | J |
| Internal energy density | $u\,[\widetilde{\mathbf{T}},\widetilde{\mathbf{V}}]$ | E | $\mathsf{L^{-1}\,M\,T^{-2}}$ | $\mathrm{J\,m^{-3}}$ |
| **K** | | | | |
| Kinetic co-energy (of a system) | $T^{*}\,[\widetilde{\mathbf{T}},\mathscr{S}]$ | E | $\mathsf{L^{2}\,M\,T^{-2}}$ | J |
| Kinetic energy (of a system) | $T\,[\widetilde{\mathbf{T}},\mathscr{S}]$ | E | $\mathsf{L^{2}\,M\,T^{-2}}$ | J |
| Kinetic energy density | $k[\widetilde{\mathbf{T}},\widetilde{\mathbf{V}}]$ | E | $\mathsf{L^{-1}\,M\,T^{-2}}$ | $\mathrm{J\,m^{-3}}$ |
| **L** | | | | |
| Lagrangian (of a system) | $L\,[\widetilde{\mathbf{T}},\mathscr{S}]$ | E | $\mathsf{L^{2}\,M\,T^{-2}}$ | J |
| Lagrangian density | $\mathcal{L}\,[\widetilde{\mathbf{T}},\widetilde{\mathbf{V}}]$ | E | $\mathsf{L^{-1}\,M\,T^{-2}}$ | $\mathrm{J\,m^{-3}}$ |
| Linear strain | $\varepsilon\,[\overline{\mathbf{I}},\mathbf{L}]$ | C | 1 | – |
| **M** | | | | |
| Magnetic charge (of a system) | $G\,[\overline{\mathbf{T}},\mathscr{S}]$ | C | $\mathsf{L^{2}\,M\,T^{-2}\,I^{-1}}$ | Wb |
| Magnetic charge density | $\check{g}\,[\overline{\mathbf{T}},\widetilde{\mathbf{V}}]$ | C | $\mathsf{L^{-1}\,M\,T^{-3}\,I^{-1}}$ | $\mathrm{Wb\,m^{-3}}$ |
| Magnetic current | $I_{\mathrm{m}}[\overline{\mathbf{T}},\widetilde{\mathbf{S}}]$ | C | $\mathsf{L^{2}\,M\,T^{-3}\,I^{-1}}$ | V |
| Magnetic current density | $\mathbf{J}_{\mathrm{m}}[\overline{\mathbf{T}},\widetilde{\mathbf{S}}]$ | C | $\mathsf{M\,T^{-3}\,I^{-1}}$ | $\mathrm{V\,m^{-2}}$ |
| Magnetic energy density | $w_{\mathrm{m}}\,[\widetilde{\mathbf{T}},\widetilde{\mathbf{V}}]$ | E | $\mathsf{L^{-1}\,M\,T^{-2}}$ | $\mathrm{J\,m^{-3}}$ |
| Magnetic field strength | $\mathbf{H}\,[\overline{\mathbf{T}},\widetilde{\mathbf{L}}]$ | S | $\mathsf{L^{-1}\,I}$ | $\mathrm{A\,m^{-1}}$ |
| Magnetic flux | $\Phi\,[\widetilde{\mathbf{I}},\overline{\mathbf{S}}]$ | C | $\mathsf{L^{2}\,M\,T^{-2}\,I^{-1}}$ | Wb |
| Magnetic flux density | $\mathbf{B}\,[\widetilde{\mathbf{I}},\overline{\mathbf{S}}]$ | C | $\mathsf{M\,T^{-2}\,I^{-1}}$ | $\mathrm{Wb\,m^{-2}}$ |
| Magnetic scalar potential | $\phi_{\mathrm{m}}\,[\overline{\mathbf{T}},\widetilde{\mathbf{P}}]$ | S | $\mathsf{I}$ | A |

| *Name* | *Time-space* | | *dim* | *SI unit* |
|---|---|---|---|---|
| Magnetic scalar potential impulse | $\varphi_\mathrm{m}\,[\overline{\mathbf{T}}, \widetilde{\mathbf{P}}]$ | S | $\mathsf{T\,I}$ | C |
| Magnetic vector potential | $\mathbf{A}\,[\widetilde{\mathbf{I}}, \overline{\mathbf{L}}]$ | C | $\mathsf{L\,M\,T^{-2}\,I^{-1}}$ | $\mathrm{Wb\,m^{-1}}$ |
| Magnetomotive force | $F_\mathrm{m}\,[\overline{\mathbf{T}}, \widetilde{\mathbf{L}}]$ | S | $\mathsf{I}$ | A |
| Magnetomotive force impulse | $\mathcal{F}_\mathrm{m}\,[\overline{\mathbf{T}}, \widetilde{\mathbf{L}}]$ | S | $\mathsf{T\,I}$ | C |
| Mass (of a system) | $M\,[\overline{\mathbf{I}}, \mathscr{S}]$ | P | $\mathsf{M}$ | kg |
| Mass content | $M^\mathrm{c}\,[\overline{\mathbf{I}}, \widetilde{\mathbf{V}}]$ | S | $\mathsf{M}$ | kg |
| Mass current | $I_\mathrm{m}\,[\overline{\mathbf{T}}, \widetilde{\mathbf{S}}]$ | S | $\mathsf{M\,T^{-1}}$ | $\mathrm{kg\,s^{-1}}$ |
| Mass current density | $\mathbf{J}_\mathrm{m}\,[\overline{\mathbf{T}}, \widetilde{\mathbf{S}}]$ | S | $\mathsf{L^{-2}\,M\,T^{-1}}$ | $\mathrm{kg\,m^{-2}\,s^{-1}}$ |
| Mass density | $\rho\,[\overline{\mathbf{I}}, \widetilde{\mathbf{V}}]$ | S | $\mathsf{L^{-3}M}$ | $\mathrm{kg\,m^{-3}}$ |
| Mass flow | $M^\mathrm{f}\,[\overline{\mathbf{T}}, \widetilde{\mathbf{S}}]$ | S | $\mathsf{M}$ | kg |
| Mass production | $M^\mathrm{p}\,[\overline{\mathbf{T}}, \widetilde{\mathbf{V}}]$ | S | $\mathsf{M}$ | kg |
| Mass source | $\sigma_\mathrm{m}\,[\overline{\mathbf{T}}, \widetilde{\mathbf{V}}]$ | S | $\mathsf{L^{-3}\,M\,T^{-1}}$ | $\mathrm{kg\,s^{-1}\,m^{-3}}$ |
| Moment of a force (on a system) | $\mathbf{M}\,[\widetilde{\mathbf{T}}, \mathscr{S}]$ | S | $\mathsf{L^{2}\,M\,T^{-2}}$ | N m |
| Momentum (of a system) | $\mathbf{P}\,[\widetilde{\mathbf{I}}, \mathscr{S}]$ | S | $\mathsf{L\,M\,T^{-1}}$ | N s |
| Momentum content | $\mathbf{P}^\mathrm{c}\,[\widetilde{\mathbf{I}}, \widetilde{\mathbf{V}}]$ | S | $\mathsf{L\,M\,T^{-1}}$ | N s |
| Momentum (of a particle) | $\mathbf{p}\,[\widetilde{\mathbf{I}}, \mathscr{P}]$ | S | $\mathsf{L\,M\,T^{-1}}$ | N s |
| Momentum current (= surface force) | $\mathbf{F}^\mathrm{s}\,[\widetilde{\mathbf{T}}, \widetilde{\mathbf{S}}]$ | S | $\mathsf{L\,M\,T^{-2}}$ | N |
| Momentum density | $\mathbf{p}\,[\widetilde{\mathbf{I}}, \widetilde{\mathbf{V}}]$ | S | $\mathsf{L^{-2}\,M\,T^{-1}}$ | $\mathrm{N\,s\,m^{-3}}$ |
| Momentum flow (= surface impulse) | $\mathbf{J}^\mathrm{s}\,[\widetilde{\mathbf{T}}, \widetilde{\mathbf{S}}]$ | S | $\mathsf{L\,M\,T^{-1}}$ | N s |
| Momentum production (= volume impulse) | $\mathbf{J}^\mathrm{v}\,[\widetilde{\mathbf{T}}, \widetilde{\mathbf{V}}]$ | S | $\mathsf{L\,M\,T^{-1}}$ | N s |
| Momentum production rate (= volume force) | $\mathbf{F}^\mathrm{v}\,[\widetilde{\mathbf{T}}, \widetilde{\mathbf{V}}]$ | S | $\mathsf{L\,M\,T^{-2}}$ | N |
| **O** | | | | |
| Optical path difference | $OPD\,[\overline{\mathbf{I}}, \overline{\mathbf{L}}]$ | C | $\mathsf{L}$ | m |
| Optical ray length | $OPL\,[\overline{\mathbf{I}}, \overline{\mathbf{L}}]$ | C | $\mathsf{L}$ | m |
| **P** | | | | |
| Particle content | $N^\mathrm{c}\,[\overline{\mathbf{I}}, \widetilde{\mathbf{V}}]$ | S | 1 | – |
| Particle content density | $n\,[\overline{\mathbf{I}}, \widetilde{\mathbf{V}}]$ | S | $\mathsf{L^{-3}}$ | $\mathrm{m^{-3}}$ |
| Particle current | $I_\mathrm{p}\,[\overline{\mathbf{T}}, \widetilde{\mathbf{S}}]$ | S | $\mathsf{T^{-1}}$ | $\mathrm{s^{-1}}$ |
| Particle current density | $\mathbf{J}_\mathrm{p}\,[\overline{\mathbf{T}}, \widetilde{\mathbf{S}}]$ | S | $\mathsf{L^{-2}\,T^{-1}}$ | $\mathrm{s^{-1}\,m^{-2}}$ |
| Particle flow | $N^\mathrm{f}\,[\overline{\mathbf{T}}, \widetilde{\mathbf{S}}]$ | S | 1 | – |

| *Name* | *Time-space* | | *dim* | *SI unit* |
|---|---|---|---|---|
| Particle number (of a system) | $N\,[\bar{\mathbf{I}},\mathscr{S}]$ | S | 1 | – |
| Phase | $\phi\,[\widetilde{\mathbf{I}},\overline{\mathbf{P}}]$ | C | 1 | rad |
| Phase difference in space | $\mathit{\Delta}_s\phi\,[\bar{\mathbf{I}},\overline{\mathbf{L}}]$ | C | 1 | rad |
| Phase difference in time | $\mathit{\Delta}_t\phi\,[\widetilde{\mathbf{T}},\overline{\mathbf{P}}]$ | C | 1 | rad |
| Position vector | $\mathbf{r}\,[\bar{\mathbf{I}},\overline{\mathbf{P}}]$ | C | $\mathsf{L}$ | m |
| Position vector (initial-) | $\mathbf{R}\,[\overline{\mathbf{P}}]$ | C | $\mathsf{L}$ | m |
| Potential energy (of a system) | $V\,[\widetilde{\mathbf{T}},\mathscr{S}]$ | E | $\mathsf{L}^2\,\mathsf{M}\,\mathsf{T}^{-2}$ | J |
| Potential energy content | $V\,[\widetilde{\mathbf{T}},\widetilde{\mathbf{V}}]$ | E | $\mathsf{L}^2\,\mathsf{M}\,\mathsf{T}^{-2}$ | J |
| Potential energy density | $v\,[\widetilde{\mathbf{T}},\widetilde{\mathbf{V}}]$ | E | $\mathsf{L}^{-1}\,\mathsf{M}\,\mathsf{T}^{-2}$ | J $m^{-3}$ |
| Power (= energy current) | $P\,[\overline{\mathbf{T}},\widetilde{\mathbf{V}}]$ | E | $\mathsf{L}^2\,\mathsf{M}\,\mathsf{T}^{-3}$ | W |
| Poynting vector | $\mathbf{S}\,[\overline{\mathbf{T}},\widetilde{\mathbf{S}}]$ | E | $\mathsf{L}^2\,\mathsf{M}\,\mathsf{T}^{-3}$ | W |
| Pressure (force/area) | $p\,[\widetilde{\mathbf{T}},\widetilde{\mathbf{S}}]$ | S | $\mathsf{L}^{-1}\,\mathsf{M}\,\mathsf{T}^{-2}$ | N $m^{-2}$ |
| Pressure (work/volume increase) | $p\,[\bar{\mathbf{I}},\widetilde{\mathbf{V}}]$ | S | $\mathsf{L}^{-1}\,\mathsf{M}\,\mathsf{T}^{-2}$ | J $m^{-3}$ |
| Probability current density vector | $\mathbf{j}\,[\overline{\mathbf{T}},\widetilde{\mathbf{S}}]$ | C | $\mathsf{L}^{-2}\,\mathsf{T}^{-1}$ | $s^{-1}m^{-2}$ |
| Probability density | $\rho\,[\bar{\mathbf{I}},\widetilde{\mathbf{V}}]$ | C | $\mathsf{L}^{-3}$ | $m^{-3}$ |
| **R** | | | | |
| Relative displacement vector | $\mathbf{h}[\bar{\mathbf{I}},\overline{\mathbf{L}}]$ | C | $\mathsf{L}$ | m |
| Relative position vector | $\mathbf{G}\,[\overline{\mathbf{L}}]$ | C | $\mathsf{L}$ | m |
| **S** | | | | |
| Scalar magnetic potential | $\phi_{\mathrm{m}}\,[\overline{\mathbf{T}},\widetilde{\mathbf{P}}]$ | S | $\mathsf{I}$ | A |
| Scalar magnetic potential impulse | $\varphi_{\mathrm{m}}\,[\overline{\mathbf{T}},\widetilde{\mathbf{P}}]$ | S | $\mathsf{I}\,\mathsf{T}$ | C |
| Shear strain | $\gamma\,[\overline{\mathbf{T}},\mathbf{L}]$ | C | 1 | – |
| Shear stress | $\tau\,[\widetilde{\mathbf{T}},\mathbf{S}]$ | S | $\mathsf{L}^{-1}\,\mathsf{M}\,\mathsf{T}^{-2}$ | N $m^{-2}$ |
| Spin tensor | $\mathsf{W}\,[\overline{\mathbf{T}},\overline{\mathbf{L}}]$ | C | $\mathsf{T}^{-1}$ | $s^{-1}$ |
| Strain deviatoric | $\boldsymbol{\varepsilon}'\,[\bar{\mathbf{I}},\mathbf{L}]$ | C | 1 | – |
| Strain rate tensor | $\mathsf{D}\,[\overline{\mathbf{T}},\mathbf{L}]$ | C | $\mathsf{T}^{-1}$ | $s^{-1}$ |
| Strain rate deviatoric | $\mathsf{D}'\,[\overline{\mathbf{T}},\mathbf{L}]$ | C | $\mathsf{T}^{-1}$ | $s^{-1}$ |
| Strain tensor | $\boldsymbol{\varepsilon}\,[\bar{\mathbf{I}},\overline{\mathbf{L}}]$ | C | 1 | – |
| Stream vector | $\boldsymbol{\psi}\,[\overline{\mathbf{T}},\widetilde{\mathbf{L}}]$ | S | $\mathsf{L}^{-1}\,\mathsf{M}\,\mathsf{T}^{-1}$ | N s $m^{-2}$ |
| Stress deviator | $\boldsymbol{\tau}'\,[\widetilde{\mathbf{T}},\mathbf{S}]$ | S | $\mathsf{L}^{-1}\,\mathsf{M}\,\mathsf{T}^{-2}$ | N $m^{-2}$ |
| Stress tensor | $\boldsymbol{\tau}\,[\widetilde{\mathbf{T}},\mathbf{S}]$ | S | $\mathsf{L}^{-1}\,\mathsf{M}\,\mathsf{T}^{-2}$ | N $m^{-2}$ |
| Stress tensor (symmetric) | $\boldsymbol{\sigma}\,[\widetilde{\mathbf{T}},\mathbf{S}]$ | S | $\mathsf{L}^{-1}\,\mathsf{M}\,\mathsf{T}^{-2}$ | N $m^{-2}$ |

| *Name* | *Time-space* | | *dim* | *SI unit* |
|---|---|---|---|---|
| Stress vector (= traction) | $\mathbf{t}\,[\tilde{\mathbf{T}},\tilde{\mathbf{S}}]$ | S | $\mathsf{LMT^{-2}}$ | N |
| Surface charge density | $\sigma\,[\bar{\mathbf{I}},\tilde{\mathbf{S}}]$ | S | $\mathsf{L^{-2}TI}$ | $\mathrm{C\,m^{-2}}$ |
| Surface force | $\mathbf{F}^{s}[\tilde{\mathbf{T}},\tilde{\mathbf{S}}]$ | S | $\mathsf{LMT^{-2}}$ | N |
| Surface force impulse | $\mathbf{J}^{s}\,[\tilde{\mathbf{T}},\tilde{\mathbf{S}}]$ | S | $\mathsf{LMT^{-1}}$ | N s |

**T**

| | | | | |
|---|---|---|---|---|
| Temperature | $T\,[\tilde{\mathbf{T}},\bar{\mathbf{P}}]$ | C | $\Theta$ | K |
| Temperature difference | $G\,[\tilde{\mathbf{T}},\bar{\mathbf{L}}]$ | C | $\Theta$ | K |
| Temperature gradient | $\mathbf{g}\,[\tilde{\mathbf{T}},\bar{\mathbf{L}}]$ | C | $\mathsf{L^{-1}}\,\Theta$ | $\mathrm{K\,m^{-1}}$ |
| Thermacy (= temperature integral) | $\mathscr{T}\,[\tilde{\mathbf{I}},\bar{\mathbf{P}}]$ | C | $\Theta\,\mathsf{T}$ | K s |
| Torque (surface) | $\mathbf{T}\,[\tilde{\mathbf{T}},\tilde{\mathbf{S}}]$ | S | $\mathsf{L^2MT^{-2}}$ | N m |
| Torque (volume-) | $\mathbf{T}\,[\tilde{\mathbf{T}},\tilde{\mathbf{V}}]$ | S | $\mathsf{L^2MT^{-2}}$ | N m |

**V**

| | | | | |
|---|---|---|---|---|
| Velocity (of a particle) | $\mathbf{v}\,[\bar{\mathbf{T}},\mathscr{P}]$ | C | $\mathsf{LT^{-1}}$ | $\mathrm{m\,s^{-1}}$ |
| Velocity ♠ (Chap. 12) | $\mathbf{v}\,[\tilde{\mathbf{I}},\bar{\mathbf{L}}]$ | C | $\mathsf{LT^{-1}}$ | $\mathrm{m\,s^{-1}}$ |
| Velocity ♠ (Chap. 12) | $\mathbf{v}\,[\bar{\mathbf{T}},\bar{\mathbf{P}}]$ | C | $\mathsf{LT^{-1}}$ | $\mathrm{m\,s^{-1}}$ |
| Velocity (relative) | $\mathbf{V}\,[\bar{\mathbf{T}},\bar{\mathbf{L}}]$ | C | $\mathsf{LT^{-1}}$ | $\mathrm{m\,s^{-1}}$ |
| Velocity circulation | $\Gamma\,[\tilde{\mathbf{I}},\bar{\mathbf{L}}]$ | C | $\mathsf{L^2T^{-1}}$ | $\mathrm{m^2\,s^{-1}}$ |
| Velocity gradient tensor | $\mathsf{L}\,[\bar{\mathbf{T}},\mathbf{L}]$ | C | $\mathsf{T^{-1}}$ | $\mathrm{s^{-1}}$ |
| Velocity potential | $\phi\,[\tilde{\mathbf{I}},\bar{\mathbf{P}}]$ | C | $\mathsf{L^2T^{-1}}$ | $\mathrm{m^2\,s^{-1}}$ |
| Virtual work | $W^{*}\,[\tilde{\mathbf{T}},\bar{\mathbf{L}}]$ | S | $\mathsf{L^2MT^{-2}}$ | J |
| Viscous force | $\mathbf{F}^{v}\,[\tilde{\mathbf{T}},\tilde{\mathbf{S}}]$ | S | $\mathsf{LMT^{-2}}$ | N |
| Viscous stress deviator | $\boldsymbol{\tau}'\,[\tilde{\mathbf{T}},\mathbf{S}]$ | S | $\mathsf{L^{-1}MT^{-2}}$ | $\mathrm{N\,m^{-2}}$ |
| Viscous stress tensor | $\boldsymbol{\tau}\,[\tilde{\mathbf{T}},\mathbf{S}]$ | S | $\mathsf{L^{-1}MT^{-2}}$ | $\mathrm{N\,m^{-2}}$ |
| Voltage | $V\,[\tilde{\mathbf{T}},\bar{\mathbf{L}}]$ | C | $\mathsf{L^2MT^{-3}I^{-1}}$ | V |
| Voltage impulse | $‘V\,[\tilde{\mathbf{T}},\bar{\mathbf{L}}]$ | C | $\mathsf{L^2MT^{-2}I^{-1}}$ | Wb |
| Volume dilatation | $\Theta\,[\bar{\mathbf{I}},\tilde{\mathbf{V}}]$ | C | 1 | - |
| Volume dilatation rate | $\theta\,[\bar{\mathbf{T}},\bar{\mathbf{L}}]$ | C | $\mathsf{T^{-1}}$ | $\mathrm{s^{-1}}$ |
| Volume force | $\mathbf{F}[\tilde{\mathbf{T}},\tilde{\mathbf{V}}]$ | S | $\mathsf{LMT^{-2}}$ | N |
| Volume force density | $\mathbf{f}[\tilde{\mathbf{T}},\tilde{\mathbf{V}}]$ | S | $\mathsf{L^{-2}MT^{-2}}$ | $\mathrm{N\,m^{-3}}$ |
| Volume (force) impulse | $\mathbf{J}^{v}[\tilde{\mathbf{T}},\tilde{\mathbf{V}}]$ | S | $\mathsf{LMT^{-1}}$ | N s |
| Vortex flux (= vortex strength) | $W\,[\tilde{\mathbf{I}},\bar{\mathbf{S}}]$ | C | $\mathsf{L^2T^{-1}}$ | $\mathrm{m^2\,s^{-1}}$ |
| Vorticity | $\mathbf{w}\,[\tilde{\mathbf{I}},\bar{\mathbf{S}}]$ | C | $\mathsf{T^{-1}}$ | $\mathrm{s^{-1}}$ |

| *Name* | *Time-space* | | *dim* | *SI unit* |
|---|---|---|---|---|
| **W** | | | | |
| Wave vector | $\mathbf{k}\,[\widetilde{\mathbf{I}},\overline{\mathbf{L}}]$ | C | $\mathsf{L}^{-1}$ | $\mathrm{m}^{-1}$ |
| Work (given to or by a system) | $W\,[\overline{\mathbf{T}},\mathscr{S}]$ | E | $\mathsf{L}^2\,\mathsf{M}\,\mathsf{T}^{-2}$ | J |
| Work (of surface forces) | $W\,[\overline{\mathbf{T}},\widetilde{\mathbf{S}}]$ | E | $\mathsf{L}^2\,\mathsf{M}\,\mathsf{T}^{-2}$ | J |
| Work (of volume forces) | $W\,[\overline{\mathbf{T}},\widetilde{\mathbf{V}}]$ | E | $\mathsf{L}^2\,\mathsf{M}\,\mathsf{T}^{-2}$ | J |

# Appendix F
# List of Symbols Used in This Book

| | |
|---|---|
| | A |
| $A$ | [*Geometry*] Area |
| $A$ | [*Field theory*] Action |
| $A_H$ | [*Mechanics*] Hamiltonian action |
| $A_L$ | [*Mechanics*] Lagrangian action |
| $A_\alpha$ | [*Electromagnetism*] Four-dimensional electromagnetic vector potential |
| $\mathscr{A}$ | [*Electromagnetism*] 1-form of magnetic vector potential |
| $\mathbf{A}$ | [*Electromagnetism*] Magnetic vector potential |
| $a$ | [*Geometry*] Area of infinitesimal surface element |
| $a$ | [*Electromagnetism*] Line integral of magnetic vector potential $\mathbf{A}$ |
| $\mathbf{a}$ | [*Mechanics*] Acceleration |
| | B |
| $\mathscr{B}$ | Body |
| $\check{\mathbf{B}}$ | [*Electromagnetism*] Magnetic flux density (pseudovector) |
| $\check{\mathbf{b}}$ | [*Solid mechanics*] Burgers vector (dislocations) (pseudovector) |
| | C |
| $C$ | [*Electromagnetism*] Capacitance |
| $C$ | [*Thermodynamics*] Thermal capacity |
| $\mathscr{C}_p(K)$ | $p$-dimensional chain space over cell complex $K$ |
| $\overline{\mathbf{C}}$ | [*Algebraic topology*] Incidence matrix faces–edges of primal complex $= [\overline{r}_{\beta\alpha}]$ |
| $\widetilde{\mathbf{C}}$ | [*Algebraic topology*] Incidence matrix faces–edges of dual complex $= [\widetilde{r}_{\beta\alpha}]$ |
| $c$ | [*Relativistic formulation*] Light speed |
| $c$ | [*Thermodynamics*] Specific heat |
| $c$ | [*Fluid mechanics*] Sound speed |
| $\mathbf{c}_\mathrm{p}$ | [*Algebraic topology*] $p$-dimensional chain |
| $c^\mathrm{p}$ | [*Algebraic topology*] $p$-dimensional cochain ≡ discrete $p$-form |

E. Tonti, *The Mathematical Structure of Classical and Relativistic Physics*, Modeling and Simulation in Science, Engineering and Technology, DOI 10.1007/978-1-4614-7422-7, 

| D | |
|---|---|
| $\mathbf{D}$ | [*Electromagnetism*] Electric displacement |
| $\overline{\mathsf{D}}$ | [*Algebraic topology*] Incidence matrix volumes–faces of primal complex= $[\overline{d}_{h\alpha}]$ |
| $\widetilde{\mathsf{D}}$ | [*Algebraic topology*] Incidence matrix volumes–faces of dual complex= $[\widetilde{d}_{h\alpha}]$ |
| $d$ | [*Interactions*] Piezoelectric modulus |
| $\bar{d}_{h\alpha}$ | [*Algebraic topology*] Incidence numbers between primal cell $\bar{\mathbf{v}}_h$ and primal face $\bar{\mathbf{s}}_\alpha$ |
| $\tilde{d}_{h\alpha}$ | [*Algebraic topology*] Incidence numbers between dual cell $\tilde{\mathbf{v}}_h$ and dual face $\tilde{\mathbf{s}}_\alpha$ |
| $\mathsf{D}$ | [*Fluid dynamics*] Strain rate tensor |
| **E** | |
| $E$ | [*Solid mechanics*] Elastic modulus |
| $E$ | [*Mechanics*] Total energy |
| $E$ | [*Electromagnetism*] Voltage |
| $E$ | [*Electromagnetism*] Electromagnetic energy |
| $E^{\mathrm{c}}$ | [*Fluid mechanics*] Energy content (spatial description) |
| $E^{\mathrm{f}}$ | [*Fluid mechanics*] Energy flow (spatial description) |
| $\mathscr{E}$ | [*Electromagnetism*] Impulse of electromotive force |
| $\mathscr{E}$ | [*Electromagnetism*] Even 1-form of electric field intensity |
| $\mathbf{E}$ | [*Electromagnetism*] Electric field strength |
| $e$ | [*Electromagnetism*] Electron charge |
| $e$ | [*Solid mechanics*] Extension |
| $\mathbf{e}$ | [*Fluid mechanics*] Relative velocity |
| $\mathbf{e}_h$ | [*Geometry*] Base vector in a vector space |
| $\mathbf{e}_{\mathrm{p}}^h$ | [*Algebraic topology*] $h$th $p$-dimensional cell |
| **F** | |
| $F$ | [*Thermodynamics*] Helmholtz free energy |
| $F_{\mathrm{m}}$ | [*Electromagnetism*] Magnetomotive force |
| $F_{\mathrm{n}}$ | [*Fluid mechanics*] Force normal to a plane surface |
| $F_{\alpha\beta}$ | [*Electromagnetism*] First electromagnetic tensor in space-time |
| $\mathscr{F}_{\mathrm{m}}$ | [*Electromagnetism*] Impulse of magnetomotive force |
| $\mathbf{F}$ | [*Mechanics*] Force |
| $\mathbf{F}^{\mathrm{v}}$ | [*Fluid mechanics*] Volume force ≡ momentum flow rate |
| $\mathbf{F}^{\mathrm{s}}$ | [*Fluid mechanics*] Surface force (also denoted by $\mathbf{T}$) |
| $\mathbf{F}_{\mathrm{e}}$ | [*Solid mechanics*] Elastic restoring force |
| $\mathbf{F}_{\mathrm{d}}$ | [*Solid mechanics*] Viscous force |
| $\mathbf{F}^{\mathrm{imp}}$ | [*Mechanics*] Impressed force |
| $\check{\mathbf{F}}$ | [*Electromagnetism*] Electric pseudovector potential |
| $\mathsf{F}$ | [*Solid mechanics*] Deformation gradient |

| | G |
|---|---|
| $\mathbf{f}$ | [*Mechanics*] Force for unit volume |
| $G$ | [*Thermodynamics*] Gibbs free energy ≡ Gibbs function |
| $G$ | [*Thermodynamics*] Temperature difference |
| $G$ | [*Fluid mechanics*] Hydraulic pressure gradient |
| $G$ | [*Gravitation*] Gravitational constant |
| $G$ | [*Solid mechanics*] Shear modulus |
| $G^{\mu\nu}$ | [*Electromagnetism*] Second electromagnetic tensor |
| $G^{\mathrm{f}}$ | [*Electromagnetism*] Magnetic charge flow |
| $\check{G}$ | [*Electromagnetism*] Magnetic charge (pseudoscalar) |
| $\check{G}^{\mathrm{c}}$ | [*Electromagnetism*] Magnetic charge content (pseudoscalar) |
| $\check{G}^{\mathrm{p}}$ | [*Electromagnetism*] Magnetic charge production (pseudoscalar) |
| $\mathcal{G}$ | [*Algebraic topology*] Additive group of discrete forms (= cochains) |
| $\mathbf{G}$ | [*Solid mechanics*] Relative position vector |
| $\overline{\mathbf{G}}$ | [*Algebraic topology*] Incidence matrix edges–vertices (primal) |
| $\widetilde{\mathbf{G}}$ | [*Algebraic topology*] Incidence matrix edges–vertices (dual) |
| $g$ | [*Classical gravitation*] Modulus of acceleration of gravity |
| $g$ | [*Tensorial notation*] Determinant of metric tensor |
| $g_k$ | [*Geometry*] Covariant natural base vectors |
| $g^k$ | [*Geometry*] Contravariant natural base vectors |
| $g_{hk}$ | [*Tensorial notation*] Covariant form of metric tensor in three-dimensional space |
| $g^{hk}$ | [*Tensorial notation*] Contravariant form of metric tensor in three-dimensional space |
| $g_{\mu\nu}$ | [*Relativity*] Space-time metric tensor |
| $\bar{g}_{\alpha h}$ | [*Algebraic topology*] Incidence number between an edge $\bar{\mathbf{l}}_\alpha$ and a vertex $\bar{\mathbf{p}}_h$ of primal complex |
| $\tilde{g}_{\alpha h}$ | [*Algebraic topology*] Incidence number between an edge $\tilde{\mathbf{l}}_\alpha$ and a vertex $\tilde{\mathbf{p}}_h$ of dual complex |
| $\mathbf{g}$ | [*Classical gravitation*] Acceleration of gravity |
| $\mathbf{g}$ | [*Thermal conduction*] Temperature gradient |

| | H |
|---|---|
| $H$ | [*Analytical mechanics*] Hamiltonian function |
| $H$ | [*Thermodynamics*] Entalpy |
| $H$ | [*Geometry*] Hypervolume (extension) |
| $\check{\mathcal{H}}$ | [*Electromagnetism*] 1-form of magnetic field intensity |
| $\mathbf{H}$ | [*Solid mechanics*] Displacement gradient matrix |
| $\check{\mathbf{H}}$ | [*Electromagnetism*] Magnetic field strength (pseudovector) |
| $\mathbf{H}$ | [*Algebraic topology*] Hypercell |
| $\overline{\mathbf{H}}$ | [*Space-time geometry*] Hypervolume (inner orientation) |
| $\widetilde{\mathbf{H}}$ | [*Space-time geometry*] Hypervolume (outer orientation) |

| | |
|---|---|
| | % H |
| $h$ | [*Mechanics*] Dumping coefficient |
| $h$ | [*Quantum mechanics*] Planck's constant |
| $\hbar$ | [*Quantum mechanics*] Planck's constant/$2\pi$ |
| $\mathbf{h}$ | [*Solid mechanics*] Relative displacement |
| $\mathbf{h}$ | [*Gravitation*] Gravitational flux density vector |
| | I |
| $I$ | [*Electromagnetism*] Electric current |
| $I_a$ | [*Mechanics*] Moment of inertia |
| $I_e$ | [*Field theories*] Energy current |
| $I_s$ | [*Irreversible thermodynamics*] Entropy current |
| $\check{I}_m$ | [*Electromagnetism*] Magnetic current |
| $\mathbf{I}$ | [*Time elements*] Instant |
| $\bar{\mathbf{I}}$ | [*Time elements*] Primal instant |
| $\tilde{\mathbf{I}}$ | [*Time elements*] Dual instant |
| $\mathbf{i}$ | [*Fluid mechanics*] Piezometric gradient |
| | J |
| $J$ | [*Solid mechanics*] Second-order moment |
| $J^\mu$ | [*Electromagnetism*] Four-dimensional electric current density |
| $J$ | [*Solid mechanics*] Moment of inertia |
| $\check{\mathscr{J}}$ | [*Electromagnetism*] 2-form of electric current density |
| $\mathbf{J}$ | [*Electromagnetism*] Electric current density |
| $\mathbf{J}$ | [*Mechanics*] Impulse |
| $\mathbf{J}_s$ | [*Thermodynamics*] Entropy current density |
| $\mathbf{J}_e$ | [*Field theories*] Energy current density |
| $\mathbf{J}_p$ | [*Quantum mechanics*] Particle current density |
| $\mathbf{J}^v$ | [*Continuum mechanics*] Impulse of volume force |
| $\mathbf{J}^s$ | [*Solid mechanics*] Impulse of surface force |
| $\check{\mathbf{J}}_m$ | [*Electromagnetism*] Magnetic current density (hypothetical) (pseudovector) |
| | K |
| $K$ | [*Solid mechanics*] Bulk modulus |
| $K_\mu$ | [*Relativistic mechanics*] Four-dimensional force (Minkowsky) |
| $\mathbf{K}$ | [*Algebraic topology*] Cell complex |
| $\bar{\mathbf{K}}$ | [*Algebraic topology*] Primal cell complex |
| $\tilde{\mathbf{K}}$ | [*Algebraic topology*] Dual cell complex |
| $\mathbf{K}^n$ | [*Algebraic topology*] $n$-dimensional cell complex |
| $k$ | [*Wave motion*] Wave number |
| $k$ | [*Elasticity*] Stiffness |
| $k$ | [*Elasticity*] Angular stiffness |
| $k_B$ | [*Thermodynamics*] Boltzmann's constant |
| $\mathbf{k}$ | [*Wave motion*] Wave vector |

| L | |
|---|---|
| $L$ | [*Mechanics*] Lagrangian function (Lagrangian) |
| $L$ | [*Electromagnetism*] Inductance |
| $\mathbf{L}$ | [*Geometry*] Line without orientation |
| $\overline{\mathbf{L}}$ | [*Geometry*] Line with inner orientation |
| $\widetilde{\mathbf{L}}$ | [*Geometry*] Line with outer orientation |
| $\check{\mathbf{L}}$ | [*Mechanics*] Angular momentum (pseudovector) |
| $\check{\mathbf{L}}^{\mathrm{f}}$ | [*Fluid mechanics*] Angular momentum flow (pseudovector) |
| $\check{\mathbf{L}}^{\mathrm{c}}$ | [*Fluid mechanics*] Angular momentum content (pseudovector) |
| $\mathsf{L}$ | [*Fluid mechanics*] Velocity gradient tensor |
| $\bar{\mathbf{l}}^{k}$ | [*Algebraic topology*] *k*th primal 1-cell |
| $\tilde{\mathbf{l}}^{k}$ | [*Algebraic topology*] *k*th dual 1-cell |
| **M** | |
| $M$ | [*Mechanics*] Mass |
| $M$ | [*Solid mechanics*] Bending moment |
| $M^{\mathrm{f}}$ | [*Fluid mechanics*] Mass flow (spatial description) |
| $M^{\mathrm{c}}$ | [*Fluid mechanics*] Mass content (spatial description) |
| $M^{\mathrm{p}}$ | [*Fluid mechanics*] Mass production |
| $\mathbf{M}$ | [*Mechanics*] Moment of a force |
| $\check{\mathbf{M}}$ | [*Electromagnetism*] Magnetization (pseudovector) |
| $\mathsf{M}$ | Matrix |
| $m$ | [*Mechanics*] Mass |
| $m$ | [*Theory of beams*] Couple/length |
| $m$ | [*Theory of beams*] Torsional couple/length |
| $m_0$ | [*Relativistic mechanics*] Rest mass of particle |
| $m_k$ | [*Algebraic topology*] Multiplicity of cell $\mathbf{e}^{k}_{p-1}$ |
| $m_{ik}$ | [*Fluid mechanics*] Couple stress tensor |
| **N** | |
| $N$ | [*Solid mechanics*] Traction |
| $N$ | [*Diffusion, chemistry*] Particle number |
| $N^{\mathrm{c}}$ | [*Diffusion, chemistry*] Particles content |
| $N^{\mathrm{f}}$ | [*Diffusion, chemistry*] Particle flow |
| $N_{\mathrm{A}}$ | [*Chemical physics*] Avogadro's number |
| $n$ | [*Thermodynamics*] Number of moles |
| $n_h$ | [*Algebraic topology*] Multiplicity of cell $\mathbf{e}^{h}_{p}$ in chain $\mathbf{c}_p$ |
| $\mathbf{n}$ | Unit vector normal to a surface (true vector = polar vector) |
| $\check{\mathbf{n}}$ | Unit vector normal to a surface (pseudovector) |
| **P** | |
| $P$ | [*Mechanics*] Power |
| $P$ | [*Thermal conduction*] Heat-generation rate |
| $P$ | [*Quantum mechanics*] Probability |
| $P^{\mathrm{c}}$ | [*Quantum mechanics*] Probability content (spatial description) |

| | |
|---|---|
| | % P |
| $P^{\mathrm{f}}$ | [*Quantum mechanics*] Probability flow (spatial description) |
| $\mathscr{P}$ | Particle |
| $\mathbf{P}$ | [*Mechanics*] Momentum |
| $\mathbf{P}^{\mathrm{f}}$ | [*Fluid mechanics*] Momentum flow (spatial description) |
| $\mathbf{P}^{\mathrm{c}}$ | [*Fluid mechanics*] Momentum content (spatial description) |
| $\mathbf{P}^{\mathrm{p}}$ | [*Fluid mechanics*] Momentum production (spatial description) |
| $\mathbf{P}$ | [*Algebraic topology*] Point without orientation |
| $\overline{\mathbf{P}}$ | [*Algebraic topology*] Point with inner orientation |
| $\widetilde{\mathbf{P}}$ | [*Algebraic topology*] Point with outer orientation |
| $p$ | [*Thermodynamics*] Pressure |
| $p^{\mathrm{th}}$ | [*Fluid mechanics*] Thermodynamic pressure |
| $p^{\mathrm{mech}}$ | [*Fluid mechanics*] Mechanical pressure |
| $p_i$ | [*Analytical mechanics*] Generalized momenta |
| $p_\beta$ | [*Relativistic mechanics*] Four-momentum |
| $\mathbf{p}$ | [*Mechanics*] Momentum of a particle |
| $\mathbf{p}$ | [*Fluid mechanics*] Momentum density |
| $\mathbf{p}^k$ | [*Algebraic topology*] *k*th primal 0-cell |
| $\tilde{\mathbf{p}}^k$ | [*Algebraic topology*] *k*th dual 0-cell |
| | Q |
| $Q$ | [*Thermodynamics*] Heat |
| $Q$ | [*Electromagnetism*] Electric charge |
| $Q^{\mathrm{c}}$ | [*Electromagnetism*] Electric charge content (spatial description) |
| $Q^{\mathrm{f}}$ | [*Electromagnetism*] Electric charge flow (spatial description) |
| $Q^{\mathrm{f}}_{\mathrm{m}}$ | [*Electromagnetism*] Magnetic charge flow (spatial description) |
| $Q^{\mathrm{p}}$ | [*Electromagnetism*] Electric charge production (hypothetical) |
| $Q^{\mathrm{s}}$ | [*Electromagnetism*] Electric charge stored |
| $Q_i$ | [*Analytical mechanics*] Generalized forces |
| $q$ | [*Electromagnetism*] Small electric charge |
| $q$ | [*Theory of beams*] Force for unit length |
| $q^i$ | [*Analytical mechanics*] Generalized coordinates |
| $\mathbf{q}$ | [*Thermal conduction*] Heat current density |
| $\mathbf{q}$ | [*Fluid mechanics*] Mass current density |
| | R |
| $R$ | [*Electromagnetism*] Resistance |
| $R$ | [*Thermodynamics*] Universal gas constant |
| $R$ | [*Relativistic gravitation*] Linear invariant of Riemann tensor |
| $R_{\mu\nu}$ | [*Relativistic gravitation*] Contracted Riemann tensor |
| $\check{\mathscr{R}}$ | [*Electromagnetism*] 3-form of electric charge density |
| $\mathscr{R}$ | Symbol for reversal of motion |
| $\mathscr{R}$ | Rigid body |
| $Re$ | [*Fluid mechanics*] Reynolds number |

| % | R |
|---|---|
| $R_H$ | [*Electromagnetism*] Hall constant |
| $\mathbf{R}$ | [*Solid mechanics*] Initial position vector |
| $\widetilde{\mathsf{R}}$ | [*Algebraic topology*] Incidence matrix face–edge of dual complex |
| $\bar{r}_{\alpha\beta}$ | [*Algebraic topology*] Incidence number face–edge |
| $\tilde{r}_{\alpha\beta}$ | [*Algebraic topology*] Incidence numbers between dual face $\tilde{\mathbf{s}}_\alpha$ and dual edge $\tilde{\mathbf{l}}_\beta$ |
| $\mathbf{r}$ | [*Solid mechanics*] Position vector |
| | **S** |
| $S$ | [*Analytical mechanics*] Hamilton's principal function |
| $S$ | [*Optics*] Eikonal function |
| $S$ | [*Thermodynamics*] Entropy |
| $S^{\mathrm{c}}$ | [*Thermodynamics*] Entropy content (spatial description) |
| $S^{\mathrm{f}}$ | [*Thermodynamics*] Entropy flow (spatial description) |
| $S^{\mathrm{p}}$ | [*Thermodynamics*] Entropy production (spatial description) |
| $S_{\mathrm{a}}$ | [*Interactions*] Seebeck coefficient |
| $\mathbf{S}$ | [*Electromagnetism*] Poynting vector |
| $\mathbf{S}$ | Surface without orientation in spatial description |
| $\mathcal{S}$ | Surface without orientation in material description |
| $\mathscr{S}$ | System (material description) |
| $\overline{\mathbf{S}}$ | Surface with inner orientation |
| $\widetilde{\mathbf{S}}$ | Surface with outer orientation |
| $s$ | [*Optics*] optical path length |
| $s$ | [*Thermodynamics*] entropy density |
| $\mathbf{s}^\beta$ | 2-cell of the primal complex |
| $\bar{\mathbf{s}}^\alpha$ | [*Algebraic topology*] 2-cell of primal complex |
| $\tilde{\mathbf{s}}^\alpha$ | [*Algebraic topology*] 2-cell of dual complex |
| | **T** |
| $T$ | [*Mechanics*] Kinetic energy; |
| $T$ | [*Thermodynamics*] Thermodynamic temperature |
| $T$ | [*Wave motion*] Period |
| $T$ | [*Time elements*] Duration of time interval |
| $T^*$ | [*Mechanics*] Kinetic co-energy |
| $T^{\mathrm{e}}$ | [*Time elements*] Duration of time interval to reach equilibrium |
| $T^{\mathrm{r}}$ | [*Time elements*] Duration of time interval needed for registration |
| $T_{\mathrm{n}}$ | [*Fluid mechanics*] Force normal to plane surface |
| $T_{\mu\nu}$ | [*Relativistic gravitation*] Stress energy momentum tensor |
| $\mathcal{T}$ | [*Thermal conduction*] Impulse of thermodynamic temperature |
| $\mathscr{T}$ | [*Thermal conduction*] Thermacy |
| $\mathbf{T}$ | [*Continuum mechanics*] Internal surface force (also denoted by $\mathbf{F}^{\mathrm{s}}$) |
| $\mathbf{T}$ | [*Time elements*] Interval without orientation |

| | |
|---|---|
| | % T |
| $\mathbf{T}$ | [*Continuum mechanics*] Resultant of external surface forces acting on boundary of volume |
| $\mathbf{T}$ | [*Continuum mechanics*] Tension of string |
| $\overline{\mathbf{T}}$ | [*Time elements*] Interval with inner orientation |
| $\widetilde{\mathbf{T}}$ | [*Time elements*] Interval with outer orientation |
| $\mathbf{t}$ | [*Continuum mechanics*] Stress vector |
| $\mathbf{t}$ | [*Geometry*] Unit tangent vector |
| $\check{\mathbf{t}}$ | [*Geometry*] Unit tangent pseudovector |
| | U |
| $U$ | [*Thermodynamics*] Internal energy |
| $U_g$ | [*Gravitation*] Gravitational potential |
| $U$ | [*Electromagnetism*] Generalized potential |
| $u$ | [*Thermodynamics*] Internal energy density |
| $u^\alpha$ | [*Relativistic mechanics*] Four-velocity |
| $u_e$ | [*Electromagnetism*] Electric energy density |
| $\mathbf{u}$ | [*Mechanics*] Displacement |
| $\mathbf{u}$ | [*Solid mechanics*] Incremental displacement |
| $\mathbf{u}_\infty$ | [*Fluid dynamics*] Asymptotic fluid velocity |
| | V |
| $V$ | [*Electromagnetism*] Voltage |
| $V$ | [*Mechanics*] Potential energy |
| $V^c$ | [*Mechanics*] Potential energy content (spatial description) |
| $V^f$ | [*Mechanics*] Potential energy flow (spatial description) |
| $V$ | [*Space elements*] Volume as measure of extension |
| $\mathbf{V}$ | [*Fluid mechanics*] Relative velocity |
| $\mathbf{V}$ | [*Space elements*] Volume without orientation (spatial description) |
| $\mathcal{V}$ | [*Space elements*] Volume unoriented in material description |
| $\overline{\mathbf{V}}$ | Volume (as space region) with inner orientation |
| $\widetilde{\mathbf{V}}$ | Volume (as space region) with outer orientation |
| $\mathbf{v}$ | [*Mechanics*] Velocity |
| $\bar{\mathbf{v}}^k$ | [*Algebraic topology*] *k*th primal 3-cell |
| $\tilde{\mathbf{v}}^k$ | [*Algebraic topology*] *k*th dual 3-cell; |
| | X |
| $x^\alpha$ | [*Relativistic mechanics*] Space-time coordinate |
| | Y |
| $y$ | [*Solid mechanics*] Transversal displacement |
| | W |
| $W$ | [*Mechanics*] Work |
| $W^*$ | [*Mechanics*] Virtual work |
| $W$ | [*Fluid mechanics*] Vortex flux |

| | |
|---|---|
| | % W |
| $W$ | [*Mechanics*] Weight |
| $\mathbf{W}$ | [*Fluid mechanics*] Spin tensor |
| $w$ | [*Mechanics*] Weight for unit length or for unit volume |
| $w_{\mathrm{e}}$ | [*Continuum mechanics*] Elastic energy density |
| $w_{\mathrm{e}}$ | [*Electromagnetism*] Electric energy density |
| $w_{\mathrm{m}}$ | [*Electromagnetism*] Magnetic energy density |
| $\check{\mathbf{w}}$ | [*Fluid mechanics*] Vorticity pseudovector ≡ vortex flux density |
| | **Greek alphabet** |
| $\alpha$ | [*Rigid body mechanics*] Angle of rotation |
| $\alpha$ | [*Solid mechanics*] Slope |
| $\alpha_l$ | [*Thermoelasticity*] Linear expansion coefficient |
| $\check{\alpha}$ | [*Rigid body mechanics*] Angular acceleration |
| $\Gamma$ | [*Fluid mechanics*] Velocity line integral |
| $\gamma$ | [*Solid mechanics*] Shear strain, shear angle |
| $\gamma$ | [*Interaction*] Gyromagnetic ratio |
| $\delta$ | [*Solid mechanics*] Lengthening |
| $\delta$ | [*Algebraic topology*] Coboundary operator |
| $\epsilon$ | [*Electromagnetism*] Permittivity |
| $\epsilon_0$ | [*Electromagnetism*] Vacuum permittivity |
| $\varepsilon$ | [*Solid mechanics*] Linear strain |
| $\varepsilon_{ij}$ | [*Solid mechanics*] Strain tensor |
| $\epsilon_{hlj}$ | [*Tensorial notation*] Permutation symbol |
| $\boldsymbol{\varepsilon}$ | [*Solid mechanics*] Symmetric strain tensor |
| $\boldsymbol{\varepsilon}'$ | [*Solid mechanics*] Strain deviator |
| $\eta$ | [*Thermodynamics*] Efficiency of thermodynamic circle |
| $\eta_{\alpha\beta}$ | [*Relativity*] Fundamental tensor in Cartesian coordinates in inertial reference system |
| $\check{\eta}^{hki}$ | [*Tensorial notation*] Levi-Civita pseudotensor = $\sqrt{g}\,\epsilon_{hki}$ |
| $\boldsymbol{\eta}$ | [*Continuum mechanics*] Displacement |
| $\Theta$ | [*Solid mechanics*] Volume dilatation ≡ cubic dilatation |
| $\Theta_p$ | [*Algebraic topology*] Null *p*-chain |
| $\theta$ | [*Fluid mechanics*] Cubic or volume dilatation rate |
| $\kappa$ | [*Geometry*] Curvature of plane line |
| $\Lambda_k^h$ | [*Tensorial notation*] Elements of inverse of transition matrix |
| $\lambda$ | [*Thermal conduction*] Thermal conductivity |
| $\lambda$ | [*Wave motion*] Wavelength |
| $\lambda$ | [*Solid mechanics*] Second Lamé constant |
| $\lambda_h^k$ | [*Tensorial notation*] Elements of transition matrix |
| $\mu$ | [*Thermodynamics*] Chemical potential |
| $\mu$ | [*Fluid mechanics*] Viscosity coefficient |
| $\mu$ | [*Electromagnetism*] Permeability |

| | |
|---|---|
| $\mu_0$ | [*Electromagnetism*] Vacuum permeability |
| $\mu$ | [*Solid mechanics*] Shear modulus $\equiv G$ |
| $\check{\boldsymbol{\mu}}$ | [*Electromagnetism*] Magnetic moment (pseudovector) |
| $v^i$ | [*Analytical mechanics*] Generalized velocities |
| $\nu$ | [*Wave motion*] Frequency |
| $\rho$ | [*Solid mechanics*] Mass density |
| $\rho$ | [*Electromagnetism*] Charge density |
| $\rho$ | [*Interactions*] Verdet constant |
| $\rho$ | [*Quantum mechanics*] Probability density |
| $\check{\rho}_{\mathrm{m}}$ | [*Electromagnetism*] Magnetic charge density (pseudoscalar) |
| $\sigma$ | [*Electromagnetism*] Surface charge density; stress |
| $\sigma$ | [*Electromagnetism*] Electric conductivity |
| $\sigma$ | [*Continuum mechanics*] Stress |
| $\sigma$ | [*Fluid mechanics*] Surface tension |
| $\sigma_{\mathrm{q}}$ | [*Thermodynamics*] Heat source = heat generation density rate |
| $\sigma_{\mathrm{s}}$ | [*Thermodynamics*] Entropy source strength |
| $\sigma_{ij}$ | [*Continuum mechanics*] Symmetric stress tensor |
| $\sigma'_{ij}$ | [*Solid mechanics*] Components of stress deviator |
| $\boldsymbol{\sigma}$ | [*Continuum mechanics*] Symmetric stress tensor |
| $\boldsymbol{\sigma}'$ | [*Solid mechanics*] Stress deviator |
| $\tau$ | [*Solid mechanics*] Tangential stress = shear stress |
| $\tau$ | [*Relativistic mechanics*] Proper time |
| $\boldsymbol{\tau}$ | [*Continuum mechanics*] Stress tensor |
| $\boldsymbol{\tau}$ | [*Continuum mechanics*] Viscous stress |
| $\Phi$ | [*Electromagnetism*] Magnetic flux |
| $\boldsymbol{\Phi}$ | [*Thermal conduction*] Heat current |
| $\boldsymbol{\Phi}$ | [*Fluid mechanics*] Mass current |
| $\phi$ | [*Electromagnetism*] Electric potential |
| $\phi$ | [*Fluid mechanics*] Velocity potential = kinetic potential |
| $\phi$ | [*Waves motion*] Phase; |
| $\check{\phi}$ | [*Solid mechanics*] Airy stress function (pseudoscalar) |
| $\check{\phi}_{\mathrm{m}}$ | [*Electromagnetism*] Pseudoscalar magnetic potential |
| $\varphi$ | [*Electromagnetism*] Electric potential impulse |
| $\check{\varphi}_{\mathrm{m}}$ | [*Electromagnetism*] Magnetic potential impulse (pseudoscalar) |
| $\chi$ | [*Electromagnetism*] Gauge function; |
| $\check{\chi}_{\mu\nu}{}^{\alpha\beta}$ | [*Electromagnetism*] Constitutive tensor of electromagnetism in space-time |
| $\Psi$ | [*Electromagnetism*] Electric flux |
| $\Psi$ | [*Quantum mechanics*] Probability amplitude |
| $\Psi$ | [*Fluid mechanics*] Stream line integral |
| $\psi$ | [*Fluid mechanics*] Stream function |
| $\check{\boldsymbol{\psi}}$ | [*Solid mechanics*] Stream vector (pseudovector) |

| | |
|---|---|
| $\Omega$ | [*Geometry*] Finite space region |
| $\boldsymbol{\Omega}$ | [*Solid mechanics*] Rotation tensor |
| $\breve{\omega}$ | [*Mechanics*] Angular velocity (pseudovector) |
| | **Mathematical Symbols** |
| $\wedge$ | Wedge ≡ exterior product |
| $\times$ | Vector product ≡ cross product |
| $\cdot$ | Scalar product ≡ dot product |
| $\otimes$ | Tensor product |
| $\mathscr{R}$ | Operation of reversal of motion |
| $D/Dt$ | Total derivative = convected derivative |
| $\mathrm{d}$ | Exterior differential (in exterior differential forms) |
| $d_s$ | Infinitesimal space increment |
| $d_t$ | Infinitesimal time increment |
| $\partial$ | [*Algebraic topology*] Boundary operator |
| $\partial$ | [*Differential formulation*] Partial derivative |
| $\partial\partial$ | [*Algebraic topology*] Boundary of boundary |
| $\partial_t$ | Time partial derivative |
| $\nabla$ | [*Vector and tensor analysis*] Nabla |
| $\nabla f$ | [*Vector and tensor analysis*] Gradient (grad) |
| $\nabla\times$ | [*Vector and tensor analysis*] Curl (≡ rot) |
| $\nabla\cdot$ | [*Vector and tensor analysis*] Divergence (div) |
| $\nabla^2$ | [*Vector and tensor analysis*] Laplacian |
| $\Delta$ | [*Tensorial notation*] Determinant of transition matrix |
| $\Delta_s$ | Space increment |
| $\Delta_t$ | Time increment |
| $\delta$ | [*Algebraic topology*] Coboundary operator |
| $\delta_i^j$ | [*Tensorial notation*] Kronecker tensor |
| $\det(\mathbf{M})$ | Determinant of $\mathbf{M}$ |
| $\mathrm{sgn}(\Delta)$ | Sign of determinant of transition matrix |

# References


1. Achenbach, J.D.: Waves Propagation in Elastic Solids. North Holland, Amsterdam (1973)
2. Alexandrov, P.: Elementary Concepts of Topology. Dover, New York (1961)
3. Alexandrov, P.S.: Combinatorial Topology. Dover, New York (1998)
4. Alexandrov, P.S.: Topologia Combinatoria. Einaudi, Turin, Italy (1957)
5. Akhiezer, A., Akhiezer, I.: Electromagnetisme et Ondes Electromagnetiques, vol. 3. Mir, Moscow (1988)
6. Aris, R.: Vectors, Tensors and the Basic Equations of fluid Mechanics. Prentice Hall, Englewood Cliffs, NJ (1962)
7. Auchmann, B., Kurz, S.: A Geometrically Defined Discrete Hodge Operator on Simplicial Cells. IEEE Trans. Magn. **42**(4), 643–646 (2006)
8. Ballentine, L.E.: Quantum Mechanics. World Scientific, Singapore (1998)
9. Bashmakova I.G., Smimova G.S.: The Literal Calculus of Viéte and Descartes. Am. Math. Mon. **106**(3), 260–263 (1999)
10. Batchelor, G.K.: An Introduction to Fluid Dynamics. Cambridge University Press, Cambridge (1977)
11. Becker, R.: Teoria dell'Elettricità, vol. I. Sansoni, Florence, Italy (1949)
12. Benvenuto, E.: An Introduction to the History of Structural Mechanics, Part I, p. 274. Springer, Dordrecht, The Netherlands (1991)
13. Bettini, A.: Elettromagnetismo. Zanichelli, Bologna, Italy (1994)
14. Crawford, F.S.: Berkeley physics course, vol. 3. Waves, New York, N.Y. McGraw-Hill (1968)
15. Billington, E.W., Tate, A.: The Physics of Deformation and Flow. McGraw-Hill, New York (1981)
16. Bird, R.B., Stewart, W.E., Lightfoot, E.N.: Transport Phenomena. Wiley, New York (1960)
17. Bohm, D., Hiley, B.J., Stuart, A.: On a New Mode of Description in Physics. Int. J. Theor. Phys. **3**(3), 171–183 (1970)
18. Born, M., Wolf, E.: Principles of Optics: Electromagnetic Theory of Propagation, Interference and Diffraction of Light. Pergamon, Oxford (1975)
19. Born, M.: Kritische Betrachtungen zur traditionellen Darstellung der Thermodynamik. Zeitschrift fur Physik. **XXII**, 218–224 (1921)
20. Born, M.: Experiment and Theory in Physics. Dover, New York (1956)
21. Bossavit, A.: Électromagnetism en vue de la modélisation. Springer, Paris (1994)
22. Bourbaki, N.: Elements d'histoire des mathématiques. Hermann, Paris (1969)
23. Bourbaki, N.: Espaces Vectoriels Topologiques. Springer, Berlin (2007)
24. Bourgin, D.G.: Modern Algebraic Topology. MacMillan, New York (1963)
25. Belluzzi, O.: Scienza delle Costruzioni. Zanichelli, Bologna (1966)



E. Tonti, *The Mathematical Structure of Classical and Relativistic Physics*,
Modeling and Simulation in Science, Engineering and Technology,
DOI 10.1007/978-1-4614-7422-7,

26. Blakely, R.J.: Potential Theory in Gravity & Magnetic Applications. Cambridge University Press (1996)
27. Berestetskii, V.B., Lifshitz, E.M., Pitaevskii, L.P.: Relativistic Quantum Theory. Pergamon Press, Oxford (1971)
28. Born, M.: Experiment and Theory in Physics. Dover, NOME CITTA' (1956)
29. Branin, F.H.Jr.: The algebraic topological basis for network analogies and the vector calculus. In: Symposium on Generalized Networks. Polytechnic Institute of Brooklin, New York (1966)
30. Brillouin, L.: Tensors in Mechanics and Elasticity. Academic, New York (1964)
31. Bruhat, G.: Cours de Physique Generale: Optique. Masson, Paris (1959)
32. Burke, W.L.: Applied Differential Geometry. Cambridge University Press, Cambridge (1985)
33. Callen, H.B.: Thermodynamics: An Introduction to the Physical Theories of Equilibrium Thermostatics and Irreversible Thermodynamics. Wiley, New York (1960)
34. Cartan, E.: Leçons sur la Géometrie des Espaces de Riemann. Gauthier-Villars, Paris (1951)
35. Castillo, P., Rieben, R., White, D.: FEMSTER: An object-oriented class library of discrete differential forms. ACM Trans. Math. Softw. **31**(4), 425–457 (2005)
36. Castillo, P., Koning, J., Rieben, R., White, D.: A discrete differential forms framework for computational electromagnetism. Comput. Model. Eng. Sci. **1**(1), 115 (2002)
37. Cavendish, J.C., Field, D.A., Frey, W.H.: An approach to automatic three-dimensional finite element mesh generation. Int. J. Numer. Methods Eng. **21**, 329–347 (1985)
38. Cebeci, T., Smith, A.M.: Analysis of Turbulent Boundary Layers. Academic, New York (1974)
39. Chadwick, P.: Continuum Mechanics, Concise Theory and Problems. Allen & Unwin, London (1976)
40. Chevalley, C.: Fundamental Concepts of Algebra. Academic, New York (1966)
41. Chorin, A.J., Marsden, J.E.: A Mathematical Introduction to Fluid Mechanics, 3rd edn. Springer, Berlin (1993)
42. Ciarlet, P.G., Lions, J.L.: Handbook of Numerical Analysis: Special volume, Computational Chemistry, vol. X, p. 123. North-Holland, Amsterdam (2003)
43. Corson, E.M.: Introduction to Tensors, Spinors and Relativistic Wave-Equations. Blackie & Son, London (1955)
44. Coxeter, H.S.: Regular Polytopes. Dover, New York (1973)
45. Dass, T., Sharma, S.K.: Mathematical Methods In Classical And Quantum Physics. Universities Press, India (1998)
46. de Broglie, L.: Elements de Theorie des Quanta et de Mecanique Ondulatoire. Gauthier-Villars, Paris (1959)
47. de Broglie, L.: La Thermodynamique de la Particule Isolée. Gauthier-Villars, Paris (1964)
48. de Broglie, L.: Physique et Microphysique. Harper & Brothers, New York (1960)
49. de Groot, S.R., Mazur, P.: Non-Equilibrium Thermodynamics. Dover, New York (1984)
50. de Broglie, L.: La theorie des particules de spin 1/2. Gauthier-Villars, Paris (1952)
51. Henry, P.G.: Darcy and Other Pioneers in Hydraulics, Philadelphia (2003)
52. De Marchi, G.: Idraulica. **1**(I/II). Hoepli, Milan, Italy (1977)
53. Deschamps, G.A.: Electromagnetics and Differential Forms. Proc. IEEE **69**(6), 676–696 (1981)
54. Demirel, T.: Nonequilibrium Thermodynamics: Transport and Rate Processes in Physical & Biological Systems. Elsevier Science, Amsterdam (2002)
55. Desbrun, M., Kanso, E., Tong, Y.: Discrete Differential Forms for Computational Modeling. In: Bobenko, I., et al. (eds.) Discrete Differential Geometry, pp. 287–323. Birkhäuser (2008)
56. Desbrun, M., Hirani, A.N., Leok, M., Marsden, J.E.: Discrete exterior calculus. `http://arxiv.org/abs/math/0508341` (2005) (unpublished)

57. Desoer, C., Kuh, E.: Basic Circuit Theory, India Higer Education (2009)
58. Dubrovin, B.A., Novikov, S.P., Fomenko, A.T.: Geometria Contemporanea, vol. 3. Editori Riuniti, Rome (Mir Edizioni, Moscow) (1989)
59. Dugdale, J.S.: Entropy and Its Physical Meaning. Taylor & Francis, London (1996)
60. Duhem, P.: The Aim and Structure of Physical Theory. Atheneum, New York (1977)
61. Dunford, N., Schwartz, J.T.: Linear Operators in Hilbert spaces, vol. II. Interscience, New York (1964)
62. Eisenhart, L.P.: Riemannian Geometry. Princeton University Press, Princeton, NJ (1960)
63. Eriksen, L.J.: Tensor Fields (Handbuch Der Physik Band III/1). Springer, Berlin (1960)
64. Eringen, A.C., Suhubi, E.S.: Elastodynamics. Academic, New York (1974)
65. Faraday, M.: Remarks on Static Induction. Proceedings of the Royal Institution, 12 Feb 1858
66. Fer, F.: Thermodynamique Macroscopique, Tome I, II. Gordon & Breach, Paris (1970)
67. Fermi, E.: Termodinamica. Bollati Boringhieri, Turin, Italy (1963)
68. Feynman, R.P.: Theory of Fundamental Processes. Benjamin (1962)
69. Feynman, R.P., Leighton, R.B., Sands, M.: Lectures on Physics, vol. I-II-III. Addison-Wesley, Reading, MA (1963)
70. Flanders, H.: Differential Forms with Applications to the Physical Sciences. Academic, New York (1963)
71. Fleury, P., Mathieu, J.P.: Fisica Generale e Sperimentale, vol. 1–8. Zanichelli, Bologna, Italy (1970) (Italian translation of Physique générale et expérimentale. Editions Eyrolles, Paris)
72. Fouillé, A.: Electrotechnique à l'Usage des Ingénieurs. Dunod, Paris (1961)
73. Fournet, G.: Electromagnétisme à partir des équations locales. Masson, Paris (1985)
74. Franz, W.: Algebraic Topology. Harrap, London (1968)
75. Frauendiener, J.: Discrete Differential Forms in General Relativity. Classical Quantum Gravity **23**, S369–S38 (2006)
76. Frey, W.H., Cavendish, J.C.: Fast Planar Mesh Generation Using the Delaunay Triangulation. General Motors Research Publication, GMR-4555, Warren (1983)
77. Fung, Y.C.: Foundation of Solid Mechanics. Prentice Hall, Englewood Cliffs, NJ (1965)
78. Godement, R.: Cours d'algebre. Hermann, Paris (1966)
79. Goldberg, S.: Unbounded Operators. Mc Graw-Hill, New York (1966)
80. Goldstein, H.: Classical Mechanics. Addison-Wesley, Reading, MA (1957)
81. Gordon, J.E., Wagley, S.: The Science of Structures and Materials, vol. 23. Scientific American Library/Scientific American Books, New York (1988)
82. Granger, R. A.: Fluid Mechanics, Dover, New York (1995)
83. Grady, L.J., Polimeni, J.R.: Discrete Calculus, Applied Analysis on Graphs for Computational Science. Springer, London (2010)
84. Gray, H.J., Isaacs, A.: A New Dictionary of Physics. Longman, London (1975)
85. Griffiths, D.J.: Introduction to Electrodynamics. Addison-Wesley (2012)
86. Guggenheim, E.A.: Termodinamica. Einaudi, Turin, Italy (1952)
87. Hallén, E.: Electromagnetic Theory. Chapman & Hall, London (1962)
88. Halmos, P.C.: Finite-Dimensional Vector Spaces. Springer, New York (1987)
89. Hehl, F.W., Obukhov, Y.N.: Foundations of Classical Electrodynamics – Charge, Flux and Metric. Birkhäuser, Boston (2003)
90. Helmholtz, H.: Über die Erhaltung der Kraft (Wissenshaftliche Abhandlungen, I). Druck und Verlag von Reimer, Berlin (1847)
91. Hilbert, D., Cohn-Vossen, S.: Geometry and the Imagination. Chelsea, New York (1952)
92. Hilton, P.J., Wylie, S.: Homology Theory: An Introduction to Algebraic Topology. Cambridge University Press, Cambridge (1967)
93. Hirani, A.N.: Discrete exterior calculus. Ph.D. dissertation, Caltech (2003) `http://thesis.library.caltech.edu/1885/`
94. Hocking, J.G., Young, G.S.: Topology. Addison-Wesley, Reading, MA (1961)

95. Houghton, E.L., Brock, A.E.: Aerodynamics for Engineering Students. Edward Arnold (Publisher) Ltd, Ch.11 (1972)
96. Hubbert, M.K.: The Theory of Ground-Water Motion. The Journal of Geology, vol. 48, no. 8 (1941)
97. Hughes, W.F.: An Introduction to Viscous Flow. McGraw-Hill, New York (1979)
98. Hunter, S.C.: Mechanics of Continuous Media. Ellis Horwood, Chichester, UK (1976)
99. Isaacson, E., Keller, H.B.: Analysis of Numerical Methods. Wiley, New York (1966)
100. Ingarden, R.S., Jamiolkowski, A.: Classical Electrodynamics. Elsevier, Amsterdam (1985)
101. IUPAP, International Union of Pure and Applied Physics: Symbols, Units and Fundamental Constants in Physics. Document I.U.P.A.P.-25 (SUNAMCO 87-1)
102. Jackson, J.D.: Classical Electrodynamics, 3rd edn. Wiley, New York (1999)
103. Jackson, J.D.: Elettrodinamica Classica, 2nd edn. Zanichelli, Bologna, Italy (2001)
104. Jammer, M.: Concepts of Force: A Study in the Foundations of Dynamics. Harvard University Press, Cambridge, MA (1957)
105. Jammer, M.: Concepts of Mass in Classical and Modern Physics. Harvard University Press, Cambridge, MA (1961)
106. Jaunzemis, W.: Continuum Mechanics. MacMillan, New York (1967)
107. Jefimenko, O.D.: Electricity and Magnetism. Appleton-Century-Crofts, New York (1966)
108. Joas, C., Lehner, C.: The classical roots of wave mechanics: Schrödinger's transformations of the optical-mechanical analogy. Studies in History and Philosophy of Modern Physics, **40**(4), 338–351 (2009)
109. Jordan, E.C., Balmain, K.G.: Electromagnetic Waves and Radiating Systems. Prentice Hall, Englewood Cliffs, NJ (1968)
110. Jouguet, M.: Traité d'électricité theorique, Tomes I, II, III. Gauthier Villars, Paris (1955)
111. Kaufman, A. A.: Geophysical Field Theory and Method: Gravitational, Electric and Magnetic Fields. Academic, New York (1994)
112. Kato, T.: Perturbation Theory for Linear Operators. Springer, Berlin (1966)
113. Kellog, O.D.: Foundation of Potential Theory. Ungar Publishing Co. (1929)
114. Klein, F.: Elementary mathematics from an advanced standpoint: Geometry. **39**(3), 201–136 (1948) [Dover, New York]
115. Kompaneyets, A. S.: Theoretical Physics. MIR, Moscow (1965) [P.P. Plane vector fields. Academic, New York 1966]
116. Kundu, P.K., Cohen, I.M., Dowling, D.R: Fluid Mechanics with Multimedia DVD. Academic, New York (2011)
117. Lai, W.M., Rubin, D., Krempl, E.: Introduction to Continuum Mechanics. Pergamon, Oxford (1974)
118. Lamarsh, J.R.: Introduction to Nuclear Engineering, 2nd edn. Addison-Wesley, Reading, MA (1977)
119. Lamb, H.: Hydrodynamics. Dover, New York (1945)
120. Lanczos, C.: The variational Principles of Mechanics. Toronto University Press, Toronto (1952)
121. Lanczos, C.: Linear Differential Operators. Van Nostrand Company Limited, New York (1961)
122. Landau, L.D., Lifshitz, E.M.: The classical theory of fields. Pergamon Press, Oxford (1962)
123. Landau, L., Lifshitz, E.: Mechanics. Pergamon, Oxford (1900)
124. Landau, L.D., Lifshitz, E.M.: Electrodynamics of Continuous Media. Pergamon, Oxford (1960)
125. Landau, L.D., Lifshitz, E.M.: Fluid Mechanics. Pergamon, Oxford (1958)
126. Langevin, P.: Sur la nature des grandeurs et les choix d'un système d'unités électriques. Bull. Soc. Fr. Phys. **164**, 493–505 (1922) [Reprinted in *Oeuvres scientifiques de Paul Langevin*. Centre National de la Recherche Scientifique, 493–505 (1950)]

127. Langevin, P.: Sur les grandeurs champ et induction. Bull. Soc. Fr. Phys. **162**, 491–492 (1921) [Reprinted in Oeuvres scientifiques de Paul Langevin. Centre National de la Recherche Scientifique, 491–492 (1950)]
128. von Laue, M.: Relativitätstheorie, vol. 1. Vieweg, Leipzig, Germany (1921)
129. Laugwitz, D.: Differential and Riemannian Geometry. Academic, New York (1965)
130. Lebesgue, H.: Leçons sur l'Integration et la Recherche des Fonctions Primitives. Chelsea, Bronx (1928)
131. Lebesgue, H.: Sur la Mesure des Grandeurs. Gauthier Villars, Paris (1956)
132. Lefschetz, S.: Algebraic Topology, vol. XXVII. Colloquium Publications American Mathematical Society, New York (1942)
133. Lefschetz, S.: Introduction to Topology. Princeton University Press, Princeton, NJ (1949)
134. Lefschetz, S.: Topology, 2nd edn. Chelsea, New York (1965)
135. Levi-Civita, T., Amaldi, U.: Compendio di Meccanica Razionale, vols. I, II. Zanichelli, Bologna, Italy (1975)
136. Levi-Civita, T.: The Absolute Differential Calculus. Blackie & Son, London (1954)
137. Lewis, G.N., Randal, M.: Thermodynamics. McGraw-Hill, New York (1923)
138. Lichnerowicz, A.: Tensor Calculus. Methuen, London (1962)
139. Lindsay, R.B.: Energy: Historical Development of the Concept. Dowden, Hutchingon & Ross, Stroudsburg (1975)
140. Love, A.E.: A treatise on the mathematical theory of elasticity. Dover (1944)
141. Locker, J.: Spectral Theory of Non-Self-Adjoint Two-Point Differential Operators, vol. 73. American Mathematical Society, Providence, RI (2000)
142. Loitsyanskii, L.G.: Mechanics of Liquids and Gases. Pergamon, Oxford (1966)
143. Lorrain, P., Corson, D.R., Lorrain, F.: Electromagnetic Fields and Waves. Freeman, New York (1996)
144. Luré, L.: Mécanique analytique, Tome I, II. Librairie Universitaire, Louvain (1968)
145. Mach, E.: The Science of Mechanics. Open Court, Chicago (1960)
146. MacFarlane, A.G.: Dynamical System Models. Harper & Brothers, New York (1970)
147. Malvern, L.E.: Introduction to the Mechanics of Continuous Medium, Prentice Hall, Englewood Cliffs, NJ (1969)
148. Mansfield E.H.: The Bending and Strectching of Plates. Cambridge University Press (1989)
149. Maugin, G.A.: Towards an Analytical Mechanics of Dissipative Materials. Rend. Sem. Mat. Univ. Pol. Torino, vol. 58 (2) (2000).
150. Mansfield, M., O'Sullivan, C.: Understanding Physics. Wiley, London (2010)
151. Maugin, G.A, Berezowsky, A.: Material formulation of finite-strain thermoelasticity and applications. In: Third International Congress on Thermal Stress, Cracow, Poland (1999)
152. Maxwell, J.C.: Remarks on the mathematical classification of physical quantities. Proc. London Math. Soc. **3**(34), 258–266 (1871) [Reprinted by Niven, W.D.: The Scientific Papers of J.C. Maxwell. Dover, New York (1965)]
153. Maxwell, J.C.: Traité Eléméntaire d'electricité. Gauthier Villars, Paris (1884)
154. Maxwell, J.C.: Theory of Heat. Dover, New York (2001)
155. Mc Leod, E.B.: Introduction to Fluid Dynamics. Pergamon, Oxford (1963)
156. McMahon, D.: Quantum Field Theory Demystified. McGraw-Hill (2008)
157. Melehy, M.A.: Foundation of the Thermodynamic Theory of Generalized Fields. Mono Book Corporation, Baltimore, MD (1973)
158. Meyer, R.E.: Introduction to Mathematical Fluid Dynamics. Dover, New York (1982)
159. Milligan, T.A.: Modern Antenna Design. Wiley, London (2005)
160. Milne-Thomson, L.M.: Theoretical Hydrodynamics. MacMillan, New York (1955)
161. Misner, C.W., Thorne, K.S., Wheeler, J.A.: Gravitation. Freeman, San Francisco (1973)
162. Möbius, A.F.: Der barycentrische calcül. Leiptzig (1827). Translated in: The barycentric calculus. Collected works, vol. I. Leipzig, Germany (1885)

163. Møller, C.: The Theory of Relativity. Oxford University Press, Oxford (1962)
164. Moore, W.J.: Physical Chemistry. Longman, London (1961)
165. Munkres, J.R.: Elements of Algebraic Topology. Addison-Wesley/Westview Press, Reading, MA (1984)
166. Naber, G.L.: Topological Methods in Eucliedean Spaces. Dover, New York (2000)
167. Naimark, M.A.: Linear Differential Operators, Part I. Ungar, New York (1967)
168. Naimark, M.A.: Linear Differential Operators, Part II. Ungar, New York (1968)
169. Novozilov, J.V., Jappa, J.A.: Elektrodynamika. Nauka, Moscow (1978)
170. Olivieri, L., Ravelli, E.: Elettrotecnica, vol. I. Cedam, Padua, Italy (1983)
171. O'Neill, M.E., Chorlton, F.: Viscous and Compressible Fluid Dynamics. Ellis Horwood, Chichester, UK (1989)
172. Paterson, A.R.: A First Course in Fluid Dynamics. Cambridge University Press, Cambridge (1997)
173. Patterson, E.M.: Topology. Oliver and Boyd, Edinburgh, UK (1966)
174. Pauli, W.: Elettrodinamica. Bollati Boringhieri, Turin, Italy (1964)
175. Penfield, P.Jr., Haus, H.: Electrodynamics of Moving Media. MIT Press, Cambridge, MA (1967)
176. Perucca, E.: Fisica Generale e Sperimentale, vols. I, II. UTET, Turin, Italy (1940)
177. Planck, M.: Scientific Autobiography and other Papers. Philosophical Library, New York (1949)
178. Pohl, R.W.: Physical Principles of Electricity and Magnetism. Blackie & Son, London (1930)
179. Pohl, R.W.: Elettrologia, vol. II. Piccin, Padova (1972)
180. Postnikov, M.: Lectures in Geometry. Semester I, Analytic Geometry. Mir, Moscow (1982)
181. Post, E.J.: Formal Structure of Electromagnetics. North-Holland, Amsterdam (1962)
182. Post, E.J.: The constitutive map and some of its ramifications. Ann. Phys. **71**, 497–518 (1972)
183. Post, E.J.: A minor or a major predicament of physical theory. Found. Phys. **7** (3/4), 255–277 (1977)
184. Post, E.J.: Kottler-Cartan-Van Dantzig (KCD) and noninertial systems. Found. Phys. **9**, 619–640 (1978)
185. Post, E.J.: Magnetic symmetry, improper symmetry and Neumann's principle. Found. Phys. **8**(3/4), 277–294 (1978)
186. Post, E.J.: The logic of time reversal. Found. Phys. **9**(1–2), 129–161 (1979)
187. Post, E.J.: Time asymmetries in classical and in nonclassical physics. Found. Phys. **9**(11/12), 831–863 (1979)
188. Prager, W.: Introduction to Mechanics of Continua. Ginn & Company, Boston (1961)
189. Preumont, A.: Mechatronics Dynamics of Electromechanical and Piezoelectric Systems. Springer, Dordrecht, The Netherlands (2006)
190. Prigogine, I.: Introduction a la Thermodynamique des Processus Irreversibles. Interscience/Wiley, New York (1974)
191. Redlich, O.:Generalized coordinates and forces. J. Phys. Chem. **66**, 585–588 (1962)
192. Rojansky, V.: Electromagnetic Fields and Waves. Dover, New York (1979)
193. Rosen, J.: Fundamental manifestations of symmetry in physics. Found. Phys. **20**(3), 283–307 (1990)
194. Rosser, W.G.: An Introduction to the Theory of Relativity. Butterworths, London (1964)
195. Roth, J.P.: An application of algebraic topology to numerical analysis: On the existence of a solution to the network problem. Proc. Natl. Acad. Sci. USA **41**(7), 518–521 (1955)
196. Roth, J.P.: Existence and uniqueness of solution to electrical network problem via homology sequences. In: Wilf, H.S., Harary, F.: Mathematical Aspects of Electrical Network Analysis. American Mathematical Society, Providence, RI (1971)

197. Rzewusky, J.: Field Theory. Part I, Classical Theory. Polish Scientific Publisher, VVarszawa (1964)
198. Sakurai, J.J.: Invariance Principles and Elementary Particles. Princeton University Press (1964)
199. Saletan, E.J., Cromer, A.H.: Theoretical Mechanics. Wiley, New York (1971)
200. Schaefer, H.H.: Topological Vector Spaces. Springer, Berlin (1971)
201. Schelkunoff, S.A.: Electromagnetic Fields. Blaisdell, New York (1963)
202. Schönberg, M.: Quantum Theory and Geometry. In: Max Planck Festchrift, pp. 321–338. Deutscher, Berlin (1958)
203. Schouten, J.A., Van Dantzig, D.: On ordinary quantities and W-quantities. Comput. Math. **7**, 447–473 (1939)
204. Schouten, J.A.: Tensor Calculus for Physicists. 2nd edn. Dover, New York (1989)
205. Schutz, B.: Geometrical Methods of Mathematical Physics. Cambridge University Press, Cambridge (1988)
206. Schwarz, A.S.; Topology for Physicists. Springer, Berlin (1996)
207. Sears, F.W.: An Introduction to Thermodynamics, the Kinetic Theory of Gases and Statistical Mechanics. Addison-Wesley, Reading, MA (1953)
208. Sears, F.W., Zemansky, M.W.: College Physics. Addison-Wesley, London (1955)
209. Shames, I.H.: Mechanics of Fluids. McGraw-Hill, New York (1992)
210. Seifert, H., Threlfall, W.: A Textbook of Topology. Academic, New York (1980)
211. Singer, I.M., Thorpe, J.A.: Lecture notes on elementary topology and geometry. Springer, New York (1967)
212. Sobrero L., Del significato meccanico della funzione di Airy, Rendiconti della R. Accademia Nazionale dei Lincei,v. XXI, serie 6$\omega$, 1$\omega$ sem., fasc. 4, 1935, pp. 264–269.
213. Sommerfeld, A.: Lectures in Theoretical Physics: Mechanics. **I**, Academic, New York (1952)
214. Sommerfeld, A.: Lectures in Theoretical Physics: Mechanics of Deformable Bodies. **2**, Academic, New York (1950)
215. Sommerfeld, A.: Lectures in Theoretical Physics. Electrodynamics. **III**, Academic, New York (1952)
216. Sommerfeld, A. Lectures in Theoretical Physics. Thermodynamics and Statistical Mechanics. Vol. V. Academic, New York (1952)
217. Spiegel, K.R.: Vector Analysis. Shaum Publ. Co. (1959)
218. Springer, G.: Introduction to Riemann Surfaces. Springer (1957). Physics, vol. I. MacMillan, New York (1967)
219. Stern A., Tong Y., Desbrun M., Marsden J.E.: Geometric computational electrodynamics with variational integrators and discrete differential forms. `http://arxiv.org/abs/0707.4470` (2009)
220. Stone, M.H.: Linear Trasformations in Hilbert Spaces, vol. XV. Colloquium Publications, American Mathematical Society, New York (1932)
221. Synge, J.L., Schild, A.: Tensor Calculus. University Press, Toronto (1956)
222. Teixeira F.L., Chew W.C., Lattice electromagnetic theory from a topological viewpoint. J. Math. Phys. **40**(1) (1999)
223. Temple, G.: An Introduction to Fluid Dynamics. Clarendon, Oxford (1958)
224. Timoshenko, S., Goodier, J.N.: Theory of Elasticity. Mc Graw Hill (1951)
225. Tolman, R.C.: The Measurable Quantities of Physics. Phys. Rev. **IX**(3), 237–253 (1917)
226. Tolman, R.C.: The Principles of Statistical Mechanics. Oxford University Press, Oxford (1962)
227. Tonti, E.: On the mathematical structure of a large class of physical theories. Rend. Acc. Lincei **LII**, 48–56 (1972)
228. Tonti, E.: A mathematical model for physical theories. Rend. Acc. Lincei **LII**, 175–181; 350–356 (1972)

229. Tonti, E. On the geometrical structure of electromagnetism. In: Ferrarese, G. (ed.) Gravitation, Electromagnetism and Geometrical Structures, for the 80th birthday of A. Lichnerowicz , pp. 281–308. Pitagora Editrice, Bologna, Italy (1995)
230. Tonti, E.: Finite Formulation of the Electromagnetic Field. PIER Prog. Electromagn. Res. **32**, 1–42 (2001) [EMW Publishing]
231. Tonti, E.: A Finite Formulation for the Wave Equation. J. Comput. Acoust. **9**(4), 1355–1382 (2001)
232. Tonti, E.: A direct discrete formulation of field laws: The cell method. CMES Comput. Model. Eng. Sci. **2**(2), 237–258 (2001)
233. Tonti, E. Finite formulation of the electromagnetic field. International Compumag Soc. Newslett. **8**(1), 5–11 (2001)
234. Tonti, E.: Finite formulation of electromagnetic field. IEEE Trans. Magn. **38**(2), 333–336 (2002)
235. Tonti, E., Nuzzo, E.: Gradiente Rotore Divergenza. Pitagora Editrice, Bologna, Italy (2007)
236. Treder, H.J.: Gravitationstheorie und Äquivalenzprinzip. Akademic, Berlin (1979)
237. Truesdell, C., Toupin, R.: The Classical Field Theories. Encycl. Phys. **III**(1) (1960) [Springer, Berlin]
238. Truesdell, C.: The physical components of vectors and tensors. ZAMM-J. Appl. Math. Mech./Zeitschrift für Angewandte Mathematik und Mechanik, **33**(10–11), 345–356 (1953)
239. Truesdell, C.: A First Course in Rational Continuum Mechanics: General Concepts, vol. 1. Academic, New York (1991)
240. Van Dantzig, D.: On the geometrical representations of elementary physical objects and the relations between geometry and physics. Nieuw Archief voor Wiskunde **3**(2), 73–89 (1954)
241. Veblen, O., Whitehead, J.H.: The foundations of Differential Geometry. Cambr. Tracts **29**, 55–56 (1932)
242. Vanderlinde, J.: Classical Electromagnetic theory, 2nd ed., Springer (2004)
243. Von Westenholz, C.: Differential Forms in Mathematical Physics. vol. I/II. North-Holland, Amsterdam (1980)
244. Wallace, A.W.: An Introduction to Algebraic Topology. Pergamon Press, Oxford (1967)
245. Wallace, A.W.: Algebraic Topology, Homology and Cohomology. W. A. Benjamin, New York (1970)
246. Warnick, K.F., Russer, P.: Two, Three and Four-Dimensional Electromagentics Using Differential Forms. Turk. J. Elec. Engin, **14**(1), (2006)
247. Weyl, H.: Space Time Matter. Dover, New York (1922)
248. Weyl, H.: Philosophy of Mathematics and Natural Science. Princeton University Press, Princeton, NJ (1949)
249. Wichmann, E.H.: Berkeley physics course, Vol. 4. Quantum Physics. MacGraw-Hill, (1971)
250. Wigner, E.P.: Group Theory and its Applications to the Quantum Mechanics of Atomic Spectra. Academic, New York (1959)
251. Williams, J.H.Jr.: Fundamentals of Applied Dynamics. Wiley, New York (1996)
252. Willmore, J.T.: An Introdution To Differential Geometry. Clarendon, Oxford (1959)
253. Whitney, H.: Geometric Integration Theory. Princeton University Press, Princeton, NJ (1957)
254. Whittaker, E.: A History of the Theories of Aether and Electricity, vol. 2. Humanities Press, New York (1973)
255. Yih, C.S.: Fluid Mechanics. McGraw-Hill, New York (1969)
256. Yilmaz, H.: Introduction to the Theory of Relativity and the Principles of Modern Physics. Blaisdell Pub. Co., New York (1965)
257. Yourgrau, W., Mandelstam, S.: Variational Principles in Dynamics and Quantum Theory. Dover, New York (1968)
258. Zund, J.D., Brown, E.: The theory of bivectors. Tensor N.S. **22**, 179–185 (1971)

# Index

algebraic formulation, 2, 4, 9, 17, 64, 65, 70, 81, 107, 111, 164, 185, 219, 433
analogy, 1, 2, 9–13, 17, 82, 248, 392, 393
association, 2, 5, 113, 116, 122, 126, 132, 138, 143, 197, 216, 221, 229, 230, 436

balance equation, 15, 17, 93, 156, 164, 170, 175, 176, 182, 224, 230, 238, 389, 432, 433, 451, 454
boundary operator, 194

cell complex, 4, 5, 7, 8, 17, 58–60, 63–68, 70, 73, 75, 76, 81, 83, 84, 93, 183–186, 191, 193, 197, 208, 213–216, 223, 241, 436
cell method, 17, 84, 111, 307
chain, 8, 190, 191, 193, 197, 306, 436
circuital equation, 8, 15, 16, 169, 171, 174, 182, 200, 230
coboundary operator, 8, 204, 207, 217–219, 231, 233, 430, 433
coboundary process, 8, 18, 194, 196, 200–204, 206, 212, 213, 219, 432, 433, 435, 436
cochain, 7, 8, 169, 183, 196, 198, 204
coface, 185, 186, 193, 200, 201, 203, 213, 215
configuration variable, 5, 99, 101, 148, 156, 160, 221, 229, 231, 237, 253, 274, 281, 290, 292, 325, 326, 330, 335, 389, 392, 440
conjugate variable, 149, 151, 442
constitutive equation, 7, 9, 148, 156, 160–163, 165, 170, 177, 224, 226, 229, 231, 233, 241, 245, 250, 274, 326, 430, 432, 448

defining equation, 155, 160, 176, 177, 223, 251, 281, 432, 433, 451, 454
differential form, 8, 24, 169, 181–184, 199, 207, 217–219, 231, 233, 415
discrete form, 8, 233, 306, 415, 430, 433, 435, 436
dual cell complex, 4, 5, 69–71, 78, 79, 84, 217, 218, 221, 229, 306, 335, 433, 435
dual instant, 59, 248
dual interval, 59, 60, 258, 327

energy variable, 99, 102, 116, 253
exterior differential, 8, 18, 169, 181, 200, 218, 219, 231, 233

field theory, 4, 10, 11, 87, 92, 95, 448
fundamental equation, 9, 136, 148, 157, 159, 182, 225, 237, 238, 326, 338, 369, 390
fundamental problem, 9, 103, 104, 224, 237, 241, 273, 325, 326, 357, 387, 389

global variable, 3, 4, 17, 36, 57, 64, 81, 106, 111, 123, 135, 138, 139, 142, 143, 156, 164, 175, 182, 198, 207, 212, 229, 230, 232, 273, 276, 295, 299, 306, 327, 337, 360, 389, 390, 436, 440

impressed source, 237, 238
induced source, 237, 238
inherited association, 135, 136
inner orientation, 22, 44, 53, 75, 85, 116, 187, 201, 221, 233, 415, 431, 432, 435, 436, 440
integral variable, 2, 105, 110, 145, 182
interaction equation, 156, 165

E. Tonti, *The Mathematical Structure of Classical and Relativistic Physics*, Modeling and Simulation in Science, Engineering and Technology,
DOI 10.1007/978-1-4614-7422-7, © Springer Science+Business Media New York 2013

irrotational, 34, 52, 156, 170, 172, 174, 177, 359

Legendre transform, 262, 271, 386

material equation, 156, 160, 233
Maxwell, 12, 30, 31, 48, 89, 147, 160, 222, 253, 274, 301, 307
Mechanical theory, 7
mechanical theory, 4, 7, 87

odd differential form, 8, 415, 420, 426
oddness principle, 59, 121, 122, 130, 133
outer orientation, 44, 45, 48, 49, 53, 56, 57, 76, 79, 83, 90, 115, 116, 122, 145, 160, 174, 175, 191, 197, 202, 204, 221, 224, 242, 279, 288, 361, 371, 431, 432, 436, 440

phenomenological equation, 156, 160, 174
primal cell complex, 5, 71, 72, 76, 84, 187, 217, 218, 221, 306, 335, 430, 432, 433, 435
primal instant, 61, 249, 334
primal interval, 58, 59, 332, 335
pseudoscalar, 145, 146, 277, 288, 363, 393, 473
pseudovector, 22, 145–147, 283, 285, 293, 393

quantum mechanics, 2, 95, 149, 157, 184, 392, 438

reversal of motion, 35, 36, 59, 130, 161, 162, 165, 175, 242, 244, 248, 249, 251, 265, 279, 289, 293, 331, 333, 361, 369, 387, 388

solenoidal, 34, 172, 177, 362, 363
source variable, 6, 8, 99, 100, 116, 156, 160, 217, 221, 229, 231, 237, 242, 245, 253, 274, 279, 281, 325, 326, 335, 389, 440, 443
space element, 2, 3, 8, 39, 40, 42, 49, 64, 66, 84, 105, 106, 111, 116, 127, 136, 142, 145, 156, 164, 174, 197, 217, 221, 222, 229, 261, 415, 436, 440
space global variable, 97, 109, 110, 112

time complex, 58, 87, 247, 248
time element, 23, 436
time global variable, 97, 125–127, 129, 134
time reversal, 35
topological equation, 7, 8, 15, 16, 156, 168–171, 174, 175, 177, 219, 230